GEODESY

Treatise on Geophysics

GEODESY

Editor-in-Chief
Professor Gerald Schubert
Department of Earth and Space Sciences and Institute of Geophysics and Planetary Physics,
University of California Los Angeles, Los Angeles, CA, USA

Volume Editor
Dr. Tom Herring
Massachusetts Institute of Technology, Cambridge, MA, USA

ELSEVIER

AMSTERDAM • BOSTON • HEIDELBERG • LONDON • NEW YORK • OXFORD
PARIS • SAN DIEGO • SAN FRANCISCO • SINGAPORE • SYDNEY • TOKYO

Elsevier B.V.
Radarweg 29, 1043 NX Amsterdam, the Netherlands

First edition 2009

British Library Cataloguing in Publication Data
A catalogue record for this book is available from the British Library

Library of Congress Control Number: 2009929981

ISBN: 978-0-444-53460-6

For information on all Elsevier publications
visit our website at elsevierdirect.com

Printed and bound in Spain

09 10 11 12 10 9 8 7 6 5 4 3 2 1

Contents

Preface

Geophysics is the physics of the Earth, the science that studies the Earth by measuring the physical consequences of its presence and activity. It is a science of extraordinary breadth, requiring 10 volumes of this treatise for its description. Only a treatise can present a science with the breadth of geophysics if, in addition to completeness of the subject matter, it is intended to discuss the material in great depth. Thus, while there are many books on geophysics dealing with its many subdivisions, a single book cannot give more than an introductory flavor of each topic. At the other extreme, a single book can cover one aspect of geophysics in great detail, as is done in each of the volumes of this treatise, but the treatise has the unique advantage of having been designed as an integrated series, an important feature of an interdisciplinary science such as geophysics. From the outset, the treatise was planned to cover each area of geophysics from the basics to the cutting edge so that the beginning student could learn the subject and the advanced researcher could have an up-to-date and thorough exposition of the state of the field. The planning of the contents of each volume was carried out with the active participation of the editors of all the volumes to insure that each subject area of the treatise benefited from the multitude of connections to other areas.

Geophysics includes the study of the Earth's fluid envelope and its near-space environment. However, in this treatise, the subject has been narrowed to the solid Earth. The *Treatise on Geophysics* discusses the atmosphere, ocean, and plasmasphere of the Earth only in connection with how these parts of the Earth affect the solid planet. While the realm of geophysics has here been narrowed to the solid Earth, it is broadened to include other planets of our solar system and the planets of other stars. Accordingly, the treatise includes a volume on the planets, although that volume deals mostly with the terrestrial planets of our own solar system. The gas and ice giant planets of the outer solar system and similar extra-solar planets are discussed in only one chapter of the treatise. Even the *Treatise on Geophysics* must be circumscribed to some extent. One could envision a future treatise on Planetary and Space Physics or a treatise on Atmospheric and Oceanic Physics.

Geophysics is fundamentally an interdisciplinary endeavor, built on the foundations of physics, mathematics, geology, astronomy, and other disciplines. Its roots therefore go far back in history, but the science has blossomed only in the last century with the explosive increase in our ability to measure the properties of the Earth and the processes going on inside the Earth and on and above its surface. The technological advances of the last century in laboratory and field instrumentation, computing, and satellite-based remote sensing are largely responsible for the explosive growth of geophysics. In addition to the enhanced ability to make crucial measurements and collect and analyze enormous amounts of data, progress in geophysics was facilitated by the acceptance of the paradigm of plate tectonics and mantle convection in the 1960s. This new view of how the Earth works enabled an understanding of earthquakes, volcanoes, mountain building, indeed all of geology, at a fundamental level. The exploration of the planets and moons of our solar system, beginning with the Apollo missions to the Moon, has invigorated geophysics and further extended its purview beyond the Earth. Today geophysics is a vital and thriving enterprise involving many thousands of scientists throughout the world. The interdisciplinarity and global nature of geophysics identifies it as one of the great unifying endeavors of humanity.

The keys to the success of an enterprise such as the *Treatise on Geophysics* are the editors of the individual volumes and the authors who have contributed chapters. The editors are leaders in their fields of expertise, as distinguished a group of geophysicists as could be assembled on the planet. They know well the topics that had to be covered to achieve the breadth and depth required by the treatise, and they know who were the best of their colleagues to write on each subject. The list of chapter authors is an impressive one, consisting of geophysicists who have made major contributions to their fields of study. The quality and coverage achieved by this group of editors and authors has insured that the treatise will be the definitive major reference work and textbook in geophysics.

Each volume of the treatise begins with an 'Overview' chapter by the volume editor. The Overviews provide the editors' perspectives of their fields, views of the past, present, and future. They also summarize the contents of their volumes and discuss important topics not addressed elsewhere in the chapters. The Overview chapters are excellent introductions to their volumes and should not be missed in the rush to read a particular chapter. The title and editors of the 10 volumes of the treatise are:

Volume 1: Seismology and Structure of the Earth

> Barbara Romanowicz
> University of California, Berkeley, CA, USA
> Adam Dziewonski
> Harvard University, Cambridge, MA, USA

Volume 2: Mineral Physics

> G. David Price
> University College London, UK

Volume 3: Geodesy

> Thomas Herring
> Massachusetts Institute of Technology, Cambridge, MA, USA

Volume 4: Earthquake Seismology

> Hiroo Kanamori
> California Institute of Technology, Pasadena, CA, USA

Volume 5: Geomagnetism

> Masaru Kono
> Okayama University, Misasa, Japan

Volume 6: Crust and Lithosphere Dynamics

> Anthony B. Watts
> University of Oxford, Oxford, UK

Volume 7: Mantle Dynamics

> David Bercovici
> Yale University, New Haven, CT, USA

Volume 8: Core Dynamics

> Peter Olson
> Johns Hopkins University, Baltimore, MD, USA

Volume 9: Evolution of the Earth

> David Stevenson
> California Institute of Technology, Pasadena, CA, USA

Volume 10: Planets and Moons

> Tilman Spohn
> Deutsches Zentrum für Luft-und Raumfahrt, GER

In addition, an eleventh volume of the treatise provides a comprehensive index.

The *Treatise on Geophysics* has the advantage of a role model to emulate, the highly successful *Treatise on Geochemistry*. Indeed, the name *Treatise on Geophysics* was decided on by the editors in analogy with the geochemistry compendium. The *Concise Oxford English Dictionary* defines treatise as "a written work dealing formally and systematically with a subject." Treatise aptly describes both the geochemistry and geophysics collections.

The *Treatise on Geophysics* was initially promoted by Casper van Dijk (Publisher at Elsevier) who persuaded the Editor-in-Chief to take on the project. Initial meetings between the two defined the scope of the treatise and led to invitations to the editors of the individual volumes to participate. Once the editors were on board, the details of the volume contents were decided and the invitations to individual chapter authors were issued. There followed a period of hard work by the editors and authors to bring the treatise to completion. Thanks are due to a number of members of the Elsevier team, Brian Ronan (Developmental Editor), Tirza Van Daalen (Books Publisher), Zoe Kruze (Senior Development Editor), Gareth Steed (Production Project Manager), and Kate Newell (Editorial Assistant).

G. Schubert
Editor-in-Chief

Contributors

D. C. Agnew
University of California San Diego, San Diego, CA, USA

G. Blewitt
University of Nevada, Reno, NV, USA

D. P. Chambers
The University of Texas at Austin, Austin, TX, USA

D. Crossley
St. Louis University, St. Louis, MO, USA

V. Dehant
Royal Observatory of Belgium, Brussels, Belgium

R. S. Gross
Jet Propulsion Laboratory, California Institute of Technology, Pasadena, CA, USA

T. A. Herring
Massachusetts Institute of Technology, Cambridge, MA, USA

J. Hinderer
Institut de Physique du Globe, Strasbourg, France

C. Jekeli
The Ohio State University, Columbus, OH, USA

K. Latychev
University of Toronto, Toronto, ON, Canada

P. M. Mathews
University of Madras, Chennai, India

J. X. Mitrovica
University of Toronto, Toronto, ON, Canada

T. Niebauer
Migro-g Solutions Inc., Erie, CO, USA

P. A. Rosen
California Institute of Technology, Pasadena, CA, USA

M. Simons
California Institute of Technology, Pasadena, CA, USA

M. E. Tamisiea
Harvard-Smithsonian Center for Astrophysics, Cambridge, MA, USA

J. M. Wahr
University of Colorado, Boulder, Co, USA

R. J. Warburton
GWR Instruments, San Diego, CA, USA

1 Overview

T. A. Herring, Massachusetts Institute of Technology, Cambridge, MA, USA

1.1 Introduction

Modern geodesy as discussed in this volume started with the development of distance measurement using propagating electromagnetic signals and the launch of Earth-orbiting satellites. With these developments, space-based geodesy allowed global measurements of positions, changes in the rotation of the Earth, and the Earth's gravity field. These three areas (positioning, Earth rotation, and gravity field) are considered the three pillars of geodesy. The accuracy of current measurement systems allows time variations to be observed in all three areas. Also the complexity of problems is such that each of the pillars interacts with each other and also with many other branches of Earth Science. This interaction is most apparent in the role that water plays in modern geodetic measurements. Every chapter in this volume mentions the role of water. It is critical because it can move rapidly and over large distances; it can exist in all three phases, gas, fluid, and solid; and modern geodetic methods are accurate enough that their measurements are sensitive to its effects. In its vapor form, its refractive properties delay microwave signals propagating through the Earth's atmosphere. For geodetic positioning, this is a noise source but it is a signal for metrological applications. In the liquid form, it forms oceans that affect both the tidal signal and the rotation of the Earth. Also in liquid form, its mass changes the gravity field as it is moved through the hydrologic cycle. In solid form, it has a gravitational and deformation signal that changes if melting of the ice unloads the surface of the Earth. The interactions between the pillars include the elastic loading effects of changing mass loads that can be seen in the gravity field and in the positions of ground stations. The movement of water to and from the oceans can be seen with altimeter satellites whose orbital information is derived from measurements from ground stations whose positions are affected by the changing mass load. In modern, time-dependent geodetic data analysis these interactions need to be accounted for. The common interface between the geodetic methods is the coordinate systems and reference frames used to analyze data.

Coordinate systems and the associated reference frames form a core theme of the chapters in this volume. Two other unifying themes are the measurement systems of geodesy that are used again throughout the volume, and interplay of errors in measurements, signals, and noise. In this chapter we examine these three themes.

1.2 Coordinate Systems

A common aspect of geodesy that permeates the literature on the subject is the coordinate system definition. With the development of space-based methods that allow global measurements to be made, the subject has been, in some senses, greatly simplified in recent years while at the same time complicated by the increased accuracy needs. The simplification comes from being able to use a Cartesian coordinate system with an origin at the center of mass of the Earth and axes aligned in a well-defined manner to the outer surface of the Earth. The complications arise from the need for a coordinate system definition that accounts for deformations from plate tectonics, tides, tectonic events, and other time-variable deformations. The measurement and understanding of these deformations are themes that are common to many of the chapters in this volume.

Historically, coordinate systems in geodesy are divided into two parts: horizontal coordinates such as latitude and longitude; and a vertical coordinate called height. These two systems are fundamentally different in that the horizontal one is geometric and based on the direction of the normal to an ellipsoidal body that represents the average shape of the Earth. The height system, called orthometric height, is based on the gravitational potential field of the Earth. In this system, surfaces of constant height are equipotential surfaces and hence fluid will not flow along these surfaces. One special height surface is called the geoid and is associated with the surface representing mean sea level. In Chapters 2 and 5, the subtle problems with these definitions – from defining an equipotential surface to the meaning of mean sea level in an ocean with currents – are discussed. In conventional geodetic systems there is a blend of potential-based and geometric-based systems. The determination of the geodetic latitude and longitude is complicated by measurements being made in the Earth's gravity field. Specifically with the development of electronic distance measurements, the distance measurement itself does not depend on the gravity field (except for small relativistic effects) but the projection of the direct measurement to a horizontal measurement does depend on the slope of the measured line. The angle between measured direction and local vertical (which depends on the local direction of gravity) can be easily measured but is strictly not the correct measurement. The angle to the local normal to the ellipsoid is needed. The difference between these directions is called the 'deflection of the vertical' and as discussed in Chapter 2 can be determined from measurements of gravity and solving the appropriate boundary-value problem. Deflection of vertical is also the difference between geodetic and astronomic latitude and longitude.

The blend of potential-based and geometry-based systems in geodetic coordinate systems is a consequence of the methods available for making measurements. The local nature of these measurements resulted in large differences between systems adopted by different countries. The reference ellipsoidal shape for the Earth could have differences in the semimajor axis of over a kilometer (e.g., Clarke-1866 6378206.0 m, and Everest 6377276.0 m (Smith, 1996)) and it was not uncommon at boundaries of

countries to have coordinate differences of several hundred meters. The advent of space-based measurements allows the determination of a best-fit ellipsoidal shape for the Earth based on global data.

There are simple relationships between the gravity of the Earth, expressed in spherical harmonics; the moments of inertia of the Earth; and an appropriate flattening of the Earth. After the central force term in the gravity field, the flattening term is the largest, being at least 1000 times larger than any other terms in the field. The leading terms in the gravitational potential field, V, of the Earth can be written in spherical harmonics as (e.g., Stacey, 1992)

$$V(r, \theta) = -\frac{GM}{r} \left(\mathcal{J}_0 P_0 - \mathcal{J}_1 \frac{a}{r} P_1(\cos\theta) - \mathcal{J}_2 \left(\frac{a}{r}\right)^2 P_2(\cos\theta) \right)$$

[1]

where G is the gravitational constant; M is the adopted mass of the Earth; θ and r are colatitude and radial distances to the point where the potential is determined; a is the equatorial radius of the Earth; $\mathcal{J}_0$, $\mathcal{J}_1$, and $\mathcal{J}_2$ are coefficients of the gravity field; and P_0, P_1, and P_2 are Legendre functions. When the adopted mass of the Earth matches that of the Earth, the $\mathcal{J}_0 P_0$ term is 1, and if the center of mass of the Earth corresponds to the origin of the coordinate system used to measure latitude, $\mathcal{J}_1$ is zero. When only the largest terms in the gravity field are considered, the Earth is axially symmetric and thus there is no longitude dependence. The second-degree harmonic term, $\mathcal{J}_2$, is related to both the moments of inertia of the Earth, which will control the Earth's rotational behavior, and the flattening of the shape of the Earth. The moments of inertia of the Earth can be expressed in terms of $\mathcal{J}_2$ through MacGullagh's formula yielding

$$\mathcal{J}_2 = (C - A)/(Ma^2)$$

[2]

where A and C are the minimum and maximum moments of inertia of the axially symmetric Earth. By setting the potential to be the same at the equator and the pole, the semimajor and semiminor axes of an ellipsoid, a and c, that match these values allows the flattening of the ellipsoid to be derived from the gravity field. Neglecting higher-order terms in the gravity field, we have (Stacey, 1992)

$$f = \frac{a-c}{c} = \frac{C-A}{Ma^2} \left(\frac{a^2}{c^2} + \frac{c}{2a} \right) + \frac{1}{2} \frac{a^2 c \omega^2}{GM} \approx \frac{3}{2}\mathcal{J}_2 + \frac{1}{2}m$$ [3]

where f is the flattening, and the centrifugal component of acceleration from the rotation of the Earth (rotation rate ω) has been included. The ratio of the equatorial gravitational to rotational force is given by m. From eqns [1]–[3], we see the first-order relationships between the gravity field of the Earth, the moments of inertia of the Earth, and the shape of the Earth. With modern space-based measurements, all of these quantities can be measured with great precision. Chapter 2 details the theory of gravitational potential, Chapters 10 and 9 examine the rotational consequences of moments on inertia, and Chapters 5 examines the detailed shape of the geoid based on altimeter measurements from space.

With the development of space-based geodetic systems, it became possible to define a purely geometric global coordinate system. These new systems are Cartesian and have their origin approximately at the center of mass of the Earth. The axes are aligned approximately along the rotation axis and the Greenwich meridian. The alignments here are only approximate because, with the accuracy of modern measurements, the center of mass and rotation axis move with respect to the figure of the Earth. The rotation axis motions are large, of order 10 m over a year, and have been known for about over a century. The center of mass movements are much smaller, of order 1–2 cm, and have been measured for the past few decades. Even now, center-of-mass variations are not fully accounted for in modern coordinate systems. The current International Terrestrial Reference Frames (ITRF) (Altamimi *et al.*, 2002; Altamimi, 2005) have an origin at the center of figure of the Earth. Mass movements on the surface of the Earth, mainly atmospheric and hydrospheric, result in movement of the center of mass relative to center of figure which needs to be accounted for an accurate representation of the gravity field. When center-of-figure reference frames are used, the gravity fields in these frames should have time-dependent first-degree spherical harmonic terms included. Currently, these terms need to be observationally determined and the routine production of such values is not yet available. Chapter 8 discusses in detail the measurement of time-variable gravity. To further complicate accounting for center-of-mass movements, the mass movements that cause the center of mass to move, also typically load the surface, and hence deform the surface of the Earth. The deconvolution of the effects of surface deformations affecting the center of figure and the mass contributions that affect center-of-mass positions is not trivial. This subject is explored in Lavallée and Blewitt (2002) and Blewitt (2003), and is discussed in Chapter 11 in this volume.

The treatment of Earth rotation variations is commonplace in modern systems because the effects are so large. However, even with this topic there are subtle problems associated with deformations of the Earth. The motion of the rotation axis with respect to the crust of the Earth is dominated by a resonance term called the Chandler wobble, named after the person to first observe the effect, and an annual signal due mainly to seasonal mass movements. The Chandler wobble has a period of about 430 days with the frequency related to the moments of inertia of the Earth, the elastic properties of the mantle, and the effects of the oceans and the fluid core. The annual signal and Chandler wobble cause motions of the pole with an amplitude of about 10 m. Superimposed on these large quasi-periodic signals are other broadband signals that arise from movements in the atmosphere, oceans, and fluid core. The details of these motions are discussed in the Chapter 9. In addition to these periodic-type variations there is a secular drift of the pole due to changes of moments of inertia resulting from glacial isostatic adjustment which is discussed in Chapter 7.

In addition to changes in the position of the rotation axis with respect to the crust of the Earth, the rotation rate also varies with both a secular term, due to energy dissipation in the Earth–Moon system, and with shorter-period variations due to momentum exchanges between the atmosphere, oceans, and fluid core, and the solid Earth. For period less than a few years, the atmosphere is the dominant source of changes in angular momentum. For longer-period terms, the fluid core is the most likely source. Rotation-rate changes are discussed in Chapters 9 and 10.

The concept of rotation changes introduces another coordinate system that is needed in modern geodesy. This additional coordinate system is one that does not rotate in inertial space. Rotation rate variations are measured relative to this nonrotating system. Historically, this external, nonrotating frame has been defined by measurements to optical stars and to bodies in our solar system. Both of these systems have problems when very accurate measurements are needed. Easily seen optical stars are quite close to us and exhibit the so-called proper motions meaning that they move relative to a nonrotating frame. Solar system measurements suffer from lack of accurate observations and their interpretation requires that the equations of motions of objects in the solar system contain all of the correct force models. So neither optical objects nor solar system dynamics provides a stable nonrotating coordinate system. Such a coordinate is provided by extragalactic radio sources whose directions can be measured using very-long-baseline interferometry (VLBI) (Ma *et al.*, 1998). Changes in rotation rate are measured relative to this nonrotating system, and time as measured by the rotation of the Earth, Universal Time 1 or UT1, is the integration of the rotation rate.

With the definitions of an inertial coordinate system and one that is attached to the Earth, instantaneous rotation angles can be defined that rotate one system into the other. The spectrum of these rotation angles shows strong peaks in the nearly diurnal band and broad spectral response outside the diurnal band. Historically, the rotation vector has been divided into two parts. When the rotation axis is viewed from the rotating Earth, the nearly diurnal retrograde motion of the rotation axis represents a motion that is slow in inertial space and is called nutation. This motion has been known about since ancient Greek times. The word 'nutation' is Greek for nodding. The nodding is cause by the time-dependent torques applied to the equatorial bulge of the Earth by the gravitational forces from the Moon, Sun, and planets. The average torque applied by these bodies causes a secular motion of the pole in inertia space called precession. The longer-period motion is called polar motion. In an approximate sense, nutation is caused by external torques applied to the Earth and pole motion due to mass and angular momentum redistributions in the Earth. The separation and interpretation of these motions are discussed in Chapters 9 and 10.

The final aspect of measuring changes in the rotation of the Earth is the definition of rotation angles. All points on the surface of the Earth move relative to the interior just as the interior itself moves due to mantle convection. To define rotation changes, the deformation component must be accounted for. For secular motions due to plate tectonics, the separation is made making the average motion have no net rotation, the so-called no-net-rotation or NNR frame. The current definitions of rotation angles for the Earth as distributed by the International Earth Rotation Service (IERS) use this definition. With modern geodetic accuracy, nonsecular motions of sites are now evident due mainly to loading effects. The gravitational consequences of the mass movements and loads are discussed in Chapter 8 but there are also consequences for Earth rotation measurements when small numbers of sites are used. Earth rotation measurements made using large number of sites allow some of these nonsecular site movements

to be averaged. However, the spatial correlation of loading effects suggests that there is an upper limit to the number of sites that are needed to effectively measure Earth rotation changes and average out the loading effects and other nonsecular motions. Ultimately, in closed system analysis, nonsecular deformations would be accounted for, for example, from the analysis of gravity changes, before the rotation angles were computed.

1.3 Geodetic Methods

Modern geodetic methods have in recent decades pushed the precision of measurements and the completeness of models needed to explain these measurements. In geodesy, it is common to talk about the fractional precision of measurement. A 1-km distance measured with a precision of 1 mm is said to be precise to 1 part per million or 1 ppm. On large scales, modern geodetic measurements are precise to better than 1 part per billion (1 ppb) in many cases. For global height measurement, 1 ppb corresponds to 6 mm uncertainties in heights. For gravity measurements, 1 ppb is 10 nanometers per second or 1 micro-Gal where Gal is the CGS unit for gravity named after Galileo. The Earth's gravity is about 980 Gals. There are a variety of geodetic systems that achieve these precisions. In this volume all the modern geodetic measurement methods are discussed. They fall into three basic classes: ground-based positioning systems that provide geometric positioning and tracking of external objects; satellite systems that sense the Earth's gravity field and/or make measurements directly from space; and ground-based instrumentation that measure the gravity field on the Earth's surface. Interpretation of geodetic measurements often requires inputs from all these systems.

1.3.1 Ground-Based Positioning Systems

Four major geodetic measurements fall into this category: VLBI; Satellite Laser Ranging (SLR); Global Positioning System (GPS); and the Doppler Orbitography and Radiopositioning Integrated by Satellite (DORIS) system. These systems use either microwave (VLBI, GPS, and DORIS) or optical (SLR) frequency signals to provide the carrier for their measurements. The basic measurement types include group and phase delay and Doppler shift measurements. Estimates of the positions and motions of locations on the Earth's surface obtained from the analysis of data from these systems are used in the formation of the ITRF. The latest version of the ITRF is ITRF 2005 (Altamimi, 2005).

VLBI is a microwave-based measurement system that measures the difference in arrival times of incoherent signals from radio sources by cross-correlation. Most commonly the radio sources are extragalactic objects but beacons from satellites have also been used. Group and phase delays along with Doppler shift are measured. The phase delays are difficult to use in geodetic measurements with this system because they are measured modulo 2π and reconstructing the number of cycles of phase is nontrivial in geodetic measurements. In the astronomical applications of VLBI, use of the phase delays is common. Individual delay measurements have precisions of 1–10 mm. Generally radio telescopes with diameters between 10 and 30 m are used for VLBI measurements and because of the separation of the telescopes (thousands of kilometers) the delay measurements are made relative to independent hydrogen maser clocks at the observatories. The radio signals propagate through the Earth's atmosphere and are delayed through the refractive index of the atmosphere's gas constituents. VLBI measurements are made at two relatively high frequencies (usually ~2.3 and 8 GHz) and are delayed by the propagation in the ionosphere (removed with

a dual-frequency correction since this medium is dispersive) and they are strongly affected by the dipole component of the refractivity of atmospheric water vapor. **Figure 1** shows the distribution of VLBI sites included in ITRF 2005. Of the sites shown, about 40 stations are currently operating. VLBI requires coordinated measurements so that distant telescopes look at the same objects at the same time. Measurements with VLBI are usually 24-h-duration observing sessions using four to eight radio telescopes around the world. VLBI activities are now coordinated through the International VLBI Service (IVS) (Schlueter *et al.*, 2002).

VLBI is the only modern technique with direct access to a stable inertial reference frame. It uniquely contributes to maintaining time as measured by the rotation of the Earth and to monitoring the motion of the Earth's rotation axis in inertial space. These latter measurements provide unique insights into the interaction of the fluid-outer and solid-inner cores and the mantle of the Earth. This topic is dicussed in detail in Chapter 10. VLBI also provides accurate position and atmospheric delay measurements.

SLR is an optical-based system that uses short-pulsed laser and accurate timing equipment to measure the round trip flight time between a ground system and a satellite equipped with special corner-cube retroreflectors. Some laser systems have enough power to make range measurements to corner-cube

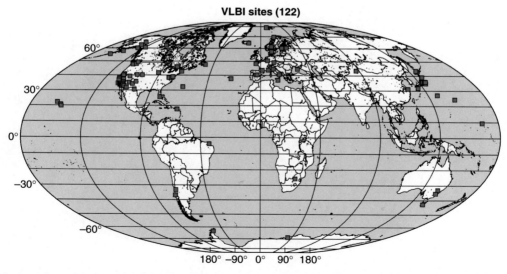

VLBI sites (122)

Figure 1 Locations of the 122 VLBI sites that are included in the ITRF 2005 reference frame. Of the sites shown, about 40 are still actively making measurements. Measurements with VLBI are organized into sessions, usually of 24-h duration, using four to eight radio telescopes.

arrays on the Moon. Specially equipped optical telescopes are used for these measurements. Measurements accuracies of 1–10 mm are common for modern systems. The optical frequencies used in this system are not affected by the Earth's ionosphere and the contribution from atmospheric water vapor is much less than for microwave systems (the optical frequencies are too high to efficiently excite the dipole water vapor resonance). **Figure 2** shows the distribution of SLR sites used in ITRF 2005. Of these sites, about 35 are currently active. SLR measurements do not need to be closely coordinated and hence SLR stations can operate semiautonomously. Some stations operate continuously while others operate on 8-h shifts. There are priorities set for which of the current 26 satellites with corner cubes should be tracked when multiple satellites are visible. The coordination of SLR measurements is provided by the International Laser Ranging Service (ILRS) (Pearlman *et al.*, 2002).

SLR provides measurements to about 26 orbiting spacecrafts and the Moon. These measurements are used to determine the positions of the SLR sites and the orbits of the spacecraft. Perturbations to the spacecraft orbits are used to study the Earth's gravity field and its temporal variations. For geodetic positioning, the most commonly observed satellites are the pair of Laser Geodynamics Satellites (LAGEOS). These are high area-to-mass ratio satellites orbiting at about 6000 km altitude. At this altitude, the orbits

are sensitive to the lower-degree terms in the Earth's gravity and measurements to LAGEOS provided the first measurements of secular changes in the lower-degree gravity field coefficients. The altitude is also high enough to be only slightly affected by atmospheric drag although there are some not well-understood electromagnetic perturbations to the orbits. The high altitude of the LAGEOS satellites also means that center of mass of the Earth can be well determined from the measurements. The primary aim of much of the SLR tracking is precise orbit determination (POD). For the altimeter missions discussed in Chapter 5, the determination of orbits with subcentimeter accuracy is critical.

The GPS is a microwave group and phase-delay measurement system using relatively low L-band signals (~1.2 and 1.5 GHz). The system was initially designed as a military navigation system that used group delays. Time tags are encoded into the transmitted signal and a ground receiver can measure the difference in time between when a signal was transmitted from a satellite based on the satellite's clock and when it was received on the ground based on the ground receiver's clock. If the clocks were synchronized, the time difference would be a measure of range. In the navigation application, accurate clocks are placed on the satellites and because the ground receivers can simultaneously make measurements to multiple satellites, the ground receiver's clock error can be

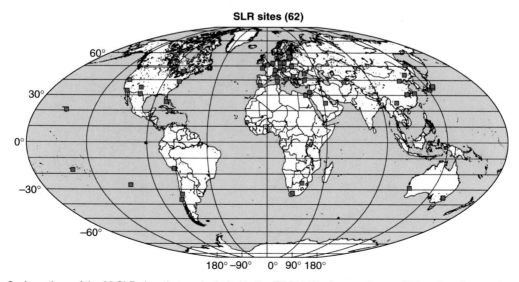

Figure 2 Locations of the 62 SLR sites that are included in the ITRF2005 reference frame. Of the sites shown, about 35 are still actively making measurements. These stations observe continuously with priorities, but not detailed scheduling, set by the International Laser Ranging Service (ILRS).

estimated or eliminated by differencing measurements. The ability of GPS receivers to make measurements simultaneously to multiple satellites and the narrow bandwidth of the transmitted signals means relatively inexpensive electronics can be used to make very precise measurements. In the geodetic applications of GPS, the phase measurements are used which is possible because multiple ground stations can see the same satellites, and either by estimation or differencing, the contributions of local oscillator phases in the receivers and transmitters can be eliminated. Phase measurements of range changes accurate to a few millimeters are possible. Dual-frequency measurements are used to eliminate the ionospheric delays but because of the low frequency, the dual-frequency corrections are not always adequate. The second-order effects from the Earth's magnetic field need to be accounted for in the most accurate analyses. Continuous tracking of signals from the satellites allows the number of cycles in the phase to be counted and only when the signal from the satellite is interrupted or when a satellite is first seen is it necessary to estimate the number of integer cycles in the phase. **Figure 3** shows the GPS sites included in the ITRF 2005 reference frame. We also show on this figure, the additional 1000 GPS sites whose data are routinely available in the international GPS data archives. GPS stations track satellites continuously, are low power,

and are suited to autonomous operation. Many thousands of GPS stations operate continuously around the world. The coordination of station standards and data analysis is through the International Global Navigation Satellite System Service (IGS) (Beutler *et al.*, 1999). In addition to GPS, the IGS also includes measurements from receivers that track the Russian GLONASS satellites and in the future will include measurements from the European Galileo systems. All these systems use similar frequency bands and their satellites are in medium Earth orbit (MEO) with altitudes near 20 000 km (and 12-h orbital periods.)

GPS and the other Global Navigation Satellite Systems (GNSS) provide inexpensive, very precise positioning capabilities on both static and moving receivers. The systems are often deployed in dense networks to monitor tectonic deformations with continuous and occasional occupations. The large numbers of stations available provide robust measurements of polar motion and short-period variations in length of day. Dense networks of GPS stations are also used to monitor atmospheric water vapor and in many cases results are used in operational weather forecasting. The high date rates available with GPS with sampling frequencies up 50 Hz make the system ideal for tracking fast-moving objects such as aircraft and for monitoring seismic events. Chapter 11 discusses many of the applications of GPS along with other space geodetic positioning systems.

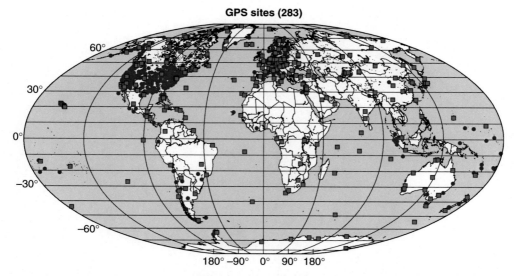

GPS sites (283)

Figure 3 Locations of the 283 GPS sites that are included in the ITRF 2005 reference frame (red squares) and the approximately 1000 additional sites (blue circles) whose data is made freely available through international data archives. These sites operate continuously.

The DORIS system is also a microwave system that uses two-way Doppler tracking. Unlike the other geodetic systems, DORIS stations have active communication with the satellites being tracked. The satellites initiate communications with the ground stations. This system allows two-way communications that can cancel errors due to different frequencies in the oscillators in the spacecraft and ground stations. DORIS is primarily used for POD and plays a critical role in tracking satellites used for radar altimetry discussed in Chapter 5. Earth orientation measurements, center-of-mass variations and station locations are also determined in the analysis of DORIS data. The activities of DORIS are coordinated under the International DORIS Service (Tavernier *et al.*, 2006). In **Figure 4**, the DORIS sites used in the ITRF 2005 are shown.

1.3.2 Satellite Systems

A wide variety of satellite systems are used in modern geodetic measurements. Those that are currently being tracked with laser ranging systems are listed in **Table 1**. The applications of satellites fall into different categories. Five of the entries are reflector arrays on the Moon, three from the Apollo program, and two from the Russian Luna program. All of the Russian GLONASS satellites have corner-cube arrays, as do two of the GPS satellites. The GPS arrays are used for accuracy evaluation. The LAGEOS, Starlette, and Stella spacecrafts are

passive, spherical bodies covered with corner-cube reflectors and are primarily used for gravity field determination and positioning. The Jason-1 and ICESAT satellites are microwave and laser altimeter satellites with Jason-1 used primarily for ocean studies and ICESAT used primarily for ice-sheet studies. Chapter 5 discusses results from these and earlier altimeter satellites that are no longer operating. The CHAMP and GRACE satellites are used for gravity field studies with GRACE being a pair of satellites with accurate range and range-rate measurements between them. These measurements provide very accurate determinations of changes in the Earth's gravity field as detailed in Chapter 8. The ERS and Enisat satellites carry synthetic aperture radars (SARs) which when used in an interferometric mode, as discussed in Chapter 12, allow high spatial resolution maps of deformation fields or topography to be made.

1.3.3 Gravimeters

Gravimeter instrumentation for precise geodesy are classed as absolute gravimeters which normally use the acceleration of an object falling in vacuum, or superconducting instruments which measure changes in gravity on a test mass that is cooled to very low-temperatures to minimize the random accelerations from thermal noise. The applications and development of these classes of gravimeters are discussed in the Chapters 3 and 4. The absolute gravimeters

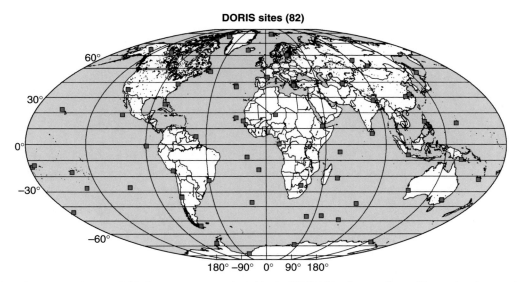

Figure 4 Locations of the 82 DORIS sites that are included in the ITRF2005 reference frame. These sites have active communication between the ground stations and the satellites that are tracked with DORIS.

Table 1 Active geodetic satellites being tracked with SLR (from International Laser Ranging Service)

Satellite name	Altitude (km)	Inclination (deg)	First data date
Ajisai	1485	50	13 Aug. 1986
ALOS	720	98	14 Aug. 2006
ANDE-RR Active	400	51.6	12 Jan. 2007
ANDE-RR Passive	400	51.6	12 Jan. 2007
Apollo11 Sea of Tranquility	356 400	5	20 Aug. 1969
Apollo14 Fra Mauro	356 400	5	07 Feb. 1971
Apollo15 Hadley Rille	356 400	5	01 Sep. 1971
Beacon-C	927	41	02 Jan. 1976
CHAMP	474	87	01 Jan. 2000
Envisat	800	98	10 Apr. 2002
ERS-2	800	99	24 Apr. 1995
Etalon-1	19 105	65	26 Jan. 1989
Etalon-2	19 135	65	13 Jul. 1989
ETS-8	803	98.62	
GFO-1	800	108	22 Apr. 1998
GIOVE-A	29 601	56	11 May 2006
GLONASS-89	19 140	65	21 Mar. 2003
GLONASS-95	19 140	65	26 Aug. 2005
GLONASS-99	19 140	65	
GPS-35	20 195	54	18 Oct. 1993
GPS-36	20 030	55	21 Apr. 1994
GRACE-A	485–500	89	18 Mar. 2002
GRACE-B	485–500	89	18 Mar. 2002
ICESat	600	94	05 Mar. 2003
Jason-1	1336	66	01 Jan. 2003
LAGEOS-1	5850	110	10 May 1976
LAGEOS-2	5625	53	24 Oct. 1992
Larets	691	98.204	04 Nov. 2003
Luna17 Sea of Rains	356 400	5	21 May 1975
Luna21 Sea of Serenity	356 400	5	16 Nov. 1973
Starlette	815	50	03 Jan. 1976
Stella	815	99	30 Sep. 1993

provide absolute measurements that are suitable for monitoring long-term changes in gravity due either to uplift or secular mass movements. These instruments are normally moved from place to place to make repeated measurements often over many years. The superconducting gravimeters provide continuous high-precision gravity measurements at observatory locations. They are suitable for tidal studies, particularly the effects of the fluid core on tidal amplitudes, seismic modes from large earthquakes, and atmospheric mass change excitations.

1.4 Error Sources, Signals, and Noise

Several common themes run through the chapters in this volume. The first of these is that geodesy is a combination of an observational and a theoretical

science. During various stages of its history, either new theories or new observations determine the direction of the science. During the last 20 years, observational accuracies have often been driving new theoretical developments. The other critical aspect of modern measurements is the errors in these measurements. At times assessing error levels has been difficult, especially when accuracy is considered, because there have been no standards to compare against. Many of the chapters in this volume discuss the error spectra of geodetic measurements in the time and space domains. In many cases, a contribution that can be considered noise for one user may be a signal for others. The effects of water vapor and liquid water fall into the category of being both noise and signal.

An example that theoretical and observational drivers have changed over the years is the study of nutation. For much of geodetic history, the motion of the rotation axis in space was computed theoretically from the ephemerides of the Moon and Sun and knowledge of the moments of inertia of the Earth. At the time that VLBI systems were being deployed worldwide and used on a regular basis, a new nutation theory was adopted by the International Astronomical Union (IAU) called the IAU1980 theory of nutation. This new theory included effects of elasticity of the Earth (derived from seismic observations) and a fluid core whose shape was computed from the assumption of hydrostatic equilibrium throughout the Earth. Just a few years after the adoption of this theory, VLBI observations revealed discrepancies between the predictions of the theory and measurements, which resulted in a new assessment of the theory. Some parts of the assessment showed that the original theories were truncated at too coarse a level for the modern VLBI accuracies and other parts showed that the theory needed to be extended to include the effects of nonhydrostatic forces, magnetic coupling, and the presence of the solid inner core. These developments are discussed in Chapter 10. Chapter 4 on superconducting gravimeters discusses a very different measurement type that can be used to address the theory. Currently, there are theoretical predictions about the conductivity structure at the core–mantle boundary that generate results consistent with VLBI measurements that need to be assessed by an independent method.

1.5 Conclusions

Modern geodesy interacts with nearly all fields of Earth Science and the chapters in this volume show these interactions. The chapters progress through the different areas of geodesy starting with the theory of the Earth's gravity field, followed by ground-based measurements of gravity and space-based altimetric measurements. The accuracy of the modern measurements allows temporal variations in gravity to be measured and this topic is discussed in three chapters dealing with tidal variations in gravity, short-term variations that can now be measured from space, and long-period (secular) variations that are related to the

viscoelastic response of the Earth to unloading of ice mass that occurred at the end of the Last Pleistocene Ice Age. Of great interest currently is the separation of this secular signal from the signal due to current melting of ice sheets and glaciers. The combination of gravity measurements and surface uplift measurements shows promise for allowing this separation. The same mass movements that effect time-dependent gravity also change the rotation of Earth on many timescales. Two chapters address Earth rotation changes in inertial space and relative to the crust of the Earth. The volume concludes with the theory, instrumentation, and applications of modern position methods, and surface deformation monitoring.

References

Altamimi Z, Sillard P, and Boucher C (2002) ITRF2000 A new release of the international terrestrial reference frame for earth science applications. *Journal of Geophysical Research* 107(B10): 2214 (doi:10.1029/2001JB000561).

Altamimi Z (2005) http://itrf.ensg.ign.fr/ITRF_solutions/2005/ITRF2005.php (accessed 8 Jan. 2007).

Beutler G, Rothacher M, Schaer S, Springer TA, Kouba J, and Neilan RE (1999) The international GPS service (IGS): An interdisciplinary service in support of Earth sciences. *Advances in Space Research* 23(4): 631–635.

Blewitt G (2003) Self-consistency in reference frames, geocenter definition, and surface loading of the solid Earth. *Journal of Geophysical Research* 108(B2): 2103 (doi:10.1029/2002JB002082).

Lavallée D and Blewitt G (2002) Degree-1 Earth deformation from very long baseline interferometry measurements. *Geophysical Research Letters* 29(20): 1967 (doi:10.1029/2002GL015883).

Ma C, Arias EF, Eubanks TM, *et al.* (1998) The International Celestial reference Frame as realized by very long baseline Interferometry. *Astronomical Journal* 116: 516–546.

Pearlman MR, Degnan JJ, and Bosworth JM (2002) The International Laser Ranging Service. *Advances in Space Research* 30(2): 135–143.

Schlueter W, Himwich E, Nothnagel A, Vandenberg N, and Whitney A (2002) IVS and its important role in the maintenance of the global reference systems. *Advances in Space Research* 30(2): 145–150.

Smith JR (1996) *Introduction to Geodesy: The History and Concepts of Modern Geodesy,* (Wiley Series in Surveying and Boundary Control) 224pp. New York: Wiley.

Stacey FD (1992) *Physics of the Earth,* 3rd edn., 513pp. Brisbane: Brookfield Press.

Tavernier G, Fagard H, and Feissel M (2006) The International DORIS Service: Genesis and early achievements. *Journal of Geodesy* 80: 8–11 403–428.

2 Potential Theory and Static Gravity Field of the Earth

C. Jekeli, The Ohio State University, Columbus, OH, USA

Glossary

deflection of the vertical Angle between direction of gravity and direction of normal gravity.

density moment Integral over the volume of a body of the product of its density and integer powers of Cartesian coordinates.

disturbing potential The difference between Earth's gravity potential and the normal potential.

eccentricity The ratio of the difference of squares of semimajor and semiminor axes to the square of the semimajor axis of an ellipsoid.

ellipsoid Surface formed by rotating an ellipse about its minor axis.

equipotential surface Surface of constant potential.

flattening The ratio of the difference between semimajor and semiminor axes to the semimajor axis of an ellipsoid.

geodetic reference system Normal ellipsoid with defined parameters adopted for general geodetic and gravimetric referencing.

geoid Surface of constant gravity potential that closely approximates mean sea level.

geoid undulation Vertical distance between the geoid and the normal ellipsoid, positive if the geoid is above the ellipsoid.

geopotential number Difference between gravity potential on the geoid and gravity potential at a point.

gravitation Attractive acceleration due to mass.

gravitational potential Potential due to gravitational acceleration.

gravity Vector sum of gravitation and centrifugal acceleration due to Earth's rotation.

gravity anomaly The difference between Earth's gravity on the geoid and normal gravity on the ellipsoid, either as a difference in vectors or a difference in magnitudes.

gravity disturbance The difference between Earth's gravity and normal gravity, either as a difference in vectors or a difference in magnitudes.

gravity potential Potential due to gravity acceleration.

harmonic function Function that satisfies Laplace's field equation.

linear eccentricity The distance from the center of an ellipsoid to either of its foci.

mean Earth ellipsoid Normal ellipsoid with parameters closest to actual corresponding parameters for the Earth.

mean tide geoid Geoid with all time-varying tidal effects removed.

multipoles Stokes coefficients.

Newtonian potential Harmonic function that approaches the potential of a point mass at infinity.

non-tide geoid Mean tide geoid with all (direct and indirect deformation) mean tide effects removed.

normal ellipsoid Earth-approximating reference ellipsoid that generates a gravity field in which it is a surface of constant normal gravity potential.

normal gravity Gravity associated with the normal ellipsoid.

normal gravity potential Gravity potential associated with the normal ellipsoid.

orthometric height Distance along the plumb line from the geoid to a point.

potential Potential energy per unit mass due to the gravitational field; always positive and zero at infinity.

sectorial harmonics Surface spherical harmonics that do not change in sign with respect to latitude.

Stokes coefficients Constants in a series expansion of the gravitational potential in terms of spherical harmonic functions.

surface spherical harmonics Basis functions defined on the unit sphere, comprising products of normalized associated Legendre functions and sinusoids.

tesseral harmonics Neither zonal nor sectorial harmonics.

zero-tide geoid Mean tide geoid with just the mean direct tidal effect removed (indirect effect due to Earth's permanent deformation is retained).

zonal harmonics Spherical harmonics that do not depend on longitude.

2.1 Introduction

Gravitational potential theory has its roots in the late Renaissance period when the position of the Earth in the cosmos was established on modern scientific (observation-based) grounds. A study of Earth's gravitational field is a study of Earth's mass, its influence on near objects, and lately its redistributing transport in time. It is also fundamentally a geodetic study of Earth's shape, described largely (70%) by the surface of the oceans. This initial section provides a historical backdrop to potential theory and introduces some concepts in physical geodesy that set the stage for later formulations.

2.1.1 Historical Notes

Gravitation is a physical phenomenon so pervasive and incidental that humankind generally has taken it for granted with scarcely a second thought. The Greek philosopher Aristotle (384–322 BC) allowed no more than to assert that gravitation is a natural property of material things that causes them to fall

(or rise, in the case of some gases), and the more material the greater the tendency to do so. It was enough of a self-evident explanation that it was not yet to receive the scrutiny of the scientific method, the beginnings of which, ironically, are credited to Aristotle. Almost 2000 years later, Galileo Galilei (1564–1642) finally took up the challenge to understand gravitation through observation and scientific investigation. His experimentally derived law of falling bodies corrected the Aristotelian view and divorced the effect of gravitation from the mass of the falling object – all bodies fall with the same acceleration. This truly monumental contribution to physics was, however, only a local explanation of how bodies behaved under gravitational influence. Johannes Kepler's (1571–1630) observations of planetary orbits pointed to other types of laws, principally an inverse-square law according to which bodies are attracted by forces that vary with the inverse square of distance. The genius of Issac Newton (1642–1727) brought it all together in his *Philosophiae Naturalis Principia Mathematica* of 1687 with a single and simple all-embracing law that in one bold stroke explained the dynamics of the entire universe (today there is more to understanding the dynamics of the cosmos, but Newton's law remarkably holds its own). The mass of a body was again an essential aspect, not as a self-attribute as Aristotle had implied, but as the source of attraction for other bodies: each material body attracts every other material body according to a very specific rule (Newton's law of gravitation; see Section 2.2). Newton regretted that he could not explain exactly why mass has this property (as one still yearns to know today within the standard models of particle and quantum theories). Even Albert Einstein (1879–1955) in developing his general theory of relativity (i.e., the theory of gravitation) could only improve on Newton's theory by incorporating and explaining action at a distance (gravitational force acts with the speed of light as a fundamental tenet of the theory). What actually mediates the gravitational attraction still intensely occupies modern physicists and cosmologists.

Gravitation since its early scientific formulation initially belonged to the domain of astronomers, at least as far as the observable universe was concerned. Theory successfully predicted the observed perturbations of planetary orbits and even the location of previously unknown new planets (Neptune's discovery in 1846 based on calculations motivated by observed perturbations in Uranus' orbit was a major triumph for Newton's law). However, it was also discovered that gravitational acceleration varies on Earth's surface, with respect to altitude and latitude. Newton's law of gravitation again provided the backdrop for the variations observed with pendulums. An early achievement for his theory came when he successfully predicted the polar flattening in Earth's shape on the basis of hydrostatic equilibrium, which was confirmed finally (after some controversy) with geodetic measurements of long triangulated arcs in 1736–37 by Pierre de Maupertuis and Alexis Clairaut. Gravitation thus also played a dominant role in geodesy, the science of determining the size and shape of the Earth, promulgated in large part by the father of modern geodesy, Friedrich R. Helmert (1843–1917).

Terrestrial gravitation through the twentieth century was considered a geodetic area of research, although, of course, its geophysical exploits should not be overlooked. But the advancement in modeling accuracy and global application was promoted mainly by geodesists who needed a well-defined reference for heights (a level surface) and whose astronomic observations of latitude and longitude needed to be corrected for the irregular direction of gravitation. Today, the modern view of a height reference is changing to that of a geometric, mathematical surface (an ellipsoid) and three-dimensional coordinates (latitude, longitude, and height) of points on the Earth's surface are readily obtained geometrically by ranging to the satellites of the Global Positioning System (GPS). The requirements of gravitation for GPS orbit determination within an Earth-centered coordinate system are now largely met with existing models. Improvements in gravitational models are motivated in geodesy primarily for rapid determination of traditional heights with respect to a level surface. These heights, for example, the orthometric heights, in this sense then become derived attributes of points, rather than their cardinal components.

Navigation and guidance exemplify a further specific niche where gravitation continues to find important relevance. While GPS also dominates this field, the vehicles requiring completely autonomous, self-contained systems must rely on inertial instruments (accelerometers and gyroscopes). These do not sense gravitation (see Section 2.6.1), yet gravitation contributes to the total definition of the vehicle trajectory, and thus the output of inertial navigation systems must be compensated for the effect of gravitation. By far the greatest emphasis in gravitation, however, has shifted to the Earth sciences, where detailed knowledge of the configuration of masses (the solid, liquid, and atmospheric components) and their transport and motion leads to improved understanding of the Earth systems (climate, hydrologic cycle, tectonics) and their interactions with life. Oceanography, in particular, also requires a detailed knowledge of a level surface (the geoid) to model surface currents using satellite altimetry. Clearly, there is an essential temporal component in these studies, and, indeed, the temporal gravitational field holds center stage in many new investigations. Moreover, Earth's dynamic behavior influences point coordinates and Earth-fixed coordinate frames, and we come back to fundamental geodetic concerns in which the gravitational field plays an essential role!

This section deals with the static gravitational field. The theory of the potential from the classical Newtonian standpoint provides the foundation for modeling the field and thus deserves the focus of the exposition. The temporal part is a natural extension that is readily achieved with the addition of the time variable (no new laws are needed, if we neglect general relativistic effects) and will not be expounded here. We are primarily concerned with gravitation on and external to the solid and liquid Earth since this is the domain of most applications. The internal field can also be modeled for specialized purposes (such as submarine navigation), but internal geophysical modeling, for example, is done usually in terms of the sources (mass density distribution), rather than the resulting field.

2.1.2 Coordinate Systems

Modeling the Earth's gravitational field depends on the choice of coordinate system. Customarily, owing to the Earth's general shape, a spherical polar coordinate system serves for most applications, and virtually all global models use these coordinates. However, the Earth is slightly flattened at the poles, and an ellipsoidal coordinate system has also been advocated for some near-Earth applications. We note that the geodetic coordinates associated with a geodetic datum (based on an ellipsoid) are never used in a foundational sense to model the field since they do not admit to a separation-of-variables solution of Laplace' differential equation (Section 2.4.1).

Spherical polar coordinates, described with the aid of **Figure 1**, comprise the spherical colatitude, θ, the longitude, λ, and the radial distance, r. Their relation to Cartesian coordinates is

$$
\begin{aligned}
x &= r \sin\theta \cos\lambda \\
y &= r \sin\theta \sin\lambda \\
z &= r \cos\theta
\end{aligned}
\qquad [1]
$$

Considering Earth's polar flattening, a better approximation, than a sphere, of its (ocean) surface is an ellipsoid of revolution. Such a surface is

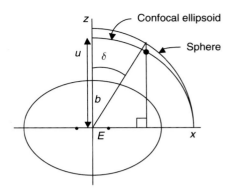

Figure 2 Ellipsoidal coordinates.

generated by rotating an ellipse about its minor axis (polar axis). The two focal points of the best-fitting, Earth-centered ellipsoid (ellipse) are located in the equator about $E = 522\,\text{km}$ from the center of the Earth. A given ellipsoid, with specified semiminor axis, b, and linear eccentricity, E, defines the set of ellipsoidal coordinates, as described in **Figure 2**. The longitude is the same as in the spherical case. The colatitude, δ, is the complement of the so-called reduced latitude; and the distance coordinate, u, is the semiminor axis of the confocal ellipsoid through the point in question. We call (δ, λ, u) ellipsoidal coordinates; they are also known as spheroidal coordinates, or Jacobi ellipsoidal coordinates. Their relation to Cartesian coordinates is given by

$$
\begin{aligned}
x &= \sqrt{u^2 + E^2} \sin\delta \cos\lambda \\
y &= \sqrt{u^2 + E^2} \sin\delta \sin\lambda \\
z &= u \cos\delta
\end{aligned}
\qquad [2]
$$

Points on the given ellipsoid all have $u = b$; and, all surfaces, $u = $ constant, are confocal ellipsoids (the analogy to the spherical case, when $E = 0$, should be evident).

2.1.3 Preliminary Definitions and Concepts

The gravitational potential, V, of the Earth is generated by its total mass density distribution. For applications on the Earth's surface it is useful to include the potential, ϕ, associated with the centrifugal acceleration due to Earth's rotation. The sum, $W = V + \phi$, is then known, in geodetic terminology, as the gravity potential, distinct from gravitational potential. It is further advantageous to define a relatively simple reference potential, or normal potential,

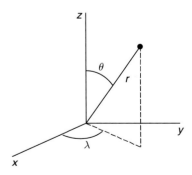

Figure 1 Spherical polar coordinates.

that accounts for the bulk of the gravity potential (Section 2.5.2). The normal gravity potential, U, is defined as a gravity potential associated with a best-fitting ellipsoid, the normal ellipsoid, which rotates with the Earth and is also a surface of constant potential in this field. The difference between the actual and the normal gravity potentials is known as the disturbing potential: $T = W - U$; it thus excludes the centrifugal potential. The normal gravity potential accounts for approximately 99.999 5% of the total potential.

The gradient of the potential is an acceleration, gravity or gravitational acceleration, depending on whether or not it includes the centrifugal acceleration. Normal gravity, γ, comprises 99.995% of the total gravity, g, although the difference in magnitudes, the gravity disturbance, δg, can be as large as several parts in 10^4. A special kind of difference, called the gravity anomaly, Δg, is defined as the difference between gravity at a point, P, and normal gravity at a corresponding point, Q, where $W_P = U_Q$, and P and Q are on the same perpendicular to the normal ellipsoid.

The surface of constant gravity potential, W_0, that closely approximates mean sea level is known as the geoid. If the constant normal gravity potential, U_0, on the normal ellipsoid is equal to the constant gravity potential of the geoid, then the gravity anomaly on the geoid is the difference between gravity on the geoid and normal gravity on the ellipsoid at respective points, P_0, Q_0, sharing the same perpendicular to the ellipsoid. The separation between the geoid and the ellipsoid is known as the geoid undulation, N, or also the geoid height (**Figure 3**). A simple Taylor expansion of the normal gravity potential along the ellipsoid perpendicular yields the following important formula:

$$N = \frac{T}{\gamma} \qquad [3]$$

This is Bruns' equation, which is accurate to a few millimeters in N, and which can be extended to $N = T/\gamma - (W_0 - U_0)/\gamma$ for the general case, $W_0 \neq U_0$. The gravity anomaly (on the geoid) is the gravity disturbance corrected for the evaluation of normal gravity on the ellipsoid instead of the geoid. This correction is $N \partial \gamma / \partial h = (\partial \gamma / \partial h)(T/\gamma)$, where h is height along the ellipsoid perpendicular. We have $\delta g = -\partial T/\partial h$, and hence

$$\Delta g = -\frac{\partial T}{\partial h} + \frac{1}{\gamma}\frac{\partial \gamma}{\partial h} T \qquad [4]$$

The slope of the geoid with respect to the ellipsoid is also the angle between the corresponding perpendiculars to these surfaces. This angle is known as the deflection of the vertical, that is, the deflection of the plumb line (perpendicular to the geoid) relative to the perpendicular to the normal ellipsoid. The deflection angle has components, ξ, η, respectively, in the north and east directions. The spherical approximations to the gravity disturbance, anomaly, and deflection of the vertical are given by

$$\delta g = -\frac{\partial T}{\partial r}, \qquad \Delta g = -\frac{\partial T}{\partial r} - \frac{2}{r} T$$
$$\xi = \frac{1}{\gamma}\frac{\partial T}{r\partial \theta}, \qquad \eta = -\frac{1}{\gamma}\frac{\partial T}{r \sin\theta \, \partial \lambda} \qquad [5]$$

where the signs on the derivatives are a matter of convention.

2.2 Newton's Law of Gravitation

In its original form, Newton's law of gravitation applies only to idealized point masses. It describes the force of attraction, **F**, experienced by two such solitary masses as being proportional to the product of the masses, m_1 and m_2; inversely proportional to the distance, ℓ, between them; and directed along the line joining them:

$$\mathbf{F} = G \frac{m_1 m_2}{\ell^2} \mathbf{n} \qquad [6]$$

G is a constant, known as Newton's gravitational constant, that takes care of the units between the left- and right-hand sides of the equation; it can be

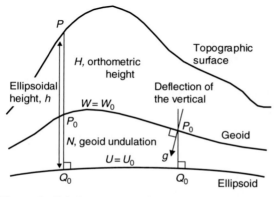

Figure 3 Relative geometry of geoid and ellipsoid.

determined by experiment and the current best value is (Groten, 2004):

$$G = (6.67259 \pm 0.00030) \times 10^{-11} \, \text{m}^3 \, \text{kg}^{-1} \, \text{s}^{-2} \quad [7]$$

The unit vector **n** in eqn [6] is directed from either point mass to the other, and thus the gravitational force is attractive and applies equally to one mass as the other. Newton's law of gravitation is universal as far as we know, requiring reformulation only in Einstein's more comprehensive theory of general relativity which describes gravitation as a characteristic curvature of the space–time continuum (Newton's formulation assumes instantaneous action and differs significantly from the general relativistic concept only when very large velocities or masses are involved).

We can ascribe a gravitational acceleration to the gravitational force, which represents the acceleration that one mass undergoes due to the gravitational attraction of the other. Specifically, from the law of gravitation, we have (for point masses) the gravitational acceleration of m_1 due to the gravitational attraction of m_2:

$$\mathbf{g} = G\frac{m_2}{\ell^2}\mathbf{n} \quad [8]$$

The vector **g** is independent of the mass, m_1, of the body being accelerated (which Galileo found by experiment).

By the law of superposition, the gravitational force, or the gravitational acceleration, due to many point masses is the vector sum of the forces or accelerations generated by the individual point masses. Manipulating vectors in this way is certainly feasible, but fortunately a more appropriate concept of gravitation as a scalar field simplifies the treatment of arbitrary mass distributions.

This more modern view of gravitation (already adopted by Gauss (1777–1855) and Green (1793–1841)) holds that it is a field having a gravitational potential. Lagrange (1736–1813) fully developed the concept of a field, and the potential, V, of the gravitational field is defined in terms of the gravitational acceleration, **g**, that a test particle would undergo in the field according to the equation

$$\nabla V = \mathbf{g} \quad [9]$$

where ∇ is the gradient operator (a vector). Further elucidation of gravitation as a field grew from Einstein's attempt to incorporate gravitation into his special theory of relativity where no reference frame has special significance above all others. It was necessary to consider that gravitational force is not a real

force (i.e., it is not an applied force, like friction or propulsion) – rather, it is known as a kinematic force, that is, one whose action is proportional to the mass on which it acts (like the centrifugal force; see Martin, 1988). Under this precept, the geometry of space is defined intrinsically by the gravitational fields contained therein.

We continue with the classical Newtonian potential, but interpret gravitation as an acceleration different from the acceleration induced by real, applied forces. This becomes especially important when considering the measurement of gravitation (Section 2.6.1). The gravitational potential, V, is a 'scalar' function, and, as defined here, V is derived directly on the basis of Newton's law of gravitation. To make it completely consistent with this law and thus declare it a Newtonian potential, we must impose the following conditions:

$$\lim_{\ell\to\infty} \ell V = Gm \quad \text{and} \quad \lim_{\ell\to\infty} V = 0 \quad [10]$$

Here, m is the attracting mass, and we say that the potential is regular at infinity. It is easy to show that the gravitational potential at any point in space due to a point mass, in order to satisfy eqns [8]–[10], must be

$$V = \frac{Gm}{\ell} \quad [11]$$

where, again, ℓ is the distance between the mass and the point at which the potential is expressed. Note that V for $\ell = 0$ does not exist in this case; that is, the field of a point mass has a singularity. We use here the convention that the potential is always positive (in contrast to physics, where it is usually defined to be negative, conceptually closer to potential energy).

Applying the law of superposition, the gravitational potential of many point masses is the sum of the potentials of the individual points (see **Figure 4**):

$$V_P = G\sum_j \frac{m_j}{\ell_j} \quad [12]$$

And, for infinitely many points in a closed, bounded region with infinitesimally small masses, dm, the summation in eqn [12] changes to an integration,

$$V_P = G\int_{\text{mass}} \frac{dm}{\ell} \quad [13]$$

or, changing variables (i.e., units), $dm = \rho \, dv$, where ρ represents density (mass per volume) and dv is a volume element, we have (**Figure 5**)

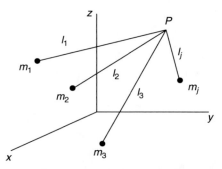

Figure 4 Discrete set of mass points (superposition principle).

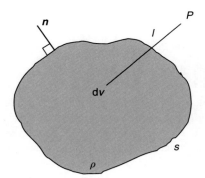

Figure 5 Continuous density distribution.

$$V_P = G \iiint_{\text{volume}} \frac{\rho}{\ell} \, dv \qquad [14]$$

In eqn [14], ℓ is the distance between the evaluation point, P, and the point of integration. In spherical polar coordinates (Section 2.1.2), these points are (θ, λ, r) and (θ', λ', r'), respectively. The volume element in this case is given by $dv = r'^2 \sin\theta' \, d\lambda' \, d\theta' \, dr'$. V and its first derivatives are continuous everywhere – even in the case that P is on the bounding surface or inside the mass distribution, where there is the apparent singularity at $\ell = 0$. In this case, by changing to a coordinate system whose origin is at P, the volume element becomes $dv = \ell^2 \sin\psi \, d\alpha \, d\psi \, d\ell$ (for some different colatitude and longitude ψ and α); and, clearly, the singularity disappears – the integral is said to be weakly singular.

Suppose the density distribution over the volume depends only on radial distance (from the center of mass): $\rho = \rho(r')$, and that P is an exterior evaluation point. The surface bounding the masses necessarily is a sphere (say, of radius, R) and because of the

spherically symmetric density we may choose the integration coordinate system so that the polar axis passes through P. Then

$$\ell = \sqrt{r'^2 + r^2 - 2r'r\cos\theta'}, \qquad d\ell = \frac{1}{\ell} r'r\sin\theta' \, d\theta'$$
$$[15]$$

It is easy to show that with this change of variables (from θ to ℓ) the integral [14] becomes simply

$$V(\theta, \lambda, r) = \frac{GM}{r}, \qquad r \geq R \qquad [16]$$

where M is the total mass bounded by the sphere. This shows that to a very good approximation the external gravitational potential of a planet such as the Earth (with concentrically layered density) is the same as that of a point mass.

Besides volumetric mass (density) distributions, it is of interest to consider surface distributions. Imagine an infinitesimally thin layer of mass on a surface, s, where the units of density in this case are those of mass per area. Then, analogous to eqn [14], the potential is

$$V_P = G \iint_s \frac{\rho}{\ell} \, ds \qquad [17]$$

In this case, V is a continuous function everywhere, but its first derivatives are discontinuous at the surface. Or, one can imagine two infinitesimally close density layers (double layer, or layer of mass dipoles), where the units of density are now those of mass per area times length. It turns out that the potential in this case is given by (see Heiskanen and Moritz, 1967, p. 8)

$$V_P = G \iint_s \rho \frac{\partial}{\partial n}\left(\frac{1}{\ell}\right) ds \qquad [18]$$

where $\partial/\partial n$ is the directional derivative along the perpendicular to the surface (**Figure 5**). Now, V itself is discontinuous at the surface, as are all its derivatives. In all cases, V is a Newtonian potential, being derived from the basic formula [11] for a point mass that follows from Newton's law of gravitation (eqn [6]).

The following properties of the gravitational potential are useful for subsequent expositions. First, consider Stokes's theorem, for a vector function, $\mathbf{f}$, defined on a surface, s:

$$\iint_s (\nabla \times \mathbf{f}) \cdot \mathbf{n} \, ds = \oint_p \mathbf{f} \cdot d\mathbf{r} \qquad [19]$$

where p is any closed path in the surface, $\mathbf{n}$ is the unit vector perpendicular to the surface, and $d\mathbf{r}$ is a differential displacement along the path. From eqn [9], we find

$$\nabla \times \mathbf{g} = \mathbf{0} \qquad [20]$$

since $\nabla \times \nabla = 0$; hence, applying Stokes's theorem, we find with $\mathbf{F} = m\mathbf{g}$ that

$$w = \oint \mathbf{F} \cdot d\mathbf{s} = 0 \qquad [21]$$

That is, the gravitational field is conservative: the work, w, expended in moving a mass around a closed path in this field vanishes. In contrast, dissipating forces (real forces!), like friction, expend work or energy, which shows again the special nature of the gravitational force.

It can be shown (Kellogg, 1953, p. 156) that the second partial derivatives of a Newtonian potential, V, satisfy the following differential equation, known as Poisson's equation:

$$\nabla^2 V = -4\pi G \rho \qquad [22]$$

where $\nabla^2 = \nabla \cdot \nabla$ formally is the scalar product of two gradient operators and is called the Laplacian operator. In Cartesian coordinates, it is given by

$$\nabla^2 = \frac{\partial^2}{\partial x^2} + \frac{\partial^2}{\partial y^2} + \frac{\partial^2}{\partial z^2} \qquad [23]$$

Note that the Laplacian is a scalar operator. Eqn [22] is a local characterization of the potential field, as opposed to the global characterization given by eqn [14]. Poisson's equation holds wherever the mass density, ρ, satisfies certain conditions similar to continuity (Hölder conditions; see Kellogg, 1953, pp. 152–153). A special case of eqn [22] applies for those points where the density vanishes (i.e., in free space); then Poisson's equation turns into Laplace' equation,

$$\nabla^2 V = 0 \qquad [24]$$

It is easily verified that the point mass potential satisfies eqn [24], that is,

$$\nabla^2 \left(\frac{1}{\ell} \right) = 0 \qquad [25]$$

where $\ell = \sqrt{(x-x')^2 + (y-y')^2 + (z-z')^2}$ and the mass point is at (x', y', z').

The solutions to Laplace' equation [24] (that is, functions that satisfy Laplace' equations) are known as harmonic functions (here, we also impose the conditions [10] on the solution, if it is a Newtonian potential and if the mass-free region includes infinity). Hence, every Newtonian potential is a harmonic function in free space. The converse is also true: every harmonic function can be represented as a Newtonian potential of a mass distribution (Section 2.3.1).

Whether as a volume or a layer density distribution, the corresponding potential is the sum or integral of the source value multiplied by the inverse distance function (or its normal derivative for the dipole layer). This function depends on both the source points and the computation point and is known as a Green's function. It is also known as the 'kernel' function of the integral representation of the potential. Functions of this type also play a dominant role in representing the potential as solutions to boundary-value problems (BVPs), as shown in subsequent sections.

2.3   Boundary-Value Problems

If the density distribution of the Earth's interior and the boundary of the volume were known, then the problem of determining the Earth's gravitational potential field is solved by the volume integral of eqn [14]. In reality, of course, we do not have access to this information, at least not the density, with sufficient detail. (The Preliminary Reference Earth Model (PREM), of Dziewonsky and Anderson (1981), still in use today by geophysicists, represents a good profile of Earth's radial density, but does not attempt to model in detail the lateral density heterogeneities.) In this section, we see how the problem of determining the exterior gravitational potential can be solved in terms of surface integrals, thus making exclusive use of accessible measurements on the surface.

2.3.1   Green's Identities

Formally, eqn [24] represents a partial differential equation for V. Solving this equation is the essence of the determination of the Earth's external gravitational potential through potential theory. Like any differential equation, a complete solution is obtained only with the application of boundary conditions, that is, imposing values on the solution that it must assume at a boundary of the region in which it is valid. In our case, the boundary is the Earth's surface and the exterior space is where eqn [24] holds (the atmosphere and other celestial bodies are neglected for the moment). In order to study the solutions to these BVPs (to show that solutions exist and are unique), we take advantage of some very important theorems and identities. It is noted that only a rather elementary introduction to BVPs is offered here with no attempt to address the much larger field of solutions to partial differential equations.

The first, seminal result is Gauss' divergence theorem (analogous to Stokes' theorem, eqn [19]),

$$\iiint_v \nabla \cdot \mathbf{f} \, dv = \iint_s f_n \, ds \qquad [26]$$

where $\mathbf{f}$ is an arbitrary (differentiable) vector function and $f_n = \mathbf{n} \cdot \mathbf{f}$ is the component of $\mathbf{f}$ along the outward unit normal vector, $\mathbf{n}$ (see **Figure 5**). The surface, s, encloses the volume, v. Equation [26] says that the sum of how much $\mathbf{f}$ changes throughout the volume, that is, the net effect, ultimately, is equivalent to the sum of its values projected orthogonally with respect to the surface. Conceptually, a volume integral thus can be replaced by a surface integral, which is important since the gravitational potential is due to a volume density distribution that we do not know, but we do have access to gravitational quantities on a surface by way of measurements.

Equation [26] applies to general vector functions that have continuous first derivatives. In particular, let U and V be two scalar functions having continuous second derivatives, and consider the vector function $\mathbf{f} = U \nabla V$. Then, since $\mathbf{n} \cdot \nabla = \partial / \partial n$, and

$$\nabla \cdot (U \nabla V) = \nabla U \cdot \nabla V + U \nabla^2 V \qquad [27]$$

we can apply Gauss' divergence theorem to get Green's first identity,

$$\iiint_v (\nabla U \cdot \nabla V + U \nabla^2 V) dv = \iint_s U \frac{\partial V}{\partial n} \, ds \qquad [28]$$

Interchanging the roles of U and V in eqn [28], one obtains a similar formula, which, when subtracted from eqn [28], yields Green's second identity,

$$\iiint_v (U \nabla^2 V - V \nabla^2 U) dv = \iint_s \left(U \frac{\partial V}{\partial n} - V \frac{\partial U}{\partial n} \right) ds \qquad [29]$$

This is valid for any U and V with continuous derivatives up to second order.

Now let $U = 1/\ell$, where ℓ is the usual distance between an integration point and an evaluation point. And, suppose that the volume, v, is the space exterior to the Earth (i.e., Gauss' divergence theorem applies to any volume, not just volumes containing a mass distribution). With reference to **Figure 6**, consider the evaluation point, P, to be inside the volume (free

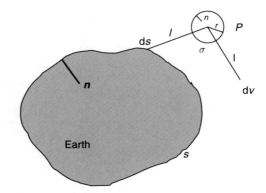

Figure 6 Geometry for special case of Green's third identity.

space) that is bounded by the surface, s; P is thus outside the Earth's surface. Let V be a solution to eqn [24], that is, it is the gravitational potential of the Earth. From the volume, v, exclude the volume bounded by a small sphere, σ, centered at P. This sphere becomes part of the surface that bounds the volume, v. Then, since U, by our definition, is a point mass potential, $\nabla^2 U = 0$ everywhere in v (which excludes the interior of the small sphere around P); and, the second identity [29] gives

$$\iint_s \left(\frac{1}{\ell} \frac{\partial V}{\partial n} - V \frac{\partial}{\partial n} \left(\frac{1}{\ell} \right) \right) ds$$
$$+ \iint_\sigma \left(\frac{1}{\ell} \frac{\partial V}{\partial n} - V \frac{\partial}{\partial n} \left(\frac{1}{\ell} \right) \right) d\sigma = 0 \qquad [30]$$

The unit vector, $\mathbf{n}$, represents the perpendicular pointing away from v. On the small sphere, $\mathbf{n}$ is opposite in direction to $\ell = r$, and the second integral becomes

$$\iint_\sigma \left(-\frac{1}{r} \frac{\partial V}{\partial r} + V \frac{\partial}{\partial r} \left(\frac{1}{r} \right) \right) d\sigma = -\iint_\Omega \frac{1}{r} \frac{\partial V}{\partial r} r^2 \, d\Omega$$
$$- \iint_\Omega V \, d\Omega$$
$$= -\iint_\Omega \frac{\partial V}{\partial r} r \, d\Omega - 4\pi \bar{V} \qquad [31]$$

where $d\sigma = r^2 d\Omega$, Ω is the solid angle, 4π, and $\bar{V}$ is an average value of V on σ. Now, in the limit as the radius of the small sphere shrinks to zero, the right-hand side of eqn [31] approaches $0 - 4\pi V_P$. Hence, eqn [30] becomes (Kellogg, 1953, p. 219)

$$V_P = \frac{1}{4\pi} \iint_s \left(\frac{1}{\ell} \frac{\partial V}{\partial n} - V \frac{\partial}{\partial n} \left(\frac{1}{\ell} \right) \right) ds \qquad [32]$$

with **n** pointing down (away from the space outside the Earth). This is a special case of Green's third identity. A change in sign of the right-hand side transforms **n** to a normal unit vector pointing into v, away from the masses, which conforms more to an Earth-centered coordinate system.

The right-hand side of eqn [32] is the sum of single- and double-layer potentials and thus shows that every harmonic function (i.e., a function that satisfies Laplace' equation) can be written as a Newtonian potential. Equation [32] is also a solution to a BVP; in this case, the boundary values are independent values of V and of its normal derivative, both on s (Cauchy problem). Below and in Section 2.4.1, we encounter another BVP in which the potential and its normal derivative are given in a specified linear combination on s. Using a similar procedure and with some extra care, it can be shown (see also Courant and Hilbert, 1962, vol. II, p. 256 (footnote)) that if P is on the surface, then

$$V_P = \frac{1}{2\pi} \int\int_s \left(\frac{1}{\ell} \frac{\partial V}{\partial n} - V \frac{\partial}{\partial n} \left(\frac{1}{\ell} \right) \right) \mathrm{d}s \qquad [33]$$

where n points into the masses. Comparing this to eqn [32], we see that V is discontinuous as one approaches the surface; this is due to the double-layer part (see eqn [18]).

Equation [32] demonstrates that a solution to a particular BVP exists. Specifically, we are able to measure the potential (up to a constant) and its derivatives (the gravitational acceleration) on the surface and thus have a formula to compute the potential anywhere in exterior space, provided we also know the surface, s. Other BVPs also have solutions under appropriate conditions; a discussion of existence theorems is beyond the present scope and may be found in (Kellogg, 1953). Equation [33] has deep geodetic significance. One objective of geodesy is to determine the shape of the Earth's surface. If we have measurements of gravitational quantities on the Earth's surface, then conceptually we are able to determine its shape from eqn [33], where it would be the only unknown quantity. This is the basis behind the work of Molodensky *et al.* (1962), to which we return briefly at the end of this section.

2.3.2   Uniqueness Theorems

Often the existence of a solution is proved simply by finding one (as illustrated above). Whether such as solution is the only one depends on a corresponding uniqueness theorem. That is, we wish to know if a certain set of boundary values will yield just one potential in space. Before considering such theorems, we classify the BVPs that are typically encountered when determining the exterior potential from measurements on a boundary. In all cases, it is an exterior BVP; that is, the gravitational potential, V, is harmonic ($\nabla^2 V = 0$) in the space exterior to a closed surface that contains all the masses. The exterior space thus contains infinity. Interior BVPs can be constructed, as well, but are not applicable to our objectives.

- *Dirichlet problem (or, BVP of the first kind)*. Solve for V in the exterior space, given its values everywhere on the boundary.
- *Neumann problem (or, BVP of the second kind)*. Solve for V in the exterior space, given values of its normal derivative everywhere on the boundary.
- *Robin problem (mixed BVP, or BVP of the third kind)*. Solve for V in the exterior space, given a linear combination of it and its normal derivative on the boundary.

Using Green's identities, we prove the following theorems for these exterior problems; similar results hold for the interior problems.

Theorem 1. If V is harmonic (hence continuously differentiable) in a closed region, v, and if V vanishes everywhere on the boundary, s, then V also vanishes everywhere in the region, v.

Proof. Since $V = 0$ on s, Green's first identity (eqn [28]) with $U = V$ gives

$$\int\int\int_v (\nabla V)^2 \, \mathrm{d}v = \int\int_s V \frac{\partial V}{\partial n} \mathrm{d}s = 0 \qquad [34]$$

The integral on the left side is therefore always zero, and the integrand is always non-negative. Hence, $\nabla V = \mathbf{0}$ everywhere in v, which implies that $V = $ constant in v. Since V is continuous in v and $V = 0$ on s, that constant must be zero; and so $V = 0$ in v.

This theorem solves the Dirichlet problem for the trivial case of zero boundary values and it enables the following uniqueness theorem for the general Dirichlet problem.

Theorem 2 (Stokes' theorem). If V is harmonic (hence continuously differentiable) in a closed region, v, then V is uniquely determined in v by its values on the boundary, s.

Proof. Suppose the determination is not unique: that is, suppose there are V_1 and V_2, both harmonic in v and both having the same boundary values on s. Then the function $V = V_2 - V_1$ is harmonic in v with all boundary values equal to zero. Hence, by Theorem 1, $V_2 - V_1 = 0$ identically in v, or $V_2 = V_1$ everywhere, which implies that any determination is unique based on the boundary values.

Theorem 3. If V is harmonic (hence continuously differentiable) in the exterior region, v, with closed boundary, s, then V is uniquely determined by the values of its normal derivative on s.

Proof. We begin with Green's first identity, eqn [28], as in the proof of Theorem 1 to show that if the normal derivative vanishes everywhere on s, then V is a constant in v. Now, suppose there are two harmonic functions in v: V_1 and V_2, with the same normal derivative values on s. Then the normal derivative values of their difference are zero; and, by the above demonstration, $V = V_2 - V_1 = $ constant in v. Since V is a Newtonian potential in the exterior space, that constant is zero, since by eqn [10], $\lim_{\ell \to \infty} V = 0$. Thus, $V_2 = V_1$, and the boundary values determine the potential uniquely.

This is a uniqueness theorem for the exterior Neumann BVP. Solutions to the interior problem are unique only up to an arbitrary constant.

Theorem 4. Suppose V is harmonic (hence continuously differentiable) in the closed region, v, with boundary, s, and, suppose the boundary values are given by

$$g = \alpha V|_s + \beta \frac{\partial V}{\partial n}\bigg|_s \qquad [35]$$

Then V is uniquely determined by these values if $\alpha/\beta > 0$.

Proof. Suppose there are two harmonic functions, V_1 and V_2, with the same boundary values, g, on s. Then $V = V_2 - V_1$ is harmonic with boundary values

$$\alpha(V_2 - V_1)|_s + \beta\left(\frac{\partial V_2}{\partial n} - \frac{\partial V_1}{\partial n}\right)\bigg|_s = 0 \qquad [36]$$

With $U = V = V_2 - V_1$, Green's first identity, eqn [28], gives

$$\iiint_v (\nabla(V_2 - V_1))^2 \, dv = \iint_s (V_2 - V_1)\frac{-\alpha}{\beta}(V_2 - V_1) \, ds \qquad [37]$$

Then

$$\iiint_v (\nabla(V_2 - V_1))^2 \, dv + \frac{\alpha}{\beta}\iint_s (V_2 - V_1)^2 \, dS = 0 \qquad [38]$$

Since $\alpha/\beta > 0$, eqn [38] implies that $\nabla(V_2 - V_1) = 0$ in v; and $V_2 - V_1 = 0$ on s. Hence $V_2 - V_1 = $ constant

in v; and $V_2 = V_1$ on s. By the continuity of V_1 and V_2, the constant must be zero, and the uniqueness is proved.

The solution to the Robin problem is unique only in certain cases. The most famous problem in physical geodesy is the determination of the disturbing potential, T, from gravity anomalies, Δg, on the geoid (Section 2.1.3). Suppose T is harmonic outside the geoid; the second of eqns [5] provides an approximate form of boundary condition, showing that this is a type of Robin problem. We find that $\alpha = -2/r$, and, recalling that when v is the exterior space the unit vector n points inward toward the masses, that is, $\partial/\partial n = -\partial/\partial r$, we get $\beta = 1$. Thus, the condition in Theorem 4 on α/β is not fulfilled and the uniqueness is not guaranteed. In fact, we will see that the solution obtained for the spherical boundary is arbitrary with respect to the coordinate origin (Section 2.4.1).

2.3.3 Solutions by Integral Equation

Green's identities show how a solution to Laplace's equation can be transformed from a volume integral, that is, an integral of source points, to a surface integral of BVPs, as demonstrated by eqn [32]. The uniqueness theorems for the BVPs suggest that the potential due to a volume density distribution can also be represented as due to a generalized density layer on the bounding surface, as long as the result is harmonic in exterior space, satisfies the boundary conditions, and is regular at infinity like a Newtonian potential. Molodensky *et al.* (1962) supposed that the disturbing potential is expressible as

$$T = \iint_s \frac{\mu}{\ell} \, ds \qquad [39]$$

where μ is a surface density to be solved using the boundary condition. With the spherical approximation for the gravity anomaly, eqn [5], one arrives at the following integral equation

$$2\pi\mu \cos\zeta - \iint_s \left(\frac{\partial}{\partial r}\left(\frac{1}{\ell}\right) + \frac{2}{r\ell}\right)\mu \, ds = \Delta g \qquad [40]$$

The first term accounts for the discontinuity at the surface of the derivative of the potential of a density layer, where ζ is the deflection angle between the normal to the surface and the direction of the (radial) derivative (Heiskanen and Moritz, 1967, p. 6; Günter, 1967, p. 69). This Fredholm integral equation of the second kind can be simplified with further approximations, and, a solution for the density, μ, ultimately leads to the solution for the disturbing potential (Moritz, 1980).

Other forms of the initial representation have also been investigated, where Green's functions other than $1/\ell$ lead to simplifications of the integral equation (e.g., Petrovskaya, 1979). Nevertheless, most practical solutions rely on approximations, such as the spherical approximation for the boundary condition, and even the formulated solutions are not strictly guaranteed to converge to the true solutions (Moritz, 1980). Further treatments of the BVP in a geodetic/mathematical setting may be found in the volume edited by Sansò and Rummel (1997).

In the next section, we consider solutions for T as surface integrals of boundary values with appropriate Green's functions. In other words, the boundary values (whether of the first, second, or third kind) may be thought of as sources, and the consequent potential is again the sum (integral) of the product of a boundary value and an appropriate Green's function (i.e., a function that depends on both the source point and the computation point in some form of inverse distance in accordance with Newtonian potential theory). Such solutions are readily obtained if the boundary is a sphere.

2.4   Solutions to the Spherical BVP

This section develops two types of solutions to standard BVPs when the boundary is a sphere: the spherical harmonic series and an integral with a Green's function. All three types of problems are solved, but emphasis is put on the third BVP since gravity anomalies are the most prevalent boundary values (on land, at least). In addition, it is shown how the Green's function integrals can be inverted to obtain, for example, gravity anomalies from values of the potential, now considered as boundary values. Not all possible inverse relationships are given, but it should be clear at the end that, in principle, virtually any gravitational quantity can be obtained in space from any other quantity on the spherical boundary.

2.4.1   Spherical Harmonics and Green's Functions

For simple boundaries, Laplace's equation [24] is relatively easy to solve provided there is an appropriate coordinate system. For the Earth, the solutions commonly rely on approximating the boundary by a sphere. This case is described in detail and a more accurate

approximation based on an ellipsoid of revolution is briefly examined in Section 2.5.2 for the normal potential. In spherical polar coordinates, (θ, λ, r), the Laplacian operator is given by Hobson (1965, p. 9)

$$\nabla^2 = \frac{1}{r^2}\frac{\partial}{\partial r}\left(r^2\frac{\partial}{\partial r}\right) + \frac{1}{r^2}\frac{1}{\sin\theta}\frac{\partial}{\partial\theta}\left(\sin\theta\frac{\partial}{\partial\theta}\right)$$
$$+ \frac{1}{r^2\sin^2\theta}\frac{\partial^2}{\partial\lambda^2} \qquad [41]$$

A solution to $\nabla^2 V = 0$ in the space outside a sphere of radius, R, with center at the coordinate origin can be found by the method of separation of variables, whereby one postulates the form of the solution, V, as

$$V(\theta, \lambda, r) = f(\theta)g(\lambda)h(r) \qquad [42]$$

Substituting this and the Laplacian above into eqn [24], the multivariate partial differential equation separates into three univariate ordinary differential equations (Hobson, 1965, p. 9; Morse and Feshbach, 1953, p. 1264). Their solutions are well-known functions, for example,

$$V(\theta, \lambda, r) = P_{nm}(\cos\theta)\sin m\lambda\frac{1}{r^{n+1}} \qquad [43a]$$

or

$$V(\theta, \lambda, r) = P_{nm}(\cos\theta)\cos m\lambda\frac{1}{r^{n+1}} \qquad [43b]$$

where $P_{nm}(t)$ is the associated Legendre function of the first kind and n, m are integers such that $0 \leq m \leq n$, $n \geq 0$. Other solutions are also possible (e.g., $g(\lambda) = e^{a\lambda}$ ($a \in \mathbb{R}$) and $h(r) = r^n$), but only eqns [43] are consistent with the problem at hand: to find a real-valued Newtonian potential for the exterior space of the Earth (regular at infinity and 2π-periodic in longitude). The general solution is a linear combination of solutions of the forms given by eqns [43] for all possible integers, n and m, and can be written compactly as

$$V(\theta, \lambda, r) = \sum_{n=0}^{\infty}\sum_{m=-n}^{n}\left(\frac{R}{r}\right)^{n+1}v_{nm}\bar{Y}_{nm}(\theta, \lambda) \qquad [44]$$

where the $\bar{Y}_{nm}$ are surface spherical harmonic functions defined as

$$\bar{Y}_{nm}(\theta, \lambda) = \bar{P}_{n|m|}(\cos\theta)\begin{cases}\cos m\lambda, & m \geq 0 \\ \sin|m|\lambda, & m < 0\end{cases} \qquad [45]$$

and $\bar{P}_{nm}$ is a normalization of P_{nm} so that the orthogonality of the spherical harmonics is simply

$$\frac{1}{4\pi} \iint_\sigma \bar{Y}_{nm}(\theta, \lambda)\, \bar{Y}_{n'm'}(\theta, \lambda)\, \mathrm{d}\sigma$$

$$= \begin{cases} 1, & n = n' \quad \text{and} \quad m = m' \\ 0, & n \neq n' \quad \text{or} \quad m \neq m' \end{cases} \qquad [46]$$

and where $\sigma = \{(\theta, \lambda)|0 \leq \theta \leq \pi, 0 \leq \lambda \leq 2\pi\}$ represents the unit sphere, with $\mathrm{d}\sigma = \sin\theta\, \mathrm{d}\theta\, \mathrm{d}\lambda$. For a complete mathematical treatment of spherical harmonics, one may refer to Müller (1966). The bounding spherical radius, R, is introduced so that all the constant coefficients, v_{nm}, also known as Stokes constants, have identical units of measure. Applying the orthogonality to the general solution [44], these coefficients can be determined if the function, V, is known on the bounding sphere (boundary condition):

$$v_{nm} = \frac{1}{4\pi} \iint_\sigma V(\theta, \lambda, R)\, \bar{Y}_{nm}(\theta, \lambda)\, \mathrm{d}\sigma \qquad [47]$$

Equation [44] is known as a spherical harmonic expansion of V and with eqn [47] it represents a solution to the Dirichlet BVP if the boundary is a sphere. The solution thus exists and is unique in the sense that these boundary values generate no other potential. We will, however, find another equivalent form of the solution.

In a more formal mathematical setting, the solution [46] is an infinite linear combination of orthogonal basis functions (eigenfunctions) and the coefficients, v_{nm}, are the corresponding eigenvalues. One may also interpret the set of coefficients as the spectrum (Legendre spectrum) of the potential on the sphere of radius, R (analogous to the Fourier spectrum of a function on the plane or line). The integers, n, m, correspond to wave numbers, and are called degree (n) and order (m), respectively. The spherical harmonics are further classified as zonal ($m = 0$), meaning that the zeros of $\bar{Y}_{n0}$ divide the sphere into latitudinal zones; sectorial ($m = n$), where the zeros of $\bar{Y}_{nm}$ divide the sphere into

longitudinal sectors; and tesseral (the zeros of $\bar{Y}_{nm}$ tessellate the sphere) (**Figure 7**).

While the spherical harmonic series has its advantages in global representations and spectral interpretations of the field, a Green's function representation provides a more local characterization of the field. Changing a boundary value anywhere on the globe changes all coefficients, v_{nm}, according to eqn [47], which poses both a numerical challenge in applications, as well as in keeping a standard model up to date. However, since the Green's function essentially depends on the inverse distance (or higher power thereof), a remote change in boundary value generally does not appreciably affect the local determination of the field.

When the boundary is a sphere, the solutions to the BVPs using a Green's function are easily derived from the spherical harmonic series representation. Moreover, it is possible to derive additional integral relationships (with appropriate Green's functions) among all the derivatives of the potential. To formalize and simultaneously simplify these derivations, consider harmonic functions, f and h, where h depends only on θ and r, and function g, defined on the sphere of radius, R. Thus let

$$f(\theta, \lambda, r) = \sum_{n=0}^{\infty} \sum_{m=-n}^{n} \left(\frac{R}{r}\right)^{n+1} f_{nm} \bar{Y}_{nm}(\theta, \lambda) \qquad [48]$$

$$h(\theta, r) = \sum_{n=0}^{\infty} (2n+1) \left(\frac{R}{r}\right)^{n+1} h_n P_n(\cos\theta) \qquad [49]$$

$$g(\theta, \lambda, R) = \sum_{n=0}^{\infty} \sum_{m=-n}^{n} g_{nm} \bar{Y}_{nm}(\theta, \lambda) \qquad [50]$$

where $P_n(\cos\theta) = \bar{P}_{n0}(\cos\theta)/\sqrt{2n+1}$ is the nth degree Legendre polynomial. Constants f_{nm} and g_{nm} are the respective harmonic coefficients of f and g when these functions are restricted to the sphere of radius, R. Then, using the decomposition formula for Legendre polynomials,

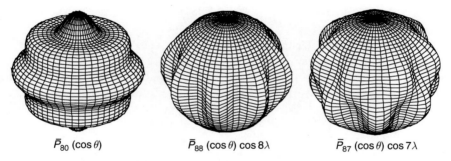

$$\bar{P}_{80}(\cos\theta) \qquad\qquad \bar{P}_{88}(\cos\theta)\cos 8\lambda \qquad\qquad \bar{P}_{87}(\cos\theta)\cos 7\lambda$$

Figure 7 Examples of zonal, sectorial, and tesseral harmonics on the sphere.

$$P_n(\cos\psi) = \frac{1}{2n+1} \sum_{m=-n}^{n} \bar{Y}_{nm}(\theta, \lambda)\,\bar{Y}_{nm}(\theta', \lambda') \qquad [51]$$

where

$$\cos\psi = \cos\theta\cos\theta' + \sin\theta\sin\theta'\cos(\lambda-\lambda') \qquad [52]$$

it is easy to prove the following theorem.

Theorem (convolution theorem in spectral analysis on the sphere).

$$f(\theta, \lambda, r) = \frac{1}{4\pi} \iint_\sigma g(\theta', \lambda', R)h(\psi, r)\mathrm{d}\sigma \qquad [53]$$
$$\text{if and only if} \quad f_{nm} = g_{nm}h_n$$

Here, and in the following, $\mathrm{d}\sigma = \sin\theta'\mathrm{d}\theta'\mathrm{d}\lambda'$. The angle, ψ, is the distance on the unit sphere between points (θ, λ) and (θ', λ').

Proof. The forward statement [53] follows directly by substituting eqns [51] and [49] into the first equation [53], together with the spherical harmonic expansion [50] for g. A comparison with the spherical harmonic expansion for f yields the result. All steps in this proof are reversible, and so the reverse statement also holds.

Consider now f to be the potential, V, outside the sphere of radius, R, and its restriction to the sphere to be the function, $g : g(\theta, \lambda) = V(\theta, \lambda, R)$. Then, clearly, $h_n = 1$, for all n, by the theorem above, we have

$$V(\theta, \lambda, r) = \frac{1}{4\pi} \iint_\sigma V(\theta', \lambda', R)U(\psi, r)\mathrm{d}\sigma \qquad [54]$$

where

$$U(\psi, r) = \sum_{n=0}^{\infty} (2n+1) \left(\frac{R}{r}\right)^{n+1} P_n(\cos\psi) \qquad [55]$$

For the distance

$$\ell = \sqrt{r^2 + R^2 - 2rR\cos\psi} \qquad [56]$$

between points (θ, λ, r) and (θ', λ', R), with $r \geq R$, the identity (the Coulomb expansion; Cushing, 1975, p. 155),

$$\frac{1}{\ell} = \frac{1}{R} \sum_{n=0}^{\infty} \left(\frac{R}{r}\right)^{n+1} P_n(\cos\psi) \qquad [57]$$

yields, after some arithmetic (based on taking the derivative on both sides with respect to r),

$$U(\psi, r) = \frac{R(r^2 - R^2)}{\ell^3} \qquad [58]$$

Solutions [44] and [54] to the Dirichlet BVP for a spherical boundary are identical (in view of the convolution theorem [53]). The integral in [54] is known as the Poisson integral and the function U is the corresponding Green's function, also known as Poisson's kernel.

For convenience, one separates Earth's gravitational potential into a reference potential (Section 2.1.3) and the disturbing potential, T. The disturbing potential is harmonic in free space and satisfies the Poisson integral if the boundary is a sphere. In deference to physical geodesy where relationships between the disturbing potential and its derivatives are routinely applied, the following derivations are developed in terms of T, but hold equally for any exterior Newtonian potential. Let

$$T(\theta, \lambda, r) = \frac{GM}{R} \sum_{n=0}^{\infty} \sum_{m=-n}^{n} \left(\frac{R}{r}\right)^{n+1} \delta C_{nm} \bar{Y}_{nm}(\theta, \lambda) \qquad [59]$$

where M is the total mass (including the atmosphere) of the Earth and the δC_{nm} are unitless harmonic coefficients, being also the difference between coefficients for the total and reference gravitational potentials (Section 2.5.2). The coefficient, δC_{00}, is zero under the assumption that the reference field accounts completely for the central part of the total field. Also note that these coefficients specifically refer to the sphere of radius, R.

The gravity disturbance is defined (in spherical approximation) to be the negative radial derivative of T, the first of eqns [5]. From eqn [59], we have

$$\delta g(\theta, \lambda, r) = -\frac{\partial}{\partial r} T(\theta, \lambda, r)$$
$$= \frac{GM}{R^2} \sum_{n=0}^{\infty} \sum_{m=-n}^{n} \left(\frac{R}{r}\right)^{n+2} (n+1)\delta C_{nm} \bar{Y}_{nm}(\theta, \lambda) \qquad [60]$$

and, applying the convolution theorem [53], we obtain

$$T(\theta, \lambda, r) = \frac{R}{4\pi} \iint_\sigma \delta g(\theta', \lambda', R)H(\psi, r)\mathrm{d}\sigma \qquad [61]$$

where with $g_{nm} = (n+1)\delta C_{nm}/R$ and $f_{nm} = \delta C_{nm}$, we have $h_n = f_{nm}/g_{nm} = R/(n+1)$, and hence (taking care to keep the Green's function unitless)

$$H(\psi, r) = \sum_{n=0}^{\infty} \frac{2n+1}{n+1} \left(\frac{R}{r}\right)^{n+1} P_n(\cos\psi) \qquad [62]$$

The integral in [61] is known as the Hotine integral, the Green's function, H, is called the Hotine kernel, and with a derivation based on equation [57], it is given by (Hotine 1969, p. 311)

$$H(\psi, r) = \frac{2R}{\ell} - \ln\left(1 + \frac{\ell}{2R\sin^2\psi/2}\right) \qquad [63]$$

Equation [61] solves the Neumann BVP when the boundary is a sphere.

The gravity anomaly (again, in spherical approximation) is defined by eqn [5]

$$\Delta g(\theta, \lambda, r) = \left(-\frac{\partial}{\partial r} - \frac{2}{r}\right) T(\theta, \lambda, r) \qquad [64]$$

or, also,

$$\Delta g(\theta, \lambda, r) = \frac{GM}{R^2} \sum_{n=0}^{\infty} \sum_{m=-n}^{n} \left(\frac{R}{r}\right)^{n+2} (n-1)\delta C_{nm} \bar{Y}_{nm}(\theta, \lambda)$$

$$[65]$$

In this case, we have $g_{nm} = (n-1)\delta C_{nm}/R$ and $h_n = R/(n-1)$. The convolution theorem in this case leads to the geodetically famous Stokes integral,

$$T(\theta, \lambda, r) = \frac{R}{4\pi} \iint_{\sigma} \Delta g(\theta', \lambda', R) S(\psi, r) \mathrm{d}\sigma \qquad [66]$$

where we define Green's function to be

$$S(\psi, r) = \sum_{n=2}^{\infty} \frac{2n+1}{n-1} \left(\frac{R}{r}\right)^{n+1} P_n(\cos \psi)$$
$$= 2\frac{R}{\ell} + \frac{R}{r} - 3\frac{R\ell}{r^2} - 5\frac{R^2}{r^2}\cos\psi$$
$$- 3\frac{R^2}{r^2}\cos\psi \ln\frac{\ell + r - R\cos\psi}{2r} \qquad [67]$$

more commonly called the Stokes kernel. Equation [66] solves the Robin BVP if the boundary is a sphere, but it includes specific constraints that ensure the solution's uniqueness – the solution by itself is not unique, in this case, as proved in Section 2.3.2. Indeed, eqn [65] shows that the gravity anomaly has no first-degree harmonics for the disturbing potential; therefore, they cannot be determined from the boundary values. Conventionally, the Stokes kernel also excludes the zero-degree harmonic, and thus the complete solution for the disturbing potential is given by

$$T(\theta, \lambda, r) = \frac{GM}{r}\delta C_{00}$$
$$+ \frac{GM}{R} \sum_{m=-1}^{1} \left(\frac{R}{r}\right)^2 \delta C_{1m} \bar{Y}_{1m}(\theta, \lambda)$$
$$+ \frac{R}{4\pi} \iint_{\sigma} \Delta g(\theta', \lambda', R) S(\psi, r) \mathrm{d}\sigma \qquad [68]$$

The central term, δC_{00}, is proportional to the difference in GM of the Earth and reference ellipsoid and is zero to high accuracy. The first-degree harmonic coefficients, δC_{1m}, are proportional to the center-of-mass coordinates and can also be set to zero with appropriate definition of the coordinate system (see Section 2.5.1). Thus, the Stokes integral [66] is the

more common expression for the disturbing potential, but it embodies hidden constraints.

We note that gravity anomalies also serve as boundary values in the harmonic series form of the solution for the disturbing potential. Applying the orthogonality of the spherical harmonics to eqn [65] yields immediately

$$\delta C_{nm} = \frac{R^2}{4\pi(n-1)GM} \iint_{\sigma} \Delta g(\theta', \lambda', R) \bar{Y}_{nm}(\theta', \lambda')\mathrm{d}\sigma,$$
$$n \geq 2 \qquad [69]$$

A similar formula holds when gravity disturbances are the boundary values ($n-1$ in the denominator changes to $n+1$). In either case, the boundary values formally are assumed to reside on a sphere of radius, R. An approximation results if they are given on the geoid, as is usually the case.

2.4.2 Inverse Stokes and Hotine Integrals

The convolution integrals above can easily be inverted by considering again the spectral relationships. For the gravity anomaly, we note that $f = r\Delta g$ is harmonic with coefficients, $f_{nm} = GM(n-1)\delta C_{nm}/R$. Letting $g_{nm} = GM\, \delta C_{nm}/R$, we find that $h_n = n-1$; from the convolution theorem, we can write

$$\Delta g(\theta, \lambda, r) = \frac{1}{4\pi R} \iint_{\sigma} T(\theta', \lambda', R) \hat{Z}(\psi, r) \mathrm{d}\sigma \qquad [70]$$

where

$$\hat{Z}(\psi, r) = \sum_{n=0}^{\infty} (2n+1)(n-1)\left(\frac{R}{r}\right)^{n+2} P_n(\cos\psi) \qquad [71]$$

The zero- and first-degree terms are included provisionally. Note that

$$\hat{Z}(\psi, r) = -R\left(\frac{\partial}{\partial r} + \frac{2}{r}\right) U(\psi, r) \qquad [72]$$

that is, we could have simply used the Dirichlet solution [54] to obtain the gravity anomaly, as given by [70], from the disturbing potential. It is convenient to separate the kernel function as follows:

$$\hat{Z}(\psi, r) = Z(\psi, r) - \sum_{n=0}^{\infty} (2n+1)\left(\frac{R}{r}\right)^{n+2} P_n(\cos\psi) \quad [73]$$

where

$$Z(\psi, r) = \sum_{n=1}^{\infty} (2n+1)n\left(\frac{R}{r}\right)^{n+2} P_n(\cos\psi) \qquad [74]$$

We find that

$$\Delta g(\theta, \lambda, r) = -\frac{1}{r} T(\theta, \lambda, r)$$
$$+ \frac{1}{4\pi R} \iint_\sigma T(\theta', \lambda', R) Z(\psi, r) \, d\sigma \quad [75]$$

Now, since Z has no zero-degree harmonics, its integral over the sphere vanishes, and one can write the numerically more convenient formula:

$$\Delta g(\theta, \lambda, r) = -\frac{1}{r} T(\theta, \lambda, r) + \frac{1}{4\pi R} \iint_\sigma (T(\theta', \lambda', R)$$
$$- T(\theta, \lambda, R)) Z(\psi, r) \, d\sigma \quad [76]$$

This is the inverse Stokes formula. Given T on the sphere of radius R (e.g., in the form of geoid undulations, $T = \gamma N$), this form is useful when the gravity anomaly is also desired on this sphere. It is one way to determine gravity anomalies on the ocean surface from satellite altimetry, where the ocean surface is approximated as a sphere. Analogously, from eqns [60] and [64], it is readily seen that the inverse Hotine formula is given by

$$\delta g(\theta, \lambda, r) = \frac{1}{r} T(\theta, \lambda, r) + \frac{1}{4\pi R} \iint_\sigma (T(\theta', \lambda', R)$$
$$- T(\theta, \lambda, R)) Z(\psi, r) \, d\sigma \quad [77]$$

Note that the difference of eqns [76] and [77] yields the approximate relationship between the gravity disturbance and the gravity anomaly inferred from eqns [5].

Finally, we realize that, for $r = R$, the series for $Z(\psi, R)$ is not uniformly convergent and special numerical procedures (that are outside the present scope) are required to approximate the corresponding integrals.

2.4.3 Vening-Meinesz Integral and Its Inverse

Other derivatives of the disturbing potential may also be determined from boundary values. We consider here only gravity anomalies, being the most prevalent data type on land areas. The solution is either in the form of a series – simply the derivative of the series [59] with coefficients given by eqn [69], or an integral with appropriate derivative of the Green's function. The horizontal derivatives of the disturbing potential are often interpreted as the deflections of the vertical to which they are proportional in spherical approximation (eqn [5]):

$$\left\{ \begin{array}{c} \xi(\theta, \lambda, r) \\ \eta(\theta, \lambda, r) \end{array} \right\} = \frac{1}{\gamma(\theta, r)} \left\{ \begin{array}{c} \frac{1}{r} \frac{\partial}{\partial \theta} \\ -\frac{1}{r \sin\theta} \frac{\partial}{\partial \lambda} \end{array} \right\} T(\theta, \lambda, r) \quad [78]$$

where ξ, η are the north and east deflection components, respectively, and γ is the normal gravity. Clearly, the derivatives can be taken directly inside the Stokes integral, and we find

$$\left\{ \begin{array}{c} \xi(\theta, \lambda, r) \\ \eta(\theta, \lambda, r) \end{array} \right\} = \frac{R}{4\pi r \, \gamma(\theta, r)} \iint_\sigma \Delta g(\theta', \lambda', R)$$
$$\times \frac{\partial}{\partial \psi} S(\psi, r) \left\{ \begin{array}{c} \cos\alpha \\ \sin\alpha \end{array} \right\} d\sigma \quad [79]$$

where

$$\left\{ \begin{array}{c} \frac{1}{r} \frac{\partial}{\partial \theta} \\ -\frac{1}{r \sin\theta} \frac{\partial}{\partial \lambda} \end{array} \right\} = \frac{1}{r} \left\{ \begin{array}{c} \frac{\partial \psi}{\partial \theta} \\ -\frac{1}{\sin\theta} \frac{\partial \psi}{\partial \lambda} \end{array} \right\} \frac{\partial}{\partial \psi} \quad [80]$$

and

$$\frac{\partial \psi}{\partial \theta} = \frac{1}{\sin\psi} (\sin\theta \cos\theta' - \cos\theta \sin\theta' \cos(\lambda' - \lambda)) = \cos\alpha$$
$$-\frac{1}{\sin\theta} \frac{\partial \psi}{\partial \lambda} = \frac{1}{\sin\psi} \sin\theta' \sin(\lambda' - \lambda) = \sin\alpha \quad [81]$$

The angle, α, is the azimuth of (θ', λ') at (θ, λ) on the unit sphere. The integrals [79] are known as the Vening-Meinesz integrals. Analogous integrals for the deflections arise when the boundary values are the gravity disturbances (the Green's functions are then derivatives of the Hotine kernel).

For the inverse Vening-Meinesz integrals, we need to make use of Green's first identity for surface functions, f and g.

$$\iint_s f \Delta^*(g) \, ds + \iint_s \nabla f \cdot \nabla g \, ds = \int_b f \, \nabla g \cdot \mathbf{n} \, db \quad [82]$$

where b is the boundary (a line) of surface, s; ∇ and Δ^* are the gradient and Laplace–Beltrami operators, which for the spherical surface are given by

$$\nabla = \left(\begin{array}{c} \frac{\partial}{\partial \theta} \\ \frac{1}{\sin\theta} \frac{\partial}{\partial \lambda} \end{array} \right), \quad \Delta^* = \frac{\partial^2}{\partial \theta^2} + \cot\theta \frac{\partial}{\partial \theta} + \frac{1}{\sin^2\theta} \frac{\partial^2}{\partial \lambda^2} \quad [83]$$

and where **n** is the unit vector normal to b. For a closed surface such as the sphere, the line integral vanishes, and we have

$$\iint_\sigma f \Delta^*(g)\, \mathrm{d}\sigma = - \iint_\sigma \nabla f \cdot \nabla g\ \mathrm{d}\sigma \qquad [84]$$

The surface spherical harmonics, $\bar{Y}_{nm}(\theta, \lambda)$, satisfy the following differential equation:

$$\Delta^* \bar{Y}_{nm}(\theta, \lambda) + n(n+1)\bar{Y}_{nm}(\theta, \lambda) = 0 \qquad [85]$$

Therefore, the harmonic coefficients of $\Delta^* T$ (θ, λ, r) on the sphere of radius, R, are

$$[\Delta^* T(\theta, \lambda, r)]_{nm} = -n(n+1)\frac{GM}{R}\delta C_{nm} \qquad [86]$$

Hence, by the convolution theorem (again, considering the harmonic function, $f = r\Delta g$),

$$\Delta g(\theta, \lambda, r) = -\frac{1}{4\pi R}\iint_\sigma \Delta^* T(\theta', \lambda', R) W(\psi, r)\, \mathrm{d}\sigma \qquad [87]$$

where

$$W(\psi, r) = \sum_{n=2}^{\infty}\frac{(2n+1)(n-1)}{n(n+1)}\left(\frac{R}{r}\right)^{n+2} P_n(\cos\psi) \qquad [88]$$

and where the zero-degree term of the gravity anomaly must be treated separately (e.g., it is set to zero in this case). Using Green's identity [84] and eqns [80] and [81], we have

$$\Delta g(\theta, \lambda, r) = \frac{1}{4\pi R}\iint_\sigma \nabla T(\theta', \lambda', R) \cdot \nabla W(\psi, r)\, \mathrm{d}\sigma$$
$$= \frac{R\gamma_0}{4\pi R}\iint_\sigma (\xi(\theta', \lambda', R)\cos\alpha$$
$$+ \eta(\theta', \lambda', R)\sin\alpha)\frac{\partial}{\partial\psi}W(\psi, r)\, \mathrm{d}\sigma \qquad [89]$$

where normal gravity on the sphere of radius, R, is approximated as a constant: $\gamma(\theta, R) \approx \gamma_0$. Equation [89] represents a second way to compute gravity anomalies from satellite altimetry, where the along-track and cross-track altimetric differences are used to approximate the deflection components (with appropriate rotation to north and east components). Employing differences in altimetric measurements benefits the estimation since systematic errors, such as orbit error, cancel out. To speed up the computations, the problem is reformulated in the spectral domain (see, e.g., Sandwell and Smith, 1996).

Clearly, following the same procedure for $f = T$, we also have the following relationship:

$$T(\theta, \lambda, r) = \frac{R\gamma_0}{4\pi}\iint_\sigma (\xi(\theta', \lambda', R)\cos\alpha$$
$$+ \eta(\theta', \lambda', R)\sin\alpha)\frac{\partial}{\partial\psi}B(\psi, r)\, \mathrm{d}\sigma \qquad [90]$$

where

$$B(\psi, r) = \sum_{n=2}^{\infty}\frac{(2n+1)}{n(n+1)}\left(\frac{R}{r}\right)^{n+1} P_n(\cos\psi) \qquad [91]$$

It is interesting to note that instead of an integral over the sphere, the inverse relationship between the disturbing potential on the sphere and the deflection of the vertical on the same sphere is also more straightforward in terms of a line integral:

$$T(\theta, \lambda, R) = T(\theta_0, \lambda_0, R) + \frac{\gamma_0}{4\pi}\int_{(\theta_0, \lambda_0)}^{(\theta, \lambda)} (\xi(\theta', \lambda', R)\, \mathrm{d}s_\theta$$
$$- \eta(\theta', \lambda', R)\, \mathrm{d}s_\lambda) \qquad [92]$$

where

$$\mathrm{d}s_\theta = R\,\mathrm{d}\theta, \quad \mathrm{d}s_\lambda = R\sin\theta\ \mathrm{d}\lambda \qquad [93]$$

2.4.4 Concluding Remarks

The spherical harmonic series, [59], represents the general solution to the exterior potential, regardless of the way the coefficients are determined. We know how to compute those coefficients exactly on the basis of a BVP, if the boundary is a sphere. More complicated boundaries would require corrections or, if these are omitted, would imply an approximation. If the coefficients are determined accurately (e.g., from satellite observations (Section 2.6.1), but not according to eqn [69]), then the spherical harmonic series model for the potential is not a spherical approximation. The spherical approximation enters when approximate relations such as eqns [5] are used and when the boundary is approximated as a sphere. However determined, the infinite series converges uniformly for all $r > R_c$, where R_c is the radius of the sphere that encloses all terrestrial masses. It may also converge below this sphere, but would represent the true potential only in free space (above the Earth's surface, where Laplace's equation holds). In practice, though, convergence is not an issue since the series must be truncated at some finite degree. Any trend toward divergence is then part of the overall model error, due mostly to truncation.

Model errors exist in all the Green's function integrals as they depend on spherical approximations in the boundary condition. In addition, the surface of integration in these formulas is formally assumed to be the geoid (where normal derivatives of the potential coincide with gravity magnitude), but it is approximated as a sphere. The spherical approximation results, in the first place, from a neglect of the ellipsoid flattening, which is about 0.3% for the Earth. When working with the disturbing potential, this level of error was easily tolerated in the past (e.g., if the geoid undulation is 30 m, the spherical approximation accounts for about 10 cm error), but today it requires attention as geoid undulation accuracy of 1 cm is pursued.

The Green's functions all have singularities when the evaluation point is on the sphere of radius, R. For points above this sphere, it is easily verified that all the Legendre series for the Green's functions converge uniformly, since $|P_n(\cos\psi)| \leq 1$. When $r = R$, the corresponding singularities of the integrals are either weak (Stokes integral) or strong (e.g., Poisson integral), requiring special definition of the integral as Cauchy principal value.

Finally, it is noted that the BVP solutions also require that no masses reside above the geoid (the boundary approximated as a sphere). To satisfy this condition, the topographic masses must be redistributed appropriately by mathematical reduction and the gravity anomalies, or disturbances measured on the Earth's surface must be reduced to the geoid. The mass redistribution must then be undone (mathematically) in order to obtain the correct potential on or above the geoid. Details of these procedures are found in Heiskanen and Moritz (1967, chapter 3). In addition, the atmosphere, having significant mass, affects gravity anomalies at different elevations. These effects may also be removed prior to using them as boundary values in the integral formulas.

2.5 Low-Degree Harmonics: Interpretation and Reference

The low-degree spherical harmonics of the Earth's gravitational potential lend themselves to interpretation with respect to the most elemental distribution of the Earth's density, which also leads to fundamental geometric characterizations, particularly for the second-degree harmonics. Let

$$C_{nm}^{(a)} = \frac{a}{GM}\left(\frac{R}{a}\right)^{n+1} v_{nm} \qquad [94]$$

be unitless coefficients that refer to a sphere of radius, a. (Recall that coefficients, v_{nm}, eqn [44], refer to a sphere of radius, R.) Relative to the central harmonic coefficient, $C_{00}^{(a)} = 1$, the next significant harmonic, $C_{20}^{(a)}$, is more than 3 orders of magnitude smaller; the remaining harmonic coefficients are at least 2–3 orders of magnitude smaller than that. The attenuation after degree 2 is much more gradual (**Table 1**), indicating that the bulk of the potential can be described by an ellipsoidal field. The normal gravitational field is such a field, but it also adheres to a geodetic definition that requires the underlying ellipsoid to be an equipotential surface in the corresponding normal gravity field. This section examines the low-degree harmonics from these two perspectives of interpretation and reference.

2.5.1 Low-Degree Harmonics as Density Moments

Returning to the general expression for the gravitational potential in terms of the Newtonian density integral (eqn [14]), and substituting the spherical harmonic series for the reciprocal distance (eqn [57] with [51]),

Table 1 Spherical harmonic coefficients of the total gravitational potential[a]

Degree (n)	Order, (m)	$C_{nm}^{(a)}$	$C_{n,-m}^{(a)}$
2	0	−4.841 70E − 04	0.0
2	1	−2.398 32E − 10	1.424 89E − 09
2	2	2.439 32E − 06	−1.400 28E − 06
3	0	9.571 89E − 07	0.0
3	1	2.030 48E − 06	2.481 72E − 07
3	2	9.048 02E − 07	−6.190 06E − 07
3	3	7.212 94E − 07	1.414 37E − 06
4	0	5.399 92E − 07	0.0
4	1	−5.361 67E − 07	−4.735 73E − 07
4	2	3.505 12E − 06	6.624 45E − 07
4	3	9.908 68E − 07	−2.009 76E − 07
4	4	−1.884 72E − 07	3.088 27E − 07

[a]GRACE model GGM02S (Tapley *et al.*, 2005).

$$V(\theta, \lambda, r) = G \iiint_{\text{volume}} \rho \frac{1}{r'} \sum_{n=0}^{\infty} \left(\frac{r'}{r}\right)^{n+1}$$

$$\times \left(\frac{1}{2n+1} \sum_{m=-n}^{n} \bar{Y}_{nm}(\theta, \lambda) \bar{Y}_{nm}(\theta', \lambda')\right) dv$$

$$= \sum_{n=0}^{\infty} \sum_{m=-n}^{n} \left(\frac{R}{r}\right)^{n+1} \left(\frac{G}{R^{n+1}(2n+1)}\right.$$

$$\times \iiint_{\text{volume}} \rho(r')^n \bar{Y}_{nm}(\theta', \lambda') dv\right)$$

$$\times \bar{Y}_{nm}(\theta, \lambda) \qquad [95]$$

yields a multipole expansion (so called from electrostatics) of the potential. The spherical harmonic (Stokes) coefficients are multipoles of the density distribution (cf. eqn [44]),

$$v_{nm} = \frac{G}{R^{n+1}(2n+1)} \iiint_v \rho(r')^n \bar{Y}_{nm}(\theta', \lambda') dv \qquad [96]$$

One may also consider the nth-order moments of density (from the statistics of distributions) defined by

$$\mu_{\alpha\beta\gamma}^{(n)} = \iiint_v (x')^{\alpha}(y')^{\beta}(z')^{\gamma} \rho \, dv, \quad n = \alpha + \beta + \gamma \qquad [97]$$

The multipoles of degree n and the moments of order n are related, though not all $(n+1)(n+2)/2$ moments of order n can be determined from the $2n+1$ multipoles of degree n, when $n \geq 2$ (clearly risking confusion, we defer to the common nomenclature of order for moments and degree for spherical harmonics). This indeterminacy is directly connected to the inability to determine the density distribution uniquely from external measurements of the potential (Chao, 2005), which is the classic geophysical inverse problem.

The zero-degree Stokes coefficient is coordinate invariant and is proportional to the total mass of the Earth:

$$v_{00} = \frac{G}{R} \iiint_{\text{volume}} \rho \, dv = \frac{GM}{R} \qquad [98]$$

It also represents a mass monopole, and it is proportional to the zeroth moment of the density, M.

The first-degree harmonic coefficients (representing dipoles) are proportional to the coordinates of the center of mass, (x_{cm}, y_{cm}, z_{cm}), which are proportional to the first-order moments of the density, as verified by recalling the definition of the first-degree spherical harmonics:

$$v_{1m} = \frac{G}{\sqrt{3}R^2} \iiint_{\text{volume}} \rho \begin{cases} r'\sin\theta'\sin\lambda', & m = -1 \\ r'\cos\theta', & m = 0 \\ r'\sin\theta'\cos\lambda', & m = 1 \end{cases} dv$$

$$= \frac{GM}{\sqrt{3}R^2} \begin{cases} y_{cm}, & m = -1 \\ z_{cm}, & m = 0 \\ x_m, & m = 1 \end{cases} \qquad [99]$$

Nowadays, by tracking satellites, we have access to the center of mass of the Earth (including its atmosphere), since it defines the center of their orbits. Ignoring the small motion of the center of mass (annual amplitude of several millimeters) due to the temporal variations in the mass distribution, we may choose the coordinate origin for the geopotential model to coincide with the center of mass, thus annihilating the first-degree coefficients.

The second-order density moments likewise are related to the second-degree harmonic coefficients (quadrupoles). They also define the inertia tensor of the body. The inertia tensor is the proportionality factor in the equation that relates the angular momentum vector, **H**, and the angular velocity, $\boldsymbol{\omega}$, of a body, like the Earth:

$$\mathbf{H} = I\omega \qquad [100]$$

and is given by

$$I = \begin{pmatrix} I_{xx} & I_{xy} & I_{xz} \\ I_{yx} & I_{yy} & I_{yz} \\ I_{zx} & I_{zy} & I_{zz} \end{pmatrix} \qquad [101]$$

It comprises the moments of inertia on the diagonal

$$I_{xx} = \iiint_{\text{volume}} \rho(y'^2 + z'^2) \, dv,$$

$$I_{yy} = \iiint_{\text{volume}} \rho(z'^2 + x'^2) \, dv, \qquad [102]$$

$$I_{zz} = \iiint_{\text{volume}} \rho(x'^2 + y'^2) \, dv$$

and the products of inertia off the diagonal

$$I_{xy} = I_{yx} = -\iiint_{\text{volume}} \rho \, x'y' \, dv,$$

$$I_{xz} = I_{zx} = -\iiint_{\text{volume}} \rho \, x'z' \, dv, \qquad [103]$$

$$I_{yz} = I_{zy} = -\iiint_{\text{volume}} \rho \, y'z' \, dv$$

Note that

$$I_{xx} = \mu_{020}^{(2)} + \mu_{002}^{(2)}, \quad I_{xy} = -\mu_{110}^{(2)}, \quad \text{etc} \quad [104]$$

and there are as many (six) independent tensor components as second-order density moments. Using the explicit expressions for the second-degree spherical harmonics, we have from eqn [96] with $n = 2$:

$$v_{2,-2} = -\frac{\sqrt{15}\,G}{5R^3}I_{xy}, \quad v_{2,-1} = -\frac{\sqrt{15}\,G}{5R^3}I_{yz}$$

$$v_{2,1} = -\frac{\sqrt{15}\,G}{5R^3}I_{xz} \quad v_{2,0} = \frac{\sqrt{5}\,G}{10R^3}\left(I_{xx} + I_{yy} - 2I_{zz}\right) \quad [105]$$

$$v_{2,2} = \frac{\sqrt{15}\,G}{10R^3}\left(I_{yy} - I_{xx}\right)$$

These are also known as MacCullagh's formulas. Not all density moments (or, moments of inertia) can be determined from the Stokes coefficients.

If the coordinate axes are chosen so as to diagonalize the inertia tensor (products of inertia are then equal to zero), then they are known as principal axes of inertia, or also 'figure' axes. For the Earth, the z-figure axis is very close to the spin axis (within several meters at the pole); both axes move with respect to each other and the Earth's surface, with combinations of various periods (daily, monthly, annually, etc.), as well as secularly in a wandering fashion. Because of these motions, the figure axis is not useful as a coordinate axis that defines a frame fixed to the (surface of the) Earth. However, because of the proximity of the figure axis to the defined reference z-axis, the second-degree, first-order harmonic coefficients of the geopotential are relatively small (**Table 1**).

The arbitrary choice of the x-axis of our Earth-fixed reference coordinate system certainly did not attempt to eliminate the product of inertia, I_{xy} (the x-axis is defined by the intersection of the Greenwich meridian with the equator, and the y-axis completes a right-handed mutually orthogonal triad). However, it is possible to determine where the x-figure axis is located by combining values of the second-degree, second-order harmonic coefficients. Let u, v, w be the axes that define a coordinate system in which the inertia tensor is diagonal, and assume that $I_{ww} = I_{zz}$. A rotation by the angle, $-\lambda_0$, about the w- (also z-) figure axis brings this ideal coordinate system back to the conventional one in which we calculate the harmonic coefficients. Tensors transform under rotation, defined by matrix, $\mathcal{R}$, according to

$$I_{xyz} = \mathcal{R}I_{uvw}\mathcal{R}^{\mathrm{T}} \quad [106]$$

With the rotation about the w-axis given by the matrix,

$$\mathcal{R} = \begin{pmatrix} \cos\lambda_0 & -\sin\lambda_0 & 0 \\ \sin\lambda_0 & \cos\lambda_0 & 0 \\ 0 & 0 & 1 \end{pmatrix} \quad [107]$$

and with eqns [105], it is straightforward to show that

$$v_{2,-2} = -\frac{\sqrt{15}G}{10R^3}\left(I_{uu} - I_{vv}\right)\sin 2\lambda_0$$

$$v_{2,2} = -\frac{\sqrt{15}G}{10R^3}\left(I_{uu} - I_{vv}\right)\cos 2\lambda_0 \quad [108]$$

Hence, we have

$$\lambda_0 = \frac{1}{2}\tan^{-1}\frac{v_{2,-2}}{v_{2,2}} \quad [109]$$

where the quadrant is determined by the signs of the harmonic coefficients. From **Table 1**, we find that $\lambda_0 = -14.929°$; that is, the u-figure axis is in the mid-Atlantic between South America and Africa.

The second-degree, second-order harmonic coefficient, $v_{2,2}$, indicates the asymmetry of the Earth's mass distribution with respect to the equator. Since $v_{2,2} > 0$ (for the Earth), equations [108] show that $I_{vv} > I_{uu}$ and thus the equator 'bulges' more in the direction of the u-figure axis; conversely, the equator is flattened in the direction of the v-figure axis. This flattening is relatively small: 1.1×10^{-5}.

Finally, consider the most important second-degree harmonic coefficient, the second zonal harmonic, $v_{2,0}$. Irrespective of the x-axis definition, it is proportional to the difference between the moment of inertia, I_{zz}, and the average of the equatorial moments, $(I_{xx} + I_{yy})/2$. Again, since $v_{2,0} < 0$, the Earth bulges more around the equator and is flattened at the poles. The second zonal harmonic coefficient is roughly 1000 times larger than the other second-degree coefficients and thus indicates a substantial polar flattening (owing to the Earth's early more-fluid state). This flattening is approximately 0.003.

2.5.2 Normal Ellipsoidal Field

Because of Earth's dominant polar flattening and the near symmetry of the equator, any meridional section of the Earth is closer to an ellipse than a circle. For this reason, ellipsoidal coordinates have often been advocated in place of the usual spherical coordinates. In fact, for geodetic positioning and geographic mapping, because of this flattening, the conventional (geodetic) latitude and longitude are coordinates that define the direction of the perpendicular to an ellipsoid, not the radial direction from the origin. These geodetic coordinates are not the ellipsoidal coordinates defined in Section 2.1 (eqn [2]) and would be rather useless in potential modeling because they do not separate Laplace' differential equation.

Harmonic series in terms of the ellipsoidal coordinates, (δ, λ, u), however, can be developed easily. They have not been adopted in most applications, perhaps in part because of the nonintuitive nature of the coordinates. Nevertheless, it is advantageous to model the normal (or reference) gravity field in terms of these ellipsoidal coordinates since it is based on an ellipsoid. Laplace' equation in these ellipsoidal coordinates can be separated, analogous to spherical coordinates, and the solution is obtained by successively solving three ordinary differential equations. Applied to the exterior gravitational potential, V, the solution is similar to the spherical harmonic series (eqn [44]) and is given by

$$V(\delta, \lambda, u) = \sum_{n=0}^{\infty} \sum_{m=-n}^{n} \frac{Q_{n|m|}(iu/E)}{Q_{n|m|}(ib/E)} v_{nm}^e \bar{Y}_{nm}(\delta, \lambda) \quad [110]$$

where E is the linear eccentricity associated with the coordinate system and Q_{nm} is the associated Legendre function of the second kind. The coefficients of the series, v_{nm}^e, refer to an ellipsoid of semiminor axis, b, and, with the series written in this way, they are all real numbers with the same units as V. An exact relationship between these and the spherical harmonic coefficients, v_{nm}, was given by Hotine (1969) and Jekeli (1988).

With this formulation of the potential, Dirichlet's BVP is solved for an ellipsoidal boundary using the orthogonality of the spherical harmonics:

$$v_{nm}^e = \frac{1}{4\pi} \iint_\sigma V(\delta, \lambda, b) \bar{Y}_{nm}(\delta, \lambda) \, d\sigma \quad [111]$$

where $d\sigma = \sin\delta \, d\delta \, d\lambda$ and $\sigma = \{(\delta, \lambda) | 0 \leq \delta \leq \pi, 0 \leq \lambda \leq 2\pi\}$. Note that while the limits of integration and the differential element, $d\sigma$, are the same as for the unit sphere, the boundary values are on the ellipsoid. Unfortunately, integral solutions with analytic forms of a Green's function do not exist in this case, because the inverse distance now depends on two surface coordinates and there is no corresponding convolution theorem. However, approximations have been formulated for all three types of BVPs (see Yu *et al.*, 2002, and references therein). Forms of ellipsoidal corrections to the classic spherical

integrals have also been developed and applied in practice (e.g., Fei and Sideris, 2000).

The simplicity of the boundary values of the normal gravitational potential allows its extension into exterior space to be expressed in closed analytic form. Analogous to the geoid in the actual gravity field, the normal ellipsoid is defined to be a level surface in the normal gravity field. In other words, the sum of the normal gravitational potential and the centrifugal potential due to Earth's rotation is a constant on the ellipsoid:

$$V^e(\delta, \lambda, b) + \phi(\delta, b) = U_0 \quad [112]$$

Hence, the normal gravitational potential on the ellipsoid, $V^e(\delta, \lambda, b)$, depends only on latitude and is symmetric with respect to the equator. Consequently, it consists of only even zonal harmonics, and because the centrifugal potential has only zero- and second-degree zonals, the corresponding ellipsoidal series is finite (up to degree 2). The solution to this Dirichlet problem is given in ellipsoidal coordinates by

$$V^e(\delta, \lambda, u) = \frac{GM}{E} \tan^{-1}\frac{E}{u} + \frac{1}{2}\omega_e^2 a^2 \frac{q}{q_0}\left(\cos^2\delta - \frac{1}{3}\right) \quad [113]$$

where a is the semimajor axis of the ellipsoid, ω_e is Earth's rate of rotation, and

$$q = \frac{1}{2}\left(\left(1 + 3\frac{u^2}{E^2}\right)\tan^{-1}\frac{E}{u} - 3\frac{u}{E}\right), \quad q_0 = q|_{u=b} \quad [114]$$

Heiskanen and Moritz (1967) and Hofmann-Wellenhof and Moritz (2005) provide details of the straightforward derivation of these and the following expressions.

The equivalent form of V^e in spherical harmonics is given by

$$V^e(\theta, \lambda, r) = \frac{GM}{r}\left(1 - \sum_{n=1}^{\infty} \mathcal{J}_{2n}\left(\frac{a}{r}\right)^{2n} P_{2n}(\cos\theta)\right) \quad [115]$$

where

$$\mathcal{J}_{2n} = (-1)^{n+1}\frac{3e^{2n}}{(2n+1)(2n+3)}\left(1 - n + \frac{5n}{e^2}\mathcal{J}_2\right), \quad n \geq 1 \quad [116]$$

and $e = E/a$ is the first eccentricity of the ellipsoid. The second zonal coefficient is given by

$$\mathcal{J}_2 = \frac{e^2}{3}\left(1 - \frac{2}{15}\frac{\omega_e^2 a^2 E}{q_0 \, GM}\right) = \frac{I_{zz}^e - I_{xx}^e}{Ma^2} \quad [117]$$

where the second equality comes directly from the last of eqns [105] and the ellipsoid's rotational symmetry ($I_{xx}^e = I_{yy}^e$). This equation also provides a direct relationship between the geometry (the eccentricity or flattening) and the mass distribution (difference of second-order moments) of the ellipsoid. Therefore, $\mathcal{J}_2$ is also known as the dynamic form factor – the flattening of the ellipsoid can be described either geometrically or dynamically in terms of a difference in density moments.

The normal gravitational potential depends solely on four adopted parameters: Earth's rotation rate, ω_e; the size and shape of the normal ellipsoid, for example, a, $\mathcal{J}_2$; and a potential scale, for example, GM. The mean Earth ellipsoid is the normal ellipsoid with parameters closest to actual corresponding parameters for the Earth. GM and $\mathcal{J}_2$ are determined by observing satellite orbits, a can be calculated by fitting the ellipsoid to mean sea level using satellite altimetry, and Earth's rotation rate comes from astronomical observations. **Table 2** gives current best values (Groten, 2004) and adopted constants for the Geodetic Reference Systems of 1967 and 1980 (GRS67, GRS80) and the World Geodetic System 1984 (WGS84).

In modeling the disturbing potential in terms of spherical harmonics, one naturally uses the form of the normal gravitational potential given by eqn [115]. Here, we have assumed that all harmonic coefficients refer to a sphere of radius, a. Corresponding coefficients for the series of $T = V - V^e$ are, therefore,

$$\delta C_{nm}^{(a)} = \begin{cases} 0, & n = 0 \\ C_{n0}^{(a)} - \dfrac{-\mathcal{J}_n}{\sqrt{2n+1}}, & n = 2, 4, 6, \ldots \\ C_{nm}^{(a)}, & \text{otherwise} \end{cases} \quad [118]$$

For coefficients referring to the sphere of radius, R, as in eqn [59], we have $\delta C_{nm} = (a/R)^n \delta C_{nm}^{(a)}$. The harmonic coefficients, $\mathcal{J}_{2n}$, attenuate rapidly due to

the factor, e^{2n}, and only harmonics up to degree 10 are significant. Normal gravity, being the gradient of the normal gravity potential, is used only in applications tied to an Earth-fixed coordinate system.

2.6 Methods of Determination

In this section, we briefly explore the basic technologies that yield measurements of the gravitational field. Even though we have reduced the problem of determining the exterior potential from a volume integral to a surface integral (e.g., either eqns [66] or [69]), it is clear that in theory we can never determine the entire field from a finite number of measurements. The integrals will always need to be approximated numerically, and/or the infinite series of spherical harmonics needs to be truncated. However, with enough effort and within the limits of computational capabilities, one can approach the ideal continuum of boundary values as closely as desired, or make the number of coefficients in the series representation as large as possible. The expended computational and measurement effort has to be balanced with the ability to account for inherent model errors (such as the spherical approximation) and the noise of the measuring device. To be useful for geodetic and geodynamic purposes, the instruments must possess a sensitivity of at least a few parts per million, and, in fact, many have a sensitivity of parts per billion. These sensitivities often come at the expense of prolonging the measurements (integration time) in order to average out random noise, thus reducing the achievable temporal resolution. This is particularly critical for moving-base instrumentation such as on an aircraft or satellite where temporal resolution translates into spatial resolution through the velocity of the vehicle.

Table 2 Defining parameters for normal ellipsoids of geodetic reference systems

Reference system	a (m)	J_2	GM (m^{3}s^{-2})	ω_e (rad s^{-2})
GRS67	6378160	1.0827E − 03	3.98603E14	7.2921151467E − 05
GRS80	6378137	1.08263E − 03	3.986005E14	7.292115E − 05
WGS84	6378137	1.08262982131E − 03	3.986004418E14	7.2921151467E − 05
Best current values[a]	6378136.7 ± 0.1 (mean-tide system)	(1.0826359 ± 0.0000001)E − 03 (zero-tide system)	(3.986004418 ± 0.000000008)E14 (includes atmosphere)	7.292115E − 5 (mean value)

[a]Groten E (2004) Fundamental parameters and current (2004) best estimates of the parameters of common relevance to astronomy, geodesy, and geodynamics. *Journal of Geodesy* 77: (10–11) (The Geodesist's Handbook pp. 724–731).

2.6.1 Measurement Systems and Techniques

Determining the gravitational field through classical measurements relies on three fundamental laws. The first two are Newton's second law of motion and his law of gravitation. Newton's law of motion states that the rate of change of linear momentum of a particle is equal to the totality of forces, $\mathbf{F}$, acting on it. Given more familiarly as $m_i \mathrm{d}^2 \mathbf{x}/\mathrm{d}t^2 = \mathbf{F}$, it involves the inertial mass, m_i; and, conceptually, the forces, $\mathbf{F}$, should be interpreted as action forces (like propulsion or friction). The gravitational field, which is part of the space we occupy, is due to the presence of masses like the Earth, Sun, Moon, and planets, and induces a different kind of force, the gravitational force. It is related to gravitational acceleration, $\mathbf{g}$, through the gravitational mass, m_g, according to the law of gravitation, abbreviated here as $m_g \mathbf{g} = \mathbf{F}_g$. Newton's law of motion must be modified to include $\mathbf{F}_g$ separately. Through the third fundamental law, Einstein's equivalence principle, which states that inertial and gravitational masses are indistinguishable, we finally get

$$\frac{\mathrm{d}^2 \mathbf{x}}{\mathrm{d}t^2} = \mathbf{a} + \mathbf{g} \qquad [119]$$

where $\mathbf{a}$ is the specific force ($\mathbf{F}/m_i$), or also the inertial acceleration, due to action forces. This equation holds in a nonrotating, freely falling frame (that is, an inertial frame), and variants of it can be derived in more complicated frames that rotate or have their own dynamic motion. However, one can always assume the existence of an inertial frame and proceed on that basis.

There exists a variety of devices that measure the motion of an inertial mass with respect to the frame of the device, and thus technically they sense $\mathbf{a}$; such devices are called accelerometers. Consider the special case that an accelerometer is resting on the Earth's surface with its sensitive axis aligned along the vertical. In an Earth-centered frame (inertial, if we ignore Earth's rotation), the free-fall motion of the accelerometer is impeded by the reaction force of the Earth's surface acting on the accelerometer. In this case, the left-hand side of eqn [119] applied to the motion of the accelerometer is zero, and the accelerometer, sensing the reaction force, indirectly measures (the negative of) gravitational acceleration. This accelerometer is given the special name, gravimeter.

Gravimeters, especially static instruments, are designed to measure acceleration at very low frequencies (i.e., averaged over longer periods of time), whereas accelerometers typically are used in navigation or other motion-sensing applications, where accelerations change rapidly. As such, gravimeters generally are more accurate. Earth-fixed gravimeters actually measure gravity (Section 2.1.3), the difference between gravitation and centrifugal acceleration due to Earth's spin (the frame is not inertial in this case). The simplest, though not the first invented, gravimeter utilizes a vertically, freely falling mass, measuring the time it takes to fall a given distance. Applying eqn [119] to the falling mass in the frame of the device, one can solve for g (assuming it is constant): $x(t) = 0.5gt^2$. This free-fall gravimeter is a special case of a more general gravimeter that constrains the fall using an attached spring or the arm of a pendulum, where other (action) forces (the tension in the arm or the spring) thus enter into the equation.

The first gravimeter, in fact, was the pendulum, systematically used for gravimetry as early as the 1730s and 1740s by P. Bouguer on a geodetic expedition to measure the size and shape of the Earth (meridian arc measurement in Peru). The pendulum served well into the twentieth century (until the early 1970s) both as an absolute device, measuring the total gravity at a point, or as a relative device indicating the difference in gravity between two points (Torge, 1989). Today, absolute gravimeters exclusively rely on a freely falling mass, where exquisitely accurate measurements of distance and time are achieved with laser interferometers and atomic clocks (Zumberge et al., 1982; Niebauer et al., 1995). Accurate relative gravimeters are much less expensive, requiring a measurement of distance change only, and because many errors that cancel between measurements need not be addressed. They rely almost exclusively on a spring-suspended test mass (Nettleton, 1976; Torge, 1989). Developed early in the twentieth century in response to oil-exploration requirements, the relative gravimeter has changed little since then. Modern instruments include electronic recording capability, as well as specialized stabilization and damping for deployment on moving vehicles such as ships and aircraft. The accuracy of absolute gravimeters is typically of the order of parts per billion, and relative devices in field deployments may be as good but more typically are at least 1 order of magnitude less precise. Laboratory relative (in time) gravimeters, based on cryogenic instruments that monitor the virtual motion of a test mass that is electromagnetically suspended using superconducting persistent currents (Goodkind, 1999), are as accurate as portable absolute devices (or more), owing to the stability of the currents and the controlled laboratory environment (*see* Chapters 3 and 4).

On a moving vehicle, particularly an aircraft, the relative gravitational acceleration can be determined only from a combination of gravimeter and kinematic positioning system. The latter is needed to derive the (vertical) kinematic acceleration (the left-hand side of eqn [119]). Today, the GPS best serves that function, yielding centimeter-level precision in relative three-dimensional position. Such combined GPS/gravimeter systems have been used successfully to determine the vertical component of gravitation over large, otherwise-inaccessible areas such as the Arctic Ocean (Kenyon and Forsberg, 2000) and over other areas that are more economically surveyed from the air for oil-exploration purposes (Hammer, 1982; Gumert, 1998). The airborne gravimeter is specially designed to damp out high-frequency noise and is usually stabilized on a level platform. Three-dimensional moving-base gravimetry has also been demonstrated using the triad of accelerometers of a high-accuracy inertial navigation system (the type that are fixed to the aircraft without special stabilizing platforms). The orientation of all accelerometers on the vehicle must be known with respect to inertial space, which is accomplished with precision gyroscopes. Again, the total acceleration vector of the vehicle, d^2x/dt^2, can be ascertained by time differentiation of the kinematic positions (from GPS). One of the most critical errors is due to the cross-coupling of the horizontal orientation error, $\delta\psi$, with the large vertical acceleration (the lift of the aircraft, essentially equal to $-g$). This is a first-order effect ($g\sin\delta\psi$) in the estimation of the horizontal gravitation components, but only a second-order effect ($g(1-\cos\delta\psi)$) on the vertical component. Details of moving-base vector gravimetry may be found in Jekeli (2000a, chapter 10) and Kwon and Jekeli (2001).

The ultimate global gravimeter is the satellite in free fall (i.e., in orbit due to sufficient forward velocity) – the satellite is the inertial mass and the 'device' is a set of reference points with known coordinates (e.g., on the Earth's surface, or another satellite whose orbit is known; *see* Chapter 5). The measuring technology is an accurate ranging system (radar or laser) that tracks the satellite as it orbits

(falls to) the Earth. Ever since *Sputnik*, the first artificial satellite, launched into the Earth orbit in 1957, Earth's gravitational field could be determined by tracking satellites from precisely known ground stations. Equation [119], with gravitational acceleration expressed as a truncated series of spherical harmonics (gradient of eqn [44]), becomes

$$\frac{d^2x}{dt^2} = \sum_{n=0}^{n_{max}} \sum_{m=-n}^{n} v_{nm} \nabla\left(\left(\frac{R}{r}\right)^{n+1} \bar{Y}_{nm}(\theta, \lambda + \omega_e t)\right) + \delta\mathbf{R}$$

[120]

where ω_e is Earth's rate of rotation and $\delta\mathbf{R}$ represents residual accelerations due to action forces (solar radiation pressure, atmospheric drag, Earth's albedo, etc.), gravitational tidal accelerations due to other bodies (Moon, Sun, planets), and all other subsequent indirect effects. The x left-hand side of eqn [120] is more explicitly $\mathbf{x}(t) = \mathbf{x}(\theta(t), \lambda(t) + \omega_e t, \mathbf{r}(t))$, and the spatial coordinates on the right-hand side are also functions of time. This makes the equation more conceptual than practical since it is numerically more convenient to transform the satellite position and velocity into Keplerian orbital elements (semi-major axis of the orbital ellipse, its eccentricity, its inclination to the equator, the angle of perigee, the right ascension of the node of the orbit, and the mean motion) all of which also change in time, but most much more slowly. This transformation was derived by Kaula (1966) (see also Seeber, 1993).

In the most general case ($n_{max} > 2$ and $\delta\mathbf{R} \neq 0$), there is no analytic solution to eqn [120] or its transformations to other types of coordinates. The positions of the satellite are observed by ranging techniques and the unknowns to be solved are the coefficients, v_{nm}. Numerical integration algorithms have been specifically adapted to this problem and extremely sophisticated models for $\delta\mathbf{R}$ are employed with additional unknown parameters to be solved in order to estimate as accurately as possible the gravitational coefficients (e.g., Cappelari *et al.*, 1976; Pavlis *et al.*, 1999). The entire procedure falls under the broad category of dynamic orbit determination, and the corresponding gravitational field modeling may be classified as the 'timewise' approach. The partial derivatives of eqn [120] with respect to unknown parameters, $\mathbf{p} = \{\ldots, v_{nm}, \ldots\}$, are integrated numerically in time, yielding estimates for $H = \partial\mathbf{x}/\partial\mathbf{p} = \{\ldots, \partial\mathbf{x}/\partial v_{nm}, \ldots\}$. These are then used in a least-squares adjustment of the linearized model relating observed positions (e.g., via ranges) to parameters

$$\delta \mathbf{x} = H \delta \mathbf{p} + \mathbf{e} \qquad [121]$$

where $\delta \mathbf{x}$ and $\delta \mathbf{p}$ are differences with respect to previous estimates, and $\mathbf{e}$ represents errors. (Tapley, 1973).

A gravimeter (or accelerometer) on a satellite does not sense the presence of a gravitational field. This is evident from the fact that the satellite is in free fall (apart from small accelerations due to action forces such as atmospheric drag) and the inertial test mass and the gravimeter, itself, are equally affected by gravitation (i.e., they are all in free fall). However, two accelerometers fixed on a satellite yield, through the difference in their outputs, a gradient in acceleration that includes the gradient of gravitation. On a nonrotating satellite, the acceleration at an arbitrary point, $\mathbf{b}$, of the satellite is given by

$$\mathbf{a}_b = \frac{d^2 \mathbf{x}}{dt^2} - \mathbf{g}(\mathbf{b}) \qquad [122]$$

in a coordinate system with origin at the center of mass of the satellite. Taking the difference (differential) of two accelerations in ratio to their separation, we obtain

$$\frac{\delta \mathbf{a}_b}{\delta \mathbf{b}} = -\frac{\delta \mathbf{g}}{\delta \mathbf{b}} \qquad [123]$$

where the ratios represent tensors of derivatives in the local satellite coordinate frame. For a rotating satellite, this equation generalizes to

$$\frac{\partial \mathbf{a}_b}{\partial \mathbf{b}} = -\frac{\partial \mathbf{g}_b}{\partial \mathbf{b}} + \Omega^2 + \frac{d}{dt}\Omega \qquad [124]$$

where Ω is a skew-symmetric matrix whose off-diagonal elements are the components of the vector that defines the rotation rate of the satellite with respect to the inertial frame. Thus, a gradiometer on a satellite (or any moving vehicle) senses a combination of gravitational gradient and angular acceleration (including a centrifugal type). Such a device is scheduled to launch for the first time in 2007 as part of the mission GOCE (Gravity Field and Steady-State Ocean Circulation Explorer; Rummel et al., 2002). If the entire tensor of gradients is measured, then, because of the symmetry of the gravitational gradient tensor and of Ω^2, and the antisymmetry of $d\Omega/dt$, the sum $\delta \mathbf{a_b}/\delta \mathbf{b}$ $+(\delta \mathbf{a_b}/\delta \mathbf{b})^T$ eliminates the latter, while the difference $\delta \mathbf{a_b}/\delta \mathbf{b} - (\delta \mathbf{a_b}/\delta \mathbf{b})^T$ can be used to infer Ω, subject to initial conditions.

When the two ends of the gradiometer are fixed to one frame, the common linear acceleration, $d^2 \mathbf{x}/dt^2$, cancels, as shown above; but if the two ends are independent, disconnected platforms moving in similar orbits, the gravitational difference depends also on their relative motion in inertial space. This is the concept for satellite-to-satellite tracking, by which one satellite precisely tracks the other and the change in the range rate between them is a consequence of a gravitational difference, a difference in action forces, and a centrifugal acceleration due to the rotation of the baseline of the satellite pair. It can be shown that the line-of-sight acceleration (the measurement) is given by

$$\frac{d^2 \rho}{dt^2} = \mathbf{e}_\rho^T (\mathbf{g}(\mathbf{x}_2) - \mathbf{g}(\mathbf{x}_1)) + \mathbf{e}_\rho^T (\mathbf{a}_2 - \mathbf{a}_1)$$
$$+ \frac{1}{\rho}\left(\left| \frac{d}{dt}\Delta \mathbf{x} \right|^2 - \left(\frac{d\rho}{dt} \right)^2 \right) \qquad [125]$$

where $\mathbf{e}_\rho$ is the unit vector along the instantaneous baseline connecting the two satellites, ρ is the baseline length (the range), and $\Delta \mathbf{x} = \mathbf{x}_2 - \mathbf{x}_1$ is the difference in position vectors of the two satellites. Clearly, only the gravitational difference projected along the baseline can be determined (similar to a single-axis gradiometer), and then only if both satellites carry accelerometers that sense the nongravitational accelerations. Also, the orbits of both satellites need to be known in order to account for the centrifugal term.

Two such satellite systems were launched recently to determine the global gravitational field. One is CHAMP (Challenging Mini-Satellite Payload) in 2000 (Reigber et al., 2002) and the other is GRACE (Gravity Recovery and Climate Experiment) in 2002 (Tapley et al., 2004a). CHAMP is a single low-orbiting satellite (400–450 km altitude) being tracked by the high-altitude GPS satellites, and it also carries a magnetometer to map the Earth's magnetic field. GRACE was more specifically dedicated to determining with extremely high accuracy the long to medium wavelengths of the gravitational field and their temporal variations. With two satellites in virtually identical low Earth orbits, one following the other, the primary data consist of intersatellite ranges observed with K-band radar. The objective is to sense changes in the gravitational field due to mass transfer on the Earth within and among the atmosphere, the hydrosphere/cryosphere, and the oceans (Tapley et al., 2004b).

An Earth-orbiting satellite is the ideal platform on which to measure the gravitational field when seeking global coverage in relatively short time. One simply designs the orbit to be polar and circular;

and, as the satellite orbits, the Earth spins underneath, offering a different section of its surface on each satellite revolution. There are also limitations. First, the satellite must have low altitude to achieve high sensitivity, since the nth-degree harmonics of the field attenuate as $(R/r)^{n+1}$. On the other hand, the lower the altitude, the shorter the life of the satellite due to atmospheric drag, which can only be countered with onboard propulsion systems. Second, because of the inherent speed of lower-orbit satellites (about $7\,\mathrm{km\,s}^{-1}$), the resolution of its measurements is limited by the integration (averaging) time of the sensor (typically 1–10 s). Higher resolution comes only with shorter integration time, which may reduce the accuracy if this depends on averaging out random noise. **Figure 8** shows the corresponding achievable resolution on the Earth's surface for different satellite instrumentation parameters, length of time in polar orbit and along-orbit integration time, or smoothing (Jekeli, 2004). In each case, the indicated level of resolution is warranted only if the noise of the sensor (after smoothing) does not overpower the signal at this resolution.

Both *CHAMP* and *GRACE* have yielded global gravitational models by utilizing traditional satellite-tracking methods and incorporating the range rate appropriately as a tracking observation (timewise approach). However, the immediate application of eqn [125] suggests that gravitational differences can be determined *in situ* and used to determine a model for the global field directly. This is classified as the spacewise approach. In fact, if the orbits are known with sufficient accuracy (from kinematic orbit determination, e.g., by GPS), this procedure utilizes a linear relationship between observations and unknown harmonic coefficients:

$$\left.\frac{\mathrm{d}^2\rho}{\mathrm{d}t^2}\right|_{x_1,\,x_2} = \sum_{n=0}^{n_{\max}}\sum_{m=-n}^{n} v_{nm}\mathbf{e}_\rho^{\mathrm{T}}\left(\nabla U_{nm}(\theta,\,\lambda,\,r)|_{x_2}\right.$$
$$\left. -\nabla U_{nm}(\theta,\,\lambda,\,r)|_{x_1}\right) + \delta a + \delta c \qquad [126]$$

where $U_{nm}(\theta,\,\lambda,\,r) = (R/r)^{n+1}\bar{Y}_{nm}(\theta,\,\lambda)$ and δa, δc are the last two terms in eqn [125]. Given the latter and a set of line-of-sight accelerations, a theoretically straightforward linear least-squares adjustment solves for the coefficients. A similar procedure can be used for gradients observed on an orbiting satellite:

$$\left.\frac{\partial\mathbf{a}}{\partial\mathbf{b}}\right|_{x} = \sum_{n=0}^{n_{\max}}\sum_{m=-n}^{n} v_{nm}\nabla\nabla^{\mathrm{T}}U_{nm}(\theta,\,\lambda,\,r)|_{x} + \psi \qquad [127]$$

where ψ comprises the rotational acceleration terms in eqn [124].

In situ measurements of line-of-site acceleration or of more local gradients would need to be reduced from the satellite orbit to a well-defined surface (such as a sphere) in order to serve as boundary values in a solution to a BVP. However, the model for the field is already in place, rooted in potential theory (truncated series solution to Laplace' equation), and one may think of the problem more in terms of fitting a three-dimensional model to a discrete set of observations. This operational approach can readily, at least conceptually, be expanded to include observations from many different satellite systems, even airborne and ground-based observations.

Recently, a rather different theory has been considered by several investigators to model the gravitational field from satellite-to-satellite tracking observations. The method, first proposed by Wolff (1969), makes use of yet another fundamental law: the law of conservation of energy. Simply, the range rate between two satellites implies an along-track velocity difference, or a difference in kinetic energy. Observing this difference leads directly, by the conservation law, to the difference in potential energy, that is, the gravitational potential and other potential energies associated with action forces and Earth's rotation. Neglecting the latter two, conservation of energy implies

$$V = \frac{1}{2}\left|\frac{\mathrm{d}}{\mathrm{d}t}x_1\right|^2 - E_0 \qquad [128]$$

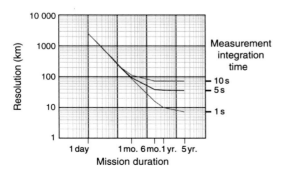

Figure 8 Spatial resolution of satellite measurements vs mission duration and integration time. The satellite altitude is 450 km (after Jekeli, 2004) mo., month. Reproduced by permission of American Geophysical Union.

where E_0 is a constant. Taking the along-track differential, we have approximately

$$V(\mathbf{x}_2) - V(\mathbf{x}_1) = \left| \frac{\mathrm{d}}{\mathrm{d}t} \mathbf{x}_1 \right| \frac{\mathrm{d}}{\mathrm{d}t} \rho \qquad [129]$$

where $|\mathrm{d}\mathbf{x}_2/\mathrm{d}t - \mathrm{d}\mathbf{x}_1/\mathrm{d}t| \approx \mathrm{d}\rho/\mathrm{d}t$. This very rough conceptual relationship between the potential difference and the range rate applies to two satellites closely following each other in similar orbits. The precise formulation is given by (Jekeli, 1999) and holds for any pair of satellites, not just two low orbiters. Mapping range rates between two polar orbiting satellites (such as *GRACE*) yields a global distribution of potential difference observations related again linearly to a set of harmonic coefficients:

$$V(\mathbf{x}_2) - V(\mathbf{x}_1) = \sum_{n=0}^{n_{max}} \sum_{m=-n}^{n} v_{nm} \left(U_{nm}(\theta, \lambda, r) \big|_{x_2} \right.$$
$$\left. - U_{nm}(\theta, \lambda, r) \big|_{x_1} \right) \qquad [130]$$

The energy-based model holds for any two vehicles in motion and equipped with the appropriate ranging and accelerometry instrumentation. For example, the energy conservation principle could also be used to determine geopotential differences between an aircraft and a satellite, such as a GPS satellite (at which location the geopotential is known quite well). The aircraft would require only a GPS receiver (to do the ranging) and a set of accelerometers (and gyros for orientation) to measure the action forces (the same system components as in airborne accelerometry discussed above). Resulting potential differences could be used directly to model the geoid using Bruns' equation.

2.6.2   Models

We have already noted the standard solution options to the BVP using terrestrial gravimetry in the form of gravity anomalies: the Green's function approach, Stokes integral, eqn [66], and the harmonic analysis of surface data, either using the integrals [69] or solving a linear system of equations (eqn [65] truncated to finite degree) to obtain the coefficients, δC_{nm}. The integrals must be evaluated using quadratures, and very fast numerical techniques have been developed when the data occupy a regular grid of coordinates on the sphere or ellipsoid (Rapp and Pavlis, 1990). Similar algorithms enable the fast solution of the linear system of eqns [65].

For a global harmonic analysis, the number of coefficients, $(n_{max} + 1)^2$, must not be greater than the number of data. A general, conservative rule of thumb for the maximum resolution (half-wavelength) of a truncated spherical harmonic series is, in angular degrees on the unit sphere,

$$\Delta\theta = \frac{180°}{n_{max}}$$

Thus, data on a $1° \times 1°$ angular grid of latitudes and longitudes would imply $n_{max} = 180$. The number of data (64 800) is amply larger than the number of coefficients (32 761). This majority suggests a least-squares adjustment of the coefficients to the data, in either method, especially because the data have errors (Rapp, 1969). As n_{max} increases, a rigorous, optimal adjustment usually is feasible, for a given computational capability, only under restrictive assumptions on the correlations among the errors of the data. Also, the obvious should nevertheless be noted that the accuracy of the model in any area depends on the quality of the data in that area. Furthermore, considering that a measurement contains all harmonics (up to the level of measurement error), the estimation of a finite number of harmonics from boundary data on a given grid is corrupted by those harmonics that are in the data but are not estimated. This phenomenon is called aliasing in spectral analysis and can be mitigated by appropriate filtering of the data (Jekeli, 1996).

The optimal spherical harmonic model combines both satellite and terrestrial data. The currently best known model is EGM96 complete to degree and order $n_{max} = 360$ (Lemoine *et al.*, 1998). It is an updated model for WGS84 based on all available satellite tracking, satellite altimetry, and land gravity (and topographic) data up to the mid-1990s. Scheduled to be revised again for 2006 using more recent data, as well as results from the satellite missions, *CHAMP* and *GRACE*, it will boast a maximum degree and order of 2160 (5' resolution). In constructing combination solutions of this type, great effort is expended to ensure the proper weighting of observations in order to extract the most appropriate information from the diverse data, pertaining to different parts of the spatial gravitational spectrum. The satellite tracking data dominate the estimation of the lower-degree harmonics, whereas the fine resolution of the terrestrial data is most amenable to modeling the higher degrees. It is beyond the present scope to delve into the numerical methodology of combination methods. Furthermore, it is an unfinished story as new *in situ* satellite measurements from *GRACE* and *GOCE* will affect the combination methods of the future.

Stokes integral is used in practice only to take advantage of local or regional data with higher resolution than was used to construct the global models. Even though the integral is a global integral, it can be truncated to a neighborhood of the computation point since the Stokes kernel attenuates rapidly as the reciprocal distance. Moreover, the corresponding truncation error may be reduced if the boundary values exclude the longer-wavelength features of the field. The latter constitute an adequate representation of the remote zone contribution and can be included separately as follows. Let $\Delta g^{(n_{max})}$ denote the gravity anomaly implied by a spherical harmonic model, such as given by eqn [65], truncated to degree, n_{max}. From the orthogonality of spherical harmonics, eqn [46], it is easy to show that

$$T^{(n_{max})}(\theta, \lambda, r) = \frac{GM}{R} \sum_{n=2}^{n_{max}} \sum_{m=-n}^{n} \left(\frac{R}{r}\right)^{n+1} \delta C_{nm} \bar{Y}_{nm}(\theta, \lambda)$$
$$= \frac{R}{4\pi} \int\int_{\sigma} \Delta g^{(n_{max})}(\theta', \lambda', R) S(\psi, r) d\sigma$$

$$[131]$$

Thus, given a spherical harmonic model $\{\delta C_{nm} | 2 \leq n \leq n_{max}, -n \leq m \leq n\}$, we first remove the model in terms of the gravity anomaly and then restore it in terms of the disturbing potential, changing Stokes formula [66] to

$$T(\theta, \lambda, r) = \frac{R}{4\pi} \int\int_{\sigma} \left(\Delta g(\theta', \lambda', R) - \Delta g^{(n_{max})}(\theta', \lambda', R)\right)$$
$$\times S(\psi, r) d\sigma + T^{(n_{max})}(\theta, \lambda, r) \quad [132]$$

In theory, if $\Delta g^{(n_{max})}$ has no errors, then the residual $\Delta g - \Delta g^{(n_{max})}$ excludes all harmonics of degree $n \leq n_{max}$, and orthogonality would also allow the exclusion of these harmonics from S. Once the integration is limited to a neighborhood of (θ, λ, R), as it must be in practice, there are a number of ways to modify the kernel so as to minimize the resulting truncation error (Sjöberg, 1991, and references therein). The removal and restoration of a global model, however, is the key aspect in all these methods.

In practical applications, the boundary values are on the geoid, being the surface that satisfies the boundary condition of the Robin BVP (i.e., we require the normal derivative on the boundary and measured gravity is indeed the derivative of the potential along the perpendicular to the geoid). The

integral in eqn [132] thus approximates the geoid by a sphere. Furthermore, it is assumed that no masses exist external to the geoid. Part of the reduction to the geoid of data measured on the Earth's surface involves redistributing the topographic masses on or below the geoid. This redistribution is undone outside the solution to the BVP (i.e., Stokes integral) in order to regain the disturbing potential for the actual Earth. Conceptually, we may write

$$T_P = \frac{R}{4\pi} \int\int_{\sigma} \left(\Delta g - \Delta g^{(n_{max})} - \delta c\right)_{P'} S_{P, P'} d\sigma$$
$$+ T_P^{(n_{max})} + \delta T_P \quad [133]$$

where δc is the gravity reduction that brings the gravity anomaly to a geoid with no external masses, and δT_P is the effect (called indirect effect) on the disturbing potential due to the inverse of this reduction. This formula holds for T anywhere on or above the geoid and thus can also be used to determine the geoid undulation according to Bruns' formula [3].

2.7 The Geoid and Heights

The traditional reference surface, or datum, for heights is the geoid (Section 2.1.3). A point at mean sea level usually serves as starting point (datum origin) and this defines the datum for vertical control over a region or country. The datum (or geoid) is the level continuation of the reference surface under the continents, and the determination of gravity potential differences from the initial point to other points on the Earth's suface, obtained by leveling and gravity measurements, yields heights with respect to that reference (or in that datum). The gravity potential difference, known as the geopotential number, at a point, P, relative to the datum origin, $\bar{P}_0$, is given by (since gravity is the negative vertical derivative of the gravity potential)

$$C_P = W_0 - W_P = \int_{\bar{P}_0}^{P} g \, dn \quad [134]$$

where g is gravity (magnitude), dn is a leveling increment along the vertical direction, and W_0 is the gravity potential at $\bar{P}_0$. By the conservative nature of the gravity potential, whatever path is taken for the integral yields a unique geopotential number for P. From these potential differences, one can define various types of height, for example, the orthometric height (**Figure 3**):

$$H_P = \frac{C_P}{\bar{g}_P} \qquad [135]$$

where

$$\bar{g}_P = \frac{1}{H_P} \int_{P_0}^{P} g\,dH \qquad [136]$$

is the average value of gravity along the plumb line from the geoid at P_0 to P. Other height systems are also in use (such as the normal and dynamic heights), but they all rely on the geopotential number (see Heiskanen and Moritz, 1967, chapter 4, for details).

For a particular height datum, there is theoretically only one datum surface (the geoid). But access to this surface is far from straightforward (other than at the defined datum origin). If $\bar{P}_0$ is defined at mean sea level, other points at mean sea level are not on the same level (datum) surface, since mean sea level, in fact, is not level. Erroneously assuming that mean sea level is an equipotential surface can cause significant distortions in the vertical control network of larger regions, as much as several decimeters. This was the case, for example, for the National Geodetic Vertical Datum of 1929 in the US for which 26 mean sea level points on the east and west coasts were assumed to lie on the same level surface. Accessibility to the geoid (once defined) at any point is achieved either with precise leveling and gravity (according to eqns [134] and [135]), or with precise geometric vertical positioning and knowledge of the gravity potential. Geometric vertical positioning, today, is obtained very accurately (centimeter accuracy or even better) with differential GPS. Suppose that an accurate gravity potential model is also available in the same coordinate system as used for GPS. Then, determining the GPS position at $\bar{P}_0$ allows the evaluation of the gravity potential, W_0, of the datum. Access to the geoid at any other point, P, or equivalently, determining the orthometric height, H_P, can be done by first determining the ellipsoidal height, h, from GPS. Then, as shown in **Figure 3**,

$$H_P = h_P - N \qquad [137]$$

where, with $T = W - U$ evaluated on the geoid, Bruns' extended equation (Section 2.1) yields

$$N = T/\gamma - (W_0 - U_0)/\gamma \qquad [138]$$

where U_0 is the normal gravity potential of the normal ellipsoid.

In a sense, the latter is a circular problem: determining N requires N in order to locate the point on the geoid where to compute T. However, the computation of T on the geoid can be done with assumptions on the density of the topographic masses above the geoid and a proper reduction to the geoid, using only an approximate height. Indeed, since the vertical gradient of the disturbing potential is the gravity disturbance, of the order of $5 \times 10^{-4}\,\mathrm{m\,s^{-2}}$, a height error of 20 m leads to an error of $10^{-2}\,\mathrm{m^2\,s^{-2}}$ in T, or just 1 mm in the geoid undulation. It should be noted that a model for the disturbing potential as a series of spherical harmonics, for example, derived from satellite observations, satisfies Laplace's equation and, therefore, does not give the correct disturbing potential at the geoid (if it lies below the Earth's surface where Laplace's equation does not hold).

The ability to derive orthometric heights (or other geopotential-related heights) from GPS has great economical advantage over the laborious leveling procedure. This has put great emphasis on obtaining an accurate geoid undulation model for land areas. Section 2.6 briefly outlined the essential methods to determine the geoid undulation from a combination of spherical harmonics and an integral of local gravity anomalies. When dealing with a height datum or the geoid, the constant $N_0 = -(W_0 - U_0)/\gamma$ requires careful attention. It can be determined by comparing the geoid undulation computed according to a model that excludes this term (such as eqn [133]) with at least one geoid undulation (usually many) determined from leveling and GPS, according to eqn [137]. Vertical control and the choice of height datum are specific to each country or continent, where a local mean sea level was the adopted datum origin. Thus, height datums around the world are 'local geoids' that have significant differences between them. Investigations and efforts have been under way for more than two decades to define a global vertical datum; however, it is still in the future, awaiting a more accurate global gravity potential model and, perhaps more crucially, a consensus on what level surface the global geoid should be.

On the oceans, the situation is somewhat less complicated. Oceanographers who compute sea-surface topography from satellite altimetry on the basis of eqn [137] depend critically on an accurate geoid undulation, or equivalently on an accurate model of T. However, no reduction of the disturbing potential from mean sea level to the geoid is necessary, the deviation being at most 2 m and causing an error in geoid undulation of less than 0.1 mm. Thus, a spherical harmonic model of T is entirely appropriate. Furthermore, it is reasonable to ensure that the constant, $W_0 - U_0$, vanishes over the oceans. That is, one may choose the geoid such that it best fits mean sea level and choose an ellipsoid that best fits this geoid. It means that the global average value of the geoid undulation should be zero (according to eqn [138]). The latter can be achieved with satellite altimetry and oceanographic models of sea surface topography (Bursa *et al.*, 1997).

Several interesting and important distinctions should be made in regard to the tidal effects on the geoid. The Sun and the Moon generate an appreciable gravitational potential near the Earth (the other planets may be neglected). In an Earth-fixed coordinate system, this extraterrestrial potential varies in time with different periods due to the relative motions of the Moon and Sun, and because of Earth's rotation (Torge, 2001, p. 88). There is also a constant part, the permanent tidal potential, representing the average over time. It is not zero, because the Earth–Sun–Moon system is approximately coplanar. For each body with mass, M_B, and distance, r_B, from the Earth's center, this permanent part is given by

$$V_c^B(\theta, r) = \frac{3}{4} GM_B \frac{r^2}{r_B^3} \left(3\cos^2\theta - 1\right) \left(\frac{1}{2}\sin^2\varepsilon - \frac{1}{3}\right) \quad [139]$$

where ε is the angle of the ecliptic relative to the equator ($\varepsilon \approx 23°\,44'$). Using nominal parameter values for the Sun and the Moon, we obtain at mean Earth radius, $R = 6371\text{km}$,

$$V_c^{s+m}(\theta, R) = -0.97\left(3\cos^2\theta - 1\right)\text{m}^2\,\text{s}^{-2} \quad [140]$$

The gravitational potential from the Sun and Moon also deforms the quasi-elastic Earth's masses with the same periods and similarly includes a constant part. These mass displacements (both ocean and solid Earth) give rise to an additional indirect change in potential, the tidal deformation potential (there are also secondary indirect effects due to loading of the ocean on the solid Earth, which can be neglected in this discussion; *see* Chapter 6). The indirect effect is modeled as a fraction of the direct effect (Lambeck, 1988, p. 254), so that the permanent part of the tidal potential including the indirect effect is given by

$$\bar{V}_c^{s+m}(\theta, R) = (1 + k_2)V_c^{s+m}(\theta, R) \quad [141]$$

where $k_2 = 0.29$ is Love's number (an empirical number based on observation). This is also called the mean tide potential.

The mean tidal potential is inherent in all our terrestrial observations (the boundary values) and cannot be averaged away; yet, the solutions to the BVP assume no external masses. Therefore, in principle, the effect of the tide potential including its mean, or permanent, part should be removed from the observations prior to applying the BVP solutions. On the other hand, the permanent indirect effect is not that well modeled and arguably should not be removed; after all, it contributes to the Earth's shape as it actually is in the mean. Three types of tidal systems have been defined to distinguish between these corrections. A mean quantity refers to the quantity with the mean tide potential retained (but time-varying parts removed); a nontidal quantity implies that all tidal effects (time-varying, permanent, direct and indirect effects) have been removed computationally; and the zero-tide quantity excludes all time-varying parts and the permanent direct effect, but it retains the indirect permanent effect.

If the geoid (an equipotential surface) is defined solely by its potential, W_0, then a change in the potential due to the tidal potential, V^{tide} (time-varying and constant parts, direct and indirect effects), implies that the W_0-equipotential surface has been displaced. The geoid is now a different surface with the same W_0. This displacement is equivalent to a change in geoid undulation, $\delta N = V^{\text{tide}}/\gamma$, with respect to some predefined ellipsoid. The permanent tidal effect (direct and indirect) on the geoid is given by

$$\delta\bar{N}(\theta) = -0.099(1 + k_2)\left(3\cos^2\theta - 1\right)\text{m} \quad [142]$$

If N represents the instantaneous geoid, then the geoid without any tidal effects, that is, the nontidal geoid, is given by

$$N_{nt} = N - \delta N \quad [143]$$

The mean geoid is defined as the geoid with all but the mean tidal effects removed:

$$\bar{N} = N - (\delta N - \delta\bar{N}) \quad [144]$$

This is the geoid that could be directly observed, for example, using satellite altimetry averaged over time. The zero-tide geoid retains the permanent indirect effect, but no other tidal effects:

$$N_z = N - \left(\delta N + 0.099k_2\left(3\cos^2\theta - 1\right)\right)\text{m} \quad [145]$$

The difference between the mean and zero-tide geoids is, therefore, the permanent component of the direct tidal potential. We note that, in principle, each of the geoids defined above, has the same potential value, W_0, in its own field. That is, with each correction, we define a new gravity field and the corresponding geoid undulation defines the equipotential surface in that field with potential value given by W_0. This is fundamentally different than what happens in the case when the geoid is defined as a vertical datum with a specified datum origin point. In this case one needs to consider also the vertical displacement of the datum point due to the tidal deformation of the Earth's surface. The potential of the datum then changes because of the direct tidal potential, the indirect effect due to mass changes, and the indirect effect due to the vertical displacement of the datum (for additional details, see Jekeli (2000b)).

References

Bursa M, Radej K, Sima Z, True S, and Vatrt V (1997) Determination of the geopotential scale factor from TOPEX/Poseidon satellite altimetry. *Studia Geophysica et Geodaetica* 41: 203–216.

Cappelari JO, Velez CE, and Fuchs AJ (eds.) (1976) *Mathematical theory of the Goddard Trajectory Determination System.* GSFC Document X-582-76-77. Greenbelt MD: Goddard Space Flight Center.

Chao BF (2005) On inversion for mass distribution form global (time-variable) gravity field. *Journal of Geodynamics* 29: 223–230.

Courant R and Hilbert D (1962) *Methods of Mathematical Physics, vol. II. New York:* Wiley.

Cushing JT (1975) *Applied Analytical Mathematics for Physical Scientists.* New York: Wiley.

Dziewonsky AD and Anderson DL (1981) Preliminary reference Earth model. *Physics of the Earth and Planetary Interiors* 25: 297.

Fei ZL and Sideris MG (2000) A new method for computing the ellipsoidal correction for Stokes's formula. *Journal of Geodesy* 74(2): 223–231.

Goodkind JM (1999) The superconducting gravimeter. *Review of Scientific Instruments* 70: 4131–4152.

Groten E (2004) Fundamental parameters and current (2004) best estimates of the parameters of common relevance to astronomy, geodesy, and geodynamics. *Journal of Geodesy* 77: (10–11) (The Geodesist's Handbook pp. 724–731).

Gumert WR (1998) An historical review of airborne gravimetry. *The Leading Edge* 17(1): 113–116.

Günter NM (1967) *Potential Theory and Its Applications to Basic Problems of Mathematical Physics.* New York: Frederick Ungar.

Hammer S (1982) Airborne gravity is here!. *Oil and Gas Journal* 80: 2 (January 11, 1982).

Heiskanen WA and Moritz H (1967) *Physical Geodesy.* San Francisco, CA: Freeman.

Helmert FR (1884) *Die Mathematischen und Physikalischen Theorien der Höheren Geodäsie,* vol. 2. Leipzig, Germany: B.G. Teubner, (reprinted in 1962 by Minerva GMBH, Frankfurt/Main, Germany).

Hobson EW (1965) *The Theory of Spherical and Ellipsoidal Harmonics.* New York: Chelsea.

Hofmann-Wellenhof B and Moritz H (2005) *Physical Geodesy.* Berlin, Germany: Springer-Verlag.

Hotine M (1969) *Mathematical Geodesy.* Washington, DC: US Department of Commerce.

Jekeli C (1988) The exact transformation between ellipsoidal and spherical harmonic expansions. *Manuscripta Geodaetica* 14: 106–113.

Jekeli C (1996) Spherical harmonic analysis, aliasing, and filtering. *Journal of Geodesy* 70: 214–223.

Jekeli C (1999) The determination of gravitational potential differences from satellite-to-satellite tracking. *Celestial Mechanics and Dynamical Astronomy* 75(2): 85–100.

Jekeli C (2000a) *Inertial Navigation Systems with Geodetic Applications.* Berlin, Germany: Walter deGruyter.

Jekeli C (2000b) Heights, the geopotential, and vertical Datums. *Report no.459, Department of Civil and Environmental Engineering and Geodetic Science, Ohio State University.* Columbus, OH: Ohio State University.

Jekeli C (2004) High-resolution gravity mapping: The next generation of sensors. In: Sparks RSJ (ed.) *Geophysical Monograph 150, IUGG, vol. 19: The State of the Planet: Frontiers and Challenges in Geophysics,* pp. 135–146. Washington, DC: American Geophysical Union.

Kaula WM (1966) *Theory of Satellite Geodesy.* London: Blaisdell.

Kellogg OD (1953) *Foundations of Potential Theory.* New York: Dover.

Kenyon S and Forsberg R (2000) Arctic gravity project. *SEG Technical Program Expanded Abstracts* 19: 410–413.

Kwon JH and Jekeli C (2001) A new approach for airborne vector gravimetry using GPS/INS. *Journal of Geodesy* 74: 690–700.

Lambeck K (1988) *Geophysical Geodesy.* Oxford: Clarendon Press.

Lemoine FG, Kenyon SC, Factor JK, et al. (1998) *The development of the joint NASA GSFC and the National Imagery and Mapping Agency (NIMA) geopotential model EGM96. NASA Technical Paper NASA/TP-1998-206861,* Greenbelt, MD: Goddard Space Flight Center.

Martin JL (1988) *General Relativity: A Guide to Its Consequences for Gravity and Cosmology.* New York: Wiley.

Molodensky MS, Eremeev VG, and Yurkina MI (1962) *Methods for Study of the External Gravitational Field and Figure of the Earth.* Jerusalem, Israel: Israel Program for Scientific Translations (Translation from Russian).

Moritz H (1980) *Advanced Physical Geodesy.* Karlsruhe, Germany: Wichmann, (reprinted in 2001 by Civil and Environmental Engineering and Geodetic Science, Ohio State University, Columbus, OH).

Morse PM and Feshbach H (1953) *Methods of Theoretical Physics, Parts I and II.* New York: McGraw-Hill.

Müller C (1966) *Spherical Harmonics.* Lecture Notes in Mathematics. Berlin, Germany: Springer-Verlag.

Nettleton LL (1976) *Gravity and Magnetics in Oil Prospecting.* New York: McGraw-Hill.

Niebauer T, Sasagawa G, Faller J, Hilt R, and Klopping F (1995) A new generation of absolute gravimeters. *Metrologia* 32(3): 159–180.

Pavlis DE, Moore D, Luo S, McCarthy JJ, and Luthcke SB (1999) *GEODYN Operations Manual, 5 vols.* Greenbelt, MD: Raytheon ITSS.

Petrovskaya MS (1979) Upward and downward continuations in the geopotential determination. *Bulletin Géodésique* 53: 259–271.

Rapp RH (1969) Analytical and numerical differences between two methods for the combination of gravimeter and satellite data. *Bollettino di Geofisica Teorica ed Applicata* XI(41–42): 108–118.

Rapp RH and Pavlis NK (1990) The development and analysis of geopotential coefficient models to spherical harmonic degree 360. *Journal of Geophysical Research* 95(B13): 21885–21911.

Reigber Ch, Balmino G, Schwintzer P, *et al.* (2002) A high quality global gravity field model from CHAMP GPS tracking data and accelerometry (EIGEN-1S). *Geophysical Research Letters* 29(14): 37–41 (doi:10.1029/2002GL015064).

Rummel R, Balmino G, Johannessen J, Visser P, and Woodworth P (2002) Dedicated gravity field missions – Principles and aims. *Journal of Geodynamics* 33: 3–20.

Sandwell DT and Smith WHF (1996) Marine gravity anomaly form GEOSAT and ERS-1 satellite altimetry. *Journal of Geophysical Research* 102(B5): 10039–10054.

Sansò F and Rummel R (eds.) (1997) Lecture Notes in Earth Sciences 65: Geodetic Boundary Value Problems in View of the One Centimeter Geoid. Berlin, Germany: Springer-Verlag.

Seeber G (1993) *Satellite Geodesy, Foundations, Methods, and Applications*. Berlin, Germany: Walter deGruyter.

Sjöberg LE (1991) Refined least squares modifications of Stokes' formula. *Manuscripta Geodaetica* 16: 367–375.

Tapley BD (1973) Statistical orbit determination theory. In: Tapley BD and Szebehely V (eds.) *Recent Advances in Dynamical Astronomy*, pp. 396–425. Dordtecht, The Netherlands: D. Reidel.

Tapley BD, Bettadpur S, Ries JC, Thompson PF, and Watkins M (2004b) GRACE measurements of mass variability in the Earth system. *Science* 305(5683): 503–505.

Tapley BD, Bettadpur S, Watkins M, and Reigber Ch (2004a) The gravity recovery and climate experiment, mission overview and early results. *Geophysical Research Letters* 31(9): 1501 (doi:10.1029/2004GL019920).

Tapley BD, Ries J, Bettadpur S, *et al.* (2005) GGM02 – An improved Earth gravity field model from GRACE. *Journal of Geodesy* 79: 137–139 (doi:10.1007/s00190-005-0480-z).

Torge W (1989) *Gravimetry*. Berlin, Germany: Walter deGruyter.

Torge W (2001): *Geodesy, 3rd edn. Berlin, Germany: Walter de Gruyter.*

Wolff M (1969) Direct measurement of the Earth's gravitational potential using a satellite pair. *Journal of Geophysical Research* 74: 5295–5300.

Yu J, Jekeli C, and Zhu M (2002) The analytical solutions of the Dirichlet and Neumann boundary value problems with ellipsoidal boundary. *Journal of Geodesy* 76(11–12): 653–667.

Zumberge MA, Rinker RL, and Faller JE (1982) A portable apparatus for absolute measurements of the Earth's gravity. *Metrologia* 18(3): 145–152.

3 Gravimetric Methods – Absolute Gravimeter: Instruments Concepts and Implementation

T. Niebauer, Migro-g Solutions Inc., Erie, CO, USA

3.1 Absolute and Relative Gravity and Gravimeters

3.1.1 Definition of Absolute and Relative Gravity and Their Practical Uses

Terrestrial gravity measurements are made to determine the magnitude of the acceleration of gravity due to the Earth along the direction of a locally freely falling body. The local magnitude of gravity, denoted by the familiar symbol 'g', is referred to as 'little g'. In most physics text books, g is assumed to be a constant value of $9.8\,\mathrm{m\,s^{-2}}$ (often rounded to $10\,\mathrm{m\,s^{-2}}$). There is typically very little discussion in introductory books that indicate that gravity actually varies from point to point on the Earth or that it can be of practical interest to measure its value. The two most common questions asked to me by college-educated persons are (1) Isn't gravity the same everywhere? and (2) Why would anyone want to measure gravity? It seems to be a well-kept secret that the measurement of gravity on the earth's surface can provide a wealth of information about the shape of the earth, the position of the continents, as well as the water and mineral contents of the ground underneath our feet.

Measurement of gravity has a long and rich history closely connected to the measurement of time. The first accurate pendulums were used to measure time and also to measure gravity. Currently, the most accurate measurements of gravity are still linked to very precise time measurements of the trajectory for a free-falling object. Gravity measurements have been used to answer questions about the Earth's gravity field since the time of Galileo (1564–1642) who purportedly dropped stones from the leaning tower of Pisa to test Aristotle's theory that heavy things fall faster than light objects. Currently very sensitive space measurements like NASA's Gravity Probe B are being used to test Einstein's theory of relativity. Gravity is also being used to answer very practical questions such as the amount of water, oil, and other minerals that are buried deep within the ground.

Real-world gravity measurements historically have been made using the CGS unit called a Galileo (or Gal) who is often considered the father of gravity. In these units, the Earth's field is given as 980 Gal (or approximately 1000 Gal), where $1\,\mathrm{Gal} = 1\,\mathrm{cm\,s^{-2}}$. A calibrated measurement of the acceleration of gravity given in absolute units of distance and time (m s^{-2} in SI units) is referred to as an 'absolute measurement of gravity'.

Terrestrial gravity measurements are higher at the poles than at the equator by several parts in 1000 (several Gal) due to a combination of a decreasing distance to the center of the Earth and less centrifugal acceleration at the poles. Thus, gravity is obviously useful for geodetic purposes such as determining the overall shape of the Earth and determining vertical crustal motion. Motion of the Earth's crust can in turn provide information about processes inside the mantle of the Earth (Lambert *et al.*, 1998). At a higher level of precision of parts per million (milli-Gal or mGal), gravity is also sensitive to subsurface density anomalies and changes such as those caused by water table, oil and gas reservoirs, mineral deposits, man-made excavations, and even magma chambers underneath active volcanoes.

Gravity surveys are conducted at one point in time over a large physical area to provide information about subsurface density anomalies. Gravity surveys over a limited area provide gravity snapshots or maps that can be useful for mineral exploration. Gravity information over very large areas provides geodetic information about the shape of the Earth. Surveys over smaller areas can be used to infer subsurface density fluctuations and are useful for oil/gas exploration. Typical gravity surveys used for oil/gas exploration require a precision of about 0.1 mGal or 100 microGal (μGal). 3D gravity surveys that are repeated over time are sometimes referred to as 'time-lapse gravity' or 4D gravity surveys (the latter term is borrowed from the seismic community where the four dimensions refer to the usual three spatial dimensions plus time). Time-lapse gravity surveys provide a much more sensitive way of looking at subsurface density changes due to fluid motion or man-made construction because large spatial gravity variations caused by deep structure that remain constant can be ignored because they are 'common-mode' and will be removed when the time-lapsed gravity surveys are subtracted. Time-lapse gravity surveys are able to provide information at the level of 1 part in 10^9 (1 μGal level or lower) even though the background static variations of gravity over the survey area are many thousands of times larger. The ability to measure slow (months or longer) nonperiodic gravity changes below 1 μGal becomes very difficult due to many environmental sources of gravity (air pressure, water table, etc.) changes which are difficult to model. Thus, the microGal (1 μGal) represents a practical limit for most terrestrial gravity measurements of slowly varying nonperiodic gravity signals.

The absolute gravity value at a given site is not usually needed to make a gravity map for commercial applications such as oil/gas exploration. In these cases, it is enough to simply measure the difference of gravity over a survey region. This is fortuitous because most gravity meters that are used to detect small changes in gravity have some difficulty in determining an accurate value for absolute gravity due to excessive drift or tares between calibrations. Gravity meters that measure sensitive gravity differences in either space or time but are not typically used to make absolute gravity measurements are referred to as 'relative gravity meters'. Gravity differences over finite spatial distances and/or times are correspondingly sometimes referred to as 'relative gravity' changes. Most commercial measurements of gravity employ relative gravity meters.

Relative gravity changes should not be confused with the subject of gravity gradients which are defined as the mathematical differential of gravity in all directions (a tensor with five independent components).

'Absolute gravity', on the other hand, refers to the exact value of gravity (connected directly to metrology standards of length and time) at any one point in space and time near the surface of the Earth. In most applications, the absolute gravity value is not needed but it has the advantage that it is by definition calibrated and accurate.

Gravity has scientific, commercial, and military applications. The scientific applications include many geophysical investigations such as determining the shape of the Earth (geodesy), Earth tides, plate tectonics (crustal motion), vulcanology, global sea-level changes, arctic ice-sheet mass balance, and water table variations among many others. Many of the practical uses of gravity are described in Torge's book called *Geodesy*. Gravity is also being used in the scientific field of metrology to help define a practical definition for the kilogram and to provide a link between the kilogram and the Watt force balance (e.g., Newell *et al.*, 1999, 1998). Commercial applications include oil/gas exploration, mining exploration, and reservoir monitoring. Military applications include navigation and detection of subsurface structures (tunnels or voids) (Edwards *et al.*, 1997).

Absolute gravity meters can be used in any of these applications but often the same results can be obtained with a smaller or less-expensive relative gravity meter. Thus, it is difficult to make a hard rule about when absolute or relative gravity meters should be used but one can make a few generalizations. Absolute gravity measurements are usually necessary for geoid (shape of the Earth) measurements or for determining absolute network points (used to calibrate relative gravity meters). Absolute gravity meters are also used for measuring the slowly varying nonperiodic variations of gravity (over the periods of years). For example, absolute gravity has been used to watch small uplift signals (1 mm yr^{-1}) in the Canadian shield due to post-glacial rebound (e.g., Lambert *et al.*, 2006) as well as for reservoir monitoring in the largest North American oil/gas reservoir in Prudhoe Bay, Alaska (e.g., Brady *et al.*, 2006). Absolute gravity meters are clearly necessary to calibrate relative gravity meters and for metrology work such as the redefinition of the kilogram in terms of the Watt balance. In most applications, however, a smaller and less-expensive relative gravity meter can achieve similar results. Until absolute gravity meters become smaller and less expensive, most practical measurements of gravity will continue to be made with relative spring gravity meters that are small, light, easy and quick to operate, and inexpensive in comparison with absolute gravity meters with equivalent precision.

3.1.2   Definition of Precision and Accuracy

Before describing absolute and relative gravity meters it is useful to discuss and define the common specifications of these gravity meters in terms of, precision, repeatability, and accuracy. Loosely speaking, 'precision' is a measure of how fine a measurement can be made at any specific time and location. Precision is often connected with the number of significant digits to which a value has been reliably measured. 'Repeatability' is a measure of the agreement between identical measurements but typically done at different times or after the measuring instrument has been moved and then brought back. Repeatability can be lower than precision if the instrument setup or the passage of time affects the reading. 'Accuracy' is how many of the digits can be trusted to be absolutely correct. The term accuracy typically relates to a reproducible standard usually kept by a national or international agency. These different terms are often used interchangeably by geoscientists (and manufacturers of instruments) when describing the quality of instruments or measurements which can create a great deal of confusion. It

is therefore helpful to go a bit beyond these simplistic definitions and put these different measures of quality on firmer mathematical ground.

3.1.2.1 *Precision*

Precision is normally the least understood term for the quality of a gravity measurement but it is one of the most important specifications for determining which instrument should be used for any given application. Precision refers to the relative uncertainty of a reported gravity measurement. It can be given in units of acceleration or as a fraction of the Earth's gravity value. For example, a 1 mGal precision is sometimes referred to as a 1 part per million (1 ppm) measurement. The precision of gravity meter can be thought of as the resolution or the number of meaningful digits that can be read from the instrument. Precision is a difficult quantity to understand because it depends upon four quite different things: the quality of the instrument, the noise of the site at which gravity is being measured, the length of time that the value is averaged, and the frequency band of the noise and/or signal. The precision can be limited by any one or by a combination of these factors.

Clearly, the quality of the instrument must be high enough to report a stable value over the time of the measurement. However, it can happen that the background noise (of a site) is higher than the fundamental or intrinsic noise of a gravity meter. In these cases, the background noise will set the limit for the precision of the measurement. It is well known that high-precision gravity measurements are much easier to obtain in a deep underground mine far away from an ocean or man-made noise than on the busy streets of a populous city. The length of time for the measurement is also a critical parameter for the precision of any measurement. Averaging data is a very common way to increase the precision of a measurement. It is therefore always important to specify the averaging time associated with any precision specification. Finally, the noise typically depends strongly upon the frequency at which the observation is made. For example, it is typically much easier to measure gravity signals that are varying over a few hours (e.g., tidal signals) than to measure very slowly varying signals (signals that vary over periods of months or years). Thus, gravity noise tends to have a red spectrum because it tends to increase at low frequencies. Often it is impossible to separate the background and instrument noise so it is important to remember that the noise typically is composed of both types of noise.

If the sum of the instrument and background noise is white (or more precisely if all individual measurements are uncorrelated), then the precision of the instrument at any given frequency will improve with the square-root of the observation time. The simplest type of noise is called 'white noise' and is characterized by a flat frequency (or amplitude) spectrum. In this case, one way to report the precision of an instrument is to include the averaging time in the specification. This is sometimes referred to as a spectral noise density. A spectral noise amplitude, η, for gravity measurements can be given in units of acceleration per square root of the smallest frequency bin (i.e., μGal $Hz^{-1/2}$). The smallest frequency bin is given by the reciprocal of the observation time. The noise spectrum is a very good way to discuss noise. It describes exactly how the noise depends upon the frequency and it also incorporates the averaging time explicitly through the units. **Figure 1** shows the noise spectrum of a portable Earth tide (PET) gravity meter created with 1 day of data sampled at 1 Hz. It is clear that the noise between 0.05 and 1 Hz (1–20 s period) is dominated by microseismic noise. This background noise is real and is caused by slow undulations of the ground at periods similar to typical ocean waves. It is interesting to note that these data were taken in Boulder, CO, very far from the ocean! At frequencies between 0.001 and 0.05 Hz (between 5 min and 20 s period), the noise level is almost constant and is less than about 2 μGal $Hz^{-1/2}$. At frequencies from 0.0001 to 0.001 (15 min to 3 h), the gravity background rises rapidly due to the gravitational effect of the tidal forces from the Sun and Moon.

Much longer gravity records (years) have been made using superconducting (and other low drift) relative gravity meters. The noise spectrums from these long records show a very nice tidal band and then the noise rises at lower and lower frequencies. Part of the rise at low frequencies is due to instrumental drift and some of it is due to very long period Earth motion (plate tectonics, water table, etc.). A good discussion of the noise spectrum at long periods is given by Van Camp *et al.* (2005). Notice that for short measurement times, the background noise is dominated by the microseismic background noise that can be as high as 100 μGal $Hz^{-1/2}$.

One can untangle the background seismic noise from the instrument noise by subtracting two different instruments. **Figure 2** shows the difference between two PET gravity meters that are recording simultaneously at 1 Hz for 1 day at the same location. Real seismic noise and Earth tides are removed in the difference signal.

The white-noise background of the difference spectrum is 3 μGal $Hz^{-1/2}$. We expect that the differential spectrum will be $\sqrt{2}$ larger than noise spectrum for each meter so that this is consistent with the background noise of 2 μGal $Hz^{-1/2}$ measured with one instrument between 0.001 and 0.05 Hz (**Figure 1**).

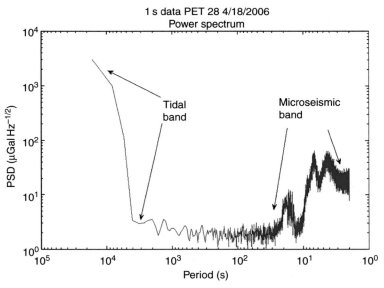

Figure 1   One day of 1 Hz measurements with a portable Earth tide gravity meter.

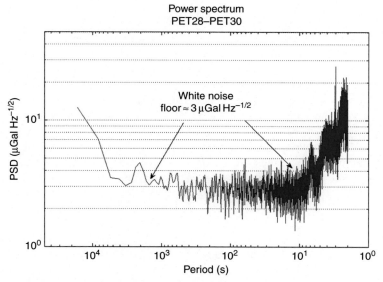

Figure 2   Difference between two PET gravity meters (1 day sampled at 1 Hz).

If the average background noise (seismic plus instrument noise) is η, the precision, P, for any observation time is then given by multiplying the spectral noise amplitude by the square-root of the smallest frequency bin. Alternatively, we can divide the spectral noise amplitude by the square root of the observation time given in seconds (i.e., $P = \eta/\sqrt{T_{\text{observation}}}$). For example, if we are taking gravity measurements that have a spectral noise amplitude dominated by seismic noise of $\eta = 100\,\mu\text{Gal}\,\text{Hz}^{-1/2}$, we will be able to obtain a precision of better than $10\,\mu\text{Gal}$ after 1 min. Higher precision can be obtained with an increasingly steep penalty of increased observation time. A 10 min average would be necessary to achieve $4\,\mu\text{Gal}$ and 2 h to achieve $1\,\mu\text{Gal}$. The spectral noise amplitude is a useful figure of merit for the precision if we can assume that the noise is time independent or stationary in time.

In general, the combined site and instrument noise is time dependent. For example, it is quite usual in many different types of measurements that noise tends to increase at low frequencies (long time averages). Sometimes this is called red or pink noise. A measurement that exhibits some sort of drift or linear variation of the measurement with time will give rise to a noise spectrum that increases at low frequencies. A drift in the instrument (or background) will cause the precision of a measurement to stop improving after the accumulation of the drift is equal to the uncorrelated noise component (that is being diminished by averaging). Equating the

random noise level with the drift allows one to calculate a critical observation time beyond which the decrease of random noise through averaging is offset by a larger drift error. Mathematically, this is given by an observation time, $T_{\text{obs}} = [\eta/d]^{2/3}$, where η and d are the spectral noise amplitude and the drift of the gravity meter. For example, if we use a gravity meter with a drift of 10 mGal per month and a background (seismic) noise of $100\,\mu\text{Gal}\,\text{Hz}^{-1/2}$, the crossover point will occur after about 15 min of observation. High-precision gravity meters (i.e., superconducting gravity meter) can have a low noise figure of $1\,\mu\text{Gal}\,\text{Hz}^{-1/2}$(in the tidal frequency band) and a drift as low as $2\,\mu\text{Gal}\,\text{yr}^{-1}$. In this case, the crossover point is about 3 years. It is therefore not an accident that most survey gravity meters use an integration time of between 3 and 15 min and many tidal studies done with superconducting gravity meters involve many years of observations. This type of calculation can provide an unbiased way to choose the type of instrument needed and the optimal integration time for any given application.

Drift is the simplest type of time behavior for the noise of a gravity measurement. Gravity meters can also exhibit nonlinear drift and even jumps (or tares) in gravity. Furthermore, one must recognize that the environmental background noise (real gravity changes that cannot be modeled) also contribute and can dominate the noise level of a measurement at any timescale. At one extreme timescale (small timescales), one can consider gravity measurements taken on a boat. Over long time periods, the ocean level is very stable but at short time intervals the wave action can produce huge accelerations (100 000 mGal or more) that will ultimately limit the precision of measurements. At long timescales (many years), the gravity noise at a specific location on the Earth may become dominated by tectonic motion or water table variations. It is often difficult to distinguish between noise that is related to instrument imperfections or background environmental sources. Indeed, one can often obtain the same level of precision using different qualities of instrument as long as the environmental noise dominates the precision.

3.1.2.2 Repeatability

Repeatability is perhaps the most useful specification for a gravity meter because it can be used as a guide to what a typical user can expect to achieve in a real-world field measurement. Repeatability does not refer to the absolute accuracy of the measurement but it does include other important sources of variation that may not be included in the reported precision of a gravity meter. In particular, gravity meters always exhibit some tendency to jump or 'tare' when they are moved. These tares can be a result of physical movement of the balancing mechanisms inside the instrument (similar to the tares found in a bathroom scale after moving). It is easy to envisage common sources of tares by considering a common spring gravity meter. Spring gravity meters are in principle very sensitive scales that balance the restoring force of a spring against gravity. The length of the spring and the mass must be controlled to an incredible level for spring gravity meters. For example, if the stretched length of the spring inside of a gravity meter is 1 cm, then the length of the spring must be controlled to 10^{-11} cm to achieve a precision of 1 μGal (1 ppm). This is 10 times smaller than a molecule of water! It is easy to imagine that the length of the spring or the clamping mechanism of the spring may change by this small amount when the instrument is moved or bumped during a survey. Similarly, a small change of the balanced mass will cause a tare in the measurement. The mass can change due to condensation caused by humidity or outgassing. Mass can also be removed or added by clamping the instrument during shipment. Mass changes will cause the value of gravity to jump or tare.

In addition to tares internal to the gravity meter, there may be setup errors that introduce a small jump or tare every time the instrument is moved. For example, a gravity meter must be leveled to about 9 s of arc (~45 μrad) in order to achieve a 1 μGal reading. The error is quadratic in the absolute error in leveling. Leveling errors will normally cause a random jump in the measurements but will always be biased toward a smaller value of gravity because of the quadratic nature of the error.

The precision of the gravity meter together with drift, tares, and possible setup errors will place a lower limit on the repeatability of any gravity meter. As always, environmental noise can also place a limit on the repeatability at any site if it is larger than the intrinsic instrument repeatability.

The repeatability for a survey gravity meter can be defined as the variation (1 sigma) in gravity that will occur if the gravity meter is repeatedly set up at a given site, then packed up, and transported some distance and finally brought back to the same site. A common way to measure repeatability with a relative spring gravity meter is to make gravity measurements at several different stations and then repeat these measurements several times to check the repeatability of the gravity meter. These data can be least-squares fit to a solution that includes a unique value of gravity at every station and a drift parameter. The residuals of the fit can be viewed as a measure of the repeatability of the instrument. It is important to scale the standard deviation of the residuals by the square root of the degrees of freedom of the least-squares fit in order to arrive at the repeatability.

Figure 3 shows data from a CG5 spring gravity meter that was taken to a survey in Northern Canada in December 2003. The survey consisted of 39 stations on two different lines across the survey area. The average separation of each station was 40 m. The first line consisted of 23 sites (Station ID #1–#23) and were repeated twice. The second line consisted of 16 stations (Station ID #24–#39) were also repeated twice. Station #1 was common for both lines and was repeated six times during the survey in order to control for drift. The survey locations are plotted in **Figure 3** with the station ID given on the right-hand vertical axis. The gravity data were then least-squares fit to a network model consisting of 39 unique gravity values and a single drift parameter. The difference between the measured gravity and the model (residuals) is also plotted in **Figure 3** using units given on the left-hand vertical axis. The residual difference has a standard deviation of 3 μGal. The repeatability obtained by scaling the residuals by

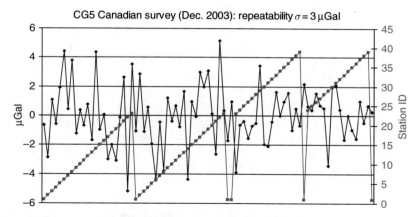

Figure 3 CG5 repeatability using data from a CG5 Survey in Northern Canada during December.

the number of degrees of freedom is 5 µGal. No tares or jumps in the gravity meter were observed during the survey. This survey is typical of data taken with many different CG5 instruments and is consistent with the body of data used to set the repeatability specification of the CG5 at 5 µGal.

3.1.2.3 Accuracy

The accuracy of a measurement is much simpler to define and to understand than the precision or the repeatability. Unfortunately, the accuracy is also the most elusive because there is no way to determine the absolute accuracy of an instrument unless a better instrument exists. Clearly, the accuracy of a gravity meter is limited by its repeatability and precision. In addition, the accuracy must take into account any systematic errors (errors that do not change) that will affect the reported value of gravity.

The accuracy refers to the correctness of the measurement in absolute terms. As we have discussed, gravity has units of acceleration given by meters per second per second (or $m s^{-2}$). These units are defined and kept by standards laboratories worldwide and are the concerns of metrology. The 'second' is defined as the duration of 9 192 631 770 periods of the radiation corresponding to the transition between the two hyperfine levels of the ground state of the cesium-133 atom at 0 K. The meter is now a derived quantity from the speed of light. It was redefined in 1983 to be length traveled by light in vacuum during 1/299 792 458 of a second. In practice, the standard meter can be realized by an iodine-stabilized laser (Chartier *et al.*, 1993) manufactured with a recipe given by the Bureau International des Poids et Mesures (BIPM), in Sevres, France.

An absolute gravity meter is defined as an instrument that directly measures the acceleration of a free particle in terms of fundamental units of space and time as defined by the standards community. Spring gravity meters (including superconducting spring gravity meters) are not considered absolute gravity meters for two fundamental reasons. First, they do not measure gravity directly but the force needed to keep a proof mass stationary. Secondly and more importantly, the measurement is not directly tied to metrological standards. The proof mass is not tied to the kilogram. The force measurements are not tied to fundamental units such as the watt, meter, or the second. Sometimes the distinction between relative and absolute instruments is misunderstood to mean precision or repeatability. It is quite possible to make an absolute gravity meter with much worse precision than a relative gravity meter. For example, we will discuss pendulum gravity meters, an early type of absolute gravity meter that have now been superseded by relative spring gravity meters because the repeatability of the spring gravimeters is much higher. At the other extreme, the relative superconducting gravity meter is the most precise gravimeter but its accuracy is limited by the ability to calibrate the instrument *in situ* and long-term drit and/or tares.

Absolute gravity meters have one very demanding requirement. In order for an absolute gravity meter to be considered a metrological standard, there must be an associated systematic error budget that takes into account all errors in the instrument that can affect the measured value of gravity. It is notoriously difficult to identify, enumerate, and estimate the size for all errors in any instrument. Many systematic errors will not affect the precision or repeatability because they will remain fixed over time but they must be included in the systematic error budget. Some of these errors may be fixed with respect to a given instrument and some may depend upon the setup conditions or site conditions. The systematic error budget (Niebauer *et al.*, 1995) for a modern-day free-fall absolute gravity meter called the FG5 provides a limit of about 1–2 µGal.

It must be noted, however, that the accuracy of an absolute gravity meter cannot be independently checked except by other absolute gravity meters of equivalent quality or by a superior instrument. Different absolute gravity meters are sometimes brought to a single location to intercompare their measured values as one type of check on the accuracy. International intercomparisons of gravimeters (Francis and Van Dam, 2003) show agreement between instruments of the same type to be consistent with the systematic error budget of the instruments. However, it is a bit disconcerting that intercomparisons between different types of absolute gravity meters that are held at the BIPM in Sevres, France, every 4 years typically easily exhibit a variance in excess of 10 µGal (Vitushkin *et al.*, 2002). Presumably these differences indicate the presence of one or more systematic errors in one or more of the different instruments but there is no simple way to determine which gravimeter has the larger systematic errors especially if the errors are constant and do not change over time. The disagreement (variation) between different absolute instruments accepted by the different standards laboratories in different countries sets how well a thing can be measured absolutely. The effort to refine absolute instruments and resolve discrepancies between different methods and absolute instruments for measuring gravity is an ongoing effort by the metrological community.

The situation where different types of absolute gravity meters are able to measure with a high precision and yet disagree with the values obtained with similar instruments is a common occurrence in gravity. Gravity is the weakest force in the universe and is 10^{40} times smaller than electrostatic forces. This means that one must eliminate the many subtle influences of the nongravitational interaction of materials to an extraordinary level in order to measure gravity. A further difficulty with gravity is that it is fundamentally tied to acceleration through the laws of general relativity. According to Einstein, one cannot distinguish the difference between a gravity meter in an accelerating elevator from one that is stationary

with a different gravity field. This means that any motion of the ground couples directly and fundamentally into the measurement. That is why gravity simply cannot be measured as well in an area with high seismic activity as it can at a very quiet site.

The difficulty associated with measuring gravity can also be seen even more dramatically in the measurement of the fundamental gravity constant, *G*, often referred to as big *G* (gravity is known as little *g*). *G* can be measured with a precision of 1 ppm or higher but apparently different methods of measuring gravity can disagree at the 1% level even when these measurements are made carefully by groups of scientists in the world's best metrology laboratories (Quinn, 2002). This situation can be frustrating for the scientist who needs to know *G* precisely but it is fruitful ground for the metrologist who wants to understand the source of systematic errors or for the scientist who is seeking for new information about the fundamental laws of physics.

It is important to realize, however, that the difficulty in determining the absolute accuracy of a gravity measurement does not hinder most geoscientist's use of gravity. For almost all real-world practical applications, absolute accuracy is not critical. As long as one knows how 'precise' and 'repeatable' their gravity measurements are, one can measure crustal motion, measure tidal signals, look for subsurface voids, or monitor earthquakes and reservoirs with a high degree of certainty!

3.2 Gravity Meters

Gravimeters are separated into two categories: absolute and relative. Absolute gravimeters can be defined as any device or instrument that measures gravity directly in terms of standard time and length units (m s^{-2}). Early absolute gravimeters made use of the physical pendulum commonly described in elementary physics books in which the period is related to the location of the center of mass and gravity. In this case, the period and the distance can be linked to standards of length and time to give an absolute gravity determination. The pendulum was eventually replaced by a free-fall absolute gravity meter that monitors the descent of a freely falling body using laser standards and atomic time standards.

Relative gravity meters vary a great deal in the construction and the type of materials that are used but basically they are always some type of mass that is being suspended by a spring. The spring material may be some sort of low-expansion metal or quartz. The spring can even be a superconducting magnetic field which produces a very low friction suspension. Relative gravimeters are distinguished from absolute gravimeters because they indirectly measure the acceleration of gravity by measuring the stretch of a spring. The stretch of the spring is proportional to a force constant (K = spring constant) that must be calibrated. The stretch is also proportional to the suspended mass.

Figure 4 shows a historical progression of the accuracy/precision of gravity meters since the 1600s. It is interesting to note that initially absolute gravity meters made with a pendulum were the most precise instrument until early in the 1900s where spring gravity meters jumped ahead of the pendulum. Free-fall gravity meters started behind the pendulum but quickly advanced to meet and finally beat spring gravity meters in the 1980s. The graph ends at 1980 largely because gravity measurements have a similar repeatability today (2006) of about 1 µGal. This limiting accuracy is mostly due to environmental sources of gravity errors and background noise level of the Earth. Higher-precision gravity measurements can be made but only for cases where the signals are periodic or have a known time signature and persist over long times so that the background noise can be averaged to obtain higher precision. For example, it is possible to measure tidal signals with a precision of nanoGal ($10^{-12}g$) if one observes

gravity continuously for long periods of time (years). Long-period measurements (DC) and/or short-step measurements tend to be limited by the background noise level of 1 µGal (Van Camp *et al.*, 2005).

3.2.1 Absolute Gravity Meters

There are only two types of methods that have been effectively used to realize an absolute gravity meter: a proof mass on a pendulum or in free fall. Both methods allow an object to fall under the Earth's gravity, although the pendulum has a radial constraint. The force of gravitational attraction of the Earth is proportional to the falling mass in both of these methods so that the resulting acceleration (given by the force divided by its mass) is independent of the size or shape of the falling mass. One might suspect that the mass enters into the pendulum through its moment of inertia but as we shall see there is a clever method to avoid this problem. The mass used in these instruments is sometimes called a test mass or proof mass to highlight the fact that any mass would follow the same path. All proof masses follow the same invisible laws of gravity in the same way that different leaves trace out the same invisible forces in a flowing stream. Free-fall absolute gravity meters can use macroscopic proof masses made of glass and metal or they can even use individual atoms if they are properly cooled with lasers before they are allowed to free fall.

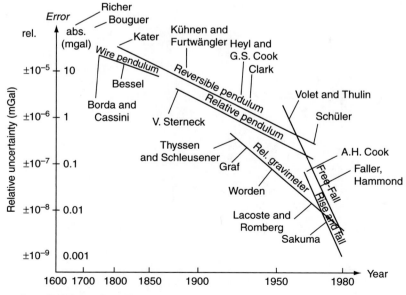

Figure 4 Accuracy of gravity meters, from Torge.

3.2.1.1 *Pendulum absolute gravimeters*

Galileo noticed that the period of a pendulum is nearly independent of the amplitude of its swing in 1602 which gave him the idea of using pendulums as clocks. Christian Huygens built the first successful clock based on the pendulum in 1657. Gravity measurements made using a pendulum were first made popular by Pierre Bouguer in 1749. The pendulum was the primary method for measuring absolute gravity for about 200 years. A very accurate absolute pendulum gravity meter was developed by Captain Henry Kater (Kater, 1818). Kater realized that a pendulum could be swung around two different axes in order to measure absolute gravity using a simple distance and time measurement.

Figure 5 shows a schematic diagram of the Kater pendulum. It has two sets of knife edges that can be used to swing the pendulum from two different points about its center of gravity. The center of gravity can be adjusted using adjustable weights in order to make the swing period for the two suspension points equal.

The equation of motion for a pendulum (with no friction) can be written as

$$\ddot{\theta} = -\frac{mgL}{I}\sin(\theta)$$

where θ is the angle of swing and m is the mass of the pendulum. The distance between the knife edge and the center of mass is given by L and the moment of inertial of the pendulum about the axis of rotation defined by the knife edges is given by I. In general the distance between the center of mass and the knife edge and the moment of inertia will be different for

the two sets of knife edges. The period is nearly independent of the swing amplitude as long as the angles are small. The period of the pendulum around each knife edge is given by

$$\Gamma_i = 2\pi\sqrt{\frac{I_i}{mgL_i}}, \quad i = 0 \text{ or } 1$$

The moment of inertia around each knife edge is related to the same moment of inertia around the center of mass by $I_i = I_{cm} + mL_i^2$. All moments of inertia can be eliminated by subtracting the two moments about the knife edges to obtain

$$m(L_2^2 - L_1^2) = \frac{mg}{4\pi^2}\left(L_2\Gamma_2^2 - L_1\Gamma_1^2\right)$$

Gravity is now given simply by

$$g = \frac{4\pi^2\left(L_2^2 - L_1^2\right)}{\left(L_2\Gamma_2^2 - L_2\Gamma_2^2\right)}$$

Notice that gravity is given only by distance and time measurements. It does not depend upon the mass of the pendulum! The expression is further simplified if one tunes the periods around each knife edge to be the same by moving the adjustable weights at the end of the pendulum. In this case, the difference of the two distance measurements is eliminated and only the sum of the two distance measurements becomes important. The expression for gravity then becomes

$$g = \frac{4\pi^2(L_1 + L_2)}{\Gamma^2} = \frac{4\pi^2 L_{\text{tot}}^2}{\Gamma^2}$$

This result is nice because one does not have to locate the center of mass because the only distance that matters is the total distance, L_{tot}, between the flexures. This equation shows that gravity now only depends upon two measurements one of which is a distance and the other a time. This is an ideal absolute gravity meter because one can obtain a calibrated result simply by measuring the distance between the knife edges with a standard meter stick and the period with a standard clock. The Kater pendulum is seen to be calibrated by its very design.

The pendulum was a very successful absolute gravity instrument but it finally reached its limiting accuracy of a few parts in 10^7 due to the difficulty in producing low-loss knife edges. The pendulum was eventually replaced with a free-fall gravity meter in which the proof mass was allowed to free fall in a

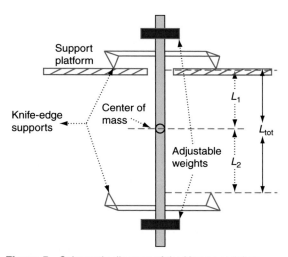

Figure 5   Schematic diagram of the Kater pendulum.

vacuum. This eliminated the frictional problem inherent in the pendulum.

3.2.1.2 Free-fall absolute gravimeters

Free-fall absolute gravimeters became technically feasible in the 1960s due to the availability of high-quality vacuum technology, electronics, and clocks. Obviously, there are many technical difficulties involved with dropping (or throwing upwards) an object and then determining the distance and time for free fall. The trajectory of the falling mass is determined by reflecting a laser beam off the falling object and comparing it with light that is reflecting from a stationary (or reference) object. The falling object must be launched vertically in order to keep the object from rotating or translating too much so that the measurement will function. One must catch the object in a nondestructive way. It is useful to automate the system so that repeated measurements can be done. A less obvious but important part of the absolute gravimeter is that it should provide isolation from ground noise or seismic surface noise that will otherwise introduce acceleration noise of about $10^{-6}g$ over a broad frequency range peaking at around 0.17 Hz (or 6 s). The absolute gravimeter must have a large dynamic range (10^9) in which all of the digits are useful and are not corrupted by systematic errors (due to air resistance or electromagnetic forces).

Many different types of absolute gravimeters can and have been built but they generally consist of three independent systems each of which addresses the tasks outlined above. The components of an absolute gravimeter and their function are outlined below:

1. Free-fall chamber

   a. A vacuum chamber to remove air resistance
   b. A test mass that will be monitored during free fall
   c. Isolation from electromagnetic influences on test mass
   d. A catch and release mechanism

2. Time and distance measurement system

   a. Laser interferometer encodes free-fall distance onto the phase of an atomic system.
   b. Wavelength-stabilized laser tied to metrological standards

      i. Mach–Zender or Michelson-type interferometer
      ii. Atom interferometer

   c. Precise time measurements of fringes made using an

      i. atomic clock standard

3. Seismic isolation

3.2.1.2.(i) The free-fall chamber The free-fall chamber contains the catch and release mechanism for the freely falling test mass. The trajectory of the test mass should be sufficiently vertical to avoid Coriolis forces arising from the rotating Earth and also so that the light will continue to reflect off the object during its descent. The test mass also should not rotate too fast so that rotational motion is confused with linear acceleration in the vertical direction.

The free-fall chamber should also shield the object from electromagnetic forces. This task is a bit daunting given that on a molecular level electrostatic forces are 10^{40} times larger than gravitational forces! The free-fall chamber is usually grounded and conductive so that it acts like a type of Faraday cage to keep outside sources of electrostatic charge from influencing the interior. The dropped object is usually fabricated to be conductive (to avoid static charges) and nonmagnetic. However, even a conductive nonmagnetic test mass will feel a magnetic eddy force when dropping through a magnetic field gradient. Care must be taken to keep magnetic field gradients minimized as well.

The free-fall chamber is evacuated to avoid forces due to air resistance. One must eliminate these forces below 10^{-9} N if we want to measure the acceleration of a 100 gm freely falling mass to 1 µGal precision. A nice experiment to demonstrate the dependence of air resistance on velocity is done with physics students at the Hockaday school. **Figure 6** shows the measured force of air resistance as a function of velocity measured by dropping 10 coffee filters (a 10 gm mass) with a large cross-sectional area. Notice that the force is 0.1 N at about 2.8 m s^{-1} in air. This plot implies that one must decrease the pressure by at least nine orders of magnitude to get the air resistance low enough to measure absolute gravity with high precision. This means that a vacuum of about 10^{-7} Torr (1 atm = 760 Torr) would be needed for a 1 µGal free-fall measurement.

A similar limit has been determined (Niebauer, 1987) for a more conventional dropped object using a calculation of air drag (Kennard, 1938) given by $F_{drag} = \rho ScV/4$, where ρ and c are the air density

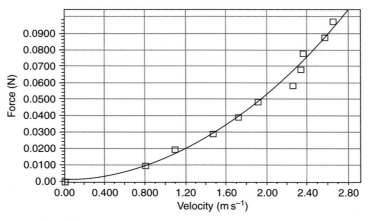

Figure 6   Air resistance vs velocity, from Hockaday.

and mean speed of the air molecules, S is the surface area of the object, and V is the velocity.

The potential sources of systematic errors for measurements of absolute gravity have been described and cataloged (Niebauer *et al.*, 1995) and are manageable at parts in $10^9 g$ and below if care is taken.

One particularly successful strategy to address all of these concerns was first introduced by Zumberge (1981). They constructed a small elevator that served as a platform to drop and catch a reflective test mass. The test mass terminated in three spherical 'feet' that rested in three V-grooves fastened to the floor of the elevator.

The elevator was gently lifted to a predetermined height where it rested momentarily before starting the drop. Once the drop was initiated, the elevator was accelerated downwards by an external motor at about $2g$ for about 20 ms until the test mass was free falling about 2 mm behind (and no longer touching) the elevator. Once the separation was achieved, the elevator reduced its acceleration to g in order to maintain a fixed separation between itself and the freely falling test mass during the measurement part of the freefall trajectory. The test mass and co-moving elevator fell about 20 cm in 0.2 s during which time the trajectory of the test mass was externally measured using a laser interferometer. Near the bottom of the free-fall chamber, the elevator acceleration was decreased until the object was caught softly (while still falling at close to g). Once the object was caught, the elevator was slowed quickly to a rest position at the bottom of the free-fall chamber. This cycle was automated and could be repeated every few seconds. The co-moving elevator solved several

problems at once. First it provided an elegant way of dropping and catching the object in a nondestructive way that produced a release mechanism that did not impart very much horizontal or rotational velocity to the test mass. It also acted as a sort of miniature Faraday cage to isolate the test mass electromagnetically from the rest of the free-fall chamber motor and vacuum pump voltages. In addition, it acted as a kind of 'drag-free' chamber to reduce air resistance on the test mass. The drag-free chamber reduced the vacuum requirement to about 10^{-6} Torr. In a very real sense the freely falling test mass and drag-free elevator became the equivalent of a typical low-orbit Earth satellite that uses the same drag-free external satellite to remove the corrupting influence of air resistance on the orbit. The only difference is that the absolute gravimeter orbit is tragically short lived (0.2 s) and must be re-launched after each encounter with the Earth. We have sometimes jokingly referred to this method as 'satellites in a can'. The advantage of using a satellites in a can to measure gravity is that it allows us to get the satellite much closer to the Earth for measurement of gravity near (or potentially below) the Earth's surface. It is also clearly much cheaper to launch satellites in a can!

An important variant of the free-fall method described above was first implemented by Sakuma (1970) in which the test mass was launched upwards and allowed to rise and fall. Measurements on rise and fall test masses are more resistant to systematic errors that tend to cancel on each half of the measurement. For example, air resistance acts as a downward force as the test mass travels upwards but then becomes an upwards force when the test mass is falling downwards. On average, this force cancels and allows accurate measurements of gravity with a lower level of vacuum. Sakuma launched his objects upwards by releasing an elastic element. This device was not automated and required a great deal of time to prepare so that only a few measurements could be made in a day. The co-moving elevator system has also been used to successfully launch a test mass repeatedly in a rise and fall manner quickly (about two times per second). While the rise and fall method continues to be a useful area for research, it has not become a practical method because of the difficulty in achieving a perfectly vertical trajectory compared to the dropping method. It is possible to make very good measurements of gravity using this method; however, only very small physical changes in the launching mechanism cause the test mass to deviate from a vertical trajectory. This problem is minimized for straight free fall because the initial velocity is close to zero. Thus, the horizontal components of the initial velocity are proportionally much smaller for free fall compared to the rise and fall method.

It is also possible to drop cooled atoms instead of macroscopic objects. Kasevich and Steven (1991) at Stanford have pioneered this free-fall method. They introduce a source of laser-cooled sodium atoms into a free-fall chamber. Atom interferometery is used to measure the free-fall acceleration of the atoms. This method has an advantage over dropping macroscopic mirrors and using laser interferometery because there are no 'moving parts' that can wear out. Currently, this technology is still under development in the laboratory but it has the potential to become the absolute gravimeter technology of choice in the future.

3.2.1.2.(ii) The distance and time measurement: Trajectory determination

Once the test mass is somehow coaxed into free fall, the absolute gravimeter must very accurately determine its trajectory in order to determine the acceleration g (gravity). This is often done with some sort of laser ranging. It is much more challenging to do this for near-Earth satellites (satellites in a can) than traditional laser ranging for satellites orbiting high above the Earth because the orbit is very short lived (0.1–0.2 s) for the free-falling test masses in the free-fall chamber. A very good method for measuring the trajectory of a test mass is to use an optical interferometer in which the free-fall distance is used to modulate the phase of light reflected from its surface. This method

was first pioneered by Faller and Dicke in 1963 even before the advent of common lasers using a white-light interferometer. **Figure 7** shows a schematic diagram of the apparatus. A monochromatic but incoherent (white) light source was collimated and fed into an optical system (interferometer) that split the light so that some of it reflected off a falling retroreflector (sometimes known as a cat's eye) and another portion of which reflected from a fixed mirror. The light was then recombined onto the photodetector to convert the intensity variations caused by interference of these two beams into an electrical signal that could be used for timing. Incoherent light will only interfere when the two path lengths are identical. When this special condition is met by the falling retroreflector, the recombined light will momentarily be bright due to constructive interference. The interferometer was designed so that this special condition of equal path lengths between the light source and the falling test mass and the reference mirror occurred at three distinct displacements of the falling test mass. **Figure 8** shows the three different cases or displacements of the falling test mass together with the corresponding fixed light path that caused a white-light interference. The time for each white-light fringe (or interference) was then measured. In this case, a fringe can be thought of as a momentary bright flash of light when the path length of the light reflecting from the falling object matches one of the three fixed light paths reflected from the stationary mirror. (It is common to use the word 'fringe' for the time-varying behavior of the recombined light intensity as the constructive or destructive interferences occur even though one usually thinks of a fringe being a spatial variation of intensity.) These three conditions

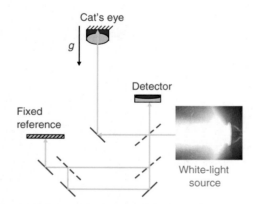

Figure 7 Schematic diagram of the Faller–Dicke absolute gravity meter (Princeton, 1963).

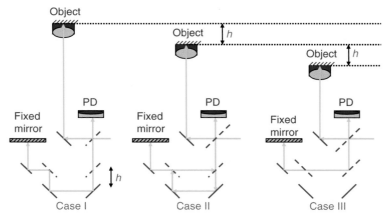

Figure 8   Three optical paths (cases I, II, and III) for interference.

provide three time and distance pairs (t, z): $[(0, 0)$, (t_1, h), $(t_2, 2h)]$ that are sufficient to solve for the initial velocity and acceleration of the object. The acceleration is independent of the initial velocity which means that the precise starting position of the drop was not critical to the measurement of gravity. The value of gravity is also independent of the composition of the dropped object because all matter experiences the same force of gravity according to Einstein's equivalence principle.

The white-light interferometer was quickly replaced with a coherent laser source once they became easily available. A coherent laser source will produce fringes when the path length of the reflected beam (the changing arm of the interferometer) differs from that of the reference path (fixed arm) by any multiple of $(1/2)\lambda$, where λ is the wavelength of the laser. The factor of $1/2$ occurs because the path length of each arm is twice its separation from the splitting/recombining objects (as the object falls $(1/2)\lambda$ the total path length change for that arm is λ which causes a new interference condition). This removes the need to have three different fixed paths and one can adopt a simpler Michelson-type interferometer arrangement such as that used in many instruments after the introduction of the laser. A simplified Michelson-type interferometer is shown in **Figure 9**.

The mirrored surfaces are meant to indicate a retroreflector commonly called a corner-cube (or cube-corner depending upon which English variant is used). The corner-cube has the important property that the reflected path length is independent of a horizontal translation or a rotation of the cube around the vertex. The corner-cube also reflects the

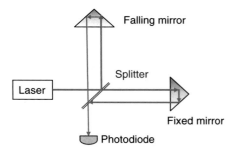

Figure 9   Michelson interferometer used to measure gravity from a falling mirror.

light parallel to the incoming beams so that the interferometer will stay aligned even if the falling corner-cube suffers small horizontal translation or rotations during its descent. These are very important practical advantages because it is not trivial to release or launch an object without imparting some initial horizontal velocity or rotation. In fact, the free-fall method would probably not be viable without a corner-cube!

The optical interferometer using a laser is an ideal method for measuring distance in a gravimeter because it has a very large dynamic range. The interferometer can be used to measure a free-fall distance of many meters or indeed many kilometers in the case of space satellites with a precision of under 1 nm. The corner-cube permits the selection of only one axis of motion with a high degree of rejection of motion in other axes (horizontal motion and rotation around the optical center). It is also possible to use other types of interferometers such as the Mach–Zender interferometer first used in the commercial FG5 absolute gravimeter. **Figure 10** shows a

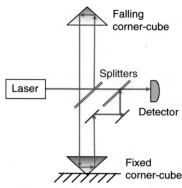

Figure 10 FG5 Mach–Zender interferometer for measuring gravity.

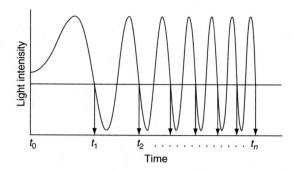

Figure 12 Two-beam interferometer signal for gravimeter with zero-crossings.

schematic diagram of a Mach–Zender arrangement where the laser is split and one beam (the reflected beam) is sent into a changing arm that reflects from both a falling mirror and a fixed reference mirror. The other arm (transmitted beam) is combined with the changing arm on a second beam splitter.

It is possible to recombine both beams on the first beam splitter but then one must consider the multiple beam interferences that occur due to light traveling multiple times in the changing arm. In this case, the interferometer behaves more like a cavity. The variation with two splitters avoids cavity effects and results in simple two-beam interference. Two-beam interferometers can be understood simply as the recombination of two waves from each arm of the interferometer as shown in **Figure 11**.

When the two wave fronts are in phase, they recombine to produce a large amplitude and is known as constructive interference. At this point the light hitting the photodiode is bright. When the two waves are completely out of phase, the interference is destructive and the intensity of the light hitting the photodiode is dark. This pattern of light and dark fringes repeats as the object falls. The time for successive cycles for each fringe becomes shorter

and shorter as the object falls faster and faster due to the acceleration of gravity. The optical signal that hits the photodiode is shown in **Figure 12**.

The times for each zero-crossing of each fringe are denoted by t_n. This figure only shows a small fraction of the fringes that occur in a real gravimeter. In reality, the fringes start at DC and sweep to about 6 MHz over about 0.2 s during which time over 600 000 cycles or fringes are observed. The laser interferometer is similar to the white-light interferometer example given first except that now it is possible to make many, more than three, time measurements during the free fall. **Figure 13** is a graph of the position and time for the first few zero-crossings.

One can choose some number of fringes (minimum of 3 and maximum of 600 000 for a 0.2 s drop using red laser light) and fit these to the usual curve for a freely falling body. If we assume that we measure the time for every M zero-crossings, the free-fall equation becomes

$$x_n = n\frac{M\lambda}{2} = x_0 + v_0 t_n + \frac{1}{2}g t_n^2$$

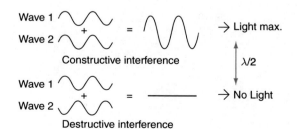

Figure 11 Two-beam interference with coherent light.

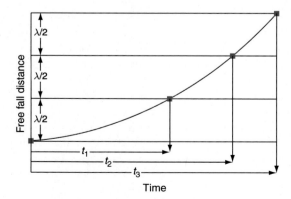

Figure 13 First four zero-crossings of the free-fall gravity meter (position vs. time).

where the initial position, velocity, and gravity are linear parameters that can be determined using a common least-squares fit. This equation is sufficient to achieve a 0.1 mGal measurement for g. For higher accuracy one must take into account the effect of the vertical gravity gradient, γ, in the equation which adds a cubic and quartic term as given by

$$n\frac{M\lambda}{2} = x_0 + v_0\left(t_n + \frac{1}{6}\gamma t_n^3\right) + \frac{1}{2}g_0\left(t_n^2 + \frac{1}{12}\gamma t_n^4\right)$$

Where, x_0, v_0, and g_0 are the position, velocity, and gravity at $t = 0$. This equation will achieve a 10 μGal measurement. In order to reach 1 μGal , one must take into account the finite speed of light. This is normally done by 'retarding' the time variable to take into account that the object position occurs at an earlier time than the fringe formation in the interferometer by an amount $t_n' = t_n - c/x_n$.

The value of gravity obtained with this type of instrument does not depend upon the initial position or velocity of the free-falling body and only depends upon the wavelength of light and the time measurements. Thus, it produces a value for gravity that is 'absolute' or intrinsically calibrated by the laser wavelength and atomic clock used to make the measurements. For highest accuracy, a primary standard laser is used for the light source and an atomic clock is used for the timing. In 2001, the BIPM passed a resolution that the ballistic absolute gravimeter described here is a primary method for measurement of gravity.

3.2.1.2.(iii) Seismic isolation Absolute gravimeters must measure the free-fall distance to about 1 nm precision in order to achieve a 1 μGal. Seismic noise varies around the earth but generally is about $10^{-6}\,\mathrm{m\,Hz}^{-1/2}$ and is found to have a peak at a period of about 6 s (Raab and Coyne). Background noise therefore places a limit on the precision for absolute gravity meters of about 100–1000 μGal Hz$^{-1/2}$ unless some type of seismic isolation is done. The FG5 absolute gravity meter accomplishes this by placing the reference arm of the interferometer on a long-period seismometer called a 'Superspring' developed by Rinker and Faller in 1980. The Superspring has a period of 60–90 s and can reduce the seismic noise so that the noise on one drop is about 3–5 μGal at a quiet location. It is also possible to measure the seismic noise using an accelerometer

and then remove the ground noise from the measured free-fall signal.

3.2.1.3 Commercial absolute gravity meters

Currently, the most accurate commercial absolute gravity meter is called the FG5. The FG5 is primarily a laboratory instrument consisting of an evacuated dropping chamber, interferometer, superspring, interferometer, and an iodine-stabilized laser (usually a WEO model 100 I$_2$ stabilized laser). The instrument can be assembled in about 1 h and is ready to take measurements once all of the components have reached thermal equilibrium. A photograph of the FG5 is shown in **Figure 14**. The instrument is about 1.5 m tall and the instrument requires about 1 m^2 of reasonably flat area for setup. At a quiet site it is capable of obtaining a 3 μGal measurement after only one drop but more typically it has a single drop scatter of 10 μGal. Different FG5 instruments are consistent with each other at the 1–2 μGal level. This value is consistent with the systematic error budget (Niebauer *et al.*, 1995) and has been verified in several different intercomparisons (Sasagawa *et al.*, 1995; Francis and Van Dam, 2006) where metrology experts check instrument alignment and maintenance and then carefully setup the instruments. The positions of the instruments are interchanged to remove uncertainties due to horizontal gradients.

Figure 14   FG5 absolute gravity meter, from MicroLacoste.

The best intercomparisons have been done at seismically quiet locations. It is interesting to note that when different types of absolute gravity meters are intercompared at a more active site and less control of the setup and maintenance of the instruments is required, errors of 10 μGal or larger are observed (Vitushkin *et al.*, 2002). In general, it is always true that exquisite care is necessary to achieve accuracies of 1 μGal with any type of gravity meter.

3.2.2 Relative Gravity Meters

There have been many different types of relative gravity meters developed over the years. Herschel proposed the use of a static spring to balance a mass against the force of gravity in 1833. Static spring gravity meters began to dominate pendulum gravity meters from about 1930 onwards. McGarva Bruckshaw (1941) provides a very detailed discussion of the many ideas that had been tried by 1941. Bruckshaw points out that the main advantage of the static gravity meters was that they could provide a measurement in a few minutes instead of an hour or two that was more typical for pendulum measurements. It is interesting that although Bruckshaw discussed many different gravity meters of his day, he did not even mention the one gravity meter invented by LaCoste and Romberge in 1932 that was to make all of them obsolete by the time he gave his talk at the Imperial college of London!

It is useful to point out the immense difficulty of measuring gravity with a spring before discussing some of the solutions that have been wrought over time. The spring extension for a simple vertically hanging spring is proportional to gravity. Thus, one must measure the extension of a spring with the same precision of the gravity measurement. For example, if we consider a simple vertical spring stretched by 0.01 m (1 cm), we need to measure its extension to about 1 Å (or 10^{-10} m) in order to measure gravity to 10 μGal. One angstrom is the size of a very small atom (hydrogen). A human hair on this scale is 1 500 000 Å! It very difficult to measure the extension of a spring to this level even with today's technology. Moreover, it is difficult to imagine that a spring extension will be stable as time passes or with changes in temperature, humidity, or pressure. For example, the temperature dependence of the spring constant of most metal springs is about 10^{-3} per °C which means that the temperature of the spring/mass system must be held to a constant temperature to about 10 μ °C if one wants to make a 10 μGal measurement with a

common metal spring. The spring gravity meter is a great example of an instrument that works much better in practice than could be expected on simple theoretical grounds!

In 1932, Lucien LaCoste was given a homework problem to design a vertical seismograph by his professor Dr. Arnold Romberg. This project led LaCoste to invent a novel beam suspension that solved many problems of spring gravity meters. In particular, his suspension amplified the stretch of the spring associated with gravity changes so that gravity could be measured with high precision before the advent of sensitive electronics. LaCoste and Romberg eventually started the famous LaCoste & Romberg (L&R) gravity company in 1939 where they revolutionized relative spring gravity instruments (Chris *et al.*, 1996).

3.2.2.1 The L&R zero-length spring suspension

The L&R suspension is shown in **Figure 15**. A mass is suspended horizontally from a beam that can rotate freely around a low-friction pivot point. The force of gravity is balanced with spring that is somewhat disadvantaged by pulling upwards at a nominal angle of about 45° (instead of pulling directly upwards on the mass). As we will show, this

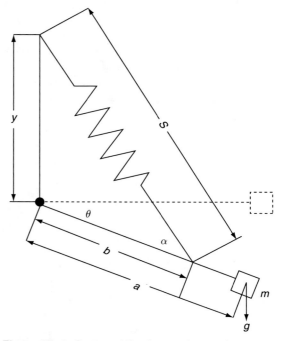

Figure 15 LaCoste and Romberg spring gravity meter suspension.

suspension produces an astatic type of gravity meter (Heiskanen and Vening Meinesz, 1958) which has no restoring torque on the beam at any angle when the system is balanced and becomes unstable when gravity is not perfectly balanced by the spring. The system is therefore very sensitive with respect to gravity changes.

The astatic nature of the suspension can be demonstrated by considering the potential energy of the system given by

$$U = \frac{1}{2}k(S - S_0)^2 - mga\sin(\theta)$$

where S is the stretched length of the spring and S_0 is the unstretched length of the spring (i.e., the stretch at which the restoring force from the spring is zero). The stretched spring length can be written as

$$S^2 = y^2 + b^2 + 2yb\sin(\theta)$$

One can see immediately that if the unstretched length is zero (i.e., $S_0 = 0$), then the potential energy can be written as

$$\begin{aligned}U &= 1/2k\left(y^2 + b^2 + 2yb\sin(\theta)\right) - mga\sin(\theta) \\ &= 1/2k\left(y^2 + b^2\right) + (kyb - mga)\sin(\theta)\end{aligned}$$

The gravity meter is balanced when $kyb = mga$. Note that the balanced beam has a potential energy with no angular dependence. It follows that there will be no torque on the system at any angle and therefore there will be no preferred angle of the beam. Typically, the gravity meter is balanced by moving the top suspension of the spring up or down (thereby changing y). Gravity is then directly proportional to the position of the top of the spring, y, and can be written as $g = kyb/ma$.

The torque on the beam, without any simplifying assumptions, is generally given by

$$\Gamma = -\frac{\partial U}{\partial \theta} = \left(mga - kyb + \frac{kS_0 yb}{\sqrt{y^2 + b^2 + 2yb\sin(\theta)}}\right)\cos\theta$$

Again we see the simplification if the unstretched length of the spring is zero ($S_0 = 0$). This removes the last term which depends upon the angle of the beam in a complicated way and gives a very simple form for the torque on the beam $\Gamma = (mga - kyb)\cos\theta$. One can imagine how happy the eager student LaCoste was to set this term to zero in his homework problem and show the simplified torque formula!

The name zero-length spring was given to the special condition where the length of an unstressed spring is zero. A simple example of a zero-length spring is a flat circular unstressed spring. When the weight is applied to the center, the spring deforms creating a force that is proportional to the stretch of the spring from its zero position. An extension spring can also be made to behave mathematically as a zero-length spring as long as its pretension is equal to the force required to stretch the spring to its length with no load attached. For example, if a spring has a force constant $k = 1\,\mathrm{N\,m^{-1}}$ and an unstretched length of 10 cm, it would need to be wound with a pretension of 10 gm in order to have zero length. The coils of this zero-length spring will just start to separate after a weight of 10 gm is placed on the end of the spring. There is some confusion in the literature surrounded by the advantages of using a zero-length spring. It is sometimes said to be the source of a long period but in fact its primary purpose in the L&R suspension is to create a beam suspension in which the restoring force is not dependent upon the beam angle.

Another common misunderstanding of the L&R suspension written in many books and papers is that the zero length provides an infinite period spring suspension resulting in high sensitivity for gravity. While it is true that a spring with an infinite period also has no torque at any angle, this is not quite the same thing as an astatic beam. The difficulty with interpreting the L&R astatic beam as having an infinite period is that one typically associates a spring period with a stable system with some restoring force that becomes smaller and smaller as the instrument is tuned to have an infinite period. The L&R suspension, however, has a torque proportional to the cosine of the angle when the beam balance is not perfect (or $mga \neq kyb$). This means that the beam will either fall down or up depending upon the relative strength of the spring and gravity forces if the beam is not perfectly balanced. In other words, the perfect L&R suspension is unstable as shown in **Figure 15**. The balancing condition is a saddle point rather than a point of stability. It is therefore mathematically incorrect to ascribe a period (even infinity) to this equation. The cosine dependence of the torque upon angle makes the system much more sensitive to small gravity changes than a spring with a truly infinite period. The likely criticism of LaCoste's professor Romberg to seeing this equation for the first time might have been to ask, "How can one use a system that is inherently unstable in a practical device?" The answer requires a reanalysis of the system where the attachment point of the spring is not directly above the pivoting hinge. This can be accomplished by

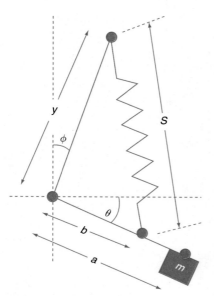

Figure 16 Tilted L&R suspension.

tilting the instrument by an angle ϕ as shown in **Figure 16**.

Recomputing the torque for a zero-length spring ($S_0 = 0$) we find

$$\Gamma = (mga - ykb \cos \phi)\cos \theta - ykb \sin \phi \sin \theta$$

The last term is proportional to the negative of the sine of the beam angle. This term provides the necessary restoring torque to stabilize the beam against mismatches in the balance of the gravity meter. This tilt is necessary for the L&R suspension. However, only a very small tilt ($\phi > 0$) is needed to stabilize the system once the beam is balanced. To first order in phi and theta, the balancing of the meter is still performed by adjusting $y = mga/kb$ and the resulting torque is then given by $\Gamma = -ykb\phi\theta$.

The dynamic equation for the beam angle is then given by the classic harmonic oscillator:

$$\ddot{\theta} = -\omega^2\theta, \quad \omega^2 = \frac{vkb\phi}{I} = \frac{mga\phi}{I}$$

If we have a massless beam, the frequency is given by

$$\omega^2 = \frac{mga\phi}{I} = \frac{g\phi}{a}$$

A tilt of $\phi \approx 10^{-4}$ will provide a period of about 20 s for a 1 cm beam ($a = 1$ cm). This is equivalent

to about 20 arcsec. Once the beam is balanced, a small change in gravity will tend to unbalance the beam and it will sag until the restoring torque from the $\sin(\theta)$ term balances the $\cos(\theta)$ term. This can be simplified to yield the equation $\phi\theta = \Delta g/g$. Note that if Δg is 10 μGal and $\phi \approx 10^{-4}$, then the beam will sag 20 arcsec (or $\theta \approx 10^{-4}$). Twenty seconds of arc is easily visible in the standard field telescopes available in the 1940s. This example shows how the L&R suspension allowed the use of a simple spring to measure gravity with high sensitivity even as early as the 1940s (before high-precision distance measurements and electronics were easily available). The dependence of the spring constant on temperature was overcome by using special low temperature coefficient spring alloys combined with a precise temperature control.

Figure 17 is a photograph of an L&R G meter. The instrument is leveled with the three black knobs. The top of the spring is moved with the silver rotary knob on the center of the top lid. The eyepiece of the telescope is also visible on the top of the lid. The telescope is used to look at the beam position. The instrument weighs 34 lbs and has dimensions of $24 \times 17 \times 23$ inch. This gravity meter has been a popular survey instrument for over 50 years because it is lightweight and can take a 10 μGal measurement in a few minutes. Surveyors can collect as many as 100 points a day with this instrument in difficult field terrain. This instrument dominated land gravity measurements from 1940 and is still widely used today.

Figure 17 L&R G meter, from MicroLacoste.

3.2.2.2 Quartz vertical spring gravity meters

The CG3 land gravity meter was developed by Hugill (1990) at Scintrex Ltd. in 1990. Further description of the gravimeter is given by Seigel *et al.*, (1993). The CG3 used a vertical quartz spring element rather than the L&R beam suspension. As discussed earlier, a simple vertically hanging spring gravimeter requires a much better measurement of the vertical position (10 000 times better) than the L&R suspension. This was done through the use of a capacitive plates that acts as both a high-precision position sensor and also as a force transducer to supplement the spring force. This technique was based upon a linear feedback system developed by Vaillant (1986). Small capacitive plates have very high sensitivity to position which allow position measurements with a noise of about $1 \, nm \, Hz^{-1/2}$ at room temperature. The sensing usually is done with a three-plate sensor where a voltage is placed on the outer plates and the differential voltage difference between each external plate and the center plate is measured. Typically this voltage is modulated and synchronously detected for noise immunity. Another important feature of the capacitive plates is that they can be used to provide a DC force to keep the position of the mass fixed. The quartz spring acts to levitate the mass while the capacitive transducer is used to offset any changes in the gravity force on the mass. The same capacitor plates can be used as both a position sensor and as a force transducer. The system is very insensitive to linearity of the spring constant because the spring is always kept at the same length by the capacitive electronics.

There are some advantages of using fused quartz over metal. Fused quartz is a very strong material which has similar tensile strength per unit mass as music wire (piano wire) used for high-quality metal springs. However, it has a much lower coefficient of thermal expansion of about $5.5 \times 10^{-7} \, cm \, cm^{-1} \, K^{-1}$ (average from 20°C to 320°C) which is about an order of magnitude lower than for music wire.

Quartz is relatively easy to work using heat. This results in a monolithic sensor that avoids clamping of critical spring elements. The spring, the frame, and all connecting elements are made from quartz.

The current state-of-the-art for quartz spring gravity meters is embodied in the CG5 commercially sold by Scintrex, LTD. It sensor is quite light and does not need to be clamped during movement. It has a monolithic quartz design that is much more resistant to tares that affect metal spring gravity meters of the same accuracy/precision.

3.2.2.3 The superconducting gravity meter

The gravity meter that used a superconducting magnetic field to levitate a mass was first developed by Prothero (1967) and Prothero and Goodkind (1968). The superconducting gravity meter consists of three parts; the levitated mass (sphere), the field coils, and the magnetic shield. The suspended mass is a 1 inch diameter hollow niobium shell manufactured with a slight mass asymmetry so that it has a preferred orientation when levitated. Two coils are used to independently adjust both the levitation force and the magnetic gradient. When current is placed in the coils, currents in the sphere are produced, which by the Faraday induction law precisely cancel magnetic flux from entering the sphere. The currents in the coils induce currents in the surface of the superconducting sphere that are perfectly stable in the absence of any ohmic losses. Interaction of the magnetic field in the coils and the sphere produces a 'spring-like' force that is used to balance the gravitational force. The position of the mass (niobium sphere) is measured by a capacitance bridge that surrounds the sphere and is sealed with a partial pressure of helium gas in a separate cavity inside the coils. A superconducting cylinder with a hemispherical closure on one end provides the primary magnetic shielding. Currents induced in the shield's outer surface cancel any changes in external magnetic field due to the Earth or other nearby magnetic material. Current in the coils induces persistent currents in the inside surface of the shield, which prevent levitation magnetic field from extending outside the shield. An additional μmetal shield is placed on the outside of the vacuum can. The entire instrument is cooled by liquid helium to 4 K.

The superconducting spring is of much higher quality than the metal or quartz springs discussed above. In principle, there is no temperature coefficient because it is held in a 4 K liquid helium bath. The principle of superconductivity produces a very uniform magnetic field free of gradients and other imperfections that would result from conventional room temperature coils. The drift from this type of gravity meter varies from 1 to 10 $\mu Gal \, yr^{-1}$. Michel van Camp *et al.* (2005) recently determined that the superconducting gravity meter has a precision of 0.1–0.3 $\mu Gal \, Hz^{-1/2}$. Thus, in most practical situations the short-term precision is limited by the environmental noise background created by seismic noise.

The major difficulty with the current superconducting gravity meters is that they suffer a large uncontrollable tare when they are moved. This limitation may be removed in the future but at the moment this problem requires them to be used in static applications.

3.2.3 Gravimeter Applications

Both types of relative gravity meters (metal/quartz) are much smaller and less expensive than either the absolute or superconducting gravity meter. Thus, relative gravity meters are used for the bulk of all gravity measurements. Both types of relative gravity meters have similar accuracy and precision but quartz gravity meters are resistant to tares at the 10–20 µGal level whereas metal spring gravity meters typically exhibit tares of this size during a normal survey. Tares can be identified and corrected if the user is willing to make many loops to get repeated data at every site. Quartz gravity meters are therefore preferred on high-accuracy land surveys due to the less frequent tares at the 10–20 µGal. Metal spring gravity meters have a sometimes important practical advantage in their ability to measure gravity without electronics due to the mechanical advantage provided by the lever arm and zero-length spring configuration. This can be helpful for surveys that are in very remote areas where an electronics failure cannot be tolerated. Therefore, metal spring gravity meters tend to be used more often in high-risk lower-accuracy land surveys where it is important that the job get done (at the 20–100 µGal level) even if the electronics may fail. However, either type of relative gravity meter can achieve a similar result if the measurements are taken carefully and enough repeat measurements are taken to identify and correct for drift and tares in the instrument.

Superconducting gravity meters are typically employed to measure Earth tides at a specific point on the Earth. It is common for them to provide amplitudes at the tidal frequencies with an incredible precision of 1 nanoGal (0.001 µGal or $10^{-12}g$) after integrating the tidal signal for 1–3 years. Park *et al.* (2005) recently used superconducting gravity meters to observe free Earth oscillations that were excited by the Sumatra–Andaman earthquake in December 2004.

Absolute gravity meters have traditionally been used for establishing base stations used for relative gravity networks. They are also used to calibrate relative gravity meters. Francis *et al.* (1998) describe a method where a superconducting gravity meter can be calibrated with an absolute gravity meter by correlating Earth-tide measurements taken with both instruments over a few days of observations. Absolute

gravity meters are particularly useful for monitoring long-term gravity changes (over 1 year or longer). The lack of drift and the good agreement between different absolute gravity meters allow measurements to be combined in a meaningful way to measure very small changes of gravity in real field conditions. Lambert *et al.* (2001) have taken absolute gravity measurements in Churchill Canada for 20 years (1987 to present) and have demonstrated that the earth's crust is rising locally at about 3 mm yr^{-1} due to post-glacial rebound. These measurements provide new bounds on the mantle viscosity as well as provide an independent way to measure vertical crustal motion (other than GPS or VLBI).

The study of gravity changes is also known as 4D gravity in the commercial survey market. Absolute gravity measurements were started in 2002 and are currently being repeated annually in order to monitor subsurface density changes due to water injection into a large oil reservoir in Prudhoe Bay, Alaska. Approximately 300 stations are located over 10 km^2 area, many of which are on top of water in Prudhoe Bay. The survey is conducted while the surface of the bay is frozen so that high-accuracy static gravity measurements can be made. The survey required a special absolute gravity meter to be fabricated that could maintain temperature stability and be automatically aligned in temperatures that could be as low as −40°C. **Figure 18** shows an absolute gravity meter deployed in Prudhoe Bay. The unit collects both gravity and location (latitude, longitude, and elevation) using GPS at every location.

The absolute gravity measurements are being used to monitor both the amount of subsurface water that is injected and the location of the flow. While the project is still in its early phases, the gravity is already improving the knowledge of the reservoir significantly over what can be measured by the sparse well information.

Figure 18 A10 deployed in Prudhoe Bay, from MicroLacoste.

References

Brady JL, Hare JL, Ferguson JF, *et al.* (2006) *Results of the World's First 4D Microgravity Surveillance of a Waterflood – Prudhoe Bay, Alaska*, SPE 101762, SPE Annual Technical Conference, San Antonio, TX.

Chartier JM, Labot J, Sasagawa G, Niebauer TM, and Hollander W (1993) A portable iodine stabilized He–Ne laser and its use in an absolute gravimeter. *IEEE Transactions on Instrumentation and Measurement* 43.

Chris H, Lucien J, and LaCoste B (1996) Portrait of a scientist–inventor. *Earth in Space* 8(9): 12–13.

Edwards AJ, Maki JT, and Peterson DG (1997) Gravity gradiometry as a tool for underground facility detection. *Journal of Environmental and Engineering Geophysics* 2: 139–143.

Francis O, Niebauer TM, Sasagawa G, Klopping F, and Gschwind J (1998) Calibration of a superconducting gravimeter by comparison with an absolute gravimeter FG5 in Boulder. *Geophysical Research Letters* 25(7): 1075–1078.

Francis O and van Dam TM (2003) Processing of the Absolute data of the ICAG01. *Cahiers du Centre Européen de Géodynamique et de Séismologie* 22: 45–48.

Francis O and van Dam T (2006) Analysis of results of the international comparison of absolute gravimeters in Walferdange (Luxembourg) of November 2003. *Cahiers du Centre Europeen de Geodynamique et de Seismologie* 26: 1–24.

Heiskanen WA and Vening Meinesz FA (1958) *The Earth and Its Gravity Field*. New York: McGraw-Hill.

Hugill A (1990) *The Scintrex CG-3M Autograv Automated Gravity Meter, Description and Field Results*, SEG Conference, San Francisco.

Kasevich M and Chu S (1991) Atomic interferometry using stimulated Raman transitions. *Physical Review Letters* 67: 181–184.

Kater H (1818) *Philosophical Transactions of the Royal Society of London* 108: 33.

Kennard EH (1938) *Kinetic Theory of Gases*. New York: McGraw-Hill.

Lambert A, Courtier N, and James TS (2006) Long-term monitoring by absolute gravimetry: Tides to postglacial rebound. *Journal of Geodynamics* 41(1–3): 307–317 (ESS Cont.# 2005056).

Lambert A, Courtier N, Sasagawa GS, *et al.* (2001) New constraints on Laurentide postglacial rebound from absolute gravity measurements. *Geophysical Research Letters* 28(10): 2109–2112.

Lambert A, James TS, and Thorleifson LH (1998) Combining geomorphological and geodetic data to determine postglacial tilting in Manitoba. *Journal of Paleolimnology* 19(3): 365–376.

McGarva Bruckshaw J (1941) Gravity meters. *Proceedings of the Physical Society* 53: 449–467.

Newell DB, Steiner RL, and Williams ER (1998) An accurate measurement of Planck's constant. *Physical Review Letters* 81: 2404–2407.

Newell DB, Steiner RL, and Williams ER (1999) *The NIST Watt Balance: Recent Results and Future Plans*. Atlanta, GE: American Physical Society, 20–26 March.

Niebauer T (1987) *New Absolute Gravity Instruments for Physics and Geophysics*. Thesis in Physiscs, University of Colorado, Boulder, CO.

Niebauer TM, (1987) PhD Thesis, University of Colorado physics pp. 24.

Niebauer TM, Sasagawa GS, Faller JE, Hilt R, and Klopping F (1995) A new generation of absolute gravimeters. *Metrologia* 32(3): 346–352.

Park J, Song TR, Tromp J, *et al.* (2005) Long-period behavior of the 26 December 2004 Sumatra-Andaman earthquake from its excitation of Earth's free oscillations. *Science* 308: 1139–1144.

Prothero WA (1967) A Cryogenic Gravimeter, PhD Thesis, University of California, San Diego.

Prothero WA and Goodkind JM (1968) A Superconducting gravimeter *Review of Scientific Instruments* 39: 1257–1262.

Quinn TJ (2002) Measuring big *G. Nature* 408: 919–921.

Raab F and Coyne D Effect of Microseismic Noise on a LIGO Interferometer, *A Portable Apparatus for Absolute Measurements of the Earth's Gravity*. Thesis in Physics, LIGO-T960187-01 - D 2/20/97

Rinker R (1980) Phd Thesis, University of Colorado.

Rinker R (1983) Super-spring – *A New Type of Low-Frequency Vibration Isolator*. Thesis in Physics, University of Colorado, Boulder, CO.

Sakuma A (1970) Recent developments in the absolute measurement of gravitational acceleration. *National Bureau of Standards (US) Special Publication* 343: 447–456.

Sasagawa G, Klopping FJ, Niebauer TM, Faller JE, and Hilt R (1995) Intracomparison tests of the FG5 absolute gravity meters. *Geophysical Research Letter* 22: 461–464.

Seigel HO, Brcic I, and Mistry P (1993) A high precision, μGal resolution, land gravimeter with worldwide range. In: Seigel HO (ed.) *A Guide to High Precision Land Gravimeter Surveys*. Concord, ON: Scintrex Ltd.

Torge W (2001) *Geodesy*. Berlin, Germany: Walter de Gruyter GmbH & Co. KG.

Vaillant HD (1986) An inherently linear electrostatic feedback method for gravity meters. *Journal of Geophysical Research* 91: 10463–10469.

Van Camp M, Williams SDP, and Francis O (2005) Uncertainty of absolute gravity measurements. *Journal of Geophysical Research* 110: B05406 (doi:10.1029/2004JB003497).

Vitushkin L, Becker M, Jiang Z, *et al.* (2002) Results of the sixth international comparison of absolute gravimeters ICAG-2001. *Metrologia* 39(5): 407–427.

Zumberge MA (1981) Internal working note for California Institute of Technology LIGO Project-MS 51–53, University of Colorado, Boulder, CO.

4 Gravimetric Methods – Superconducting Gravity Meters

J. Hinderer, Institut de Physique du Globe, Strasbourg, France

D. Crossley, St. Louis University, St. Louis, MO, USA

R. J. Warburton, GWR Instruments, San Diego, CA, USA

4.1 The Superconducting Gravimeter

4.1.1 Historical

4.1.1.1 Early years at UCSD

The superconducting gravimeter (SG) was first introduced by Prothero and Goodkind (1968) as part of Prothero's (1967) thesis work on the design and development of the instrument at University of California at San Diego (UCSD). The SG broke new ground in geophysics instrumentation, and was an elegant realization of the principles of superconductivity. Although the basic sensor configuration has remained unchanged for nearly 40 years, continuous improvements in all other aspects of the original design have successfully converted the SG from a prototype laboratory instrument to a reliable research tool. (Note that in this article, the traditional gravity abbreviations are frequently used: 1 microgal = 1 μGal = 10 mm s^{-2}, 1 nanogal = 1 nGal = 0.01 nm s^{-2}, and cpd = cycles per (solar) day.)

In 1970, Richard Warburton became a postdoctoral student with John Goodkind, and this collaboration was the foundation for the eventual line of commercial SGs. Richard Reineman, an undergraduate laboratory assistant working with William Prothero in 1969, was integral to the effort as a development technician with Goodkind and Warburton. Prothero and Goodkind (1972) published the first observations taken over a 4 month period and obtained precise tidal amplitude and phases, new information on ocean tide loading, and a recording of seismic normal modes following the 7.1-magnitude Kamchatka earthquake from 1969. Within a few years, the UCSD group generated significant papers using SG data on ocean tide loading (Warburton et al., 1975) and the effects of barometric pressure on gravity (Warburton and Goodkind, 1977). This phase of the SG research culminated with a detailed tidal analysis of 18 months of data that included the first estimate of the effect of the nearly diurnal wobble on the resonant amplification of small diurnal tidal phases (Warburton and Goodkind, 1978). These papers are still recommended reading for those interested in the basic issues concerning the treatment of gravity data.

4.1.1.2 Early commercial model TT instruments (1981–94)

The early publications as well as presentations at various conferences caught the attention of Paul Melchior (Royal Observatory of Brussels, Belgium, ROB) and Rudolf Brien and Bernd Richter (Institut für Angewandte Geodäsie, Germany, IfAG; now known as Bundesamt für Kartographie und Geodäsie, BKG), who contacted Goodkind about the possibility of using SGs to expand their previous work based on LaCoste Romberg (LCR) gravity meters. As a result, the commercial venture GWR Instruments (Goodkind, Warburton, and Reineman) was formed in 1979 to manufacture two SGs, one for ROB and one for IfAG. From this point on, two different design streams continued: Goodkind refined the original UCSD instruments and used them to develop new areas of geophysical research, and GWR began the manufacture of instruments for other scientific groups from their San Diego facilities. Eric Brinton joined GWR in 1986 to continue development of refrigeration, electronics, and data acquisition systems.

Melchior purchased a GWR Model TT30 dewar similar to those used at UCSD. The SG sensor was simply cooled by insertion through the 5 inch diameter neck of a dewar with the internal 200 l volume ('belly') filled with liquid helium. In a typical dewar, the belly is surrounded by a vacuum space, which contains two radiation shields with many thin layers of aluminized Mylar ('superinsulation') placed on the surface of the belly and shields. Hold time depends critically on efficiently using the cooling power of the gas (enthalpy) as it flows past the shields and neck before exhausting at room temperature. The dewar was suspended from a large concrete pier by 2 μm and a rear fixed point, which were used for leveling.

In 1981, the Model TT30 was installed in the basement vault at ROB. Visiting scientists who were familiar with modern SG installations would have been greeted by an eerie silence – there was no compressor noise, not even in an adjacent room. Silence had its disadvantages, however, as almost 200 l of liquid helium had to be replenished every 3 weeks or so, and each of these refills caused unpleasant disturbances to the data stream. Despite many problems that originated with a helium leak in the TT30 vacuum can lid (described in detail in De Meyer and Ducarme, 1989), this instrument was to continue

without major interruptions for nearly 18 years until it was decommissioned in 2000. Early tidal results from the Brussels SG can be found in Ducarme (1983).

Helium was very expensive in Germany, so IFAG/BKG asked GWR to develop a refrigerated dewar system for their instrument. These systems use a cryogenic refrigerator (coldhead and compressor) to intercept and reduce the flow of heat via radiation and conduction from the outside of the dewar to its belly. This reduces the rate of boil-off so that the 'hold time' of the liquid helium is lengthened. Hold time depends on the cooling power of the refrigerator being used and how well it can be thermally coupled to the neck and radiation shields. On the first TT40 refrigerated dewar, the coldhead was bolted onto the top of the dewar wall and penetrated into the vacuum space through a special port sealed with an O-ring. The coldhead's two cooling stages (at 65 and 11 K) were connected directly to the outer and inner radiation shields using copper braid. The TT40 design was extremely successful with a holdtime of well over 400 days versus 50 days unrefrigerated. This project was beginning of a collaboration between GWR Instruments and Bernd Richter for developing new and improved SG models that has continued to the present day.

The provision of a commercial instrument proved to be a landmark opportunity for the geodetic and gravity community. In 1981, SG Model TT40 was installed in the basement of a castle at Bad Homburg, near Frankfurt Germany. At the International Union of Geodesy and Geophysics (IUGG) meeting in Hamburg, Richter (1983) presented a paper on data from the TT40 that showed the first gravitational determination of the 14 month Chandler component of polar motion, with amplitude of $\pm 5\,\mu$Gal ($1\,\mu$Gal $= 10\,$nm s^{-2}). To say the least, this took the audience by surprise and convincingly demonstrated the capabilities of the new gravimeter.

At the same time, Goodkind (1983) repeated his determination of the nearly diurnal wobble parameters from the tidal amplitudes using data from 1978, but the problem of accurately computing the ocean tidal loading of small waves still was a limiting factor. A few years later, Richter (1985) reported on the extension of his data set to 3 years of successful SG operation.

In 1985, Richter installed a second instrument – a new model TT60 – at Bad Homburg, and for over a year until the beginning of 1987 he obtained parallel recording with the original TT40. His thesis (Richter, 1987) contained many interesting insights into the operation of the instrument and its data, but being in German it was not widely read. A significant result was that the gravity residuals from both instruments were highly correlated at the sub-μGal level (Richter, 1990), indicating that significant geophysical signals were still left in the data at this level. At the time, the cause was ascribed to the atmosphere, but this was before the environmental influence of hydrology on gravity became widely appreciated. Following the experimental work of Richter and the theoretical speculations of Melchior and Ducarme (1986), new refrigerated Model TT70's were installed in Wuhan, China, in 1986 (Hsu *et al.*, 1989), and in Strasbourg, France, in 1987 (Hinderer and Legros, 1989). The Model TT70 introduced the use of internal tiltmeters and thermal levelers to automatically keep the SG aligned with the plumbline at its tilt minimum.

As with the TT40 and TT60, the first two TT70's were manufactured with the coldhead bolted into the top of the dewar and penetrating the vacuum space. Although very efficient, this design made it difficult to service the coldhead without warming the dewar to room temperature. Later, TT70 models were manufactured with the coldhead supported by a separate frame and inserted through the neck of the dewar. Cooling power was coupled to the neck and radiation shields only via helium gas. In the new TT70 design, the coldhead could easily be removed for servicing or removal of any ice that may build up between the coldhead and the gravity sensor unit (Warburton and Brinton, 1995). New TT70's were soon installed at the National Astronomical Observatory in Mizusawa, Japan, in 1988, and two were located side by side at Kyoto University in the same year (Tsubokawa, 1991). Meanwhile, in 1989, Richter moved the TT60 from Bad Homburg to Wettzell, a fiducial station of the German geodetic network, and a TT70 was installed at Cantley, Canada in the same year (Bower *et al.*, 1991). Approximately 12 TT70 SGs were manufactured between 1986 and 1994, and many of these instruments are still operating as part of the Global Geodynamics Project (GGP) network (T007 Esashi, T009 Kyoto, T011 Kakioka, T012 Cantley, T015 Brasimone, T016 Kamioka, and T020 Metsahovi) – see Crossley (2004). TT70 dewars are 150 cm tall, have an 80 cm diameter, and weigh 150 kg. They require an annual transfer of 200 l of liquid helium and servicing the coldheads at 1–2 year intervals. In 1993, it was found that the TT70 SG was less susceptible to horizontal noise sources when the dewar was mounted from the bottom (Warburton and Brinton, 1995). **Figure 1** shows the TT70 operating at Cantley, Canada, after modification to the bottom-mounted configuration. The large

Figure 1 An example of the TT70 type instrument installed at Cantley, Canada, in 1989. The rack on the far left contains the chart recorders and noncritical electronics; the more sensitive components are in the enclosed temperature-controlled rack next to it. The 200 l dewar sits on three feet placed on small granite blocks. The front two feet with the X and Y thermal levelers attached to the dewar bottom are visible. The coldhead is supported from a frame resting on the top of a concrete pier. Normally, the entire SG is enclosed by thermal insulation.

concrete pier, previously used to support the dewar, now only supports the coldhead.

4.1.1.3 The Compact SG (1994–2002)

In 1993, GWR produced a much smaller 125 l Compact Dewar designed to operate on a 1 m^2 pier or platform, so that it could be easily operated at many preexisting geodetic installations. The compact SG is 104 cm high, 66 cm wide, and weighs 90 kg. The SG sensor is built into the dewar belly, which allows the neck and radiation shields to be custom designed to mirror the dimensions of the coldhead. The Compact Dewar uses the same APD Cryogenics DE202 coldhead and HC-2 helium compressor as used in Models TT60 and TT70. However, the smaller neck diameter and volume dramatically reduces the heat load on the outer radiation shield, and the improved neck/coldhead interface allows much more efficient use of the coldhead cooling power. As a result, the dewar efficiency is doubled and less than 100 l of liquid helium is used annually.

Figure 2 shows Compact C023 operating in Medicina, Italy (Romagnoli *et al.*, 2003). The APD HC-2 compressor and its water chiller are on the left side (rear), with flexible stainless tubes connecting the compressed helium gas to the coldhead. The two thermal levelers and a third fixed point are attached to an aluminum band wrapped around the circumference of the dewar. These are supported by three feet that rest on small granite blocks placed upon the floor.

Figure 2 A compact instrument C023 installed at Medicina, Italy.

The coldhead is supported and centered in the neck using a metal tripod support frame. The neck–coldhead interface is sealed with a rubber gasket, which prevents air from entering, and provides vibration isolation between the coldhead and the SG sensor. The addition of stiff internal spokes placed between the inner belly and the outer dewar wall makes the compact SG less sensitive to horizontal noise. This structural change produces lower noise levels than observed in previous SGs (Boy *et al.*, 2000).

The first compact SG, C021, was tested next to T002 at the Royal Observatory Brussels before being moved in 1995 to a seismic station in Membach, Belgium, where it is still operating. The SG is installed in a separate room at the end of a 100-m-long tunnel, and because it is close to long-period seismometers, care was taken to minimize and measure vibrations produced by the cryogenic refrigeration system (Van Camp, 1995). Approximately 13 compact SGs were manufactured from 1994 to 2002 and are installed at over half the GGP stations.

4.1.2 Basic Principles of Operation

4.1.2.1 Superconducting components
Seismometers and relative gravimeters are based on a test mass suspended by a spring that is attached to the instrument support. A change in gravity or motion of the ground generates a voltage that becomes the output signal (velocity or acceleration). This system works well in many modified forms for seismometers and is still used successfully in the LCR and Scintrex models of field gravimeters. The major problem at periods longer than the seismic normal-mode range, for example, at 4 h and longer for the tides, is that (even in a thermally well-regulated environment) the mechanical aspects of a spring suspension cause erratic

drift that is difficult to remove by postprocessing. Field gravimeters repeatedly occupy a reference station to monitor this drift, and observatory spring instruments have to be rezeroed when the signal exceeds the range of the voltmeter. Since the 1980s, spring gravimeters have incorporated electrostatic feedback that considerably improves their linearity and drift performance (e.g., Larson and Harrison, 1986).

The SG almost completely solves the drift problem by replacing the mechanical spring with the levitation of a test mass using a magnetic suspension. **Figure 3** shows a diagram of the GSU; the three major superconducting elements are the levitated mass (sphere), the field coils, and the magnetic shield. The displacement transducer is formed by a capacitance bridge that surrounds the sphere and is sealed with a partial pressure of helium gas in a separate cavity inside the coils. The field is generated by two niobium wire coils (superconducting below a temperature of 9.2 K) that carry, in principle, perfectly stable superconducting persistent currents to provide an extremely stable magnetic field. The stability depends on the zero-resistance property of superconductors – after the currents are 'trapped', no resistive (ohmic) losses are present to cause them to decay in time. The test mass is a small 2.54–cm-diameter sphere, also made of niobium, that weighs about 5 g. The coils are axially aligned; one just below the center of the sphere and one displaced about 2.5 cm below the sphere. When current flows in the coils, secondary currents are induced on the surface of the sphere, which by the Faraday induction law precisely cancel magnetic flux from entering the sphere. As with the currents in the coils, the induced currents are perfectly stable in the absence of any ohmic losses. The levitation force is produced by the interaction between the magnetic field from the coils and the currents induced on the surface of the superconducting sphere. **Figure 4** shows a schematic of the sphere, coils, capacitance bridge, and magnetic flux lines.

The use of two coils allows the operator to independently adjust both the levitation force and the magnetic gradient. The upward levitation force is mainly produced by the lower coil. Its current can be precisely adjusted to balance the time-averaged downward force of gravity on the sphere at the center of the displacement transducer. The upper coil is used to adjust the magnetic force gradient ('spring constant'), which can be made very weak. As a result, a very small change in gravity (acceleration) gives a large displacement of the test mass, generating an instrument of very high sensitivity.

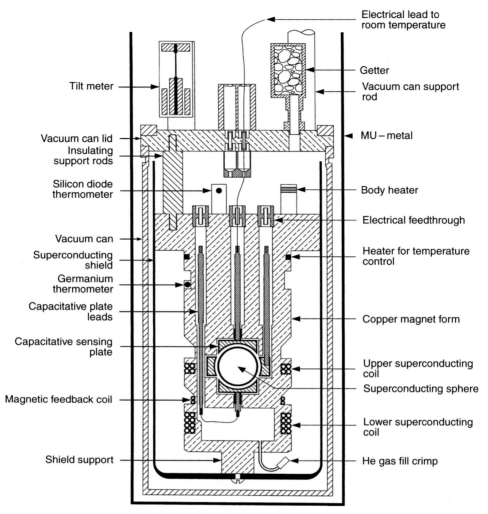

Tilt meter

Vacuum can lid
Insulating
support rods

Silicon diode
thermometer

Vacuum can
Superconducting
shield
Germanium
thermometer
Capacitative plate
leads

Capacitative sensing
plate

Magnetic feedback coil

Shield support

Electrical lead to
room temperature

Getter
Vacuum can support
rod

MU – metal

Body heater

Electrical feedthrough

Heater for temperature
control

Copper magnet form

Upper superconducting
coil
Superconducting sphere

Lower superconducting
coil

He gas fill crimp

Figure 3 Schematic of SG sensor showing arrangement of the sphere, coils, vacuum can, and shielding.

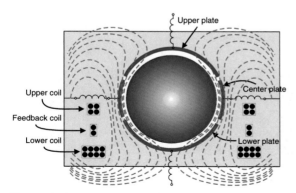

Upper plate

Upper coil

Feedback coil

Lower coil

Center plate

Lower plate

Figure 4 Schematic diagram of the Nb sphere, coils, plates, and general pattern of magnetic flux lines. Flux is excluded from the sphere and is confined externally by the Nb shield.

Because the levitation is magnetic, changes in the Earth's magnetic field would seriously degrade the stability of an unshielded SG. A superconducting cylinder with a hemispherical closure on one end surrounds the sphere and levitation coils and is attached to the bottom of the copper magnet form. This provides the primary magnetic shielding from changes in the Earth's magnetic fields, which in its absence would seriously degrade the stability of the magnetic levitation force. When the magnetic coils are turned on, persistent currents also are induced in the inside surface of the shield, which prevents the levitation magnetic field from penetrating the shield. An additional μ-metal shield is placed on the outside of the vacuum can. During the initialization process, this shield reduces the Earth's magnetic field by a factor of about 100 before the superconducting components cool through their transition temperature. This process minimizes any trapped flux in the sphere, coils, or shield that could produce instability in the sensor.

4.1.2.2 Displacement transducer and feedback system

Relative motion between the ground (to which the coil assembly is attached) and the sphere, or any other perturbation of the gravity potential, moves the sphere from its equilibrium position. The position of the sphere is detected by using a phase-sensitive lock-in amplifier in conjunction with a capacitance bridge. Three capacitor plates surround the sphere with 1 mm clearance (**Figure 3**). The upper and lower plates are hemispherical caps that surround the upper and lower portions of the sphere. The center plate is a spherical ring around the equator of the sphere (**Figure 4**). A 10 kHz reference signal from the lock-in amplifier drives the primary of a carefully shielded transformer. The two balanced secondary windings of the transformer apply equal and opposite voltages to the upper and lower capacitor plates. The AC signal from the center ring plate is proportional to the displacement of the sphere from the center of the bridge. The sensor is operated in feedback to take advantage of the increased linear dynamic range and rapid response compared to open-loop operation. The AC signal is amplified, demodulated, filtered, and applied to an integrator network. The DC output is connected to a precision resistor in series with a five-turn coil wound on the copper magnetic form below the sphere. The resulting feedback force is proportional to the product of the feedback current and the current on the surface of the sphere. This force is given by $F = CI_F(I_{IC} + I_{IF})$, where I_F is the feedback current, I_{IC} is the current induced on the surface of the sphere by the levitation field, I_{IF} is the current induced on the surface of the sphere by the feedback field, and C is a constant. Because I_{IC} is proportional to g and I_{IC} is atmost the maximum amplitude of the tides, the maximum nonlinearity is $(I_{IF}/I_{IC})_{MAX} \sim 10^{-7}$. Therefore, the sensor is extremely linear. The gain (scale factor) of the sensor is determined by the geometry, the resistor size, the number of turns on the coils, and the mass of the sphere.

4.1.2.3 Temperature control

The sensor and superconducting shield are located inside a vacuum can surrounded by the liquid helium bath at about 4.2 K. In response to atmospheric pressure, the boiling point varies by about $1\,mK\,mb^{-1}$, and during storms may change as much as 100 mK.

Therefore, the sensor must be temperature regulated (Goodkind, 1999). A germanium thermometer measures temperature and forms one arm of a Wheatstone bridge that has its null point preset to 4.5 K. The bridge output supplies feedback power of a few milliwatts, which is applied to a heater attached to the copper magnetic form. Variations in control power almost perfectly follow the inverse of atmospheric pressure. With small bath-temperature variations, high vacuum isolation, and high thermal conductivity of materials, it is relatively easy to regulate to a few microkelvins at cryogenic temperatures. As a result, the SG is almost completely isolated from environmental effects caused by changes in external temperature, humidity, and barometric pressure. This is a major advantage over mechanical gravimeters that operate near room temperature.

4.1.2.4 Tilt compensation system

To measure the magnitude of the gravitational acceleration g, the gravimeter must be aligned with a plumbline along g. For most types of gravimeter, the test mass is constrained to move only along its axis of measurement. Therefore, when its axis is tilted with respect to the vertical plumbline, it measures only the component of g along its axis of measurement. The measured magnitude is $g\cos\theta$, where θ is the angle between the vertical and axis of the instrument. For small angles, the apparent decrease in gravity produced by tilts is $\delta(\theta) = (g\cos\theta - g) \approx g(\theta^2/2)$. The LCR is aligned along g by tilting it systematically along two orthogonal directions and setting it at the maximum value of g.

When an SG is tilted, the component of gravity along its axis of measurement decreases as $g\cos\theta$. The magnetic force gradient perpendicular to its axis of measurement is relatively weak, however, so the sphere moves off axis in response to the force component $g\sin\theta$. Because the magnetic levitation force supporting the sphere decreases off axis, the sphere position moves downward. This apparent increase in gravity has the same angular dependence as the equation above but its magnitude is about 2 times larger. As a result, the SG tilt dependence becomes $\delta(\theta)_{SG} \approx -g_{SG}(\theta^2/2)$, where the magnitude of $|g_{SG}| \approx |g|$. Therefore, the SG is aligned along g by tilting systematically along two orthogonal directions and setting it at the minimum value of g (not at the maximum, as for other gravimeters). This phenomenon explains why cultural noise such as nearby trains or automobiles will cause downward spikes on the SG signal. The horizontal accelerations move the sphere off axis where the magnetic support force is weaker. This effect was first observed in 1981 when the TT30 was installed in the cellar vault in the ROB. In contrast, trains did not affect operation of the LCR gravity meter operating in a nearby vault.

The SG is supplied with an automatic leveling system consisting of two tiltmeters mounted orthogonally on top of the gravimeter vacuum can and two thermally activated levelers that are placed under two of the dewar support points. After tilt-minimizing the SG sensor, the tiltmeters are 'aligned' to the same null by electronically setting their output voltages to zero. In feedback, the tiltmeters continuously adjust the power controlling the expansion of the levelers to keep alignment better than 1 μrad. This leveling precision is essential in gravity studies where apparent tilt-induced gravity changes must be kept less than 1 nGal (1 nGal = 0.01 nm s^{-2}). The tilt minimum adjustment is made on initial installation and checked every year or so by the operator. A recent study (Iwano and Fukuda, 2004) on SG data from the Syowa station shows the clear advantage of the tilt compensation system in reducing the noise in gravity, especially in the tidal range.

4.1.2.5   *Sphere and sphere resonance*
The sphere is a hollow superconducting shell that is manufactured with a slight mass asymmetry so that it has a preferred orientation when levitated. Various manufacturing processes are discussed in Warburton and Brinton (1995). A small hole is drilled on the top of the sphere to allow the helium gas to enter and to prevent a differential pressure from developing when it is cooled to 4 K. Just as important, the volume of the shell displaces 15 times less helium gas than the volume of the sphere; so the hole reduces buoyancy forces that result from changes in the surrounding helium gas pressure.

When the gravimeter is tilted, particularly impulsively, the horizontal displacement of the sphere turns into an orbital motion (precession) with an associated vertical component in the feedback output. This mode appears as a sphere 'resonance' that has a period of 60–120 s depending on the particular instrument. In the absence of trapped magnetic fields and helium gas in the chamber, the Q of this mode is several thousand, so it is always excited making the instrument not usable. Slow damping of the mode is provided by adding helium gas to the chamber, but the resonance remains underdamped and is clearly visible in some of the instruments' data. By comparison, the vertical resonance of the sphere is heavily

damped with a period close to 1 s. Further technical details on the instrument design can be found in Goodkind (1991, 1999).

4.1.3   Development of the Dual-Sphere Design

In the early commercial SGs manufactured up to 1990, offsets (or 'tares') occurred in the SG gravity records that affected both long-term stability and measurement of tidal factors (Seama *et al.*, 1993; Harnisch and Harnisch, 1995; Hinderer *et al.*, 1994). The offsets could be quite large (up to 100 μGal) if caused by mechanical shock from transferring liquid helium, power failures, or earthquakes. Small instrumental offsets less than 5 μGal could occur at random intervals that were not associated with outside disturbances. Rapid offsets larger than 0.2 μGal and occurring in less than 1 min could be easily detected and corrected (Merriam *et al.*, 2001). However, there was a concern that the residual data would depend arbitrarily on the threshold value chosen in automatic offset detection programs (Harnisch and Harnisch, 1997). When two SGs were operated side by side, the difference in recordings provided a much clearer determination of the occurrence of offsets (Klopping *et al.*, 1995).

On the basis that random offsets will seldom occur in two sensors simultaneously, a dual-sphere SG was manufactured to solve the instrument offset problem (Richter and Warburton, 1998). The two spheres are mounted one above the other and separated by about 20 cm. The lower sensor is manufactured exactly like previous single-sphere sensors, and the temperature and tilt control remain the same. Small differences in the sphere masses, superconducting shield, coil windings, and machining tolerances produce magnetic asymmetries that are not identical in the two sensors. These asymmetries produce slightly different tilt minima and require more complicated electronics to align the tilt minimum of the upper sensor with the lower sensor.

The complications of a dual-sphere system are justified by providing a built-in instrumental offset detector. Because the outputs are treated as signals from two different gravimeters, the user can combine the processed data sets, select the least disturbed sphere output for any one time period, or convert the two signals into a gravity gradient by using the known vertical separation. CD029 was the first dual-sphere SG produced and it was tested at Bad Homburg beginning July 1998 before being moved permanently to Wettzell in November 1998 (Harnisch *et al.*, 2000). Results at Bad Homburg showed that careful calibration of both amplitude and phase was required to minimize the difference signal and that indeed offsets of a few tenths of a µGal could easily be detected. Surprisingly, after moving to Wettzell, no spontaneous random offset exceeding about 0.1 µGal was observed. Larger offsets that occurred were due to other causes, that is, failures in the cooling system or during extensive maintenance procedures (coldhead maintenance, He refills, or removal of ice from the neck).

More recent data from all four dual-sphere SGs support the early observations with CD029 (Kroner *et al.*, 2004, 2005). No offsets have occurred in CD029 at Wettzell or D034 at Moxa, and only one or two offsets per year are observed in CD028 at Bad Homburg and in CD037 in Sutherland. From these data, GWR concludes that the changes in its manufacturing process, in particular improvements to the Nb spheres and shields, have greatly reduced the incidence of random offsets. With this success, one might argue that the dual-sphere SG is no longer needed. However, as gravity changes are examined with higher and higher resolution, they may still be used to discriminate sub-µGal observations of instrumental origin (see figure 3 of Meurers, 2004), or illuminate more subtle instrumental effects that need improvement (Kroner *et al.*, 2005). More importantly, the gradient signal itself may yet prove useful in modeling environmental gravity variations.

4.1.4 Instrument Performance

4.1.4.1 Instrument drift
One SG design goal was to produce an instrument that is stable to a few µGal per year. Meeting this goal with a commercial instrument took approximately a decade. Drift in the early model TT70s could be more than 100 µGal yr^{-1} during the first year of operation but decreased steadily with time to less than 10–20 µGal yr^{-1} after 5 years of operation. Drifts were modeled as the sum of one or two exponentials, a low-degree polynomial, or a combination of both (Boy *et al.*, 2000; Seama *et al.*, 1993), but uncertainty in the exact functional form decreased the precision with which long-period tides and polar motion could be determined from the data.

A dramatic decrease in drift was made in the early 1990s owing to a variety of design changes described in Warburton and Brinton (1995). At present, drifts are

characterized by a small initial exponential followed by a small linear term. The exponential component decays in 4–6 months, after which it is negligible. For example, the initial drift for C023 manufactured in 1995 is $d(t) = -16.0\exp(-t/31)\,\mu$Gal, where t is in days (Schwahn *et al.*, 2000). Long-term linear drift rates (including real secular changes) reported for the nine SGs operating in Europe vary from 1.6 to 4.9 µGal yr^{-1} (Crossley *et al.*, 2004). An 8-year comparison of an FG5 absolute gravimeter (AG) next to SG C023 in Membach confirmed that, after removal of the linear term (4.2 µGal yr^{-1}), the SG has an identical spectra to the AG for frequencies less than 1 cpd (cycles per (solar) day) (Van Camp *et al.*, 2005). This confirms that the SG drift is restricted to DC (very low frequency), and that it provides a continuous low noise record of all gravity variations.

The only reliable method to determine instrumental drift is to compare the SG with collocated measurements made with an AG at regular intervals. The SG provides a complete time history of gravity at the site, which is invaluable for correlating with other geophysical variables, whereas the AG provides an absolute reference from which the drift and secular changes can be inferred. Precise drift measurement is complicated by real gravity variations due to hydrology, crustal uplift, annual terms, seasonal terms, or signals of unknown origin (e.g., Zerbini *et al.*, 2002).

4.1.4.2 Calibration stability
The question of calibration will be covered in more detail later in this report, but it is worth noting here that recalibration of single sensor instrument using AGs over time periods of 5 years or more have rarely shown differences that exceed the calibration error (between 0.1% and 0.01%). For dual-sphere sensors, the constancy of calibration is verified to 10^{-3}% by least-squares fitting the time series of the upper sphere to the lower sphere (Kroner *et al.*, 2005). This constancy means that the analysis of very long records by tidal analysis, for example, can be done without fear of any evolution in the calibration constants.

4.1.4.3 Instrumental noise and precision
Most of the GGP data acquisition systems record and digitize the full output of the SG electronics board. A voltmeter with 7.5 digits of resolution has the equivalent of 22 bits; when applied to a signal that includes the full range of tides (300 µGal), this translates into a smallest significant change of about

0.1 nGal. This is in effect the quantization noise of the SG + data acquisition system, but not necessarily its precision. There is no reference gravity more accurate than the SG itself, so the precision of the SG can only be obtained by inference. In the frequency domain for studies of tides or normal modes, it is common for the SG to measure small periodic tidal signals and long-period seismic signals with a sensitivity of 1 nGal and better. Therefore, 1 nGal is generally referred to as the nominal precision, or sensitivity, of the SG.

The determination of the instrumental noise of the SG is complicated by the fact that sources of the Earth noise (signals from the atmosphere, oceans, and hydrology) are generally much larger than instrumental noise for frequencies ranging from 3×10^{-8} Hz (1 cycle per year) to 1 Hz. The SG noise is higher than Earth noise only in the small subseismic band between 1 and 20 mHz where the noise level is typically from 1 to $3 \text{ nm s}^{-2} \text{ Hz}^{-1/2}$ (0.1–0.3 µGal $\text{Hz}^{-1/2}$) for most SG stations (Rosat *et al.*, 2004). This is more than 2 orders of magnitude lower than the noise level of the AG as reported by Crossley *et al.* (2001) and Van Camp *et al.* (2005).

It is common practice to filter and decimate data to 1 min samples to look at small temporal gravity variations; assuming white noise, the above noise level translates into a precision of 0.01–0.03 µGal. This level is consistent with common experience. For example, Meurers (2000) easily observed gravity signals of magnitude 0.1–0.3 µGal over 10–30 min intervals and Imanishi *et al.* (2004) identified reliable coseismic offsets of a similar magnitude.

The best way to determine real-world accuracy in the temporal domain is to compare different instruments operating side by side. Two comparisons have been done – one in Miami (Richter, 1990; Klopping *et al.*, 1995, Harnisch and Harnisch, 1995) and one in Boulder (Harnisch *et al.*, 1998), with the result that different pairs of instruments agree to 0.1 µGal, or better. For a dual-sphere instrument, the data streams from the two spheres are largely independent, and the difference signals between two sensors of the four dual-sphere SGs are typically 1 µGal over record lengths of years. Additionally, the residual curves from the dual SG in Moxa agree within a few tenths of a µGal to the polar motion modeled using data provided by the International Earth Rotation Service (IERS). These results suggest that 0.1 µGal is the time-domain accuracy of the SG for long periods.

4.1.5 Recent Developments

4.1.5.1 Ultralong hold time dewar

Although Compact Dewars were extremely successful, the goal remained to further decrease helium consumption and annual disturbances from helium transfers. By 1997, the first commercial (and somewhat) practical 4 K refrigeration systems became available. The Leybold Vacuum Products KelKool 4.2 GM coldhead produced cooling power of 50 W at 50 K at its upper stage and 0.5 W at 4 K at its lower stage. It cooled well below the 4.2 K liquefaction temperature of helium and produced more than 5 times the cooling power of the APD DE202. Soon afterward, GWR produced an ultralong hold time dewar (ULHD) based on the 125 l Compact Dewar design with its neck modified to accommodate the much larger 4.2 GM coldhead. This is a closed-cycle system, because the helium gas condenses in the neck on the lower stage and drips back into the storage volume of the dewar (Richter and Warburton, 1998). This success pointed to the future in which SGs could operate indefinitely without transferring liquid helium or consumption of liquid helium. The first ULHD system is shown in **Figure 5**.

Figure 5 One of the first dual-sphere instruments (CD029, now at Wettzell, Germany) with a ULHD using a Leybold KelKool 4.2 K GM coldhead. The support cranes used to insert and remove the coldhead are shown in the background.

Two dual SGs using the KelKool 4.2 GM coldhead (CD029 and CD030) have been operating continuously since June 1999 at Wettzell and at Bad Homburg, Germany. The coldheads are extremely reliable and require maintenance only at approximately 3 year intervals. However, the coldhead is too heavy for one person to handle and requires a support crane for insertion and removal. Also, the combination of compressor and water chiller requires 7 kW of power versus 2 kW power for the DE202 system.

Two more ULHD systems were manufactured, R038 operating in Concepción, Chile, since December 2002, and C043, which replaced T016 at Syowa station, Antarctica, in March 2003. Neither of these stations could supply 7 kW, so these ULHDs were designed to use a Leybold 4.2 Lab coldhead, which with water cooling reduced the power load to 3.5 kW (Warburton *et al.*, 2000).

4.1.5.2 Observatory dewar

Within a year of shipping C043, Leybold stopped manufacturing the 4.2 Lab coldhead and the 5 year effort to develop a closed-cycle system was threatened. Fortunately, in the same year, Sumitomo Heavy Industries, who had extensive experience with large 4 K cryocoolers, entered the market with a new smaller refrigeration system – the SHI RDK-101 coldhead and CAN-11 compressor. Physically, the RDK-101 coldhead is about the same size as the APD 202, but it uses small 16–mm-diameter flex hoses and is easy for one person to handle. The CAN-11 compressor uses only 1.3 kW power and is air cooled. As a result of the smaller size and power, the RDK-101 has less than half the cooling and liquefaction power of the Leybold 4.2 Lab, but it has excellent prospects for continued future production in Japan.

Over the next 2 years, GWR re-engineered and tested several coldhead/dewar interfaces to fully utilize all the cooling power of the RDK-101. Small dewars have several attractive features: they are visually appealing, lighter and easier to move and install, and, most importantly, the input heat load decreases with surface area. Larger dewars provide longer hold times in the event of coldhead failure. After experimenting with dewar volumes as small as 10 l, GWR chose a 35 l capacity for its Observatory Dewar design. This compromise allows 20 days operation in failure mode: either with the power off, or the coldhead inoperative, and enough excess cooling capacity to liquefy helium gas at a rate greater than 1 l per day.

Figure 6 The newest Observatory Model SG and SHI RDK-101 coldhead. All of the control and data acquisition electronics are contained inside one temperature-regulated electronics enclosure. During installation, the front panel is removed and all the controls are accessed by the local keyboard and computer screen normally stored inside the enclosure. After installation, all functions and data retrieval are remotely accessed through an Internet connection. The dewar is 42 cm in diameter and the combined height of the dewar and coldhead is 130 cm.

The first OSG O040 was installed in Walferdange in December 2003, the second in South Korea in March 2005, with two more installed in Taiwan in March 2006. **Figure 6** shows OSG O049 SG dewar system, with all its control electronics and data acquisition system.

4.1.5.3 Data acquisition system and remote control

In the early years, GWR supplied analog electronics (Gravimeter Electronics Package – GEP) for controlling the gravity, temperature, and tilt functions; and a current supply (Dual Power Supply and Heater Pulser – DPS) for sphere levitation. Analog filters were copies of those used for the International Deployment of Accelerometers (IDA) network of LCR gravity meters and took the approach of dividing the signal between a low-passed tidal gravity output that was the main system output, and a short-period high-passed output that displayed signals such as earthquakes and disturbances. The high-frequency signal was initially recorded only on a strip chart recorder, and atmospheric pressure was also sampled at the station. Each user provided their own data acquisition system.

As more users began to acquire SGs for a variety of purposes, it became standard practice to sample at high rate (1–10 s) the full signal (tides + seismic frequencies). Nevertheless, it took some time for the gravimeter community to achieve the goal of a common set of standards for the acquisition and exchange of SG data. As part of this goal, GWR provided a replacement gravity card (GGP gravity card) with a filter designed for 1 s sampling and additional circuitry for measuring the phase response of the gravimeter system (GGP newsletters #2 and #3).

The first GWR data acquisition system manufactured in 1995 used CSGI software running on a PC with a QNX operating system. Soon, however, uncertainty in the future of the QNX operating system and software maintenance convinced GWR to develop a Windows-based system. In 1999, GWR and BKG reported on an ambitious project to control the SG remotely (Warburton *et al.*, 2000). The prototype Remote SG R038 has been operating at Concepción, Chile, since December 2002, and almost all goals for remote operation have since been implemented. In 2004, GWR decided that all new observatory SGs should be provided with a GWR data acquisition system with remote control capabilities. This is required not only to further standardize GGP data but also to enable GWR to remotely diagnose problems as they arise and to solve them without requiring travel to the site of operation.

4.1.6 User Requirements

4.1.6.1 Operation and maintenance

The SG sensing unit contains only one moving mechanical part, the niobium sphere. It is therefore virtually free of any maintenance requirements, and this has been verified by field installations of 15 years. SG support equipment, however, does need periodic maintenance to assure proper operation and can fail unexpectedly as the result of a lightning strike or other natural catastrophes. Many of the major gaps in SG data have been caused by power supply failure during major storms, or failure of the data acquisition systems. Planning for failure of either electronics or refrigeration is necessary to minimize interruptions in long (decadal) gravity records. It is most important to keep the dewar at least partially filled with liquid helium so that the sensor and superconductors remain below 4.5 K. Upon complete helium loss, the sensor will start warming up to room temperature. Although no damage occurs to the sensor, it requires that the sphere be relevitated, which reactivates the

initial drift discussed in Section 4.1.4.1. In practice, therefore, operators are very careful to make sure the liquid helium volume is kept above a minimum level, so that in the case of a power or coldhead failure there is enough time to either transfer more liquid helium or to fix the source of the failure. With the coldhead off, the maximum hold time for a Compact Dewar is about 60 days. Prudently, most operators do not let the liquid He fall below about one-third full. Therefore, even under severe interruptions, such as the fire at Mt. Stromlo, Australia, in January 2003, the operator has at least 20 days to solve the resulting problems without warming the sensor up. It is also important to follow the manufacturers' and GWR's instructions for maintenance of the coldhead, compressor, and water chiller to prevent equipment failure. Many operators keep a backup refrigeration system available for immediate replacement.

At most of the GGP stations, operators check weekly that the refrigeration system and data acquisition system are functioning properly and ensure general site integrity. When problems develop, they will be observed either in the support status variables that monitor operation of the support equipment (temperature control, tilt-leveling control, and the refrigeration system), or will cause an increase in the instrumental gravity noise. For example, refrigeration problems cause immediate increase in helium boil-off rate and warming of dewar neck thermometers. Ice buildup around the coldhead that touches the inside of the dewar neck will cause an immediate increase in noise observed through the mode filter and on the gravity residual. Problems with the leveling system will be observed on the tilt X and Y balance signals and as gravity noise on the mode filter and gravity residual.

The new GWR data acquisition system (DDAS) allows the operator to monitor about 30 status variables remotely. In addition, alarm levels can be set to automatically generate warnings and alert the operator by e-mail to initiate investigation and repair. After collection and analysis of 1 month data, the operator can enter a calibration factor, tidal parameters, and barometric pressure admittance, and the DDAS will automatically generate a theoretical tide and display the gravity residual signal in real time. This allows visual examination of the gravity noise at the sub-μGal scale. Changes in noise level are immediately observable and with some experience can be identified as those of geophysical origin (atmosphere, ocean, or earthquakes) or due to possible equipment problems. In the latter case,

GWR can consult online with the operator to analyze the problem and provide a rapid solution. Remote access should reduce the frequency of data gaps and ensure higher quality of overall long-term data.

4.1.6.2 Site location

The SG measures an extremely wide bandwidth of signals from periods of seconds to years and the origin of signal sources ranges from local to global. The scientific objective is to determine the signals of most importance and site selection is paramount to achieving these goals. In practice, a site may be chosen to maximize gravity signals of interest and to minimize signals that are of less interest that will be considered as 'noise'. For example, a site must be near a volcano if one wishes to measure signals from magma intrusion, or it must be close to the ocean to measure nonlinear ocean tides or sea-level changes. In contrast, if the goal is to measure short-period signals, such as seismic waves, normal modes, or tides, the site needs to be as quiet as possible (see below).

Site selection may be restricted to the country providing the research funds and by the goals of the funding agency; however, some sites are cooperative efforts chosen to expand the geographic distribution of the GGP network. Examples of the latter include Ny-Alesund, Norway (Sato *et al.*, 2006a), Sutherland, South Africa (Neumeyer and Stobie, 2000), and Concepción, Chile (Wilmes *et al.*, 2006). Neumeyer and Stobie (2000) discuss both geographic and practical criteria used for choosing the Sutherland site. The SG has been used in a wide variety of field situations including harsh conditions such as a salt mine in Asse, Germany (Jentzsch *et al.*, 1995), and on the edge of the Antarctic continent at Syowa station (Sato *et al.*, 1991). Most of the sites, however, are more instrument friendly and have been chosen to use existing facilities where other geophysical instrumentation is already operating and to share infrastructure (such as power, telephone, or satellite communications) and staff.

4.1.6.3 Site noise

The SG is an extremely low-noise instrument and requires a quiet site for optimum operation. A quiet site is one that is removed, by distances of at least several kilometers, from nearby cultural environments such as a city, town, railroad, or major highway. These are strong noise-generating environments that not only increase short-period disturbances but also introduce ground tilts and loadings that will be seen in the residual gravity. If at all possible, it is highly recommended that potential sites be pretested for ambient Earth noise using a STS-1 VBB long-period seismometer. To detect low-noise signals, the site noise must approach the new low-noise model of Peterson (1993) and the best IRIS Global Seismographic Network (GSN) stations (Widmer-Schnidrig, 2003; Berger *et al.*, 2004).

Note that a site may be isolated from cultural influences, but may still be part of a scientific research station containing large (and massive) instruments such as a VLBI antenna. These fiducial stations are of great interest from geodetic and geophysical points of view because of the advantages of combining gravity and geodetic measurements. If, however, an SG is housed within a busy scientific building that is visited constantly or is running other machinery, the quality of the data will clearly suffer. Ideally, the SG should be housed by itself some distance (100 m or more) from other instruments or heavy traffic areas and should pass the seismometer test proposed above.

4.1.6.4 Site stability

Monitoring of long-term crustal deformation using gravity and space techniques requires careful integration of GPS, AG, SG, environmental sensors, and ocean gauges (Zerbini *et al.*, 2001; Zerbini *et al.*, 2002; Richter *et al.*, 2004). All piers must be constructed carefully, since differences in pier construction and separation of piers (non-collocation) may cause subtle and spurious signals of non-geophysical origin. The deformation characteristics of the ground immediately below and around the instrument are obviously important. Ideally, if an instrument can be placed on a concrete pad that is anchored directly to nearby nonfractured bedrock, deformations at the SG will reflect those over a much larger surrounding area. Frequently, SGs are located in an underground setting together with other instrumentation (seismometers and tiltmeters) that is normally on bedrock. There is no problem in siting an SG within such a complex, provided an environment is created around the instrument to protect it and the associated electronics from excessive humidity and temperature changes. The coldhead and compressor may be too noisy to be placed near other instruments and these need to be sited with some care.

For above-ground situations, it may not be possible to find bedrock and the gravimeter will have to sit on a concrete pad secured to unconsolidated material. The type of material (clay, gravel, sand, etc.) will play an important factor in the interpretation of meteorological and other seasonal effects. For example, it has been found that porous material can compress and deform under loading and thus generate an unwanted signal. Whatever the height of the SG with respect to the local ground level, an important factor effect is the soil moisture content of the local and regional area from 1 to 100 m around the gravimeter. Soil moisture resides largely in a layer no more than 1 or 2 m thick, but the effect on an SG can be significant. Installation of groundwater sensors, soil moisture probes, rainfall gauge, and other meteorological sensors are required for developing advanced hydrological models at the sub-μGal level.

The requirements of location, site noise, and site preparation need to be assessed very carefully by potential new users. Considerable experience lies both with the manufacturer (GWR) and with many experienced SG owners, who have maintained their stations for a decade or longer. Although there is no central funding source for establishing new SG sites, potential new users are encouraged to contact existing SG groups for advice, particularly through the GGP website, GGP workshops, and by visiting operating GGP sites.

4.2   SG Data Analysis

As with all modern geodetic techniques, data from SGs require specialized processing before it can be used to its fullest advantage. There is a large amount of GGP data available from GGP-ISDC (Information System Data Center) online, hosted by International Center for Earth Tides (ICET), but it may be of limited use to scientists unfamiliar with these kinds of data. The purpose of this section is to review the common procedures in analyzing the data for different end uses. A summary of the GGP project is given in Crossley et al. (1999), and station names and details can be found in Crossley (2004).

4.2.1   Preprocessing

4.2.1.1   Second data sampled from an SG

SG data acquisition systems usually record two types of signals: the gravity feedback signal at high accuracy (0.1 nGal) and at precise times, and many auxiliary channels such as environmental data (e.g., atmospheric pressure, room temperature) and instrument parameters (e.g., tilts). The gravity is typically recorded at 1, 2, or 5 s

intervals, whereas the auxiliary signals are often sampled at lower rates, for example, 1 min.

Examples of raw SG data are shown in **Figure 7**. **Figure 7(a)** shows data from the Cantley and Boulder instruments at 1 and 5 s sampling, respectively. There is a noticeable difference in high-frequency noise between the recordings due to the different sampling rates and antialiasing filters. Most of the noise is microseismic, that is, the incessant propagation of surface waves generated by wind and ocean turbulence, between 1 and 10 s, in the period range where most SG data are sampled. **Figure 7(b)** shows data from the Strasbourg instrument for the day after the Sumatra–Andaman earthquake (26 December 2004). The earthquake clearly dominates the tides, but SGs are not gain-ranging instruments, so the surface waves from large events are frequently clipped.

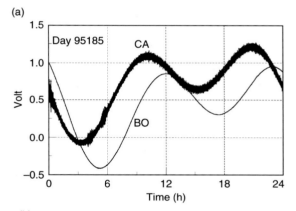

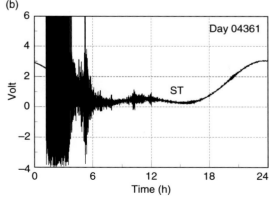

Figure 7   Examples of 1 day of raw SG data (a) from stations CA (Cantley, Canada) (1 s sampling) and BO (Boulder, USA) 5 s sampling and (b) from station ST (Strasbourg, France) (2 s sampling) that includes the $M_w = 9.3$ Sumatra–Andaman earthquake.

The 1 or 2 s data are usually not sent to ICET, although there is a mechanism for receiving it. Initially, the issue was one of file size, but also the high rate data are not generally interesting from a geodetic point of view. Some of the raw data are transferred to the Incorporated Research Institutions for Seismology (IRIS) database, where it is of interest in the study of seismic normal modes of the Earth (Widmer-Schnidrig, 2003).

4.2.1.2 Minute data from the GGP database (ICET)

Most of the data available at ICET/GFZ are 1 min data sent by each station on a regular basis (one file per month). The requirements of GGP are that operators must apply a digital decimation filter to their raw data and resample it at 1 min using a zero-phase digital filter. The local air pressure is also decimated to 1 min, and the two series are combined in a simple ASCII file; each sample is date and time stamped and the file has appropriate header information. A sample of data from the Strasbourg station is shown in **Table 1** in the PRETERNA format that is part of the ETERNA Earth tide analysis program (Wenzel, 1996b). This data is uncorrected, that is, data disturbances such as spikes, offsets, and other problems still remain, inevitably smoothed by the decimation filter.

A display of minute data from ICET is shown in **Figure 8**; this is from station Boulder for the month of September 1997. The month is chosen for illustration purposes because it has data gaps, spikes, and (not visible at this resolution) signals that seem to be offsets.

Generally, the data at ICET vary from bad months like this to completely trouble-free data; a typical file may contain one or two problems to be fixed. Each monthly file is about 1.5 MB uncompressed.

Two different philosophies have been used to solve problems in the data: either leave gaps (i.e., simply ignore bad segments), or remove disturbances and offsets and fill the gaps with a synthetic signal from a model. The former (leave gaps) requires that all processing steps have to maintain the integrity of the gaps and this reduces the flexibility of using most time series analysis algorithms in their standard form. The latter (fix gaps) requires further choices about what level of disturbances to treat and what to leave alone, but it is the preferred approach within the gravity community. The main reason is the convenience of dealing with continuously sampled data rather than data in sequences of blocks and gaps. Computer algorithms are also easier to implement for continuous data than data organized into blocks, but this is not the only reason to repair bad data.

Some geophysicists would argue, with justification, that one should not 'invent' data for the convenience of processing. Pagiatakis (2000) proposed the treatment of SG data using a least-squares inversion for all possible constituents at once, as is commonly done for other complex data sets, for example, very long baseline interferometry (VLBI). In passing over 'bad' data, the amplitudes of offsets are included as variables in the inversion; so they are removed simultaneously with tides, pressure, drift, and other modeled effects. This obviously requires a large computational effort,

---

Table 1   Sample GGP 1-min data file

Filename	:	ST041200.GGP
Station	:	Strasbourg, France
Instrument	:	GWR C026
Phase Lag (sec)	:	17.1800 0.0100 measured
N Latitude (deg)	:	48.6217 0.0001 measured
E Longitude (deg)	:	7.6838 0.0001 measured
Height (m)	:	180.0000 1.0000 estimated
Gravity Cal (nm.s-2/V)	:	−792.0000 1.0000 measured
Pressure Cal (hPa/V)	:	22.2222 0.1000 estimated
Author	:	jhinderer@eost.u-strasbg.fr

```
yyyymmdd  hhmmss  gravity(V)  pressure(V)
C**************************************************************************************
77777777
20041201  000000  2.448085   4.914420
20041201  000100  2.452300   4.912670
20041201  000200  2.456466   4.910337
20041201  000300  2.460378   4.908314
20041201  000400  2.464132   4.906599
```

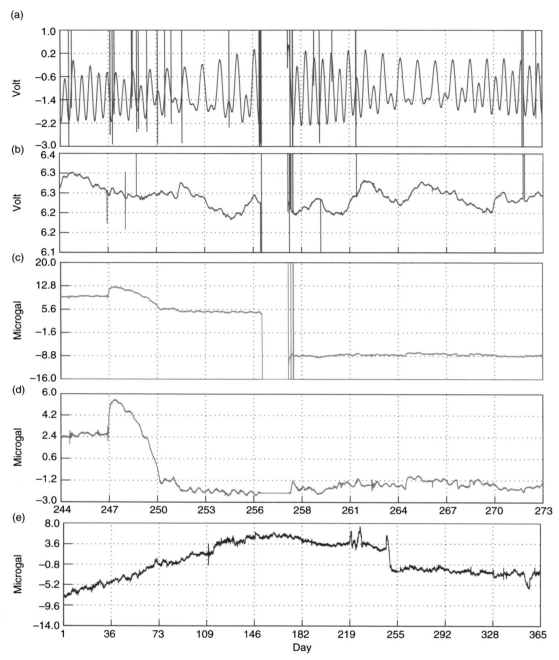

Figure 8 Preprocessing of 1 min data for station BO (Sep. 1997): (a) original gravity; (b) original pressure; (c) after small gaps (<10 s) are filled and removal of tides and nominal pressure; (d) after a large gap with offset is corrected; and (e) the residual gravity for the whole of 1997 after gaps and offsets are removed.

particularly when using 1 s data (as opposed to 1 min data) and when the data spans many years or even decades. The inversion also requires interpretation of the resulting parameter covariance matrix to see what tradeoffs exist in the model. This approach does not give an intuitive feel for the effects of various processing steps and there are many different corrections to be included. For the rest of the article, we turn to the gap-filling approach and assume that the goal is to produce and analyze a uniformly sampled residual gravity series.

A simple model will suffice for most of the following discussion.

g (observed) = g (disturbances)

    (of instrument and station origin)

+ g (tides)

    (solid Earth, ocean)

+ g (nontidal loading)

    (atmosphere, ocean currents)

+ g (polar)

    (annual and Chandler polar motion)

+ g (drift)

    (instrument drift function)

+ g (hydro)

    (rainfall, soil moisture, groundwater)

+ g (residual)

    (ocean currents, other signals,

    deformation, tectonics,

    slow eartquakes, etc.)   [1]

The purpose of data processing and analysis is to remove the effects that are not the subject of investigation and model the effects that remain. We note at this point that the instrument amplitude calibration factors are required in the next step in order to convert the observed data (g, p) in volts to their equivalent gravity and pressure values:

$$g\,(\mu Gal) = g\,(volt)^*GCAL$$

$$p\,(hPa) = p\,(volt)^*PCAL$$

Calibration issues are discussed in detail later in this chapter.

4.2.1.3 Remove–restore technique

The most widely used approach is to first make nominal corrections for some of the largest influences such as tides and pressure (remove), then fix the problems in the residual signal and put back the removed signals (restore). The data can then be used for whatever processing and analysis is desired. Before any corrections are made to gravity, the first step is therefore to deal with problems in the atmospheric pressure, because this will be used in the remove–restore phase. **Figures 8(a)** and **8(b)** show the residual gravity and pressure for 1 month; clearly, the pressure has disturbances and gaps at the same time as in the gravity channel, indicating problems in the data acquisition system.

With pressure data, it is difficult to generate data within gaps, short of running computationally

intensive weather forecasting codes or interpolating data from meteorological stations surrounding the gravimeter. Meteorological services give hourly pressure data that is available by request or through the Internet. For simplicity, a linear extrapolation between the beginning and end points of the gap, even for gaps extending for days or more, may be sufficient. Pressure variations have only a small seasonal signal, and no significant daily variation, so this interpolation is at least reasonable, if not ideal. For longer gaps of several days or more, an auxiliary source of data as mentioned above should be considered, especially where the pressure channel has failed but the gravity data are still good.

Pressure sensors are usually factory calibrated (PCAL), but this calibration factor needs to be monitored by the user. From time to time, the sensors need recalibration for drift or other problems, or even replacement, and this can introduce a dilemma in the processing. Having detected a problem, should one allow for instrument drift over a previous period of time, or simply assume an offset? No one solution is ideal in all cases. We assume that the pressure record has been made continuous through gaps, and been cleaned of spikes and other disturbances. Depending on the circumstances, this might have required ancillary data, or perhaps linear interpolation.

The gravity signal is not so easily dealt with. The data in **Figure 8(a)** are dominated by the tidal signal that varies usually between 100 and 300 μGal in amplitude, depending mainly on station latitude and the phase of the lunisolar cycle. For very short gaps or spikes (10 s or less), linear interpolation will work without much problem even without removing the tides. This is especially effective if applied to the original raw data (at second sampling), because spikes or gaps are then not smeared into the record by the second-to-minute decimation filter. For longer gaps, we need a version of the remove–restore philosophy.

4.2.1.4 Treatment of gaps in gravity data

To repair gravity data, we normally construct a preliminary model based on the tides and nominal atmospheric pressure. For longer gaps, it may be appropriate to add instrument drift and perhaps polar motion. The tides are usually modeled by summing wave groups with specific gravimetric factors (δ, κ) determined in some prior tidal analysis at the station (see below). The preliminary model is then subtracted from the observed data to reveal hidden problems that need to be fixed. We call the residual gravity g (temp):

$$g(\text{temp}) = g(\text{observed}) - g(\text{removed}) \qquad [2a]$$

where

$$\begin{aligned} g(\text{removed}) = \; & g(\text{tides}) + g(\text{nominal pressure}) \\ & + g(\text{drift}) + g(\text{polar motion}) \end{aligned} \qquad [2b]$$

The instrument calibration factor already enters eqn [2]. Note that a linear function is normally used to fill in a residual gap, so we are interested only in modeling signals that are not linear during a gap. Polar motion is very well defined by IERS data, but at a level of $5\,\mu\text{Gal yr}^{-1}$ (or $0.01\,\mu\text{Gal day}^{-1}$) it is not important for gaps of less than a week or so. Instrument drift over the time period of most gaps can usually be taken as a simple linear function of time. Other components of the signal such as hydrology or nontidal ocean loading are, however, too complex to be included as part of the removed function, even though they can have an effect of several μGal over time spans of days.

Figure 8(c) shows g (temp) for the case when g (removed) consists of a local tide model and a nominal pressure correction. All the short spikes were previously replaced by a simple linear interpolation on the original data, as discussed above. Because the pressure is included in the g (removed) model, it is corrected before the gravity in order to avoid introducing spurious signals in the gravity. The gravity series still has a major problem between days 256–258, but otherwise looks 'better'.

At ICET/GFZ, the basic files for all stations have uncorrected minute data of the form of **Figures 8(a)** and **8(b)**, that is, raw data decimated to 1 min but with no data repair prior to decimation. ICET produces corrected minute data in which data repair has been done on this already-decimated minute data. Some users also send data that have been repaired at the second sampling rate prior to decimation.

4.2.1.5   Disturbances and offsets

To make further progress, we now work with the residual series g (temp), as in **Figure 8(c)**. The major disturbance at day 256 contains an offset and there is a suspicious signal at day 247 that requires consideration. Repairing (fixing) the offsets in **Figure 8(c)** is one of the most critical processing steps, requiring choices as to which problems to leave alone, and which to fix. We consider an offset (or tare) to be any apparent change in the instrument

base level. This could be a small offset ($0.5\,\mu$Gal) that takes place within one time step (e.g., 1 s) due to an instrumental or electronic event. Alternatively, an offset may occur during helium refilling (maybe 10–$20\,\mu$Gal) or a major electrical strike that causes level changes of $100\,\mu$Gal or more.

Here we choose to fill the gap between days 256 and 258 with a straight line and to absorb the apparent level change into an offset that is $13.816\,\mu$Gal. We have no means of knowing the actual offset during this time (the gap), so we simply join the good data before and after with a straight line and compute the offset needed to minimize the residual signal. This is a large offset that is clearly associated with something that happened to the instrument, involving also the pressure; the instrument base level was changed by a non-geophysical event. Unfortunately, the data log is unavailable for this time period at Boulder, but at most stations there should be a written explanation of such events (stored in the GGP database as 'log' files). Some disturbances will not cause an obvious offset in the data, but clear offsets should be removed.

The 'fixed' residual is shown in **Figure 8(d)** (that interpolates the gap), but there is still an anomaly at day 247 with a jump of $2\,\mu$Gal followed by a gradual decrease of $4\,\mu$Gal. We know that rapid changes of gravity can be caused by physical processes such as heavy rainfall (Crossley and Xu, 1998; Klopping *et al.*, 1995), but there is no reason at this point to adjust the data, even though the disturbance causes an apparent offset. Even though a log file helps to indicate a non-geophysical cause for an offset, it usually cannot help to decide what the size of a particular offset may be.

After going through the whole year 1997, we find a total of 29 problems that need to be fixed, three of which were readily apparent offsets. The residual curve for the whole year is shown in **Figure 8(e)**. Day 247 still shows up as a possible nonphysical anomaly, as there is no apparent subsequent recovery from the level jump. Without the 'offset', the data would show a smooth annual variation that arises predominantly from polar motion. Although it might be tempting to treat more aggressively the level change as an offset, doing so will affect all the subsequent data, and it is clearly risky to do so without good reason.

4.2.1.6   Automatic procedures

Ideally, each day's raw data record should be scanned visually for potential disturbances. Where this is not possible, simple numerical algorithms can scan the data for the rapid detection of anomalies. One

Table 2 Maximum slew rates for some geophysical signals

	(μGal min^{-1})
Solid earth tides	0.95
Atmospheric pressure	0.20
Groundwater variations	0.02

method uses the slew rate (Crossley *et al.*, 1993) that computes the simple forward data derivative and flags data points that exceed a certain threshold. We indicate from typical records the maximum slew rates in **Table 2**, although these may be exceeded in some situations. With the solid tides in the gravity signal, slew rates can reach up to 1 μGal min^{-1}, and this rate is important when trying to determine the phase calibration of the instrument, as discussed later. The next largest slew is from the atmosphere, and this is an order of magnitude larger than the slews from groundwater, rainfall, and soil moisture. Disturbances and offsets can have much larger slew rates, but sometimes they are small slews that cannot be distinguished from the real signals above. Most users would use slew rate as a diagnostic but not as a corrective procedure. Another approach is to compare each data point with a predicted value based on the accumulated statistics of previous data.

Two widely used software packages, TSOFT (Van Camp and Vauterin, 2005) and PREGRED (part of the ETERNA package; Wenzel, 1996b), deal especially with the preprocessing of gravity data, provide some automation of data analysis, and incorporate many other processing options. TSOFT is probably the most widely used package for GGP data.

To give further insight into the effect of processing choices, we show some examples from Hinderer *et al.* (2002a). In each case, the same 6 months of data (March–December 1997) were given to different scientists to repair, and the results were compared. **Figure 9(a)** shows how four different operators approached the problem of a moderate-sized earthquake that was removed. The results ranged from removal of just the large surface waves (SR) to removal of most of the disturbed record (JPB). For a transient disturbance without offset, four different choices were made about which parts of the signal to repair (**Figure 9(b)**); depending on the choice, the final and starting levels of the different options obviously diverge. Finally, a highly disturbed portion of the record due to a helium refill (**Figure 9(c)**) was an opportunity for two operators (JPB, DC2) to remove

an apparent offset, and two others chose to leave it in. Some of this processing was done using TSOFT and some using other algorithms and procedures.

4.2.1.7 Processing for different purposes

For several end purposes, the long-term trends in the data are unimportant. This is the case for tidal processing, that is, for finding gravimetric tidal factors, ocean loading, and related signals such as the resonance effect of the free core nutation (FCN). Other possibilities are signals in the 1–24 h range (e.g., the possible Slichter triplet) and, of course, high-frequency seismic modes. In these cases, a more aggressive treatment of disturbances and offsets might be beneficial. **Figure 10(a)** shows the effect of the different processing options from **Figure 9** in the frequency domain, that is, the power spectral density (PSD) function. There are five different options ranging from a very minimal treatment (DC1 – not shown in **Figure 9**) to a comprehensive removal of problems (JPB). The latter has effectively reduced the noise floor in the long-period seismic band (up to about 5×10^{-5} Hz or 5.5 h) by more than an order of magnitude. This is mainly due to the removal of offsets and disturbances that contain high frequencies.

When the goal is to preserve the long-term evolution of the gravity residual rather than search for short-period signals, offsets must be examined carefully. If offsets occurred randomly in time and with random amplitudes, their accumulated impact is similar to a brown noise or random walk type of process. The mean value drifts from the true gravity, and this would be detectable only by comparison with absolute gravity measurements. Large offsets (5–10 μGal or more) need to be corrected, especially when they are logged as due to specific causes. Sometimes, smaller offsets happen without apparent cause, but if not corrected a bias can easily accumulate. Even with appropriate software this can be a time-consuming process, as already discussed.

We show in **Figure 10(b)** the effect of different options in the time domain, over the 6 month test period referred to earlier. All the traces coincide at the beginning but diverge by up to 1.5 μGal after 6 months, the largest part of this difference being due to the different treatments of the disturbance in **Figure 9(c)**. As indicated above, the use of the AG measurements can be very useful in identifying offsets at the level of a few μGal. Note that in **Figure 10(b)** instrument drift plays no role in the processing differences.

(a)

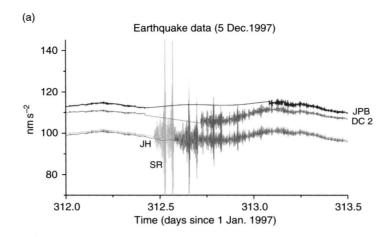

(b)

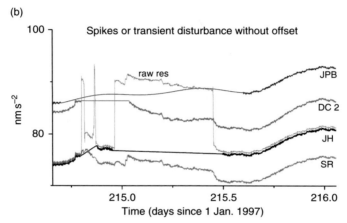

(c)

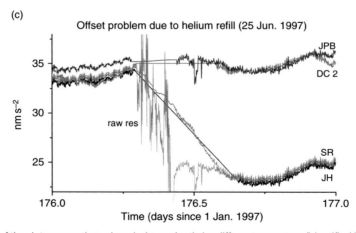

Figure 9 Examples of the data corrections done independently by different operators (identified by initials) for the removal/interpolation of (a) an earthquake, (b) disturbances without offsets, and (c) an offset caused by a helium refill. From Hinderer J, Rosat S, Crossley D, Amalvict M, Boy J-P, and Gegout P (2002a) Influence of different processing methods on the retrieval of gravity signals from GGP data. *Bulletin d'Informations des Marées Terrestres* 123: 9278–9301.

Long gaps in a record make determination of offsets difficult, if not impossible. There are two saving strategies – either to rely on the co-location of AG measurement to anchor the SG record at the ends of a long gap, or possibly use a fit to the polar motion when there is a suspected large offset within the

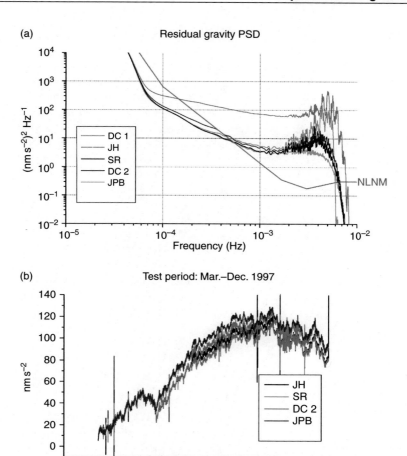

(a) Residual gravity PSD

(b) Test period: Mar.–Dec. 1997

Figure 10 The cumulative effect of data corrections for a test period of 6 months of data from different treatments (a) in the time domain, showing a 1.5 μGal offset between treatments and (b) in the frequency domain – NLNM is the new low noise model of Hinderer *et al.* (2002a).

downtime of the gravimeter. Long gaps in the record are generally due to serious problems in the instrument or data acquisition system, so an offset is certainly possible. Unfortunately, other instruments (e.g., spring gravimeters), even when available, have too much drift to provide a good interpolation signal to fill in an SG record for more than a few hours.

4.2.1.8 Restoring the signal

When all problems have been removed from g (temp), it is easy to recover the observed gravity from [2], so that

$$g\,(\text{fixed}) = g\,(\text{temp}) + g\,(\text{removed}) \qquad [3]$$

is now the observed gravity with the problem segments repaired. Referring to eqn [1], we note that the fixed data have a signal in the gaps that do not

contain the more complex effects such as hydrology. In all other respects, however, the data in the good portions are identical to the observations, and the fixed gravity can now be analyzed for the components in eqn [1]. By subtracting g (fixed) from g (observed), it is always possible to see exactly what has been removed during this stage of the processing.

4.2.2 Solid Earth and Ocean Tides

4.2.2.1 Tide-generating potential

The basis for computing a theoretical gravity tide is the tide-generating potential (TGP). Using the orbital and rotational data for the Earth as forced by the Sun and Moon, this potential was first given as a catalog by Doodson (1921), who included only terms of degree 1–3 (i.e., 24, 12, and 8 h periods). The tides occur in the rotating frame of the Earth's mantle, but they also lead to forced nutations of the Earth in the space-based (nonrotating) coordinate system (e.g., Melchior and Georis, 1968). There is one tidal wave associated with each component of the Earth's nutation. The most extensive tidal developments now include the perturbation effects of all the major planets and terms up to degree 6 for the moon (period 4 h) as well as terms allowing for the nonspherical shape of the major bodies. The most recent references on the tidal potential for gravity work are Hartmann and Wenzel (1995a, 1995b), Roosbeek (1996), and Wenzel (1996a).

There are two approaches to cataloging the TGP. First, the ephemeris – a catalog of apparent positions of the bodies in the solar system as seen from a position and time on the Earth; each body is defined by a longitude, latitude, right ascension, and obliquity. In this method, an ephemeris tide program (e.g., Merriam, 1992a) generates a time series from the TGP on the surface of a rigid Earth. The accuracy of this TGP is directly related both to the accuracy of the ephemeris itself and the analytic theory used to describe the motions of the planetary bodies. Merriam estimated for GTIDE an rms time-domain accuracy of 0.25 nGal, with a maximum error of 0.8 nGal.

Wenzel (1996a) compared three tidal ephemeris programs: GTIDE as updated in Merriam (1993a), and two JPL programs DE200T and DE403T that he used to generate six benchmark gravity tides series. These series were gravity at 1 h spacing for two example epochs (1987–93; 2017–23) at the Black Forest Observatory (BFO) in Germany (chosen because of the availability of gravimeter data at BFO). The internal time-domain accuracy of the benchmark series derived from the JPL ephemerides is better than 0.1 nGal, and in the frequency domain the rms errors are about 100 times smaller.

The second approach, which is a more widely used method for the TGP, is to sum together a number of harmonic terms, each of which represents a partial tide. The frequencies and amplitudes of these waves can be derived from the ephemerides by constructing a time series and then filtering it, as in the CTE potential (Cartwright and Taylor, 1971; Cartwright and Edden, 1973), or using analytic means as for the Xi89 potential (Xi, 1987, 1989), and the potential RATGP95 (Roosbeek, 1996). Alternatively, the waves can be found by fitting coefficients to a series based on an ephemeris – the spectral method – as used in the Tamura87 and Tamura93 potentials (Tamura, 1987, 1993) and in HW95 (Hartmann and Wenzel, 1995a, 1995b). These TGPs (except for CTE that has now been superseded) shown in **Table 3**, are those most commonly used in high-precision gravity work. They include tables (catalogs) specifying the frequencies and amplitudes of the waves (actually wave groups) that must be summed. By selecting different subsets of waves, the speed of computation can be adjusted according to the accuracy required.

Merriam (1993a) compared the catalogs of Tamura (1987) and Xi (1989) with GTIDE and concluded that although their differences should be detectable using SGs, in practice either of the catalogs could be used for SG analysis. Wenzel (1996a) compared all the above series and catalogs and concluded that the HW95 was the most accurate for high-precision work. Roosbeek (1996) notes this is expected because HW95 is derived from one of the benchmark series itself and its only error should be computational. Tamura93 is widely used as a compromise between speed and accuracy.

4.2.2.2 Elastic response of the Earth

Once the tidal forcing series has been accurately computed for a precise latitude and longitude, it is necessary to allow for the deformation of the Earth to compute an approximation to the observed solid tide.

Table 3 Tide-generating potential used in SG data reduction

Catalog	# waves	# coeffs.	Degree	Time (ngal)	Frequency (ngal)
CTE	505	1010	3	38.44	0.565
Tamura87	1200	1326	4	8.34	0.118
Xi89	2934	2934	4	6.42	0.090
Tamura93	2060	3046	4	3.08	0.046
RATGP95	6499	7202	6	2.00	0.026
HW95	12935	19271	6	0.14	0.002

The last two columns refer to rms accuracies in the time and frequency domains respectively.
Modified from Wenzel H-G (1996a) Accuracy assessment for tidal potential catalogues. *Bulletin d'Informations des Marées Terrestres* 124: 9394–9416.

This is usually done within the tide program through the use of elastic load Love numbers (h_n, k_n, l_n) for each of the harmonics (denoted by n) in the TGP. The Love numbers are computed by solving the gravitoelastic equations of motion for the Earth and finding the surface displacement $\mathbf{u}$ and potential Ψ for any kind of forced deformation (e.g., tides or loading) (e.g., Wang, 1997):

$$\mathbf{u} = h_n\boldsymbol{\psi}^1/\mathbf{e}_r + \ell_n\nabla_1\boldsymbol{\psi}^1/g_0$$

$$\boldsymbol{\psi} = k_n\boldsymbol{\psi}^1 \qquad [4]$$

$$\delta_n = 1 + 2h_n - (n+1)k_n/n$$

Here, $\boldsymbol{\psi}^1$ is the perturbation in the gravity potential, $\mathbf{e}_r$ is the radial vector, ∇_1 is the surface gradient operator, and g_0 the surface gravity. The Love numbers completely describe any kind of deformation, elastic or inelastic, and therefore contain all the complexity of the actual Earth, that is, resonances for all the Earth's normal modes, anelasticity and frequency dependency (e.g., Dickman, 2005). Love numbers have been frequently computed for seismic Earth models such as PREM (Dziewonski and Anderson, 1981) and given in a number of different forms; they are in principle complex numbers because of the Earth's anelasticity (e.g., Mathews, 2001). Of interest here are the real gravimetric tidal factors (δ_n, κ_n), where κ_n defines the phase of the tidal wave as observed (or computed, note it is 0 for an elastic model) with respect to the TGP and δ_n is found from a combination involving h_n and k_n as above (ℓ_n is not used in gravity). Typical elastic values are h_n = 0.602, 0.291, and 0.175, and k_n = 0.298, 0.093, and 0.043, yielding δ = 1.155, 1.167, and 1.121 for n = 2, 3, and 4, respectively. A nominal pair of values is taken as (δ = 1.16, κ = 0).

Gravimetric factors can be used to construct the synthetic tide at a station and are thus the simplest way to 'de-tide' a gravity record (especially a short one where tidal fitting would be problematic), and of course are perfectly suited to constructing the 'removal' signal discussed above. They can be found empirically from a tidal analysis at a station, noting that there should be no phase lag entered into the program (e.g., Crossley and Xu, 1998). When the synthetic tide is reconstructed from the empirical gravimetric factors, both the ocean loading, considered in the next subsection, and the system phase lag will be automatically included along with the solid tide.

4.2.2.3 Ocean tides and loading

Ocean tides have the same frequencies as solid-Earth tides, but they derive from the vertical and (small) horizontal components of the TGP that generates water movement in shallow areas such as the continental shelves. This variable water depth loads the crust and modifies the amplitudes of the bodily tides as seen on land stations. One problem in computing ocean tides is their variability, due primarily to oceanic weather systems. A further difficulty is the need to define the coastal topography and bathymetry, particularly in remote and icebound locations such as Antarctica.

Ocean tidal loading is almost an invisible effect when doing tidal analysis, because the gravimetric factors are adjusted to the total tide signal that includes ocean loading. There are several problems for which ocean tides and their loading effects need to be studied in detail, and in these cases the ocean loading must be computed separately. Depending on the location, ocean tide loading varies between 1% and 10% of the body tide.

As in atmospheric loading (see Section 4.2.3), the ocean load function is generally expressed as the convolution of a Green's function with a suitable data set representing ocean heights. Most ocean loading programs consider the tide heights to be given, for each wave, within cells that follow the coastlines, but the height variability due to weather systems is ignored. Ocean tide heights are now determined almost exclusively by satellite altimetry (TOPEX/ POSEIDON), but this still leads to a large number of ocean tide models constructed with differing assumptions and constraints.

Ocean tide loading is an essential element of the complex problem that includes mean sea levels and global warming. It is therefore an important topic in gravimetry that has greatly benefited from the availability of high-quality SG data. Baker and Bos (2003) and Boy et al. (2003) have discussed the issue of whether SG data can discriminate between competing ocean tide models, and they emphasize the need for accurate SG calibration. Several computer programs can do the appropriate calculations. Three of the most widely used are SPOTL (Agnew, 1997), OLFG by Scherneck (1991), and GOTIC2 by Matsumoto et al. (2001). Boy et al. (2003) showed significant differences between these programs and tide models that left the role of calibration unclear. On the other hand, Bos and Baker (2005) concluded that with care the computational errors can be resolved and the programs show almost identical results for all programs and for all tide models. This suggests that there are still calibration issues for SGs. They offer a new program CARGA that incorporates most of the features of the other programs.

Ocean loading corrections are frequently geared to the need for precise geodetic information in particular areas such as Antarctica (Bos *et al.*, 2000) or the Pacific Northwest (Lambert *et al.*, 2003).

4.2.2.4   Tidal analysis

Tidal analysis consists of determining the amplitudes and phases of tidal waves of specific frequencies from observational data. How many waves can be determined and to what accuracy depends on the length of the record used and on the noise characteristics of the site. In most approaches, the ocean tides are grouped in with the solid tides of the same frequency, and they cannot be separated by a simple fit of data to known tidal frequencies. Ocean tide variability is reflected in the time dependence of the gravimetric tidal factors at a particular site.

Three programs are used within the SG community for tidal analysis. The most widespread program is ETERNA, developed over many years by Wenzel (1996b) and now in its final form (version 3.3). It consists of a suite of several programs that deal with all the common aspects of processing gravimeter data and it can be adapted to a variety of different data sets; it can also be used for analysis of strain and tilt data as well as gravity. ETERNA is a harmonic method that does a direct least-squares fitting of the (δ, κ) factors for various wave groups to gravity data sampled at fixed intervals, for example, 1 min or 1 h. The program can handle different sub-blocks of data if there are gaps. The user has the choice of simultaneously fitting an admittance function to pressure or other auxiliary data. It includes polar motion and different assumptions regarding the instrument drift function. The program is well documented and available through ICET.

The second approach is a program called BAYTAP-G that was developed by Japanese geophysicists during the 1980s (Tamura *et al.*, 1991). The approach is a hybrid method using a combination of harmonic series and the response method (Lambert, 1974) to estimate the various components of a gravity record (i.e., the tidal parameters, a pressure perturbation effect, a 'drift' function, and irregular noise). One of the distinguishing aspects of the program is a focus on a statistical description of the 'drift', which in this context means the entire long-term gravity signal and not just the instrument drift, and of the random noise Using Bayesian estimates, the procedure involves the nonlinear estimation of a tradeoff parameter between the tidal harmonic series and the residual gravity.

A third program is available, called VAV, which was described most recently by Venedikov and Viera (2004). The origins of this approach go back 50 years or more when gravimeter records were processed by the application of suitable filters of various lengths to account for the tides. Comparisons exist between VAV and ETERNA but the former has not been as widely adopted by the SG community as the other programs. Dierks and Neumeyer (2002) compared all three programs using both synthetic data and a 1 year observed SG data set from station Sutherland (SA). They found the performance of the three programs to be similar, but with different treatments of the statistics between signals (tides, pressure, and drift) and residual gravity. The spectra of the final residuals were noticeably different between the programs, and the reader should consult their paper for details (also for valuable tips on how to set some of the parameters in the program inputs).

A few words should be said about the response method in SG tidal analysis, as discussed by Merriam (2000). Rather than allocating for all waves within a group the same gravimetric factor, the response method selectively interpolates the gravimetric factors using only waves that seem unperturbed by noise (i.e., that are close to their theoretical expectations). Waves that seem anomalous are thus treated independently, requiring a more hands-on treatment, but one that has some practical advantages. The additional flexibility allows a greater reduction in tidal residuals, because it can accommodate the more variable ocean loading.

4.2.3   Atmospheric Pressure Effects

The atmosphere provides a significant gravity effect (up to 10% of the tidal signal) with a transfer function (or admittance factor) that approximates $-0.3 \,\mu\text{Gal}$ hPa^{-1} for a typical continental station. The effect is a combination of gravitational attraction by atmospheric density anomalies with a loading that vertically deforms the crust and mantle. For a positive atmospheric density anomaly, simple theory gives about $-0.4 \,\mu\text{Gal} \,\text{hPa}^{-1}$ for the upward attraction and $+0.1 \,\mu\text{Gal} \,\text{hPa}^{-1}$ for the loading. A number of well-studied empirical and physical methods exist for making a pressure correction to the gravity data, but even with the most sophisticated treatments it is not possible to completely remove the atmospheric pressure effect.

4.2.3.1 Single admittance factors

Atmospheric effects on gravity became an important consideration with the higher precision and lower noise of the SG compared to previous instruments. Warburton and Goodkind (1977; henceforth called WG77) anticipated most of the issues that were revisited in many subsequent papers (e.g. Spratt, 1982; Müller and Zürn, 1983; Rabbell and Zschau, 1985; and van Dam and Wahr, 1987). A useful review of the role of local, regional, and global effects was given by Merriam (1992b), from which we restate the following conclusions:

1. Approximately 90% of the atmospheric effect comes from the local zone, defined as within 0.5° (or 50 km) of the station, and 90% of this effect is from the direct Newtonian attraction of the atmosphere; the remaining 10% is from loading and deformation of the crust.
2. The regional zone extends from 0.5° to ∼3°, or from 50 to 100–500 km (depending on the topography surrounding the station), and in this zone the atmospheric correction is small (a few percent) and primarily comes from loading.
3. The rest of the atmosphere, at distances >3°, is the global effect and contributes only a few percentage of the total atmospheric gravity.

Merriam found a local admittance factor of $-0.356\,\mu$Gal hPa^{-1}, and with a variable regional correction, a combined admittance that varied between -0.27 and $-0.43\,\mu$Gal hPa^{-1}. This is consistent with the factor of -0.30 found by WG77 for frequencies <1 cpd, a factor that is widely applied for the quick 'correction' of gravity due to atmospheric pressure. Note that this factor applies to relative changes in pressure and gravity that have to be propagated from one data sample to the next by maintaining appropriate reference levels.

4.2.3.2 Frequency-dependent admittance

WG77 first recognized that a single admittance factor was not appropriate at all frequencies, partly because of the separate effect of atmospheric tides at harmonics of a solar day (e.g., Elstner and Schwahn, 1997). In addition, however, the atmosphere shows more coherency toward higher frequencies (2–8 cpd) and the admittance factor from data at a single station is larger than at lower frequencies (<1 cpd) when regional scales become important. WG77 also introduced the cross spectrum between gravity and pressure to find an admittance function in the frequency domain, and this was extended by Crossley $et\ al.$ (1995) and Neumeyer (1995), who showed the advantages that could be realized under certain conditions. Frequency dependence is equivalent to allowing the admittance function to vary in time, as would arise in the passage of weather systems (e.g., Müller and Zürn, 1983).

To give some insight into these points, we show in **Table 4** a comparison of different processing methods, from Crossley $et\ al.$ (1995). Four different corrections are given with the notation TD = time domain; FD = frequency domain; δg = gravity correction in the TD; Δg = Fourier transform of δg; and $p(t)$ = pressure.

Note that fitted values of the admittance α (cases 2 and 3) are smaller than the nominal correction (case 1), because they respond to the lower coherence in the atmospheric signal at long periods and at frequencies that are harmonics of 1 cpd. This is clear in **Figure 11** that shows the amplitude component of the admittance from 2 years of gravity and pressure data from Cantley (CA). The smooth background obtained by binning segments of the spectra uses an averaging window of 1.5 cpd and represents the increasing coherence with frequency of the atmosphere as sampled from a single location. Note that the asymptotic value is consistent with the values of $0.356\,\mu$Gal hPa^{-1} quoted by Merriam. The sharp negative lines are computed separately from a narrow window of 0.01 cpd centered on the harmonics of a day and reflects the low coherence between gravity at a single location and the global atmospheric tides. There are issues with using such a function, however, because the choice of averaging window has a large effect on how much signal is removed from the gravity residuals.

Table 4 Comparison of atmospheric pressure corrections

Case	Type	Formula	α-value	rms
1	Nominal admittance in the TD	$\delta g(t) = -0.3\,p(t)$	0.30	
2	Fitted admittance in the TD	$\delta g(t) = -\alpha\,p(t)$	0.257	1.147
3	Fitted admittance in the FD	$\Delta g(\omega) = -\alpha\,P(\omega)$	0.255	1.142
4	General FD admittance	$\Delta g(\omega) = -\alpha(\omega)\,P(\omega)$	Fig. 11	0.672

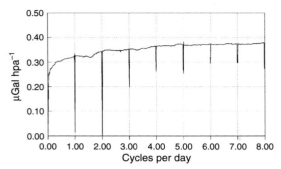

Figure 11   Amplitude of frequency-dependent barometric admittance in **Table 4**, case 4. Reproduced from Crossley D, Jensen O, and Hinderer J (1995) Effective barometric admittance and gravity residuals. *Physics of the Earth and Planetary Interiors* 90: 221–241, with permission from Elsevier.

In the time domain, we compare the signals in **Figure 12**. The upper trace is the uncorrected gravity, the middle trace is corrected with a single admittance (case 3), and the lower trace is for a frequency-dependent correction (case 4). As noted by Kroner and Jentzsch (1999), one should not assume that a smaller residual necessarily means a 'better' correction because the atmosphere is not the only source of gravity variations. In the period range 1–8 cpd, it is known that rainfall and hydrology also contribute significantly to gravity and this extra 'noise' considerably complicates the determination of an accurate admittance. Except for specialized studies, a frequency-dependent admittance is rarely done as part of regular processing. Aside from ease of use, a nominal value of $-0.30\,\mu\text{Gal hPa}^{-1}$ has the advantage that such a correction can be quickly 'restored' to gravity residuals if required.

We note that the atmospheric admittance effects in coastal areas depart from continental areas (discussed above) because water responds to surface pressure as a fluid rather than elastically as for the solid Earth. This inverted-barometer effect permits the water column to adjust isostatically to ocean surface pressure, and the ocean bottom pressure does not contribute to atmospheric loading. The local admittance for near-coastal stations may therefore depart from that represented in **Figure 11**.

4.2.3.3   Green's functions and nonlocal pressure corrections

The admittance approach is suited to single pressure series taken at the station but it can approximate only the local part of the atmospheric effect. For regional and global corrections, it is necessary to compute gravity directly from spatial meteorological data. All discussions on loading start with Farrell (1972), who showed how to convolve an observed distribution of surface pressure (data) with a Green's function (or kernel) that represents the effect of a point load for a given source–station separation. The method is quite general and can be used for any kind of surface load as well as local and nonlocal corrections for any kind of atmospheric model.

Two-dimensional loading calculations can take one of two forms: either the surface pressure alone is used or the surface temperature is included which allows the incorporation of thermal changes in the air column using a simple gas law (Merriam, 1992b). The former option is easier to implement because pressure data are widely available from meteorological

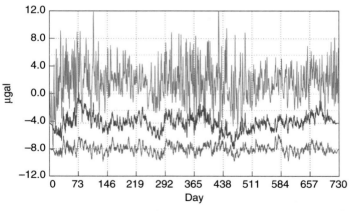

Figure 12   Pressure corrections using an empirical scalar admittance function. The upper trace (green) shows uncorrected gravity residuals (tides, polar motion, and drift removed), the middle trace is after correction using a single admittance (blue), and the lower trace the correction using an empirical frequency dependent admittance. Reproduced from Crossley D, Jensen O, and Hinderer J (1995) Effective barometric admittance and gravity residuals. *Physics of the Earth and Planetary Interiors* 90: 221–241, with permission from Elsevier.

centers, though sampled usually only at 6 h intervals. The thermal admittance given by Merriam is $\delta g = +0.013 \ (T_c - 15°) \ \mu\text{Gal} \, °\text{C}^{-1}$ which combines the effects from local and regional zones. In the passage of a cold front, for example, with pressure and temperature changes of 3 hPa and 10°C, respectively (Müller and Zürn, 1983), the gravity effect from temperature is only 10% (i.e., 0.1 μGal) of the pressure effect and may be reasonably ignored.

Numerical computations using global surface meteorological data need to be done for a spherical Earth (Merriam, 1992b), so the computational task becomes nontrivial, especially when using the high-resolution pressure data. Some notable results were found by Mukai *et al.* (1995) and Sun (1995). Boy *et al.* (2002) considered the improvement to be expected at low frequencies when the regional and global loading is done explicitly using various assumptions for the vertical structure of the atmosphere and different meteorological data sets (ECMWF at 1.125° vs NCEP at 2.5°). In addition, they demonstrated that the inverted-barometer assumption is appropriate for periods longer than 1 week. We show in **Figures 13(a)** and **13(b)** their results for two different corrections in zone 2 (the region >0.5°) that provide a reduced gravity residual compared to the local admittance. Note that the use of a simple gas law model of the atmosphere (pseudostratified loading) reduces the size of the nonlocal effect compared to the loading that uses only the surface pressure. As in Merriam (1992b), the atmospheric correction due to temperature and water vapor has been assumed to be small.

Finally, we mention the importance of including topography when making atmospheric corrections in hilly or mountainous terrain (Boy *et al.*, 2002).

4.2.3.4 3-D atmospheric corrections

In recent years, and of particular importance to the correction of atmospheric effects in the *GRACE* satellite data (Boy and Chao, 2005), it has been recognized that seasonal mass changes within the atmosphere are not reflected in surface pressure data. This means that an additional 3-D Newtonian attraction needs to be modeled when high-quality atmospheric corrections are required. This has been known for some time, and a correction for this effect can take the form of a simple annual sinusoidal term of amplitude about 0.8 μGal, as reported in Zerbini *et al.* (2001, 2002). A more complete treatment of the effect and the computations required can be found in Neumeyer *et al.* (2004a). Here we show in **Figure 13(c)** sample calculations from this study for four stations in central

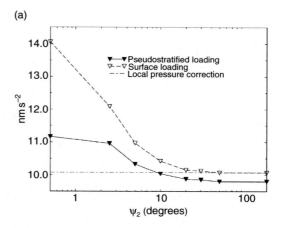

(a)

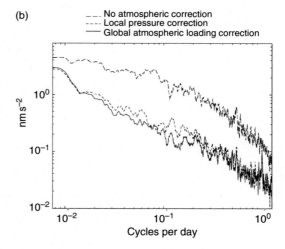

(b)

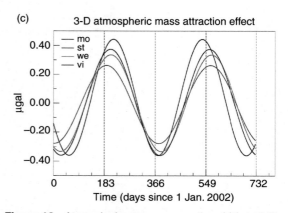

(c)

Figure 13 Atmospheric pressure corrections (a) from 0.5° to 180° using 2-D Green's functions compared to a local scalar admittance, (b) the reduction of noise in the frequency domain, and (c) seasonal air mass effect for atmospheric attraction at four GGP stations. (a) and (b) Reproduced from Boy J-P, Gegout P, and Hinderer J (2001) Gravity variations and global pressure loading. *Journal of Geodetic Society of Japan* 47: 1, 267–272, with permission from Geodetic society of Japan.

Europe over a 2 year period (only the annual fitted function is shown). In the summer months, the air masses within the atmosphere rise and the gravitational attraction effect decreases (upward); this causes a net positive gravity effect at the ground that is about 1 μGal larger than that in winter months. Neumeyer *et al.* (2004a) showed that the effect is not the smooth function shown in **Figure 13(c)**, but there are sudden changes that occur over periods of days. The computational effort to include this term is not trivial and it is being done routinely only at a few SG stations; one limitation is the need to interpolate the meteorological data that is available only at 6 h sampling and 0.5° (50 km) spacing. This is also a factor for the data reduction in *GRACE* because the atmospheric pressure corrections are done on the inter-satellite distance timescales that are aliased by 6 h sampling.

4.2.4 Calibration Issues

4.2.4.1 Basics
SGs are relative instruments and no 'factory calibration' is provided by the manufacturer. The instrument must be calibrated to convert its output feedback voltage, as recorded by a digitizing voltmeter, to units of acceleration. The amplitude calibration factor (GCAL) is frequently called the scale factor of the instrument and is expressed in $\mu Gal\ V^{-1}$ or $nm\ s^{-2}\ V^{-1}$, with a typical value of about $-80\ \mu Gal\ V^{-1}$ (**Table 5**). In addition, the system transfer function of the complete system (sensor and electronics) must be measured to determine the frequency dependence of both the amplitude and phase. Although it is easy to find the approximate calibration (to a few % in amplitude or phase), it is nontrivial to improve the calibration to be better than 0.1%. There are three applications for which accurate calibration is most important: (1) ocean tide loading or solid Earth tidal deformation including the FCN resonance, (2) the subtraction of a synthetic model for tides and the atmosphere whenever residuals are required to high accuracy, and (3) for determining precise spectral amplitudes in normal-mode studies.

The SG output has a very large bandwidth, from ~1 s (the raw sampling) to periods as long as the data length (years to decades). It is common to refer to the long period limit as 'DC' (vaguely meaning beyond

Table 5 Representative SG calibration experiments using an absolute gravimeter

Station	Instrument	AG or method	#drops	Time	SF ($\mu Gal\ V^{-1}$)	(%)
BH[a]	CD030_L	FG5 #101 platform	18 000	2 yr	-73.690 ± 0.088	0.12
					-73.971 ± 0.023	0.03
	CD030_U	FG5 #101 platform			-67.626 ± 0.084	0.12
					-67.922 ± 0.041	0.06
BO[b]	C024	FG5 #205 platform	20 800	9 d	-80.281 ± 0.063	0.08
					-80.341 ± 0.009	0.01
CA[c]	T012	JILA-2	NA	3 yr	-78.3 ± 0.1	0.13
CB[d]	C031	FG5 #206	15 778	6 d	-76.098 ± 0.169	0.22
			46 560	12 d	-75.920 ± 0.061	0.08
MA[e]	T011	FG5 #210	100 000	27 d	-92.801 ± 0.034	0.04
					$-92.851 \pm 0.049A$	0.06
					$-92.879 \pm 0.036B$	0.04
MB[f]	C021	FG5 #202	275 468	47 d	-78.457 ± 0.001	0.06
MC[g]	C023	FG5 #101, 103, 206 platform	18 000	4 yr	-74.822 ± 0.137	0.18
					-74.824 ± 0.013	0.02
ST[h]	TT05	JILA-5	5600	1 d	-76.05 ± 0.55	0.72
ST[i]	C026	FG5 #206	412 244	3 yr	-79.19 ± 0.05	0.06
ST[j]	C026	FG5 #206	450 000	4 yr	-79.40 ± 0.03	0.04
SY[k]	T016	FG5 #203	55 743	15 d	-58.168 ± 0.061	0.10

[a]Falk *et al.* (2001);
[b]Francis *et al.* (1998);
[c]Merriam *et al.* (2001);
[d]Amalvict *et al.* (2001b);
[e]Imanishi *et al.* (2002);
[f]Francis (1997);
[g]Hinderer *et al.* (1991);
[h]Amalvict *et al.* (2001a);
[i]Amalvict *et al.* (2002);
[j]Iwano *et al.* (2003).
Scale factors (SF) are by direct regression except (a) tidal analysis, (b) modified least squares. Station codes, etc., are given in Crossley (2004).

tidal periods), and some of the calibration methods effectively measure this DC amplitude calibration constant, whereas other techniques attempt to find other parts of the transfer function. Even in well-controlled laboratory experiments, it is difficult to provide a suitable acceleration at periods beyond a few hours. This is not an issue for seismometers because they are not designed to measure periods longer than 1 h. For gravimeters at tidal periods, however, one has to either extrapolate the results from shorter periods or extract the calibration from the data under normal operating conditions.

As noted by Richter (1991), some of the methods used to calibrate other types of gravimeters are inappropriate for SGs. One of these is the use of calibration 'lines' where an instrument is repeatedly moved to precise locations arranged in a horizontal or vertical line; due to the instrument size this is impractical for the SG. The application of electromagnetic or electrostatic forces is also difficult due to the need for a precise knowledge of the reference force.

For a relative calibration, an SG is compared over a period of time with another instrument whose calibration is known (including calibration by an AG). In this case, the two sets of measurements can be compared indirectly, for example, by comparing the elastic gravimetric factors from a tidal analysis, or by a direct regression of the data from one instrument to the other. For absolute calibration, two methods have been used – either a test mass is moved with respect to the gravimeter and the Newtonian attraction is measured directly, or the instrument itself is subject to a known acceleration provided by an acceleration platform.

4.2.4.2 Amplitude calibration, relative methods
4.2.4.2.(i) Calibration using a theoretical tide
The best theoretical solid Earth tidal models are accurate to about 1 nGal in frequency-domain amplitude. This is about 10^{-5} of the full tidal amplitude ($300 \, \mu$Gal), so it would seem a straightforward task to calibrate an instrument using the tidal signal. In practice, however, the observed tidal amplitudes are a combination of theoretical amplitudes, elastic response of the Earth, and ocean tidal loading. Variability in ocean crustal loading of about 1 μGal limits the accuracy of tidal calibration to about 0.3% accuracy.

Several authors have argued that the tidal analysis of an SG record is a useful calibration tool (e.g., Goodkind, 1996). Of equal importance is the relative stability of the amplitude calibration, and Merriam (1993b) has inferred this for the model TT70 at

Cantley using the tidal admittances of the M2, O1, and K1 waves from 3 years of data. He concluded that the amplitude calibration was stable at the 0.013% level and the phase calibration to within 0.01° for M2. The use of a theoretical tide (solid tide + elastic yielding + ocean tide loading) for calibration is still, however, model-based, and therefore of a different character to the other methods listed below.

4.2.4.2.(ii) Calibration using spring gravimeters
Many early SG calibrations came from the side-by-side comparison of data from spring gravimeters that had already been calibrated either in the laboratory or on calibration lines. Due to limitations in the original calibrations, it is unlikely that the amplitude factors can be trusted below about the 0.5% level (Richter, 1991).

4.2.4.2.(iii) Calibration using AGs
For several reasons, the use of an AG has become the most widely used method in recent years. First, the SG is not disturbed during the calibration because the AG measurements are done on a nearby pier or adjoining room. The instrument separation needs to be kept small, however, to prevent horizontal gradients due to local gravity variations (e.g., hydrology), from affecting the instruments differently. Second, even though the AG has a repeatability of 1–2 μGal over a typical measurement campaign, with care a precision of about 0.05% can be achieved (**Table 5**).

Another advantage is that the processing of the data from the two instruments is straightforward and no complicated physical effects have to be accounted for because the instruments measure the same temporal changes whatever their origin. Finally, the co-location of the two instruments also enables the long-term drift of the SG to be estimated and this provides obvious advantages in cross-checking the AG measurements.

Hinderer *et al.* (1991) reported an early calibration of a model TT70 SG using a JILA-5 AG in Strasbourg. They introduced the direct regression technique whereby all SGs and AGs are used without correction, save for allowing different linear drifts between the two lasers of the JILA instrument. When tested against theoretical tides for O_1 (a common method for European stations because the solid tide O_1 is large and the ocean tide loading is small), agreement was found at about the 0.1% level. The full-size T012 at Cantley was calibrated a number of times using AGs. Bower *et al.* (1991) obtained 0.4% from tidal gravimetric factors using a JILA-2, and

Merriam *et al.* (2001) reported on a series of nine calibrations to achieve a precision of 0.13%.

Several authors have engaged in extensive campaigns to intercompare the instruments. Francis (1997) reported on 47 days of recording involving a large number of AG drops. A precision of about 0.1% was achieved both by linear regression and by comparing tidal amplitude factors. A significant finding of this study was that the precision of calibration was not significantly improved for times longer than 5 days, suggesting that this is an optimum period for such a comparison. Similar results were found for the C024 instrument in Boulder (Francis *et al.*, 1998) that agreed very well at a level of 0.08% with the absolute calibration using an acceleration platform.

Okubo *et al.* (1997) performed tests using FG5 #107 at two of the Japanese SG sites, T007 in Esashi and T011 in Matsushiro. Their results pointed out that the internal consistency of the SG was 0.3 and 0.62 μGal for Esashi and Matsushiro, respectively, even better than the error in the AG residuals at the two sites (1.55 and 1.52 μGal); the overall precision was 0.2%.

A sampling of other results of AG calibrations is given in **Table 5**. Amalvict *et al.* (2001a, 2002) showed a series of 28 different calibration (totaling almost half a million drops) runs over a 4 year period for ST, with an average length of 5 days each. The regression method gave an overall precision of 0.04% with a repeatability of 0.1%. Imanishi *et al.* (2002) presented a study of the various ways to cross-calibrate the AG and SG using a simple regression, tidal factors, and a combined method (**Table 5**). They found that the SG is not the only instrument subject to 'drift', and the AG showed a bias of several μGal over the 1 month of the experiment. Nevertheless, they were able to extract a very precise calibration of about 0.04%, which approaches the best calibrations by an acceleration platform method. Finally, Falk *et al.* (2001) and Harnisch *et al.* (2002) have given useful results on a number of SGs and AGs, together with some acceleration platform calibration results.

Note that for BH the two spheres have different calibration factors – this is true of all dual-sphere instruments. Also, MC is one of many stations that have been visited by several different AGs. The 'errors' are generally the precision of a least-squares fit, and are known to underestimate uncertainties obtained from differing sets of measurements. Thus, some SGs have been DC-calibrated to better than 0.05%, but others are probably closer to 0.15%.

Francis and Hendrickx (2001) turned the calibration issue around and used an SG as the reference signal in the calibration of a spring gravimeter. This is possible due to the extremely low drift and high precision of the SGs and produces good results, especially when the variable drift of the spring gravimeter is modeled along with the calibration, as in Meurers (2002).

4.2.4.3 Amplitude calibration, absolute methods
4.2.4.3.(i) Calibration using a moving mass
Warburton *et al.* (1975) first described the use of a heavy mass calibration experiment. They rolled a 321 kg hollow steel sphere filled with mercury to a fixed position under the SG, thus generating a 10 μGal test signal. Within the accuracy of the various parameters, the most critical being the exact distance between the center of the test sphere and the niobium sphere, they found a precision of 0.2%. In such an experiment, other factors limiting accuracy are the homogeneity of the sphere, the exact geometry of the position of the sphere with respect to the niobium sphere, as well as loading and tilting effects that are induced by the heavy mass itself.

A similar calibration was done by Goodkind *et al.* (1993), who used a moving mass system to test the inverse square law of Newtonian gravity. A spherical mass was moved vertically under the SG in a specially designed chamber; an accuracy of 0.09% was achieved for two different types of sphere materials. The ultimate limit in this type of calibration is the uncertainty in the gravitational constant, though at 0.01% it is below the precision of the above calibrations.

Achilli *et al.* (1995) implemented a unique moving mass system for the full-sized T015 SG in Brasimone as part of a test for constancy of the gravitational constant. Their ring, weighing 272 kg, is raised and lowered around the SG from strong supports in the roof, but a limited range of about 1 m reduced the mass effect to 6.7 μGal. During two calibrations, they found a precision of 0.2% and a repeatability of 0.3%.

4.2.4.3.(ii) Calibration using a moving platform
The idea of using an acceleration platform for the calibration of spring gravimeters was used successfully for LCR gravimeters by Van Ruymbeke (1989). The idea was applied to SGs by Richter (1991), who pioneered the development of what is now known as 'The Frankfurt Calibration Platform'. The SG instrument design, in which the dewar and coldhead are mounted separately, limits the maximum vertical range through which the gravimeter can be lifted. For a

Table 6 Maximum accelerations available from a moving platform, amplitude ±5 mm

Period (s)	Acceleration (μGal)	Output (V)
300	219.30	3.96042
480	85.60	1.50698
600	54.81	0.95092
900	24.36	0.43222
1200	13.70	0.25497
2400	3.42	0.07782
3600	1.52	0.02585
7200	0.38	0.00065

Modified from Richter (1991).

sinusoidal signal of amplitude 5 mm, **Table 6** shows the maximum accelerations that are possible. Note that for periods longer than 2 h, the amplitude is smaller than atmospheric and hydrological effects. Other limitation to this technique arises from the gravity gradient effect, mechanical stresses deforming the support systems, and the need to avoid tilting the gravimeter, which causes the horizontal resonance of the sphere to be excited.

This type of calibration also allows the phase response to be measured through the phase offset between the instrument response and the acceleration function (Richter *et al.*, 1995a). A calibration precision of better than 0.02% in amplitude and 0.1 s in time delay has been achieved using this system (**Table 6**).

4.2.4.4 Phase calibration
4.2.4.4.(i) Response of the analog filter
The ideal amplitude gain of the electronic analog antialiasing filter should be constant from DC up to the corner frequency of the antialiasing filter and the ideal phase response should be linear with frequency. These aspects are generally met in most SG recordings, but depart from the ideal at high frequencies, which for gravity data analysis means periods shorter than about 1 h. The GGP filter, used on the GWR gravity card, is an eight-pole Bessel filter designed to achieve the desired linearity.

The phase response at longer periods can be well represented by $\phi(\omega) = -\alpha\omega$, where α is a constant, ϕ is measured in degrees, and ω is in cpd. Given $\phi(\omega)$, it is simple to find the group delay $\tau(\omega) = -d\phi/d\omega$, usually measured in seconds. For a linear phase filter, one has (with α in degress per cpd)

$$\tau(\omega) = 240\,\alpha = \text{constant} \qquad [5]$$

For example, for the IDA filter, $\alpha = 0.15°\text{cpd}^{-1}$ so $\tau = 36.0\,\text{s}$. For the GGP filter board at 1 s sampling,

$\alpha = 0.035°\text{cpd}^{-1}$ so $\tau = 8.4\,\text{s}$. In order to usefully measure the nonelastic response of the Earth to tidal waves, accuracies of 0.002° are required in the tidal wave phases (Wenzel, 1994). The phase lag of the recording system therefore has to be known at least to the same accuracy, that is, 0.5 s. An electronic calibration method is necessary to achieve this precision.

4.2.4.4.(ii) System response
The antialiasing filter is not the only component in the data acquisition system that causes time delays and phase shifts. They can occur throughout the electronics, including the feedback sensor, the digitizing voltmeter, and the time stamping of the signal onto the hard disk. Therefore, the overall phase response is best determined *in situ*, and not assumed from the design characteristics of only the antialiasing filter.

4.2.4.4.(iii) The step response method
The amplitude and phase response (the transfer function) of the complete system can be measured by injecting a known voltage to the feedback coil and recording the output voltage of the feedback network. The transfer function completely describes the electronic response of the system, but is not a substitute for determining the instrument calibration, that is, the conversion from voltage to acceleration. The injected voltage can be either a step function, which gives both amplitude and time responses but is sensitive to noise, or a series of sine waves that suppresses noise but requires a longer measurement time (Van Camp *et al.*, 2000). A third possibility is to inject a randomized input signal to the instrument and find the transfer function by cross-spectral analysis, but Richter and Wenzel (1991) conclude that this method is best suited to instruments with electrostatic or electromagnetic feedback, such as the spring gravimeters used in the IDA network.

The initial experiment to determine instrumental phase lag was by Richter and Wenzel (1991) in which they outlined the method and gave results for several spring gravimeters and one SG, the TT60 at Wettzell. Results for the latter showed an average time lag of 38.73 ± 0.14 s, and a phase determination of 0.002° for diurnal tidal waves, at the target accuracy level noted above. Wenzel (1995) later determined even better results for an LCR meter with a group delay accuracy of ± 0.004 s, far superior to the nominal value determined from the electrical components themselves.

Van Camp *et al.* (2000) have discussed in the utmost detail the phase response measurement for a compact SG, in this case the C012 at Membach. The electronic step was applied to various combinations of filter type and time intervals (1–4 min), and the output was Fourier-analyzed to determine the amplitude and phase response. The time lag results showed an accuracy of ± 0.003 s for the GGP1 filter and ± 0.075 s for the raw gravity signal output. The sine wave method gave very similar results, confirming the integrity of the method. The amplitude gain of the electronics system was determined to be flat from 500 to 2000 s, the longest period determined by either method. Further details are given on the GGP website.

4.2.5 Other Corrections to Residual Gravity

4.2.5.1 Polar motion

One important signal contained in an SG record is the 14 month (435 day) oscillation of the rotation pole, or Chandler wobble. As noted previously, Richter's (1983) first observation of this signal, with only 5 µGal amplitude, was a turning point in the refinement of gravity residuals. Since then, every SG station has recorded data that when suitably processed show the polar motion (e.g., Harnisch and Harnisch, 2006a). Naturally, some records are very clear and others not so clear, depending on the epoch (amplitude of the motion) and the quality of the instrument and site.

With most data sets, it is not difficult to see the polar motion once tides and atmospheric pressure are subtracted. Indeed, for the most part, the highly accurate space geodetic series for the polar orientation that is given on the IERS website can be used directly at any of the gravity stations. A simple conversion is usually made between the (x, y) amplitudes of polar motion (m_1, m_2) in radians and the gravity effect δg in µGal (e.g., Wahr, 1985):

$$\delta g = 3.90 \times 10^{-9} \sin 2\theta \left[\cos(m_1 \lambda) - \sin(m_2 \lambda) \right] \quad [6]$$

where (θ, λ) are station latitude and longitude. The numerical factor includes the nominal value of 1.16 for the gravimetric delta factor, whereas fitted solutions are usually closer to $\delta = 1.18$ (e.g., Loyer et al., 1999; Harnisch and Harnisch, 2006b). In many studies, the polar motion is considered as a quantity very accurately determined from space geodetic data, but it is interesting to observe its signature in either an individual or collective gravity series.

4.2.5.2 Instrument drift

From a historical viewpoint, particularly for spatial gravity surveys, the term 'drift' has been applied to almost any unwanted time-varying gravity signal. This usage persists in the SG literature and sometimes causes confusion. Here we use drift as an instrument characteristic, because all other gravity variations have specific causes.

From an instrumental point of view (e.g., Goodkind, 1999), drift is likely to be either a linear or exponential function of time, but its size is not easy to predict. The exponential behavior can be reset after a loss of levitation or other magnetic changes within the sensor. Under normal operation, the user can generally assume that drift will level off after installation and gradually become more linear as time progresses. Representative values of instrument drift are frequently less than $4 \, \mu\text{Gal yr}^{-1}$ where these have been checked carefully with AGs (see above).

From a processing point of view, other functions such as polynomials have been used as a model for instrument drift but there is no physical reason to prefer such a choice. We recommend that an exponential drift be assumed from some initialization event for the instrument and later this may be replaced with a simple linear function. Drift is not to be confused with a secular change of gravity, even though the two cannot be separated except using combined SG–AG observations.

4.2.5.3 Hydrology

Hydrology is perhaps the most complex of the intermediate scale (hour–year) variations in gravity (e.g., Harnisch and Harnisch, 2006a). This is due to two factors. The first is its variability, due largely to the local water storage balance at the station that involves many components (rain and snowfall, soil moisture, evapotranspiration, and runoff). Rainfall is relatively easy to assess, as is the groundwater level, which is usually measured in a nearby well. The direct measurement of soil moisture is not easy and has not been done at most SG stations.

The second problem is one of length scales. The connectedness (permeability) of the soils and groundwater system is inhomogeneous at the local length scales (meter to kilometers), and so an assessment of the amount of moisture surrounding an SG involves extensive measurements. This quantity predominantly affects the attraction term rather than the loading term.

As a result, even though groundwater variations are extremely useful, they are not entirely reliable for the purpose of determining an admittance factor for 'correcting' gravity residuals. The simplest case is a horizontal layer (Bouguer slab) of moisture of thickness h and fractional porosity ϕ, yielding a gravity perturbation (e.g., Crossley *et al.*, 1998),

$$\delta g = 2\pi G\rho\phi h = 0.42\phi\,\mu\text{Gal cm}^{-1} \qquad [7]$$

This is similar in application to the single admittance factor for atmospheric pressure, but the porosity (and permeability) can usually be estimated only very approximately, thus limiting the accuracy of the correction. Adding to the complications are SG stations located underground where a soil moisture layer may be present both above and below the instrument. In this case, the hydrology correction becomes quite problematic and is currently pursued as a research problem rather than as a correction that is part of normal processing.

4.2.5.4 *Residual gravity*
Having completed the above processing steps and corrections, one finally arrives at a residual gravity series that represents the unknowns including effects such as ocean currents, secular changes in elevation due to tectonics, gravity changes due to slow and silent earthquakes, or eigenmodes of the Earth (perhaps from the Earth's core). In addition, of course, all the subtle effects of mismodeling will be applied, which may include wrong assumptions regarding disturbances, offsets, drift, loading, and hydrology.

On one conclusion there can be little doubt – the ability of SGs to reliably measure effects at the 0.1 μGal level has opened up many interesting scientific possibilities, as well as posed many challenging issues that will be discussed in the following sections of this chapter.

4.3 Scientific Achievements Using SGs

4.3.1 The Global Geodynamics Project
GGP is an international research effort that was launched as a SEDI (Study of the Earth's Interior) initiative at the Vienna IUGG General Assembly in 1991. In 2003, it changed to become an Inter-Commission Project of International Association of Geodesy (IAG) and reports to Commission 2 (The Gravity Field) and Commission 3 (Earth Rotation and Geodynamics). It consists of a worldwide network of SGs, currently about 20 instruments, run by independent national groups. The groups agreed to provide vertical gravity acceleration data in a standard form, basically untouched raw data decimated to 1 min samples, and sent at the end of every month to a database. As indicated in **Figure 14**, GGP stations are sparsely distributed worldwide, with only two regional clusters of instruments, one in Europe and a smaller one in Japan. The coverage in the Southern Hemisphere is still weak, despite the effort of installing stations in Australia, Indonesia, and

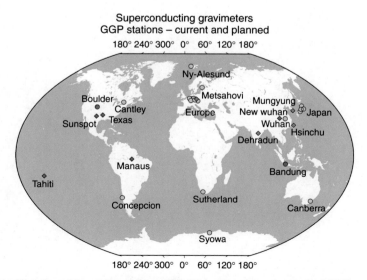

Figure 14 Geographical location of the stations of the GGP (Global Geodynamics Project). Yellow stations are active, red are currently not operating, and green are recently installed.

Antarctica by Japanese colleagues, in South Africa by GFZ Potsdam (Germany), and very recently in South America by BKG (Germany) – the TIGO project. Phase I of GGP was the period 1997–2003 and we are currently in phase II (2003–07).

The scientific objectives of the GGP cover geophysical phenomena throughout the wide period range of the instruments (from 1 s to several years), covering topics such as normal modes, mantle rheology, tides, solid Earth–oceans–atmosphere interactions, hydrology, and Earth rotation. **Figure 15** represents schematically the gravity spectrum that is observable by SGs ranging from seconds (ocean noise) to several years (secular changes). We refer the reader to the EOS article by Crossley *et al.* (1999), where a full description is provided. Other review papers on SGs have also appeared (e.g., Goodkind, 1999; Hinderer and Crossley, 2000; Meurers, 2001a; Hinderer and Crossley, 2004). We will show below some of the most interesting results which owe their existence to the collection of the worldwide GGP data of high quality.

The wide spectrum of geophysical phenomena that are observable with SGs is evident in **Figure 15**. Basically the range of observable periods (or characteristic time constants) covers 8 orders of

magnitude from 1 s to several years. The highest frequency detectable by SGs is ~1 s, because of the feedback system limitation, and on the left the figure shows background noise mainly caused by ocean noise with two dominant peaks at 5–6 s and 10–15 s. At slightly longer periods we have the seismology region including the normal modes generated by earthquakes – periods up to 54 min which is the gravest period of the Earth elastic normal modes. Between 150 and 500 s (2–7 mHz), these modes form the incessant oscillations ('hum') unrelated to earthquakes but rather of atmospheric and/or oceanic origin. At periods longer than about 6 h (depending on the core stability profile), another class of eigenmodes are the gravity-inertial modes (also called core modes) predominantly confined to the liquid core. A particularly interesting and isolated long-period oscillation is the Slichter mode (actually a triplet due to rotation and ellipticity) arising from translation of the solid inner core. Its period, between 4 and 8 h, depends primarily on the density jump of the inner-core boundary (ICB).

From 4 h up to 18.6 years, there are many spectral lines due to lunisolar tides, the most important of which are semi-diurnal and diurnal. The study of diurnal gravity tides includes a resonance effect due

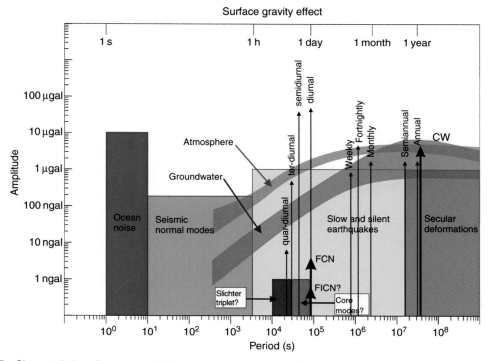

Figure 15 Characteristics of geophysical phenomena observable by SGs.

to the existence of wobble modes for an Earth model with a fluid outer core (FCN) and a solid inner core (free inner core nutation). Toward longer periods, there are many signals (atmospheric pressure, snow, soil water content, ocean circulation, Earth's rotation) that are related to the seasonal solar cycle which lead to contribute to a strong annual term in gravity. CW stands for Chandler wobble, which is another rotational mode of the Earth with a 435-day period; its observation provides valuable information on mantle rheology at long periods.

On the right of the figure we have long-period secular deformations; for instance, gravity changes caused by post-glacial rebound (PGR) or tectonics. Also indicated in shaded bands are the effects due to atmospheric pressure and hydrology (groundwater). In contrast to periodic phenomena that have sharp spectral lines, these effects are broadband and cover a large part of the spectrum, atmospheric effects being roughly 1 order of magnitude larger than hydrologic effects. **Figure 15** is a normalized amplitude spectrum where a sinusoidal wave with unit amplitude will always appear with unit amplitude regardless of the period of observation. On this type of spectrum,

noise would decrease as a function of record length. We have chosen to represent the typical variability of hydrology and atmospheric pressure induced effects in gravity (equivalent time standard deviation) as time independent. The two bands are derived from the analysis of water table level and air pressure records of several years length available at some of the GGP stations.

An example of the different contributions to the time-variable gravity signal is shown in **Figure 16** in Strasbourg, France. The largest contribution is of course the solid Earth tides with several hundreds of nanometers per square second and multiple periods (semi-diurnal, diurnal, fortnightly, monthly, semi-annual, annual). Ocean loading is by far smaller of the order of $1\,\mathrm{nm\,s^{-2}}$ (this term can be larger for a station near the oceans). Nontidal ocean loading is slightly larger and exhibits a strong seasonal feature. Atmospheric loading is broadband and can easily cause gravity changes of several tens of nanometers per square second. Hydrology (soil moisture + snow) is predominantly seasonal and can also lead to gravity perturbations of similar amplitude. Finally, the Earth's rotation changes (polar motion and length of day)

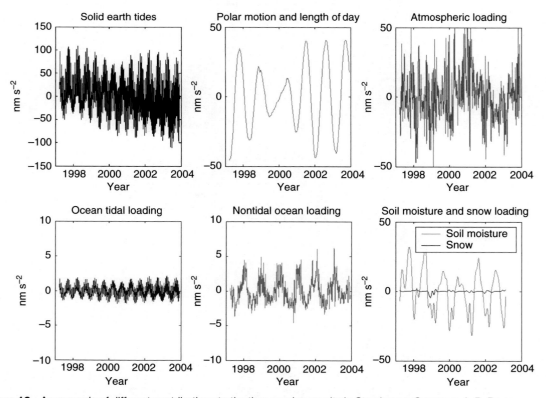

Figure 16 An example of different contributions to the time-varying gravity in Strasbourg. Courtesy: J.-P. Boy.

induce a gravity signal of several tens of nanometers per square second with a dominant beating between an annual term and the 435 day Chandler wobble.

We now review some scientific studies in the various period ranges, as discussed in **Figure 15**. Section 4.3.2 deals with the short-period seismic phenomena and other normal modes up to a 1 day period. In Section 4.3.3, atmospheric effects on gravity are described while tidal contributions (fluid core resonance effect, linear and nonlinear ocean loading) are discussed in Section 4.3.4. Section 4.3.5 deals with nontidal ocean loading, hydrology contributions are covered in Section 4.3.6, and Section 4.3.7 covers Earth rotation and polar motion effects. Section 4.3.8 deals with tectonics, and Section 4.3.9 considers the problem of the calibration/validation of gravity satellite data with SG ground observations. Finally, Section 4.3.10 suggests new projects, especially using SGs in regional arrays.

4.3.2 Seismic and Subseismic Signals

Investigation with SGs of the seismic normal modes excited by large earthquakes has led to some new and impressive results, due primarily to the low noise levels of SGs at periods longer than 500 s, as indicated in **Figure 17** (Rosat *et al.*, 2003a). At periods

longer than 1000 s, the best SGs have lower noise than the worldwide seismometer limit NLNM (new low noise model, Peterson, 1993); noisier SG stations cross the NLNM at longer periods (4.2 h for the station Be, for instance). This extensive compilation of all the GGP stations extends earlier results from just a few SG stations but which already were convincing in terms of low noise (Banka and Crossley, 1999; Van Camp, 1999). In addition, metrological comparisons between SGs and broadband seismometers (Freybourger *et al.*, 1997; Hinderer *et al.*, 2002b), as well as between AGs and SGs over a large spectral range (Crossley *et al.*, 2001; Francis *et al.*, 2004), have demonstrated the excellent characteristics of SGs.

Of recent large earthquakes, the 2001 Peru event of magnitude M = 8.4 strongly excited the long-period seismic modes, and significant observations were made by the GGP network. In particular, the fully split $_0S_2$ multiplet (with five individual singlets), which has been rarely visible on a single instrument, could be fully analyzed from the Strasbourg C026 instrument (Rosat *et al.*, 2003a) but was also present at other SG stations.

The most important new result was the detection of the overtone $_2S_1$ (see **Figure 18**), which is an elastic mode, unlike the Slichter triplet $_1S_1$ whose

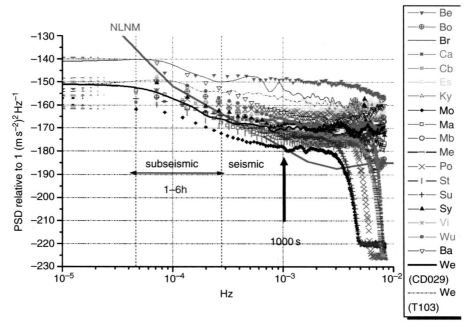

Figure 17 Noise levels of the SGs from the GGP network. Reproduced from Rosat S, Hinderer J, Crossley D, and Rivera L (2003a) The search for the Slichter Mode: Comparison of noise levels of superconducting gravimeters and investigation of a stacking method. *Physics of the Earth and Planetary Interiors* 140: 183–202, with permission from Elsevier.

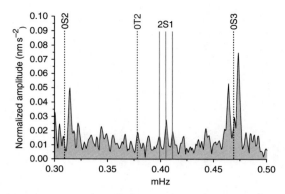

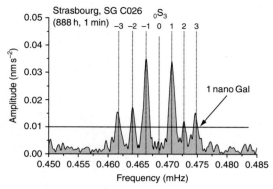

Figure 18 Normalized amplitude spectrum (in nm s^{-2}) in the frequency band 0.3–0.5 mHz after the $M_w = 8.4$ Peru event exhibiting the triplet of the $_2S_1$ mode. Reproduced from Rosat S, Rogister Y, Crossley D, and Hinderer J (2006) A search for the Slichter triplet with superconducting gravimeters: Impact on the density jump at the inner core boundary. *Journal of Geodynamics* 41: 296–306, with permission from Elsevier.

Figure 19 Observation of the well-resolved singlets of $_0S_3$ at Strasbourg using 888 h of data after the 2004 Sumatra earthquake of magnitude M = 9.3. The singlet $m = 0$ was not excited at Strasbourg. Reproduced from Rosat S, Sato T, Imanishi Y, *et al.* (2005) High-resolution analysis of the gravest seismic normal modes after the 2004 $M_w = 9$ Sumatra earthquake using superconducting gravimeter data. *Geophysical Research Letters* 32: L13304 (doi:10.1029/2005GL023128), with permission from Elsevier.

period is determined primarily by a gravitational restoring force. The eigenfunctions of the two modes share some similarities (Rosat *et al.*, 2003b). The detection of this mode strongly benefited from the stacking method proposed by Courtier *et al.* (2000), also used by Guo *et al.* (2006), that is applicable to all degree 1 modes with their associated surface gravity changes.

The $M_w = 9.3$ Sumatra huge earthquake on 26 December 2004 even more strongly excited the low-frequency seismic modes and provided a unique opportunity to improve the determination of the period and Q of the gravest seismic modes. An example of the strong signal-to-noise ratio is shown in **Figure 19**, where all singlets of the spheroidal mode $_0S_3$ are visible on the Strasbourg SG (Rosat *et al.*, 2005).

We also show the background normal modes of the Earth (frequently referred to as the 'hum') that were first discovered in the Syowa, Antarctica, SG record (Nawa *et al.*, 1998), and later seen on other SG records (Nawa *et al.*, 2000; Rosat *et al.*, 2003b; Widmer-Schnidrig, 2003). **Figure 20** shows the power spectral density in the frequency band 1.5–5.5 mHz of the gravity data observed in Strasbourg during the period 1997–2001 with SG C026. The spectral peaks of the fundamental spheroidal modes are clearly visible. Rather than selecting periods without earthquakes above a specific magnitude, we used a statistical approach by computing the quartiles of all available gravity data in period of study.

The discovery of the hum has generated numerous studies in observational seismology and also led

to theoretical arguments suggesting that the preferred excitation mechanism is not earthquakes, but rather the atmosphere (Lognonné *et al.*, 1998; Kobayashi and Nishida, 1998; Nishida and Kobayashi, 1999; Nishida *et al.*, 2000, 2002; Suda *et al.*, 1998; Tanimoto *et al.*, 1998; Tanimoto and Um, 1999; Roult and Crawford, 2000). Two recent array-based studies in seismology, however, suggest an alternative explanation in terms of interactions between the atmosphere, the oceans, and the seafloor in case of stormy weather (Rhie and Romanowitz, 2004; Tanimoto, 2005).

One major goal of GGP was to detect the translational motion of the solid inner core (the Slichter triplet $_1S_1$), because knowing its period(s) would bring a new constraint on the density contrast at the ICB (Rosat *et al.*, 2006) and possibly also on viscosity just above the ICB (Smylie *et al.*, 2001). We do not review here previous studies and controversies on the theoretical issues in modeling this eigenmode, nor different claims for detection and counter studies; instead, we refer the reader to the review in Hinderer *et al.* (1995).

It is, however, worth showing the current status of the search. For example, in **Figure 21** (Rosat *et al.*, 2004), we see that there is obviously no clear observational evidence in this stacking of SG data of any of the theoretically predicted triplets. Evidently, many spectral peaks may emerge slightly above the background level of 0.01 nm s^{-2}, but as pointed out by

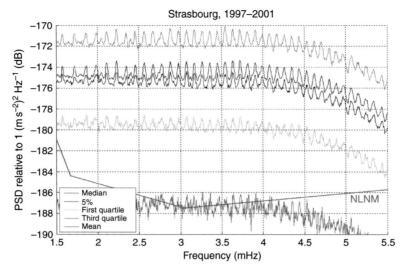

Figure 20 Power spectral density of the gravity data observed in Strasbourg during the period 1997–2001 clearly exhibiting the presence of the 'hum'. Courtesy: S. Rosat.

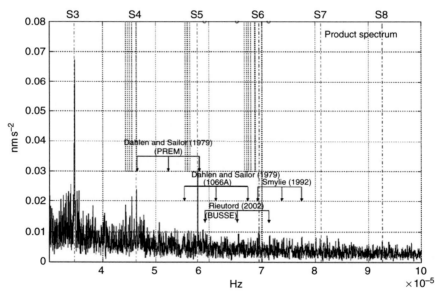

Figure 21 Product spectrum resulting from the multistation analysis method applied to five SGs. The three frequencies computed according to the splitting formula of Dahlen and Sailor (1979) for the Earth models 1066A and PREM, predicted by Rieutord (2002) based on the Earth's model by Busse (1974), and the peaks that Smylie (1992) identified in SG observations to be the inner-core translation, are indicated by vertical arrows. The modulations of the diurnal harmonic S_1 of the atmosphere from S_3 to S_8 (vertical dashed and dotted lines) and nonlinear tides (vertical dotted lines) around 4 and 6 h are also shown. Reproduced from Rosat S, Hinderer J, Crossley D, and Boy J-P (2004) Performance of superconducting gravimeters from long-period seismology to tides. *Journal of Geodynamics* 38: 461–476, with permission from Elsevier.

Florsch *et al.* (1995a), these are no more significant than statistical fluctuations. Some claims have been made (e.g., Guo *et al.*, 2006), but they are (as yet) generally unsupported by other observations and by the SG community.

Although not yet observed, the identification of $_1S_1$ remains a unique topic for the GGP network and interest remains high. This has led to the development of techniques for stacking the SG records, all based on the original idea of Cummins *et al.* (1991) as developed in Courtier *et al.* (2000). Despite low noise levels in the band from 3 to 6 h, and despite the methodological attempts to enhance any global signature of degree 1 triplet (see Rosat *et al.* 2003a), the problem probably remains one of excitation, as discussed by Crossley (1993). Even for the largest modern earthquake (Chile 1960, magnitude 9.5), the seismic moment yields a gravity signal of no more than 1 nGal, and to be observed this amplitude needs to be maintained for some time beyond the earthquake. An alternative possibility, first used by Smylie (1992), is that the excitation could be intermittent or randomly caused perhaps by fluid motions in the outer core.

4.3.3 Atmospheric Loading

As discussed earlier, several methods of correcting for the atmosphere have been proposed. We now show a specific example of the 2-D atmospheric pressure corrections done by Boy *et al.* (1998, 2001, 2002), and other groups in Japan (Mukai *et al.*, 1995), using 2-D pressure data originating from the weather prediction centers such as NCEP (National Center for Environmental Predictions) or ECMWF (European Centre for Medium-Range Weather Forecasts). **Figure 22** shows the computations for the stations Canberra in Australia and Strasbourg in France, where the residual gravity level (in nm s^{-2}) is plotted as a function of the solid angle in degrees around the station.

Except for small angles (close to the station), the 2-D correction performs better than the local barometric admittance close to 3 nm s^{-2} hPa^{-1} (straight horizontal line – note this is the same as $-0.3\,\mu$Gal mb^{-1} quoted earlier). Because of the known correlation between length and timescales in the atmospheric processes (Green, 1999), long-period pressure effects (exceeding a few days) require large-scale loading computations to be adequately represented. The results from a coastal station from the GGP network (CB station in **Figure 22**) clearly discriminated against the use of the NIB (noninverted barometer) ocean reaction to air pressure changes.

As we previously discussed, in order to have the most accurate atmospheric corrections for the new gravity satellite missions, *CHAMP* and *GRACE*, 2-D pressure loading computations have been extended to 3-D modeling where pressure, temperature, and humidity data varying with height are also taken into account in addition to the surface data (see Svenson and Wahr, 2002; Boy and Chao, 2005; Neumeyer *et al.*, 2004a). It was recently shown for the station Medicina, with the help of balloon radio sounding, that there is a seasonal vertical air mass change in the atmosphere without ground pressure changes which is caused by warming and cooling (Simon, 2002) and that leads by attraction to a non-negligible gravity effect of the order of 1 μGal.

In an attempt to model the atmosphere not just as a source of noise, but to investigate actual meteorological effects in gravity, Meurers (1999, 2000, 2001b) analyzed SG measurements from the Vienna station and found an interesting new phenomenon. It appears that during some weather disturbances, rainfall occurs without generating a large ground pressure change but causing a significant gravity drop. Meurers (1999) model suggests that vertical

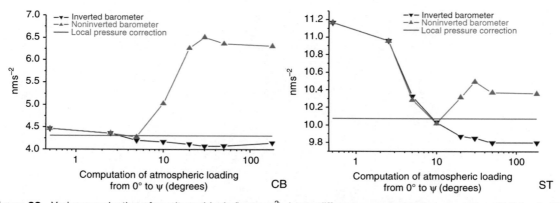

Figure 22 Variance reduction of gravity residuals (in nm s^{-2}) due to different pressure loading corrections (2-D load with inverted or noninverted barometer response, local admittance); left is for Canberra (CB) station in Australia and right for Strasbourg (ST) station in France. Reproduced from Boy J-P, Gegout P, and Hinderer J (2001) Gravity variations and global pressure loading. *Journal of Geodetic Society of Japan* 47(1): 267–272, with permission from Geodetic Society of Japan.

convective air motion (air mass exchange or water transport) does not alter the ground pressure (total air column mass unchanged) but does modify gravity through Newtonian attraction. This is an example where gravity could be of indirect use to meteorologists to indicate air movements without detectable ground pressure signature.

4.3.4 Tides and Nearly Diurnal Earth Wobbles

As discussed previously, Earth and ocean tides are by far the largest components of surface gravity changes. The SGs have brought two areas of improvement to tidal studies. First, the high sensitivity of these meters which enables them to detect small-amplitude tidal signals previously hidden in noise (e.g., nonlinear ocean tides) and to retrieve with better precision larger tidal signals (see Ducarme et al., 2002). With this high precision, Xu et al. (2004a) revisited the question of the possible latitude dependence of tidal gravimetric factors. Using 19 GGP stations they found that the discrepancy of the four principal waves (O_1 and K_1 in the diurnal band, M_2 and S_2 in the semi-diurnal band) between observations and theoretical models (Dehant et al., 1999; Mathews, 2001) is less than 0.2%. This means that there is no significant latitude dependence. Second, the much lower instrumental drift of SGs versus mechanical spring meters permits more precise studies of long period tides (M_f, M_m, S_{Sa}, and S_a) (Sato et al., 1997a; Hinderer et al., 1998; Mukai et al., 2001; Ducarme et al., 2004; Boy et al., 2006).

4.3.4.1 Resonance effects in diurnal tides

Of the tides themselves, the largest components occur in both semi-diurnal and diurnal frequency bands. Within the diurnal band, some waves are affected by a resonance that occurs due to the nearly diurnal free wobble (NDFW), also called FCN, which is a differential rotation of the fluid outer core with respect to the mantle. In a co-rotating reference frame, the FCN period is of course nearly diurnal; in an absolute frame, it is approximately 430 days. The observation of this resonance requires precise amplitude and phase measurements of the diurnal tidal waves that are close in frequency to the eigenfrequency. In particular, the small-amplitude waves ψ_1 and φ_1 are critical in the retrieval of the FCN parameters, that is, the period and damping of this resonance mode. A very clear example of this can be found in Hinderer et al. (2000), where there is a noticeable improvement in the FCN adjustment when using data from the compact SG (C026) compared to earlier data from the TT70 in Strasbourg (see their figures 4 and 6).

We do not discuss here the numerous papers using GGP data to derive the FCN parameters and refer the reader to previous reviews on this subject (Hinderer and Crossley, 2000, 2004). A study combining the analysis of six SGs by Sun et al. (2002) leads to an FCN period of 429.0 (424.3–433.7) sidereal days and a Q factor of 9500 (6400, 18 700). A more global analysis by Xu et al. (2004), who used simultaneously tidal gravity observations at 19 GGP stations, leads to an FCN period of 429.9 (427.2–432.7) sidereal days and a Q value about 20 000 (12 000–72 000). The FCN period is in good agreement with space geodetic studies (Herring et al., 1986; Neuberg et al., 1987; Merriam, 1994; Defraigne et al., 1994). However, most of the time, the gravity-derived Q factor is much smaller than that obtained using VLBI (Herring et al., 1986). An artifact of the analysis is sometimes even a negative Q (see table 1 in Florsch and Hinderer (2000); Sun et al., 2002, 2004). The Q discrepancy between gravimeters and VLBI measurements is not due to imprecision in the gravity observations, but is rather of a methodological nature. The classical least-squares method usually applied to the determination of the resonance parameters (obviously, a linearized form applied to a nonlinear problem) is inadequate because it implies Gaussian statistics (Sato et al., 2002; Sun et al., 2004) that are not correct for this problem. Florsch and Hinderer (2000) have demonstrated that an appropriate Bayesian method is required to solve for the FCN parameters due to the nonstandard form of the probability distribution for Q.

Figure 23 shows the Bayesian approach applied to Canberra (CB) SG data by plotting the joint probability distribution for the eigenperiod T and quality factor Q. Taking a vertical slice through the distribution shows that the shape is somewhat Gaussian for the period T, leading to values found in previous studies based on the least-squares method. This is no longer the case for a horizontal slice in which the distribution of Q shows a preferred range of high values exceeding 10^4 (including infinity which means no damping at all) in agreement with estimates from lunisolar nutation observations by VLBI.

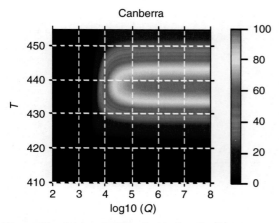

Figure 23 Joint probability density function for eigenperiod T and quality factor Q using SG data from Canberra (CB) station. Courtesy: N. Florsch.

4.3.4.2 Ocean loading

4.3.4.2.(i) Semi-diurnal and diurnal tides
We have previously discussed the ocean loading correction to gravity and referred to the articles of Baker and Bos (2003), Boy *et al.* (2003), and Bos and Baker (2005). Clearly, observations from the GGP network can be compared to computations from different global ocean tide models, some of them being partly hydrodynamic, including the often-quoted classical Schwiderski (1980) model, or assimilating data derived directly from satellite altimetry.

The present situation, from the comprehensive treatment by Bos and Baker (2005) is that the class of recent ocean tide models, specifically GOT00 (Ray, 1999), NAO99b (Matsumoto *et al.*, 2000), and FES99 (Lefévre *et al.*, 2002), is in better agreement with SG observations than earlier models such as Schwiderski (1980). When differences are carefully examined, most models are in agreement and there is no 'best' model valid in all areas of the world. One must mention that the spread along the real (in-phase) axis of the tidal residuals is in general larger than along the imaginary axis (out-of-phase) related to possible amplitude calibration problems. By contrast, the out-of-phase component seems to be more reliable because of a better determination of the instrumental phase lag (Van Camp *et al.*, 2000) and is hence a strong validation tool for ocean tides.

A corollary to this investigation is the limits we can place on Earth's inelasticity from the gravimetric amplitude factors and phase delays after correction for ocean tidal loading. This was pointed out by Baker and Bos (2003), where they compared for some waves the gravimetric amplitude factor and phase delay to some reference elastic or slightly inelastic values (Dehant *et al.*, 1999). In particular, they used the small value ($\leq 0.2\,\mu$Gal) of ocean loading for wave O1 in Europe for testing inelasticity in the Earth's tidal response with European SG stations. These observations could help in rejecting inelasticity models exhibiting an increase in amplitude larger than 0.3%. The small phase lag of a few hundredths of a degree is consistent with the Mathews (2001) inelastic body tide model for 0_1.

4.3.4.2.(ii) Long-period tides
We have pointed out that the SGs have a very small instrumental drift compared to classical spring meters and this is why the investigation of long-period tides is particularly suited to SG data. In a recent paper, Boy *et al.* (2006) analyzed long series from 18 GGP stations to estimate the ocean loading for the monthly (M_m) and fortnightly (M_f) tides. The available models were an equilibrium tide (Agnew and Farrell, 1978), Schwiderski (1980), and three recent hydrodynamical models with satellite altimeter data assimilation – NAO99b (Matsumoto *et al.*, 2000), FES99 (Lefévre *et al.*, 2002), and TPXO.6 (Egbert and Erofeeva, 2002). They concluded that the uncertainty for M_m is still too large to discriminate between the newer models. On the other hand, the hydrodynamical models for M_f are clearly closer to the SG observations than the equilibrium tidal model or the older model proposed by Schwiderski (1980) (see **Figure 24**).

4.3.4.2.(iii) Nonlinear tides
Almost all tidal theory and associated ocean tide loading follow the response method (reviewed earlier and classically used in oceanography). Merriam (1995; 2000) shows that nonlinear tides can be seen in SG records. At CA (Cantley) in Canada, it is possible to see clearly these small abnormal waves originating from nonlinearities in the ocean tidal response at the Bay of Fundy. The sensitivity inferred from the SG measurements is equivalent to 1 mm of open ocean tide, which means that these observations are a unique tool for validating these small ocean tidal waves that exist only near the coastlines (e.g., Sinha and Pingree, 1997).

Some years ago, Florsch *et al.* (1995b) identified small signals of a few nGal in the SG residuals with periods around 6 h from Strasbourg and Cantley stations. A more systematic study by Boy *et al.* (2004) on various GGP stations demonstrated that these signals are indeed due to nonlinear tides in the quar-diurnal frequency band.

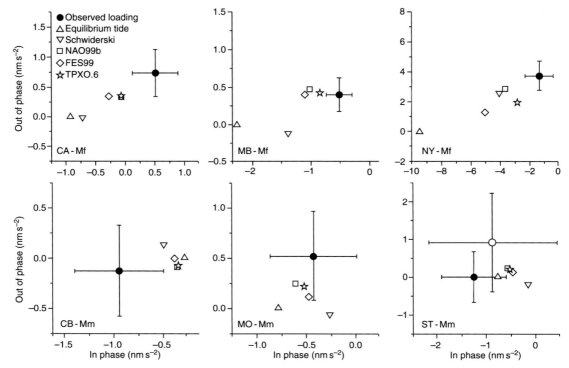

Figure 24 Observed tidal loading for M_f (top) and M_m (bottom) waves and estimated tidal loading for six SGs and five ocean models. Reproduced from Boy J-P, Llubes M, Ray R, Hinderer J, and Florsch N (2006) Validation of long-period oceanic tidal models with superconducting gravimeters. *Journal of Geodynamics* 41: 112–118, with permission from Elsevier.

Figure 25 shows the fair agreement for the M_4 nonlinear tide both in amplitude and phase between SG observations and the predicted contributions from recent ocean tidal models (Flather (1976), Mog2D, and Pingree and Griffiths (1980, 1981)). Mog2D (2-D gravity wave model) is a barotropic, nonlinear model from Lynch and Gray (1979), that was later developed for tidal- and atmospheric-driven applications both at coastal and global scales.

Due to the integrative properties of the gravity-loading Green's functions, inland SG observations act as a complementary large-scale validation tool to point-like tide gauge observations of nonlinear ocean tides. The latter are generated only in shallow water near the coasts and hence mostly escape detection by satellite altimeters like TOPEX/POSEIDON or JASON.

4.3.5 Nontidal Ocean Circulation

In addition to tidal oceanic contributions, nontidal effects related to the general oceanic circulation are also detectable in SG measurements. Virtanen and Mäkinen (2003) investigated the loading effect of the

Baltic Sea on the Metsahovi (Finland) instrument (T020), which is located only 15 km away from the open sea. They found a good correlation between SG residuals and sea-level changes from the nearby Helsinki tide gauge. Short-period variations are mostly driven by wind stress moving water only locally, whereas long-term variations are caused by water exchange through the Danish Straits. It is therefore useful to combine gravity observations with tide gauge measurements (and with precise positioning measurements) to better test the loading from Baltic Sea.

Sato *et al.* (2001) made another important contribution to nontidal effects using SG records from Esashi, Canberra, and Syowa to investigate the nontidal annual contribution from sea-level changes (see also Fukuda and Sato, 1997). They demonstrated the importance of the steric correction to sea-surface height (SSH) change – this is the coefficient used to compensate the thermal expansion of the oceans due to sea-surface temperature (SST) change. The steric part does not involve any additional mass change and hence does not alter gravity by loading; thus, inland SG measurements are a unique tool to distinguish between steric and nonsteric SSH components.

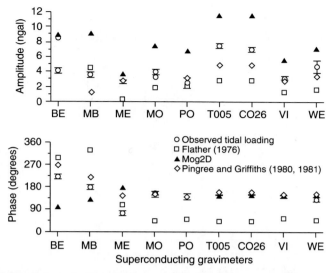

Figure 25   Amplitude (nGal) and phase (in degree with respect to Greenwich) of M_4 observed tidal loading and loading predictions according to recent tidal ocean models. Reproduced from Boy J-P, Llubes M, Ray R, *et al*. (2004) Non-linear oceanic tides observed by superconducting gravimeters in Europe. *Journal of Geodynamics* 38: 391–405, with permission from Elsevier.

4.3.6   Hydrology

Early studies in gravity focused mainly on tides, oceanic and atmospheric effects – starting with the references already made to Warburton and colleagues in the 1970s. In the last decade, however, more and more attention has been paid to the hydrology signature in gravity, especially in SG data where the drift is not a large issue. Most of the studies are restricted to modeling local effects by trying to find correlations between gravity residuals and a relevant hydrological parameter such as water table level, rainfall, or soil moisture (see, e.g., Crossley and Xu, 1998; Virtanen, 2001; Kroner, 2001; Kroner *et al.*, 2004; Takemoto *et al.*, 2002; Ijpelaar *et al.*, 2002; Harnisch and Harnisch, 2002). A typical data set over a 2 week period is shown in **Figure 26**.

Most authors use an admittance (in μGal per millimeter of water) that depends on local porosity and permeability (neither of which is well characterized by spot measurements). The difficulty is in separating local effects from a regional or even continental hydrology signal (see van Dam *et al.*, 2001a, 2001b) knowing that both will share a similar seasonal variation because the meteorological forcing is similar. For this reason, active experiments, where known amounts of water were added to specific areas in the gravimeter vicinity, were conducted at Moxa Observatory and helped significantly in the validation and parametrization of hydrology models (Kroner and Jahr, 2006).

More recently, a study was devoted to the seasonal changes in SGs (Boy and Hinderer, 2006). These changes can be linked to global hydrology models such as LadWorld (Milly and Shmakin, 2002) or GLDAS (Rodell *et al.*, 2004), as shown in **Figure 27**. For Cantley there is a strong snow contribution in winter, adding to the soil humidity, while in Wettzell it is much smaller. For both stations, there is a good agreement between the gravity residuals and the estimated continental water storage loading effects. In fact, for more than half of the 20 analyzed SGs, there is such a good correlation, as shown in **Figure 28**. For the other stations, the discrepancies may be associated with local hydrology effects, especially when a station is partly underground like Moxa or Strasbourg.

Finally, a series of papers (Zerbini *et al.*, 2001, 2002; Romagnoli *et al.*, 2003; Richter *et al.*, 2004) carefully examined both height and gravity changes from continuous GPS and SG (C023) observations at the station Medicina (Italy). These studies, from which we show an example in **Figure 29**, provide a convincing interpretation of the seasonal signal from the combined loading contributions of air pressure, ocean circulation, and hydrology (surficial water table), illustrating once more the need to have simultaneous height and gravity measurements in addition to the monitoring of various environmental parameters in the close vicinity of the station.

(a)

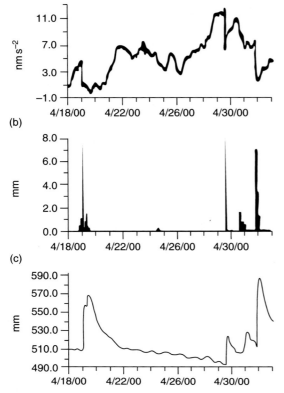

4.3.7 Earth Rotation

At the long period end of the observable spectrum of gravity (**Figure 15**), there are two isolated theoretical periods of the Earth's normal-mode spectrum. One is the inner-core wobble (ICW), whose period is of the order of years. For model PREM, wobble programs yield a value of about 700 days, but the theory is hardly suitable for such a long-period motion. Mathews *et al.* (2002) quote a value of 6.6 yr that is derived from their theory and with VLBI observations and models, but at either period the predicted small amplitude makes this a difficult target (e.g., Guo *et al.*, 2006).

The other mode is much more accessible, and is of course the CW, with a period of about 435 days in the mantle reference frame. This is one of the two components of polar motion normally seen in gravity studies, the other being the smaller amplitude forced annual wobble that is seen in combination with the annual tide and other seasonal effects. Note that at much longer periods there is an 18.6 lunisolar year tide that will be extremely difficult to identify in gravity. Discussion of Richter's (1983) observation of the CW has been mentioned earlier and almost every SG in the GGP network has reported a clear signal of the CW – see, for example, Richter *et al.* (1995b) and Sato *et al.* (1997b) for the Japanese Antarctic station. Here, we review only a few results.

Loyer *et al.* (1999) showed the importance of using a long data set when trying to infer the transfer function of the polar motion; clearly more than

Figure 26 Gravity residual observations (a), accumulated precipitation (per hour) (b), and groundwater table level (c) at Moxa, Germany, from 18 Apr. to 2 May 2000. Reproduced from Kroner C, Jahr T, and Jentzsch G (2004) Results from 44 months of observations with a superconducting gravimeter at Moxa/Germany. *Journal of Geodynamics* 38(3–5): 263–280, with permission from Elsevier.

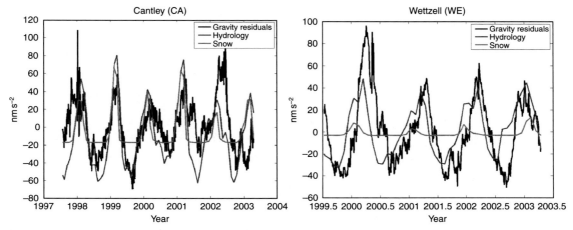

Figure 27 Gravity residuals, hydrology (soil moisture + snow), and snow-modeled contributions at stations Cantley (Canada) and Wettzell (Germany). Reproduced from Boy J-P and Hinderer J (2006) Study of the seasonal gravity signal in superconducting gravimeter data. *Journal of Geodynamics* 41: 227–233, with permission from Elsevier.

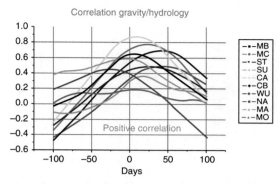

Figure 28 Correlation between gravity and hydrology (continental soil moisture + snow) for some GGP stations; there is a clear positive value with a time lag depending on the station location.

6.5 years of continuous data are required to separate the annual component (whatever its origin) from the 435 day term. They found a gravimetric amplitude factor $\delta = 1.18 \pm 0.10$ and a phase delay of a few degrees using only the Strasbourg station. Harnisch and Harnisch (2006b) generalized this study by using data from 12 GGP stations with lengths varying between 4 and 18 years to investigate the polar motion contribution to gravity. They found that in general the gravimetric amplitude factors for the Chandler wobble are close to the nominal value of 1.16 and phase lags of a few degrees. Xu *et al.* (2004b) also made a similar study on five SGs from the GGP network, and we show in **Figure 30** the typically

good agreement between gravity residuals and polar motion for the stations Membach and Potsdam. Combining the results leads to a δ factor of 1.16 ± 0.07 and to a phase delay of $-1.30° \pm 1.33°$. Finally, a data set of nine SGs was analyzed for the polar motion response by Ducarme *et al.* (2006), who found a mean δ factor of 1.179 ± 0.004, similar to Loyer *et al.* (1999).

A value slightly larger than the nominal value of 1.16 is to be expected when including mantle inelasticity and/or ocean pole tide contribution. The variability on the phase delays between different stations is however still large and unexplainable in terms of Earth's rheological properties.

4.3.8 Tectonic Effects

The final class of long-term gravity changes is not periodic, but secular (i.e., aperiodic) and primarily due to active tectonics or PGR. Until recently, most studies have been done using AGs (e.g., Niebauer *et al.*, 1995), rather than SGs. One reason for this is that an SG needs a much longer residence time at a station to get a good result, due to instrumental considerations. Thus, there is rarely an SG located close enough to the zone of interest for a particular study.

A second reason is that the SGs have a small instrument drift that must be accounted for as part of any estimate of the secular gravity trend. The potential of

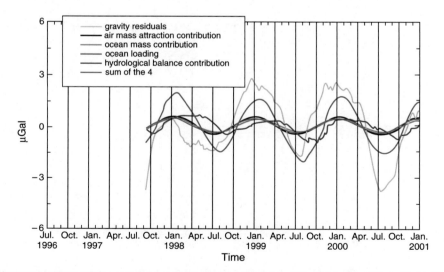

Figure 29 Observed and modeled seasonal gravity changes at Medicina, Italy. Reproduced from Zerbini S, Negusini M, Romagnoli C, Domenichini F, Richter B, and Simon D (2002) Multi-parameter continuous observations to detect ground deformation and to study environmental variability impacts. *Global and Planetary Changes* 34: 37–58, with permission from Elsevier.

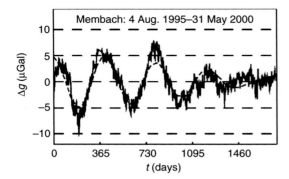

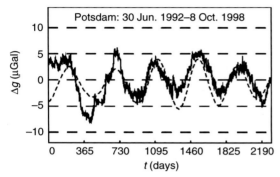

Figure 30 Comparison between gravity residuals (solid line) and polar motion gravity prediction (dashed line) at stations Membach (Belgium) and Potsdam (Germany). Reproduced from Xu J-Q, Sun H-P, and Yang X-F (2004b) A study of gravity variations caused by polar motion using superconducting gravimeter data from the GGP network. *Journal of Geodesy* 78: 201–209 (doi: 10.1007/s00190-004-0386-1), with permission from Springer.

long-term gravity measurements to solve tectonic problems is however significant, as clearly demonstrated by a recent study (Lambert *et al.*, 2006) devoted to PGR in North America. In particular, they could show that the admittance of gravity to height changes (in μGal per millimeter) deduced from combined GPS and AG measurements is indeed close to the theoretically predicted value by Wahr *et al.* (1995) and by James and Ivins (1998) for a viscoelastic model.

In fact, combining gravity and height changes measurements is very efficient in discriminating PGR effects from present-day ice melting contributions as shown by a recent investigation of the long-term gravity changes in Svalbard (Sato *et al.*, 2006a, 2006b). Another study based on 7 years of collocated gravity measurements of SG and AG at Membach (Belgium) by Francis *et al.* (2004) indicates that there is a small decrease in gravity connected to uplift seen by GPS at the same location. A longer data set is required to determine if this gravity decrease is due

to PGR or active tectonics in the Ardenne mountains (see also Van Camp *et al.*, 2002).

Such comparisons of AG and GPS measurements are much easier to do when there are continuous SG measurements at the same location. In fact, the continuity of the SG record is often important to check the integrity of the AG measurements (discussed earlier in connection with the SG calibration) and to model effects such as hydrology that increase the AG scatter. Amalvict *et al.* (2004, 2006) analyzed the long-term gravity changes of the Strasbourg SG with regular AG measurements and collocated GPS and interpreted the results in terms of hydrology and tectonics of the Rhine graben. **Figure 31** shows the gravity trend which seems to be present in an 8 year data set (1997–2004) and how hydrological contributions partly explain some of the gravity features.

4.3.9 Ground/Satellite Gravity Field Comparison

There is a major international effort in the present decade to measure variations in the Earth's global gravity field using low-orbit satellites. The first satellite *CHAMP* was launched in 2000 and was followed 2 years later by *GRACE*. In the near future, there will be a third mission called *GOCE* that will orbit even closer to the ground and hence be even more sensitive to smaller-scale gravity changes. The primary goal of these missions is to use the temporal changes of the Earth's gravity field to infer changes in regional and continental water storage, and ocean circulation (see Wahr *et al.*, 1998; Rodell and Famiglietti, 1999).

A major concern with satellite measurements of time-varying gravity is how to calibrate and validate such observations. In addition to comparisons with models (the primary technique used to date), there are several possibilities using actual measurements at the ground (GPS, gravity) or at the ocean bottom (water pressure). This problem is important because if successful, it would produce an independent method of validation that does not rely solely on modeling. We will show hereafter that surface measurements from the GGP network provide a useful additional constraint on space gravity data. The validation signal is related to seasonal gravity changes that are coherent on length scales appropriate to satellite altitudes (typically a few hundreds of kilometers).

A first study directly comparing *CHAMP* data to six SG ground observations was done by Neumeyer *et al.* (2004b) and has led to satisfactory results for all the stations in the 1 year analysis period (from

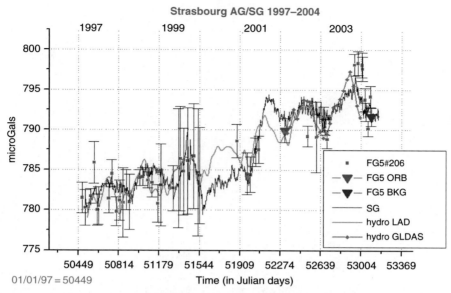

Figure 31 An example of SG/AG superposition at Strasbourg (France) from 1997 to 2004. The gravity observations are corrected for tides (solid + ocean load), air pressure, and polar motion; the instrumental drift of the SG has been removed and hydrological contributions added. Reproduced from Amalvict M, Hinderer J, and Rozsa S (2006) Crustal vertical motion along a profile crossing the Rhine graben from the Vosges to the Black Forest Mountains: Results from absolute gravity, GPS and levelling observations. *Journal of Geodynamics* 41: 358–368, with permission from Elsevier.

December 2000 to December 2001). The superposition of the monthly gravity mean values from the SG residuals (after correction for solid tides, ocean and atmospheric loadings, and polar motion) with the *CHAMP*-reconstructed values at the SG sites is rather good. Neumeyer *et al.* (2006) recently extended this study to *GRACE* data, pointing out again the partial agreement between surface and satellite-derived gravity at specific locations. Before a detailed comparison can be made, however, one has to remember that ground gravity measurements include necessarily a contribution from the vertical motion of the instrument through the ambient gravity field. This signal does not affect the orbiting satellite, and hence there is a difference in the gravity changes as seen at (moving) ground level and by the satellite (Hinderer *et al.*, 2006).

We note in these studies that the comparison of single station results with the large-scale satellite solutions is problematic due to the completely different error budgets involved. *GRACE* data, for example, are good to 1 µGal only over length scales longer than 500–1000 km, whereas SGs are good to the same accuracy (or better) at a single point. In order to average SG measurements and reduce local effects, there have been attempts to assemble a network solution from nearby SG stations rather than

doing the above single-station comparison. Within the existing rather sparse GGP network, Europe is obviously the best place to try such an approach.

The approach was first initiated using 1 year of SG data by Crossley and Hinderer (2002) and Crossley *et al.* (2003) and extended to longer data sets by Crossley *et al.* (2003, 2004, 2005). This approach was further extended to a 21 month time interval to intercompare surface data (GGP European subnetwork), satellite data from *GRACE*, and theoretical predictions for two hydrology models (LAD and GLDAS) (Andersen *et al.*, 2005a; Hinderer *et al.*, 2006). The results show the existence of an annual signal that is coherent over Europe with an amplitude of a few µGal, mostly due to the seasonal loading from continental hydrology (soil moisture + snow) according to recent models such as LadWorld (Milly and Shmakin, 2002) or GLDAS (Rodell *et al.*, 2004). There is even a possibility to detect in *GRACE* data interannual signals (Andersen and Hinderer, 2005) and, in particular, there is a clear evidence that *GRACE* has been affected by the heat wave that occurred in summer 2003 in Europe (Andersen *et al.*, 2005b). The Wettzell (Germany) and the Medicina (Italy) SG data seem to confirm this point as shown by **Figure 32**.

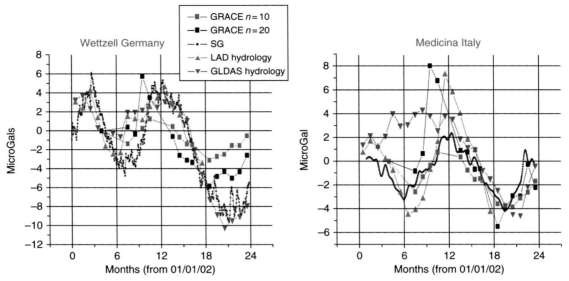

Figure 32   An example of the superposition of gravity changes seen by *GRACE* (in red for $n = 10$ and in black for $n = 20$) and the superconducting gravimeter in Wettzell (Germany) and Medicina (Italy) (in blue) and contributions from continental hydrology models like GLDAS (in green) and LADWorld (in purple).

4.3.10   Future Possibilities

Several recent SG results point the way to future studies involving gravity. One of these is the study by Imanishi *et al.* (2004) of coseismic displacements that accompanied a large $M_w = 8.0$ earthquake in Japan. Using three SGs in a linear array (**Figure 33**), they showed that gravity offsets

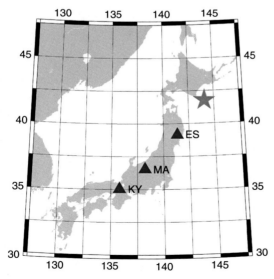

Figure 33   Japan SG array (blue stations) that observed gravity offsets due to a local large earthquake (red).

occurred that matched the theoretical prediction of static displacements that accompany a fault rupture process. The offsets were 0.58, 0.10, and 0.07 µGal at stations ES, MA, and KY respectively, in line with the epicenter at Tokachi-oki (near Hokkaido). Such small offsets (especially close to 0.1 µGal) would hardly be identified in an individual record, but this is the first time a gravity array has performed at this level of precision.

The significance of this result can be transferred to another context, that of subduction-induced silent earthquakes that have been identified off the coast of Japan and in the Cascadia subduction zone off Vancouver Island. In the latter region, episodic tremor that accompanies the silent slip events can last for several days, but there is no identifiable earthquake. Such events are identified primarily as horizontal displacements on GPS arrays, but Lambert *et al.* (2006) have shown (**Figure 34**) that AG measurements also reveal coincident offsets in gravity of several µGal, in between which there is a secular drift (in this case a negative trend). Imagine how precisely this signal would be measured by an inline SG array of the same type that exists in Japan!

It needs hardly be said that hydrology will continue to increase in importance as one of the important targets for future geodetic and geophysical measurements. At the present time, an SG is part of a project to monitor the state of underground water

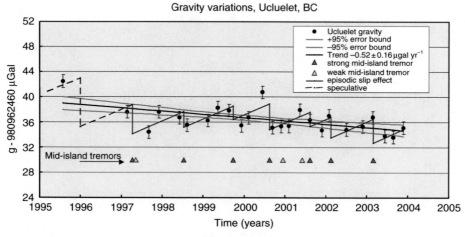

Figure 34 Variations in AG at Ucluelet (western coast of Vancouver Island) showing some concordance with the episodic slip and seismic tremor activity above the Cascadia subduction zone. Courtesy: T. Lambert.

storage at the Texas Hydrological Observatory, that is, surface watersheds and underlying aquifers (C. Wilson, U. Texas), and there have been some proposals to use gravity to monitor active subsurface tectonics in the US (Harry, 1997).

The question of monitoring tectonic processes from *GRACE* data has been raised (Mikhailov *et al.*, 2004), but of course a combination with SG measurements would give a much stronger data set. Also of recent interest is the proposal to use AG measurements in conjunction with *GRACE* data to monitor Fennoscandian post-glacial uplift, and such a project would benefit considerably from one or two stations that have an SG for continuous monitoring. Along the same lines, a proposal is underway to combine an SG with AG and portable Scintrex-type gravimeters to calibrate/validate *GRACE* satellite measurements in West Africa (Hinderer *et al.*, 2005). One profile covers the extremely dry regions from the central Sahara desert down to the coast of Benin, and another is a triplet of stations in the monsoon region of very high rainfall in the Cameroon region.

Our final suggestion for the future deployment of SGs is for the monitoring of volcanoes – both their slow deformation and their explosive activity. Carbone *et al.* (2003) reported several years of gravity along an NS profile on the flank of Mt. Etna that showed variations of more than 50 mGal using Scintrex portable instruments and a continuously operating LCR base station. Had the base station been equipped with an SG, the results would have been even more impressive.

As has been shown by Rymer and Locke (1995), the combination of gravity measurements with surface elevation changes is capable of distinguishing between several different modes of volcanic behavior: surface lowering can accompany either loss of magma or magma injection, and surface elevation (inflation) again associated with either loss or gain of magma. Both gravity and GPS measurements were done in the study of Furuya *et al.* (2003) – shown in **Figure 35** – and this allows a more detailed interpretation of the subsurface mass changes than using either method alone. The gravity signals are very large (~100 μGal) compared to those discussed elsewhere in this chapter, and there would be no question of their detection by an SG, even without detailed AG backup measurements.

4.4 Conclusions

We now summarize the main points of the use of SGs in geodetic measurements:

1. Over the 40 years since it was first developed, the SG has proved to be an extremely reliable instrument for determining gravity variations from 1 s to periods of several years, and several instruments have been operated for more than a decade without interruption.

2. Calibration changes in SGs are virtually nonexistent, and drift rates are at the level of a few μGal per year, making them ideal for long-term monitoring of the gravity field (tectonics, seasonal changes, and polar motion).

3. The precision of 0.1 nGal in frequency domain and an accuracy of better than 0.1 μGal in the time

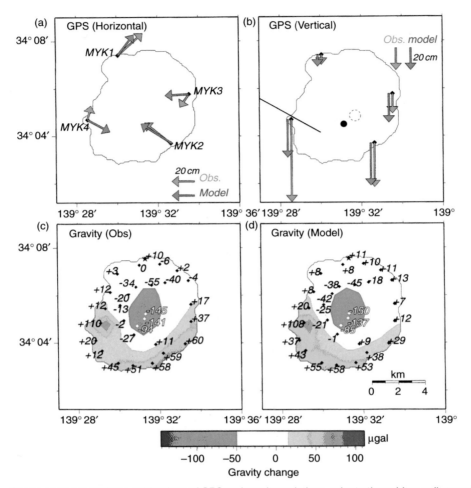

Figure 35 Observed and modeled combinations of GPS and gravity variations, prior to the caldera collapse at Miyakejima Volcano, Japan, in 2000. Reproduced from Furuya M, Okubo S, Sun W, Tanake Y, Oikawa J, and Watanabe H (2003) Spatio-temporal gravity changes at Miyakejima Volcano, Japan: Caldera collapse, explosive eruptions, and magma movement. *Journal of Geophysical Research* 108(B4): 2219, with permission from American Geophysical Union.

domain have contributed to major improvements in tidal analysis, ocean loading computations, and atmospheric effects that were not possible with other types of gravimeter.

4. With SGs, it is now possible to seriously model the gravity effects in hydrology due to rainfall and to quantify these effects in soil moisture, groundwater variations, and atmospheric mass changes.

5. The SG has proven to be very useful, in the same way that spring gravimeters were, in the study of the Earth's free oscillations, particularly at periods longer than 500 s.

6. The combination of SGs and satellite measurements of the gravity field is proving to be consistent and complementary indicators of regional hydrology.

7. New possibilities exist for the use of SGs in hydrology, tectonics, and the monitoring of volcanoes.

8. Future studies in gravity would be much enhanced if SGs can be deployed in array form and combined, where appropriate, with AGs, portable gravimeters, and GPS receivers.

Acknowledgments

Much of the development of the SG has been carried out at GWR by contracts initiated by Bernd Richter and funded by the Institut für Angewandte Geodäsie (IFAG, Frankfurt, Germany) and later by the Bundesamt für Kartographie und Geodäsie (BKG, Frankfurt). Without this involvement, the commercial venture would have been difficult to maintain. We also acknowledge the support of the US Department of Energy under Grant No. DE-FG03-95ER81979 and the US National Science Foundation under Grants No. 9529827 and EAR Award No. 0409381.

References

Achilli V, Baldi P, Casula G, et al. (1995) A calibration system for superconducting gravimeters. *Bulletin Geodesique* 69: 73–80.

Agnew DC (1997) NLOADF: A program for computing ocean-tide loading. *Journal of Geophysics* 102: 5109–5110.

Agnew DC and Farrell WE (1978) Self-consistent equilibrium ocean tides. *Geophysical Journal of the Royal Astronomical Society* 55: 171–181.

Amalvict M, Hinderer J, Boy J-P, and Gegout P (2001a) A three year comparison between a superconducting gravimeter (GWR C026) and an absolute gravimeter (FG5#206) in Strasbourg (France). *Journal of Geodetic Society of Japan* 47: 410–416.

Amalvict M, McQueen H, and Govind R (2001b) Absolute gravity measurements and calibration of SG-CT031 at Canberra, 1999–2000. *Journal of Geodetic Society of Japan* 47: 334–340.

Amalvict M, Hinderer J, Gegout P, Rosat S, and Crossley D (2002) On the use of AG data to calibrate SG instruments in the GGP network. *Bulletin d'Informations des Marées Terrestres* 135: 10621–10626.

Amalvict M, Hinderer J, Makinen J, Rosat S, and Rogister Y (2004) Long-term and seasonal gravity changes and their relation to crustal deformation and hydrology. *Journal of Geodynamics* 38: 343–353.

Amalvict M, Hinderer J, and Rozsa S (2006) Crustal vertical motion along a profile crossing the Rhine graben from the Vosges to the Black Forest Mountains: Results from absolute gravity, GPS and levelling observations. *Journal of Geodynamics* 41: 358–368.

Andersen O and Hinderer J (2005) Global inter-annual gravity changes from GRACE: Early results. *Geophysical Research Letters* 32: L01402 (doi:10.1029/2004GL020948).

Andersen O, Hinderer J, and Lemoine FG (2005a) Seasonal gravity field variations from GRACE and hydrological models. In: Jekeli C, Bastos L, and Fernandes J (eds.) *Gravity, Geoid and Space Missions. IAG Symposia*, vol. 129, pp. 316–321. Berlin: Springer Verlag.

Andersen O, Seneviratne S, Hinderer J, and Viterbo P (2005b) GRACE-derived terrestrial water storage depletion associated with the 2003 European heat wave. *Geophysical Research Letters* 32: L18405 (doi:10.1029/2005GL023574).

Baker T and Bos M (2003) Validating Earth and ocean tide models using tidal gravity measurements. *Geophysical Journal International* 152: 468–485.

Banka D and Crossley D (1999) Noise levels of superconducting gravimeters at seismic frequencies. *Geophysical Journal International* 139: 87–97.

Berger J, Davis P, and Ekström G (2004) Ambient Earth noise: A survey of the Global Seismographic Network. *Journal of Geophysical Research* 109: B11307 (doi:10.1029/204JB003408).

Bos MS and Baker TF (2005) An estimate of the errors in gravity ocean tide loading computations. *Journal of Geodesy* 79: 50–63.

Bos M, Baker TF, Lyard FH, Zürn W, and Rydelek PA (2000) Long-period lunar tides at the geographic South Pole and recent models of ocean tides. *Geophysical Journal International* 143: 490–494.

Bower DR, Liard J, Crossley D, and Bastien R (1991) Preliminary calibration and drift assessment of the superconducting gravimeter GWR12 through comparison with the absolute gravimeter JILA 2, Proceedings of the Workshop: Non Tidal Gravity Changes Intercomparison Between Absolute and Superconducting Gravimeters, Conseil de l'Europe, *Cahiers du Centre Europeen de Geodynamique et de Seismologie* 3, Luxembourg, pp. 129–142.

Boy J-P and Chao BF (2005) Precise evaluation of atmospheric loading effects on Earth's time-variable gravity field. *Journal of Geophysical Research* 110: B08412 (doi:10.1029/2002JB002333).

Boy J-P and Hinderer J (2006) Study of the seasonal gravity signal in superconducting gravimeter data. *Journal of Geodynamics* 41: 227–233.

Boy J-P, Hinderer J, and Gegout P (1998) Global atmospheric loading and gravity. *Physics of the Earth and Planetary Interiors* 109: 161–177.

Boy J-P, Hinderer J, Amalvict M, and Calais E (2000) On the use of long records of superconducting and absolute gravity observations with special application to the Strasbourg station, France, Proceedings of the Workshop: High-Precision Gravity Measurements with Application to Geodynamics and Second GGP Workshop. *Cahiers du Centre Européen de Géodynamique et de Séismologie*, 17, Luxembourg, pp. 67–83.

Boy J-P, Gegout P, and Hinderer J (2001) Gravity variations and global pressure loading. *Journal of Geodetic Society of Japan* 47: 1, 267–272.

Boy J-P, Gegout P, and Hinderer J (2002) Reduction of surface gravity data from global atmospheric pressure loading. *Geophysical Journal International* 149: 534–545.

Boy J-P, Llubes M, Hinderer J, and Florsch N (2003) A comparison of tidal ocean loading models using superconducting gravimeter data. *Journal of Geophysical Research* 108(B4): 2193 (doi:10.1029/2002JB002050).

Boy J-P, Llubes M, Ray R, et al. (2004) Non-linear oceanic tides observed by superconducting gravimeters in Europe. *Journal of Geodynamics* 38: 391–405.

Boy J-P, Llubes M, Ray R, Hinderer J, and Florsch N (2006) Validation of long-period oceanic tidal models with superconducting gravimeters. *Journal of Geodynamics* 41: 112–118.

Busse FH (1974) On the free oscillation of the Earth's inner core. *Journal of Geophysical Research* 79: 753–757.

Carbone D, Greco F, and Budetta G (2003) Combined discrete and continuous gravity observations at Mount Etna. *Journal of the Volcanology and Geothermal Research* 123(1–2): 123–135.

Cartwright DE and Taylor RJ (1971) New computations of the tide generating potential. *Geophysical Journal of the Royal Astronomical Society* 23(1): 45–74.

Cartwright DE and Edden CA (1973) Corrected tables of spherical harmonics. *Geophysical Journal of the Royal Astronomical Society* 33(3): 253–264.

Courtier N, Ducarme B, Goodkind J, et al. (2000) Global superconducting gravimeter observations and the search for the translational modes of the inner core. *Physics of the Earth and Planetary Interiors* 117: 3–20.

Crossley D (1993) Core modes and Slichter modes – fact and fancy. *Bulletin d'Informations des Marées Terrestres* 117: 8628–8638.

Crossley D (2004) Preface to the global geodynamics project. *Journal of Geodynamics* 38: 225–236.

Crossley D and Xu X (1998) Analysis of superconducting gravimeter data, from Table Mountain, Colorado. *Geophysical Journal International* 135: 835–844.

Crossley D and Hinderer J (2002) GGP Ground truth for satellite gravity missions. *Bulletin d'Informations des Marées Terrestres* 136: 10735–10742.

Crossley D and Hinderer J (2005) Using SG arrays for hydrology in comparison with GRACE satellite data, with extension to seismic and volcanic hazards. *Korean Journal of Remote Sensing* 21(1): 31–49.

Crossley D, Jensen O, Xu H, and Hinderer J (1993) A slew rate detection criterion applied to SG data processing. *Bulletin d'Informations des Marées Terrestres* 117: 8675–8704.

Crossley D, Jensen O, and Hinderer J (1995) Effective barometric admittance and gravity residuals. *Physics of the Earth and Planetary Interiors* 90: 221–241.

Crossley D, Xu H, and Van Dam T (1998) Comprehensive analysis of 2 years of data from Table Mountain, Colorado. In: Ducarme B and Paquet P (eds.) *Proceedings of the 13th International Symposium on Earth Tides*, pp. 659–668. Brussels: Observatoire Royal de Belgique.

Crossley D, Hinderer J, Casula G, et al. (1999) Network of superconducting gravimeters benefits a number of disciplines. *EOS, Transactions, American Geophysical Union* 80: 121–126.

Crossley D, Hinderer J, and Amalvict M (2001) A spectral comparison of absolute and superconducting gravimeter data. *Journal of Geodetic Society of Japan* 47: 373–379.

Crossley D, Hinderer J, Llubes M, and Florsch N (2003) The potential of ground gravity measurements to validate GRACE data. *Advances in Geosciences* 1: 1–7.

Crossley D, Hinderer J, and Boy J-P (2004) Regional gravity variations in Europe from superconducting gravimeters. *Journal of Geodynamics* 38: 325–342.

Crossley D, Hinderer J, and Boy J-P (2005) Time variation of the European gravity field from superconducting gravimeters. *Geophysical Journal International* 161: 257–264.

Cummins P, Wahr J, Agnew D, and Tamura Y (1991) Constraining core undertones using stacked IDA gravity records. *Geophysical Journal International* 106: 189198.

Dahlen FA and Sailor RV (1979) Rotational and elliptical splitting of the free oscillations of the Earth. *Geophysical Journal of the Royal Astronomical Society* 58: 609–623.

Defraigne P, Dehant V, and Hinderer J (1994) Stacking gravity tide measurements and nutation observations in order to determine the complex eigenfrequency of the nearly diurnal free wobble. *Journal of Geophysical Research* 99(B5): 9203–9213.

Dehant V, Defraigne P, and Wahr J (1999) Tides for a convective Earth. *Journal of Geophysical Research* 104: 1035–1058.

De Meyer F and Ducarme B (1989) Non-tidal gravity changes observed with a superconducting gravimeter. In: Helsinki E (ed.) *Proceedings of 11th International Symposium on Earth Tides*, pp. 167–184. Stuttgart: Schweizerbart'sche Verlagsbuchhandlung.

Dickman SR (2005) Rotationally consistent Love numbers. *Geophysical Journal International* 161: 31–40.

Dierks O and Neumeyer J (2002) Comparison of Earth Tides Analysis Programs. *Bulletin d'Informations des Marées Terrestres* 135: 10669–10688.

Doodson AT (1921) The harmonic development of the tide generating potential. *Proceedings of the Royal Society of London* A 100: 306–328.

Dragert H, Wang K, and James T (2001) A silent slip event on the deeper Cascadia subduction interface. *Science* 292: 1525–1528.

Ducarme B (1983) Tidal gravity parameters at Brussels reconfirmed by a superconducting gravimeter. *Physics of Earth and Planetary Interiors* 32: 1–3.

Ducarme B, Sun H-P, and Xu J-Q (2002) New investigation of tidal gravity results from the GGP network. *Bulletin d'Informations des Marées Terrestres* 136: 10761–10776.

Ducarme B, Venedikov A, Arnoso J, and Vieira R (2004) Determination of the long period tidal waves in the GGP superconducting gravity data. *Journal of Geodynamics* 38: 307–324.

Ducarme B, Venedikov A, Arnoso J, Chen X-D, Sun H-P, and Vieira R (2006) Global analysis of the GGP superconducting gravimeters network for the estimation of the pole tide gravimetric amplitude factor. *Journal of Geodynamics* 41: 334–344.

Dziewonski AM and Anderson DL (1981) Preliminary reference Earth model. *Physics of Earth and Planetary Interiors* 25: 297–356.

Egbert GD and S Erofeeva E (2002) Efficient inverse modeling of barotropic ocean tides. *Journal of Atmospheric and Oceanic Technology* 19: 183–204.

Elstner C and Schwahn W (1997) Precise mean parameters for daily and subdaily persistent air pressure waves at Potsdam for the period 1893–1992. In: Ducarme B and Paquet P (eds.) *Proceedings of the 13th International Symposium on Earth Tides*, pp. 469–476. Brussels: Observatoire Royal de Belgique.

Falk R, Harnisch M, Harnisch G, Nowak I, Richter B, and Wolf P (2001) Calibration of superconducting gravimeters SG103, C023, CD029, and CD030. *Journal of Geodetic Society of Japan* 47(1): 22–27.

Farrell WE (1972) Deformation of the Earth by surface loads. *Review of Geophysics. Space Physics* 10(3): 751–797.

Flather RA (1976) A tidal model of the North-West European continental shelf. *Mémoires de la Société Royale des Sciences de Liège* 9: 141–164.

Florsch N and Hinderer J (2000) Bayesian estimation of the Free Core Nutation parameters from the analysis of precise tidal gravity data. *Physics of Earth and Planetary Interiors* 117: 21–35.

Florsch N, Hinderer J, and Legros H (1995b) Identification of quarter diurnal waves in superconducting gravimeter data. *Bulletin d'Informations des Marées Terrestres* 122: 9189–9198.

Florsch N, Legros H, and Hinderer J (1995a) The search for weak harmonic signals in a spectrum with application to gravity data. *Physics of Earth and Planetary Interiors* 90: 197–210.

Francis O (1997) Calibration of the C021 superconducting gravimeter in Membach (Belgium) using 47 days of absolute gravity measurements, in *Gravity, Geoid and Marine Geodesy*, Tokyo, Japan, IAG Symposium 117, Springer, pp. 212–219.

Francis O and Hendrickx M (2001) Calibration of the LaCoste-Romberg 906 by comparison with the spring gravimeter C021 in Membach (Belgium). *Journal of Geodetic Society of Japan* 47(1): 16–21.

Francis O, Niebauer T, Sasagawa G, Klopping F, and Gschwind J (1998) Calibration of a superconducting gravimeter by comparison with an absolute gravimeter FG5 in Boulder. *Geophysical Research Letters* 25(7): 1075–1078.

Francis O, Van Camp M, van Dam T, Warnant R, and Hendrickx M (2004) Indication of the uplift of the Ardenne in long-term gravity variations in Membach (Belgium). *Geophysical Journal International* 158: 346–352.

Freybourger M, Hinderer J, and Trampert J (1997) Comparative study of superconducting gravimeters and broadband seismometers STS-1/Z in subseismic frequency bands. *Physics of the Earth and Planetary Interiors* 101: 203–217.

Fukuda Y and Sato T (1997) Gravity effects of sea level variation at the Superconducting Gravimeter sites, estimated from ERS-1 and Topex-Poseidon altimeter data. In: Segawa, et al. (eds.) *IAG Symposia vol. 117, Gravity, Geoid, and Marine Geodesy*, pp. 107–114. Berlin: Springer-Verlag.

Furuya M, Okubo S, Sun W, Tanake Y, Oikawa J, and Watanabe H (2003) Spatio-temporal gravity changes at Miyakejima Volcano, Japan: caldera collapse, explosive eruptions, and magma movement. *Journal of Geophysical Research* 108(B4): 2219.

Goodkind JM (1983) Q of the nearly diurnal free wobble. In: Kuo JT (ed.) *Proceedings of The 9th International Symposium on Earth Tides*, pp. 569–575.. New York: E. Schweizerbartsche Verlagsbuchhandlung.

Goodkind JM (1991) The superconducting gravimeter: principles of operation, current performance and future prospects, Proc. Workshop: Non Tidal Gravity Changes Intercomparison Between Absolute and Superconducting Gravimeters, Conseil de l'Europe, Cahiers du Centre Europeen de Geodynamique et de Seismologie, 3, Luxembourg, pp. 81–90.

Goodkind JM (1996) Test of theoretical solid earth and ocean gravity tides. *Geophysical Journal International* 125: 106–114.

Goodkind JM (1999) The superconducting gravimeter. *Review of Scientific Instruments* 70(11): 4131–4152.

Goodkind JM, Czipott PV, Mills AP, *et al.* (1993) Test of the gravitational square law at 0.4 to 1.4 m mass separation. *Physical Review D* 47(4): 1290–1297.

Green J (1999) *Atmospheric Dynamics*. New York: Academic Press.

Guo J-Y, Dierks O, Neumeyer J, and Shum CK (2006) Weighting algorithms to stack superconducting gravimeter data for the potential detection of the Slichter modes. *Journal of Geodynamics* 41: 326–333.

Harnisch M and Harnisch G (1995) Processing of the data from two superconducting gravimeters, recorded in 1990–1991 at Richmond (Miami, Florida). Some problems and results. *Bulletin d'Informations des Marées Terrestres* 122: 9141–9147.

Harnisch M and Harnisch G (1997) Long time behaviour of superconducting gravimeters derived from observed time series. *Bulletin d'Informations des Marées Terrestres* 127: 9796–9805.

Harnisch M and Harnisch G (2002) Seasonal variations of hydrological influences on gravity measurements at Wettzell. *Bulletin d'Informations des Marées Terrestres* 137: 10.

Harnisch G and Harnisch M (2006a) Hydrological influences in long gravimetric data series. *Journal of Geodynamics* 41: 1–3, 276–287.

Harnisch M and Harnisch G (2006b) Study of long-term gravity variations, based on data of the GGP co-operation. *Journal of Geodynamics* 41(1–3): 318–325.

Harnisch M, Harnisch G, Richter B, and Schwahn W (1998) Estimation of polar motion effects from time series recorded by superconducting gravimeters. In: Ducarme B and Paquet P (eds.) *Proceedings of the 13th International Symposium. Earth Tides, Brussels 1997*, pp. 511–518. Brussels: Observatoire Royal de Belgique.

Harnisch M, Harnisch G, Nowak I, Richter B, and Wolf P (2000) The dual sphere superconducting gravimeter C029 at Frankfurt a.M. and Wettzell. First results and calibration. Proc. of the Workshop: High-Precision Gravity Measurements with Application to Geodynamics and Second GGP Workshop, Cahiers du Centre Européen de Géodynamique et de Séismologie, 17, Luxembourg, pp. 39–56.

Harnisch M, Harnisch G, and Falk R (2002) Improved scale factors of the BKG superconducting gravimeters, derived from comparisons with absolute gravity measurements. *Bulletin d'Informations des Marées Terrestres* 135: 10627–10642.

Harry DL (1997) The use of GPS and microgal absolute gravimetry to constrain active tectonics in the subsurface: Preliminary report, AGU Chapman Conference on Microgal Gravimetry, St. Augustine, Fl.

Hartmann T and Wenzel H-G (1995a) The HW95 tidal potential catalogue. *Geophysical Research Letters* 22, 24: 3553–3556.

Hartmann T and Wenzel H-G (1995b) Catalog HW95 of the tide generating potential. *Bulletin d'Informations des Marées Terrestres* 123: 9278–9301.

Herring TA, Gwinn CR, and Shapiro II (1986) Geodesy by radioInterferometry: Studies of the forced nutations of the Earth: I. Data analysis. *Journal of Geophysical Research* 91(B5): 4745–4754.

Hinderer J and Legros H (1989) Gravity perturbations of annual period. In: Helsinki E (ed.) *Proceedings of the 11th International Symposium. Earth Tides*, pp. 425–429. Stuttgart: Schweizerbart'sche Verlagsbuchhandlung.

Hinderer J and Crossley D (2000) Time variations in gravity and inferences on the Earth's structure and dynamics. *Surveys in Geophysics* 21: 1–45.

Hinderer J and Crossley D (2004) Scientific achievements from the first phase (1997–2003) of the Global Geodynamics Project using a worldwide network of superconducting gravimeters. *Journal of Geodynamics* 38: 237–262.

Hinderer J, Florsch N, Mäkinen J, Legros H, and Faller JE (1991) On the calibration of a superconducting gravimeter using absolute gravity measurements. *Geophysical Journal International* 106: 491–497.

Hinderer J, Crossley D, and Xu H (1994) A two year comparison between the French and Canadian superconducting gravimeter data. *Geophysical Journal International* 116: 252–266.

Hinderer J, Crossley D, and Jensen O (1995) A search for the Slichter triplet in superconducting gravimeter data. *Physics of Earth and Planetary Interiors* 90: 183–195.

Hinderer J, Boy JP, and Legros H (1998) A 3000 day registration of the superconducting gravimeter GWR T005 in Strasbourg, France. In: Ducarme B and Paquet P (eds.) *Proceedings of the 13th International Symposium on Earth Tides*, pp. 617–624. Brussels: Observatoire Royal de Belgique.

Hinderer J, Boy JP, Gegout P, Defraigne P, Roosbeek F, and Dehant V (2000) Are the free core nutation parameters variable in time? *Physics of Earth and Planetary Interiors* 117: 37–49.

Hinderer J, Rosat S, Crossley D, Amalvict M, Boy J-P, and Gegout P (2002a) Influence of different processing methods on the retrieval of gravity signals from GGP data. *Bulletin d'Informations des Marées Terrestres* 123: 9278–9301.

Hinderer J, Amalvict M, Crossley D, Rivera L, Leveque J-J, and Luck B (2002b) Tides, earthquakes and ground noise as seen by an absolute gravimeter and its superspring; a comparison with a broadband seismometer and a superconducting gravimeter. *Metrologia* 39: 495–501.

Hinderer J, de Linage C, and Boy J-P (2005) Issues in the ground validation of satellite-derived gravity measurements: A proposal to validate GRACE in Africa from the Sahara to the equatorial monsoon. *EOS Transactions*, *American Geophysical Union* 85(52): *Fall Meeting Supplement* Abstract G33A-0028.

Hinderer J, Andersen O, Lemoine F, Crossley D, and Boy J-P (2006) Seasonal changes in the european gravity field from GRACE: A comparison with superconducting gravimeters and hydrology model predictions. *Journal of Geodynamics* 41: 59–68.

Hsu H-T, Tao GQ, Song XL, Baker TF, Edge RJ, and Jeffries G (1989) Gravity tidal datum at Wuchang of China. In: Helsinki E (ed.) *Proceedings of the 11th International Symposium on Earth Tides*, pp. 187–195. Stuttgart: Schweizerbart'sche Verlagsbuchhandlung.

Ijpelaar R, Troch P, Warderdam P, Stricker H, and Ducarme B (2002) Detecting hydrological signals in time series of *in-situ* gravity measurements. *Bulletin d'Informations des Marées Terrestres* 135: 10837–10838.

Imanishi Y (2005) On the possible cause of long period instrumental noise (parasitic mode) of a superconducting gravimeter. *Journal of Geodesy* 78(11–12): 683–690.

Imanishi Y, Higashi T, and Fukuda Y (2002) Calibration of the superconducting gravimeter T011 by parallel observation with the absolute gravimeter FG5 210 – a Bayesian approach. *Geophysical Journal International* 151: 867–878.

Imanishi Y, Sato T, Higashi T, Sun W, and Okubo S (2004) A network of superconducting gravimeters detects submicrogal coseismic gravity changes. *Science* 306: 476–478.

Iwano S and Fukuda Y (2004) Superconducting gravimeter observations without a tilt compensation system. *Physics of the Earth and Planetary Interiors* 147(4): 343–351.

Iwano S, Kimura I, and Fukuda Y (2003) Calibration of the superconducting gravimeter TT70 #016 at Syowa station by parallel observation with the absolute gravimeter FG5 #203. *Polar Geoscience* 16: 22–28.

James TS and Ivins ER (1998) Predictions of Antarctic crustal motions driven by present-day ice sheet evolution and by isostatic memory of the Last Glacial Maximum. *Journal of Geophysical Research* 103: 4993–5017.

Jentzsch G, Kroner C, Flach D, and Gommlich G (1995) Long and aperiodic effects in the recording of the superconducting gravimeter in the Asse salt mine in Northern Germany, Proc. 2nd Workshop: Non Tidal Gravity Changes Intercomparison Between Absolute and Superconducting Gravimeters, Conseil de l'Europe, Cahiers du Centre Européen de Géodynamique et de Séismologie, 11, Luxembourg, pp. 187–189.

Klopping FJ, Peter G, Berstis KA, Carter WE, Goodkind JM, and Richter BD (1995) Analysis of two 525 day long data sets obtained with two side-by-side, simultaneously recording superconducting gravimeters at Richmond, Florida, U.S.A., Proc. 2nd Workshop: Non Tidal Gravity Changes Intercomparison Between Absolute and Superconducting Gravimeters, Conseil de l'Europe, Cahiers du Centre Europeen de Geodynamique et de Seismologie, 11, Luxembourg, pp. 57–69.

Kobayashi N and Nishida K (1998) Continuous excitation of planetary free oscillations by atmospheric disturbances. *Nature* 395: 357–360.

Kroner C (2001) Hydrological effects on gravity at the Geodynamic Observatory Moxa. *Journal of the Geodetic Society of Japan* 47(1): 353–358.

Kroner C and Jentzsch G (1999) Comparison of different pressure reductions for gravity data and resulting consequences. *Physics of the Earth and Planetary Interiors* 115: 205–218.

Kroner C and Jahr T (2006) Hydrological experiments around the superconducting gravimeter at Moxa Observatory. *Journal of Geodynamics* 41: 268–275.

Kroner C, Jahr T, and Jentzsch G (2004) Results from 44 months of observations with a superconducting gravimeter at Moxa/Germany. *Journal of Geodynamics* 38(3-5): 263–280.

Kroner C, Dierks O, Neumeyer J, and Wilmes H (2005) Analysis of observations with dual sensor superconducting gravimeters. *Physics of the Earth and Planetary Interiors* 153: 210–219.

Lambert A (1974) Earth tide analysis and prediction by the response method. *Journal of Geophysical Research* 79(32): 4952–4960.

Lambert A, Pagiatakis SD, Billyard AP, and Dragert H (2003) Improved ocean tidal loading corrections for gravity and displacement: Canada and northern United States. *Journal of Geophysical Research* 103(B12): 30231–30244.

Lambert A, Courtier N, and James TS (2006) Long-term monitoring by absolute gravimetry:Tides to postglacial rebound. *Journal of Geodynamics* 41: 307–317.

Larson JV and Harrison JC (1986) An improved analysis of the electrostatic feedback of LaCoste and Romberg gravity meters. In: *Proceedings of the 10th International Symposium on Earth Tides, Madrid 1985*, pp. 1–8. Madrid: Consejo Superior de Investigaciones Cientificas.

Lefèvre F, Lyard F, Le Provost C, and Schrama EJO (2002) FES99: A global tide finite element solution assimilating tide gauge and altimetric information. *Journal of the Atmospheric and Oceanic Technology* 19: 1345–1356.

Lognonné P, Clevede E, and Kanamori H (1998) Computation of seismograms and atmospheric oscillations by normal-mode summation for a spherical earth model with realistic atmosphere. *Geophysical Journal International* 135: 388–406.

Loyer S, Hinderer J, and Boy JP (1999) Determination of the gravimetric factor at the Chandler period from Earth's orientation data and superconducting gravimetry observations. *Geophysical Journal International* 136: 1–7.

Lynch DR and Gray WG (1979) A wave equation model for finite element tidal computations. *Computers and Fluids* 7: 207–228.

Matsumoto K, Takanezawa T, and Ooe M (2000) Ocean tide models developed by assimilating TOPEX/Poseidon altimeter data into hydro-dynamical model: A global model and a regional model around Japan. *Journal of Oceanography* 56: 567–581.

Matsumoto K, Sato T, Takanezawa T, and Ooe M (2001) GOTIC2: A program for the computation of ocean tidal loading effect. *Journal of Geodetic Society of Japan* 47: 243–248.

Mathews PM (2001) Love numbers and gravimetric factor for diurnal tides. *Journal of Geodetic Society of Japan* 47: 231–236.

Mathews PM, Herring TA, and Buffett B (2002) Modeling of nutation and precession: new nutation series for nonrigid Earth and insights into the Earth's interior. *Journal of Geophysical Research* 107(B4): 10.1029/2001JB000390.

Melchior P and Georis B (1968) Earth tides, precession-nutations and the secular retardation of Earth's rotation. *Physics of Earth and Planetary Interiors* 1(4): 267–287.

Melchior P and Ducarme B (1986) Detection of inertial gravity oscillations in the Earth's core with a superconducting gravimeter at Brussels. *Physics of Earth and Planetary Interiors* 42: 129–134.

Merriam J (1992a) An ephemeris for gravity tide predictions at the 1 ngal level. *Geophysical Journal International* 108: 415–422.

Merriam J (1992b) Atmospheric pressure and gravity. *Geophysical Journal International* 109: 488–500.

Merriam J (1993a) A comparison of recent tide catalogues. *Bulletin d'Informations des Marées Terrestres* 115: 8515–8535.

Merriam J (1993b) Calibration, phase stability, and a search for non-linear effects in data from the superconducting gravimeter at Cantley, Quebec. In: Hsu H-T (ed.) *Proceedings of the 12th International Symposium on Earth Tides, Beijing, China*, pp. 12. Beijing: Science Press.

Merriam J (1994) The nearly diurnal free wobble resonance in gravity measured at Cantley, Quebec. *Geophysical Journal International* 119: 369–380.

Merriam J (1995) Non-linear tides observed with the superconducting gravimeter at Cantley, Quebec. *Geophysical Journal International* 123: 529–540.

Merriam J (2000) The response method applied to the analysis of superconducting gravimeter data. *Physics of the Earth and Planetary Interiors* 121(3–4): 289–299.

Merriam J, Pagiatakis S, and Liard J (2001) Reference level stability of the Canadian superconducting gravimeter installation. *Journal of Geodetic Society of Japan* 47(1): 417–423.

Meurers B (1999) Air pressure signatures in the SG data of Vienna. *Bulletin d'Informations des Marées Terrestres* 131: 10195–10200.

Meurers B (2000) Gravitational effects of atmospheric processes in SG gravity data. In: Ducarme B and Barthelemy J (eds.) *Proceedings of the Workshop: 'High Precision Gravity Measurements with Application to Geodynamics and Second GGP Workshop'*, vol. 17, pp. 57–65. Luxembourg: 1999, ECGS Cahiers.

Meurers B (2001a) Superconducting gravimetry in Geophysical research today. *Journal of Geodetic Society of Japan* 47(1): 300–307.

Meurers B (2001b) Tidal and non-tidal gravity variations in Vienna – A Five Years' SG Record. J. *Journal of Geodetic Society of Japan* 47(1): 392–1397.

Meurers B (2002) Aspects of gravimeter calibration by time domain comparison of gravity records. *Bulletin d'Informations des Marées Terrestres* 135: 10643–10650.

Meurers B (2004) Investigation of temporal gravity variations in SG-records. *Journal of Geodynamics* 38: 423–435.

Mikhailov V, Tikhotsky S, Diament M, Panet I, and Ballu V (2004) Can tectonic processes be recovered from new gravity satellite data? *Earth and Planetary Science Letters* 228(3-4): 281–297.

Milly C and Shmakin A (2002) Global modeling of land water and energy balances. Part I: The land dynamics (LaD) model. *Journal of Hydrometeorology* 3: 283–299.

Mukai A, Higashi T, Takemoto S, Nakagawa I, and Naito I (1995) Accurate estimation of atmospheric effects on gravity observations made with a superconducting gravimeter at Kyoto. *Physics of the Earth and Planetary Interiors* 91: 149–159.

Mukai A, Takemoto S, Higashi T, and Fukuda Y (2001) Oceanic tidal loadings estimated from gravity observations in Kyoto and Bandung. *Journal of Geodetic Society of Japan* 47(1): 261–266.

Müller T and Zürn W (1983) Observation of gravity changes during the passage of cold fronts. *Journal of Geophysics* 53: 155–162.

Nawa K, Suda N, Fukao Y, Sato T, Aoyama Y, and Shibuya K (1998) Incessant excitation of the Earth's free oscillations. *Earth Planets Space* 50: 3–8.

Nawa K, Suda N, Fukao Y, et al. (2000) Incessant excitation of the Earth's free oscillations: Global comparison of superconducting gravimeter records. *Physics of the Earth and Planetary Interiors* 120: 289–297.

Neuberg J, Hinderer J, and Zürn W (1987) Stacking gravity tide observations in Central Europe for the retrieval of the complex eigen-frequency of the nearly diurnal free wobble. *Geophysical Journal of the Royal Astronomical Society* 91: 853–868.

Neumeyer J (1995) Frequency-dependent atmospheric pressure correction on gravity variations by means of cross-spectral analysis. *Bulletin d'Informations des Marées Terrestres* 122: 9212–9220.

Neumeyer J and Stobie B (2000) The new Superconducting Gravimeter Site at the South African Geodynamic Observatory Sutherland (SAGOS), Proceedings of the Workshop: High-Precision Gravity Measurements with Application to Geodynamics and Second GGP Workshop, Cahiers du Centre Européen de Géodynamique et de Séismologie, 17, Luxembourg, 85-96.

Neumeyer J, Hagedoorn J, Leitloff, and Schmidt R (2004a) Gravity reduction with three-dimensional atmospheric pressure data for precise ground gravity measurements. *Journal of Geodynamics* 38: 437–450.

Neumeyer J, Schwintzer P, Barthelmes F, et al. (2004b) Comparison of superconducting gravimeter and CHAMP satellite derived temporal gravity variations. In: Reigber Ch, Lühr H, Schwintzer P, and Wickert J (eds.) *Earth Observations with CHAMP Results from Three Years in Orbit,* pp. 31–36.

Neumeyer J, Barthelmes F, Dierks O, et al. (2006) Combination of temporal gravity variations resulting from Superconducting Gravimeter (SG) recordings, GRACE satellite observations and global hydrology models. *Journal of Geodesy* (DOI:10.1007/s00190-005-0014-8).

Niebauer T, Sasagawa G, Faller J, Hilt R, and Klopping F (1995) A new generation of absolute gravimeters. *Metrologia* 32: 159–180.

Nishida K and Kobayashi N (1999) Statistical features of Earth's continuous free oscillations. *Journal of Geophysical Research* 104: 28741–28750.

Nishida K, Kobayashi N, and Fukao Y (2000) Resonant oscillations between the solid Earth and the atmosphere. *Science* 287: 2244–2246.

Nishida K, Kobayashi N, and Fukao Y (2002) Origin of Earth's ground noise from 2 to 20 mHz. *Geophysical Research Letters* 29: 52-1–52-4.

Okubo S, Yoshida S, Sato T, Tamura Y, and Imanishi Y (1997) Verifying the precision of an absolute gravimeter FG5: Comparison with superconducting gravimeters and detection of oceanic tidal loading. *Geophysical Research Letters* 24: 489–492.

Pagiatakis SD (2000) Superconducting gravimeter data treatment and analysis. *Cahiers du Centre Européen de Géodynamique et Séismologie* 17: 103–113.

Peterson J (1993) Observations and modelling of seismic background noise, Open-File Report 93-332, U.S. Department of Interior, Geological Survey, New Mexico: Albuquerque.

Pingree RD and Griffiths KD (1980) Currents driven by a steady uniform windstress on the shelf seas around the British Isles. *Oceanologica Acta* 3: 227–235.

Pingree RD and Griffiths KD (1981) S2 tidal simulations on the North-West European Shelf. *Journal of the Marine Biological Association of the United Kingdom* UK 61: 609–616.

Prothero WA (1967) A cryogenic gravimeter, PhD Thesis, University of California at San Diego, La Jolla.

Prothero WA and Goodkind JM (1968) A superconducting gravimeter. *Review of Scientific Instruments* 39: 1257–1262.

Prothero WA and Goodkind JM (1972) Earth tide measurements with the superconducting gravimeter. *Journal of Geophysical Research* 77: 926–932.

Rabbell W and Zschau J (1985) Static deformations and gravity changes at the earth's surface due to atmospheric loading. *Journal of Geophysics* 56: 81–99.

Ray RD (1999) A global ocean tide model from TOPEX/Poseidon altimeter: GOT99.2, NASA Tech. Memo., TM-209478, 58 pp.

Rhie J and Romanowitz B (2004) Excitation of Earth's continuous free oscillations by atmosphere–ocean–seafloor coupling. *Nature* 431: (doi:10.1038/nature02942).

Richter B (1983) The long-period tides in the Earth tide spectrum. In: Proceedings of XVIII Gen. Ass. IAG, Hamburg, 1, pp. 204–216. Columbus, Ohio: Ohio State University Press, 1984: Ohio State University Press.

Richter B (1985) Three years of registration with the superconducting gravimeter. *Bulletin d'Informations des Marées Terrestres* 94: 6344–6352.

Richter B (1987) Das supraleitende Gravimeter, Deutsche Geodät. Komm., Reihe C, 329, Frankfurt am Main, 124pp.

Richter B (1990) In: McCarthy D and Carter W (eds.) *IUGG Geophysical Monograph No. 59 (9): The Long Period Elastic Behavior of the Earth,* pp. 21–25.

Richter B (1991) Calibration of superconducting gravimeters, Proceedings of the Workshop: Non Tidal Gravity Changes Intercomparison Between Absolute and Superconducting Gravimeters, Conseil de l'Europe, Cahiers du Centre Europeen de Geodynamique et de Seismologie 3, Luxembourg, pp. 99–107.

Richter B and Wenzel H-G (1991) Precise instrumental phase lag determination by the step response method. *Bulletin d'Informations des Marées Terrestres* 111: 8032–8052.

Richter B and Warburton RJ (1998) A new generation of superconducting gravimeters. In: *Proceedings of the 13th International Symposium on Earth Tides, Brussels 1997,* pp. 545–555. Brussels: Observatoire Royal de Belgique.

Richter B, Wilmes H, and Nowak I (1995a) The Frankfurt calivration system for relative gravimeters. *Metrologia* 32: 217–223.

Richter B, Wenzel H-G, Zürn W, and Klopping F (1995b) From Chandler wobble to free oscillations: Comparison of cryogenic gravimeters and other instruments over a wide period range. *Physics of the Earth and Planetary Interiors* 91: 131–148.

Richter B, Zerbini S, Matonti F, and Simon D (2004) Long-term crustal deformation monitored by gravity and space techniques at Medicina, Italy and Wettzell, Germany. *Journal of Geodynamics* 38: 281–292.

Rieutord M (2002) Slichter modes of the Earth revisited. *Physics of the Earth and Planetary Interiors* 131: 269–278.

Rodell M and Famiglietti J (1999) Detectability of variations in continental water storage from satellite observations of the time dependent gravity field. *Water Resources Research* 35(9): 2705–2723.

Rodell M, Houser PR, Jambor U, et al. (2004) The global land data assimilation system. *Bulletin of the American Meteorological Society* 85(3): 381–394.

Romagnoli C, Zerbini S, Lago L, et al. (2003) Influence of soil consolidation and thermal expansion effects on height and gravity variations. *Journal of Geodynamics* 521–539.

Roosbeek F (1996) RATGP95: A harmonic development of the tide-generating potential using an analytic method. *Geophysical Journal International* 126: 197–204.

Rosat S, Hinderer J, Crossley D, and Rivera L (2003a) The search for the Slichter Mode: Comparison of noise levels of superconducting gravimeters and investigation of a stacking method. *Physics of the Earth and Planetary Interiors* 140: 183–202.

Rosat S, Hinderer J, and Rivera L (2003b) First observation of $_2S_1$ and study of the splitting of the football mode $_0S_2$ after the June 2001 Peru event of magnitude 8.4. *Geophysical Research Letters* 30: 21 2111 (doi:10.1029/2003GL018304).

Rosat S, Hinderer J, Crossley D, and Boy J-P (2004) Performance of superconducting gravimeters from long-period seismology to tides. *Journal of Geodynamics* 38: 461–476.

Rosat S, Sato T, Imanishi Y, et al. (2005) High-resolution analysis of the gravest seismic normal modes after the 2004 $M_w = 9$ Sumatra earthquake using superconducting gravimeter data. *Geophysical Research Letters* 32: L13304 (doi:10.1029/2005GL023128).

Rosat S, Rogister Y, Crossley D, and Hinderer J (2006) A search for the Slichter triplet with superconducting gravimeters: Impact on the density jump at the inner core boundary. *Journal of Geodynamics* 41: 296–306.

Roult G and Crawford W (2000) Analysis of 'background' free oscillations and how to improve resolution by subtracting the atmospheric pressure signal. *Physics of the Earth and Planetary Interiors* 121: 325–338.

Rymer H and Locke C (1995) Microgravity and ground deformation precursors to eruption: A review, In: Proceedings of the Workshop: New Challenges for Geodesy in Volcano Monitoring. Cah. Cent. Europ. Géodyn. et Séism., Luxembourg 8, pp. 21–39.

Sato T, Shibuya K, Ooe M, et al. (1991) Long term stability of the superconducting gravimeter installed at Syowa station, Antarctica, Proceedings of the 2nd Workshop: Non Tidal Gravity Changes Intercomparison Between Absolute and Superconducting Gravimeters, Conseil de l'Europe, Cahiers du Centre Européen de Géodynamique et de Séismologie, Luxembourg, 11, pp. 71–75.

Sato T, Ooe M, Nawa K, Shibuya K, Tamura Y, and Kaminuma K (1997a) Long-period tides observed with a superconducting gravimeter at Syowa station, Antarctica, and their implication to global ocean tide modeling. *Physics of the Earth and Planetary Interiors* 103: 39–53.

Sato T, Nawa K, Shibuya K, et al. (1997b) Polar motion effect on gravity observed with a superconducting gravimeter at Syowa station, Antarctica. In: IAG Symposia. Segawa et al. (eds.) *Gravity, Geoid, and Marine Geodesy,* vol. 117, pp. 99–106. Berlin: Springer-Verlag.

Sato T, Fukuda Y, Aoyama Y, et al. (2001) On the observed annual gravity variation and the effect of sea surface height variations. *Physics of the Earth and Planetary Interiors* 123: 45–63.

Sato T, Tamura Y, Matsumoto K, Imanishi Y, and McQueen H (2002) Parameters of the fluid core resonance estimated from superconducting gravimeter data. *Bulletin d'Informations Marées Terrestres* 136: 10751–10760.

Sato T, Boy J-P, Tamura Y, et al. (2006a) Gravity tide and seasonal gravity variation at Ny-Ålesund, Svalbard in Arctic. *Journal of Geodynamics* 41: 234–241.

Sato T, Okuno J, Hinderer J, et al. (2006b) A geophysical interpretation of the secular displacement and gravity rates observed at Ny-Alesund, Svalbard in the Arctic – Effects of the post-glacial rebound and present-day ice melting. *Geophysical Journal International* 165: 729–743, doi: 10.1111/1365-246X.2006.02992.x.

Scherneck H-G (1991) A parameterized Earth tide observation model and ocean tide loading effects for precise geodetic measurements. *Geophysical Journal International* 106: 677–695.

Schwahn W, Baker T, Falk R, et al. (2000) Long-Term Increase of Gravity at the Medicina Station (Northern Italy) Confirmed by Absolute and Superconducting Gravimetric Time Series, Proceedings of the Workshop: High-Precision Gravity Measurements with Application to Geodynamics and Second GGP Workshop, Cahiers du Centre Européen de Géodynamique et de Séismologie, 17, Luxembourg, pp. 145–168.

Schwiderski EW (1980) On charting global ocean tides. *Rev. Geophys.* 18: 243–268.

Seama N, Fukuda Y, and Segawa J (1993) Superconducting gravimeter observations at Kakioka, Japan. *Journal of Geomagnetism and Geoelectricity* 45: 1383–1394.

Simon D (2002) Modeling of the field of gravity variations induced by the seasonal air mass warming during 1998–2000. *Bulletin d'Informations Marées Terrestres* 136: 10821–10836.

Sinha B and Pingree RD (1997) The principal lunar semidiurnal tide and its harmonics: Baseline solutions for M2 and M4 constituents on the North-West European continental shelf. *Continental Shelf Research* 17: 1321–1365.

Smylie DE (1992) The inner core translational triplet and the density near Earth's center. *Science* 255: 1678–1682.

Smylie D, Francis O, and Merriam J (2001) Beyond tides-Determination of core properties from superconducting gravimeter observations. *Journal of the Geodetic Society of Japan* 47(1): 364–372.

Spratt RS (1982) Modelling the effects of the atmospheric pressure variations on gravity. *Geophysical Journal of the Royal Astronomical Society* 71: 173–186.

Suda N, Nawa K, and Fukao Y (1998) Earth's background free oscillations. *Science* 279: 2089–2091.

Sun H-P (1995) Static deformation and gravity changes at the Earth's surface due to atmospheric pressure, PhD thesis, Obs. Roy. de Belgique, Bruxelles.

Sun H-P, Hsu H-T, Jentzsch G, and Xu J-Q (2002) Tidal gravity observations obtained with a superconducting gravimeter at Wuhan/China and its application to geodynamics. *Journal of Geodynamics* 33: 187–198.

Sun H-P, Jentzsch G, Xu J-Q, Hsu H-Z, Chen X-D, and Zhou J-C (2004) Earth's free core nutation determined using C032 superconducting gravimeter at station Wuhan/China. *Journal of Geodynamics* 38: 451–460.

Svenson S and Wahr J (2002) Estimated effects of the vertical structure of atmospheric mass on the time-variable geoid. *Journal of Geophysical Research* 107(B9): 2194 (doi:10.1029/2000JB000024).

Takemoto S, Fukuda Y, Higashi T, *et al.* (2002) Effect of groundwater changes on SG observations in Kyoto and Bandung. *Bulletin d'Informations Marées Terrestres* 135: 10839–10844.

Tamura Y (1987) A harmonic development of the tide generating potential. *Bulletin d'Informations des Marées Terrestres* 99: 6813–6855.

Tamura Y (1993) *Additional terms to the tidal harmonic tables, Proceedings of 12th International Symposium. Earth Tides, Beijing 1993*, pp. 345–350, Beijing/New York: Science Press.

Tamura Y, Sato T, Ooe M, and Ishiguro M (1991) A procedure for tidal analysis with a Bayesian information criterion. *Geophysical Journal International* 104: 507–516.

Tanimoto T (2005) The oceanic excitation hypothesis for the continuous oscillations of the Earth. *Geophysical Journal International* 160: 276–288.

Tanimoto T, Um J, Nishida K, and Kobayashi N (1998) Earth's continuous oscillations observed on seismically quiet days. *Geophysical Research Letters* 25: 1553–1556.

Tanimoto T and Um J (1999) Cause of continuous oscillations of the Earth. *Journal of Geophysical Research* 104: 28723–28739.

Tsubokawa T (1991) Absolute and superconducting gravimetry in Japan, Proceedings of 2nd Workshop: Non Tidal Gravity Changes Intercomparison Between Absolute and Superconducting Gravimeters, Conseil de l'Europe, *Cahiers du Centre Europeen de Geodynamique et de Seismologie* 11, Luxembourg, pp. 47–71.

Van Camp M (1995) Noise induced by the refrigerating device of a superconducting gravimeter in the seismological station of Membach (Belgium). *Bulletin d'Informations des Marées Terrestres* 123: 9302–9314.

Van Camp M (1999) Measuring seismic normal modes with the GWR C021 superconducting gravimeter. *Physics of the Earth and Planetary Interiors* 116: 81–92.

Van Camp M and Vauterin P (2005) Tsoft: graphical and interactive software for the analysis of time series and Earth tides. *Computers in Geosciences* 31(5): 631–640.

Van Camp M, Wenzel H-G, Schott P, Vauterin P, and Francis O (2000) Accurate transfer function determination for superconducting gravimeters. *Geophysical Research Letters* 27: 1, 37–40.

Van Camp M, Camelbeeck T, and Francis O (2002) Crustal motions across the Ardenne and the Roer Graben (north western Europe) using absoluite gravity measurements. *Metrologia* 39: 503–508.

Van Camp M, Williams DP, and Francis O (2005) Uncertainty of absolute gravity measurements. *Journal of Geophysical Research* 110: B05406 (doi:10.1029/2004JB003497).

Van Dam TM and Wahr JM (1987) Displacements of the Earth's surface due to atmospheric loading: Effects on gravity and baseline measurements. *Journal of Geophysical Research* 92(82): 1281–1286.

Van Dam T, Wahr J, Milly P, *et al.* (2001a) Crustal displacements due to continental water storage. *Geophysical Research Letters 28,* 4: 651–654.

Van Dam T, Wahr J, Milly P, and Francis O (2001b) Gravity changes due to continental water storage. *Journal of Geodetic Society of Japan* 47(1): 249–1254.

Van Ruymbeke M (1989) A calibration system for gravimeters using a sinusoidal acceleration resulting from a periodic movement. *Bulletin Géodesique* 63: 223–235.

Venedikov A and Viera R (2004) Guidebook for the practical use of the computer program VAV – Version 2003. *Bulletin d'Informations des Marées Terrestres* 121: 11037–11103.

Virtanen H (2001) Hydrological studies at the gravity station Metsahovi, Finland. *Journal of Geodetic Society of Japan* 47(1): 328–333.

Virtanen H and Mäkinen J (2003) The effect of the Baltic Sea level on gravity at the Metsähovi station. *Journal of Geodynamics, 35,* 4–5: 553–565.

Warburton RJ and Goodkind JM (1977) The influence of barometric-pressure variations on gravity. *Geophysical Journal of the Royal Astronomical Society* 48: 281–292.

Warburton RJ and Goodkind JM (1978) Detailed gravity-tide spectrum between one and four cycles per day. *Geophysical Journal of the Royal Astronomical Society* 52: 117–136.

Warburton RJ and Brinton EW (1995) Recent developments in GWR Instruments' superconducting gravimeters, Proc. 2nd Workshop: Non-tidal gravity changes Intercomparison between absolute and superconducting gravimeters, Cahiers du Centre Européen de Géodynamique et de Séismologie, Luxembourg, 11, pp. 3–56.

Warburton RJ, Beaumont C, and Goodkind JM (1975) The effect of ocean tide loading on tides of the solid earth observed with the superconducting gravimeter. *Geophysical Journal of the Royal Astronomical Society* 43: 707–720.

Warburton RJ, Brinton EW, Reineman R, and Richter B (2000) Remote operation of superconducting gravimeters, Proceeding of the Workshop: High-Precision Gravity Measurements with Application to Geodynamics and Second GGP Workshop, Cahiers du Centre Européen de Géodynamique et de Séismologie, 17, Luxembourg, pp. 125–136.

Wahr JM (1985) Deformation induced by polar motion. *Journal of Geophysical Research* 90(B11): 9363–9368.

Wahr J, Han D, and Trupin A (1995) Predictions of vertical uplift caused by changing polar ice volumes on a viscoelastic Earth. *Geophysical Research Letters* 22: 977–980.

Wahr J, Molenaar M, and Bryan F (1998) Time variability of the Earth's gravity field: hydrological and oceanic effects and their possible detection using GRACE. *Journal of Geophysical Research* 103(B12): 30205–30229.

Wang R (1997) In: Helmut W, Zürn W, and Wenzel H-G (eds.) Tidal response of the Earth, in Tidal Phenomena, Lecture Notes in Earth Sciences, pp. 27–57. Berlin: Springer.

Wenzel H-G (1994) Accurate instrumental phase lag determination for feedback gravimeters. *Bulletin d'Informations des Marées Terrestres* 118: 8735–8751.

Wenzel H-G (1995) Accurate Instrumental Phase Lag Determination for Feedback Gravimeters. In: Hsu HT (ed.) *Proceedings of the 12th International Symposium. Earth Tides*, pp. 191–198. Beijing, New York: Science Press.

Wenzel H-G (1996a) Accuracy assessment for tidal potential catalogues. *Bulletin d'Informations des Marées Terrestres* 124: 9394–9416.

Wenzel H-G (1996b) The nanogal software: Earth tide processing package ETERNA 3.30. *Bulletin d'Informations des Marées Terrestres* 124: 9425–9439.

Widmer-Schnidrig R (2003) What can superconducting gravimeters contribute to normal mode seismology. *Bulletin of the Seismological Society of America* 93(3): 1370–1380.

Wilmes H, Boer A, Richter B, *et al.* (2006) A new data series observed with the remote superconducting gravimeter GWR R038 at the geodetic fundamental station TIGO in Concepción (Chile). *Journal of Geodynamics* 41: 5–13.

Xi Q-W (1987) A new complete development of the tide generating potential for the epoch J200.0. *Bulletin d'Informations des Marées Terrestres* 99: 6766–6681.

Xi Q-W (1989) The precision of the development of the tidal generating potential and some explanatory notes. *Bulletin d'Informations des Marées Terrestres* 105: 7396–7404.

Xu J-Q, Sun H-P, and Ducarme B (2004a) A global experimental model for gravity tides of the Earth. *Journal of Geodynamics* 38: 293–306.

Xu J-Q, Sun H-P, and Yang X-F (2004b) A study of gravity variations caused by polar motion using superconducting gravimeter data from the GGP network. *Journal of Geodesy* 78: 201–209 (doi:10.1007/s00190-004-0386-1).

Zerbini S, Richter B, Negusini M, *et al.* (2001) Height and gravity various by continuous GPS, gravity and environmental parameter observations in the southern Po Plain, near Bologna, Italy. *Earth and Planetary Science Letters* 192: 267–279.

Zerbini S, Negusini M, Romagnoli C, Domenichini F, Richter B, and Simon D (2002) Multi-parameter continuous observations to detect ground deformation and to study environmental variability impacts. *Global and Planetary Changes* 37–58.

Relevant Websites

http://www.eas.slu.edu – GGP Home Page, Department of Earth & Atmospheric Sciences, Saint Louis University.

http://ggp.gfz-potsdam.de – Global Geodynamic Project Information System and Data Center.

5 Gravimetric Methods – Spacecraft Altimeter Measurements

D. P. Chambers, The University of Texas at Austin, Austin, TX, USA

Nomenclature

a	semi-major axis of satellite orbit
c	speed of light in a vacuum
f	radar altimeter frequency
i	inclination of satellite orbit
t	time
CM	center of mass
E	integrated density of free electrons along altimeter path
H_{SAT}	normal height of satellite above a reference ellipsoid
J_2	second zonal harmonic of the earth's gravity field
P	satellite orbit period
P_R	exact repeat period
R	true satellite range (distance between satellite altimeter and ocean surface)
R_E	earth's equatorial radius
R_P	earth polar radius
R_{meas}	raw range measured by an altimeter, including all biases and errors
λ	longitude
μ_E	gravitational constant of the earth
τ_{tide}	alias period of ocean tide
ϕ	latitude
ω_E	rotation rate of earth-fixed reference frame about inertial reference frame
Δx	spacing between adjacent ascending or descending passes on a satellite ground track

ΔR	corrections to range to remove biases and errors	ΔSSH_{IB}	inverted barometer correction to SSH
ΔR_{dry}	correction to range due to dry gases in atmosphere	$\dot{\Omega}$	rate at which ascending node of satellite orbit moves relative to inertial axis
ΔR_{iono}	correction to range due to free electrons in the ionosphere		

Glossary

crossover point The geographical point where an ascending and descending ground track pass cross.

dynamic ocean topography DOT; also known as sea surface topography and dynamic topography. The elevation of the sea surface height above the geoid, caused by surface geostrophic currents. Variations range from ± 3 m.

exact-repeat orbit A special type of orbit used for satellite altimetry. The ground track will repeat its pattern on the Earth's surface after a certain period of time.

geoid The equipotential surface that corresponds to mean sea level of an ocean at rest which includes both the gravitational potential and the rotational potential.

geoid height The height of the geoid above a reference ellipsoid.

ground track Trace of the satellite nadir point on the surface of the Earth as the satellite orbits and the Earth rotates under it.

ground track pass One segment of the ground track, either moving south to north (an ascending pass) or north to south (descending pass). Passes are ordered sequentially in time, starting from the beginning of the repeat period (see repeat period). Altimeter data are normally distributed as pass files.

mean sea surface MSS; the time-average of the sea surface height. This can be computed for a single satellite along a particular ground track pass, or as a grid from a combination of multiple satellites.

nadir point Point where the vector from the satellite normal to the geoid intercepts the geoid.

nodal day Time required for the Earth to make one rotation relative to the satellite orbit's line of nodes.

reference ellipsoid First-order approximation of the geoid, which models the dominant equatorial bulge. Generally defined by a equatorial radius and a flattening coefficient. SSH is computed relative to a reference ellipsoid.

repeat period Amount of time before a ground track repeats itself (see exact-repeat orbit). The stated repeat period is normally given as the closest integer number of days, instead of the exact time; for example, a 17.0505-day repeat is typically referred to as a 17-day repeat period.

sea state bias SSB; a bias in altimeter range measurement that is related to surface waves and the surface wind state.

sea surface height SSH; the elevation of the sea surface above a reference ellipsoid ($= H_{SAT} - R$). Variations range from ± 100 m.

sea surface height anomaly SSHA; deviation of SSH away from the time-average mean surface. Variations range from ±40 cm in most of the ocean to ±1 m for eddies.

significant wave height SWH; average height of the 1/3 highest waves in the field of view.

sun-synchronous orbit A special type of exact-repeat orbit where the orbital node rate equals the rate of revolution of the Earth around the sun. This means that the angle between the orbital plane and the vector between the Earth and the Sun is constant.

5.1   Introduction

Satellite altimetry was first proposed in the 1960s as a way to obtain global, synoptic measurements of the sea surface. Previously, one could measure variations in the sea surface regularly in time from tide gauges, but only along coastlines. In the deep ocean, one could measure water properties such as temperature, salinity, and currents from a ship, but it took many months to obtain the data along a single transect, and the transect was often not repeated for years. There were also many regions of the ocean that were never observed, due to the difficulty in reaching all the vast areas of the oceans with surface vessels.

Satellites, on the other hand, orbit the Earth approximately once every 90 min. Because the Earth rotates beneath them, the satellite's path traces out a ground track so that after a matter of days it has covered a significant percentage of the Earth's surface (**Figure 1**). Thus, a satellite is the perfect platform for an instrument to measure the Earth's sea surface height (SSH), or sea level (SL). The measurement can be made at regular intervals along the ground track – a 1 s interval is common which corresponds to a spacing of about 6 km because of the satellite's motion. The spacing between adjacent ground tracks can also be adjusted by changing the period needed for the ground track to repeat (Section 5.3.1). Longer repeat periods lead to smaller distance between adjacent ground tracks at the equator. Thus, a satellite's orbit parameters can be adjusted to give a ground track that provides whatever repeat interval or ground track spacing is desired. In essence, an altimeter can be considered as a global distribution of tide gauges that sample the SL regularly at intervals of several days instead of every minute or hour. One sacrifices the high temporal sampling of the tide gauge for a much denser spatial sampling.

Measuring SSH from space is conceptually simple. Imagine a satellite orbiting the Earth and that its position in some well-defined reference frame is known accurately at any time. A vector from the center of mass (CM) of the satellite normal to a mathematical model of the Earth's surface defines the location of the subsatellite point longitude (λ) and latitude (ϕ) at some time, t, as well as the height of the satellite above that surface along the vector (H_{SAT}). For altimetry, this surface is generally modeled as an ellipsoid and not a sphere to account for the Earth's bulge at the equator. A commonly used ellipsoid is the one defined for the TOPEX/Poseidon (T/P) mission: equatorial radius = 6 378 136.3 m, flattening = 1/298.257. Then, if the satellite measures the distance between it and the ocean surface below (R) via a ranging instrument, or altimeter, along this same nadir vector, then one can determine the SSH above the reference ellipsoid as a scalar quantity

$$\text{SSH}(\phi, \lambda, t) = H_{SAT}(\phi, \lambda, t) - R(\phi, \lambda, t) \qquad [1]$$

In practice the measurement is not so simple. The orbit of the satellite is not constant, but changes due to perturbations from drag, the Earth's nonuniform gravity, solar and Earth radiation pressure, and even the ocean tides. In order to determine H_{SAT} accurately, the satellite has to be continuously tracked, and its orbit estimated from the tracking data (e.g., Tapley *et al.*, 2004a). Determining the range

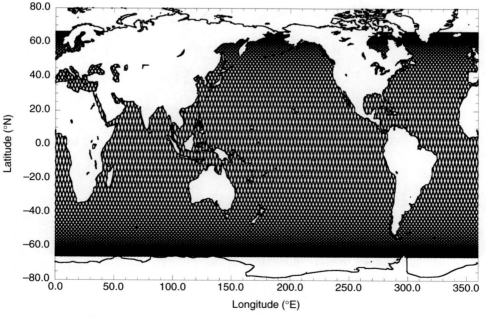

Figure 1 Ground track for TOPEX/Poseidon.

accurately is also not trivial. A radar altimeter consists of an antenna that both transmits a microwave pulse of known power and receives the partial power pulse reflected back from the ocean surface. To first order, the measured range is computed as

$$R_{\text{meas}}(\phi, \lambda, t) = c\frac{\Delta t}{2} \qquad [2]$$

where Δt is the two-way travel time from initial transmission of the pulse to the arrival of the reflection, and c is the speed of light in a vacuum. The travel time of the pulse needs to be known to better than 10^{-10} s in order to ensure a range precision of 3 cm for a typical satellite altitude. Additionally, because the microwave pulse is traveling through the atmosphere and not a vacuum, corrections for attenuation need to be applied. The effective footprint of the pulse at the ocean surface is also of the order of 10 km, and the energy is reflected differently by wave crests than from troughs, which will bias the range estimate. Thus, the true range is

$$R(\phi, \lambda, t) = R_{\text{meas}}(\phi, \lambda, t) + \sum_i \Delta R(\phi, \lambda, t) \qquad [3]$$

where ΔR is a set of corrections to account for the path delay and the surface scattering effects. Ignoring any of these corrections can lead to errors of the order of tens of centimeter in range or more. Finally, the orbit height is defined to the center of mass (CM) of the satellite, while the range is measured relative to the phase center of the radar antenna, which is attached to the body of the satellite some distance from the CM. In order to recover SSH as accurately as possible, the vector from the antenna center to the CM needs to be known precisely, including any change caused by burning fuel to maintain the orbit.

Since the original development of the satellite altimeter concept, several missions have been flown, each progressively returning more accurate measurements of SSH as experience has been gained. The GEOS-3 mission was launched by the National Aeronautics and Space Administration (NASA) in April 1973, with an altimeter that measured range with a 25 cm precision for 1-s averages (Stanley, 1979). The initial orbit accuracy was about 5 m for the radial component (H_{SAT}). Although this meant that GEOS-3 could not observe the smallest variations in SSH, it demonstrated the capability and provided initial information on the approximately ± 100 m deviations of SSH relative to the ellipsoid. The satellite continued to operate for nearly 4 years.

The Seasat mission was launched in July 1978 (also by NASA) with a significantly more precise altimeter (5 cm) as well as instruments to measure some of the atmosphere path delays (Lame and Born, 1982). The orbit determination had also improved, although initial orbits (and SSH) still had errors of 1 m or more. Although it failed in 3 months due to a short circuit in the power system, Seasat demonstrated the usefulness of altimetry for applications beyond mapping the mean shape of the SSH.

Based on the hint of groundbreaking science suggested by Seasat, NASA began developing another altimeter mission. It was soon combined with a similar mission proposed by the French space agency, Centre Nationale d'Etudes Spatiales (CNES) and became known as TOPEX/Poseidon (the combination of the two original mission names). The goals of the T/P mission were to improve the altimeter range measurement (including both precision and corrections) as well as the orbit accuracy in order to reach a goal for SSH accuracy of 13 cm (Stewart *et al.*, 1986).

The importance of having altimeter data for study before the launch of the new mission was recognized, especially with only 3 months of Seasat data. During the T/P development phase, the US Navy launched the Geosat altimeter mission on 12 March 1985, with a primary goal to map the marine geoid at very small wavelengths (McConathy and Kilgus, 1987). Several groups, especially those headed by Jim Mitchell and Bob Cheney, convinced the Navy to maneuver Geosat from its classified orbit into the unclassified Seasat orbit at the end of its primary mission (Born *et al.*, 1987). It operated until December 1989 and provided a wealth of data for planning the new mission – most notably in efforts to improve orbit determination.

As noted, the goal of the T/P mission was to measure SSH to an accuracy of 13 cm root mean square (RMS), which was nearly an order of magnitude better than that had been accomplished up to that time. To meet this ambitious goal, the satellite was instrumented with more science equipment than any previous altimeter mission. This included two separate altimeters (one dual frequency) which could share the single antenna, a microwave radiometer to measure the delay due to water vapor along the path, and three independent tracking systems: a reflector for satellite laser ranging (SLR), a doppler orbitography and radiopositioning integrated by satellite (DORIS) tracking receiver, and an experimental global positioning system (GPS) receiver (Fu *et al.*, 1994). The primary altimeter (TOPEX) was based on the Seasat and Geosat altimeters with extensive modifications. These included adding another frequency to estimate ionosphere path delay directly (Section 5.2.1), increased precision ($\sim$2 cm RMS), and redundant circuits in order to prolong life (i.e., the primary TOPEX_A and a backup TOPEX_B altimeter). The secondary altimeter (POSEIDON) was an experimental, solid-state altimeter that operated approximately 10% of the time in order to test the applicability of a low-power, low-mass altimeter for future missions.

One of the more important aspects of the T/P mission is often overlooked. A Science Working Team (SWT) was created 5 years before the anticipated launch date, and tasked with the development of algorithms and models to meet the SSH accuracy goal. The team far exceeded the aim, and within 2 years of the launch in August 1992, SSH accuracy was estimated to be better than 4 cm RMS (Fu *et al.*, 1994). One of the main reasons for the improved accuracy was better precise orbit determination (POD), mainly due to improvement in Earth gravity field models (Nerem *et al.*, 1994a; Tapley *et al.*, 1996).

At the same time NASA and CNES were developing the T/P mission, the European Space Agency (ESA) was developing its first Earth Remote Sensing (ERS-1) satellite. Although ERS-1 included a radar altimeter, the goals of the mission were significantly different than T/P. T/P was designed as a dedicated radar altimeter mission in order to obtain the most accurate measurement of SSH possible for scientific applications. The ERS-1 altimeter, on the other hand, was just one of many remote sensing instruments on the platform. This led to compromises in the mission due to requirements of the other instruments, mainly that the satellite was in a sun-synchronous orbit at a relatively low altitude compared to T/P, so drag forces were much higher. In addition, one of the tracking systems (PRARE) failed shortly after launch in July 1991, so orbit accuracy was significantly poorer than what was available for T/P. In addition, the ERS-1 altimeter operated at only a single frequency, so could not directly measure the path delay in the ionosphere.

Although ERS-1 was launched over a year before T/P, data from its altimeter were not used widely until years later, unlike the data from the T/P mission, which many scientists embraced early on. This was partly due to the ease of use of the T/P data products, which were designed to be used by a wide variety of scientists without having to be an expert in radar altimetry. The ERS-1 data products, on the other hand, required more expertise to use. In addition, the range and orbits for the early T/P mission were significantly more accurate than those for ERS-1, and so were more useful at studying much smaller SSH signals. Finally, updates to important algorithms to improve the accuracy of T/P were incorporated directly into the processing stream and data products were re-issued. ERS-1 algorithms were essentially fixed, and users had to collect algorithm changes from various published (and unpublished) sources and correct the data themselves (Scharroo, 2002).

Since the launches of ERS-1 and T/P, ESA has launched ERS-2 in April 1995 and Envisat in March 2002. ERS-2 carried essentially the same single-frequency altimeter as ERS-1, but the PRARE tracking system worked so that orbits were slightly better. Envisat carries both an improved two-frequency radar altimeter as well as a DORIS tracking receiver. Between the launches of ERS-1 and ERS-2, the US Navy launched a Geosat follow-on (GFO) mission into the Seasat/Geosat ERM ground track in February 1998. The GFO altimeter was again only a single-frequency instrument, but unlike Geosat, the satellite had an SLR reflector array and a GPS receiver. Unfortunately, the GPS receiver failed early in the mission, and hence orbit accuracy has suffered. NASA and CNES launched a follow-on mission to T/P in December 2001: Jason-1. Jason-1 was designed to provide as accurate an SSH measurement as T/P in the same orbit, but using less power from a solid-state altimeter derived from the POSEIDON instrument.

It is not the intent of this chapter to provide every detail of how a radar altimeter works, how orbits are determined precisely, or to document all the science that has been accomplished with every altimeter. Such detail is beyond the scope of a single chapter. For more details, the reader is pointed to the excellent book edited by Fu and Cazenave (2001), *Satellite Altimetry and Earth Sciences*, especially the first chapter by Chelton *et al.* (2001). Instead, we will summarize the important corrections necessary to obtain the most accurate altimeter range measurement, provide some background on types of altimeter orbits and how they are changed depending on the application, how the measurement is calibrated and validated, and how the data are used to study tides, SL change, and the shape of the geoid. More importantly, we will comment on various corrections that are needed for the raw T/P and Jason-1 altimeter data that have been discussed extensively in Science Team meetings and mentioned in some literature, but have not been summarized completely. Although these corrections are applied to analysis products distributed by NASA, CNES, and other groups, as well as some releases of the geophysical data records (GDRs), there are still some versions of the GDRs that have none or only a few of these corrections. Thus, we believe it is important to summarize them in one location. We also tend to examine results of the T/P and Jason-1 missions, as these data are more widely used by scientists and the author is more familiar with them.

5.1.1   Altimeter Jargon

Before proceeding, it is worthwhile to note some jargon that is used in altimetry. Since these terms are often used in the current literature without definition, it is important to understand what they mean, and will enable an easier discussion.

The 'nadir point' of the orbiting satellite is the point where the vector from the satellite normal to the reference ellipsoid intercepts the ellipsoid. The nadir point will trace out a 'ground track' on the surface of the Earth (**Figure 1**). Satellite altimeter missions are often in a special orbit that will repeat the ground track after a certain amount of time, the 'repeat period'. This type of orbit is known as an 'exact-repeat orbit'. A particular repeat period is often referred to as a 'cycle'. Thus, Cycle 1 will refer to the first repeat period (or cycle) of the mission, while Cycle 100 will refer to the 100th repeat period. Different mission will have different repeat periods. For example, Geosat had a 17-day repeat period (exactly 17.0505 days), while T/P and Jason-1 have 10-day repeats (9.915625 days). See Section 5.3.1 for more details on how exact-repeat orbits are determined.

In satellite altimetry, the data are often distributed as files along a particular ground track, called a 'pass'. The numbering of passes is determined by the time the pass crosses the equator relative to the time of the start of the repeat cycle. The first pass crosses the equator immediately after the start of the cycle, while the last pass crosses the equator just before the start of the next cycle. Passes are coincident in time, but not in location. Pass 2 will never be adjacent geographically to Pass 1, only adjacent in time. Passes are also often referenced by the direction the satellite is moving. For an 'ascending pass', the satellite is moving from south to north as it crosses the equatorial plane (**Figure 2**). A satellite is moving north to south on a 'descending pass'. The geographical point where an

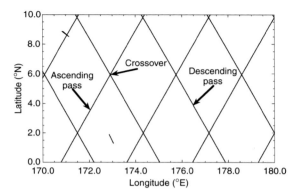

Figure 2   Crossover point for T/P, showing ascending and descending passes.

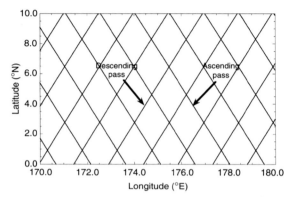

Figure 3   Crossover point for Geosat, showing ascending and descending passes.

ascending and descending pass cross is referred to as a 'crossover' point (**Figure 2**). One can determine crossovers from a single satellite ground track (called a 'single-satellite crossover'), or between ground tracks of different altimeter satellites (called a 'dual-satellite crossover').

Note that ascending passes do not always slope south-to-north from west-to-east, while descending passes slope north-to-south from west-to-east. This pattern is only valid for satellites with an inclination less than 90°. If the inclination is greater than 90°, then ascending passes slope south-to-north from east-to-west, and similarly for descending passes (**Figure 3**).

5.2   Measuring Range

A radar altimeter consists of a radio transmitter that emits very sharp pulses of microwave energy, a receiver to record the pulse after it has been reflected from the ocean surface, and a clock to record the time between transmission and reception. Several thousand pulses are transmitted each second. Many of these pulses must be averaged together to obtain a smooth estimate of the pulse shape because the instantaneous pulse is the superposition of the scatter from many wave facets on the sea surface. In reality, the processing of the altimeter signal is far more complicated than this, as the pulse shape is changed because of the reflection and becomes more of a stretched waveform, and this is what the time of transit is estimated from. The precision of the range will also be dependent on how many waveforms are averaged before determining the transit time. For this discussion, we will assume that a two-way travel time has been determined from the waveforms at a 1-s (1 Hz) spacing, which is typical for radar altimeters. For more details on how the transit time is actually calculated, the reader is directed to Chelton *et al.* (2001).

As discussed before (eqn [2]), the raw range measurement is then calculated from the two-way transit time assuming that the speed of light is constant. However, this is only true in a vacuum, and moving through a medium like the atmosphere will slow down the electromagnetic wave. Thus, the raw range measurement will be too long, with the error depending on the properties of the atmosphere. It must be corrected in order to determine an accurate range. In altimetry, the corrections are given in terms of the negative path delay that need to be added to the range in order to determine the correct value (eqn [3]).

5.2.1 Ionosphere Path Delay

Free electrons in the Earth's ionosphere will refract the radar pulse and delay the path, ΔR_{iono}, by an amount proportional to the density of free electrons along the path (E) and inversely proportional to the square of the frequency (f) of the microwave pulse:

$$\Delta R_{iono}(f) = -\frac{40.25E}{f^2} \qquad [4]$$

where E is the integrated density of free electrons (electrons m^{-2}) along the path and f is the radar frequency in hertz. The units of the path delay will be in meters, and the negative sign is following the convention set by T/P that all path delay corrections are added to the raw range. Thus, the effect of the correction is to shorten the raw range measurement.

The electron count varies over the day (more free electrons in the day), over the seasons (more free electrons in the winter), over the 11-year solar cycle (more during the solar maximum), and meridionally (maximum at the magnetic equator). Typical values of E range from 10^{16} to 10^{18} electrons m^{-2}, which implies, for a 13 GHz altimeter, errors are in the range of 2–20 cm. If E is known precisely, this range error could be removed completely. However, before T/P, altimeters relied on models of the electron density to determine the path delay.

One way to determine the path correction directly is to operate a two-frequency altimeter. The path delay will be different for each frequency band, as will the raw range measurement, but the corrected range measurements should be identical, that is,

$$R = R_1 + \Delta R(f_1)$$
$$R = R_2 + \Delta R(f_2) \qquad [5]$$

where the subscript 1 denotes the primary frequency for science, and 2 denotes the secondary frequency. If these two equations are differenced, and the path delay equation (eqn [4]) is substituted, then

$$R_1 - R_2 = -40.25E\left[\frac{1}{f_2^2} - \frac{1}{f_1^2}\right]$$
$$R_1 - R_2 = -40.25E\left[\frac{f_1^2 - f_2^2}{f_1^2 f_2^2}\right] \qquad [6]$$
$$\Delta R(f_1) = \frac{f_2^2}{f_1^2 - f_2^2}(R_1 - R_2)$$

Thus, given the measured ranges at the two known frequency bands, one can obtain the ionosphere path delay correction directly. Generally, the only correction applied to the raw ranges before combining is the sea state bias (SSB) (Section 5.2.4), as this is the only other frequency-dependent range correction.

For T/P and Jason-1, the primary science frequency is in the Ku-band (13.6 GHz), while the secondary band is in the C-band (5.3 GHz). The reason for this is that the Ku-band is affected less by the ionosphere than the C-band (eqn [4]) for a particular electron density, and the instrumental errors for the Ku-band are less (Chelton *et al.*, 2001). The only other dual-frequency altimeter that has operated is Envisat, using Ku-band (13.575 GHz) and S-band (3.2 GHz) frequencies. The estimated error in the ionosphere delay correction derived from the dual-frequency TOPEX has been estimated to be less than 1 cm RMS (Imel, 1994; Zlotnicki, 1994). The Jason-1 correction agrees with that of TOPEX to 1.3 cm

RMS, indicating an accuracy comparable to TOPEX or even slightly better (Chambers *et al.*, 2003a).

Single-frequency altimeters rely on models of the ionosphere electron density and eqn [4] in order to estimate a correction. Original models for Seasat and Geosat had errors in the electron density as large as 50% (Cheney *et al.*, 1991), but subsequent models (e.g., Bilitza, 1997; Daniell *et al.*, 1995) have reduced the errors to 20% or less. However, this is still equivalent to path delay errors as large as 3 cm RMS and these models will never match the accuracy of the correction from a dual-frequency altimeter.

Although the POSEIDON altimeter is also a single-frequency altimeter and cannot measure the electron content directly, T/P carries a DORIS tracking receiver for POD (see Section 5.3.3). DORIS beacons at approximately 50 locations around the world transmit at two frequencies (0.4 and 2.0 GHz) in order to correct for the ionosphere path delay in the measurement between the satellite and the station. This allows for the calculation of electron content along a path between T/P and the DORIS ground stations. In order to be used by an altimeter, though, the slant measurement has to be interpolated to the radar beam path, generally by using an empirical model (Fleury *et al.*, 1991). Although the results comparing the DORIS ionosphere corrections with those obtained from TOPEX for several periods when the two operated within the same 10-day cycle indicate that the DORIS model is better than the empirical models, the global RMS error is still about 2 cm (Imel, 1994; Zlotnicki, 1994). More troubling (Zlotnicki, 1994) showed that the signal was correlated over very large distances, indicating long-wavelength errors.

Another method for determining total electron content (TEC) is from analyzing the two frequencies transmitted by GPS satellites and detected by ground receivers (e.g., Mannucci *et al.*, 1999). As with the determination from DORIS, this technique will only determine TEC accurately for vectors from ground stations to the GPS satellites. To produce global maps, the data have to be interpolated. Comparisons with ionosphere path delay derived from these TEC maps with those derived from the dual-frequency TOPEX altimeter indicate that while differences near GPS ground stations are small (less than 1 cm RMS), the difference away from ground stations can be up to 10 times larger (e.g., Imel, 1994; Schreiner *et al.*, 1997). Thus, although models of global electron density are improving, none can

meet the accuracy of 1 cm RMS that can be achieved with a dual-frequency altimeter.

5.2.2   Dry Troposphere Path Delay

The radar pulse is also slowed by gases in the Earth's troposphere. The amount of 'dry' gases is nearly constant, and produces height errors with a magnitude of about 2 m. However, this path delay does not change appreciably over time. The path delay (ΔR_{dry}) can be related to the sea level pressure (SLP) (P_0) and latitude (e.g., Chelton *et al.*, 2001) as

$$\Delta R_{dry} \approx -0.2277 P_0 (1 + 0.0026 \cos 2\phi) \qquad [7]$$

which has units of cm if P_0 is in mbars.

The only practical way to obtain global SLP data is from numerical weather prediction models that assimilate available SLP measurements such as those from the National Center for Environmental Prediction (NCEP) or the European Center for Medium-range Weather Forecasting (ECMWF). The data from the ECMWF model is used in most modern altimeters such as T/P, Jason-1, and Envisat, as its SLP data are widely considered more accurate than those from NCEP (e.g., Trenberth and Olson, 1988). Although SLP pressures are typically high (about 950–1050 mbars), meaning that the total range correction is about 2.3 m, the correction is not very sensitive to errors in SLP. For example, a 1 mbar error in SLP is equivalent to a 2 mm error in the path delay.

An approximate measure of the SLP error is a comparison of the output of the ECMWF and NCEP models, although this will not account for similar errors that cancel. Typical RMS differences are 2–3 mbars equatorward of 40° latitude, and 4–7 mbars poleward of 40° latitude (Chelton *et al.*, 2001), which suggests errors in the path delay of between 5 mm RMS in the tropics and sometimes larger than 1 cm at higher latitudes.

5.2.3   Wet Troposphere Path Delay

The path delay caused by both water vapor and cloud liquid water can be determined using a passive microwave radiometer. As mentioned earlier, Geosat did not carry a microwave radiometer, so had to rely on data from models or other satellite radiometers. The wet troposphere correction for Geosat was originally based on the Fleet Numerical Oceanography Center (FNOC) operational analysis of humidity.

This was shown to significantly degrade Geosat ranges in the tropics where the FNOC model consistently underestimated water vapor (Emery et al., 1990; Zimbelman and Busalacchi, 1990). Later updates to the data used a correction based on interpolating water vapor measurements determined from other satellites (SSMI and TOVS) to the Geosat ground track (Cheney et al., 1991), but the error was still of the order of 4 cm RMS (Shum et al., 1993).

A two-channel radiometer, operating at two frequencies near 20 GHz, is the minimum required to determine the path delay. The two frequencies are necessary to remove factors affecting the microwave brightness temperatures other than water vapor. However, it has been shown that operating a downward-looking satellite radiometer with a third frequency allows one to correct for effects of wind-induced changes to sea-water emissivity (e.g., Keihm et al., 1995). Both the TOPEX Microwave Radiometer and the Jason-1 Microwave radiometer operate at three frequencies: 18.0, 21.0, and 37.0 GHz. Overall, the estimated accuracy of the T/P path delays are estimated to be about 1 cm (Ruf et al., 1994) with estimates for Jason-1 being even lower (e.g., Brown et al., 2004). Keihm et al. (1995) estimated that the error in wet troposphere path delay would be about 1–2 cm higher if the 37 GHz channel were not used, which is consistent with the finding for the ERS-1 and ERS-2 missions which use a dual-frequency radiometer at 23.8 and 36.5 GHz. Eymard et al. (1996) estimated that the error in path delay for ERS-1 was 2 cm.

One problem with microwave radiometers is that they have historically suffered from calibration drifts. Although there was considerable work undertaken to minimize such drifts in the TMR, a drift became apparent in the 18.0 GHz channel over the course of the mission. This effect was a nearly linear drift in path delay of approximately 1.5 mm yr^{-1} over the first four years of the mission (Keihm et al., 2000; Scharroo et al., 2004), and was first observed in the comparison of SSH with tide gauges (Chambers et al., 1998a; Mitchum, 1998; see also Section 5.4.2). The sense of the error was to make the measured SSH lower than it actually was. More recent analysis suggests that the drift lasted for about 6.5 years before leveling off (Scharroo et al., 2004). We should note that not every T/P GDR or data product that is available to the public has been corrected for this error. Therefore, the user needs to verify if this correction has been applied to T/P data, or corrected for it (e.g., Ruf, 2002; Scharroo et al., 2004).

There is also a smaller change in the TMR wet path delay that varies in time. It was not noticed until the TMR and JMR path delays were compared early in the T/P-Jason calibration phase (Chambers et al., 2003b). It was observed that the two path delays changed abruptly by nearly a centimeter at several periods during the 210-day period before returning to normal. It was quickly established that the error was in the TMR measurement, and in fact had existed for the entire T/P mission (Chambers et al., 2003a; Zlotnicki and Desai, 2004). The timing was eventually linked to transition from one yaw state to another and has also been shown to exist in the JMR path delays, but to a smaller extent (Brown et al., 2004; Desai and Haines, 2004; Obligas et al., 2004; Zlotnicki and Desai, 2004). The cause for the correlation with the transition from one yaw state to another was linked to higher than normal temperatures on the MR feedhorn that had not been simulated when determining the MR calibration coefficients (Brown et al., 2004). New calibration coefficients have been determined to remove this effect in the JMR data (Brown et al., 2004) and were used to generate the first-release Jason-1 GDRs. Although a similar calibration has not been applied to the TMR data in the T/P GDRs, a new calibration exists and will be applied to a planned re-processing (P. Callahan, personal communication, 2004).

In addition to the yaw-mode path delay biases, the JMR has also suffered two additional biases in the path delay. The first occurred in Nov. 2002, when the path delay correction became biased by about 3–4 mm (Desai and Haines, 2004; Zlotnicki and Desai, 2004). Approximately a year later, it became biased again by a further 7–8 mm in the same direction. The changes can be seen clearly in **Figure 4**, which shows the mean difference with path delay estimated from ECMWF water vapor averaged for each Jason-1 repeat cycle. The cause has been linked to thermal shock to the instrument, and the error will bias the SSH measurement globally by the same amount as the path delay bias. The original release Jason-1 GDRs will be affected by this problem, although re-processed GDRs are currently being made to correct for this.

The ERS and Envisat radiometers are also not immune to changing biases in the path delay. The ERS-2 radiometer suffered a far larger change in the 23.8 GHz channel than has been observed in either the TMR or JMR channels, resulting in a 10 K change in brightness temperature after 16 June 1996 (Scharroo et al., 2004). The Envisat radiometer also

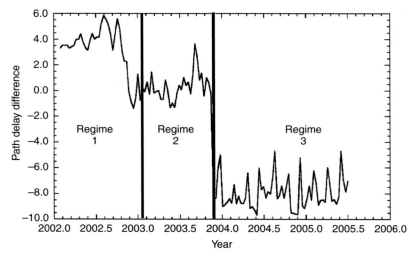

Figure 4 Path delay from JMR minus path delay estimated from ECMWF model stored on GDRs, averaged globally for each repeat cycle. Note that path delays are stored as negative values on GDRs, but that here we have differenced the positive magnitudes.

appears to have undergone a 1 K change in brightness temperature for both channels starting in 2003 (Scharroo *et al.*, 2004). Corrections for these drifts have to be applied by the user to the released data records (e.g., Scharroo *et al.*, 2004).

5.2.4　Sea State Bias

As mentioned briefly before, the range is also biased due to surface waves in the footprint of the radar pulse. The bias is composed of three components. The first, known as the electromagnetic bias (EMB), arises because wave troughs backscatter more power per unit area than wave crests. This creates a bias in the range toward the troughs and away from the mean sea level (MSL), which is desired. The second effect is known as the skewness and is caused by the fact that the onboard tracker determines the two-way travel time from the half-power point of the waveform that corresponds to the median scattering surface and not the mean scattering surface (Chelton *et al.*, 2001). The third component relates to how the waveforms are tracked and range is estimated, so is referred to as the tracker bias. Although these biases are distinctly different and related to different instrumental effects, they are difficult to separate completely as they are all related to the size of waves in the footprint. Several studies have attempted to determine a theoretical value for the EMB portion alone, based on assumptions about the wave number spectrum

(Srokosz, 1986; Rodriguez *et al.*, 1992; Glazman *et al.*, 1996). However, as more research has been conducted, it has become more and more apparent that the EMB also tends to be affected by the radar frequency as well as wind speed (Melville *et al.*, 1991; Walsh *et al.*, 1991), and so varies from one instrument to another. Because a part of the skewness and tracker biases are also linearly dependent on significant wave height (SWH) (e.g., Hayne and Hancock, 1982; Srokosz, 1986), any empirical estimate of the EMB from altimeter data will include the portion of the skewness and tracker bias that is related to SWH. Because of this, one usually considers the combined effect as the SSB. Although theoretical studies of the EMB portion continue (Elfouhaily *et al.*, 1999, 2000), they still cannot adequately explain the full SSB in altimetry. Therefore, in practice, the combined SSB is estimated empirically from each altimeter's data so as to reduce the variance of the SSH data from that altimeter.

SSB is expressed as a linear function of SWH, which is the average height of 1/3 highest waves in the field of view

$$SSB = -b\text{SWH} \qquad [8]$$

where b is determined empirically. Wave conditions cause changes in the slope of the leading edge of the radar altimeter waveform. High slopes indicate small waves, while shallow slopes indicate large waves, very nearly with a linear relationship. The power of the returned waveform is also related to the roughness of the surface, caused by the wind speed. Chelton et al. (2001) describe completely the history and process for determining SWH and wind speed from the waveforms. In this discussion, we will assume that a measure of SWH and wind speed is returned at the same time as altimeter range and SSH.

Determining an empirical estimate of SSB is relatively straightforward. First, one determines SSH with all corrections except SSB at a particular location along a ground track (x) and at a particular time (t), which to first order should be the true SSH (SSH_{true}) corrupted by the SSB (bSWH), or

$$SSH(x, t) = SSH_{true}(x, t) + bSWH(x, t) \quad [9]$$

If the SSH was measured in the same location again some Δt later, then the residual (ΔSSH) will be

$$\Delta SSH = SSH(x, t + \Delta t) - SSH(x, t) \\ = b(SWH(x, t + \Delta t) - SWH(x, t)) + \varepsilon \quad [10]$$

assuming that the difference in the true SSH is small from one time to another. In reality, this is not necessarily true, but with a large-enough number of observations, it should be in a mean sense, provided there are no biases. The factor ε accounts for any errors in the data and unmodeled variations. Thus, one seeks to find the value of b to minimize the sum of ε^2 over all observations

$$\sum_{x,t} \varepsilon^2 = \sum_{x,t} [\Delta SSH - b\Delta SWH]^2 \quad [11]$$

Differences in the values estimated for b arise from how SSH is corrected, particularly if orbit error corrections are applied (Chelton et al., 2001), how many residuals are used, and what types of residuals are used. For example, early estimates for Seasat and Geosat used residuals based on differencing SSH along ground track passes from adjacent repeat cycles. One problem with this method is that there were fairly large geoid errors that did not completely cancel out because the observations were not made at exactly the same location. Altimeter crossovers, which are derived from interpolating the SSH to the point where an ascending and descending ground track cross do a better job of removing the geoid error, but also results in far fewer points – there are a few thousand crossovers for a 10-day T/P cycle compared to nearly 400 000 collinear points. However, as Gaspar et al. (1994) pointed out, collinear points are usually correlated over fairly large distances and so are not completely independent,

while crossovers are. Also, the time gap between crossovers is always smaller than the repeat cycle. For T/P, it is on average 3.5 days. This will reduce effects of the unmodeled SL variations that are ignored in the estimation process.

Another factor that affects the result is how b is modeled. The earliest models assumed b was constant everywhere over the ocean; that is, the SSB was just a percentage of SWH. However, Witter and Chelton (1991) found evidence in the Geosat data that b was not uniform and Fu and Glazman (1991) proposed another model that was a nonlinear function of both SWH and wind speed. This led to the model adopted by the T/P project before launch, which was empirically derived from aircraft measurements (Hevizi et al., 1993)

$$b = a_0 + a_1 U + a_2 U^2 \quad [12]$$

where U is the wind speed and the parameters a_0, a_1, and a_2 depend on the frequency of the altimeter. The pre-launch algorithm was evaluated in several studies (Chelton, 1994; Gaspar et al., 1994; Rodriguez and Martin, 1994). It was found that a function of the form

$$b = a_0 + a_1 U + a_2 U^2 + a_3 SWH \quad [13]$$

resulted in approximately the same variance of residuals within $\pm 40°$ as the original parametrization (eqn [12]), but significantly lowered the variance poleward of $40°$, where wave heights and wind speeds tend to be higher. This four-parameter model (eqn [13]) of Gaspar et al. (1994) was adopted for re-processing of the T/P data.

Ideally, the SSB would not assume an *a priori* parametric model, but would be determined as a map of SSB versus SWH and U that could be used as a lookup table. Such nonparametric models were originally studied by Witter and Chelton (1991), Chelton (1994), and Rodriguez and Martin (1994), but did not show significant improvement over a parametric model. In preparation for the Jason-1 mission, Gaspar and Florens (1998) proposed a new method to calculate a nonparameteric SSB model and demonstrated that their model further reduced the variance in mid- and high latitudes by about $1\,cm^2$. This SSB model was selected for the Jason-1 mission.

Errors in the SSB are difficult to quantify. Chelton (1994) found that estimating a parametric model with different data sets amounted to differences of about 1% of SWH, which he took as one estimate of the error. One thing to note is that the error will have a time-variable component as well as a bias component, as there will always be some waves in the ocean on average. This bias is also not constant, since the mean wave state varies from point to point. Assuming the 1% of SWH value is a reasonable upper bound on the error, then the maximum potential bias and variable errors are plotted in **Figure 5**. The time-variable portion of the error is probably smaller than 2 cm RMS except for areas with the highest waves. More troubling are the potential of systematic bias errors as large as 3–4 cm. This is especially a problem for studies of MSL change which depend on linking data from two different altimeters, as discussed in the following paragraph. However, for studies that remove a local mean of the SSH in order to observe the time-varying SSH, any bias in the SSB model will be removed, provided means are computed for each set of altimeter data.

Several investigators on the T/P SWT noticed significant changes in the observed SWH starting in mid-1997. It was confirmed via comparison with buoys that the TOPEX SWH measurement was in error. The size of the change in SWH was about

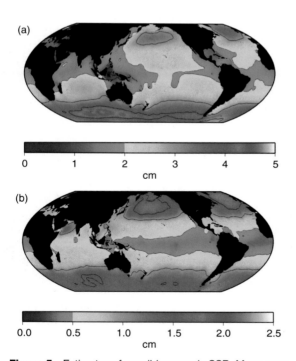

Figure 5 Estimates of possible errors in SSB. Mean error (a) is computed assuming error is 1% of mean SWH (Chelton, 1994). The variable error (b) is computed assuming error is 1% of the RMS of SWH. Note that the scale of the variable portion is half the size of the mean portion.

50 cm over 2 years and would have caused a change in SSB of about 1 cm if there had not been a negating factor. Instrument engineers found that the source of the problem was a noticeable degradation in the TOPEX altimeter point-target-response (PTR), which would affect not only the SWH, but the range measurement as well (Fu, 1998), and did so in such a way as to nearly cancel the SSB effect. However, the PTR degradation was only expected to increase, and since there was a redundant altimeter onboard that had never operated (TOPEX_B) it was recommended that the original primary science altimeter (TOPEX_A) be turned off and TOPEX_B be turned on for at least a few cycles to see if this corrected the problem. This was done in February 1999. All tests indicated TOPEX_B performed as well as early TOPEX_A data, so the decision was made to keep TOPEX_B on as the primary science instrument for the remainder of the mission. The last T/P 10-day repeat cycle to use TOPEX_A was Cycle 235, ending on 9 February 1999. TOPEX_B was turned on at the beginning of Cycle 236 on the same day.

Other than a few initial calibrations based on the first few months of data, no significant changes were made to the data algorithms, in particular the SSB algorithm, or to the tracker. Shortly after the switch, it was noticed that there was a slight offset in the TOPEX_B data relative to tide gauge data (see Section 5.4.2 for discussion), but the nature of the error was not clear until Jason-1 was launched and investigators began comparing Jason-1 data with TOPEX_B data. Chambers *et al.* (2003b) noticed that there was a residual relationship between the TOPEX_B and Jason-1 data as a function of SWH, even though coefficients to a SSB model had been adjusted for the new Jason-1 altimeter. They found that the relationship was reduced only if the SSB model for TOPEX_B was also adjusted, suggesting an error in the model. Upon further investigation, they demonstrated that the offset between TOPEX_B and TOPEX_A SSH that had been observed in the tide gauge calibration was removed when separate SSB models were used for TOPEX_A and TOPEX_B. An important consequence of the error in the TOPEX SSB models was that the step function was not globally uniform (**Figure 6**), as suggested by the tide gauge calibration. The change varied from 5 to 6 mm in the tropics (where most of the tide gauges for the calibration were located) to more than 2 cm in the far Southern Ocean. The effect of the erroneous SSB for TOPEX_B on the

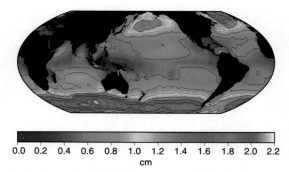

Figure 6 Mean bias correction to TOPEX_B data by switching from old SSB model to new model.

determination of global mean sea level (GMSL) rate was to decrease the real measurement by more than 1 mm yr^{-1} (Chambers *et al.*, 2003b).

This highlights why empirical estimation of altimeter SSB models is still so important. Theoretically, the TOPEX_A and TOPEX_B instruments were identical, with the exact same operating frequencies. However, there were clearly differences in the tracker bias, and this leaked into the SSB. Work on non-parametric models has continued, and recently models have been developed for TOPEX_A, TOPEX_B, Jason-1, and other altimeters (Labroue *et al.*, 2004). However, SSB models still have errors of up to several centimeters at some wave heights, and also can have biases of several centimeters between nonparametric models and parametric models based on the same data. Thus, the SSB is perhaps the largest current contributor to SSH error and is still a topic of ongoing research.

5.2.5 Inverted Barometer Correction

The inverted barometer (IB) correction is not a correction to the raw range measurement *per se*, but a correction to the SSH measurement to account for SL variations caused by changes in the atmosphere pressure. A 1 mbar increase in the atmospheric pressure causes a decrease of approximately 1 cm in the sea surface. Because there is no net pressure gradient at the sea surface due to this atmospheric loading, the signal must be removed from the SSH data before using it to compute geostrophic currents so as not to corrupt the calculation. The size of the variation is also large in some regions, up to 15 cm or more in the Southern Ocean. Because height variations are large and not due to internal ocean dynamics, they are frequently removed using a model that assumes an

instantaneous, static response to surface atmospheric pressure (P_{atm})

$$\Delta SSH_{IB} = \alpha(P_{atm} - P_{ref}) \qquad [14]$$

The correction is subtracted from the SSH measurement. Different scaling factors (α) have been used for different altimeters, based on interpretations of various static IB studies (e.g., Wunsch, 1972; Dickman, 1988). For example, the Jason-1 GDRs use a value -9.948 mm mbar^{-1}, while the T/P and Geosat GDR handbooks recommend a value of -10.1.

More problematic is the different recommendations for the reference pressure in the GDR handbooks. Even though Ponte *et al.* (1991) noted that the reference pressure should be the average pressure over the ocean as a function of time (see also Dorandeu and Le Traon, 1999), the reference pressure used for the Geosat and T/P GDRs was a fixed mean pressure based on the global average (1013.3 mbar), not the time-mean over only the ocean ($\sim$1010.9 mbars). It also did not include the time variation, which has a seasonal amplitude of about 0.6 mbar, and intraseasonal variations of nearly the same magnitude (Dorandeu and Le Traon, 1999). One artifact of this was that the GMSL calculated from T/P data with the IB correction in the GDRs applied had an erroneous seasonal variation (Dorandeu and Le Traon, 1999). It also means that there is an approximately 24 mm bias difference between SSH computed with the different reference pressures.

Jason-1 GDRs use the correct IB response, calculated relative to the instantaneous mean ocean pressure derived from the ECMWF atmospheric model, that is,

$$\Delta SSH_{IB} = -9.948(P_{atm} - P_{ref}(t)) \qquad [15]$$

5.3 Satellite Orbit

5.3.1 Orbit and Ground Track

The subsatellite point of any object orbiting the Earth will trace out a path referred to as the ground track when the orbit is projected from an inertial reference frame (i.e., one with a fixed orientation in space) to an Earth fixed reference frame (i.e., one with a fixed orientation relative to the Earth). To first order, the two frames can be considered to have the same center and z-axis along the pole, but that the Earth-fixed frame rotates about the inertial frame at a rate ω_E where

$$\omega_E = \frac{2\pi}{86\,164\,s} \qquad [16]$$

Note that the period of rotation is equivalent to the length of a sidereal day (expressed in solar seconds), not a solar day. Because of this, after every revolution of the satellite in its orbit, the subsatellite point will cross the equator westward by an amount equivalent to the angle that the Earth has rotated during that time:

$$\Delta_E = \omega_E P \qquad [17]$$

For a circular orbit, the orbital period (P) is given by

$$P = 2\pi \left(\frac{a^3}{\mu_E}\right)^{1/2} \left[1 - \frac{3}{2}\mathcal{J}_2 \left(\frac{R_E}{a}\right)^2 \left(4\cos^2 i - 1\right)\right] \qquad [18]$$

and is dependent on orbit's inclination relative to the equatorial plane (i), the mean radius of the orbit (a), the gravitational constant of the Earth (μ_E), the Earth's equatorial radius (R_E), and the second zonal harmonic of the Earth's gravity field ($\mathcal{J}_2$) which perturbs the period slightly from that predicted by a point mass assumption. For typical satellite altimeter orbits, the orbital period will be about 90 min, which means that the subsatellite point will cross the equator on the next revolution approximately 22.5° west of the previous point.

We have already included the effect of the Earth's equatorial bulge ($\mathcal{J}_2$) in the calculation of the orbital period. However, the same perturbation causes the angular momentum vector of the orbit to precess, like that of a spinning top. The effect of this is to cause the node of the orbit, or the point here it crosses the inertial equatorial plane, to move either east or west. The node rate ($\dot{\Omega}$) for a circular orbit is determined from

$$\dot{\Omega} = -\frac{3}{2}\mathcal{J}_2 \left(\frac{\mu_E}{a^3}\right)^{1/2} \left(\frac{R_E}{a}\right)^2 \cos i \qquad [19]$$

For altimeter satellite orbits, the value for $\dot{\Omega}$ is $\pm 3°$ per solar day. The effect of the node rate acts opposite of the axis rotation for inclinations less than 90°, and in the same direction for inclinations greater than 90° so the total ground track shift (Δ) is

$$\Delta = \left(\omega_E - \dot{\Omega}\right) P \qquad [20]$$

and positive values mean a shift of the ground track westward (**Figure 7**).

Many different types of ground tracks can be found by slightly adjusting either the radius of the orbit or the inclination to cause a change in Δ. One special type of ground track that is commonly used for satellite altimeter missions is an exact-repeat round track, or orbit. An exact-repeat orbit has a ground track that repeats itself (to within a very small difference) after a certain period, P_R. For altimetric studies, this is desirable for several reasons. For one, the satellite will repeat measurements at a consistent interval, which is important for examining

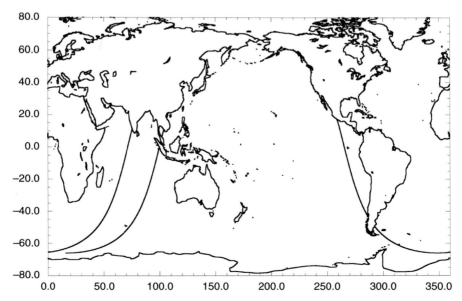

Figure 7 Plot for T/P ground track for first three passes, showing westward shift of ground track in time.

the time-varying SSH. Second, by repeating measurements over the same area, one can obtain a precise average of the SSH that is not affected by large time-variable SSH signals such as tides and ocean eddies, which can have variations in SSH of more than a meter. Finally, there is also a need to make measurements in an area as small as possible. Otherwise, the sampling of stationary geoid gradients (of order 5–10 cm km^{-1}) in time could be interpreted as real SSH variation even if there has been no change.

One major drawback to an exact-repeat orbit is that portions of the ocean are still never measured due to gaps in the ground track coverage (**Figure 1**). Given enough time, a satellite altimeter in a nonrepeating ground track will eventually observe the SSH at a scale approaching the along-track sampling rate, but perhaps only once for any particular area. Although there is the potential to bias the measurement due to large time-variable features like an eddy, the error introduced may not be significant compared to the size of the desired signal. This was one of the reasons that the original Geosat mission was in a nonrepeating ground track, as the US Navy was more interested in obtaining any measurement of the SSH in areas that had never been observed in order to detect seamounts, trenches, and undersea ridges, than the most accurate mean measurement.

A sun-synchronous orbit is a special type of exact-repeat orbit (Mazzega, 1989). This type of orbit has a node rate $\dot{\Omega} = 2\pi/365.2422$ days, which means that the angle between the orbital plane and the vector between the Earth and the Sun will maintain a constant angle. This is beneficial in that the satellite will fly over a region at roughly the same local time for every ascending or descending pass and can be useful for maintaining a fixed orientation of the solar panels with the Sun for charging. Although this type of orbit is usually used for satellites that require a consistent viewing angle relative to the Earth–Sun line, there have been three altimeter satellites in sun-synchronous orbits: ERS-1, ERS-2, and Envisat. These satellites were hosts to instruments other than the radar altimeter, and these required the sun-synchronous orbit. There are certain aspects of sun-synchronous orbits that will be discussed further in Section 5.3.2 that make them problematic for some applications of altimetry.

For an exact-repeat orbit there are an integer number of orbital revolutions (N) that are equivalent to D nodal days. A nodal day is the time required for the Earth to make one rotation relative to the satellite

orbit's line of nodes. Thus, to find the exact-repeat orbit, one needs to solve the equation

$$P_R = NP = \frac{2\pi D}{(\omega_E - \dot{\Omega})} \qquad [21]$$

to find the orbit radius (assuming a circular orbit) and inclination that gives one a 'D-day' repeat orbit. Eqn [21] is nonlinear so there is no analytical solution. Typically, one starts with an approximate orbital radius and inclination, calculates an initial P and $\dot{\Omega}$, then assumes a value for D and calculates N (either rounding or truncating to make it an integer). Now that an approximation of N is known, one can recompute the orbital period, adjust either the radius or inclination (usually radius) to match the period, then iterate until the solution converges. For sun-synchronous orbits, P_R is exactly an integer number of days equal to D. For other exact-repeat orbits, P_R is a value slightly shorter or larger than D. For example, T/P (with an inclination of ~66°) is called a '10-day repeat orbit' (the value of D), but the actual repeat period is 9.915 625 solar days. Geosat has a '17-day repeat period' but the actual period is 17.050 5 days.

The spacing (Δx) between adjacent ascending or descending passes of the ground track at the equator is determined from the number of revolutions in the repeat period by

$$\Delta x = R_E \left(\frac{2\pi}{N} \right) \qquad [22]$$

Thus, longer repeat periods (and hence higher N) will have shorter distances between ground tracks. T/P has 127 revolutions in the 10-day repeat period, with a ground track spacing of ~315 km, while Envisat has 501 revolutions in a 35-day repeat and a spacing of ~80 km. Note that ground tracks that are adjacent in space on the Earth's surface are not adjacent in time. For instance, although the spacing of ascending tracks for T/P is ~2.8°, the track immediately after the previous one in time is 28.3° westward at the equator (e.g., **Figure 7**).

The ground track will not remain stable due to nonconservative forces such as drag that removes energy from the orbit. The effect of drag is to reduce the semi-major axis of the orbit. This means that the orbital period will also decrease (eqn [18]) and that the ground track shifts (Δ) will also decrease (eqn [17]), so that the ground track will drift eastward relative to its nominal location (**Figure 8**). To stop the drift and reverse it, the semimajor axis must be raised via a pair of orbital maintenance maneuvers in order to increase

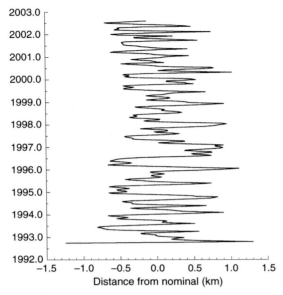

Figure 8 Plot of T/P ground track variation around nominal ground track at equator for duration of primary mission until it was moved to its current ground track. Positive excursions indicate the position is east of nominal, while negative values indicate the position is west of nominal.

the period and cause a drift back to the west. In order to maintain an exact-repeat ground track but also not expend too much fuel, the drift is allowed within a certain band, generally ±1 km from nominal for satellite altimeter missions. With some knowledge of the atmospheric density and drag properties of the spacecraft, one can effectively steer the ground track to stay within the ±1 km range (**Figure 8**).

Note that the inclination of the orbit also determines approximately the maximum latitude of the ground track

$$\phi_{max} \approx i \ \text{for} \ i < 90° \\ \phi_{max} \approx 180° - i \ \text{for} \ i > 90° \qquad [23]$$

5.3.2   Aliasing

One consequence of the relatively long intervals between successive samples of the same location due to the satellite orbit is aliasing of signals with periods smaller than the exact-repeat period into signals with periods significantly longer than repeat period. This is demonstrated in **Figure 9**, where a variation with a frequency of exactly 1 cycle per day (like an ocean tide), is sampled every 9.9 days. Here, the altimeter sees the tidal variation as a low-frequency fluctuation (period = 99 days), although with the same amplitude. This means that if the high-frequency variation is not modeled and removed from the data properly, errors will leak into the SSH measurement as long-period signals that could be interpreted as real SL variations.

Ocean tides are one such high-frequency signal. The dominant tidal frequencies occur in bands near one (diurnal) and two (semi-diurnal) cycles per day. Most of the periods are slightly different from 12 h and 24 h (**Table 1**), except for the S2 tide (caused by solar perturbations), which has a period of exactly 12 h. This has an important consequence for a sun-synchronous orbit. Because the exact-repeat period for a sun-synchronous orbit is exactly an integer

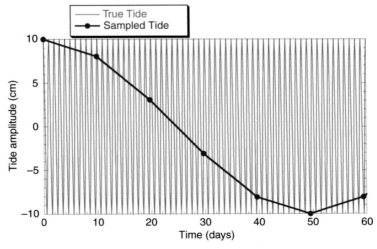

Figure 9   Aliasing of a 1 cycle-per-day signal (red) by sampling every 9.9 days (blue).

Table 1 Periods of largest tidal constituents along with altimeter alias periods (in days)

Tide symbol	Period (h)	Equilibrium amplitude (cm)	Source	T/P 10-day Alias	Geosat 17-day Alias	Envisat 35-day Alias
M_2	12.420601	16.83	L	62.1	317.1	94.5
K_1	23.934470	9.83	C	173.2	175.5	365.2
S_2	12.000000	7.83	S	58.7	168.8	$\propto$
O_1	25.819342	6.99	L	45.7	112.9	75.1
N_2	12.658348	3.26	L	49.5	52.1	97.4
P_1	24.065890	3.25	S	88.9	4466.7	365.2
K_2	11.967235	2.13	C	86.6	87.7	182.6
Q_1	26.868357	1.35	L	69.4	74.1	132.8

Source: L, lunar; S, solar; C, combined.

number of N solar days, it will always sample the S2 tide at exactly the same phase every day. Thus, the tide (or tide error) will alias into a bias, and not a long-period variation. The alias period may also be so long that it is essentially a bias, depending on the length of the mission.

The theory of tidal aliasing for altimetry is described by Parke et al. (1987). Given the exact-repeat period, P_R, one computes the tidal phase change over one repeat period by

$$\Delta\vartheta_{\text{tide}} = \frac{2\pi P_R}{P_{\text{tide}}} \left(\text{anglebetween} - \pi \text{ to } \pi\right) \quad [24]$$

where P_{tide} is the period of the particular tide (**Table 1**). Both periods (P_R and P_{tide}) must be in the same units. The resulting principal alias period is

$$\tau_{\text{tide}} = \text{abs}\left(\frac{2\pi P_R}{\Delta\vartheta_{\text{tide}}}\right) \quad [25]$$

Table 1 gives the alias periods at the major tidal constituents for the various altimeter exact-repeat orbits. Note that in addition to the bias at S2 for the sun-synchronous orbits, the periods of the K2, K1, P1 tides are close enough to 12 h and 24 h that they alias into very long periods near the semi-annual period (K2) and annual periods (K1 and P1). The P1 tide for Geosat orbits aliases into a 12-year fluctuation, which is nearly a bias over short periods. The alias periods for T/P are all less than 180 days, by design (Parke et al., 1987), in order that the data could be used to improve tide models.

Aliasing would not be important if the tidal variations were known accurately. However, this was not the case, even as recently as a decade ago. Tide models developed with Geosat data (e.g., Cartwright and Ray, 1990) had errors greater than 5 cm RMS in the open ocean. Although this was

large, it was still better than errors in older tide models based only on tide gauge data or hydrodynamic models. However, with only a year of T/P altimeter data, a reduction in error to the level of 2–3 cm was possible, mainly due to the short alias periods (e.g., Ma et al., 1994; Schrama and Ray, 1994). Although tide errors are now significantly less than they were 10 years ago, any remaining error in the model for a certain constituent will still alias into a small, undesired long-period signal. For more discussion of the contribution of altimetry to tide models, please see Section 5.5.2 or Le Provost (2001).

In addition to temporal aliasing, tides will be spatially aliased as well (Schlax and Chelton, 1994; Parke et al., 1998). Because tidal signals propagate across the ocean as long waves and there is a several-day interval between adjacent ground tracks, tidal errors will also propagate as waves. This can cause spurious signals that may be interpreted as real Kelvin or Rossby waves at certain latitudes. This was observed in Geosat data with the Cartwright and Ray (1990) tide model (Jacobs et al., 1992) and also in T/P before tide models were improved (Schlax and Chelton, 1994). With current tide models, the data from the T/P ground track now show no significant spatial aliasing and have been used to show for the first time Rossby waves at most latitudes (Chelton and Schlax, 1996).

There are SSH variations other than tides with periods of less than 10-days, which will also alias into long-period errors. More problematic is that these variations are not regular like tides; both the amplitudes and phases will shift in time. Fukamori et al. (1998) used a general ocean circulation model forced by realistic winds and heat fluxes to demonstrate that half the energy at periods of less than a year in some regions was due to barotropic variations with periods of less than 20 days. Such variations are not properly sampled by altimeters (unless they have very short repeat periods) and so will alias into long-period errors. The barotropic variations are predominantly at higher latitudes, and have amplitudes as large as 5 cm.

Fukamori *et al.* (1998) suggested that barotropic variations from a model could be used to de-alias the altimetry data, and this has been tested (e.g., Stammer *et al.*, 2000; Tierney *et al.*, 2000). The variance of the T/P SSH was reduced by about 10–20% in regions of high barotropic variations, suggesting that the model was removing the signal from the altimetry. However, the variance increased in other regions, particularly the tropics, suggesting that using the model added erroneous signals to the altimetry. Because of this, barotropic de-aliasing of altimetry is still very much a topic of research and is not currently being applied operationally to the data.

5.3.3 Precise Orbit Determination

Determining the position of a satellite accurately at any time depends on having an accurate dynamical model in order to propagate the orbit forward in time. Generally, the orbit will be in error for several reasons. The initial starting position may be wrong. Parameters in the dynamical models – such as the Earth gravitational acceleration, atmospheric drag, Earth and solar radiation pressure, etc. – may be incorrect. If there are independent measures of the satellite's position or velocity, then conceptually one can use the difference between the observed position and the computed position to estimate corrections to the dynamical models and initial position in order to reduce the differences. In practice, there is no way to measure the satellite position or velocity directly, but instead one measures either the range from another site, or the change in rate of the range. For a complete analysis of POD techniques, the reader is referred to Tapley *et al.* (2004a), or as they apply to altimetry, Chelton *et al.* (2001). Here, we wish to summarize the major improvements in POD that have led to more accurate measures of SSH from altimetry, and how orbit errors appear in the SSH. Since the errors in SSH are directly proportional to errors in the nadir altitude of the satellite (eqn [1]), one is most concerned with errors in what is referred to as the radial component of the orbit. Other components of the orbit (defined in a satellite fixed frame) are the transverse component (in the direction of satellite motion) and the normal component

(normal to the radial and transverse in a right-hand sense). Errors in the transverse and normal component affect the SSH much less than errors in the radial component, although they will affect the accuracy of positioning the subsatellite point, but not significantly compared to the width of the radar footprint. In the literature (as here), when altimeter orbit accuracy is discussed, radial orbit accuracy is implied unless otherwise stated.

Initial orbits for both Seasat and Geosat had errors of several meters (Marsh and Williamson, 1980; Schutz and Tapley, 1980; Born *et al.*, 1988). This was significantly larger than the total SSH accuracy goal for the T/P mission (13 cm), a value that was still higher than the 5 cm or better value requested by oceanographic community (Stewart *et al.*, 1986). However, the dominant signature of orbit error is that of a very long-wavelength once-per-orbital revolution sinusoid, mainly due to dynamical errors in the orbit introduced by errors in the gravity field model (e.g., Tapley *et al.*, 2004a). There are also two components of the orbit error. One varies in time so that the error is different from one pass to another. The second component is phase locked to a specific geographic location and will tend to be the same over time, and is known as geographically correlated orbit errors (Tapley and Rosborough, 1985).

Geographically correlated orbit error is aliased directly into any calculation of the mean sea surface (Section 5.5.1). However, one can effectively remove this correlated orbit error from the altimeter data by estimating and removing the mean SSH from the data and examining only the time-variable component. The time-variable component of orbit error will remain, however. Several methods have been devised to reduce the time-variable portion of the orbit error in the data. Because it is a long-wavelength signal, a bias and trend was often fit over portions of each pass of SSH anomalies and the residuals were examined. Although this meant one could only study wavelengths shorter than the pass length utilized, it did allow for a reduction of the orbit error from order 1 m to order 10 cm (Tapley et al., 1982). Tai (1988) used a sinusoid with a frequency of 1 cycle-per-revolution (cpr), and estimated the amplitude and phase over each satellite orbital revolution (two passes). Other studies have shown similar results using complex demodulation (Francis and Bergé, 1993), a Fourier series (Douglas *et al.*, 1987), estimating a sinusoid over each pass (half-revolution) (Chelton and Schlax, 1993), or an adaptive Kalman filter (Urban, 2000).

The one problem with all these techniques is that they will absorb some of the real long-wavelength ocean variability in addition to the orbit error (Van Gysen *et al.*, 1992). This is more likely when the SL variation is of the same order of magnitude as the orbit error. One of the more important long-wavelength ocean signals that may be removed is the annual variability in the SL due to heating/cooling of the upper layers of the ocean at mid-latitudes and wind forcing in the tropics. In general, when the sea surface in the Northern Hemisphere is higher than the time average, the sea surface in the Southern Hemisphere will be lower than average (**Figure 10**). The difference from peak to trough over a pass can be as large as 15–20 cm, and since the satellite circles the Earth once every revolution, this real ocean signal will tend to alias into a 1 cpr signal in the spectra of the satellite SSH anomalies (Wagner and Tai, 1994). Furthermore, if a sinusoid is fit over longer arcs (i.e., multiple revolutions), it will not remove the total orbit error and could absorb signals from other sources, since the amplitude and phase of the residual orbit error change in time. Thus, for many oceanographic applications, orbit errors in the range of 10–20 cm are more problematic than errors larger than 50 cm, since this is similar to the amplitude of real ocean variations at a similar wavelength.

(a)

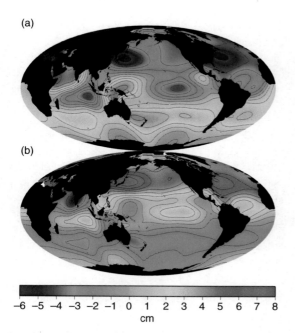

(b)

Figure 10 SSH anomalies for average Mar. (a) and average Oct. (b) based on 12 years of T/P and Jason-1 maps that have been smoothed with a 1000-km radius Gaussian.

As noted previously, the largest contributor to orbit error are errors in the Earth gravity model, followed closely by errors in drag and solar radiation pressure. The T/P orbit was designed to have a greater semi-major axis than either Geosat or Seasat in order to reduce both the effects of drag and errors in the Earth gravity field model at the time, as gravity errors (like the signal) are attenuated with distance. However, predicted orbit errors from the best gravity field models at the time, like the Goddard Earth Model (GEM) GEM-10B (Lerch *et al.*, 1981), were still estimated to be of the order of 50 cm (Nerem *et al.*, 1994a). Thus, significant work was done in order to improve the gravity field models, using in part altimeter data from the Geosat mission. These efforts led to the development of the Joint Gravity Model-1 (JGM-1) by members of the T/P SWT, which was the pre-launch model for T/P (Nerem *et al.*, 1994a). The orbit accuracy using JGM-1 along with other model improvements (Tapley *et al.*, 1994a) significantly exceeded the goal of 13 cm, reaching an estimated radial accuracy of about 5 cm (Nerem *et al.*, 1994a). However, further improvements to the gravity field model (JGM-2) were made, mainly from the inclusion of T/P tracking data (Nerem *et al.*, 1994a). The RMS of the orbit error using JGM-2 was estimated to be about 2 cm (Tapley *et al.*, 1994a), with total orbit errors (including errors in models other than gravity) of about 3 cm RMS. A further iteration of the JGM model resulted in JGM-3 (Tapley *et al.*, 1996). Total orbit errors for T/P using JGM-3 are estimated to be 2 cm RMS or less (Tapley *et al.*, 1996). Although improvements in the gravity models had the largest effect in improving T/P orbits, improvements in other models also contributed to the overall orbit accuracy. These included a model for the satellites complex shape along with its thermal and reflective properties, better solid and ocean tide models, and improved empirical parametrization (e.g., Tapley *et al.*, 1994a).

It was still difficult to match this accuracy with other satellites such as ERS-1, ERS-2, and GFO until recently because they are at a lower altitude and hence have higher drag and are sensitive to gravity signals not completely modeled by JGM-3. They have also generally had less tracking data than T/P; the PRARE system failed on ERS-1, and only limited SLR tracking was available. Although both PRARE and SLR tracking were available for ERS-2, PRARE provides significantly less tracking data than the DORIS system on T/P. The GPS system failed on GFO and it too had to rely only on SLR tracking.

However, by including tracking data from ERS-1 and ERS-2 and computing new gravity models (Tapley *et al.*, 1997; Scharroo and Visser, 1998), orbit accuracy for these low-altitude satellites were improved, although still not to the level of T/P. For example, the fit of SLR residuals to the precise orbit, which is one indicator of orbit improvement, was reduced from order 7–8 cm RMS to 3–4 cm RMS using these new gravity models (Tapley *et al.*, 1997; Scharroo and Visser, 1998). Typical values for T/P are 1–2 cm RMS.

A significant improvement in orbit determination of low-altitude satellites is now possible because of an improved gravity model determined from the Gravity Recovery and Climate Experiment (Tapley *et al.*, 2004b). For example, Lemoine *et al.* (2003) computed orbits for the low-altitude Envisat satellite using both SLR and DORIS tracking and a variety of gravity models. They found that the SLR residuals for orbits computed with the EGM96 gravity field model had an RMS of 4.8 cm, but when an early GRACE gravity model was substituted, the residual RMS was reduced to 2 cm, which is comparable to the best orbits for T/P.

Accuracy for Jason-1, which is in the same orbit as T/P, is generally even better than that of T/P, having estimated errors of 1 cm RMS or less (Luthcke *et al.*, 2003; Choi *et al.*, 2004). This is primarily due to the use of GPS tracking data in addition to SLR and DORIS. The use of new GRACE gravity models does not significantly alter the time-variable orbit error for T/P or Jason-1, although it does reduce the geographically correlated portion (Beckley *et al.*, 2004; Choi *et al.*, 2004).

The difference between a 2 cm orbit and a 1 cm orbit may not seem large, but the effects in terms of SSH error are significant for long wavelengths. Recall that the error is predominantly at 1 cpr. As Chelton *et al.* (2001) point out, a 2 cm orbit is equivalent to an erroneous rise (or fall) in SSH of about 4 cm over a distance of 10 000 km, the length scale of a typical ocean basin. This can still be a significant fraction of the basin-scale SSH variation (especially that related to the seasonal heat cycle). On the other hand, a 1-cm orbit is equivalent to an erroneous rise of only 2 cm, which is closer to the estimated accuracy of the long-wavelength-smoothed SSH anomalies.

5.4   Calibration and Validation

It is vital to calibrate and validate the SSH measurement after all the corrections discussed in

Section 5.2 are applied. The term calibration/validation (cal/val) is often used in the literature, but one should understand that each term is very distinct. The altimeter SSH anomaly (variation about a mean value) is often validated against independent measures of the SSH variation, typically from tide gauges but more recently from other satellite altimeters. By comparing SSH anomalies (SSHAs), one is not concerned with either biases in the altimeter or the independent data. Typically, validation consists of determining if the frequency and amplitudes of the observed SSHAs are similar, usually by computing the correlation and standard deviation (or RMS) between the two time series after means have been removed. High correlation and low RMS are interpreted as meaning that the altimeter is accurately measuring SSH. Often, the altimeter accuracy in measuring SSH is given from the RMS of the comparison (or mean RMS averaged over several sites) assuming that the tide gauge has no errors, or as $RMS/\sqrt{2}$ assuming that the error in the tide gauge is comparable to that of the altimeter and there are no correlated errors.

Calibration, on the other hand, refers to the comparison of absolute SSH, not anomalies. In this case, one assumes that the SSH at an independent site has been measured in the same reference frame as used by the altimeter (i.e., relative to the same local ellipsoid) and is free of biases. In this case, one can measure the mean offset between the altimeter SSH measurement and the independent 'truth' measurement and interpret it as a bias in the altimeter SSH (or range). The change in the bias over time can also be tracked using this method, although it can also be tracked using comparisons to tide gauges that are not in a known reference frame (see Section 5.4.2).

In the following sections, we will discuss the methods of altimetry cal/val and comment on the strengths and weakness of each method. We will also discuss the results of the first direct calibration between two altimeters in the same orbit and at about the same time, when T/P and Jason-1 were observing the same SSH separated in time by only 70 s as part of the Jason-1 cal/val phase. In particular, we will comment on several insights that this unique study has provided on cal/val methods.

5.4.1   Calibration at Single Site, Instrumented Platforms

The oldest cal/val method used in altimetry is based on comparing the measurement directly to an independent measure of SSH made at a site that the satellite directly overflies. This was done with both GEOS-3 and Seasat (Martin and Kolenkiewicz, 1981; Kolenkiewicz and Martin, 1982), and was successful even in the presence of meter-level orbit errors because the over flight site at Bermuda also had a NASA laser ranging station. Because of this, the orbital altitude could be determined very precisely at the calibration point. The altimeter bias is determined by comparing the SSH with that measured by one or more tide gauge at the calibration site. However, tide gauges normally measure SSH variation relative to some arbitrary level that is not defined in the same reference frame as the altimeter. Thus, one first needs to find the tide gauge reference level relative to the measurement frame used by the satellite (i.e., relative to the same reference ellipsoid). If that is done accurately, then the 'true' absolute SSH at the tide gauge can be obtained. Then the altimeter SSH measurement can be compared to the 'truth', and the difference is interpreted as the bias. Generally, multiple observations are made over a period of months or years to average out random errors.

One problem with this method was that neither Seasat nor GEOS-3 flew directly over the tide gauge, which meant that the altimeter SSH measurement had to be interpolated to the tide gauge. In addition, the survey of the tide gauge relied on traditional leveling techniques from a more distant benchmark, which may have introduced systematic errors at the several-centimeter level. However, the method was sufficient for bias determination in the presence of meter-level SSH accuracy.

Geosat had no SLR reflector and so no direct calibration of the measurement was undertaken, since the primary mission was more interested in SSH gradients than absolute SSH. This is one significant problem in trying to link Geosat and T/P data in order to determine longer time series of GMSL change (Section 5.5.3), although it has been attempted using uninstrumented tide gauge sites (Guman, 1997; Urban, 2000). However, the uncertainty in such calibrations is of the order of several centimeters.

ERS-1 SSH was calibrated over a nearly a 2-month period in 1991 using a method similar to that used for GEOS-3 and Seasat (Scharroo, 2002). There were several significant improvements in the campaign. First, the site was located on a small scientific platform in the Adriatic Sea off the coast of Venice. This meant that there was no loss of signal due to land as was the case with Bermuda. In addition, the ground track for ERS-1 was selected to fly directly over the platform. The average distance between the nearest 1 s altimeter point and the tide gauge was only 600 m. Finally, and perhaps more importantly, GPS was available to tie the tide gauge to the ERS-1 reference frame. This had the potential to significantly reduce systematic bias errors in the 'truth' measurement. The initial estimate indicated an altimeter range bias of +4.5 m (Scharroo, 2002). However, after 2 years of analysis and many corrections to the data (including 4 changes to the reference level of the tide gauge), the final range bias estimate was −41.5 cm, based on 10 overflights with values ranging from −46.9 to −21.6 cm (Scharroo, 2002).

The T/P mission used two instrumented calibration sites, the main one off the coast of California on the Harvest oil platform (Christensen et al., 1994), and the other in the Mediterranean Sea on Lampedusa Island (Ménard et al., 1994). The tide gauges were also leveled to the same reference frame as T/P using GPS. Like the ERS-1 calibration site, the Harvest platform was specifically chosen because it was very nearly under one pass of the T/P ground track so that interpolation was minimized. Unlike the ERS-1 site, the Harvest platform was designed to be a continuing calibration site, and has continued to do so to this day (Haines et al., 2003). Because of this, the initial calibration at Harvest was based on nearly a year's worth of overflights. The scatter of the bias estimates was also smaller than those for ERS-1, with a standard deviation of about 2 cm. The Lampedusa site, on the other hand, was designed for a limited campaign, and in fact only estimated a TOPEX bias based on two overflights (Ménard et al., 1994). The initial bias estimate for TOPEX was about 15 cm (Christensen et al., 1994; Ménard et al., 1994). It was eventually discovered that an oscillator correction had been improperly applied to the TOPEX data (e.g., Nerem et al., 1997). When this was corrected, the apparent TOPEX bias disappeared.

With the advent of relatively inexpensive GPS receivers, a new type of calibration instrument has been developed: the GPS buoy. The idea behind a GPS buoy is relatively simple. Data from a GPS receiver mounted on a floating buoy can be used to determine the local SSH in the same frame as the altimeter. If the buoy is deployed along a ground track at the same time as the satellite altimeter is flying overhead, one can compare the two SSH measurements and determine the bias. Born et al. (1994) demonstrated the concept using a buoy at the Harvest platform. They found that the bias determined from the GPS buoy for a single over flight agreed with the value determined at the Harvest platform to within 1 cm.

Based on the success of both the Harvest tide gauge and GPS buoy campaigns, several other calibrations sites have been devised for the Jason-1 mission, many of them using a combination of GPS-referenced tide gauges and GPS buoys. In addition to the Harvest platform, they range from a network of tide gauge sites in the UK (Woodworth et al., 2004) to sites at several islands in the Mediterranean Sea (Bonnefond et al., 2003; Jan et al., 2004; Martinez-Benjamin et al., 2004; Pavlis and Mertikas, 2004) and one site in Tasmania (Watson et al., 2003).

Although the goal of these calibrations is to quantify the absolute altimeter range bias of Jason-1, realistically only the local bias in SSH is estimated. The bias estimate will be corrupted by biases due to geographically correlated orbit errors, SSB error, or the water vapor correction, which tends to have errors correlated with the amount of water in the path. These types of biases are not constant, but vary significantly from one place to another. The large number of absolute calibrations sites for Jason-1 allows one to quantify better some sense of the true error in this type of absolute calibration (**Table 2**). Although formal errors on all the estimates (based on scatter and number of observations) are of the order of a few millimeters, the absolute calibration estimates differ by up to several centimeters. This result suggests that

realistically, it may be difficult to determine the absolute bias to better than the 1–2 cm level using these single-site calibration techniques.

5.4.2 Cal/Val from Comparison to Tide Gauges

One can compute SSH anomalies from both altimeter and tide gauge measurements (even if they have not been consistently referenced), and compare the results. This is typically done by comparing the closest SSHA along altimeter passes that bound a tide gauge to the daily average tide gauge measurement (e.g., Mitchum, 1994) or by averaging altimeter data from multiple passes around the site over some period (typically one repeat cycle or a month) and comparing it to the tide gauge data averaged over the same time period (e.g., Cheney et al., 1994; Chambers et al., 1998a). The former can be thought of as a measure of the error in the 1-s SSH measurement, while the latter is an estimate of some smoothed SSHA.

Wyrtki and Mitchum (1990) found RMS differences of 3–10 cm between tide gauge data and Geosat data averaged into 2° latitude by 8° longitude grids. Cheney and Miller (1990) and Shum et al. (1994) obtained similar results using different orbits and processing. Mitchum (1994) compared TOPEX and tide gauge data early on in the mission, and found that the two agreed with each other on the order of 5 cm RMS, which is more representative of the error in a single 1 s value because of the methodology.

Table 2   Bias estimates for Jason-1

Site/Technique	Absolute bias	Relative bias with TOPEX_B	Reference
Harvest	13.8	11.9[a]	Haines et al., 2003
Corsica	12.0	10.1[a]	Bonnefond et al., 2003
Bass Strait	13.1	11.2[a]	Watson et al., 2004
Corsica	11.8	9.9[a]	Jan et al., 2004
Gavdos, Crete	14.5	12.6[a]	Pavlis et al., 2004
United Kingdom	12.9	11.0[a]	Woodworth et al., 2004
Ibiza	12.0	10.1[a]	Martinez-Benjamin et al., 2004
Mean (std. dev.)	12.9(1.0)	11.0(1.0)	
Global residuals	—	15.3	Beckley et al., 2004
Global residuals	—	14.4	Chambers et al., 2003a
Global residuals	—	14.1	Dorandeu et al., 2004
Global mean sea level	—	15.4	Leuliette et al., 2004
Global residuals	—	14.6	Shum et al., 2003
Mean (std. dev.)		14.8 (0.6)	

[a] Relative bias computed relative to TOPEX_B absolute bias of 1.9 cm measured at Harvest (Haines et al., 2003). Bias in cm of SSH, computed so $SSH_{true} = SSH_{meas} - bias$.

Other authors who used temporal and spatial averaging found RMS errors in the range of 2–4 cm (Cheney et al., 1994; Nerem et al., 1994b; Tapley et al., 1994b). Chambers et al. (1998a) analyzed 3 years of TOPEX data around 40 tide gauges in the Pacific and Indian Oceans and found a mean RMS of 3.5 cm and an average correlation of 0.87. During the Jason-1 cal/val phase, Chambers et al. (2003a) used the unique fact that T/P, Jason-1, and the tide gauges were observing the SL variations at roughly the same time in order to quantify the mean error in each instrument, based on an assumption that errors between the three data sets were uncorrelated. They estimated errors in the TOPEX, Jason-1, and tide gauge SSHAs at 3.1, 3.7, and 1.6 cm, respectively. The larger error in Jason-1 is largely attributable to the fact that the on-board tracker for Jason-1 does less smoothing to the waveforms than the TOPEX hardware, so the range measurement is slightly noisier than that of TOPEX.

Although one cannot determine the 'absolute' altimeter bias with the general tide gauge network, one can use the data to determine changes in the bias. The strength of this is that averaging altimeter–tide gauge residuals at many sites will effectively reduce the formal error in the determination, compared to the error at a single site such as the Harvest platform. This means that one has the potential to detect even small changes in the bias, of the order of several millimeters. Although one can simply average altimeter–tide gauge residuals to reduce the noise (e.g., Chambers et al., 1998a), the variance can be reduced even more if one accounts for both land motion at the tide gauges and phase differences between the altimeter and tide gauge measurements due to propagating ocean waves (Mitchum, 1998, 2000). For example, Haines et al. (2003) found that the TOPEX residuals at the Harvest calibration site are about 3–4 cm RMS, while Mitchum (1998) found that the RMS of the residuals averaged over 55 tide gauges was less than 1 cm.

The first significant use of this method was to detect a large, quadratic trend in the first release of the TOPEX SSH data, which was causing TOPEX SL to differ from that measured by tide gauges by nearly 7 mm yr^{-1} (Mitchum, 1998). The source of the error was eventually traced to an improperly applied oscillator correction, which caused not only the drift but also the order 15 cm bias first detected in the TOPEX data (Christensen et al., 1994; Ménard et al., 1994). However, even after the oscillator error was corrected, the tide gauge calibration continued to

suggest a drift in the TOPEX measurement, of the order of 1–2 mm yr^{-1}, with the sense being that the SL rate measured by TOPEX was not as large as that measured by the tide gauges (Chambers et al., 1998a; Mitchum, 1998). The error was eventually linked to drifts in the radiometer (Keihm et al., 2000; Section 5.2.3). Note, however, that the official T/P GDRs have still not been corrected for this error, and users have to either apply an ad hoc correction based on the comparison of the average TMR path delay with that from other radiometers (Section 5.2.3) or to use a separate correction product that has been distributed with the GDRs.

After correcting for the TMR drift, another apparent bias change in the TOPEX data was noted shortly after the redundant TOPEX_B altimeter was turned on because of degradation in the original TOPEX_A altimeter. The change was small, of the order of 5 mm, and it took some time to acquire enough data to ensure that the change was significant (Mitchum, personal communication, 2002). Shortly afterwards, it was shown that the bias was caused by an error in the SSB model used for the TOPEX_B data (Chambers et al., 2003b; Section 5.2.4). After applying a new TOPEX_B SSB model, there are no statistically significant drifts or changes in bias in the TOPEX data (Chambers et al., 2003b; Leuliette et al., 2004).

A similar tide gauge calibration has been performed on data from the first release Jason-1 GDRs and found a significant drift between Jason-1 and the tide gauges of more than 5 mm yr^{-1} (Leuliette et al., 2004). However, when a different set of precise orbits were utilized, the drift was reduced to 2.5 mm yr^{-1} (Leuliette et al., 2004), and the remainder of the error has been attributed to the bias jumps in the JMR (Section 5.2.3). When new precise orbits and corrections for the JMR jumps are utilized, the apparent drift in the tide gauge calibration is not statistically different than 0 (G. Mitchum, personal communication, 2005).

Thus, comparing altimeter SSH measurements with tide gauge data has proved to be a very accurate method to detect even small changes in the bias of the altimeter. In all the cases discussed, the bias change was first observed in the tide gauge calibration and then later linked to a specific error. Although calibration at the level of less than 1 mm yr^{-1} is not necessary for many applications, it is vital for determining changes in GMSL, where the signal is only about 2–3 mm yr^{-1} (Section 5.5.3).

5.4.3 The TOPEX/Poseidon – Jason-1 Calibration Phase

Although ERS-1 and ERS-2 flew along the same ground track for 1 year as part of a calibration phase, the time separation between the two measurements was 1 day, which meant that the satellites were not observing exactly the SSH. Although the data were used to calibrate and validate both ERS-1 and ERS-2 measurements (Scharroo, 2002), the results were affected by real changes in the true ocean during the 24 h time difference.

The Jason-1 calibration phase was planned so that the satellite flew over each point along the ground track approximately 70 s before T/P. Since the ocean and atmospheric signals change much more slowly than 70 s, the TOPEX and Jason-1 science instruments effectively observed identical SSHs in this configuration. This meant that any significant differences (either biases or large variances) between the Jason-1 and TOPEX measurements would be attributable to an error in either one, or both, of the instruments.

One strength of this cal/val mission was the shear number of points available for computing statistics, as there are typically more than 400 000 observations considered good for each 10 day repeat cycle. The cal/val phase lasted for 210 days (21 complete repeat cycles), at which time T/P was maneuvered into a new orbit with a ground track half-way between the original ground track. When the cal/val phase was planned, it was assumed that TOPEX was properly calibrated and that any large bias or variance would be attributable to an error in the Jason-1 system. Contrary to expectations, however, early analysis pointed to small, yet evident problems in the TOPEX data and not the Jason-1 observations. One of these problems relates to the SSB model used for the TOPEX measurements (Chambers et al., 2003b) and has been discussed in Section 5.2.4 Another problem relates to the wet path delay measured by the TOPEX Microwave Radiometer, which was found to change during yaw transitions (see Section 5.2.3). Both of these problems were relatively small and had gone undetected for several years before they became readily apparent in the data residuals between Jason-1 and TOPEX during the cal/val phase.

More importantly, this type of cross-calibration has proved to be invaluable for linking the SSH time series from two different satellites by determining a relative bias. Chambers et al. (2003a) pointed out that

local relative biases between Jason-1 and T/P computed with coincident data differ by several centimeters from the global mean, with slightly larger differences in coastal waters and the Mediterranean Sea, where the largest number of absolute calibration sites are located, and in areas of high SWH due to mean errors in SSB models. Some of the difference has been linked to residual geographical orbit error from the JGM-3 gravity field (Beckley et al., 2004). These results give some explanation for the differences in the bias determined from various *in situ* calibration sites (**Table 2**).

In contrast to the absolute bias computations, the relative bias (Jason relative to TOPEX) estimated from the global SSH residuals during the cal/val phase are far more consistent (**Table 2**). More significantly, there is nearly a 4 cm systematic bias between the mean of the point calibrations and the 'global' calibrations. If one considers the global calibration to be the truth for linking the altimeter data from two different missions to compute a long record of GMSL variation, then relying on an estimate from a single site could be in error by up to 4 cm. While this is not a problem where missions overlap (like T/P and Jason-1), it could be disastrous if there is a long gap between altimeters.

In summary, the results of the Jason-1 and T/P cal/val phase have pointed out several important things. First, residual analysis can detect very small errors that may have gone undetected using other methods. This has proved to be especially true for calibration of the water vapor radiometers. Second, the idea of a global bias in SSH measurements – meaning that the bias is the same everywhere – is a convenient fiction. The SSH measurement depends on a number of measurements and corrections, and a lot of the errors in these constituents can have biases that differ depending on where the measurement is made – that is, SSB errors that are dependent on the mean SWH state, or wet troposphere errors that may be consistently larger in regions where there is more water vapor like coastal areas. This means that the global average bias can differ significantly from the local bias. Even considering only the longest record from Harvest (Haines et al., 2003), the difference between a single local bias estimate and the global average can differ by nearly 3 cm. This is not a trivial difference for continuing long records of GSML change, which have variations of the order of less than 1 cm and trends of only 2–3 mm yr^{-1}.

5.5 Applications to Geophysics

5.5.1 Mean Sea Surface, Gravity, and Bathymetry

For hundreds of years, the ill-defined term 'mean sea level' has been used as a geodetic reference. At best, MSL is based on a benchmark near a tide gauge that represents a time average of the local SSH. At worst, the reference is extrapolated between tide gauges or defined by some arbitrary barometric pressure value. Even if one could measure the MSL around coastlines with millions of tide gauges, such a reference will not be constant over time, since SL is rising for a variety of reasons (Section 5.5.3). In addition, the land that tide gauges are affixed to may also be lifting or subsiding, either due to elastic rebound from the ice loading during the last glaciation, volcanic activity, or pumping of water and oil from underground reservoirs. In our modern age, with satellite altimeters and GPS measuring heights to a few centimeters, a more definitive, stable reference is needed.

Once such reference is the geoid, which is a mathematical surface that represents a specific equipotential surface related to the Earth's gravitational and rotational potential (e.g., Tapley and Kim, 2001). Modern geoids are based on global Earth gravity models determined from satellite tracking data (e.g., Nerem *et al.*, 1994a; Tapley *et al.*, 1996) – most recently on data from the Gravity Recovery and Climate Experiment (GRACE) (Tapley *et al.*, 2004b). Geoids will differ depending on what gravity model is used to determine the gravitational potential, as well as what equipotential value is assumed. In recent years, an equipotential determined from an altimetric mean SSH is used.

To first order, the geoid is approximated by an ellipsoid of revolution, defined by an equatorial and polar radius (R_E and R_P), or an equatorial radius and flattening coefficient, $f = ((R_E - R_P)/R_E)$. There are many reference ellipsoids in the literature; the one most used in altimetry was defined early in the T/P mission to represent a surface with a zero mean to the measured SSH (e.g., Tapley *et al.*, 1994a). It has values of $R_E = 6\,378\,136.3$ m and $f = 1/298.257$. The geoid still has significant normal deviations above or below the ellipsoid, of order 100 m. In the literature these geoid undulations are often referred to as 'geoid height,' and should not be confused with the full geoid (which includes the portion represented by the ellipsoid).

Ninety-nine percent or more of the signal in the SSH measured by an altimeter with respect to the ellipsoid is the geoid height. The next largest quasi-permanent signal is related to the mean ocean geostrophic circulation, which causes slopes in the mean dynamic ocean topography (DOT) (e.g., Wunsch and Gaposchkin, 1980; Tapley *et al.*, 2003). The DOT is the difference between the measured SSH and the geoid, and has typical variations of ±2 m. **Figure 11** shows the mean SSH and geoid along a particular T/P pass, along with the difference. Note that the geoid pictured does not model gravity variations with wavelengths smaller than 100 km wavelength, even though these signals are in the mean SSH. These residual geoid variations appear in the raw DOT measurement. **Figure 12** shows the global DOT filtered in order to suppress these residual geoid signals. The topography slopes result in horizontal pressure gradients that are equally balanced by the Coriollis force caused by ocean currents. This causes a difference in the relationship between the topography gradient and current in the Northern Hemisphere versus the Southern. The topography is higher to the right of a current and lower to the left in the Northern Hemisphere, while the direction of the slope is reversed in the Southern Hemisphere, with higher topography to the left of the current.

One has to be careful about how the permanent lunar and solar tidal effects are represented in the geoid if it is combined with altimetry (Rapp, 1989). The altimeter measures the SSH including the mean permanent tide. The geoid can be represented with or without the permanent tide, and the difference is 20 cm over a wavelength of 20 000 km.

The remaining signals in the SSH are caused by time-variable ocean signals such as tides, planetary waves, eddies, fluctuations in the circulation, as well as exchanges of heat with the atmosphere and water mass with the atmosphere and land. For many oceanographic applications, it is the time-variable component of SSH that is of most interest, and one is most concerned with accurately removing the mean SSH (or mean sea surface, MSS) from the data. This is not trivial for two reasons. First, the altimeter does not exactly sample the same location every repeat cycle, although the subsatellite points fall roughly within a region of 7 km along the ground track and ±1 km on either side of the nominal track at the equator (**Figure 13**). Second, the MSS (geoid) is not constant within this area, or bin, but can have fairly steep gradients in both the along-track and cross-track directions. Brenner *et al.* (1990) estimated

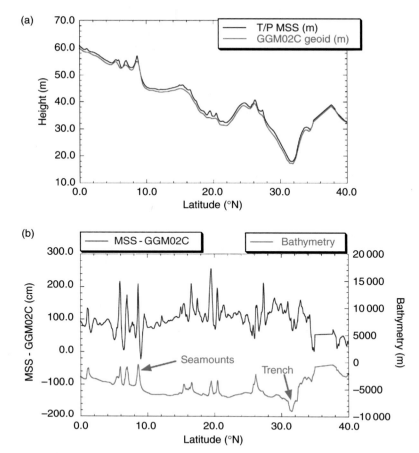

Figure 11   (a) Average SSH along a T/P pass in the western Pacific (blue), along with the GGM02C geoid (red) (Tapley *et al.*, 2005). (b) The difference (blue) shows the long-wavelength dynamic topography and higher-resolution geoid that is not modeled. The corresponding bathymetry is shown in red.

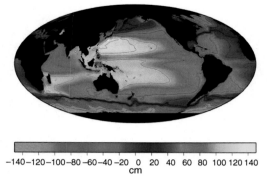

Figure 12   Global DOT from GSFCMSS00 (Wang, 2000) – GGM02C (Tapley *et al.*, 2005), filtered to suppress variations with wavelengths smaller than 444 km (e.g., Tapley *et al.*, 2003).

that the average gradient over the ocean was about 2 cm km^{-1}, which means that the difference in MSS from one end of the sampling region to the other would be as much as 14 cm. If one simply averaged the SSH points in the region and removed the mean from the data to compute the SSH anomalies without accounting for the gradients, the geoid slope would appear to be a time-variable signal since the altimeter samples the region in time (e.g., Chambers *et al.*, 1998b). To make matters worse, the geoid slopes are even larger near trenches or seamounts, with values up to 20 cm km^{-1} (Brenner *et al.*, 1990).

One can estimate the along-track gradients by averaging SSH along each ground track to create an MSS profile. This will give the average gradient over ~6–7 km (determined by the sampling rate), which one can then use to correct for the along-track error. However, this approach will not determine the cross-track geoid gradient, which could account for up to 4 cm of erroneous SSH variations over most of the ocean, but up to 40 cm near trenches. To account for this, MSS models have been developed based on

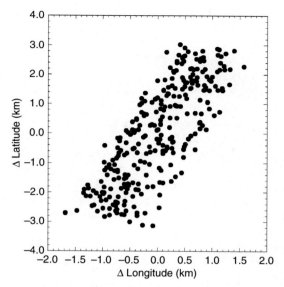

Figure 13   Sub-satellite points in one T/P bin for the entire mission.

gridding the MSS profiles from multiple altimeter missions. Then, one can determine and correct for the geoid gradients from these models. Many such models have been produced over the years, differing mainly by altimeter data used, time span of the data, and method used to map the data. Tapley and Kim (2001) discuss the principal methods that have been utilized for mapping the data. The earliest models were based on combinations of Seasat and GEOS-3 data (e.g, Marsh and Martin, 1982; Rapp, 1986; Marsh *et al.*, 1992). The Marsh *et al.* (1992) model had a spatial sampling of 1/8° (∼15 km at the equator) which was only possible because GEOS-3 had a much denser ground track than Seasat. Data from as far away as 100 km of the grid center were used in the gridding process, albeit downweighted as a function of distance. Later, data from the Geosat exact repeat mission were included (Basic and Rapp, 1992), but this did not significantly improve the resolution because it was in the same ground track as Seasat. Arguably, the errors along the Seasat–Geosat ground track were reduced because of more averaging. The largest improvement in MSS models came when T/P, ERS-1, ERS-2, and Geosat geodetic mission data were included. By the beginning of the Jason-1 mission several MSS models with resolutions of 1/30° (<4 km at the equator) were available (Cazenave *et al.*, 1996; Andersen and Knudsen, 1998; Tapley and Kim, 2001; Wang, 2000).

Although all modern global MSS models have formal errors of a few centimeters or less, one has to be careful not to interpret this as the absolute error in the MSS height. One significant problem in MSS mapping is that systematic biases in the SSH measurement will go directly into the MSS and these are not included in the formal error statistics. These errors are caused by residual biases in the range measurement, geographically correlated orbit errors, SSB error, and aliased solar tide error for the sun-synchronous altimeters. Another source of errors is the use of one-time profiles from the Geosat Geodetic Mission and ERS-1 168-day repeat mission. These profiles have not averaged out the ocean variability such as mesoscale eddies which can cause deviations of up to a meter or more. Modelers have attempted to reduce such errors by referencing the MSS profiles from other altimeters to those of T/P (e.g., Tapley and Kim, 2001), but this will not remove systematic errors in T/P, nor will it remove all of the systematic error from the other altimeters at the shortest wavelengths. If one compares two modern MSS models (**Figure 14**), one observes differences of several tens of centimeters in some areas. Although some of the difference is due to different averaging intervals, a significant portion can only be attributed to errors in one or both of the MSS models.

Another approach to estimating geoid gradients directly along the ground track of a particular satellite was introduced by Chambers *et al.* (1998b). They estimated the parameters to a two-dimensional plane for a series of bins defined along a nominal ground track based on several years of SSH measurements for the T/P mission. They demonstrated that this

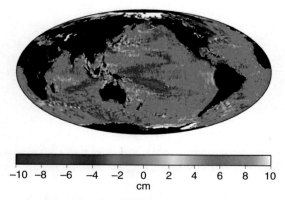

Figure 14   Difference between GSFCMSS00 (Wang, 2000) and CSRMSS98 (Tapley and Kim, 2001), averaged in 0.25° grids.

local along-track MSS model reduced the variance of SSH anomalies better than available gridded MSS models because it was tuned to the particular satellite and had the highest possible resolution, and that it could be updated quickly as more data was available. One problem with the approach was the aliasing of true SL variability into the estimation of the gradients, although later refinements have reduced the problem by either including harmonic parameters to account for the signals (Chambers, 2002), or by first removing gridded maps of long-wave SL variability (Dorandeu et al., 2003).

The MSS can be used as a direct measure of the geoid height with an error equivalent to about 1% of the signal, arising from the nongravitational dynamic topography signal. Often, a topography based on historical measurements (e.g., Levitus, 1982) has been subtracted from the MSS model and the residual is used as a measure of geoid height. However, the ocean topography model is not perfect, nor will it represent the smallest-scale gradients related to the circulation, such as those near intense jets like the Gulf Stream and Kuroshio. More importantly, systematic errors caused by the altimeter measurements in the MSS will remain.

Such proxies of geoid height over the ocean from altimetry are still vital in the determination of global geoid models in combination with tracking data from satellites (e.g., Lemoine et al., 1998; Tapley and Kim, 2001), since they are the only direct measure of short-wavelength (7–500 km) variations of ocean gravity. There have been problems in previous global models, though, arising from the difficulty in separating gravity from oceanographic currents in the MSS. Until recently, satellite tracking data alone has not been able to accurately resolve the geoid at the wavelengths where this is important (e.g., Tapley et al., 1994b). Tapley et al. (2003) demonstrated that the long-wavelength portion of the EGM96 geoid (Lemoine et al., 1998) actually contained significant oceanographic signals from the combination of the altimetry and satellite tracking data in the solution, which meant it could not be combined with altimetry to determine the dynamic ocean topography. This has become less of a problem when tracking data from the Gravity Recovery and Climate Experiment (GRACE) are utilized in the combination (Tapley et al., 2003), as the GRACE data alone can be used to determine the long-wavelength geoid to an accuracy of a few centimeters (Tapley et al., 2005).

Other methods have been used to isolate only the short-wavelength gravity information from the altimetry. Haxby et al. (1983) first showed how to differentiate a gridded MSS in order to obtain a map of gravity anomalies. One benefit of differentiating the MSS measurement is the reduction in some systematic errors, especially those with very long wavelength, although it may not eliminate all errors where data from multiple altimeter have been used. Because of this, another method to map gravity anomalies from altimetry was developed (e.g., Sandwell, 1992; Sandwell and Smith, 1997). Instead of differentiating a pre-gridded MSS model, SSH gradients along each satellite pass were computed and averaged, then converted to gravity anomalies and gridded. By doing this, more of the systematic long-wavelength errors for each particular satellite were eliminated before the data were mapped.

Much has been learned about the geophysics of the sea floor from such altimetric geoid or gravity anomaly maps. The primary cause of change in the geoid over short spatial scales is variations in the sea floor topography. Thus, by studying the altimetric geoid or gravity anomaly maps one can infer information about seamounts, trenches, spreading ridges, or subduction zones (e.g., Cazenave and Royer, 2001). The information can also be used to infer the seafloor topography, or baythmetry (e.g., Smith and Sandwell, 1994; Sandwell and Smith, 2001) although there is one important limitation to this. The gravity to topography transfer function is related to the ocean depth, and causes attenuation of the signal in SSH (Sandwell and Smith, 2001). For example, if two topography features are very close together (on the order of the ocean depth or less), they will appear in the SSH as a single feature. Very sharp changes in the topography are also spread out in the SSH. Thus, a step function in the bathymetry would appear as a more gradual slope in the SSH. Still the gravitational effect of the bathymetry is quite evident in the SSH (**Figure 11**).

5.5.2 Tides

Tides represent the largest time-variable geophysical variation in the SSH. The magnitude of ocean tides range from a few centimeters in the deep ocean to more than 10 m in some coastal areas, like the Bay of Fundy in Canada. Although explanations of ocean tides can be traced to Newton's gravitational theory in 1687, the ability to predict tides accurately has been developed only more recently. One hundred years after Newton, Laplace formulated tidal equations that were impossible to solve analytically without precise knowledge of the shape of the ocean basin. Darwin (1886) made a significant advance by introducing harmonic techniques based on fitting data from tide gauges to known frequencies dependent on solar and lunar motion. However, this still only allowed the prediction of tides at discrete points where a tide gauge was present. As Le Provost (2001) pointed out, there are less than a few hundred tide gauge sites around the globe which have records of longer than 1 year that are also more recent than 1950.

In the 1960s when large mainframe computers became commonly available, scientists began to create hydrodynamic models that attempted to solve Laplace's tidal equations numerically (e.g., Le Provost, 2001). However, these were still not able to predict tides as well as harmonic fits to tide gauge data because the models were limited in resolution, and there were still significant uncertainties in ocean boundaries, bathymetry, and tidal energy dissipation due to friction. Schwiderski (1980) attempted to reduce this error by constraining the models to fit tide gauge observations where they were available. Although these types of solutions were better than those from purely hydrodynamic models, they still did not adequately predict tides where there had been no direct observations (Woodworth, 1985).

The data from satellite altimetry, on the other hand, can be thought of as hundreds of thousands of tide gauges spread more or less uniformly over the ocean. The one significant drawback is that the altimeter does not measure the tide at a rate shorter than the tidal period, which leads to aliasing (Section 5.3.2). However, if the alias period is not too long, then one can extract the tidal parameters from the altimeter measurements of SSH with a long-enough record. Details of the methods used to determine tides from altimetry is beyond the scope of this chapter; see Le Provost (2001) for a more complete summary.

Data from the first several years of the Geosat mission were used to empirically solve for ocean tidal parameters (Cartwright and Ray, 1990). Although better than the previous Schwiderski (1980) estimate, this Geosat model still had errors over the open ocean of order 5 cm RMS, particularly in the twice-per-day lunar (M2) and solar (S2) constituents due to the aliasing of the Geosat repeat period. As noted in Section 5.3.2, the T/P mission was specifically designed to reduce the aliased period of the major tidal constituents to be less than 90 days

(Parke et al., 1987) with the goal to determine new ocean tide models.

Within 1 year of the start of the T/P mission, several empirical tide models had been calculated from the T/P data, all with estimated errors of less than 4 cm (Egbert et al., 1994; Ma et al., 1994; Schrama and Ray, 1994). With a few more years of data, even more models were available, all generally agreeing with each other within 1 cm RMS in the deep ocean (Shum et al., 1997) and agreeing within 2–3 cm RSS with tide gauges (Le Provost, 2001).

There have been incremental improvements in tide models since 1997. There are two updated models that are included on the Jason-1 GDRs. The Goddard Ocean Tide model 1999.2 (GOT99.2) (Ray, 1999) is an empirical update to an earlier model, while the FES99 model is a hydrodynamic model that assimilates T/P altimetry (Le Provost, 2001). The differences between the newer models and the older ones are relatively small over the open ocean, but the new ones do show some improvement in shallow water areas. One reason is that the nearly 10-year long record from T/P has allowed better determination of the tides in the presence of larger errors in these regions. In addition, FES99 is a finite-element hydrodynamic model, and its grid has been increased in shallow water as computers have become faster, smaller, and more affordable. Also, because the K1 tidal constituent is aliased to 173.2 days in T/P, a longer time span has allowed better separation from the semi-annual period and hence continued improvement for this constituent.

Although the problem of tide prediction may now appear to be solved, there are still areas for improvement. Most altimeter tide models have been based on binned data over relatively large areas (generally 3°, the approximate ground track spacing of T/P) in order to obtain a global solution. This is fine over the open ocean where wavelengths of tides are of the order of 1000 km or longer. However, the tidal wavelength shortens significantly with decreasing depth, as much as an order of magnitude or more. The ground track spacing of a single altimeter like T/P will not adequately support the resolution of tides in shallow water. One can estimate tides at the resolution of sampling (about 7 km) for specific T/P ground tracks (e.g., Tierney et al., 1998) but this still does not help in costal regions between ground tracks. For this, one either needs to use a hydrodynamic model (e.g., FES99) or combine data from multiple altimeters like Geosat, ERS-1, ERS-2, and

Envisat. Even though there are significant aliasing problems for all these satellites, this has been attempted by Andersen and Knudsen (1997) using T/P, Geosat, and ERS-1 data. They found, however, that including the data from the other altimeters degraded the solution, based on comparison to tide gauges. This is possibly as much due to higher errors in the Geosat and ERS-1 data as to the aliasing issues. Some improvement may be possible in the near future using data from the new T/P ground track (**Figure 15**), which was moved half-way between the old T/P and Jason-1 ground track. This pattern allows binning at a smaller resolution ($\sim$1.5° at the equator) and so will be able to resolve smaller tidal wavelengths in shallow water. In addition, Envisat errors are significantly lower than ERS-1, and so it may be able to contribute information for tidal constituents that are not aliased. In addition to the problem in shallow water, the tide models based on T/P altimetry are relatively poor poleward of $\pm$66° latitude, since T/P does not fly over these regions. At the moment, tide predictions are based purely on the hydrodynamic models (and whatever tide gauge data are available in this area). Envisat data may help for certain constituents, but only up to 82°.

Even though there is still room for improving coastal tide models, altimetry data have provided exciting new insights into other facets of tidal dynamics. One of these is the detection of internal tides (e.g., Ray and Mitchum, 1996). An internal tide is a tide-like oscillation of isotherms within the ocean. The dominant ocean tide is a barotropic response of the ocean, so there is no relative motion between isotherms (layers of constant temperature);

in other words, the entire water column moves up and down as water mass is redistributed by the barotropic tidal wave. An internal tide, on the other hand, causes vertical relative motion between the isotherms. The variations can be of the order of hundreds of meters of amplitude in the isotherms and have periods similar to oceanic tides. However, the surface displacement that would be seen by an altimeter is only a few centimeters (Le Provost, 2001), and it will have a wavelength of only a few hundred kilometers. Internal tides are generated when the barotropic ocean tides interact with sharp changes in the bathymetry. Thus, they tend to occur near seamounts, ridges, and continental shelves (Morozov, 1995).

Although theoretically detectable by altimetry, most direct observations of internal tides have been based on subsurface temperature measurements, until recently. They were not identified in altimetry until Ray and Mitchum (1996) analyzed several T/P ground tracks off the coast of Hawaii. Because of the short wavelength of internal tides, they first estimated specific tidal constituents (such as M2) along each T/P pass near the Hawaiian Islands over very small bins (5.75 km). This estimate will include both the contribution from the long-wavelength ocean tide and the shorter-wavelength internal tide. By subtracting off a model of the long-period tide (based on global T/P empirical models), the internal tide signature was isolated. The internal tide was shown to propagate away from the Hawaiian Ridge (Ray and Mitchum, 1996). Further investigation showed measurable internal tides in other regions where they had been predicted: off Tahiti, over the mid-Atlantic Ridge between Brazil and Africa, and along the European continental shelf in the North Atlantic (Ray and Mitchum, 1997). Kantha and Tierney (1997) have performed a similar analysis globally for the M2 constituent; their maps show high correlation between the amplitude of the internal tide and the bathymetry over much of the world's oceans.

5.5.3 Sea Level Change

One of the major benefits of satellite altimetry over the ocean is the ability to measure SL variations in the deep ocean away from coasts. Although local SL change has been measured for decades using tide gauges, it was not possible to measure in the open ocean before the advent of satellite altimeters. Additionally, tide gauges measure SL relative to the

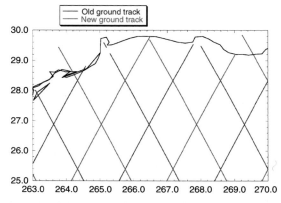

Figure 15 Ground tracks for original T/P (and current Jason-1) orbit (blue) and new T/P orbit (red) in the Gulf of Mexico.

local land motion, which may be rising or falling due to post-glacial rebound (PGR), volcanic activity, or pumping water and oil from underground reservoirs, and the tide gauges are not tied to a common reference frame. A satellite altimeter, on the other hand, is tied to a consistent reference frame and so measures the total (not relative) SL change to the best of our capability (e.g., Nerem *et al.*, 1998).

Although tide gauges with very long records show a reasonably consistent rate of SL change over the last 100 years of about 1.8 mm yr^{-1} when corrected for PGR (Douglas, 1991, 2001), there has been some question whether this truly represents the rate of GMSL, since all but a handful of the 24 gauges are on the east coast of North America and the west coast of Europe. If the GMSL rate is mainly due to redistribution of water mass from ice sheets in Greenland and Antarctica to the ocean, then the pattern of GMSL rise would be expected to be nearly uniform. If it is due to changes in ocean heating, then it would not. Because of its nearly global sampling, satellite altimetry was expected to resolve some of these questions.

Computing GMSL from altimetry is relatively straightforward, and most analyses use a procedure similar to that described in more detail by Nerem (1995) with only a few modifications. Essentially, the SSH measurements along each ground track pass are reduced to SSH anomalies (SSHAs) about the mean SSH using either a mean profile, a global MSS model, or satellite-specific along-track MSS model (Section 5.5.1). The SSHAs for each repeat cycle are then averaged, accounting for the fact that there are more observations in the high latitudes because of the ground track spacing (Nerem, 1995). From this, one obtains a point representing the GMSL for each repeat period. With a long-enough time series, the rate of GMSL change can be obtained. Note that although GMSL is often used in the literature, it is not meant to suggest that the entire ocean is rising or falling by this amount. GMSL is really just an average of local SL variations, which are generally higher than or lower than the global average. However, GMSL is often used to study small, subtle changes related to climate change that may or may not be readily apparent in the local measurements.

The Seasat mission did not last long enough to make regular measurements of GMSL, although a determination was made using a month's data (Born *et al.*, 1986). At best its data represent a single measurement of GMSL in 1978 with a fairly large error bar due to the limitations in a single-site calibration,

especially one without a large number of observations over time (Section 5.4.1). The Geosat exact repeat mission lasted nearly 3 years from 1986–89, but the GMSL measurement was corrupted by errors in the orbit and range corrections (Wagner and Cheney, 1992). The determination of the GMSL rate from Geosat ranged from 0 mm yr^{-1} (Tapley *et al.*, 1992) to −12 mm yr^{-1} (Wagner and Cheney, 1992), and because the altimeter was never calibrated like Seasat, it was difficult to tie the two measurements together.

Although ERS-1 was launched over a year before T/P, it too suffered from large errors in the orbit, as well as problems in the instrument and atmospheric corrections, which limited the ability to determine GMSL. It was not until several years of reprocessing and corrections had been made that GMSL could be determined from ERS-1 data (e.g., Anzenhofer and Gruber, 1998), and by this point a much longer and more accurate time series was being determined regularly from T/P (e.g., Nerem *et al.*, 1997).

The earliest estimates of GMSL rate from T/P (Nerem, 1995; Minster *et al.*, 1995) were unfortunately corrupted by the error in the oscillator correction that caused the estimate to be nearly 7 mm yr^{-1} too high (Nerem *et al.*, 1997; Section 5.4.2). Intermediate results were affected by a smaller drift in the TMR brightness temperatures (Section 5.2.3), but this was soon corrected and reasonable estimates of GMSL from T/P were being calculated (e.g., Nerem *et al.*, 1999). Finally, the change from TOPEX_A to TOPEX_B caused another bias in T/P GMSL of nearly 1 cm after February 1999 that was corrected when the SSB model was updated (Chambers *et al.*, 2003b; Section 5.2.4). **Figure 16** shows the latest time series of GMSL from corrected T/P and Jason-1 data. The trend over the 12-year period is 3 mm yr^{-1}. Leuliette *et al.* (2004) estimate that the error in this determination is ±0.4 mm yr^{-1}, based on a combination of the formal error in the fit and the accuracy of calibration relative to tide gauges.

Note that the RMS for the difference between the T/P and Jason-1 measurements during the 210-day overlap is 4 mm, which suggests an error in any 10-day estimate of less than 3 mm RMS. This indicates that the variations about the long-term linear trend are not noise, but represent real GMSL variations at shorter periods. Two components have been linked to seasonal transport of water mass between the hydrosphere and ocean and seasonal changes in the ocean heating (Chen *et al.*, 1998; Minster *et al.*, 1999). Each of

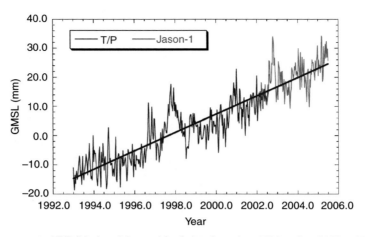

Figure 16   GMSL from corrected T/P (blue) and Jason-1 (red) data from Jan. 1993 to Jun. 2005 at 10-day intervals. The T/P data have been corrected for the oscillator error (Section 5.4.2), the drift in TMR (Section 2.3), and the different TOPEX_A and TOPEX_B SSB models (Section 5.2.4), while the Jason-1 data have had a relative bias applied (Section 5.4.1) and have been corrected for the biases in the JMR wet path delays (Section 5.2.3). The long-term linear trend is shown with a thick blue line.

these signals have amplitudes of 8–10 mm, but they are out of phase so that the signal in GMSL is of order 4 mm. Although this is a very small signal, it is detectable in the altimeter measurements.

An even larger variation is apparent in **Figure 16** during 1997, when GMSL rose steadily by nearly 25 mm before peaking at the end of the year and then falling again. This was linked to the 1997/1998 El Niño event by Nerem *et al.* (1999), who showed that there was a reduction in the global mean Outgoing Longwave Radiation at the same time, which implied that the ocean was storing more heat during this period. Additionally, they demonstrated that a model that assimilated *in situ* temperature data also reproduced the signal. The discovery that GMSL was affected to such a large amount by inter-annual variations in the ocean like El Niño has led to some concern that part of the difference between the long-term rates measured by tide gauges over 100 years and altimetry over 10 years may be due to decadal variations in the ocean. Nerem *et al.* (1999) attempted to estimate this effect using the southern oscillation index (SOI) scaled to GMSL and found that at least 10 years of continuous data were needed in order to reduce error from El Niño-period variability to less than 0.5 mm yr^{-1}.

However, this analysis did not include variations with periods of longer than approximately 7 years. At the time, large decadal variations in the ocean had not been observed with any confidence. Recently though, several studies examining variations in heat storage over the ocean suggest a significant change

shortly before T/P began making measurements. One component of SL change is the so-called steric component, which is caused by changes in the water density as temperature and salinity of the water column vary. The larger portion is the temperature, or thermosteric, component. Levitus *et al.* (2000) have shown that the temperature throughout the water column for all ocean basins has been warming over the last 50 years, and that this is equivalent to a GMSL rate of 0.55 mm yr^{-1} (Levitus *et al.*, 2005). Willis *et al.* (2004) have demonstrated that the mean thermosteric SL rate during the T/P period is closer to 1.6 mm yr^{-1}, which suggests an increase of 1 mm yr^{-1} in the thermosteric component during the 1990s compared to the longer-term rate. This is very close to the level at which the T/P and tide gauge estimates differ, which suggests that the difference is due to a change in ocean heat storage over the last 10–15 years. Whether this is an acceleration in heat storage (and hence GMSL) or a decadal variation is still unknown at the moment, as there are no signs of a reduction in GMSL or mean ocean heat storage as of yet.

Altimetry also gives a unique perspective on change in local rates of SL, which is impossible from tide gauge data without significant extrapolation. Local rates (**Figure 17**) determined over the same period of time as the GMSL curve in **Figure 16** show that there are only a few locations where the local rate is the same as the GMSL rate. In most places it is either significantly higher (as much as 2–3 times larger) or actually falling. We know that

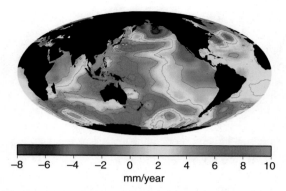

Figure 17 Map of local SL rates determined from T/P and Jason-1 data from Jan. 1993 to Dec. 2004. Data were mapped using a Gaussian with a 1000-km radius, before computing trends. All data were corrected as for **Figure 16**.

the patterns change significantly during El Niño years in the tropics when there are large local changes in the heat storage. Several studies have shown high correlation between local thermosteric and total SL rates (Cabanes *et al.*, 2000; Willis *et al.*, 2004) although there are still significant differences in the magnitude. Some of these differences may be caused by long-term changes in the salinity (Antonov *et al.*, 2002) or ocean mass variations. This is still a topic that needs much investigation.

In addition to studying the patterns of SL rate, authors have used maps of SL anomalies determined from altimeters to study other patterns of SL variability, using empirical orthogonal functions (EOFs) (e.g., Hendricks *et al.*, 1996; Chambers *et al.*, 2002). EOF analysis is used to find correlated patterns in a data set, where the patterns are represented by a spatial mode (EOFs) and an associated time series (principal components). The modes are sorted by the amount of variance of the total signal explained. Thus, the first mode will explain the largest percentage of variability and the last modes will explain the least. Chambers *et al.* (2002) used such spatial patterns estimated from T/P data in order to reconstruct global maps of SL variability from tide gauge data, using an idea first used with sea surface temperature data (Smith *et al.*, 1996). The idea is to estimate the associated time series from the tide gauge data using the altimeter patterns as a basis function. Chambers *et al.* (2002) computed only the interannual variations after removing all long-term trends from tide gauge data, and then derived a GMSL time series from the reconstructed fields back to 1950. They found significant El Niño variability throughout the time

period, and verified the results in the 1980s using Geosat data and a proxy for SL derived from sea surface temperature records. Church *et al.* (2004) recently extended this idea to better determine long-term GMSL rates from tide gauges by using the EOF patterns from T/P to model interannual variations. As longer altimeter records become available, spatial patterns associated with decadal variations will become detectable, and this will allow for the study of even more lower-frequency GMSL variations in the historical tide gauge record.

5.6 Conclusions and Future Prospects

Satellite radar altimetry has become a powerful tool for measuring both geophysical and oceanographic signals with high precision. The accuracy of the SSH measurement is now approaching that of a tide gauge, but no comparable instrument can match the satellite altimeter for the sheer density of measurements in such a short period. For the first time, we are truly obtaining global measurements of SL on a regular basis. Data from current missions like Jason-1, Envisat, and GFO are already being used in operational data-assimilating models for tracking ocean eddies, current variations, and predicting El Niño and hurricane intensification. Future radar altimeter missions like the NASA/CNES Jason-1 follow-on are planned, but will not significantly advance the capability of the measurement (merely continue the length of the data records). Several new types of altimeter systems that have significantly different resolution or precision over a radar altimeter have been either proposed or tested. In concluding this chapter, we would like to briefly describe some of these, and comment on their possible use for advancing the science of geophysics.

5.6.1 Laser Altimetry

Two laser altimeters have already been operated from satellite platforms. The Mars Orbiter Laser Altimeter (MOLA) was launched in 1996 aboard the Mars Global Surveyor mission, and conducted a multiyear mapping mission of the red planet beginning in early 1999 (Smith *et al.*, 2001). An earlier version of the MOLA instrument had been on the Mars Observer mission that never reached orbit. The Ice, Cloud, and Land Elevation Satellite (ICESat) was launched into Earth orbit January 2003, and carries the Geoscience Laser Altimeter System (GLAS), with a primary mission to measure changes in the Earth's ice sheet elevation over time (Zwally et al., 2002). Although radar altimeter data have been used to measure ice sheets for some time (e.g., Zwally and Brenner, 2001), the accuracy of the radar measurement over the ice is much worse than over the ocean, because ice sheet elevation changes so much over the footprint of the radar pulse.

The footprint of a laser is significantly smaller than that of a microwave pulse, typically 70 m compared to the effective footprint for a radar altimeter of order 10 km. This allows for the along-track separation of unique data points to be much closer, about 200 m or less compared to 7 km for the radar, and hence the laser altimeter can detect steep changes in the surface elevation with higher accuracy than a radar altimeter. However, the quality that makes the laser altimeter so good over ice sheets degrades its performance over the ocean compared to a radar altimeter. Because the along-track separation and footprint size of the laser altimeter is smaller than the wavelength of typical waves, the laser altimeter measures the wave surface of the ocean, not the mean surface like a radar altimeter. With sufficient averaging of profiles, though, one may be able to extract measurements of the mean sea surface along the ICESat ground track at a higher resolution than is possible with any radar altimeter system (Zwally et al., 2002).

Although ICESat has been operating for nearly 3 years, its data are still being calibrated and validated (e.g., Luthke et al., 2005) and so not all of the possible science applications have been explored. Additionally, the first laser failed after only 36 days of operation. There are three separate lasers on ICESat, and each was designed to operate for at least 1 year in order to meet a mission lifetime of 3 years. After an independent review (Kichak, 2003) it was theorized that the most likely cause was a short in the diode array, and that the same failure could occur in the other two lasers. It was recommended that a reduced science mode be planned, operating the lasers at a lower temperature and not continuously in order to extend the life-time as long as possible. Since then, GLAS-2 has operated for three periods during 2003 and 2004, and GLAS-3 has operated for 4 periods during 2004 and 2005. The typical period of operation has been Feb.–Mar., May–Jun., and Oct.–Nov. of each year.

Although ICESat has not met all of its science goals, much like Seasat it has demonstrated the

capability of the laser altimeter for Earth geoscience applications, and we can hope that future satellite laser altimeter missions are conducted.

5.6.2   Delay-Doppler Altimeter

A Delay-Doppler (DD) radar altimeter measures the phase change of the return signal using techniques developed for synthetic aperture radars instead of the travel time, which leads to improved precision by a factor of 2 or more at the sampling rate of a standard radar altimeter (Raney, 1998). This means that a DD altimeter can makes samples more frequently than a radar altimeter with comparable precision. The Delay-Doppler altimeter also has a smaller effective footprint than a standard radar altimeter, which means it will be more accurate in places with steep changes in the topography, such as over ice sheets.

Johns Hopkins University Applied Physics Laboratory (APL) has built, tested, and demonstrated DD altimeters on aircraft for several years now (APL, 2005). The first space-based Delay-Doppler altimeter was installed on the Cryosat satellite to measure changes in the Earth's ice sheets. Unfortunately, Cryosat crashed into the Arctic Ocean due to a launch failure on 8 Oct. 2005. As of now, there are no other Delay-Doppler altimeter missions planned, although we expect a replacement Cryosat mission to be proposed.

5.6.3   Wide Swath Altimeter

Another technique has been proposed to measure the SSH with an even higher spatial density than is even possible with a Delay-Doppler Altimeter, based on a radar interferometer (Rodriguez and Martin, 1992; Rodriguez and Pollard, 2001), and known as a wide swath altimeter (WSA). The details of the instrument are beyond the scope of this chapter, but the basic concept is that by placing two interferometric antennas on the ends of several meter booms attached to a satellite bus with a nadir radar altimeter, one could obtain a swath of SSH measurements several hundred kilometers wide, instead of the 2-km wide swath with a typical radar altimeter (Rodriguez and Pollard, 2001). The nadir instrument would be used for estimation of ionosphere and tropospheric path delays, as well as calibration of the interferometer altimeter data.

Such an instrument was being seriously considered for inclusion in the follow-on mission to Jason-1 – the Ocean Surface Topography Mission (OSTM). It was estimated that the WSA on OSTM would provide 10-day SSH anomaly maps with approximately 15 km resolution, with only a few gaps between crossovers (Rodriguez and Pollard, 2001). Although the accuracy of the off-nadir points was predicted to be not as good as the nadir points, the difference was not too great (Rodriguez and Pollard, 2001). Unfortunately, even though work on the WSA had progressed significantly beyond a simple conceptual model, budget limitations and concerns about a slippage of the planned launch date forced its removal from the final OSTM design. There are no current plans for a WSA mission, but work is ongoing on the concept.

References

Andersen OB and Knudsen (1997) Multi-satellite ocean tide modeling – the K1 constituent. *Progress in Oceanography* 40: 197–216.

Andersen OB and Knudsen P (1998) Global marine gravity field from the ERS-1 and Geosat geodetic mission altimetry. *Journal of Geophysical Research* 103: 8129–8137.

Antonov JI, Levitus S, and Boyer TP (2002) Steric sea level variations during 1957–1994: Importance of salinity. *Journal of Geophysical Research* 107: 8013 (doi:10.1029/2001JC000964).

Anzenhofer M and Gruber T (1998) Fully reprocessed ERS-1 altimeter data from 1992 to 1995: Feasibility of the detection of long-term sea level change. *Journal of Geophysical Research* 103: 8089–8112.

Basic T and Rapp RH (1992) Oceanwide prediction of gravity anomalies and sea surface heights using GEOS 3, Seasat, and Geosat altimeter data and ETOPO5U bathymetric data. *DGSS Report 416*, 89 pp. Columbus, OH: Ohio State University.

Beckley BD, Zelensky NP, Luthcke SB, and Callahan PS (2004) Towards a seamless transition from TOPEX/Poseidon to Jason – 1. *Marine Geodesy* 27: 373–390.

Bent RB, Llewellyn SK, Nesterczuk G, and Schmid PE (1976) The development of a highly successful worldwide empirical ionospheric model. In: Goodman J (ed.) *Effects of the Ionosphere on Space Systems and Communications*, pp. 13–28. Springfield, VA: National Technical Information Service.

Bilitza D (1997) International reference ionosphere – Status 1995/96. *Advances in Space Research* 20: 1755–1759.

Bonnefond P, Exertier P, Laurain O, *et al.* (2003) Absolute calibration of Jason-1 and TOPEX /Poseidon altimeters in Corsica. *Marine Geodesy* 26: 261–284.

Born GH, Mitchell JL, and Heyler GA (1987) Geosat ERM - Mission design. *Journal of Atronautical Science* 35: 119–134.

Born GH, Lemoine FG, and Crawford MJ (1988) Geosat ERM – Orbit determination. *Advances in Atronautical Science* 65: 65–81.

Born GH, Tapley BD, Ries JC, and Stewart RH (1986) Accurate measurement of mean sea level changes by altimetric satellites. *Journal of Geophysical Research* 91: 11775–11782.

Born GH, *et al.* (1994) Calibration of the TOPEX altimeter using a GPS buoy. *Journal of Geophysical Research* 99: 24517–24526.

Brenner AC, Koblinsky CJ, and Beckley BD (1990) A preliminary estimate of geoid-induced variations in repeat orbit satellite altimeter observations. *Journal of Geophysical Research* 95: 3033–3040.

Brown S, Ruf C, Keihm S, and Kitiyakara K (2004) Jason-1 microwave radiometer performance and on-orbit calibration. *Marine Geodesy* 27: 199–220.

Cabanes C, Cazenave A, and Le Provost C (2000) Sea level rise during past 40 years determined from satellite and *in situ* observations. *Science* 294: 840–842.

Cartwright DE and Ray RD (1990) Oceanic tides from GEOSAT altimetry. *Journal of Geophysical Research* 95: 3069–3090.

Cazenave A, Schaeffer P, Berge M, Brossier C, Dominh K, and Gennero MC (1996) High-resolution mean sea surface computed with altimeter data of ERS-1 (geodetic mission) and Topex-Poseidon. *Geophysical Journal International* 126: 696–704.

Cazenave A and Royer JV (2001) Applications to marine geophysics. In: Fu L-L and Cazenave A (eds.) *Satellite Altimetry and Earth Science,* International Geophysics Series, vol. 69, ch. 11, pp. 407–439. San Diego, CA: Academic Press.

Chambers DP (2002) Effect of sea level variability on the estimation of mean sea surface gradients. *Marine Geodesy* 25: 273–288.

Chambers DP, Ries JC, Shum CK, and Tapley BD (1998a) On the use of tide gauges to calibrate altimeter drift. *Journal of Geophysical Research* 103: 12885–12890.

Chambers DP, Tapley BD, and Stewart RH (1998b) Reduction of geoid gradient error in ocean variability from satellite altimetry. *Marine Geodesy* 21: 25–39.

Chambers DP, Mehlhaff CA, Urban TJ, Fujii D, and Nerem RS (2002) Low frequency variations in global mean sea level: 1950–2000. *Journal of Geophysical Research* 107: (10.1029/2001JC001089).

Chambers DP, Ries JC, and Urban TJ (2003a) Calibration and verification of Jason-1 using global along-track residuals with TOPEX. *Marine Geodesy* 26: 305–318.

Chambers DP, Hayes SA, Ries JC, and Urban TJ (2003b) New TOPEX sea state bias models and their effect on global mean sea level. *Journal of Geophysical Research* 108(C10): 3305 (10.1029/2003JC001839).

Chelton DB (1994) The sea-state bias in altimeter estimates of sea level from collinear analysis of TOPEX data. *Journal of Geophysical Research* 99: 24995–25008.

Chelton DB and Schlax MG (1993) Spectral characteristics of time-dependent orbit errors in altimeter height measurements. *Journal of Geophysical Research* 98: 12579–12600.

Chelton DB and Schlax MG (1996) Global observations of oceanic Rossby waves. *Science* 272: 234–238.

Chelton DB, Ries JC, Haines BJ, Fu L-L, and Callahan PS (2001) Satellite altimetry. In: Fu L-L and Cazenave A (eds.) *Satellite Altimetry and Earth Science,* International Geophysics Series, vol. 69, ch. 1, pp. 1–131. San Diego, CA: Academic Press.

Chen JL, Wilson CR, Chambers DP, Nerem RS, and Tapley BD (1998) Seasonal global water mass balance and mean sea level variations. *Geophysical Research Letters* 25: 3555–3558.

Cheney RE, Douglas BC, and Miller L (1989) Evaluation of GEOSAT altimeter data with application to tropical Pacific sea level variability. *Journal of Geophysical Research* 94: 4737–4747.

Cheney R, Miller L, Agreen R, Doyle N, and Lillibridge J (1994) TOPEX/POSEIDON: The 2-cm solution. *Journal of Geophysical Research* 99: 24555–24563.

Cheney RE, Doyle NS, Douglas BC, *et al.* (1991) *The Complete Geosat Altimeter GDR Handbook*, NOAA Manual NOS NGS 7. Rockville, MD.

Choi K-R, Ries JC, and Tapley BD (2004) Jason - 1 precision orbit determination by combining SLR and DORIS with GPS tracking data. *Marine Geodesy* 27: 319–332.

Christensen EJ, Haines BJ, Keihm SJ, *et al.* (1994) Calibration of TOPEX/POSEIDON at Platform Harvest. *Journal of Geophysical Research* 99: 24465–24485.

Church JA, White NJ, Coleman R, Lambeck K, and Mitrovica JX (2004) Estimates of the regional distribution of sea-level rise over the 1950 to 2000 period. *Journal of Climate* 17: 2609–2625.

Daniell RE, Brown LD, Anderson DN, *et al.* (1995) Parameterized ionospheric model: A global ionospheric parameterization based on first principles models. *Radio Science* 30: 1499–1510.

Darwin GH (1886) On the dynamical theory of the tides of long period. *Proceedings of the Royal Society, Series A* 41: 319–336.

Desai SD and Haines BJ (2004) Monitoring measurements from the Jason-1 microwave radiometer and independent validation with GPS. *Marine Geodesy* 27: 221–240.

Dickman SR (1988) Theoretical investigation of the oceanic inverted barometer response. *Journal of Geophysical Research* 93: 14941–14946.

Dorandeu J, Ablain M, Faugere Y, Mertz F, Soussi B, and Vincent P (2004) Jason - 1 global statistical evaluation and performance assessment: Calibration and cross-calibration results. *Marine Geodesy* 27: 345–372.

Dorandeu J, Ablain M, and Le Traon P-Y (2003) Reducing cross-track geoid gradient errors around TOPEX /Poseidon and Jason - 1 nominal tracks: Application to calculation of sea level anomalies. *Journal of Atmospheric and Oceanic Technology* 20: 1826–1838.

Dorandeu J and Le Traon P-Y (1999) Effects of global mean atmospheric pressure variations on mean sea level changes from TOPEX/Poseidon. *Journal of Atmospheric and Oceanic Technology* 16: 1279–1283.

Douglas BC (1991) Global sea level rise. *Journal of Geophysical Research* 96: 6981–6992.

Douglas BC (2001) Sea level change in the era of the recording tide gauge. In: Douglas BC, Kearney M, and Leatherman S (eds.) *Sea Level Rise: History and Consequences*, pp. 37–64. San Diego, CA: Academic Press.

Douglas BC, McAdoo DC, and Cheney RE (1987) Oceanographic and geographical applications of satellite altimetry. *Reviews of Geophysics* 25: 875–880.

Egbert GD, Bennet AF, and Foreman MGG (1994) TOPEX/Poseidon tides estimated using a global inverse model. *Journal of Geophysical Research* 99: 24821–24852.

Elfouhaily T, Thompson DR, Vandemark D, and Chapron B (1999) Weakly nonlinear theory and sea state bias estimations. *Journal of Geophysical Research* 104: 7641–7647.

Elfouhaily T, Thompson DR, Chapron B, and Vandemark D (2000) Improved electromagnetic bias theory. *Journal of Geophysical Research* 105: 1299–1310.

Emery WJ, Born GH, Baldwin DG, and Norris CL (1990) Satellite-derived water vapor corrections for Geosat altimetry. *Journal of Geophysical Research* 95: 2953–2964.

Eymard L, Tabary L, Gerard E, Boukabara S-A, and Le Cornec A (1996) The microwave radiometer aboard ERS-1. Part II: Validation of the geophysical products. *IEEE Transactions on Geoscience and Remote Sensing* 34: 291–303.

Fenoglio-Marc L, Dietz C, and Groten E (2004) Vertical land motion in the Mediterranean sea from altimetry and tide gauge stations. *Marine Geodesy* 27: 683–702.

Fleury R, Foucher F, and Lassudrie-Duchesne P (1991) Global TEC measurement capabilities of the DORIS system. *Advances in Space Research* 11: 51–54.

Francis O and Bergé M (1993) Estimate of the radial orbit error by complex demodulation. *Journal of Geophysical Research* 98: 16083–16094.

Fu L-L, (ed.) (1998) Minutes of the joint TOPEX/POSEIDON Jason-1 science working team meeting. *JPL Reports D-16608*, pp 32–33. Pasadena, California: Jet Propulsion Laboratory.

Fu L-L, (ed.) (2004) Minutes of the ocean surface topography science team meeting. *JPL Reports D-31211*, pp 16–17. Pasadena, CA: Jet Propulsion Laboratory.

Fu L-L and Cazanave A (eds.) (2001) *Satellite Altimetry and Earth Science*, International Geophysics Series, vol. 69. San Diego, CA: Academic Press.

Fu L-L, Christensen EJ, Lefebvre M, Ménard Y, Dorrer M, and Escudier P (1994) TOPEX/POSEIDON mission overview. *Journal of Geophysical Research* 99: 24369–24381.

Fu L-L and Glazman R (1991) The effect of the degree of wave development on the sea-state bias in radar altimetry measurement. *Journal of Geophysical Research* 96: 829–834.

Fukamori I, Raghunath R, and Fu L-L (1998) Nature of global large-scale sea level variability in relation to atmospheric forcing: A modeling study. *Journal of Geophysical Research* 103: 5493–5512.

Gaspar P and Florens JP (1998) Estimation of the sea state bias in radar altimeter measurements of sea level: Results from a new nonparametric method. *Journal of Geophysical Research* 103: 15803–15814.

Gaspar P, Ogor F, Le Traon P-Y, and Zanife OZ (1994) Estimating the sea state bias of the TOPEX and POSEIDON altimeters from crossover differences. *Journal of Geophysical Research* 99: 24981–24994.

Gaspar P, Labroue S, Ogor F, Lafitte G, Marchal L, and Rafanel M (2002) Improving nonparameteric estimates of the sea state bias in radar altimeter measurements of sea level. *Journal of Atmospheric and Oceanic Technology* 19: 1690–1707.

Glazman R, Fabrikant A, and Srokosc M (1996) Numerical analysis of the sea state bias for satellite altimetry. *Journal of Geophysical Research* 101: 3789–3799.

Guman MD (1997) *Determination of Global Mean Sea Level Variations Using Multi-Satellite Altimetry*, 195 pp. PhD Thesis, Dissertation, Department of Aerospace Engineering, The University of Texas at Austin.

Haines BJ, Dong D, Born GH, and Gill SK (2003) The Harvest experiment: Monitoring Jason–1 and TOPEX/Poseidon from a California offshore platform. *Marine Geodesy* 26: 239–260.

Haxby WF, Karner GD, LaBrecque JL, and Weissel JK (1983) Digital images of combined oceanic and continental data sets and their use in tectonic studies. *EOS Transactions of the American Geophysical Union* 64: 995–1004.

Hayne GS and Hancock DW (1982) Sea-state related altitude errors in the SEASAT radar altimeter. *Journal of Geophysical Research* 87: 3227–3231.

Hendricks JR, Leben RR, Born GH, and Koblinsky CJ (1996) Empirical orthogonal function analysis of global TOPEX/POSEIDON altimeter data and implications for detection of global sea level rise. *Journal of Geophysical Research* 101: 14131–14146.

Hevizi L, Walsh E, MacIntosh R, *et al.* (1993) Electromagnetic bias in sea surface range measurements at frequencies of the TOPEX/Poseidon satellite. *IEEE Transactions on Geoscience and Remote Sensing* 31: 367–388.

Imel DA (1994) Evaluation of the TOPEX/POSEDION dual-frequency ionosphere correction. *Journal of Geophysical Research* 99: 24895–24906.

Jacobs GA, Born GH, Parke ME, and Allen PC (1992) The global structure of the annual and semiannual sea surface height variability from the Geosat altimeter data. *Journal of Geophysical Research* 97: 17813–17828.

Jan G, Menard Y, Faillot M, Lyard F, Jeansou E, and Bonnefond P (2004) Offshore absolute calibration of space-borne radar altimeters. *Marine Geodesy* 27: 615–630.

Kantha LH and Tierney CC (1997) Global baroclinic tides. *Progress in Oceanography* 40: 163–178.

Keihm SJ, Jansses MA, and Ruf CS (1995) TOPEX/POSEIDON microwave radiometer (TMR) III: Wet tropospheric range correction algorithm and pre-launch error budget. *IEEE Transactions on Geoscience and Remote Sensing* 33: 147–161.

Keihm SJ, Zlotnicki V, and Ruf CS (2000) TOPEX microwave radiometer performance evaluation. *IEEE Transactions on Geoscience and Remote Sensing* 38: 1379–1386.

Kichak RA (2003) Independent GLAS Anomaly Review Board Executive Summary 4 Nov. 2003. http://icesat.gsfc.nasa.-gov/docs/IGARB.pdf (accessed Oct. 2006).

Kolenkiewicz R and Martin C (1982) Seasat altimeter height calibration. *Journal of Geophysical Research* 87: 3189–3197.

Kuo CY, Shum CK, Braun A, and Mitrovica JX (2004) Vertical crustal motion determined by satellite altimetry and tide gauge data in Fennoscandia. *Geophysical Research Letters* 31 (doi:10.1029/2003GL019106, L01608).

Labroue S, Gaspar P, Dorandeu J, et al. (2004) Nonparameteric estimates of the sea state bias for the Jason-1 radar altimeter. *Marine Geodesy* 27: 453–483.

Lame DB and Born GH (1982) SEASAT measurement system evaluation: Achievements and limitations. *Journal of Geophysical Research* 87: 3175–3178.

Lemoine FG, Kenyon SC, Factor JK, et al. (1998) The development of the joint NASA GSFC and NIMA geopotential model EGM96. *NASA TM-1998-206861*. Greebelt, MD: Goddard Space Flight Center.

Lemoine FG, Luthke SB, Zelensky NP, et al. (2003) An evaluation of recent gravity models with respect to altimeter satellite missions. *Presented at TOPEX/Poseidon and Jason-1 Science Working Team Meeting*. San Diego, CA: Arles, France.

Le Provost C (2001) Ocean tides. In: Fu L-L and Cazanave A (eds.) *Satellite Altimetry and Earth Science*,, International Geophysics Series, vol. 69, ch. 6, pp. 267–303. San Diego, CA: Academic Press.

Lerch FJ, Putney BH, Wagner CA, and Klosko SM (1981) Goddard Earth models for oceanographic applications (GEM-10B and GEM-10C). *Marine Geodesy* 5: 145–187.

Leuliette EW, Nerem RS, and Mitchum GT (2004) Calibration of TOPEX/Poseidon and Jason altimeter data to construct a continuous record of mean sea level change. *Marine Geodesy* 27: 79–94.

Levitus S (1982) Climatalogical atlas of the world ocean. *NOAA Professional Report 13*. Rockville, MD: US Department of Commerce.

Levitus S, Antonov J, and Boyer T (2005) Warming of the world ocean, 1955–2003. *Geophysical Research Letters* 32: L02604, doi:10.1029/2004GL021592.

Levitus S, Antonov JL, Boyer TP, and Stephens C (2000) Warming of the world ocean. *Science* 287: 2225–2229.

Luthcke SB, Zelensky NP, Rowlands DD, Lemoine FG, and Williams TA (2003) The 1-centimeter orbit: Jason - 1 precision orbit determination using GPS, SLR, DORIS, and altimeter data. *Marine Geodesy* 26: 399–421.

Luthke SB, Rowlands DD, Williams TA, and Sirota M (2005) Reduction of ICESat systematic geolocation errors and the impact on ice sheet elevation change detection. *Geophysical Research Letters* 32: L21S05, doi:10.1029/2005GL023689.

Ma XC, Shum CK, Eanes RJ, and Tapley BD (1994) Determination of ocean tides from the first year of TOPEX/POSEIDON altimeter measurements. *Journal of Geophysical Research* 99: 24809–24820.

Mannucci AJ, Iijima BA, Linqwister UJ, Pi X, Sparks L, and Wilson BD (1999) GPS and ionosphere. In: Stone WR (ed.) *Review of Radio Science*,, pp. 622–655. Oxford: Oxford University Press.

Marsh JG, Koblinsky CJ, Zwally HJ, Brenner AC, and Beckley BD (1992) A global mean sea surface based upon GEOS-3 and Seasat altimeter data. *Journal of Geophysical Research* 97: 4915–4921.

Marsh JG and Martin TV (1982) The Seasat altimeter mean sea surface. *Journal of Geophysical Research* 87: 3269–3280.

Marsh JG and Williamson RG (1980) Precise orbit analysis in support of the Seasat altimeter experiment. *Journal of the Astronautical Sciences* 27: 345–369.

Martin C and Kolenkiewicz R (1981) Calibration validation of the GEOS-3 altimeter. *Journal of Geophysical Research* 86: 6369–6381.

Martinez-Benjamin JJ, Martinez-Garcia M, Lopez SG, et al. (2004) Ibiza absolute calibration experiment: Survey and preliminary results. *Marine Geodesy* 27: 657–682.

Mazzega P (1989) The solar tides and the sun-synchronism of satellite altimetry. *Geophysical Research Letters* 16: 507–510.

McConathy DR and Kilgus CC (1987) The Navy geosat mission: And overview. *Johns Hopkins APL Techincal Digest* 8(2), 170–175.

Melville WK, Stewart RH, Keller WC, et al. (1991) Measurements of electromagnetic bias in radar altimetry. *Journal of Geophysical Research* 96: 4915–4924.

Ménard Y, Jeansou E, and Vincent P (1994) Calibration of TOPEX/POSEIDON altimeters at Lampedusa: Additional results at Harvest. *Journal of Geophysical Research* 99: 24487–24504.

Minster JF, Brossier C, and Rogel P (1995) Variation of the mean sea level from TOPEX/Poseidon data. *Journal of Geophysical Research* 100: 25153–25162.

Minster JF, Cazenave A, Serafini YV, Mercier F, Gennero MC, and Rogel P (1999) Annual cycle in mean sea level from TOPEX-Poseidon and ERS-1: Inference on the global hydrological cycle. *Global and Planetary Change* 20: 57–66.

Mitchum GT (1994) Comparison of TOPEX sea surface heights and tide gauge sea levels. *Journal of Geophysical Research* 99: 24541–24553.

Mitchum GT (1998) Monitoring the stability of satellite altimeters with tide gauges. *Journal of Atmospheric and Oceanic Technology* 15: 721–730.

Mitchum GT (2000) An improved calibration of satellite altimetric heights using tide gauge sea levels with adjustment for land motion. *Marine Geodesy* 23: 145–166.

Morozov EG (1995) Semidiurnal internal wave global field. *Deep Sea Research* 42: 135–148.

Nerem RS (1995) Measuring global mean sea level variations using TOPEX/POSEIDON altimeter data. *Journal of Geophysical Research* 100: 25135–25151.

Nerem RS, Chambers DP, Leuliette E, Mitchum GT, and Giese BS (1999) Variations in global mean sea level during the 1997–98 ENSO event. *Geophysical Research Letters* 26: 3005–3008.

Nerem RS, Eanes RJ, Ries JC, and Mitchum GT (1998) The use of a precise reference frame for sea level studies. In: Rummel R, Drewes H, Bosch W, and Hornik H (eds.) *Towards an Integrated Global Geodetic Observing System*. Munich: International Association of Geodesy.

Nerem RS, Haines BJ, Hendricks J, Minster JF, Mitchum GT, and White WB (1997) Improved determination of global mean sea level variations using TOPEX/POSEIDON altimeter data. *Geophysical Research Letters* 24: 1331–1334.

Nerem RS, Lerch FJ, Marshall JA, *et al.* (1994a) Gravity model development for TOPEX/POSEIDON: Joint gravity model-1 and 2. *Journal of Geophysical Research* 99: 24421–24447.

Nerem RS, Schrama EJ, Koblinsky CJ, and Beckley BD (1994b) A preliminary evaluation of ocean topography from the TOPEX/POSEIDON mission. *Journal of Geophysical Research* 99: 24565–24583.

Obligas E, Tran N, and Eymard L (2004) An assessment of Jason-1 microwave radiometer measurements and products. *Marine Geodesy* 27: 255–277.

Parke ME, Born G, Leben R, McLaughlin C, and Tierney C (1998) Altimeter sampling characteristics using a single satellite. *Journal of Geophysical Research* 103: 10513–10526.

Parke ME, Stewart RH, Farless DL, and Cartwright DE (1987) On the choice of orbits for an altimetric satellite to study ocean circulation and tides. *Journal of Geophysical Research* 92: 11,693–11,707.

Pavlis EC and Mertikas SP (2004) The GAVDOS mean sea level and altimeter calibration facility: Results for Jason – 1. *Marine Geodesy* 27: 631–656.

Ponte RM, Salstein DA, and Rosen RD (1991) Sea level response to pressure forcing in a barotropic numerical model. *Journal of Physical Oceanography* 21: 1043–1057.

Raney RK (1998) The delay Doppler radar altimeter. *IEEE Transactions on Geoscience and Remote Sensing* 36: 1578–1588.

Rapp RH (1989) The treatment of permanent tidal effects in the analysis of satellite altimeter data for sea surface topography. *Manuscripta Geodaetica* 14: 368–372.

Rapp RH (1986) Gravity anomalies and sea surface heights derived from a combined GEOS-3/Seasat altimeter data set. *Journal of Geophysical Research* 91: 4867–4876.

Ray RD (1999) A global ocean tide model from TOPEX/POSEIDON altimetry: GOT99.2. *NASA Technical Memorandum 1999-209478*. Greenbelt, MD: Goddard Space Flight Center.

Ray RD and Mitchum GT (1996) Surface manifestation of internal tides near Hawaii. *Geophysical Research Letters* 23: 2101–2104.

Ray RD and Mitchum GT (1997) Surface manifestation of internal tides in the deep ocean. *Progress in Oceanography* 40: 135–162.

Rodriguez E, Kim Y, and Martin YM (1992) The effect of small-wave modulation on the electromagnetic bias. *Journal of Geophysical Research* 97: 2379–2389.

Rodriguez E and Martin YM (1992) Theory and designa of interferometric synthetic aperture radars. *IEEE Proceedings-F Radar and Signal Processing* 139: 147–159.

Rodriguez E and Martin YM (1994) Estimation of the electromagnetic bias from retracked TOPEX data. *Journal of Geophysical Research* 99: 24971–24979.

Rodriguez E and Pollard BD (2001) The measurement capabilities of wide-swath ocean altimeters. In: Chelton DB (ed.) *Report of the High-Resolution Ocean Topography Science Working Group Meeting*, pp. 190–215. Corvallis, OR: College of Oceanic and Atmospheric Sciences, Oregon State University.

Ruf CS (2002) Characterization and correction of a drift in calibration of the TOPEX microwave radiometer. *IEEE Transtions on Geoscience and Remote Sensings* 40: 509–511.

Ruf C, Keihm S, Subramanya B, and Jansses M (1994) TOPEX/POSEIDON microwave radiometer performance and in-flight calibration. *Journal of Geophysical Research* 99: 24915–24926.

Sandwell DT (1992) Antarctic marine gravity field from high-density satellite altimetry. *Reviews of Geophysics, Supplement* 29: 132–137.

Sandwell DT and Smith WHF (1997) Marine gravity anomaly from Geosat and ERS-1 satellite altimetry. *Journal of Geophysical Research* 102: 10039–10054.

Sandwell DT and Smith WHF (2001) Bathymetric estimation. In: Fu L-L and Cazanave A (eds.) *Satellite Altimetry and Earth Science*, International Geophysics Series, vol. 69, ch. 12, pp. 441–457. San Diego, CA: Academic Press.

Scharroo R (2002) *A Decade of ERS Satellite Orbits and Altimetry*, 195 pp. PhD Thesis, Dissertation, Department of Aerospace Engineering, Delft University of Technology, The Netherlands.

Scharroo R, Lillibridge JL, Smith WHF, and Schrama EJO (2004) Cross-calibration and long-term monitoring of the microwave radiometers of ERS, TOPEX, GFO, Jason, and Envisat. *Marine Geodesy* 27: 279–297.

Scharroo R and Visser PNAM (1998) Precise orbit determination and gravity field improvement for the ERS satellites. *Journal of Geophysical Research* 103: 8113–8127.

Schutz BE and Tapley BD (1980) Orbit accuracy assessment for Seasat. *Journal of the Astronautical Sciences* 27: 371–390.

Schlax MG and Chelton DB (1994) Aliased tidal errors in TOPEX/POSEIDON sea surface height data. *Journal of Geophysical Research* 99: 24,761–24,725.

Schrama EJO and Ray RD (1994) A preliminary tidal analysis of TOPEX/POSEIDON altimetry. *Journal of Geophysical Research* 99: 24799–24808.

Schreiner WS, Markin RE, and Born GH (1997) Correction of signal frequency altimeter measurements for ionospheric delay. *IEEE Transactions on Geoscience and Remote Sensing* 35: 271–277.

Schwiderski EW (1980) Ocean tides. Part II: A hydrodynamical interpolation model. *Marine Geodesy* 3: 161–218.

Shum CK, Chambers DP, Ries JC, Yuan DN, and Tapley BD (1994) The determination of large-scale sea surface topography and its variations using Geosat altimetry. In: Schutz BE (ed.) *Gravimetry and Space Techniques Applied to Geodynamics and Ocean Dynamics*, Geophysical Monograph 82, IUGG Volume 17.

Shum CK, Tapley BD, and Ries JC (1993) Satellite altimetry: its applications and accuracy assessment. *Advances in Space Research* 13(11): 315–324.

Shum CK, Woodworth PL, Andersen OB, *et al.* (1997) Accuracy assessment of recent ocean tide models. *Journal of Geophysical Research* 102: 25173–25194.

Shum C, Yi Y, Cheng K, *et al.* (2003) Calibration of JASON - 1 altimeter over Lake Erie. *Marine Geodesy* 26: 335–354.

Smith WHF and Sandwell DT (1994) Bathymetric prediction from dense satellite altimetry and sparse shipboard bathymetry. *Journal of Geophysical Research* 99: 21803–21824.

Smith TM, Reynolds RW, Livezey RE, and Stokes DC (1996) Reconstruction of historical sea surface temperatures using empirical orthogonal functions. *Journal of Climate* 9: 1403–1420.

Smith DE, Zuber MT, Frey HV, *et al.* (2001) Mars orbiter laser altimeter: Experiment summary after the first year of global mapping of Mars. *Journal of Geophysical Research* 106: 23689–23722.

Srokosz MA (1986) On the joint distribution of surface elevations and slopes for a nonlinear random sea, with an application for radar altimetry. *Journal of Geophysical Research* 91: 995–1006.

Stammer D, Wunsch C, and Ponte RM (2000) De-aliasing of global high frequency barotropic motions in altimeter measurements. *Geophysical Research Letters* 27: 1175–1178.

Stanley HR (1979) The GEOS-3 project. *Journal of Geophysical Research* 84(B8): 3779–3783.

Stewart RH, Fu L-L, and Lefebvre M (1986) Science opportunities from the TOPEX/Poseidon mission. *Jet Propulsion Laboratory Publication* 86-18: 58.

Tai CK (1988) Geosat crossover analysis in the tropical pacific 1. Constrained sinsoidal crossover adjustment. *Journal of Geophysical Research* 93: 10621–10629.

Tapley BD, Bettadpur S, Watkins M, and Riegber C (2004b) The gravity recovery and climate experiment: Mission overview and early results. *Geophysical Research Letters* 31: 1–4, p L09607.

Tapley BD, Born GH, and Parke ME (1982) The Seasat altimeter data and its accuracy assessment. *Journal of Geophysical Research* 87: 3179–3188.

Tapley BD, Chambers DP, Bettadpur S, and Ries JC (2003) Large scale ocean circulation from the GRACE GGM01 geoid. *Geophysical Research Letters* 30 (doi:10.1029/2003GL018622).

Tapley BD, Chambers DP, Shum CK, Eanes RJ, Ries JC, and Stewart RH (1994b) Accuracy assessment of the large-scale dynamic ocean topography from TOPEX/POSEIDON altimetry. *Journal of Geophysical Research,* 99: 24605–24618.

Tapley BD and Kim MC (2001) Applications to geodesy. In: Fu L-L and Cazanave A (eds.) *Satellite Altimetry and Earth Science*, International Geophysics Series, vol. 69, ch. 10, pp. 371–406. San Diego, CA: Academic Press.

Tapley BD and Rosborough GW (1985) Geographically correlated orbit errors and its effect on satellite altimetry missions. *Journal of Geophysical Research* 90: 11817–11831.

Tapley BD, Schutz BE, and Born GH (2004a) *Statistical Orbit Determination*. Burlington, MA: Academic Press.

Tapley BD, Shum CK, Ries JC, Suter R, and Schutz BE (1992) Monitoring of changes in global mean sea level using geosat altimeter. In: *Sea Level Changes: Determination and Effects*, IUGG Geophysical Monographs Washington, DC., vol. 69, pp. 167–180. Washington, DC.

Tapley BD, Shum CK, Ries JC, et al. (1997) The TEG-3 geopotential model. In: Seagwa J, Fujimoto H, and Okubo S (eds.) *Proceedings of the International Association of Geodesy International Symposium No. 117, Gravity, Geoid and Marine Geodesy*, pp. 453–460. New York: Springer.

Tapley BD, Ries JC, Davis GW, et al. (1994a) Precision orbit determination for TOPEX/POSEIDON. *Journal of Geophysical Research* 99(C12): 24383–24404.

Tapley BD, Watkins MM, Ries JC, et al. (1996) The JGM-3 geopotential model. *Journal of Geophysical Research* 101(B12): 28029–28049.

Tapley B, Ries J, Bettadpur S, et al. (2005) GGM02 – An improved Earth gravity field model from GRACE. *Journal of Geodesy* doi: 10.1007/s00190-005-0480-z.

Tierney CC, Parke ME, and Born GH (1998) Ocean tides from along track altimetry. *Journal of Geophysical Research* 103: 10273–10287.

Tierney C, Wahr J, Bryan F, and Zlotniki V (2000) Short-period oceanic circulation: Implications for satellite altimetry. *Geophysical Research Letters* 27: 1255–1258.

Trenberth KE and Olson JG (1988) An evaluation and intercomparison of global analyses from the national meteorological center and the European center for medium-range weather forecasts. *Bulletin of the American Meteorological Society* 69: 1047–1057.

Urban T (2000) *The Integration and Application of Multi-Satellite Radar Altimetry*, 273 pp. Phd Thesis, Dissertation. Department of Aerospace Engineering, The University of Texas at Austin.

van Gysen H, Coleman R, Morrow R, Hirsch B, and Rizos C (1992) Analysis of collinear passes of satellite altimeter data. *Journal of Geophysical Research* 97: 2265–2277.

Wagner CA and Cheney RE (1992) Global sea level changes from satellite altimetry. *Journal of Geophysical Research* 97: 15607–15615.

Wagner CA and Tai CK (1994) Degradation of ocean signals in satellite altimetry due to orbit error removal processes. *Journal of Geophysical Research* 99: 16255–16267.

Walsh EJ, Jackson FC, Hines DE, et al. (1991) Observations on electromagnetic bias in radar altimeter surface range measurements. *Journal of Geophysical Research* 96: 20571–20583.

Wang YM (2000) The satellite altimeter data derived mean sea surface GSFC98. *Geophysical Research Letters* 27: 701–704.

Watson C, Coleman R, White N, Church J, and Govind R (2003) Absolute calibration of TOPEX/Poseidon and Jason-1 using GPS buoys in Bass Strait, Australia. *Marine Geodesy* 26: 285–304.

Willis JK, Roemmich D, and Cornuelle B (2004) Interannual variability in upper ocean heat content, temperature, and thermosteric expansion on global scales. *Journal of Geophysical Research* 109: C12036, doi:10.1029/2003JC002260.

Witter DL and Chelton DB (1991) An apparent wave height dependence in the sea-state bias in GEOSAT altimeter range measurements. *Journal of Geophysical Research* 96: 8861–8867.

Woodworth PL (1985) Accuracy of existing ocean tide models. In: Hunt JJ (ed.) *Proceedings of a Conference on the use of Satellite Data in Climate Models, Alpbach, Austria, 10–12 June, 1985*, pp. 95–98, ESA Publication SP-244, Noordwijk, The Netherlands.

Woodworth P, Moore P, Dong X, and Bingley R (2004) Absolute calibration of the Jason-1 altimeter using UK tide gauges. *Marine Geodesy* 27: 95–106.

Wunsch C (1972) Bermuda sea level in relation to tides, weather and baroclinic fluctuations. *Reviews of Geophysics and Space Research* 10: 1–49.

Wunsch C and Gaposchkin EM (1980) On using satellite altimetry to determine the general circulation of the oceans with application to geoid improvement. *Reviews of Geophysics* 18: 725–745.

Wyrtki K and Mitchum G (1990) Interannual differences of geosat altimeter heights and sea level: The importance of a datum. *Journal of Geophysical Research* 95: 2969–2975.

Zimbelman DF and Busalacchi AJ (1990) The wet tropospheric range correction: Product intercomparisons and the simulated effect for tropical Pacific altimeter retrievals. *Journal of Geophysical Research* 95: 2899–2922.

Zlotnicki V (1994) Correlated environmental corrections for TOPEX/POSEIDON, with a note of ionospheric accuracy. *Journal of Geophysical Research* 99: 24907–24914.

Zlotnicki V and Desai SD (2004) Assessment of Jason microwave radiometer's measurement of wet tropspheric path delay using comparisons with SSM/I and TMI. *Marine Geodesy* 27: 241–253.

Zwally HJ and Brenner AC (2001) Ice sheet dynamics and mass balance. In: Fu L-L and Cazanave A (eds.) *Satellite Altimetry and Earth Science*, International Geophysics Series, vol. 69, ch. 9, pp. 351–369. San Diego, CA: Academic Press.

Zwally HJ, Schutz B, Abdalati W, et al. (2002) ICESat's laser measurements of polar ice, atmosphere, ocean, and land. *Journal of Geodynamics* 34: 405–445.

Relevant Website

http://fermi.jhuapl.edu/d2p/ – Johns Hopkins University Applied Physics Laboratory.

6 Earth Tides

D. C. Agnew, University of California San Diego, San Diego, CA, USA

6.1 Introduction

The Earth tides are the motions induced in the solid Earth, and the changes in its gravitational potential, induced by the tidal forces from external bodies. (These forces, acting on the rotating Earth, also induce motions of its spin axis, treated in Chapter 10). Tidal fluctuations have three roles in geophysics: measurements of them can provide information about the Earth; models of them can be used to remove tidal variations from measurements of something else; and the same models can be used to examine tidal influence on some phenomenon. An example of the first role would be measuring the nearly diurnal resonance in the gravity tide to estimate the flattening of the core–mantle boundary (CMB); of the second, computing the expected tidal displacements at a point so we can better estimate its position with the Global Positioning System (GPS); and of the third, finding the tidal stresses to see if they trigger earthquakes. The last two activities are possible because the Earth tides are relatively easy to model quite accurately; in particular, these are much easier to model than the ocean tides are, both because the Earth is more rigid than water, and because the geometry of the problem is much simpler.

For modeling the tides it is an advantage that they can be computed accurately, but an unavoidable consequence of such accuracy is that it is difficult to use Earth-tide measurements to find out about the Earth. The Earth's response to the tides can be described well with only a few parameters; even knowing those few parameters very well does not provide much information. This was not always true; in particular, in 1922 Jeffreys used tidal data to show that the average rigidity of the Earth was much less than that of the mantle, indicating that the core must be of very low rigidity (Brush, 1996). But subsequently, seismology has determined Earth structure in much more detail than could be found with tides. Recently, Earth tides have become more important in geodesy, as the increasing precision of measurements has required corrections for tidal effects that could previously be ignored, This chapter therefore focuses on the theory needed to compute tidal effects as accurately as possible, with less attention to tidal data or measurement techniques; though of course the theory is equally useful for interpreting tidal data, or phenomena possibly influenced by tides.

There are a number of reviews of Earth tides available; the best short introduction remains that of Baker (1984). Melchior (1983) describes the subject fully (and with a very complete bibliography), but is now somewhat out of date, and should be used with caution by the newcomer to the field. The volume of articles edited by Wilhelm *et al.* (1997) is a better reflection of the current state of the subject, as are the quadrennial *International Symposia on Earth Tides* (e.g., Jentzsch, 2006). Harrison (1985) reprints a number of important papers, with very thoughtful commentary; Cartwright (1999) is a history of tidal science (mostly the ocean tides) that also provides an interesting introduction to some aspects of the field – which, as one of the older parts of geophysics, has a terminology sometimes overly affected by history.

6.1.1 An Overview

Figure 1 is a simple flowchart to indicate what goes into a tidal signal. We usually take the tidal forcing to be completely known, but it is computed using a particular theory of gravity, and it is actually the case that Earth-tide measurements provide some of the best evidence available for general relativity as opposed to some other alternative theories (Warburton and Goodkind, 1976). The large box labeled Geophysics/oceanography includes the response of the Earth and ocean to the forcing, with the arrow going around it to show that some tides would be observed even if the Earth were oceanless and rigid. Finally, measurements of Earth tides can detect other environmental and tectonic signals.

At this point it is useful to introduce some terminology. The 'theoretical tides' could be called the modeled tides, since they are computed from a set of models. The first model is the tidal forcing, or 'equilibrium tidal potential', produced by external bodies; this is computed from gravitational and astronomical theory, and is the tide at point E in **Figure 1**. The next two models are those that describe how the Earth and ocean respond to this forcing; in **Figure 1** these are boxes inside the large dashed box. The solid-Earth model gives what are called the 'body tides', which are what would be observed on an oceanless but otherwise realistic Earth. The ocean model (which includes both the oceans and the elastic Earth) gives the 'load tides', which are changes in the solid Earth caused by the shifting mass of water associated with the ocean tides. These two responses sum to give the total tide

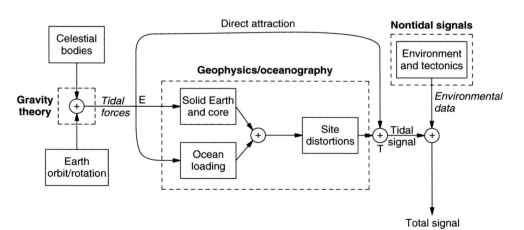

Figure 1 Tidal flowchart. Entries in italics represent things we know (or think we know) to very high accuracy; entries in boldface (over the dashed boxes) represent things we can learn about using tidal data. See text for details.

caused by the nonrigidity of the Earth; the final model, labeled 'site distortions', may be used to describe how local departures from idealized models affect the result (Section 6.6.1). This nonrigid contribution is summed with the tide from direct attraction to give the total theoretical tide, at point T in the flowchart.

Mathematically, we can describe the processes shown on this flowchart in terms of linear systems, something first applied to tidal theory by Munk and Cartwright (1966). The total signal $y(t)$ is represented by

$$y(t) = \int x_T(t-\tau)w_T(\tau)d\tau + n(t) \qquad [1]$$

where $x_T(t)$ is the tidal forcing and $n(t)$ is the noise (nontidal energy, from whatever source). The function $w(t)$ is the impulse response of the system to the tidal forcing. Fourier transforming eqn [1], and disregarding the noise, gives $Y(f) = W_T(f)X_T(f)$: $W(f)$ is the 'tidal admittance', which turns out to be more useful than $w(t)$, partly because of the bandlimited nature of X_T, but also because (with one exception) $W(f)$ turns out to be a fairly smooth function of frequency. To predict the tides we assume $W(f)$ (perhaps guided by previous measurements); to analyze them, we determine $W(f)$.

We describe the tidal forcing first, in some detail because the nature of this forcing governs the response and how tidal measurements are analyzed. We next consider how the solid Earth responds to the tidal forcing, and what effects this produces. After this we discuss the load tides, completing what we need to know to produce the full theoretical tides. We conclude with brief descriptions of analysis methods appropriate to Earth-tide data, and instruments for measuring Earth tides.

6.2 The Tidal Forces

The tidal forces arise from the gravitational attraction of bodies external to the Earth. As noted above, computing them requires only some gravitational potential theory and astronomy, with almost no geophysics. The extraordinarily high accuracy of astronomical theory makes it easy to describe the tidal forcing to much more precision than can be measured: perhaps as a result, in this part of the subject the romance of the next decimal place has exerted a somewhat excessive pull.

Our formal derivation of the tidal forcing will use potential theory, but it is useful to start by considering the gravitational forces exerted on one body (the Earth, in this case) by another. As usual in discussing gravitation, we work in terms of accelerations. Put most simply, the tidal acceleration at a point on or in the Earth is the difference between the acceleration caused by the attraction of the external body and the orbital acceleration – which is to say, the acceleration which the Earth undergoes as a whole. This result is valid whatever the nature of the orbit – it would hold just as well if the Earth were falling directly toward another body. For a spherically symmetric Earth the orbital acceleration is the acceleration caused by the attraction of the other body at the Earth's center of mass, making the tidal force the difference between the attraction at the center of mass, and that at the point of observation.

Figure 2 shows the resulting force field. At the point directly under the attracting body (the 'sub-body point'), and at its antipode, the tidal force is oppositely directed in space, though in the same way (up) viewed from the Earth. It is in fact larger at the sub-body point than at its antipode, though if the ratio a/R is small (1/60 for the forces plotted in **Figure 2**) the difference is also small.

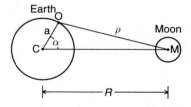 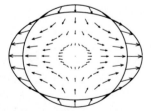

Figure 2 Tidal forcing. On the left is the geometry of the problem for computing the tidal force at a point O on the Earth, given an external body M. The right plot shows the field of forces (accelerations) for the actual Earth–Moon separation; the scale of the largest arrow is 1.14 μm s^{-2} for the Moon, and 0.51 μm s^{-2} for the Sun. The elliptical line shows the equipotential surface under tidal forcing, greatly exaggerated.

6.2.1 The Tidal Potential

We now derive an expression for the tidal force – or rather, for the more useful 'tidal potential', following the development in Munk and Cartwright (1966). If M_{ext} is the mass of the external body, the gravitational potential, V_{tot}, from it at O is

$$V_{tot} = \frac{GM_{ext}}{\rho} = \frac{GM_{ext}}{R} \frac{1}{\sqrt{1 + (a/R)^2 - 2(a/R)\cos\alpha}}$$

using the cosine rule from trigonometry. The variables are as shown in **Figure 2**: a is the distance of O from C, ρ the distance from O to M, and α the angular distance between O and the sub-body point of M. We can write the square-root term as a sum of Legendre polynomials, using the generating-function expression for these, which yields

$$V_{tot} = \frac{GM_{ext}}{R} \sum_{n=0}^{\infty} \left(\frac{a}{R}\right)^n P_n(\cos\alpha) \qquad [2]$$

where $P_2(x) = (1/2)(3x^2 - 1)$ and $P_3(x) = (1/2)(5x^3 - 3x)$.

The $n = 0$ term is constant in space, so its gradient (the force) is zero, and it can be discarded. The $n = 1$ term is

$$\frac{GM_{ext}}{R^2} a \cos\alpha = \frac{GM_{ext}}{R^2} x_1 \qquad [3]$$

where x_1 is the Cartesian coordinate along the C–M axis. The gradient of this is a constant, corresponding to a constant force along the direction to M; but this is just the orbital force at C, which we subtract to get the tidal force. Thus, the tidal potential is eqn [2] with the two lowest terms removed:

$$V_{tid}(t) = \frac{GM_{ext}}{R(t)} \sum_{n=2}^{\infty} \left(\frac{a}{R(t)}\right)^n P_n[\cos\alpha(t)] \qquad [4]$$

where we have made R and α, as they actually are, functions of time t – which makes V such a function as well.

We can now put in some numbers appropriate to the Earth and relevant external bodies to get a sense of the magnitudes of different terms. If r is the radius of the Earth, $a/R = 1/60$ for the Moon, so that the size of terms in the sum [4] decreases fairly rapidly with increasing n, in practice, we need to only consider $n = 2$ and $n = 3$, and perhaps $n = 4$ for the highest precision; the $n = 4$ tides are just detectable in very low noise gravimeters. These different values of n are referred to as the degree-n tides. For the Sun, $r/R = 1/23\,000$, so the degree-2 solar tides completely dominate.

If we consider $n = 2$, the magnitude of V_{tid} is proportional to GM_{ext}/R^3. If we normalize this quantity to make the value for the Moon equal to 1, the value for the Sun is 0.46, for Venus 5×10^{-5}, and for Jupiter 6×10^{-6}, and even less for all other planets. So the 'lunisolar' tides dominate, and are probably the only ones visible in actual measurements – though, as we will see, some expansions of the tidal potential include planetary tides.

At very high precision, we also need to consider another small effect: the acceleration of the Earth is exactly equal to the attraction of the external body at the center of mass only for a spherically symmetric Earth. For the real Earth, the C_{20} term in the gravitational potential makes the acceleration of the Moon by the Earth (and hence the acceleration of the Earth by the Moon) depend on more than just eqn [3]. The resulting Earth-flattening tides (Wilhelm, 1983; Dahlen, 1993) are however small.

We can get further insight on the behavior of the tidal forces if we use geographical coordinates, rather than angular distance from the sub-body point. Suppose our observation point O is at colatitude θ and east longitude ϕ (which are fixed) and that the sub-body point of M is at colatitude $\theta'(t)$ and east longitude $\phi'(t)$. Then we may apply the addition theorem for spherical harmonics to get, instead of [4],

$$V_{tid} = \frac{GM_{ext}}{R(t)} \sum_{n=2}^{\infty} \left(\frac{a}{R(t)}\right)^n \frac{4\pi}{2n+1}$$
$$\times \sum_{m=-n}^{n} Y_{nm}^*(\theta'(t), \phi'(t)) Y_{nm}(\theta, \phi) \qquad [5]$$

where we have used the fully normalized complex spherical harmonics defined by

$$Y_{nm}(\theta, \phi) = N_n^m P_n^m(\cos\theta) e^{im\phi}$$

where

$$N_n^m = (-1)^m \left[\frac{2n+1}{4\pi} \frac{(n-m)!}{(n+m)!}\right]^{1/2}$$

is the normalizing factor and P_n^m is the associated Legendre polynomial of degree n and order m (**Table 1**).

Table 1   Associated Legendre functions

$P_2^0(\theta) = \frac{1}{2}(3\cos^2\theta - 1)$	$P_3^0(\theta) = \frac{1}{2}(5\cos^3\theta - 3\cos\theta)$
$P_2^1(\theta) = 3\sin\theta\cos\theta$	$P_3^1(\theta) = \frac{3}{2}(5\cos^2\theta - 1)$
$P_2^2(\theta) = 3\sin^2\theta$	$P_3^2(\theta) = 15\sin^2\theta\cos\theta$
	$P_3^3(\theta) = 15\sin^3\theta$

As is conventional, we express the tidal potential as V_{tid}/g, where g is the Earth's gravitational acceleration; this combination has the dimension of length, and can easily be interpreted as the change in elevation of the geoid, or of an equilibrium surface such as an ideal ocean (hence its name, the 'equilibrium potential'). Part of the convention is to take g to have its value on the Earth's equatorial radius a_{eq}; if we hold r fixed at that radius in [5], we get

$$
\begin{aligned}
\frac{V_{\text{tid}}}{g} &= a_{\text{eq}} \frac{M_{\text{ext}}}{M_{\text{E}}} \sum_{n=2}^{\infty} \frac{4\pi}{2n+1} \left(\frac{a_{\text{eq}}}{R}\right)^{n+1} \\
&\quad \times \sum_{m=-n}^{n} Y_{nm}^{*}(\theta', \phi') Y_{nm}(\theta, \phi) \\
&= \sum_{n=2}^{\infty} K_n \frac{4\pi}{2n+1} \xi^{n+1} \\
&\quad \times \sum_{m=-n}^{n} Y_{nm}^{*}(\theta', \phi') Y_{nm}(\theta, \phi)
\end{aligned}
\qquad [6]
$$

where the constant K includes all the physical quantities:

$$
K_n = a_{\text{eq}} \frac{M_{\text{ext}}}{M_{\text{E}}} \left(\frac{a_{\text{eq}}}{\overline{R}}\right)^{n+1}
$$

where M_{E} is the mass of the Earth and $\overline{R}$ is the mean distance of the body; the quantity $\xi = \overline{R}/R$ expresses the normalized change in distance. For the Moon, K_2 is 0.35837 m, and for the Sun, 0.16458 m.

In both [5] and [6], we have been thinking of θ and ϕ as giving the location of a particular place of observation; but if we consider them to be variables, the $Y_{nm}(\theta, \phi)$ describes the geographical distribution of V/g on the Earth. The time dependence of the tidal potential comes from time variations in R, θ', and ϕ'. The first two change relatively slowly because of the orbital motion of M around the Earth; ϕ' varies approximately daily as the Earth rotates beneath M. The individual terms in the sum over m in [6] thus separate the tidal potential of degree n into parts, called 'tidal species', that vary with frequencies around 0, 1, 2, ..., n times per day; for the largest tides ($n = 2$), there are three such species. The diurnal tidal potential varies once per day, and with colatitude as $\sin\theta \cos\theta$: it is largest at mid-latitudes and vanishes at the equator and the poles. The semidiurnal part (twice per day) varies as $\sin^2\theta$ and so is largest at the equator and vanishes at the poles. The long-period tide varies as $3\cos^2\theta - 1$, and so is large at the pole and (with reversed sign) at the equator. As we will see, these spatial dependences do not carry over to those tides, such as strain and tilt, that depend on horizontal gradients of the potential.

To proceed further it is useful to separate the time-dependent and space-dependent parts a bit more explicitly. We adopt the approach of Cartwright and Tayler (1971) who produced what was, for a long time, the standard harmonic expansion of the tidal potential. We can write [6] as

$$
\begin{aligned}
\frac{V_{\text{tid}}}{g} &= \sum_{n=2}^{\infty} K_n \xi^{n+1} \frac{4\pi}{2n+1} \Bigg[Y_{n0}(\theta', \phi') Y_{n0}(\theta, \phi) \\
&\quad + \sum_{m=1}^{n} Y_{n-m}^{*}(\theta', \phi') Y_{n-m}(\theta, \phi) \\
&\quad + Y_{nm}^{*}(\theta', \phi') Y_{nm}(\theta, \phi) \Bigg] \\
&= \sum_{n=2}^{\infty} K_n \xi^{n+1} \frac{4\pi}{2n+1} \Bigg[Y_{n0}(\theta', \phi') Y_{no}(\theta, \phi) \\
&\quad + \sum_{m=1}^{n} 2\Re [Y_{nm}^{*}(\theta', \phi') Y_{nm}(\theta, \phi)] \Bigg]
\end{aligned}
$$

Now define complex (and time-varying) coefficients $T_{nm}(t) = a_n^m(t) + \mathrm{i} b_n^m(t)$ such that

$$
\frac{V_{\text{tid}}}{g} = \Re \left[\sum_{n=2}^{\infty} \sum_{m=0}^{n} T_{nm}^{*}(t) Y_{nm}(\theta, \phi) \right]
\qquad [7]
$$

$$
= \sum_{n=2}^{n=\infty} \sum_{m=0}^{n} N_n^m P_n^m(\cos\theta) [a_n^m(t) \cos m\phi + b_n^m(t) \sin m\phi]
\qquad [8]
$$

Then the T_{nm} coefficients are, for m equal to 0,

$$
T_{n0} = \left(\frac{4\pi}{2n+1}\right)^{1/2} K_n \xi^{n+1} P_n^0(\cos\theta')
\qquad [9]
$$

and, for m not equal to 0,

$$
T_{nm} = (-1)^m \frac{8\pi}{2n+1} K_n \xi^{n+1} N_n^m P_n^m(\theta') \mathrm{e}^{\mathrm{i}\phi'}
\qquad [10]
$$

from which we can find the real-valued, time-varying quantities $a_n^m(t)$ and $b_n^m(t)$, which we will use below in computing the response of the Earth.

6.2.2 Computing the Tides: Direct Computation

Equations [7]–[10] suggest a straightforward way to compute the tidal potential (and, as we will see, other theoretical tides). First, use a description of the location of the Moon and Sun in celestial coordinates (an ephemeris); other planets can be included if we wish. Then convert this celestial location to the geographical coordinates θ' and ϕ' of the sub-body point, and the distance R, using standard transformations (McCarthy and Pétit, 2004, chapter 4). Finally, use eqns [9] and [10] to get $T_{nm}(t)$. Once we have the T_{nm}, we can combine these with the spatial factors in [7] to get V_{tid}/g, either for a specific location or as a distribution over the whole Earth. As we will see, we can vary the spatial factors to find, not just the potential, but other observables, including tilt and strain, all with no changes to the T_{nm}; we need to do the astronomy only once.

A direct computation has the advantage, compared with the harmonic methods (discussed below) of being limited in accuracy only by the accuracy of the ephemeris. If we take derivatives of [10] with respect to R, θ', and ϕ', we find that relative errors of 10^{-4} in V_{tid}/g would be caused by errors of 7×10^{-5} rad (14″) in θ' and ϕ', and 3×10^{-5} in ξ. (We pick this level of error because it usually exceeds the accuracy with which the tides can be measured, either because of noise or because of instrument calibration.) The errors in the angular quantities correspond to errors of about 400 m in the location of the sub-body point, so our model of Earth rotation, and our station location, needs to be good to this level – which requires 1 s accuracy in the timing of the data.

Two types of ephemerides are available: analytical, which provide a closed-form algebraic description of the motion of the body; and the much more precise numerical ephemerides, computed from numerical integration of the equations of motion, with parameters chosen to best fit astronomical data. While numerical ephemerides are more accurate, they are less convenient for most users, being available only as tables; analytical ephemerides are or can be made available as computer code.

The first tidal-computation program based directly on an ephemeris was that of Longman (1959), still in use for making rough tidal corrections for gravity surveys. Longman's program, like some others, computed accelerations directly, thus somewhat obscuring the utility of an ephemeris-based approach to all tidal computations. Munk and Cartwright (1966) applied this method for the tidal potential. Subsequent programs such as those of Harrison (1971), Broucke *et al.* (1972), Tamura (1982), and Merriam (1992) used even more precise ephemerides based on subsets of Brown's lunar theory.

Numerical ephemerides have been used primarily to produce reference time series, rather than for general-purpose programs, although the current IERS standards use such a method for computing tidal potentials and displacements (with corrections described in Section 6.3.2.2). Most precise calculations (e.g., Hartmann and Wenzel, 1995) have relied on the numerical ephemerides produced by JPL

(Standish *et al.*, 1992). The resulting tidal series form the basis for a harmonic expansion of the tidal potential, a standard method to which we now turn.

6.2.3   Computing the Tides (I): Harmonic Decompositions

Since the work of Thomson and Darwin in the 1870s and 1880s, the most common method of analyzing and predicting the tides, and of expressing tidal behavior, has been through a 'harmonic expansion' of the tidal potential. In this, we express the T_{nm} as a sum of sinusoids, whose frequencies are related to combinations of astronomical frequencies and whose amplitudes are determined from the expressions in the ephemerides for R, θ', and ϕ'. In such an expansion, we write the complex T_{nm}'s as

$$T_{nm}(t) = \sum_{k=1}^{K_{nm}} A_{knm} \exp[i(2\pi f_{knm} t + \varphi_{knm})] \qquad [11]$$

where, for each degree and order we sum K_{nm} sinusoids with specified real amplitudes, frequencies, and phases A, f, and φ. The individual sinusoids are called 'tidal harmonics' (not the same as the spherical harmonics of Section 6.2.1).

This method has the conceptual advantage of decoupling the tidal potential from the details of astronomy, and the practical advantage that a table of harmonic amplitudes and frequencies, once produced, is valid over a long time. Such an expansion also implicitly puts the description into the frequency domain, something as useful here as in other parts of geophysics. We can use the same frequencies for any tidal phenomenon, provided that it comes from a linear response to the driving potential – which is essentially true for the Earth tides. So, while this expansion was first used for ocean tides (for which it remains the standard), it works just as well for Earth tides of any type.

To get the flavor of this approach, and also introduce some terminology, we consider tides from a very simple situation: a body moving at constant speed β in a circular orbit, the orbital plane being inclined at an angle ε to the Earth's equator. The angular distance from the ascending node (where the orbit plane and the equatorial plane intersect) is βt. The rotation of the Earth, at rate Ω, causes the terrestrial longitude of the ascending node to be Ωt; since the ascending node is fixed in space, Ω corresponds to one revolution per sidereal day. We further assume that at $t = 0$ the body is at the ascending node and longitude $0°$ is under it. Finally, we take just the real part of [6], and do not worry about signs.

With these simplifications we consider first the diurnal degree-2 tides ($n = 2$, $m = 1$). After some tedious spherical trigonometry and algebra, we find that

$$V/g = K_2 \left(\frac{6\pi}{5}\right) [\sin\varepsilon\cos\varepsilon\sin\Omega t \\ + \frac{1}{2}\sin\varepsilon(1 + \cos\varepsilon)\sin(\Omega - 2\beta)t \\ + \frac{1}{2}\sin\varepsilon(1 - \cos\varepsilon)\sin(\Omega + 2\beta)t]$$

This shows that the harmonic decomposition includes three harmonics, with arguments (of time) Ω, $\Omega - 2\beta$, and $\Omega + 2\beta$; their amplitudes depend on ε, the inclination of the orbital plane. If ε were zero, there would be no diurnal tides at all. For our simple model, a reasonable value of ε is $23.44°$, the inclination of the Sun's orbital plane, and the mean inclination of the Moon's. These numbers produce the harmonics given in **Table 2**, in which the frequencies are given in cycles per solar day (cpd). Both the Moon and Sun produce a harmonic at 1 cycle per sidereal day. For the Moon, β corresponds to a period of 27.32 days (the tropical month) and for the Sun 365.242 days (one year), so the other harmonics are at ± 2 cycles per month, or ± 2 cycles per year, from this. Note that there is not a harmonic at 1 cycle per lunar (or solar) day – this is not unexpected, given the degree-2 nature of the tidal potential.

It is convenient to have a shorthand way of referring to these harmonics; unfortunately the standard naming system, now totally entrenched, was begun by Thomson for a few tides, and then extended by Darwin in a somewhat *ad hoc* manner.

The result is a series of conventional names that simply have to be learned as is (though only the ones for the largest tides are really important). For the Moon, the three harmonics have the Darwin symbols K_1, O_1, and OO_1; for the Sun they are K_1 (again, since this has the same frequency for any body), P_1, and ϕ_1.

For the semidiurnal ($m = 2$ case), the result is

$$V/g = K_2 \left(\frac{24\pi}{5}\right) [(1 - \cos^2\varepsilon)\cos 2\Omega t + \frac{1}{2}(1 + \cos\varepsilon)^2 \\ \times \cos(2\Omega - 2\beta)t + \frac{1}{2}(1 - \cos\varepsilon)^2\cos(2\Omega + 2\beta)t]$$

again giving three harmonics, though for ε equal to $23.44°$, the third one is very small. Ignoring the last term, we have two harmonics, also listed in **Table 2**. The Darwin symbol for the first argument is K_2; again, this frequency is the same for the Sun and the Moon, so these combine to make a lunisolar tide. The second argument gives the largest tides: for the Moon, M_2 (for the Moon) or S_2 (for the Sun), at precisely 2 cycles per lunar (or solar) day, respectively.

Finally, the $m = 0$, or long-period, case has

$$V/g = K_2 \left(\frac{\pi}{5}\right) [(1.5\sin^2\varepsilon - 1) - 1.5\sin^2\varepsilon\cos 2\beta t]$$

which gives one harmonic at zero frequency (the so-called 'permanent tide'), and another with an argument of 2β, making tides with frequencies of 2 cycles per month (Mf, the fortnightly tide, from the Moon) and 2 cycles per year (Ssa, the semiannual tide, from the Sun).

This simple model demonstrates another important attribute of the tides, arising from the dependence on the orbital inclination ε. For the Sun this is nearly invariant, but for the Moon it varies

Table 2   Tidal constituents (simple model)

Argument	Moon		Sun	
	Freq. (cpd)	Amp. (m)	Freq. (cpd)	Amp. (m)
Long-period tides				
	0.000000	0.217	0.000000	0.100
2β	0.073202	0.066	0.005476	0.030
Diurnal tides				
Ω	1.002738	0.254	1.002738	0.117
$\Omega - 2\beta$	0.929536	0.265	0.997262	0.122
$\Omega + 2\beta$	1.075940	0.011	1.008214	0.005
Semidiurnal tides				
2Ω	2.005476	0.055	2.005476	0.025
$2\Omega - 2\beta$	1.932274	0.640	2.000000	0.294

by $\pm 5.13°$ from the mean, with a period of 18.61 years. This produces a variation in amplitude in all the lunar tides, which is called the 'nodal modulation'. The simple expressions show that the resulting variation is $\pm 18\%$ for O_1, and $\pm 3\%$ for M_2. Such a modulated sinusoid can be written as $\cos \omega_0 t(1 + A \cos \omega_m t)$, with $\omega_0 \gg \omega_m$; this is equal to

$$\cos \omega_0 t + \frac{1}{2} A \cos[(\omega_0 + \omega_m)t] + \frac{1}{2} A \cos[(\omega_0 - \omega_m)t]$$

so we can retain a development purely in terms of sinusoids, but with three harmonics, one at the central frequency and two smaller ones (called 'satellite harmonics') separated from it by 1 cycle in 18.61 years.

An accurate ephemeris would include the ellipticity of the orbits, and all the periodic variations in ε and other orbital parameters, leading to many harmonics; for a detailed description, see Bartels (1957/ 1985). The first full expansion, including satellite harmonics, was by Doodson (1921), done algebraically from an analytical ephemeris; the result had 378 harmonics. Doodson needed a nomenclature for these tides, and introduced one that relies on the fact that, as our simple ephemeris suggests, the frequency of any harmonic is the sum of multiples of a few basic frequencies. For any (n, m), we can write the argument of the exponent in [11] as

$$2\pi f_k t + \phi_k = \left(\sum_{l=1}^{6} D_{lk} 2\pi f_l\right) t + \sum_{l=1}^{6} D_{lk} \varphi_l$$

where the f_l's are the frequencies corresponding to various astronomical periods, and the φ_l's are the phases of these at some suitable epoch; **Table 3** gives a list. (Recent tabulations extend this notation with up to five more arguments to describe the motions of the planets. As the tides from these are small we ignore them here.) The $l=1$ frequency is chosen to be one cycle per lunar day exactly, so for the M_2 tide the D_l's are $2, 0, 0, 0, 0, 0$. This makes the solar tide, S_2, have the

D_l's $2, 2, 2, 0, 0, 0$. In practice, all but the smallest tides have D_{lk} ranging from -5 to 5 for $l > 1$. Doodson therefore added 5 to these numbers to make a compact code, so that M_2 becomes $255 \cdot 555$ and S_2 $273 \cdot 555$. This is called the Doodson number; the numbers without 5 added are sometimes called Cartwright–Tayler codes (**Table 4**).

Figure 3 shows the full spectrum of amplitude coefficients, from the recent expansion of Hartmann and Wenzel (1995). The top panel shows all harmonics on a linear scale, making it clear that only a few are large, and the separation into different species around 0, 1, and 2 cycles/day: these are referred to as the long-period, diurnal, and semidiurnal tidal bands. The two lower panels show an expanded view of the diurnal and semidiurnal bands, using a log scale of amplitude to include the smaller harmonics. What is apparent from these is that each tidal species is split into a set of bands, separated by 1 cycle/month; these are referred to as 'groups': in each group the first two digits of the Doodson number are the same. All harmonics with the same first three digits of the Doodson number are in clusters separated by 1 cycle/year; these clusters are called 'constituents', though this name is also sometimes used for the individual harmonics. As a practical matter this is usually the finest frequency resolution attainable; on the scale of this plot finer frequency separations, such as the nodal modulation, are visible only as a thickening of some of the lines. All this fine-scale structure poses a challenge to tidal analysis methods (Section 6.5.1).

Since Doodson provided the tidal potential to more than adequate accuracy for studying ocean tides, further developments did not take place for the next 50 years, until Cartwright and Tayler (1971) revisited the subject. Using eqn [6], they computed the potential from a more modern lunar ephemeris, and then applied special Fourier methods to analyze, numerically, the resulting series, and get amplitudes for the various harmonics. The result was

Table 3   Fundamental tidal frequencies

l	Symbol	Frequency (cycles/day)	Period	What
1	τ	0.9661368	24 h 50 m 28.3 s	Lunar day
2	s	0.0366011	27.3216 d	Moon's longitude: tropical month
3	h	0.0027379	365.2422 d	Sun's longitude: solar year
4	p	0.0003095	8.847 yr	Lunar perigee
5	N'	0.0001471	18.613 yr	Lunar node
6	p_s	0.0000001	20941 yr	Solar perigee

Longitude refers to celestial longitude, measured along the ecliptic.

Table 4 Largest tidal harmonics, for $n = 2$, sorted by size for each species

Amplitude (m)	Doodson number	Frequency (cpd)	Darwin symbol
Long-period tides			
−0.31459	055.555	0.0000000	M_0, S_0
−0.06661	075.555	0.0732022	Mf
−0.03518	065.455	0.0362916	Mm
−0.03099	057.555	0.0054758	Ssa
0.02793	055.565	0.0001471	*N*
−0.02762	075.565	0.0733493	
−0.01275	085.455	0.1094938	Mtm
−0.00673	063.655	0.0314347	MSm
−0.00584	073.555	0.0677264	MSf
−0.00529	085.465	0.1096409	
Diurnal tides			
0.36864	165.555	1.0027379	K_1
−0.26223	145.555	0.9295357	O_1
−0.12199	163.555	0.9972621	P_1
−0.05021	135.655	0.8932441	Q_1
0.05003	165.565	1.0028850	
−0.04947	145.545	0.9293886	
0.02062	175.455	1.0390296	J_1
0.02061	155.655	0.9664463	M_1
0.01128	185.555	1.0759401	OO_1
−0.00953	137.455	0.8981010	ρ_1
−0.00947	135.645	0.8930970	
−0.00801	127.555	0.8618093	σ_1
0.00741	155.455	0.9658274	
−0.00730	165.545	1.0025908	
0.00723	185.565	1.0760872	
−0.00713	162.556	0.9945243	π_1
−0.00664	125.755	0.8569524	$2Q_1$
0.00525	167.555	1.0082137	ϕ_1
Semidiurnal tides			
0.63221	255.555	1.9322736	M_2
0.29411	273.555	2.0000000	S_2
0.12105	245.655	1.8959820	N_2
0.07991	275.555	2.0054758	K_2
0.02382	275.565	2.0056229	
−0.02359	255.545	1.9321265	
0.02299	247.455	1.9008389	ν_2
0.01933	237.555	1.8645472	μ_2
−0.01787	265.455	1.9685653	L_2
0.01719	272.556	1.9972622	T_2
0.01602	235.755	1.8596903	$2N_2$
0.00467	227.655	1.8282556	ε_2
−0.00466	263.655	1.9637084	λ_2

a compendium of 505 harmonics, which (with errors corrected by Cartwright and Edden (1973)) soon became the standard under the usual name of the CTE representation. (A few small harmonics at the edges of each band, included by Doodson but omitted by Cartwright, are sometimes added to make a CTED list with 524 harmonics.)

More extensive computations of the tidal potential and its harmonic decomposition have been driven by the very high precision available from the ephemerides and the desire for more precision for analyzing some tidal data (gravity tides from superconducting gravimeters). Particular expansions are those of Bullesfeld (1985), Tamura (1987), Xi (1987), Hartmann and Wenzel (1995), and Roosbeek (1995) . The latest is that of Kudryavtsev (2004) , with 27 000 harmonics. **Figure 4** shows the amplitude versus number of harmonics; to get very

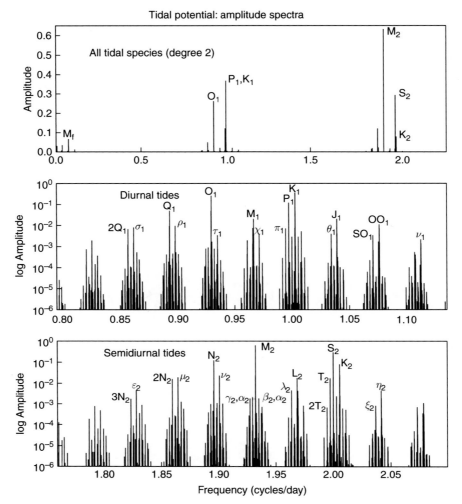

Figure 3   The spectrum of the tidal potential. Since all variations are purely sinusoidal, the spectrum is given by the amplitudes of the tidal harmonics, taken from Hartmann and Wenzel (1995), though normalized according to the convention of Cartwright and Tayler (1971). The Darwin symbols are shown for the larger harmonics (top) and all named diurnal and semidiurnal harmonics, except for a few that are shown only in **Figure 5**.

high accuracy demands a very large number. But not many are needed for a close approximation; the CTED expansion is good to about 0.1% of the total tide.

6.2.4 The Pole Tide

Both in our elementary discussion and in our mathematical development of the tidal forcing, we treated the Earth's rotation only as a source of motion of the sub-body point. But changes in this rotation also cause spatial variations in the gravitational potential, and since these have the same effects as the attraction of external bodies, they can also be regarded as tides.

The only significant one is the 'pole tide', which is caused by changes in the direction of the Earth's spin axis relative to a point fixed in the Earth. The spin produces a centrifugal force, which depends on the angular distance between the spin axis and a location. As the spin axis moves, this distance, and the centrifugal force, changes.

Mathematically, the potential at a location $\mathbf{r}$ from a spin vector $\mathbf{\Omega}$ is

$$V = \frac{1}{2}\left[|\mathbf{\Omega}|^2|\mathbf{r}|^2 - |\mathbf{\Omega}\cdot\mathbf{r}|^2\right] \qquad [12]$$

We assume that the rotation vector is nearly along the 3-axis, so that we have $\mathbf{\Omega} = \Omega(m_1\hat{x}_1 + m_2\hat{x}_2 + \hat{x}_3)$,

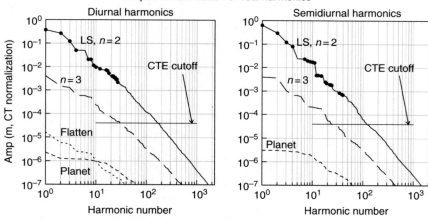

Amplitude distribution of tidal harmonics

Figure 4 Distribution of harmonic amplitudes for the catalog of Hartmann and Wenzel (1995), normalized according to Cartwright and Tayler (1971). The line 'LS, $n = 2$' refers to lunisolar harmonics of degree 2; those with large dots have Darwin symbols associated with them. Constituents of degree 3, from other planets, and from earth flattening are shown as separate distributions. The horizontal line shows the approximate cutoff level of the Cartwright and Tayler (1971) list.

with m_1 and m_2 both much less than 1. If we put this expression into [12], and subtract V for m_1 and m_2 both zero, the potential height for the pole tide is

$$\frac{V}{g} = -\frac{\Omega^2}{2g}[2(m_1 r_1 r_3 + m_2 r_2 r_3)]$$
$$= -\frac{\Omega^2 a^2}{g} \sin\theta \cos\theta (m_1 \cos\phi + m_2 \sin\phi)$$

This is a degree-2 change in the potential, of the same form as for the diurnal tides. However, the periods involved are very different, since the largest pole tides come from the largest polar motions, at periods of 14 months (the Chandler wobble) and 1 year. The maximum range of potential height is a few cm, small but not negligible; pole-tide signals have been observed in sea-level data and in very precise gravity measurements. This 'tide' is now usually allowed for in displacement measurements, being computed from the observed polar motions (Wahr, 1985). The accompanying ocean tide is marginally observable (Haubrich and Munk, 1959; Miller and Wunsch, 1973; Trupin and Wahr, 1990; Desai, 2002).

6.2.5 Radiational Tides

The harmonic treatment used for the gravitational tides can also be useful for the various phenomena associated with solar heating. The actual heating is complicated, but a first approximation is to the input radiation, which is roughly proportional to the cosine of the Sun's elevation during the day and zero at night; Munk and Cartwright (1966) called this the 'radiational tide'. The day–night asymmetry produces harmonics of degrees 1 and 2; these have been tabulated by Cartwright and Tayler (1971) and are shown in **Figure 5** as crosses (for both degrees), along with the tidal potential harmonics shown as in **Figure 3**. The unit for the radiational tides is S, the solar constant, which is 1366 W m^{-2}.

These changes in solar irradiation drive changes in ground temperature fairly directly, and changes in air temperature and pressure in very complicated ways. Ground-temperature changes cause thermoelastic deformations with tidal periods (Berger, 1975; Harrison and Herbst, 1977; Mueller, 1977). Air pressure changes, usually known as 'atmospheric tides' (Chapman and Lindzen, 1970), load the Earth enough to cause deformations and changes in gravity. Such effects are usually treated as noise, but the availability of better models of some of the atmospheric tides (Ray, 2001; Ray and Ponte, 2003; Ray and Poulose, 2005) and their inclusion in ocean-tide models (Ray and Egbert, 2004) has allowed their effects to be compared with gravity observations (Boy et al., 2006b).

That some of these thermal tidal lines coincide with lines in the tidal potential poses a real difficulty for precise analysis of the latter. Strictly speaking, if we have the sum of two harmonics with the same frequency, it will be impossible to tell how much each part contributes. The only way to resolve this is to make additional assumptions about how the response to these behaves at other frequencies. Even

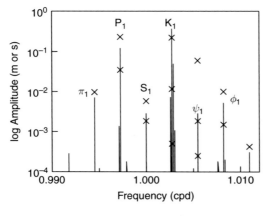

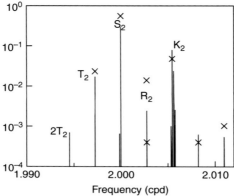

Figure 5 Radiational tides. The crosses show the amplitudes of the radiational tidal harmonics (degree 1 and 2) from Cartwright and Tayler (1971); the amplitudes are for the solar constant S being taken to be 1.0. The lines are harmonics of the gravitational tides, as in **Figure 3**.

when this is done, there is a strong likelihood that estimates of these tides will have large systematic errors – which is why, for example, the large K_1 tide is less used in estimating tidal responses than the smaller O_1 tide is.

6.3   Tidal Response of the Solid Earth

Having described the tidal forces, we next turn to the response of the solid Earth – which, as is conventional, we assume to be oceanless, putting in the effect of the ocean tides at a later step. We start with the usual approximation of a spherical Earth in order to introduce a number of concepts, many of them adequate for all but the most precise modeling of the tides. We then describe what effects a better approximation has, in enough detail to enable computation; the relevant

theory is beyond the scope of this treatment, though outlined in Chapter 10.

6.3.1   Tidal Response of a SNREI Earth

To a good approximation, we can model the tidal response of an oceanless Earth by assuming a SNREI Earth model: that is, one that is **S**pherical, **N**on-**R**otating, **E**lastic, and **I**sotropic. As in normal-mode seismology (from which this acronym comes), this means that the only variation of elastic properties is with depth. In addition to these restrictions on the Earth model, we add one more about the tidal forcing: that it has a much longer period than any normal modes of oscillation of the Earth so that we can use a quasi-static theory, taking the response to be an equilibrium one. Since the longest-period normal modes for such an Earth have periods of less than an hour, this is a good approximation.

It is simple to describe the response of a SNREI Earth to the tidal potential (Jeffreys, 1976). Because of symmetry, only the degree n is relevant. If the potential height at a point on the surface is $V(\theta, \phi)/g$, the distortion of the Earth from tidal forces produces an additional gravitational potential $k_n V(\theta, \phi)$, a vertical (i.e., radial) displacement $h_n V(\theta, \phi)/g$, and a horizontal displacement $l_n(\nabla_1 V(\theta, \phi)/g)$, where ∇_1 is the horizontal gradient operator on the sphere. So defined, k_n, h_n, and l_n are dimensionless; they are called Love numbers, after A. E. H. Love (though the parameter l_n was actually introduced by T. Shida). For a standard modern earth model (PREM) $h_2 = 0.6032$, $k_2 = 0.2980$, and $l_2 = 0.0839$. For comparison, the values for the much older Gutenberg–Bullen Earth model are 0.6114, 0.3040, and 0.0832 – not very different. In this section we adopt values for a and g that correspond to a spherical Earth: 6.3707×10^6 m and 9.821 m s^{-2}, respectively.

6.3.1.1   Some combinations of Love numbers (I): gravity and tilt

Until there were data from space geodesy, neither the potential nor the displacements could be measured; what could be measured were ocean tides, tilt, changes in gravity, and local deformation (strain), each of which possessed its own expression in terms of Love numbers – which we now derive. Since the first three of these would exist even on a rigid Earth, it is common to describe them using the ratio between what they are on an elastic Earth (or on the real Earth) and what they would be on a rigid Earth.

The simplest case is that of the effective tide-raising potential: that is, the one relevant to the ocean tide. The total tide-raising potential height is $(1 + k_n)V/g$, but the solid Earth (on which a tide gauge sits) goes up by $h_n V/g$, so the effective tide-raising potential is $(1 + k_n - h_n)V/g$, sometimes written as $\gamma_n V/g$, γ_n being called the 'diminishing factor'. For the PREM model $\gamma_2 = 0.6948$. Since tilt is just the change in slope of an equipotential surface, again relative to the deforming solid Earth, it scales in the same way that the potential does: the tilt on a SNREI Earth is γ_n times the tilt on a rigid Earth. The NS tilt is, using eqn [8] and expressions for the derivatives of Legendre functions,

$$\lambda_N = \frac{-\gamma_n}{ga} \frac{\partial V}{\partial \theta}$$
$$= \frac{-1}{a \sin \theta} \sum_{n=2}^{n=\infty} \gamma_n \sum_{m=0}^{n} N_n^m [n \cos \theta P_n^m (\cos \theta) - (n+m)$$
$$\times P_{n-1}^m (\cos \theta)][a_n^m(t)\cos m\phi + b_n^m(t)\sin m\phi] \quad [13]$$

where the sign is chosen such that a positive tilt to the North would cause a plumb line to move in that direction. The East tilt is

$$\lambda_E = \frac{\gamma_n}{ga \sin \theta} \frac{\partial V}{\partial \phi} = \frac{-1}{a \sin \theta} \sum_{n=2}^{n=\infty} \gamma_n \sum_{m=0}^{n} m N_n^m P_n^m (\cos \theta)$$
$$\times [b_n^m(t)\cos m\phi - a_n^m(t)\sin m\phi] \quad [14]$$

with the different combinations of a and b with the ϕ dependence showing that this tilt is phase-shifted relative to the potential, by 90° if we use a harmonic decomposition.

Tidal variations in gravity were for a long time the commonest type of Earth-tide data. For a spherical Earth, the tidal potential is, for degree n,

$$V_n \left(\frac{r}{a}\right)^n + k_n V_n \left(\frac{a}{r}\right)^{n+1}$$

where the first term is the potential caused by the tidal forcing (and for which we have absorbed all nonradial dependence into V_n), and the second is the additional potential induced by the Earth's deformation. The corresponding change in local gravitational acceleration is the radial derivative of the potential:

$$\frac{\partial}{\partial r} \left[V_n \left(\left(\frac{r}{a}\right)^n + k_n \left(\frac{a}{r}\right)^{n+1} \right) \right]_{r=a} = V_n \left[\frac{n}{a} - (n+1)\frac{k_n}{a} \right] \quad [15]$$

In addition to this change in gravity from the change in the potential, there is a change from the gravimeter being moved up by an amount $h_n V_n/g$. The change in gravity is this displacement times

the gradient of g, $2g/a$, plus the displacement times $-\omega^2$, where ω is the radian frequency of the tidal motion – that is, the inertial acceleration. (We adopt the Earth-tide convention that a decrease in g is positive.) If we ignore this last part (which is at most 1.5% of the gravity-gradient part), we get a total change of

$$V_n \left[\frac{n}{a} - \left(\frac{n+1}{a}\right) k_n + \frac{2h_n}{a} \right] = \frac{nV_n}{a} \left[1 - \left(\frac{n+1}{n}\right) k_n + \frac{2}{n} h_n \right] \quad [16]$$

The nV_n/a term is the change in g that would be observed on a rigid Earth (with h and k zero); the term which this is multiplied by, namely

$$\delta_n = 1 + \frac{2h_n}{n} - \left(\frac{n+1}{n}\right) k_n$$

is called the 'gravimetric factor'. For the PREM model, $\delta_2 = 1.1562$: the gravity tides are only about 16% larger than they would be on a completely rigid Earth, so that most of the tidal gravity signal shows only that the Moon and Sun exist, but does not provide any information about the Earth. The expression for the gravity tide is of course very similar to eqn [8]:

$$\delta g = \frac{g}{a} \sum_{n=2}^{n=\infty} \delta_n \sum_{m=0}^{n} N_n^m P_n^m (\cos \theta)[a_n^m(t)\cos m\phi + b_n^m(t)\sin m\phi] \quad [17]$$

6.3.1.2 Combinations of Love numbers (II): displacement and strain tides

For a tidal potential of degree n, the displacements at the surface of the Earth ($r = a$) will be, by the definitions of the Love numbers l_n and h_n,

$$u_r = \frac{h_n V}{g}, \qquad u_\theta = \frac{l_n}{g} \frac{\partial V}{\partial \theta}, \qquad u_\phi = \frac{l_n}{g \sin \theta} \frac{\partial V}{\partial \phi} \quad [18]$$

in spherical coordinates. Comparing these with [17], [13], and [14], we see that the vertical displacement is exactly proportional to changes in gravity, with the scaling constant being $h_n a/2\delta_n g = 1.692 \times 10^5 \text{ s}^2$; and that the horizontal displacements are exactly proportional to tilts, with the scaling constant being $l_n a/\gamma_n = 7.692 \times 10^5 \text{ m}$; we can thus use eqns [17], [13], and [14], suitably scaled, to find tidal displacements.

Taking the derivatives of [18], we find the tensor components of the surface strain are

$$e_{\theta\theta} = \frac{1}{ga}\left(h_n V + l_n \frac{\partial^2 V}{\partial^2 \theta}\right)$$

$$e_{\phi\phi} = \frac{1}{ga}\left(h_n V + l_n \cot\theta \frac{\partial V}{\partial \theta} + \frac{l_n}{\sin\theta}\frac{\partial^2 V}{\partial^2 \phi}\right)$$

$$e_{\theta\phi} = \frac{l_n}{ga\sin\theta}\left(\frac{\partial^2 V}{\partial\theta\partial\phi} - \cot\theta \frac{\partial V}{\partial \phi}\right)$$

We again use [8] for the tidal potential, and get the following expressions that give the formulas for the three components of surface strain for a particular n and m; to compute the total strain these should be summed over all $n \geq 2$ and all m from 0 to n (though in practice the strain tides with $n > 3$ or $m = 0$ are unobservable).

$$e_{\theta\theta} = \frac{N_n^m}{a\sin^2\theta}\left[\left(h_n \sin^2\theta + l_n\left(n^2\cos^2\theta - n\right)\right)P_n^m(\cos\theta)\right.$$

$$\left. - 2l_n(n-1)(n+m)\cos\theta P_{n-1}^m(\cos\theta)\right.$$

$$\left. + l_n(n+m)(n+m-1)P_{n-2}^m(\cos\theta)\right]$$

$$\times \left[a_n^m(t)\cos m\phi + b_n^m(t)\sin m\phi\right]$$

$$e_{\phi\phi} = \frac{N_n^m}{a\sin^2\theta}\left[\left(h_n \sin^2\theta + l_n\left(n\cos^2\theta - m^2\right)\right)P_n^m(\cos\theta)\right.$$

$$\left. - l_n(n+m)\cos\theta P_{n-1}^m(\cos\theta)\right]$$

$$\times \left[a_n^m(t)\cos m\phi + b_n^m(t)\sin m\phi\right]$$

$$e_{\theta\phi} = \frac{mN_n^m l_n}{a\sin^2\theta}\left[(n-1)\cos\theta P_n^m(\cos\theta) - (n+m)P_{n-1}^m(\cos\theta)\right]$$

$$\times \left[b_n^m(t)\cos m\phi - a_n^m(t)\sin m\phi\right]$$

Note that the combination of the longitude factors with the $a_n^m(t)$ and $b_n^m(t)$ means that $e_{\theta\theta}$ and $e_{\phi\phi}$ are in phase with the potential, while $e_{\theta\phi}$ is not.

One consequence of these expressions is that, for $n = 2$ and m equal to either 1 or 2, the areal strain, $(1/2)(e_{\theta\theta} + e_{\phi\phi})$, is equal to $V(h_2 - 3l_2)/ga$: areal strain, vertical displacement, the potential, and gravity are all scaled versions of each other. Close to the surface, the free-surface condition means that deformation is nearly that of plane stress, so vertical and volume strains are also proportional to areal strain, and likewise just a scaled version of the potential.

If we combine these expressions for spatial variation with the known amplitudes of the tidal forces, we can see how the rms amplitude of the body tides varies with latitude and direction (**Figure 6**). There are some complications in the latitude dependence; for example, the EW semidiurnal strain tides go to

zero at 52.4° latitude. Note that while the tilt tides are larger than strain tides, most of this signal is from the direct attractions of the Sun and Moon; the purely deformational part of the tilt is about the same size as the strain.

6.3.2   Response of a Rotating Earth

We now turn to models for tides on an oceanless and isotropic Earth, still with properties that depend on depth only, but add rotation and slightly inelastic behavior. Such models have three consequences for the tides:

1. The ellipticity of the CMB and the rotation of the Earth combine to produce a free oscillation in which the fluid core (restrained by pressure forces) and solid mantle precess around each other. This is known as the 'nearly diurnal free wobble' (NDFW) or 'free core nutation'. Its frequency falls within the band of the diurnal tides, which causes a resonant response in the Love numbers near 1 cycle/day. The diurnal tides also cause changes in the direction of the Earth's spin axis (the astronomical precessions and nutations), and the NDFW affects these as well, so that some of the best data on it come from astronomy (Herring *et al.*, 2002).

2. Ellipticity and rotation couple the response to forcing of degree n to spherical harmonics of other degrees, and spheroidal to toroidal modes of deformation. As a result, the Love numbers become slightly latitude dependent, and additional terms appear for horizontal displacement.

3. The imperfect elasticity of the mantle (finite Q) modifies the Love numbers in two ways: they become complex, with small imaginary parts; and they become weakly frequency dependent because of anelastic dispersion.

The full theory for these effects, especially the first two, is quite complicated. Love (1911) provided some theory for the effects of ellipticity and rotation, and Jeffreys and Vincente (1957) and Molodenskii (1961) for the NDFW, but the modern approach for these theories was described by Wahr (1981a, 1981b); a simplified version is given by Neuberg *et al.* (1987) and Zürn (1997). For more recent developments, see the article by Dehant and Mathews in this volume, and Mathews *et al.* (1995a, 1995b, 1997, 2002), Wang (1997), Dehant *et al.* (1999), and Mathews and Guo (2005).

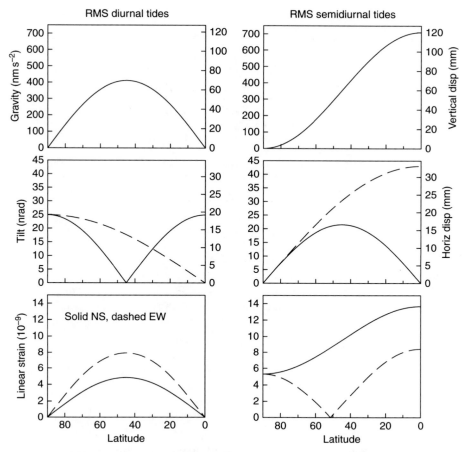

Figure 6 RMS tides. The left plots show the rms tides in the diurnal band, and the right plots the rms in the semidiurnal band. The uppermost frame shows gravity and vertical displacement (with scales for each) which have the same latitude dependence; the next horizontal displacement and tilt; and the bottom linear strain. In all plots but the top, dashed is for EW measurements, solid for NS.

To obtain the full accuracy of these theories, particularly for the NDFW correction, requires the use of tabulated values of the Love numbers for specific harmonics, but analytical approximations are also available. The next three sections outline these, using values from the IERS standards (McCarthy and Pétit, 2004) for the Love numbers and from Dehant *et al.* (1999) for the gravimetric factor.

6.3.2.1 NDFW resonance

The most important result to come out of the combination of improved theoretical development and observations has been that the period of the NDFW, both in the Earth tides and in the nutation, is significantly different from that originally predicted. The NDFW period is in part controlled by the ellipticity of the CMB, which was initially assumed to be that for a hydrostatic Earth. The

observed period difference implies that the ellipticity of the CMB departs from a hydrostatic value by about 5%, the equivalent of a 500 m difference in radius, an amount not detectable using seismic data. This departure is generally thought to reflect distortion of the CMB by mantle convection.

The resonant behavior of the Love numbers from the NDFW is confined to the diurnal band; within that band it can be approximated by an expansion in terms of the resonant frequencies:

$$L(f) = S_z + \sum_{k=1}^{2} \frac{S_k}{f - f_k} \qquad [19]$$

where $L(f)$ is the frequency-dependent Love number (of whatever type) for frequency f in cycles per solar day; The expansion in use for the IERS standards includes three resonances: the Chandler wobble, the NDFW, and the free inner core nutation

Table 5   Coefficients (real and imaginary parts) used in eqn [19] to find the frequency dependence of the Love numbers (including corrections for ellipticity) in the diurnal tidal band

	$R(S_2)$	$I(S_2)$	$R(S_1)$	$I(S_1)$	$R(S_2)$	$I(S_2)$
δ_0	1.15802	0.0	-2.871×10^{-3}	0.0	4.732×10^{-5}	0.0
$k^{(0)}$	0.29954	-1.412×10^{-3}	-7.811×10^{-4}	-3.721×10^{-5}	9.121×10^{-5}	-2.971×10^{-6}
$h^{(0)}$	0.60671	-2.420×10^{-3}	-1.582×10^{-3}	-7.651×10^{-5}	1.810×10^{-4}	-6.309×10^{-6}
$l^{(0)}$	0.08496	-7.395×10^{-4}	-2.217×10^{-4}	-9.672×10^{-6}	-5.486×10^{-6}	-2.998×10^{-7}
δ_+	1.270×10^{-4}	0.0	-2.364×10^{-5}	0.0	1.564×10^{-6}	0.0
k^+	-8.040×10^{-4}	2.370×10^{-6}	2.090×10^{-6}	$1.030 \times 10-7$	-1.820×10^{-7}	6.500×10^{-9}
$h^{(2)}$	-6.150×10^{-4}	-1.220×10^{-5}	1.604×10^{-6}	1.163×10^{-7}	2.016×10^{-7}	2.798×10^{-9}
$l^{(2)}$	1.933×10^{-4}	-3.819×10^{-6}	-5.047×10^{-7}	-1.643×10^{-8}	-6.664×10^{-9}	5.090×10^{-10}
$l^{(1)}$	1.210×10^{-3}	1.360×10^{-7}	-3.169×10^{-6}	-1.665×10^{-7}	2.727×10^{-7}	-8.603×10^{-9}
l^P	-2.210×10^{-4}	-4.740×10^{-8}	5.776×10^{-7}	3.038×10^{-8}	1.284×10^{-7}	-3.790×10^{-9}

(FICN); to a good approximation (better than 1%), the last can be ignored. **Table 5** gives the values of the S's according to the IERS standards (and to Dehant *et al.* (1999) for the gravimetric factors), scaled for f in cycles per solar day; in these units the resonance frequencies are

$$f_1 = -2.60812 \times 10^{-3} - 1.365 \times 10^{-4}i$$
$$f_2 = 1.0050624 + 2.5 \times 10^{-5}i$$

and the FICN frequency (not used in [19]) is $1.00176124 + 7.82 \times 10^{-4}i$. Dehant *et al.* (1999) use purely real-valued frequencies, with $f_1 = -2.492 \times 10^{-3}$ and $f_2 = 1.0050623$, as well as purely real values of the S's.

Figure 7 shows the NDFW resonance, for a signal (areal strain) relatively sensitive to it. Unfortunately,

the tidal harmonics do not sample the resonance very well; the largest effect is for the small ψ_1 harmonic, which is also affected by radiational tides. While tidal measurements (Zürn, 1997) have confirmed the frequency shift seen in the nutation data, the latter at this time seem to give more precise estimates of the resonant behavior.

One consequence of the NDFW resonance is that we cannot use equations of the form [17] to compute the theoretical diurnal tides, since the factor for them varies with frequency. If we construct the $a_n^m(t)$ and $b_n^m(t)$ using [11], it is easy to adjust the harmonic amplitudes and phases appropriately. Alternatively, if we find $a_n^m(t)$ and $b_n^m(t)$ using an ephemeris, we can compute the diurnal tides assuming a frequency-independent factor, and then apply corrections for

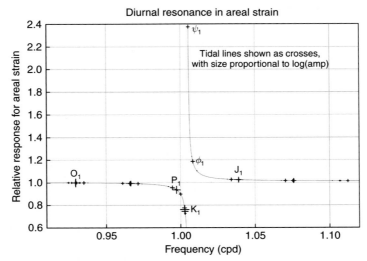

Figure 7   NDFW response, for the combination of Love numbers that gives the areal strain, normalized to 1 for the O_1 tide. The crosses show the locations of tidal harmonics, with size of symbol proportional to the logarithm of the amplitude of the harmonic.

the few harmonics that are both large and affected by the resonance; Mathews *et al.* (1997) and McCarthy and Pétit (2004) describe such a procedure for displacements and for the induced potential.

6.3.2.2 Coupling to other modes

The other effect of rotation and ellipticity is to couple the spheroidal deformation of degree n, driven by the potential, to spheroidal deformations of degree $n \pm 2$ and toroidal deformations of degree $n \pm 1$. Thus, the response to the degree-2 part of the tidal potential contains a small component of degrees 0 and 4. If we generalize the Love numbers, describing the response as a ratio between the response and the potential, the result will be a ratio that depends on latitude: we say that the Love number has become latitude dependent.

Such a generalization raises issues of normalization; unlike the spherical case, the potential and the response may be evaluated on different surfaces. This has been a source of some confusion. The normalization of Mathews *et al.* (1995b) is the one generally used for displacements: it uses the response in displacement on the surface of the ellipsoid, but takes these to be relative to the potential evaluated on a sphere with the Earth's equatorial radius. Because of the inclusion of such effects as the inertial and Coriolis forces, the gravimetric factor is no longer the combination of potential and displacement Love numbers, but an independent ratio, defined as the ratio of changes in gravity on the ellipsoid, to the direct attraction at the same point. Both quantities are evaluated along the normal to the ellipsoid, as a good approximation to the local vertical. Wahr (1981a) used the radius vector instead, producing a much larger apparent latitude effect.

The standard expression for the gravimetric factor is given by Dehant *et al.* (1999):

$$\delta(\theta) = \delta_0 + \delta_+ \frac{Y_{n+2}^m}{Y_n^m} + \delta_- \frac{Y_{n-2}^m}{Y_n^m} \qquad [20]$$

By the definition of the Y_n^m's, $\delta_- = 0$ except for $m \le n - 2$, so for the $n = 2$ tides we have for the diurnal tides

$$\delta(\theta) = \delta_0 + \delta_+ \frac{\sqrt{3}}{2\sqrt{2}} \left(7 \cos^2 \theta - 3 \right)$$

and for the semidiurnal tides

$$\delta(\theta) = \delta_0 + \delta_+ \frac{\sqrt{3}}{2} \left(7 \cos^2 \theta - 1 \right)$$

The expression for the induced potential (Wahr, 1981a) is similar, namely that the potential is gotten by replacing the term $Y_{nm}(\theta, \phi)$ in eqn [7] with

$$k_0 \left(\frac{a_e}{r} \right)^{n+1} Y_{nm}(\theta, \phi) + k_+ \left(\frac{a_e}{r} \right)^{n+3} Y_{n+2}^m(\theta, \phi) \qquad [21]$$

which of course recovers the conventional Love number for $k_+ = 0$.

The expressions for displacements are more complicated, partly because this is a vector quantity, but also because the horizontal displacements include spheroidal–toroidal coupling, which affects neither the vertical, the potential, nor gravity. For the degree-2 tides, the effect of coupling to the degree-4 deformation is allowed for by defining

$$h(\theta) = h^{(0)} + \frac{1}{2} h^{(2)} \left(3 \cos^2 \theta - 1 \right),$$
$$l(\theta) = l^{(0)} + \frac{1}{2} l^{(2)} \left(3 \cos^2 \theta - 1 \right) \qquad [22]$$

Then to get the vertical displacement, replace the term $Y_{nm}(\theta, \phi)$ in eqn [7] with

$$h(\theta) Y_2^m(\theta, \phi) + \frac{\delta_{m0} h^P}{N_2^0} \qquad [23]$$

where the δ_{m0} is the Kronecker delta, since h^P (usually called h' in the literature) only applies for $m = 0$; the N_2^0 factor arises from the way in which h^P was defined by Mathews *et al.* (1995b) .

The displacement in the $\hat{\theta}$ (North) direction is gotten by replacing the term $Y_{nm}(\theta, \phi)$ in eqn [7] with

$$l(\theta) \frac{\partial Y_2^m(\theta, \phi)}{\partial \theta} - \frac{m l^1 \cos \theta}{\sin \theta} Y_2^m(\theta, \phi) + \frac{\delta_{m1} l^P}{N_2^1} e^{i\phi} \qquad [24]$$

where again the l^P (usually called l') applies only for the particular value of $m = 1$, for which it applies a correction such that there is no net rotation of the Earth. Finally, to get the displacement in the $\hat{\phi}$ (East) direction, we replace the $Y_{nm}(\theta, \phi)$ term in [7] with

$$i \left[\frac{m l(\theta)}{\sin \theta} Y_2^m(\theta, \phi) + l^1 \cos \theta \frac{\partial Y_2^m(\theta, \phi)}{\partial \theta} + \frac{\delta_{m0} \sin \theta l^P}{N_2^0} \right] \qquad [25]$$

where again the l^P term applies, for $m = 0$, a no-net-rotation correction. The multiplication by i means that when this is applied to [7] and the real part taken, the time dependence will be $b_n^m(t) \cos m\phi - a_n^m(t) \sin m\phi$ instead of $a_n^m(t) \cos m\phi + b_n^m(t) \sin m\phi$.

Table 6 gives the generalized Love numbers for selected tides, including ellipticity, rotation, the NDFW, and anelasticity (which we discuss below). The values for the diurnal tides are from exact computations rather than the resonance approximations

Table 6   Love numbers for an Earth that includes ellipticity, rotation, anelasticity, and the NDFW resonance

	Ssa	Mf	O_1	P_1	K_1	ψ_1	M_2	M_3
δ_0	1.15884	1.15767	1.15424	1.14915	1.13489	1.26977	1.16172	1.07338
δ_+	0.00013	0.00013	0.00008	−0.00010	−0.00057	0.00388	0.00010	0.00006
δ_-	−0.00119	−0.00118						
$\Re[k^{(0)}]$	0.3059	0.3017	0.2975	0.2869	0.2575	0.5262	0.3010	0.093
$\Im[k^{(0)}]$	−0.0032	−0.0021	−0.0014	−0.0007	0.0012	0.0021	−0.0013	
k^+	−0.0009	−0.0009	−0.0008	−0.0008	−0.0007	−0.0011	−0.0006	
$\Re[h^{(0)}]$	0.6182	0.6109	0.6028	0.5817	0.5236	1.0569	0.6078	0.2920
$\Im[h^{(0)}]$	−0.0054	−0.0037	−0.0006	−0.0006	−0.0006	−0.0020	−0.0022	
$\Re[l^{(0)}]$	0.0886	0.0864	0.0846	0.0853	0.0870	0.0710	0.0847	0.0015
$\Im[l^{(0)}]$	−0.0016	−0.0011	−0.0006	−0.0006	−0.0006	−0.0001	−0.0006	
$h^{(2)}$	−0.0006	−0.0006	−0.0006	−0.0006	−0.0007	−0.0001	−0.0006	
$l^{(2)}$	0.0002	0.0002	0.0002	0.0002	0.0002	0.0002	0.0002	
$l^{(1)}$			0.0012	0.0012	0.0011	0.0019	0.0024	
l^P			−0.0002	−0.0002	−0.0003	−0.0001		

Values for the gravimetric factors are from Dehant *et al.* (1999) and for the other Love numbers from the IERS standards (McCarthy and Pétit, 2004).

given above. It is evident that the latitude dependence ranges from small to extremely small, the latter applying to the gravimetric factor, which varies by only 4×10^{-4} from the equator to 60° N. For the displacements, Mathews *et al.* (1997) show that the various coupling effects are at most 1 mm; the latitude dependence of $h(\theta)$ changes the predicted displacements by 0.4 mm out of 300.

6.3.2.3 Anelastic effects

All modifications to the Love numbers discussed so far apply to an Earth model that is perfectly elastic. However, the materials of the real Earth are slightly dissipative (anelastic), with a finite Q. Measurements of the Q of Earth tides were long of interest because of their possible relevance to the problem of tidal evolution of the Earth–Moon system (Cartwright, 1999); though it is now clear that almost all of the dissipation of tidal energy occurs in the oceans (Ray *et al.*, 2001), anelastic effects on tides remain of interest because tidal data (along with the Chandler wobble) provide the only information on Q at frequencies below about 10^{-3} Hz.

Over the seismic band (approximately 10^{-3} to 1 Hz), Q appears to be approximately independent of frequency. A general model for frequency dependence is

$$Q = Q_0 \left(\frac{f}{f_0}\right)^{\alpha} \quad [26]$$

where f_0 and Q_0 are reference values. In general, in a dissipative material the elastic modulus μ will in

general also be a function of frequency, with $\mu(f)$ and $Q(f)$ connected by the Kramers–Kronig relation (Dahlen and Tromp, 1998, chapter 6). (We use μ because this usually denotes the shear modulus; in pure compression Q is very high and the Earth can be treated as elastic.) This frequency dependence is usually termed 'anelastic dispersion'. For the frequency dependence of Q given by [26], and α small, the modulus varies as

$$\mu(f) = \mu_0 \left[1 + \frac{1}{Q_0}\left\{\frac{2}{\alpha\pi}\left[1 - \left(\frac{f_0}{f}\right)^{\alpha}\right] + i\left(\frac{f_0}{f}\right)^{\alpha}\right\}\right] \quad [27]$$

so there is a slight variation in the modulus with frequency, and the modulus becomes complex, introducing a phase lag into its response to sinusoidal forcing. In the limit as α approaches zero (constant Q), the real part has a logarithmic frequency dependence. Including a power-law variation [26] at frequencies below a constant-Q seismic band, the frequency dependence becomes

$$\mu(f) = \mu_0 \left[1 + \frac{1}{Q_0}\left\{\frac{2}{\pi}\ln\left(\frac{f_0}{f}\right)^{\alpha} + i\right\}\right], \quad f > f_m$$

$$\mu(f) = \mu_0 \left[1 + \frac{1}{Q_0}\left\{\frac{2}{\alpha\pi}\left[\alpha\ln\left(\frac{f_m}{f_0}\right)\right.\right.\right.$$
$$\left.\left.\left. + 1 - \left(\frac{f_m}{f}\right)^{\alpha}\right] + i\left(\frac{f_m}{f}\right)^{\alpha}\right\}\right], \quad f < f_m \quad [28]$$

where f_m is the frequency of transition between the two Q models.

Adding anelasticity to an Earth model has three effects on the computed Love numbers:

1. Anelastic dispersion means that the elastic constants of an Earth model found from seismology must be adjusted slightly to be appropriate for tidal frequencies. As an example, Dehant *et al.* (1999) find that for an elastic Earth model the gravimetric factor δ_0 is 1.16030 for the M_2 tide; an anelastic model gives 1.16172; for $h^{(0)}$ the corresponding values are 0.60175 and 0.61042.

2. Dispersion also means that the Love numbers vary within the tidal bands. For the semidiurnal and diurnal tides, the effect is small, especially compared to the NDFW resonance; in the long-period bands, it is significant as f approaches zero. The usual formulation for this (McCarthy and Pétit, 2004) is based on a slightly different form of [28], from Smith and Dahlen (1981) and Wahr and Bergen (1986) ; the Love numbers vary in the long-period band as

$$L(f) = A - B \left\{ \cot \frac{\alpha \pi}{2} \left[1 - \left(\frac{f_m}{f} \right)^\alpha \right] + i \left(\frac{f_m}{f} \right)^\alpha \right\} \quad [29]$$

where A and B are constants for each Love number. For the IERS standards, $\alpha = 0.15$ and $f_m = 432$ cpd (a period of 200 s). A and B are 0.29525 and -5.796×10^{-4} for k^0, 0.5998 and -9.96×10^{-4} for $h^{(0)}$, and 0.0831 and -3.01×10^{-4} for $l^{(0)}$.

3. As eqn [29] shows, the Love numbers also become complex-valued, introducing small phase lags into the tides. There are additional causes for this; in particular, the NDFW frequency can have a small imaginary part because of dissipative core–mantle coupling, and this will produce complex-valued Love numbers even in an elastic Earth. Complex-valued Love numbers can be used in extensions of eqns [7] and [8]; for example if the elastic Love-number combination introduces no phase shift (as for gravity) itself in phase, the real part is multiplied by $[a_n^m(t)\cos m\phi + b_n^m(t)\sin m\phi]$ and the imaginary part by $[b_n^m(t)\cos m\phi - a_n^m(t)\sin m\phi]$.

The most recent examination of tidal data for anelastic effects (Benjamin *et al.*, 2006) combined data from diurnal tides (in the potential, as measured by satellites), the Chandler wobble, and the 19-year nodal tide. They find a good fit for α between 0.2 and 0.3, with $f_m = 26.7$ cpd; using the IERS value of f_m gave a better fit for α between 0.15 and 0.25.

6.4 Tidal Loading

A major barrier to using Earth tides to find out about the solid Earth is that they contain signals caused by the ocean tides – which may be signal or noise depending on what is being studied. The redistribution of mass in the ocean tides would cause signals even on a rigid Earth, from the attraction of the water; on the real Earth they also cause the Earth to distort, which causes additional changes. These induced signals are called the 'load tides', which combine with the body tide to make up the total tide (**Figure 1**).

6.4.1 Computing Loads I: Spherical Harmonic Sums

To compute the load, we start with a description of the ocean tides, almost always as a complex-valued function $H(\theta', \phi')$, giving the amplitude and phase of a particular constituent over the ocean; we discuss such ocean-tide models in more detail in Section 6.4.3. The loads can then be computed in two ways: using a sum of spherical harmonics, or as a convolution of the tide height with a Green function.

In the first approach, we expand the tidal elevation in spherical harmonics:

$$H(\theta', \phi') = \sum_{n=0}^{\infty} \sum_{m=-n}^{n} H_{nm} Y_{nm}(\theta', \phi') \quad [30]$$

where the Y_{nm} are as in the section on tidal forcing, and the H_{nm} would be found from

$$H_{nm} = \int_0^\pi \sin \theta' d\theta' \int_0^{2\pi} d\phi H(\theta', \phi') Y_{nm}^*$$
$$\equiv \int_\Omega H(\theta', \phi') Y_{nm}^* d\Omega \quad [31]$$

where we use Ω for the surface of the sphere. Note that there will be significant high-order spherical-harmonic terms in H_{nm} if only because the tidal height goes to zero over land: any function with a step behavior will decay only gradually with increasing degree.

The mass distribution H causes a gravitational potential on the surface of the Earth, which we call V^L. This potential is given by the integral over the surface of the potential function times H; the potential function is proportional to r^{-1}, where r is the linear distance from the location (θ, ϕ) to the mass at (θ', ϕ'), making the integral

$$V^L(\theta, \phi) = G\rho_w a^2 \int_\Omega \frac{H(\theta', \phi')}{r} d\Omega \quad [32]$$

where ρ_w is the density of seawater, and G and a are as in Section 6.2.1. We can write the r^{-1} in terms of angular distance Δ:

$$\frac{1}{r} = \frac{1}{2a\sin(\Delta/2)} = \frac{1}{a}\sum_{n=0}^{\infty} P_n(\cos\Delta)$$

$$= \frac{1}{a}\sum_{n=0}^{\infty}\sum_{m=-n}^{n} \frac{4\pi}{2n+1} Y_{nm}(\theta', \phi') Y_{nm}^*(\theta, \phi) \quad [33]$$

where we have again used the addition theorem [5]. Combining the last expression in [33] with the spherical harmonic expansion [30] and the expression for the potential [32] gives the potential in terms of spherical harmonics:

$$V^L = G\rho_w a \sum_{n=0}^{\infty}\sum_{m=-n}^{n} \frac{4\pi}{2n+1} H_{nm} Y_{nm}(\theta, \phi) \quad [34]$$

We have found the potential produced by the load because this potential is used, like the tidal potential, in the specification of the Earth's response to the load. Specifically, we define the load Love numbers k'_n, h'_n, and l'_n such that, for a potential V^L of degree n, we have

$$u_n^z = h'_n\frac{V_n^L}{g}, \qquad u_n^b = l'_n\frac{\nabla_1 V_n^L}{g}, \qquad V_n = k'_n V_n^L \quad [35]$$

where u_n^z is the vertical displacement (also of degree n), u_n^b is the horizontal displacement, and V_n is the additional potential produced by the deformation of the Earth. These load Love numbers, like the Love numbers for the tidal potential, are found by integrating the differential equations for the deformation of the Earth, but with a different boundary condition at the surface: a normal stress from the load, rather than zero stress. For a spherical Earth, these load numbers depend only on the degree n of the spherical harmonic.

To compute the loads, we combine the definition of the load Love numbers [35] with the expression [34], using whichever combination is appropriate for some observable. For example, for vertical displacement u^z this procedure gives

$$u^z(\theta, \phi) = \frac{G\rho_w a}{g}\sum_{n=0}^{\infty}\sum_{m=-n}^{n} \frac{4\pi}{2n+1} h'_n H_{nm} Y_{nm}(\theta, \phi)$$

$$= \frac{\rho_w}{\rho_E}\sum_{n=0}^{\infty}\sum_{m=-n}^{n} \frac{3h'_n}{2n+1} H_{nm} Y_{nm}(\theta, \phi) \quad [36]$$

where ρ_E is the mean density of the Earth. A similar expression applies for the induced potential, with k'_n replacing h'_n; for the effective tide-raising potential,

sometimes called the 'self-attraction loading' or SAL (Ray, 1998), we would use $1 + k'_n - h'_n$.

Many terms are needed for a sum in [36] to converge, but such a sum provides the response over the whole Earth. Ray and Sanchez (1989) used this method to compute radial displacement over the whole Earth, with $n = 256$, and special methods to speed the computation of the H_{nm} coefficients in eqn [31]. In any method that sums harmonics, there is always room for concern about the effects of Gibbs' phenomenon (Hewitt and Hewitt, 1979) near discontinuities, but no such effect was observed in the displacements computed near coastlines. Mitrovica et al. (1994) independently developed the same method, extended it to the more complicated case of horizontal displacements, and were able to make calculations with $n = 2048$.

Given a global ocean-tide model and a need to find loads over the entire surface, this summation technique requires much less computation than the convolution methods to be discussed in the next section. For gravity the contributions from the Earth's response are, from eqn [16], $-(n+1)k'_n V_n^L/a$ from the induced potential and $2h'_n V_n^L/a$ from the displacement, making the sum

$$\delta g(\theta, \phi) = \frac{3g\rho_w}{a\rho_E}\sum_{n=0}^{\infty}\sum_{m=-n}^{n} \frac{2h'_n-(n+1)k'_n}{2n+1} H_{nm} Y_{nm}(\theta, \phi)$$

$$[37]$$

While this sum might appear to converge more slowly than eqn [36] because of the $n+1$ multiplying k'_n, the convergence is similar because for large n, nk'_n approaches a constant value, which we term k'_∞. All three load Love numbers have such asymptotic limits for large n:

$$\text{As } n \Rightarrow \infty \ \ h'_n \Rightarrow h'_\infty \ \ nk'_n \Rightarrow k'_\infty \ \ nl'_n \Rightarrow l'_\infty \quad [38]$$

so that the sum [37] converges reasonably well. A similar sum can be used to get the gravity from the direct attraction of the water:

$$\frac{3g\rho_w}{a\rho_E}\sum_{n=0}^{\infty}\sum_{m=-n}^{n} \frac{1}{4n+2} H_{nm} Y_{nm}(\theta, \phi)$$

(Merriam, 1980; Agnew, 1983).

However, summation over harmonics is not well suited to quantities that involve spatial derivatives, such as tilt or strain. To find the load tides for these, we need instead to employ convolution methods, which we now turn to.

6.4.2 Computing Loads II: Integration Using Green Functions

If we only want the loads at a few places, the most efficient approach is to multiply the tide model by a Green function which gives the response to a point load, and integrate over the area that is loaded by the tides. That is, we work in the spatial domain rather than, as in the previous section, in the wave number domain; there is a strict analogy with Fourier theory, in which convolution of two functions is the same as multiplying their Fourier transforms. Convolution methods have other advantages, such as the ability to combine different ocean-tide models easily, include more detail close to the point of observation, and handle any of the Earth-tide observables. The standard reference on the procedure remains the classic paper of Farrell (1972); Jentzsch (1997) is a more recent summary.

More formally, we find the integral over the sphere (in practice over the oceans)

$$\int_0^\pi r\sin\theta' d\theta' \int_0^{2\pi} r\, d\phi'\, G_L(\theta, \phi, \theta', \phi')\rho_w H(\theta', \phi')$$

$$[39]$$

where G_L is the Green function for an effect (of whatever type) at (θ, ϕ) from a point mass (δ-function) at (θ', ϕ'); $\rho_w g H r^2 \sin\theta\, d\theta\, d\phi$ is the applied force.

The Green functions are found, not directly, but by forming sums of combinations of the load Love numbers. The first step is to find the potential from a point mass. Take $H = \rho_w a^2 \delta(\theta', \phi')$, (where δ is the Dirac delta-function). Substitute this into eqn [32], using the sum in $P_n(\cos\Delta)$ in eqn [33]. This gives the potential as

$$V^L(\theta, \phi) = G\rho_w a \int_\Omega H(\theta', \phi') \sum_{n=0}^\infty P_n(\cos\Delta)d\Omega$$

$$= \frac{ga}{M_E} \sum_{n=0}^\infty P_n(\cos\Delta) \qquad [40]$$

which shows that the degree-n part of the potential is $V_n^L = ga/M_E$, independent of n. So, to compute vertical displacement we would apply this potential to the load Love number h_n' getting the displacement

$$u^z = \frac{a}{M_E} \sum_{n=0}^\infty h_n' \frac{V_n^L}{g} = \frac{a}{M_E} \sum_{n=0}^\infty h_n' P_n(\cos\Delta) = G_z(\Delta) \qquad [41]$$

which is thus the loading Green function for vertical displacement. Some insight into the behavior of this function can be gotten by using the asymptotic value of h_n' to write

$$G_z(\Delta) = \frac{a}{M_E} \sum_{n=0}^\infty h_\infty' P_n(\cos\Delta)$$

$$+ \frac{a}{M_E} \sum_{n=0}^\infty (h_n' - h_\infty') P_n(\cos\Delta)$$

$$= \frac{ah_\infty'}{2M_E \sin\Delta/2} + \frac{a}{M_E} \sum_{n=0}^\infty (h_n' - h_\infty') P_n(\cos\Delta)$$

$$[42]$$

where we have made use of [33]. The new sum will converge much more rapidly; in practice, it needs to include only enough terms for h_n' to have approached h_∞' to adequate precision. For Δ small, the sum approaches zero, so the analytic part shows that, for loads nearby, G_z varies as Δ^{-1}. This is the vertical displacement seen for a point load on an elastic half-space, in what is called the Boussinesq solution; in the limit of short distance, the loading problem reduces to this, which provides a useful check on numerical computations.

For the horizontal displacement the Green function is

$$u^b = \frac{ga}{M_E} \sum_{n=0}^\infty \frac{l_n'}{g} \frac{\partial V_n^L}{\partial\Delta} = \frac{a}{M_E} \sum_{n=0}^\infty l_n' \frac{\partial P_n(\cos\Delta)}{\partial\Delta} = G_b(\Delta)$$

$$[43]$$

which may again have the asymptotic part nl_n' removed and replaced by an analytic expression

$$G_b(\Delta) = \frac{al_\infty'}{M_E} \sum_{n=0}^\infty \frac{1}{n} \frac{\partial P_n(\cos\Delta)}{\partial\Delta}$$

$$+ \frac{a}{M_E} \sum_{n=0}^\infty (l_n' - l_\infty') \frac{\partial P_n(\cos\Delta)}{\partial\Delta}$$

$$= -\frac{al_\infty'}{M_E} \frac{\cos(\Delta/2)[1 + 2\sin\Delta/2]}{2\sin(\Delta/2)[1 + \sin\Delta/2]}$$

$$+ \frac{a}{M_E} \sum_{n=0}^\infty (nl_n' - l_\infty') \frac{1}{n} \frac{\partial P_n(\cos\Delta)}{\partial\Delta} \qquad [44]$$

which shows the same dependence on Δ for small distances.

For gravity, there are two parts to the loading: the direct attraction of the water mass (often called the Newtonian part), and the change caused by elastic deformation of the Earth. The first part can be found analytically by using the inverse square law and computing the vertical part of the attraction. If the elevation of our point of observation is εa, with ε small, this Green function is

$$G_{gn}(\Delta) = -\frac{g}{M_E} \left[\frac{\varepsilon + 2\sin^2\Delta/2}{(4(1+\varepsilon)\sin^2\Delta/2 + \varepsilon^2)^{3/2}} \right] \qquad [45]$$

The elastic part of the Green function follows from the harmonic expression [37]:

$$G_{ge} = \frac{g}{M_E} \sum_{n=0}^{\infty} (2b'_n - (n+1)k'_n) P_n(\cos \Delta)$$

$$= \frac{g(2b'_\infty - k'_\infty)}{2M_E \sin \Delta/2} + \frac{a}{M_E} \sum_{n=0}^{\infty} (2(b'_n - b'_\infty)$$

$$- ((n+1)k'_n - k'_\infty)) P_n(\cos \Delta) \quad [46]$$

which shows, again, a Δ^{-1} singularity for Δ small.

Likewise, the Green function for the tide-raising potential makes use of the combination $1 - k'_n + b'_n$:

$$G_{pot} = \frac{a}{M_E} \sum_{n=0}^{\infty} (1 + k'_n - b'_n) P_n(\cos \Delta)$$

$$= \frac{a(1 - b'_\infty)}{2M_E \sin \Delta/2} + \sum_{n=0}^{\infty} (k'_n - (b'_n - b'_\infty)) P_n(\cos \Delta) \quad [47]$$

and the Green function for tilt uses the same combination of Love numbers, but with the derivative of the P_n's:

$$G_t = \frac{-1}{M_E} \sum_{n=0}^{\infty} (1 + k'_n - b'_n) \frac{\partial P_n(\cos \Delta)}{\partial \Delta}$$

$$= \frac{(1 - b'_\infty)\cos(\Delta/2)}{4M_E \sin^2 \Delta/2} - \sum_{n=0}^{\infty} (k'_n - (b'_n - b'_\infty)) \frac{\partial P_n(\cos \Delta)}{\partial \Delta}$$

$$[48]$$

which has a Δ^{-2} singularity for Δ small. The tilt is thus much more sensitive to local loads than the other observables we have so far discussed.

The remaining Green function is that for strain, specifically for the strain in the direction of the load

$$e_{\Delta\Delta} = \frac{1}{a} \frac{\partial u^b}{\partial \Delta} + \frac{u^z}{a}$$

from which the Green function is, from [41] and [43],

$$G_{\Delta\Delta} = \frac{1}{M_E} \sum_{n=0}^{\infty} b'_n P_n(\cos \Delta) + \frac{1}{M_E} \sum_{n=0}^{\infty} l'_n \frac{\partial^2 P_n(\cos \Delta)}{\partial \Delta^2}$$

$$= \frac{b'_\infty}{2M_E \sin \Delta/2} + \frac{1}{M_E} \sum_{n=0}^{\infty} (b'_n - b'_\infty) P_n(\cos \Delta)$$

$$+ \frac{l'_\infty}{M_E} \sum_{n=0}^{\infty} \frac{1}{n} \frac{\partial^2 P_n(\cos \Delta)}{\partial \Delta^2} + \frac{1}{M_E} \sum_{n=0}^{\infty} (l'_n - l'_\infty)$$

$$\times \frac{\partial^2 P_n(\cos \Delta)}{\partial \Delta^2}$$

$$= \frac{b'_\infty}{2M_E \sin \Delta/2} + \frac{1}{M_E} \sum_{n=0}^{\infty} (b'_n - b'_\infty) P_n(\cos \Delta)$$

$$+ \frac{l'_\infty}{M_E} \frac{1 + \sin \Delta/2 + \sin^2 \Delta/2}{4 \sin^2 \Delta/2 [1 + \sin \Delta/2]}$$

$$+ \frac{1}{M_E} \sum_{n=0}^{\infty} (l'_n - l'_\infty) \frac{\partial^2 P_n(\cos \Delta)}{\partial \Delta^2} \quad [49]$$

which again shows a near-field singularity of Δ^{-2}. This behavior also holds for the strain perpendicular to the direction to the load; since this is given by

$$\frac{u^z}{a} + \frac{\cos \Delta}{\sin \Delta} \frac{u^b}{a} \quad [50]$$

there is no need to compute a separate Green function for it. The Green function for linear strain that would be used in [39] is

$$G_L(\Delta, \zeta) = G_{\Delta\Delta}(\Delta)\cos^2 \zeta$$

$$+ \left[\frac{G_z(\Delta)}{a} + \cot \Delta \frac{G_b(\Delta)}{a} \right] \sin^2 \zeta \quad [51]$$

where ζ is the azimuth of the load relative to the direction of extension. Areal strain has a complicated dependence on distance, because it is zero for a point load on a halfspace, except right at the load.

All of the Green functions are computed by finding the load Love numbers for a range of n, and forming the various sums. Farrell (1972) formed the sums up to $n = 10\,000$; the Love numbers can be computed at values of n spaced logarithmically and interpolated to intermediate values. Several numerical methods to accelerate the convergence of the sums are described by Farrell (1972) and Francis and Dehant (1987). The Green functions tabulated by Farrell (1972) (with the addition of the potential function by Farrell (1973)) are still widely used; Jentzsch (1997) tabulates a set for the PREM model. Kamigaichi (1998) has discussed the variations in the strain and tilt Green functions at shallow depths, forming sums up to $n = 4 \times 10^6$; the results show a smooth transition between the surface functions given here, and the Boussinesq results for the response of a halfspace at depth. Such burial does however eliminate the singularity in the strain Green functions. Examinations of the extent to which local structure, particularly lateral variations, affects computed load tides, have not be plentiful – perhaps mostly because the data most sensitive to such effects, strain and tilt, are affected by other local distortions (Section 6.6.1). As with the Love numbers for the body tides, the load Love numbers will be affected by rotation, ellipticity, anisotropy, and anelasticity. The first two produce, again, a resonant response from the NDFW, though only for loads of degree 2 and order 1 (Wahr and Sasao, 1981); Pagiatakis (1990) has examined effects from the others.

It is also possible to define load Love numbers, and Green functions derived for them, for transverse rather than normal stress, to describe the deformation of the Earth by wind stress or ocean currents, including tidal currents; see Merriam (1985, 1986) and Wilhelm (1986).

6.4.3 Ocean Tide Models

Of course, to compute loads, we need a description of the ocean tides. Producing such models globally is a difficult task that has been pursued for some time (Cartwright, 1977, 1999). The intractability of the relevant equations, the complexity of the geometry, and the sensitivity of the results to details in the models long precluded numerical solutions, one difficulty being that the oceans have barotropic modes of oscillation with periods close to the diurnal and semidiurnal tidal bands. At the same time, it was very difficult to measure tides in deep water. All this meant that until recently there were no good tidal models for computing loads.

From the Earth-tide standpoint what is important is that increasing computational power has finally rendered numerical solutions possible for realistic geometries, and that satellite altimetry has provided data with global coverage (Le Provost et al., 1995, 1998; Andersen et al., 1995; Shum et al., 1997; Desai et al., 1997). The ocean models now available are often adequate to produce estimated loads that are as accurate as available Earth-tide measurements.

Perhaps the biggest difficulty in modeling the tides is the need to represent the bathymetry in adequate detail in the model, especially in shallow water, where the wavelengths are short. This need, and the relatively coarse spacing of the altimetry data, has meant that tidal models are still divided into two groups: global and local. Global models are computed on a relatively coarse mesh (say 0.5°), and rely heavily on altimetry data (e.g., Egbert and Erofeeva, 2002); they often cannot adequately model the resonances that occur in some bodies of water (such as the Bay of Fundy), for which local models are required: these use a finer mesh, and often rely more on local tide-gauge data. Obviously, a local model is not important for computing loads unless the data are collected close by.

Most tidal models are given for particular tidal constituents, usually at least one diurnal and one semidiurnal. Unless a local resonance is present, the loads for other harmonics can be found by scaling using the ratios of the amplitudes in the equilibrium tide (Le Provost et al., 1991).

6.4.4 Computational Methods

Essentially all load programs perform the convolution [39] directly, either over the grid of ocean cells

(perhaps more finely divided near the load) or over a radial grid. Two that are generally available are GOTIC (Matsumoto et al., 2001) and SPOTL (Agnew, 1996). Bos and Baker (2005) have recently compared the results from four programs, albeit only for gravity, which is least sensitive to local loads. They found variations of a few percent because of different computational assumptions, and different coastline models. Most global ocean-tide models do not represent coastlines more accurately than their rather coarse mesh size, so some local refinement is needed. Fortunately, this problem has essentially been solved by the global coastal representations made available by Wessel and Smith (1996) – except in the Antarctic, where their coastline is (in places) the ice shelves, beneath which the tides are still present.

Figure 8 shows the computed loads for a region (Northwest Europe) with large and complex local tides. The vertical displacement and gravity loads have roughly similar forms, but the tilt and linear strain have a very different pattern, being much more concentrated near the coast, as might be expected from the different near-field behavior of their Green functions.

6.5 Analyzing and Predicting Earth Tides

As noted in Section 6.1, the tidal forces can be described to extremely high precision and accuracy, and the body tides and tidal loading can often be modeled to an accuracy that exceeds that of tidal measurements. Such measurements do however provide a check on these models, and in some cases allow them to be improved, so we briefly describe how tidal parameters are extracted from the data, and how the data are obtained. This is important whether we aim to measure the tides, use modeled tides as a calibration signal, or predict the tides to high accuracy to check the quality of ongoing measurements.

6.5.1 Tidal Analysis and Prediction

As noted in Section 6.1, the analysis of time series for tidal response is just a special case of finding the transfer function, or admittance, of a linear system, a concept first introduced into tidal analysis by Munk and Cartwright (1966). Because the tides are very band-limited, we can find the 'tidal admittance', $W_T(f)$, only for frequencies at which $X_T(f)$ contains significant energy. For ocean tides it is most

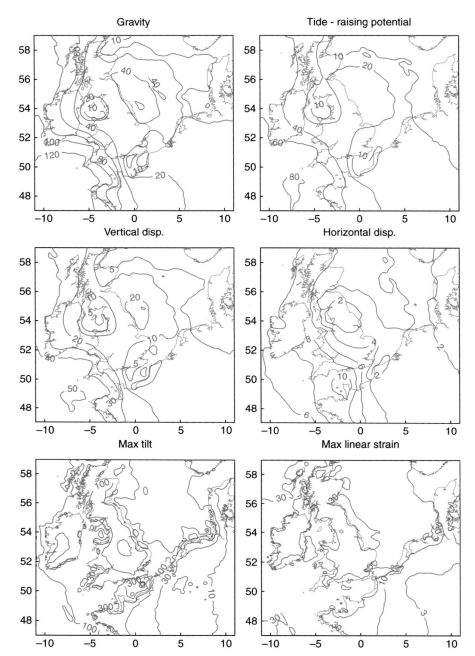

Figure 8 Loads for the M_2 tide, computed by the Green function method for the TPXO6.2 tidal model of Egbert and Erofeeva (2002), combined with a detailed model of the North Sea tides by the same group. Only the amplitude of the complex-valued quantities is shown; for tilt, displacement, and strain the value is taken along the azimuth that maximizes the amplitude. The units are nm s^{-2} for gravity, mm for the potential height and the displacements, and 10^{-9} for tilt and strain; for clarity the contour interval is logarithmic for the last two.

meaningful to take $x_T(t)$ to be the local value of the tide-raising potential, or for some analyses the tide computed for a nearby site (Cartwright *et al.*, 1969). In Earth-tide studies it may be more convenient to take as reference the tides expected for an oceanless, but otherwise realistic, earth model, so that any departure of $W(f)$ from unity will then reflect the effect of ocean loads or the inadequacy of the model.

The theory described in Section 6.3.2 shows that, except for the NDFW resonance, $W(f)$ for an ocean-less Earth varies only very slightly with frequency. The ocean tides show more variability, but only in limited areas do they have resonances within the tidal bands, so that in general the ocean load also varies smoothly with frequency (Garrett and Munk, 1971). Even the local resonances in certain bays and gulfs have a low enough Q that the response is smoothly varying over (say) the entire semidiurnal band (Ku et al., 1985). So, the more closely spaced two frequencies are, the closer the corresponding values of $W(f)$ will be, an assumption Munk and Cartwright (1966) dubbed the 'credo of smoothness'.

A naive way to find the tidal response is to take the Fourier transform of the data (using a fast Fourier transform), and use the amplitudes and phases of the result. This is a poor choice for two reasons. One problem is that the frequencies computed by the usual definition of the discrete Fourier transform usually do not coincide with the frequencies of the tidal harmonics – especially if the length of the transform, N, is chosen to work well with a fast Fourier transform algorithm. In addition, any noise will bias the amplitudes of the coefficients to be larger than the true values.

If spectral analysis is to be used, a much better method is the cross-spectral technique described by Munk and Cartwright (1966) . This method has lower-frequency resolution than others to be discussed, but makes the fewest assumptions about the form of $W(f)$, and also provides estimates of the noise as a function of frequency. This is useful because many methods assume the noise to be the same at all frequencies, and it may not be; in particular, the noise is sometimes observed to rise sharply in the tidal bands, a phenomenon called 'tidal cusping' (Munk et al., 1965; Ponchaut et al., 2001; Colosi and Munk, 2006). The cross-spectral method does however require large amounts of data to perform reliably. The procedure is described in full by Munk and Cartwright (1966) ; it depends on finding the cross-spectrum between a noise-free reference series and the data, using a slow Fourier transform to make the Fourier frequencies match the tidal frequencies relatively well, windowing to reduce bias from spectral leakage, and averaging to get a statistically consistent estimate.

By far the commonest approach to tidal analysis is least-squares fitting of a set of sinusoids with known frequencies – chosen, of course, to match the frequencies of the largest tidal constituents. That is, we aim to minimize the sum of squares of residuals:

$$\sum_{n=0}^{N}\left[y_n - \sum_{l=1}^{L}(A_l\cos(2\pi f_l t_n) + B_l\sin(2\pi f_l t_n))\right]^2 \quad [52]$$

which expresses the fitting of L sine–cosine pairs with frequencies f_l to the N data y_n, the f's being fixed to the tidal harmonic frequencies and the A's and B's being solved for.

The usual assumption behind a least-squares analysis is that the residual after fitting the sinusoids will be statistically independent random variables; but this is valid only if the noise spectrum is white, which is usually not so. One departure from whiteness is the presence of increased long-period noise outside the tidal bands; this can be removed by filtering the data before analyzing it. A more difficult problem is the tidal cusping just referred to. If the noise spectrum rises to a higher level inside the tidal bands, perhaps very much higher around the frequencies of the radiational tides, this needs to be allowed for in fitting the tides, and in finding the errors in the final tidal parameters. In particular, the relative error for a harmonic of amplitude A analyzed over a total time span T is approximately $2P(f)/A^2 T$, where $P(f)$ is the noise power spectral density at the frequency of that harmonic (Munk and Hasselmann, 1964; Cartwright and Amin, 1986). If excess energy in the tidal bands is not allowed for, the errors can be underestimated by significant amounts.

The main problem with using [52] directly for tidal analysis comes from the fine-scale frequency structure of the tidal forcing, particularly the nodal modulations. Leaving such variations out of [52], and only solving for a few large harmonics, will be inaccurate. But the simplest way of including nodal and other modulations, namely by including the satellite harmonics in [52], is not possible because the solution will be unstable unless we have 19 years of data. This instability is general, and applies whenever we try to solve for the amplitudes of harmonics separated in frequency by less than $1/T$, where T is the record length (Munk and Hasselmann, 1964). This problem is not restricted to the nodal modulation; for example, with only a month of data, we cannot get reliable results for the P_1 and K_1 lines, since they are separated by only 0.15 cycles/month.

All least-squares tidal analysis thus has to include assumptions about tidal harmonics closely spaced in frequency – which comes to an implicit assumption about the smoothness of the admittance. Usually, the admittance is assumed to be constant over frequency ranges of width $1/T$ around the main constituents, summing all harmonics within each such range to form (slowly varying) sinusoidal functions to replace the sines and cosines of [52]. Of course, if we then wish to assign the resulting amplitude to a particular harmonic (say M_2), we need to correct the amplitudes found by the ratio of this sinusoidal function to the single harmonic. All this adds complexity to the existing analysis programs (Tamura *et al.*, 1991; Wenzel, 1996; Pawlowicz *et al.*, 2002; Foreman, 2004; Van Camp and Vauterin, 2005).

A quite different approach to tidal analysis is the 'response method', also introduced by Munk and Cartwright (1966) . This does not use an expansion of the tidal potential into harmonics, but rather treats it as a time series to be fit to the data, using a set of weights to express the admittance. Lambert (1974) and Merriam (2000) have applied this method to Earth tides, and it is standard in the estimation of tides from satellite altimetry.

The basic approach is to find the tides as a weighted sum over the time variations of each spherical harmonic (not harmonics in time):

$$y(t) = \sum_{n=2}^{\infty} \sum_{m=-n}^{n} \sum_{l=-L_{nm}}^{L_{nm}} w_{nl}^m [a_n^m(t-l\Delta)] + ib_n^m(t-l\Delta)] \quad [53]$$

where the $a_n^m(t)$ and $b_n^m(t)$ are the time-varying functions that sum to give the potential in [8]. The complex-valued weights w_{nl}^m are called 'response weights'; their Fourier transform gives the admittance $W(f)$. So, for example, a single complex weight for each n and m (i.e., setting $L_{nm} = 0$) amounts to assuming a constant W for each degree and order – though even one complex weight can express both amplitude and phase response. Including more weights, with time lags (the sum over l), allows the admittance to vary with frequency, smoothly, across each tidal band. The lag interval is usually chosen to be 2 days, which makes the admittance smooth over frequencies of greater than 0.5 cpd; note that the lags can include the potential at future times because the admittance is being fit over only a narrow frequency band.

6.5.1.1 Predicting tides

All tidal predictions, other than those based on the response method, use a harmonic expansion similar to eqn [11]:

$$x(t) = \sum_{k=1}^{K} A_k \cos[2\pi f_k(t-t_0) + \phi_k^0(t_0) + \phi_k] \quad [54]$$

where the A_k's and ϕ_k's are amplitudes and phases (the 'harmonic constants') for whatever is being predicted. The f_k's are the frequencies of the different harmonics, and the ϕ_k^0's are the phases of these at a reference time t_0.

Any user of tidal constants should be aware of two pitfalls relating to the conventions for phase. One is a sign convention: whether positive phases represent lags (true in much of the older literature) or leads. The other is the reference time used. The 'local phase' is one choice, in which zero phase (for each harmonic) is at a time at which the potential from that harmonic is locally a maximum. For ocean tides this phase is usually denoted as κ (with positive phases for lags). For Earth tides local phase is convenient because on a SNREI Earth it is zero for gravity, NS tilt, vertical displacement, and areal strain. The other choice is the Greenwich phase G, in which the phase is taken to be zero (for each harmonic) at a time at which its potential would be a maximum at 0 longitude. If given for a number of places, this phase provides a 'snapshot' of the distribution of the tides at a particular instant; this phase is the norm in ocean-tide models. Since the time between maximum at Greenwich and maximum at a local place depends only on the spherical harmonic order m and on the Earth rotating 360° every 24 hs, the relationship between k and G is simple, and depends only on the tidal species number m and the longitude ϕ_W; the frequency of the harmonic is not involved. The relationship is conventionally written as

$$G = \kappa - m\phi_W$$

where for both phases a lag is taken to be positive, and longitude ϕ_W to be positive going West.

The primary complications in predicting the tides come, once again, from the various long-term modulations, notably the nodal modulations discussed in Section 6.2.3. Classical prediction methods, which used only a few constituents to minimize computation, applied nodal corrections to the A_k's and ϕ_k's of these few harmonics to produce new values that would be valid for (say) each year; a complication, since the corrections themselves change with time.

A more computationally intensive but conceptually simpler approach uses a large number of harmonics in the sum [54], including all satellite harmonics, thus automatically producing the modulations. The amplitudes and phases of a few harmonics are interpolated to give those of all harmonics, again on the assumption that the admittance is smooth; for example, through a spline interpolation of the real and imaginary parts of $W(f)$ (Le Provost et al., 1991).

6.6   Earth-Tide Instrumentation

Because the Earth tides are so small, building instruments to detect them has long been a challenge, though an easier one over time, as sensitive transducers and digital recording have become more readily available.

The earliest measurements were of tidal tilts, and over the years a wide variety of tiltmeters have been designed; Agnew (1986) describes many of them, and the designs have changed little since then. They generally fall into two classes: small instruments that sense the motion of a pendulum or of a bubble in a fluid, and larger systems that measure the motion of a fluid in a long pipe. The former are usually referred to as short-base tiltmeters, and are now generally installed in boreholes in order to obtain adequate thermal stability and relatively low rates of drift. The latter, called long-base systems, are usually installed in tunnels a few tens of meters long or longer (e.g., d'Oreye and Zürn, 2005), though a very few instruments, several hundred meters in length, have been installed near the ground surface. A similar division exists in strainmeters (Agnew, 1986): there are very short-base systems, many installed in boreholes, longer instruments installed in tunnels, and a few very long instruments, using laser light rather than physical length standards, some at the surface and others underground. One other class of instrument sensitive to tidal deformations is the ring-laser gyroscope (Schreiber et al., 2003), which detects tilts that alter the orientation of the instrument relative to the Earth's spin axis.

Historically, the second type of Earth tide to be detected was changes in gravity, and many such measurements have been made with a variety of types of gravimeters (Torge, 1989). The bulk of these used metallic springs, arranged so that changes in gravity caused extension of the spring, leading to significant phase lags and hysteresis. Such measurements are however significant in showing the widespread effects of load tides (Llubes and Mazzega, 1997), and provided early evidence of the effect of the NDFW on tidal data (see, e.g., Neuberg et al. (1987)). These uncertain phase lags are reduced by applying feedback, which was first done in the instrument of LaCoste and Romberg (Harrison and

LaCoste, 1978); this, supplied with electronic feedback, remains useful for tidal measurements. The lowest-noise tidal gravimeter is the superconducting gravimeter (Goodkind, 1999), in which a superconducting sphere is suspended in a magnetic field at liquid-helium temperatures. This gives a system with very low noise (especially at long periods) and little drift. At the periods of the semidiurnal tides, the noise ranges from −130 to −140 dB (relative to $1 \, m^2 \, s^{-4}$), with the noise being about 5 dB higher in the diurnal band (Rosat et al., 2004); these levels are about 5 dB below those of spring gravimeters (Cummins et al., 1991). Because the noise is so low, small tidal signals can be measured with great precision; recent examples include the detection of loading from small nonlinear ocean tides (Merriam, 1995; Boy et al., 2004) and the discrimination between the loads from equilibrium and dynamic models for the long-period ocean tides (Iwano et al., 2005; Boy et al., 2006a). The superconducting gravimeter also can provide accurate measurements of the NDFW resonance (Zürn et al., 1986; Florsch et al., 1994; Sato et al., 1994, 2004). However, unlike the spring gravimeters, the superconducting instrument is not portable.

Comparing the size of the load tides in **Figure 8** with the rms body tides in **Figure 6** shows that even the largest loads are but a few percent of the total gravity tide. So, to get accurate measurements of the load tides, the gravimeter must be calibrated with extreme accuracy, at least to a part in 10^3. An even higher level of accuracy is needed if tidal gravity measurements are to discriminate between the predictions of different Earth models. Calibration to this level has proved to be difficult. One method, usable only with spring gravimeters, is to make gravity measurements over a wider range and interpolate to the small range of the tides. For superconducting gravimeters, which cannot be moved, one method is to sense the response to known moving masses (Achilli et al., 1995); this is not in general possible with spring gravimeters because of their more complicated mass distribution. Another method, applicable to both types, is to place the instrument on a platform that can be moved vertically by known amounts to produce small accelerations of the same size as the tides (Richter et al., 1995). The most common approach is now to operate an absolute gravimeter next to a tidal system for several days and find the scale factor by direct comparison (Hinderer et al., 1991; Francis et al., 1998; Tamura et al., 2005). For absolute systems the scale is set by atomic standards known to much higher accuracy than is required, and comparison tests (Francis and van Dam, 2002) confirm that this method can provide calibrations to 10^{-3}. Baker and Bos (2003) summarize tidal measurements from both spring and superconducting gravimeters; from the greater scatter for the in-phase components of the residual tide they conclude that systematic errors in the calibration remain, with errors up to 4×10^{-3} in some cases. These authors, and Boy *et al.* (2003), find too much scatter in the results to distinguish between different Earth models, or to detect any latitude dependence.

The newest procedures for measuring Earth tides are the techniques of space geodesy. Since the induced potential affects satellite orbits, modeling of these provides constraints on k_2; a particularly notable result was that of Ray *et al.* (2001), who were able to measure the phase lag of the solid-Earth component of the M_2 tide as $0.204° \pm 0.047°$. Positioning techniques such Very Long Baseline Interferometry (VLBI) and GPS are sensitive to all displacements, including tides. VLBI data have been used to observe the body tides, and now have sufficient precision to be sensitive to load tides as well (Sovers, 1994; Haas and Schuh, 1998; Petrov and Ma, 2003); the currently available series can provide the amplitudes of tidal constituents to better than 1 mm. However, VLBI data are available only at a few places, whose number is not increasing.

Continuous GPS data, in contrast, are available from many locations; and it is important that the tidal displacements of such locations be accurately modeled, since any inaccuracy (especially in the vertical) will bias GPS estimates of zenith delay and hence of water vapor (Dragert *et al.*, 2000; Dach and Dietrich, 2000); for the standard 1-day processing, unmodeled tidal displacements may produce, through aliasing, spurious long-period signals (Penna and Stewart, 2003).

Tidal motions can be found with GPS in two ways. One is to process data over short time spans (say, every 3 h) to produce a time series that can then be analyzed for tidal motions. The other is to include the complex amplitude of some of the larger tidal constituents (in all three directions) as unknowns in the GPS solution, solving for any unmodeled tides. Baker *et al.* (1995) and Hatanaka *et al.* (2001) took the first approach for two local areas, finding good agreement between observed and predicted loads. Khan and Scherneck (2003) also used this method, finding that the best approach was to estimate zenith delays over the same (hourly) time span as the displacement

was found for. Schenewerk *et al.* (2001) used the second method for a global set of stations, taking data from every third day over 3 years; they found generally good agreement with predicted loads except at some high-latitude sites, probably because of using an older ocean-tide model; Allinson *et al.* (2004) applied this method to tidal loading in Britain. King (2006) compared the two methods for the GPS site at the South Pole (for which only load tides are present), using about 5 years of data. He found that the second method gave better results in the vertical; the two methods had comparable errors in the horizontal, with the precision being somewhat less than 1 mm. The K_1 and K_2 tides give poor results because their frequency is very close to the repeat time of the GPS constellation and its first harmonic. King and Padman (2005) used GPS to validate different ocean-tide models around Antarctica, with models specifically designed for this region predicting the loads better than the older global models.

6.6.1 Local Distortion of the Tides

King and Bilham (1973) and Baker and Lennon (1973) introduced an important concept into Earth-tide studies by suggesting that much of the scatter in measurements of tidal tilts was caused by strain tides which were coupled into tilts by the presence of a free surface – for example, the tunnel in which such instruments were often housed. This 'cavity effect' (Harrison, 1976) has indeed turned out to be important; any strain and tilt measurements not made with surface-mounted longbase instruments require that a cavity be created, and any such inhomogeneity in an elastic material will produce local deformations, which can produce rotations and strains that modify the tidal strains and tilts for an Earth with only radial variations. Other irregularities will have similar effects; for example, topography, which is to say an irregular free surface, will also create departures from what would be seen on a smooth Earth.

The way in which these departures are described is through the use of coupling matrices (Berger and Beaumont, 1976; King *et al.*, 1976). We can divide the actual displacement field in the Earth, **u**, into two parts, $\mathbf{u}_0$, which is the displacement that would occur with no cavity (or other inhomogeneity), and $\delta\mathbf{u}$, the difference between this ideal and the actual displacement. We can perform the same decomposition on the strain tensor **E** and the local rotation vector $\boldsymbol{\Omega}$:

$$\mathbf{E} = \mathbf{E}_0 + \delta\mathbf{E}, \qquad \boldsymbol{\Omega} = \boldsymbol{\Omega}_0 + \delta\boldsymbol{\Omega}$$

The additional deformations δE and $\delta\Omega$ will depend on the details of the inhomogeneity (e.g., the shape of the cavity) and on E_0, but not on Ω_0 because it is a rigid rotation. If we suppose E_0 to be homogeneous strain, then δE and $\delta\Omega$ at any position are linear functions of E_0, and can be related to it by a fourth-order strain–strain coupling tensor C_E and a third-order strain-rotation coupling tensor C_Ω:

$$\delta E = E_E E_0, \qquad \delta\Omega = C_\Omega E_0 \qquad [55]$$

In general, strain–tilt coupling involves both C_E and C_Ω (Agnew, 1986). Near the surface of the Earth there are only three independent components to E_0; this fact, and the symmetries of the strain tensors, mean that both C_E and C_Ω have nine independent components.

An analytical solution for C_E and C_Ω exists for an ellipsoidal cavity (Eshelby, 1957); Harrison (1976) has described results for the case where one axis of the ellipsoid is vertical: the coupled strains and tilts are largest when measured in the direction along which the cavity is smallest, whether the external strain is in that direction or not. Strain measured along a long narrow cavity is not amplified very much. The limit of this is an infinite horizontal circular tunnel, for which the strains along the tunnel (or shear strains across it) are unaltered, but both vertical and horizontal strains are amplified, as has been observed in cross-tunnel measurements of strain tides (Itseuli *et al.*, 1975; Beavan *et al.*, 1979; Takemoto, 1981). Finite element modeling (Berger and Beaumont, 1976) shows that slight departures from circularity do not much alter the strain from wall to wall, but in a square tunnel the strains concentrate near the corners; strains along a finite tunnel are undistorted if they are measured more than one tunnel diameter away from an end. Tiltmeters mounted on tunnel walls, or placed on a ledge or near a crack, will be affected by strain-coupled tilts.

Inhomogeneities (including cavities) create distortion but do not add noise. They thus mean that precise comparison between data and models is difficult at best, with the exception of frequency-dependent effects such as the NDFW. There are analytical methods for approximating topographic effects (Meertens and Wahr, 1986), and while it is always possible to build a finite-element model of any inhomogeneity (Emter and Zürn, 1985; Sato and Harrison, 1990), not enough detail of the elastic constants is usually known to make results from such a model more than a rough guide. The measurements by Baker (1980), on a pier in the middle of a tunnel, which showed a 5% change in tilts measured 0.5 m apart, imply that such modeling has to be extremely detailed.

One possible way to reduce the importance of coupling is to arrange the geometry to minimize it. Measuring strain along a tunnel is one example; another is measuring tilt in a borehole: since a horizontal strain does not cause rotation of the side of the borehole, attaching the tiltmeter to this should eliminate the cavity effect. An array of such tiltmeters installed to use observed tides to map crustal inhomogeneities (Levine *et al.*, 1989; Meertens *et al.*, 1989) still produced widely scattered results. A subsequent test using closely spaced instruments in nominally uniform geology (Kohl and Levine, 1995) also showed significant differences between boreholes, and in one borehole even between two different positions. These results suggest that any short-base measurement of tilt may be significantly affected by local inhomogeneities, and so cannot be accurately compared with theoretical tidal models.

Sometimes the aim is to use the tides to determine the effects of the inhomogeneity, so that other signals seen can be corrected. The most notable case is borehole strainmeters, in which the instrument, hole, and grout cause the strain inside the instrument wall to be significantly different from that in the far field. A simplified model that assumes axial symmetry (Gladwin and Hart, 1985) provides two coupling coefficients, one for areal and one for shear strain. This can be checked using the tides (Hart *et al.*, 1996), which also allow estimation of a full coupling tensor, along the lines of eqn [55].

References

Achilli V, Baldi P, Casula G, Errani M, *et al.* (1995) A calibration system for superconducting gravimeters. *Bulletin Geodesique* 69: 73–80.

Agnew DC (1983) Conservation of mass in tidal loading computations. *Geophysical Journal of the Royal Astronomical Society* 72: 321–325.

Agnew DC (1986) Strainmeters and tiltmeters. *Reviews of Geophysics* 24: 579–624.

Agnew DC (1996) SPOTL: Some Programs for Ocean-Tide Loading, *SIO Reference Series 96-8*, Scripps Institution of Oceanography.

Allinson CR, Clarke PJ, Edwards SJ, *et al.* (2004) Stability of direct GPS estimates of ocean tide loading. *Geophysical Research Letters* 31(15): L15603, doi: 10.1029/2004GL020588.

Andersen OB, Woodworth PL, and Flather RA (1995) Intercomparison of recent ocean tide models. *Journal of Geophysical Research* 100: 25,261–25,282.

Baker TF (1980) Tidal tilt at Llanrwst, north Wales: Tidal loading and earth structure. *Geophysical Journal of the Royal Astronomical Society* 62: 269–290.

Baker TF (1984) Tidal deformations of the Earth. *Science Progress Oxford* 69: 197–233.

Baker TF and Bos MS (2003) Validating Earth and ocean tide models using tidal gravity measurements. *Geophysical Journal International* 152: 468–485.

Baker TF, Curtis DJ, and Dodson AH (1995) Ocean tide loading and GPS. *GPS World* 6(5): 54–59.

Baker TF and Lennon GW (1973) Tidal tilt anomalies. *Nature* 243: 75–76.

Bartels J (1957/1985) Tidal forces (English translation). In: Harrison JC (eds.) *Earth Tides*, pp. 25–63. New York: Van Nostrand Reinhold.

Beavan J, Bilham R, Emter D, and King G (1979) Observations of strain enhancement across a fissure. *Veröffentlichungen der Deutschen Geodätischen Kommission, Reihe B.* 231: 47–58.

Benjamin D, Wahr JM, Ray RD, Egbert GD, and Desai SD (2006) Constraints on mantle anelasticity from geodetic observations, and implications for the J_2 anomaly. *Geophysical Journal International* 165: 3–16, doi:10.1111/j.1365-246X.2006.02915.x.

Berger J (1975) A note on thermoelastic strains and tilts. *Journal of Geophysical Research* 80: 274–277.

Berger J and Beaumont C (1976) An analysis of tidal strain observations from the United States of America: II. the inhomogeneous tide. *Bulletin of the Seismological Society of America* 66: 1821–1846.

Bos MS and Baker TF (2005) An estimate of the errors in gravity ocean tide loading computations. *Journal of Geodesy* 79: 50–63.

Boy J-P, Llubes M, Hinderer J, and Florsch N (2003) A comparison of tidal ocean loading models using superconducting gravimeter data. *Journal of Geophysical Research* 108: ETG 6-1–6–17, doi:10.1029/2002JB002050.

Boy J-P, Llubes M, and Ray R, et al. (2004) Non-linear oceanic tides observed by superconducting gravimeters in Europe. *Journal of Geodynamics* 38: 391–405.

Boy J-P, Llubes M, Ray R, Hinderer J, and Florsch N (2006a) Validation of long-period oceanic tidal models with superconducting gravimeters. *Journal of Geodynamics* 41: 112–118.

Boy J-P, Ray R, and Hinderer J (2006b) Diurnal atmospheric tide and induced gravity variations. *Journal of Geodynamics* 41: 253–258.

Broucke RA, Zurn W, and Slichter LB (1972) Lunar tidal acceleration on a rigid Earth. *AGU Geophysical Monograph* 16: 319–324.

Brush SG (1996) *Nebulous Earth: The Origin of the Solar System and the Core of the Earth from Laplace to Jeffreys.* Cambridge: Cambridge University Press.

Bullesfeld F-J (1985) Ein Beitrag Zur Harmonischen Darstellung Des Gezeitenzeugenden Potentials. *Veröffentlichungen der Deutschen Geodätischen Kommission, Reihe C.* 31: 3–103.

Cartwright D, Munk W, and Zetler B (1969) Pelagic tidal measurements: A suggested procedure for analysis. *EOS Transactions. American Geophysical Union* 50: 472–477.

Cartwright DE (1977) Oceanic tides. *Reports on Progress in Physics* 40: 665–708.

Cartwright DE (1999) *Tides: A Scientific History.* New York: Cambridge University Press.

Cartwright DE and Amin M (1986) The variances of tidal harmonics. *Deutsche Hydrographische Zeitschrift.* 39: 235–253.

Cartwright DE and Edden AC (1973) Corrected tables of tidal harmonics. *Geophysical Journal of the Royal Astronomical Society* 33: 253–264.

Cartwright DE and Tayler RJ (1971) New computations of the tide-generating potential. *Geophysical Journal of the Royal Astronomical Society* 23: 45–74.

Chapman S and Lindzen RS (1970) *Atmospheric Tides: Gravitational and Thermal.* New York: Gordon and Breach.

Colosi JA and Munk WH (2006) Tales of the venerable Honolulu tide gauge. *Journal of Physical Oceanography* 36: 967–996.

Cummins P, Wahr JM, Agnew DC, and Tamura Y (1991) Constraining core undertones using stacked IDA gravity records. *Geophysical Journal International* 106: 189–198.

Dach R and Dietrich R (2000) Influence of the ocean loading effect on GPS derived precipitable water vapor. *Geophysical Research Letters* 27: 2953–2956, doi:10.1029/1999GL010970.

Dahlen FA (1993) Effect of the Earth's ellipticity on the lunar potential. *Geophysical Journal International* 113: 250–251.

Dahlen FA and Tromp J (1998) *Theoretical Global Seismology.* Princeton, NJ: Princeton University Press.

Dehant V, Defraigne P, and Wahr JM (1999) Tides for a convective Earth. *Journal of Geophysical Research* 104: 1035–1058.

Desai SD (2002) Observing the pole tide with satellite altimetry. *Journal of Geophysical Research* 107(C11): 3186, doi:10.1029/2001JC001224.

Desai SD, Wahr JM, and Chao Y (1997) Error analysis of empirical ocean tide models estimated from Topex/Posecidon altimetry. *Journal of Geophysical Research* 102: 25,157–25,172.

Doodson AT (1921) The harmonic development of the tide generating potential. *Proceedings of the Royal Society Series A* 100: 305–329.

d'Oreye NF and Zürn W (2005) Very high resolution long-baseline water-tube tiltmeter to record small signals from Earth free oscillations up to secular tilts. *Review of Scientific Instruments* 76: 024,501.

Dragert H, James TS, and Lambert A (2000) Ocean loading corrections for continuous GPS: A case study at the Canadian coastal site Holberg. *Geophysical Research Letters* 27: 2045–2048.

Egbert GD and Erofeeva SY (2002) Efficient inverse modeling of barotropic ocean tides. *Journal of Atmospheric and Oceanic Technology* 19: 183–204.

Emter D and Zürn W (1985) Observations of local elastic effects on earth tide tilts and strains. In: Harrison JC (ed.) *Earth Tides*, pp. 309–327. New York: Van Nostrand Reinhold.

Eshelby JD (1957) The determination of the elastic field of an ellipsoidal inclusion and related problems. *Proceedings of the Royal Society Series A* 241: 376–396.

Farrell WE (1972) Deformation of the earth by surface loads. *Reviews of Geophysics* 10: 761–797.

Farrell WE (1973) Earth tides, ocean tides, and tidal loading. *Philosophical Transactions of the Royal Society Series A* 272: 253–259.

Florsch N, Chambat F, Hinderer J, and Legros H (1994) A simple method to retrieve the complex eigenfrequency of the Earth's nearly diurnal free wobble: Application to the Strasbourg superconducting gravimeter data. *Geophysical Journal International* 116: 53–63.

Foreman MGG (2004) Manual for Tidal Heights Analysis and Prediction., *Pacific Marine Science Report 77-10.* Institute of Ocean Sciences, Patricia Bay, Sidney, BC.

Francis O and Dehant V (1987) Recomputaiton of the Green's functions for tidal loading estimations. *Bulletin d'Information des Marées Terrestres* 100: 6962–6986.

Francis O and van Dam T (2002) Evaluation of the precision of using absolute gravimeters to calibrate superconducting gravimeters. *Metrologia* 39: 485–488.

Francis O, Niebauer TM, Sasagawa G, Klopping F, and Gschwind J (1998) Calibration of a superconducting gravimeter by comparison with an absolute gravimeter FG5 in Boulder. *Geophysical Research Letters* 25: 1075–1078.

Garrett C and Munk WH (1971) The age of the tide and the Q of the oceans. *Deep-Sea Research* 18: 493–503.

Gladwin MT and Hart R (1985) Design parameters for borehole strain instrumentation. *Pure and Applied Geophysics* 123: 59–80.

Goodkind JM (1999) The superconducting gravimeter. *Review of Scientific Instruments* 70: 4131–4152.

Haas R and Schuh H (1998) Ocean loading observed by geodetic VLBI. In: Ducarme B and Paquet P (eds.) *Proceedings of the 13th International Symposium on Earth Tides*, pp. 111–120. Brussels: Observatoire Royal de Belgique.

Harrison JC (1971) New Programs for the Computation of Earth Tides, *Internal Technical Report*. CIRES, University of Colorado.

Harrison JC (1976) Cavity and topographic effects in tilt and strain measurement. *Journal of Geophysical Research* 81: 319–328.

Harrison JC (1985) *Earth Tides*, New York: Van Nostrand Reinhold.

Harrison JC and Herbst K (1977) Thermoelastic strains and tilts revisited. *Geophysical Research Letters* 4: 535–537.

Harrison JC and LaCoste LJB (1978) The measurement of surface gravity. In Mueller II (ed.) *Applications of Geodesy to Geodynamics: Ninth GEOP Conference*, pp. 239–243, Columbus, OH: Ohio State University. rep. 280.

Hart RHG, Gladwin MT, Gwyther RL, Agnew DC, and Wyatt FK (1996) Tidal calibration of borehole strainmeters: Removing the effects of local inhomogeneity. *Journal of Geophysical Research* 101: 25,553–25,571.

Hartmann T and Wenzel H-G (1995) The HW95 tidal potential catalogue. *Geophysical Research Letters* 22: 3553–3556.

Hatanaka Y, Sengoku A, Sato T, Johnson JM, Rocken C, and Meertens C (2001) Detection of tidal loading signals from GPS permanent array of GSI Japan. *Journal of the Geodetic Society of Japan* 47: 187–192.

Haubrich RA and Munk WH (1959) The pole tide. *Journal of Geophysical Research* 64: 2373–2388.

Herring TA, Mathews PM, and Buffett BA (2002) Modeling of nutation-precession: Very long baseline interferometry results. *Journal of Geophysical Research* 107(B4): 2069, doi:10.1029/2001JB000165.

Hewitt E and Hewitt R (1979) The Gibbs-Wilbraham phenomenon: An episode in Fourier analysis. *Archives for the History of the Exact Sciences* 21: 129–169.

Hinderer J, Florsch N, Maekinen J, Legros H, and Faller JE (1991) On the calibration of a superconducting gravimeter using absolute gravity measurements. *Geophysical Journal International* 106: 491–497.

Itsueli UJ, Bilham R, Goulty NR, and King GCP (1975) Tidal strain enhancement observed across a tunnel. *Geophysical Journal of the Royal Astronomical Society* 42: 555–564.

Iwano S, Fukuda Y, Sato T, Tamura Y, Matsumoto K, and Shibuya K (2005) Long-period tidal factors at Antarctica Syowa Station determined from 10 years of superconducting gravimeter data. *Journal of Geophysical Research* 110: B10,403, doi:10.1029/2004JB003551.

Jeffreys H (1976) *The Earth: Its Origin, History and Physical Constitution*. Cambridge: Cambridge University Press.

Jeffreys H and Vincente RO (1957) The theory of nutation and the variation of latitude. *Monthly Notices of the Royal Astronomical Society* 117: 142–161.

Jentzsch G (1997) Earth tides and Ocean tidal loading. In: Wilhelm H, Zürn W, and Wenzel HG (eds.) *Tidal Phenomena*, pp. 145–171. Berlin: Springer-Verlag.

Jentzsch G (2006) Proceedings of the 15th International Symposium on Earth Tides. *Journal of Geodynamics* 41: 1–4.

Kamigaichi O (1998) Green functions for the earth at borehole installation sensor depths for surface point load. *Papers in Meteorology and Geophysics* 48: 89–100.

Khan SA and Scherneck HG (2003) The M_2 ocean tide loading wave in Alaska: Vertical and horizontal displacements, modelled and observed. *Journal of Geodesy* 77: 117–127, doi:10.1007/s00190-003-0312-y.

King GCP and Bilham R (1973) Tidal tilt measurement in Europe. *Nature* 243: 74–75.

King GCP, Zürn W, Evans R, and Emter D (1976) Site correction for long-period seismometers, tiltmeters, and strainmeters. *Geophysical Journal of the Royal Astronomical Society* 44: 405–411.

King M (2006) Kinematic and static GPS techniques for estimating tidal displacements with application to Antarctica. *Journal of Geodynamics* 41: 77–86.

King MA and Padman L (2005) Accuracy assessment of ocean tide models around Antarctica. *Geophysical Research Letters* 32: L23,608, doi:10.1029/2005GL023901.

Kohl ML and Levine J (1995) Measurement and interpretation of tidal tilts in a small array. *Journal of Geophysical Research* 100: 3929–41.

Ku LF, Greenberg DA, Garrett C, and Dobson FW (1985) The nodal modulation of the M_2 tide in the Bay of Fundy and Gulf of Maine. *Science* 230: 69–71.

Kudryavtsev SM (2004) Improved harmonic development of the Earth tide-generating potential. *Journal of Geodesy* 77: 829–838.

Lambert A (1974) Earth tide analysis and prediction by the response method. *Journal of Geophysical Research* 79: 4952–4960.

Le Provost C, Bennett AF, and Cartwright DE (1995) Ocean tides for and from TOPEX / POSEIDON. *Science* 267: 639–642.

Le Provost C, Lyard F, and Molines J (1991) Improving ocean tide predictions by using additional semidiurnal constituents from spline interpolation in the frequency domain. *Geophysical Research Letters* 18: 845–848.

Le Provost C, Lyard F, Molines JM, and Genco ML (1998) A hydrodynamic ocean tide model improved by assimilating a satellite altimeter-derived data set. *Journal of Geophysical Research* 103: 5513–5529.

Levine J, Meertens C, and Busby R (1989) Tilt observations using borehole tiltmeters: 1. Analysis of tidal and secular tilt. *Journal of Geophysical Research* 94: 574–586.

Llubes M and Mazzega P (1997) Testing recent global ocean tide models with loading gravimetric data. *Progress in Oceanography* 40: 369–383.

Longman IM (1959) Formulas for computing the tidal accelerations due to the Moon and the Sun. *Journal of Geophysical Research* 64: 2351–2355.

Love AEH (1911) *Some Problems of Geodynamics*. Cambridge: Cambridge University Press.

Mathews PM and Guo JY (2005) Viscoelectromagnetic coupling in precession-nutation theory. *Journal of Geophysical Research* 110(B2): B02402, doi:10.1029/2003JB002915.

Mathews PM, Buffett BA, and Shapiro II (1995a) Love numbers for diurnal tides: Relation to wobble admittances and resonance expansion. *Journal of Geophysical Research* 100: 9935–9948.

Mathews PM, Buffett BA, and Shapiro II (1995b) Love numbers for a rotating spheroidal Earth: New definitions and numerical values. *Geophysical Research Letters* 22: 579–582, doi:10.1029/95GL00161.

Mathews PM, Dehant V, and Gipson JM (1997) Tidal station displacements. *Journal of Geophysical Research* 102: 20,469–20,477.

Mathews PM, Herring TA, and Buffett BA (2002) Modeling of nutation and precession: New nutation series for nonrigid

Earth and insights into the Earth's interior. *Journal of Geophysical Research* 107(B4): 2068, doi:10.1029/2001JB000390.

Matsumoto K, Sato T, Takanezawa T, and Ooe M (2001) GOTIC2: A program for computation of oceanic tidal loading effect. *Publications of the International Latitude Observatory, Mizusawa* 47: 243–248.

McCarthy DD and Pétit G (2004) IERS Conventions (2003) *IERS Technical Note 32*. Frankfurt am Main: Verlag des Bundesamts für Kartographic und Geodäsic.

Meertens C, Levine J, and Busby R (1989) Tilt observations using borehole tiltmeters: 2, analysis of data from Yellowstone National Park. *Journal of Geophysical Research* 94: 587–602.

Meertens CM and Wahr JM (1986) Topographic effect on tilt, strain, and displacement measurements. *Journal of Geophysical Research* 91: 14057–14062.

Melchior P (1983) *The Tides of the Planet Earth*. New York: Pergamon.

Merriam JB (1980) The series computation of the gravitational perturbation due to an ocean tide. *Physics of the Earth and Planetary Interiors* 23: 81–86.

Merriam JB (1985) Toroidal Love numbers and transverse stress at the Earth's surface. *Journal of Geophysical Research* 90: 7795–7802.

Merriam JB (1986) Transverse stress Green's functions. *Journal of Geophysical Research* 91: 13,903–13,913.

Merriam JB (1992) An ephemeris for gravity tide predictions at the nanogal level. *Geophysical Journal International* 108: 415–422.

Merriam JB (1995) Non-linear tides observed with the superconducting gravimeter. *Geophysical Journal International* 123: 529–540.

Merriam JB (2000) The response method applied to the analysis of superconducting gravimeter data. *Physics of the Earth and Planetary Interiors* 121: 289–299.

Miller SP and Wunsch C (1973) The pole tide. *Nature Physical Science* 246: 98–102.

Mitrovica JX, Davis JL, and Shapiro II (1994) A spectral formalism for computing three-dimensional deformations due to surface loads, 1, theory. *Journal of Geophysical Research* 99: 7057–7074.

Molodenskii MS (1961) The theory of nutations and diurnal earth tides. *Communications de l' Observatoire Royal de Belgique, Series Geophysique* 58: 25–56.

Mueller G (1977) Thermoelastic deformations of a half-space: A Green's function approach. *Journal of Geophysics* 43: 761–769.

Munk WH and Cartwright DE (1966) Tidal spectroscopy and prediction. *Philosophical Transactions of the Royal Society Series A* 259: 533–581.

Munk WH and Hasselmann K (1964) Super-resolution of tides. In: Yoshida K (ed.) *Studies on Oceanography; A Collection of Papers Dedicated to Koji Hidaka in Commemeration of his Sixtieth Birthday*, pp. 339–344. Tokyo: University of Tokyo Press.

Munk WH, Zetler BD, and Groves GW (1965) Tidal cusps. *Geophysical Journal* 10: 211–219.

Neuberg J, Hinderer J, and Zurn W (1987) Stacking gravity tide observations in central Europe for the retrieval of the complex eigenfrequency of the Nearly Diurnal Free Wobble. *Geophysical Journal of the Royal Astronomical Society* 91: 853–868.

Pagiatakis SD (1990) The response of a realistic Earth to ocean tide loading. *Geophysical Journal International* 103: 541–560.

Pawlowicz R, Beardsley B, and Lentz S (2002) Classical tidal harmonic analysis including error estimates in MATLAB using T_TIDE, *Computers and Geosciences* 28: 929–937.

Penna NT and Stewart MP (2003) Aliased tidal signatures in continuous GPS height time series. *Geophysical Research Letters* 30: SDE 1–1, doi:10.1029/2003GL018828.

Petrov L and Ma C (2003) Study of harmonic site position variations determined by very long baseline interferometry. *Journal of Geophysical Research* 108: ETG 5–1, doi:10.1029/2002JB001801.

Ponchaut F, Lyard F, and LeProvost C (2001) An analysis of the tidal signal in the WOCE sea level dataset. *Journal of Atmospheric and Oceanic Technology* 18: 77–91.

Ray RD (1998) Ocean self-attraction and loading in numerical tidal models. *Marine Geodesy* 21: 181–192.

Ray RD (2001) Resonant third-degree diurnal tides in the seas off Western Europe. *Journal of Physical Oceanography* 31: 3581–3586.

Ray RD and Egbert GD (2004) The global S_1 tide. *Journal of Physical Oceanography* 34: 1922–1935.

Ray RD and Ponte RM (2003) Barometric tides from ECMWF operational analyses. *Annales Geophysicae.* 21: 1897–1910.

Ray RD and Poulose S (2005) Terdiurnal surface-pressure oscillations over the continental United States. *Monthly Weather Review* 133: 2526–2534.

Ray RD and Sanchez BV (1989) Radial deformation of the earth by oceanic tidal loading. *NASA Technical Memorandum*, TM-100743, Goddard Space Flight Center, Greenbelt, MD.

Ray RD, Eanes RJ, and Lemoine FG (2001) Constraints on energy dissipation in the Earth's body tide from satellite tracking and altimetry. *Geophysical Journal International* 144: 471–480.

Richter B, Wilmes H, and Nowak I (1995) The Frankfurt calibration system for relative gravimeters. *Metrologia* 32: 217–223.

Roosbeek F (1995) RATGP95: a harmonic development of the tide-generating potential using an analytical method. *Geophysical Journal International* 126: 197–204.

Rosat S, Hinderer J, Crossley D, and Boy JP (2004) Performance of superconducting gravimeters from long-period seismology to tides. *Journal of Geodynamics* 38: 461–476.

Sato T and Harrison JC (1990) Local effects on tidal strain measurements at Esashi, Japan. *Geophysical Journal International* 102: 513–526.

Sato T, Tamura Y, Higashi T, et al. (1994) Resonance parameters of the free core nutation measured from three superconducting gravimeters in Japan. *Journal of Geomagnetism and Geoelectricity* 46: 571–586.

Sato T, Tamura Y, Matsumoto K, Imanishi Y, and McQueen H (2004) Parameters of the fluid core resonance inferred from superconducting gravimeter data. *Journal of Geodynamics* 38: 375–389.

Schenewerk MS, Marshall J, and Dillinger W (2001) Vertical ocean-loading deformations derived from a global GPS network. 47: 237–242.

Schreiber KU, Klugel T, and Stedman GE (2003) Earth tide and tilt detection by a ring laser gyroscope. *Journal of Geophysical Research* 108: doi:10.1029/2001JB000569.

Shum CK, Woodworth PL, Andersen GD, et al. (1997) Accuracy assessment of recent ocean tide models. *Journal of Geophysical Research* 102: 25,173–25,194.

Smith ML and Dahlen FA (1981) The period and Q of the Chandler wobble. *Geophysical Journal of the Royal Astronomical Society* 64: 223–281.

Sovers OJ (1994) Vertical ocean loading amplitudes from VLBI measurements. *Geophysical Research Letters* 21: 357–360.

Standish EM, Newhall XX, Williams JG, and Yeomans DK (1992) Orbital ephemerides of the Sun, Moon and planets. In: Seidelmann PK (eds.) *Explanatory Supplement to the Astronomical Almanac*, pp. 279–323. Sausalito, California: University Science Books.

Takemoto S (1981) Effects of local inhomogeneities on tidal strain measurements. *Bulletin of the Disaster Prevention Research Institute of Kyoto University* 31: 211–237.

Tamura Y (1982) A computer program for calculating the tide generating force. *Publications of the International Lattitude Observatory, Mizusawa* 16: 1–20.

Tamura Y (1987) A harmonic development of the tide-generating potential. *Bulletin d'Information des Marées Terrestres* 99: 6813–6855.

Tamura Y, Sato T, Fukuda Y, and Higashi T (2005) Scale factor calibration of a superconducting gravimeter at Esashi Station, Japan, using absolute gravity measurements. *Journal of Geodesy* 78: 481–488.

Tamura Y, Sato T, Ooe M, and Ishiguro M (1991) A procedure for tidal analysis with a Bayesian information criterion. *Geophysical Journal International* 104: 507–516.

Torge W (1989) *Gravimetry*. Berlin: Walter de Gruyter Verlag.

Trupin A and Wahr J (1990) Spectroscopic analysis of global tide gauge sea level data. *Geophysical Journal International* 100: 441–453.

Van Camp M and Vauterin P (2005) Tsoft: Graphical and inter-active software for the analysis of time series and Earth tides. *Computers and Geosciences* 31: 631–640.

Wahr JM (1981a) Body tides on an elliptical, rotating, elastic and oceanless Earth. *Geophysical Journal of the Royal Astronomical Society* 64: 677–703.

Wahr JM (1981b) A normal mode expansion for the forced response of a rotating Earth. *Geophysical Journal of the Royal Astronomical Society* 64: 651–675.

Wahr JM (1985) Deformation induced by polar motion. *Journal of Geophysical Research* 90: 9363–9368.

Wahr JM and Bergen Z (1986) The effects of mantle and ane-lasticity on nutations, earth tides, and tidal variations in rotation rate. *Geophysical Journal* 87: 633–668.

Wahr JM and Sasao T (1981) A diurnal resonance in the ocean tide and in the Earth's load response due to the resonant free 'core nutation'. *Geophysical Journal of the Royal Astronomical Society* 64: 747–765.

Wang R (1997) Tidal response of the solid Earth. In: Wilhelm H, Zürn W, and Wenzel HG (eds.) *Tidal Phenomena*, pp. 27–57. Berlin: Springer-Verlag.

Warburton RJ and Goodkind JM (1976) Search for evidence of a preferred reference frame. *Astrophysical Journal* 208: 881–886.

Wenzel H-G (1996) The nanogal software: earth tide data pro-cessing package ETERNA 3.3, *Bulletin d'Information des MaréesTerrestres* 124 : 9425 − −9439.

Wessel P and Smith WHF (1996) A global, self-consistent, hierarchical, high-resolution shoreline database. *Journal of Geophysical Research* 101: 8741–8743.

Wilhelm H (1983) Earth's flattening effect on the tidal forcing field. *Journal of Geophysics* 52: 131–135.

Wilhelm H (1986) Spheroidal and torsional stress coefficients. *Journal of Geophysics* 55: 423–432.

Wilhelm H, Zürn W, and Wenzel HG (1997) *Tidal Phenomena*. Berlin: Springer-Verlag.

Xi Q (1987) A new complete development of the tide-generating potential for the epoch J2000.0. *Bulletin d'Information des Marées Terrestres* 99: 6766–6812.

Zürn W (1997) The nearly-diurnal free wobble-resonance. In: Wilhelm H, Zürn W, and Wenzel HG (eds.) *Tidal Phenomena*, pp. 95–109. Berlin: Springer-Verlag.

Zürn W, Rydelek PA, and Richter B (1986) The core-reso-nance effect in the record from the superconducting gravimeter at Bad Homburg. In: Viera R (eds.) *Proceedings of the 10th International Symposium on Earth Tides*, pp. 141–147. Madrid: Consejo Superior de Investigaciones Cientificas.

7 Time Variable Gravity: Glacial Isostatic Adjustment

J. X. Mitrovica and K. Latychev, University of Toronto, Toronto, ON, Canada

M. E. Tamisiea, Harvard-Smithsonian Center for Astrophysics, Cambridge, MA, USA

7.1 Introduction

The adjustment of the Earth in response to the glacial cycles of the Late Pleistocene ice age is reflected in a wide range of geophysical observables, including sea-level variations and anomalies in the gravitational field and rotational state of the planet. Early analyses of glacial isostatic adjustment (GIA), dating up to the early 1980s, involved geological records of Holocene sea-level trends (McConnell, 1968; Cathles, 1975; Peltier and Andrews, 1976; Wu and Peltier, 1983), regional maps of static gravity anomalies obtained by land-based surveys in previously glaciated regions such as Laurentia and Fennoscandia (Walcott, 1972; Cathles, 1975; Wu and Peltier, 1983), and astronomical measurements of secular trends in the magnitude and orientation of the rotation vector (Nakiboglu and Lambeck, 1980; Sabadini and Peltier, 1981; Wu and Peltier, 1984).

Over the last two decades, observational constraints on the GIA process obtained by space-geodetic techniques have become increasingly common and important. These observations include estimates of three-dimensional (3-D) crustal velocities using very-long-baseline-interferometry data (James and Lambert, 1993; Mitrovica et al., 1993, 1994; Peltier, 1995; Argus, 1996) and surveying using the global positioning system (Milne et al., 2001; Johansson et al., 2002). In addition, land-based (e.g., Lambert et al., 2006) and satellite-derived (e.g., Yoder et al., 1983; Mitrovica and Peltier, 1993; Ivins et al., 1993; Sabadini et al., 2002) constraints on the Earth's time-varying gravitational field have also been adopted in GIA studies.

The last of these data types, and in particular the modeling of the ice age contribution to long-wavelength components of the Earth's geopotential, the so-called low-degree Stoke's coefficients, serves as the focus of this chapter. We begin with a discussion of 'early' GIA studies of this kind, with a particular emphasis on the evolving goals of these analyses. We then move on to recent advances in such analysis, including the incorporation of rotational effects and 3-D variations in viscosity into predictions that have traditionally assumed nonrotating, spherically symmetric, viscoelastic Earth models. It seems an appropriate time for such a review given that the *GRACE* satellite gravity mission is on the threshold of providing unprecedented constraints on secular trends to much higher degree and order than has previously been possible (Tamisiea et al., 2007; Chapter 8).

7.2 $\dot{J}_2$: A Review of Early GIA Research

We begin by reproducing the usual spherical harmonic decomposition of the secular trend in the geoid height anomaly, $\dot{G}$

$$\dot{G}(\theta, \phi, t_p) = a \sum_{l=0}^{\infty} \sum_{m=0}^{l} \bar{P}_{lm}(\cos \theta)[\dot{C}_{lm}(t_p)\cos m\phi + \dot{S}_{lm}(t_p)\sin m\phi] \quad [1]$$

where θ and ϕ are the colatitude and east-longitude, respectively, t_p is the time at present day, a is the Earth's radius, and $\bar{P}_{lm}$ is the normalized

Legendre polynomial of degree l and order m. The constants $\dot{C}_{lm}$ and $\dot{S}_{lm}$ are the Stoke's coefficients. The zonal harmonics, $\dot{J}_l$, are related to the coefficients $\dot{C}_{l,m=0}$ by

$$\dot{J}_l(t_p) = -\sqrt{2l+1}\,\dot{C}_{l,m=0}(t_p) \qquad [2]$$

We note that the $\dot{J}_2$ datum is linearly proportional to the rate of change of the rotation rate, $\dot{\Omega}$ ($\dot{J}_2 \sim -0.5\dot{\Omega}/\Omega$; Wu and Peltier, 1984).

Since the pioneering study of Yoder et al. (1983), who estimated the $\dot{J}_2$ coefficient using LAGEOS satellite data, the geophysical goals of GIA analyses associated with these coefficients have changed. The earliest studies assumed that GIA was the sole contributor to the $\dot{J}_2$ signal, and the observation was used to constrain the bulk lower-mantle viscosity of the Earth model (e.g., Yoder et al., 1983; Alexander, 1983; Peltier, 1983, 1985; Rubincam, 1984; Wu and Peltier, 1984; Yuen and Sabadini, 1985). This application is illustrated in **Figure 1**(a), where we compare numerical predictions of $\dot{J}_2$ (or alternatively, $\dot{\Omega}/\Omega$) due to ongoing GIA with the observational constraint cited by Nerem and Klosko (1996). (The latter is consistent with the earlier Yoder et al. estimate of -3×10^{-11} yr^{-1}.)

The predictions are based on standard GIA methods for spherically symmetric (i.e., depth-varying), self-gravitating, Maxwell viscoelastic Earth models (see Mitrovica and Peltier (1989) for details, including a description of the Love number formulation of the theory), and they include a global ice history modified from the ICE-3G deglaciation geometry (Tushingham and Peltier, 1991) as well as a complementary water load. The calculations assume a PREM elastic structure (Dziewonski and Anderson, 1981), lithospheric thickness of 100 km, and isoviscous upper- and lower-mantle regions (henceforth ν_{UM} and ν_{LM}, respectively, with the boundary between the two taken to be 670 km depth); the former is fixed to a value of 5×10^{20} Pa s, while the latter is a free parameter of the modeling that is varied over two orders of magnitude (as given by the abscissa scale). The solid line on the figure has the classic 'inverted parabola' form common to many predictions of ongoing GIA trends as a function of ν_{LM}. In the case of a weak lower-mantle viscosity ($\nu_{LM} < 4 \times 10^{21}$ Pa s), the planetary model would have relaxed close to equilibrium since the end of the ICE-3G deglaciation (5 ky BP) and thus the predicted present-day rates are small. Models with a stiff lower mantle ($\nu_{LM} > 4 \times 10^{22}$ Pa s) are

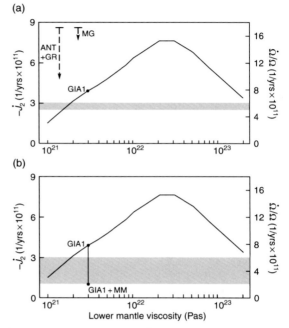

Figure 1 (a) Predictions of $\dot{J}_2$ due to GIA as a function of the lower-mantle viscosity of the Earth model (abscissa). The right ordinate scale refers to the normalized rate of change of the rotation rate, $\dot{\Omega}/\Omega$. The shaded region represents the observational constraint on $\dot{J}_2$ derived from satellite data by Nerem and Klosko (1996). The arrows at top left indicate the $\dot{J}_2$ signal associated with a net twentieth-century melting from the Antarctic and Greenland ice complexes of 1 mm yr^{-1} equivalent eustatic sea-level rise (ANT + GR) and the melting of small ice sheets and mountain glaciers tabulated by Meier (1984) (MG; ESL = 0.4 mm yr^{-1}). The result labeled GIA1 is the $\dot{J}_2$ prediction obtained with a lower-mantle viscosity of 3×10^{21} Pa s. (b) As in frame (a), except the observational constraint is based on a new analysis of uncertainties associated with the 18.6-year tide (see text). The prediction labeled GIA1 + MM is the total signal associated with the model GIA1 and a twentieth-century melt history composed of Meier's (1984) sources plus 0.6 mm yr^{-1} of ESL rise from polar ice sheets (total ESL = 1 mm yr^{-1}). All GIA calculations adopt an elastic lithospheric thickness of 100 km, and an upper-mantle viscosity of 5×10^{20} Pa s.

characterized by a slow rate of adjustment throughout the glacial cycle. Hence peak rates are predicted for models intermediate to these extremes.

According to these results, the observational constraint is reconciled by GIA predictions based on models with lower-mantle viscosity of either 2×10^{21} Pa s or $\sim 2 \times 10^{23}$ Pa s. In analyses by Peltier and colleagues (Peltier, 1983, 1985; Wu and Peltier, 1984) the higher viscosity root was ruled out by invoking independent GIA constraints associated with relative sea-level records and/or regional gravity trends over Laurentia and Fennoscandia, and it was concluded that the $\dot{J}_2$ datum 'can be used to constrain strongly the viscosity of the deep mantle' (Peltier, 1983, p. 434) to values close to 10^{21} Pa s.

Inferences of mantle viscosity based on GIA observables are generally sensitive to uncertainties in the ice load model. In this particular regard, the $\dot{J}_2$ datum is considered to be relatively robust. The degree-two zonal spherical harmonic has the same sign in high-latitude north and south polar regions (Mitrovica and Peltier, 1993; Ivins et al., 1993; Chapter 8) and thus predictions of $\dot{J}_2$ are sensitive to the total excess ice volume at last glacial maximum (LGM); that is, they are insensitive to local ice geometries or the partitioning of excess ice volumes into the Southern and Northern hemisphere. This point was made explicit in early GIA studies (e.g., Peltier, 1983), where predictions such as those shown in **Figure 1(a)** were shown for progressively larger subsets of the Late Pleistocene ice sheets (e.g., Laurentia, Laurentia plus Fennoscandia, Laurentia plus Fennoscandia plus Antarctica, etc.)

To what extent does the simple Earth model parametrization sampled in **Figure 1(a)** capture the sensitivity of the predictions of $\dot{J}_2$ to variations in the radial viscosity profile? Forward analyses indicate that the predictions are insensitive to variations in lithospheric thickness. However, these predictions demonstrate that the double root structure in the figure largely disappears (at least for the range of deep-mantle viscosity considered on the figure) when the upper-mantle viscosity is reduced to 10^{20} Pa s or when the boundary above which the viscosity is fixed to 10^{21} Pa s is moved to 1200 km depth (Mitrovica and Peltier, 1993). In these cases the predictions above a deep-mantle viscosity of 2×10^{22} Pa s tend to plateau or diminish more gradually as this viscosity is increased. The origin of this change was explored by Mitrovica and Peltier (1993) who computed radially varying Frechet kernels that measure the detailed depth-dependent sensitivity of the predictions to perturbations in the radial viscosity profile. They showed that the trend near the low-viscosity 'solutions' in **Figure 1(a)** reflects a sensitivity to structure near the base of the mantle, above the CMB. In contrast, the trend for $\nu_{LM} \sim 10^{23}$ Pa s results from a sensitivity to viscosity in the top few hundred kilometers of the lower mantle. Thus, if the upper mantle is weakened to 10^{20} Pa s, or the boundary is moved to 1200 km, a prediction of $\dot{J}_2$ will be insensitive to variations in the deep-mantle viscosity above values of $\sim 3 \times 10^{22}$ Pa s.

Since the predictions for $\nu_{LM} < 4 \times 10^{21}$ Pa s in **Figure 1(a)** appear to be robust relative to changes in both the ice history and the shallow structure of the radial Earth model, can one conclude that these predictions reflect a robust constraint on deep-mantle viscosity? For a variety of reasons, the answer is no.

First, the early suggestion that the high-viscosity solution on the figure may be rejected on the basis of independent constraints from sea-level histories and regional gravity observations has been weakened by subsequent results. The mathematical expression for the free-air gravity used in these early GIA analysis was in error (Mitrovica and Peltier, 1989), and, more importantly, it is now clear that mantle convective flow beneath Laurentia and Fennoscandia contributes significantly to the static gravity anomaly in these regions (Peltier et al., 1992; Tamisiea et al., 2007; Simons and Hager, 1997). Moreover, because of the presence of lateral variations in mantle viscosity, and the distinct resolving power of the data sets (Mitrovica and Peltier, 1991, 1993) a constraint on viscosity based upon regional sea-level data sets may not be representative of the bulk average mantle viscosity to which the $\dot{J}_2$ datum is sensitive. One may add, in this regard, that a suite of GIA-based inferences prefer a deep-mantle viscosity significantly higher than the lower root in **Figure 1(a)** (Nakada and Lambeck, 1989; Mitrovica and Forte, 1997, 2004).

The growing recognition that a number of geophysical processes other than GIA contribute to the $\dot{J}_2$ datum provides a second, and perhaps more fundamental reason to discount inferences of deep-mantle viscosity based on analyses of the kind shown in **Figure 1(a)**. Soon after the Yoder et al. (1983) analysis of *LAGEOS* data, a series of studies demonstrated that ongoing mass flux from polar ice sheets (Greenland, Antarctica) and mountain glaciers would have a potentially large impact on the low-degree Stoke's coefficients (Yoder and Ivins, 1985; Gasperini et al., 1986; Yuen et al., 1987; Peltier, 1988; Sabadini et al., 1988; Mitrovica and Peltier, 1993; Ivins et al., 1993; James and Ivins, 1995, 1997). In **Figure 1(a)** we show at top left the size of the $\dot{J}_2$ signal associated with a net mass loss from the Greenland and Antarctic ice sheets equivalent to 1 mm yr^{-1} of eustatic (globally uniform) sea-level (ESL) rise (as discussed above, this degree-two datum is only sensitive to the net change in polar ice mass), as well as the signal associated with melting from Meier's (1984) tabulation of small glaciers

(ESL ~ 0.4 mm yr^{-1}). It is clear from the figure that the signal from ongoing melting may be comparable to the variation with ν_{LM} of the GIA predictions.

In addition to ongoing melting, other analyses have discussed the potential contribution to the $\dot{J}_2$ coefficient from coupling processes at the CMB (Fang et al., 1996; Dumberry and Bloxham, 2006), mountain building (Vermeersen et al., 1994), anthropogenic water storage (Chao, 1995), and earthquakes (Chao and Gross, 1987).

7.3 New Approaches to Long-Wavelength Gravity Trends

Since a variety of processes other than GIA contribute non-negligibly to secular trends in the long-wavelength gravity coefficients, the analysis of these harmonics can take several forms. For example, if mantle viscosity is assumed known, then the trends, after correction for GIA, may be used to place bounds on recent melting from ice reservoirs. The latter is a scheme advocated, for example, by Mitrovica and Peltier (1993), who demonstrated that the $\dot{J}_2$ signal due to ongoing polar melting (as well as other low-degree even harmonics; see below) is insensitive to the geometry of the assumed melting and is linearly proportional to the equivalent ESL rise associated with the mass flux. As an alternative approach, if one could correct for non-GIA contributions, then the residual could be used to infer deep-mantle viscosity. For example, Johnston and Lambeck (1999) argued that a recent nonsteric sea-level rise of 1 mm yr^{-1} would imply a lower-mantle viscosity close to 10^{22} Pa s. This argument is consistent with the predictions shown in **Figure 1(a)**. Specifically, summing the signals associated with a GIA prediction based on $\nu_{LM} = 10^{22}$ Pa s, melting from Meier's (1984) sources (ESL = 0.4 mm yr^{-1}), and 0.6 mm yr^{-1} ESL rise from ongoing melting of polar ice sheets, yields a total $\dot{J}_2$ prediction of $(-6.0 + 0.9 + 4.0 \times 0.6 = -2.7) \times 10^{-11}$ yr^{-1}, in accord with the observational constraint.

Either of these approaches, or indeed a simultaneous estimate of viscosity and remnant melting, can also incorporate a suite of long-wavelength harmonics inferred from satellite gravity missions (e.g., Cheng et al., 1989, 1997; Nerem and Klosko, 1996; Cazenave et al., 1996). In this regard, James and Ivins (1997) used secular trends in the zonal harmonics up to degree four, as well as present-day true polar wander (TPW) rates, to map out a relationship between the inferred

deep-mantle viscosity and ESL associated with ongoing melting. They noted, as in the earlier study by Mitrovica and Peltier (1993), both an insensitivity of the low-degree $\dot{J}_l$ predictions to the geometry of ongoing mass flux and a linearity between this signal and the equivalent ESL value. They inferred a pair of viscosity solutions which tended to merge as the ESL value rose. This result also has a simple explanation in reference to **Figure 1(a)**. We can interpret the solid line in **Figure 1(a)** as the total signal associated with GIA (versus ν_{LM}) plus a present-day melting from polar ice sheets of ESL = 0.0 mm yr^{-1}. If one increases the melting rate above zero, the total signal will be the same inverted parabola, but shifted downward. The shift will increase as the ESL rise associated with the present-day melting increases, and thus the two (low and high) viscosity roots that satisfy the observational constraint will eventually merge (for an ESL rise of ~ 1.0–1.5 mm yr^{-1}) at $\sim 2 \times 10^{22}$ Pa s.

In a related effort, Devoti et al. (2001) (see also Vermeersen et al., 1998) used low-degree zonal harmonics ($\dot{J}_l$, for $l < 7$) they estimated from *LAGEOS*, *Starlette*, and *Stella* satellite data to infer upper-mantle viscosity and lithosphere thickness. They argued that residuals between the satellite-derived observations and best-fit GIA predictions were suggestive of ongoing melting from polar ice sheets. Their analysis was extended by Sabadini et al. (2002) to include the TPW datum, who inferred an increase in viscosity from the upper to lower mantle (the latter with a value generally $\sim 10^{22}$ Pa s) and an ongoing melting of the Antarctic ice sheet equivalent to a ESL rise of 0.7 mm yr^{-1}.

The extension to include higher-order harmonics comes at the price of an increasing sensitivity to errors in the Late Pleistocene and present-day ice melt geometries. The solid lines in **Figure 2** show predictions of $\dot{J}_l$ due to GIA, for $l = 2$–4, as a function of lower-mantle viscosity, computed as in **Figure 1(a)**. The dotted lines are based on a simplified disk load representation of the ICE-3G deglaciation model, as discussed in Mitrovica and Peltier (1993). The even-degree harmonics show larger amplitudes than the odd harmonic $\dot{J}_3$ since the former have the same sign in northern and southern high latitudes (as discussed above in the case of degree two), while the latter does not. Therefore the contributions to $\dot{J}_3$ from the response to loads within these two regions will tend to cancel out (Mitrovica and Peltier, 1993; James and Ivins, 1997). Accordingly, small changes in the geometry of the loads can lead to relatively large changes in the

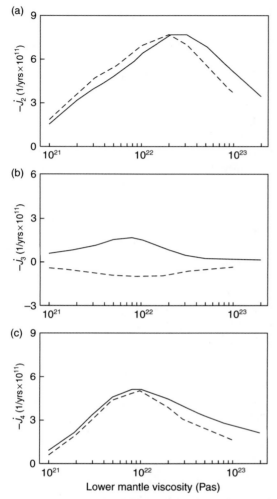

Figure 2 Solid lines – predictions of (a) $\dot{J}_2$, (b) $\dot{J}_3$, and (c) $\dot{J}_4$ due to GIA as a function of the lower-mantle viscosity (bottom abscissa scale) of the Earth model. As in **Figure 1**, the predictions use a global Late Pleistocene ice history based on the ICE-3G model (Tushingham and Peltier, 1991). The dashed line on each frame is the analogous prediction based on the disk load ice model discussed in the text.

predictions, including the change in sign evident in **Figure 2(b)** as one moves from the ICE-3G to disk model. One should note also that the sensitivity to changes in the load also increases as one moves from degree two to degree four, as one would expect since the latter samples at a higher spatial resolution.

The consideration of harmonics other than $\dot{J}_2$ can also include trends in the low-degree nonzonal, or tesseral harmonics of the Earth's gravity field ($\dot{C}_{lm}$ and $\dot{S}_{lm}$ in eqn [1]). However, the lack of observational constraints on these coefficients has meant that the potential GIA signal in these harmonics has

received much less attention. Exceptions include studies by Yuen *et al.* (1987), Ivins *et al.* (1993), and, most recently, Tamisiea *et al.* (2002). The latter two studies demonstrated that these harmonics, which reflect an azimuthal dependence in the gravity trends, are particularly sensitive to uncertainties in the geometry of the Late Pleistocene ice sheets.

7.4 Degree-Two Trends in Gravity Revisited

While the effort to consider higher-degree harmonics of the Earth's geopotential has moved forward, recent insights into both the analysis of the satellite gravity data and the theoretical framework for performing GIA predictions have led to renewed interest in the degree-two trends. In this section we discuss these issues in turn, beginning with questions that have arisen concerning the error bounds on the observed $\dot{J}_2$.

7.4.1 A Problem with the $\dot{J}_2$ Constraint?

One of the primary sources of error in the estimate of the secular trend in $\dot{J}_2$, as determined from satellite gravity data, is the uncertainty in the signal associated with the amplitude and phase of the 18.6-year tide. Cox *et al.* (2003) have suggested that the latter has been significantly underestimated, and that the error bars cited for $\dot{J}_2$ (e.g., the width of the shaded region in **Figure 1(a)** do not reflect the true uncertainty in the long-term trend. Mitrovica *et al.* (2006) examined this issue using monthly J_2 values estimated from satellite laser ranging (SLR) data by either Cox and Chao (2002) or Cheng and Tapley (2004). In particular, they corrected these two time series using bounds on the 18.6-year tide signal determined by Benjamin *et al.* (2006) from the same SLR data sets. They then estimated secular trends from the residual J_2 values over a range of time windows, all beginning in 1979 and ending between 1994 and ~2004. The $\dot{J}_2$ estimates obtained by Mitrovica *et al.* (2006) did not converge and they were different for the Cox and Chao (2002) and Cheng and Tapley (2004) time series, suggesting that the time series were still too short to accurately correct for decadal fluctuations in J_2. Their results suggested that a more realistic bound on $\dot{J}_2$ was -1.0 to -3.0×10^{-11} yr^{-1}, in accord with the conclusion of Cox *et al.* (2003). This revised uncertainty is shown in **Figure 1(b)**.

The new lower bound for the amplitude of $\dot{J}_2$ clearly permits a higher contribution to the datum from ongoing melting. As an example, consider a model GIA1, defined by a uniform lower mantle of 3×10^{21} Pa s. If this model is accurate, then the observational constraint in **Figure 1(a)** would not permit any significant ongoing melting of ice reservoirs beyond the mass flux (ESL = 0.4 mm yr^{-1}) from Meier's (1984) sources. In contrast, the constraint in **Figure 1(b)** would allow an additional ~0.6 mm yr^{-1} ESL rise from (net) melting of the Antarctic and Greenland ice mass loss (see 'GIA1 + MM' on **Figure 1(b)**). Similar arguments could be made for Earth models with $\nu_{LM} \sim 10^{23}$ Pa s (Mitrovica *et al.*, 2006).

The weakening of the $\dot{J}_2$ constraint provides a potential route to reconciling the so-called 'enigma' of global sea-level rise described by Munk (2002). Munk (2002) recognized that the traditional constraint on $\dot{J}_2$ ($\sim -3 \times 10^{-11}$ yr^{-1}; **Figure 1(a)**) was consistent with the change in the rotation rate over the last 2–3 millennia inferred from ancient eclipse observations (Stephenson and Morrison, 1995), and in this case there could be no anomalous melting of ice reservoirs beginning within the twentieth century. The new lower bound on the $\dot{J}_2$ amplitude introduces an inconsistency with the eclipse inference, and allows for the possibility of a significant melting event that onset within the last century (Mitrovica *et al.*, 2006).

The independent constraint on rotation implied by the eclipse records also has implications for previous efforts to reconcile the old $\dot{J}_2$ constraint using GIA models with $\nu_{LM} \sim 10^{22}$ Pa s and ongoing melting of order 1 mm yr^{-1} (e.g., James and Ivins, 1997; Johnston and Lambeck, 1999). As discussed, these models predict a net (GIA plus ongoing melt) $\dot{J}_2$ signal of $\sim -3 \times 10^{-11}$ yr^{-1}, in accord with **Figure 1(a)**. However, the same models predict a GIA signal of $\sim -6 \times 10^{-11}$ yr^{-1}; (**Figure 1(a)** for $\nu_{LM} \sim 10^{22}$ Pa s) that would misfit the integrated time shift over the last 2–3 ky inferred from the eclipse records.

7.4.2 Geoid Height Change: The Influence of Rotational Feedback

As discussed above, changes in the surface mass load associated with the Late Pleistocene glacial cycles perturb the magnitude and orientation of the rotation vector. The former is proportional to the $\dot{J}_2$

datum, while we have denoted the latter as TPW. There has been a growing appreciation within the GIA literature (Han and Wahr, 1989; Milne and Mitrovica, 1996, 1998; Peltier, 1998, 2004; Mitrovica *et al.*, 2001) that the time-varying centrifugal potential implied by TPW will act as a load that will perturb the gravitational and deformational response of the planet. A cartoon illustrating the geometry of this so-called rotational feedback is shown in **Figure 3**.

Figure 3 shows the situation for a pole moving counter-clockwise. In the case of ongoing GIA, the TPW would be in the direction of Hudson Bay (Sabadini and Peltier, 1981; Wu and Peltier, 1984). The centrifugal potential associated with the initial and subsequent pole position are denoted by the dashed and solid ellipses, respectively, and their difference is known as the rotational driving potential. In this 2-D cross-section along the TPW path, the driving load is negative in the two quadrants toward which the north and south rotation pole are moving, and it is positive in the remaining two quadrants. In each case, the amplitude of the perturbation is a maximum 45° from the pole position, and it is zero along the equator. The rotational driving potential is defined, in three dimensions, by the difference

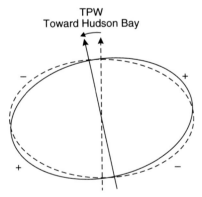

Figure 3 Schematic illustration of the geometry of the GIA-induced rotational feedback signal. The initial rotation pole and associated centrifugal potential are given by the dashed lines. A shift in the rotation pole counter-clockwise (or, as labeled on the figure, toward Hudson Bay – as expected in consequence of GIA), leads to a new orientation for both the pole and the centrifugal potential (solid lines). The so-called rotational driving potential is given by the difference between the centrifugal potential in the two states (solid minus dashed). The direct gravitational effect and deformation driven by this perturbation in the potential will act to bring the geoid (or sea surface) down in the quadrants labeled with '−' and up in quadrants labeled '+'.

between two ellipsoids of revolution (or oblate spheroids) and the magnitude is also zero along a great circle that passes through the poles and that is 90° from the great circle associated with the TPW path (i.e., a great circle perpendicular to the page through the rotation axis).

Since GIA perturbs the pole position by $\sim1°$ or less, the 'quadrential geometry' that defines the rotational driving potential is strongly dominated by a degree-two, order-one harmonic. Accordingly, the deformational and gravitational response to the driving potential is also dominated by the same harmonic (with the exception of the associated horizontal crustal motions, which have a geometry given by the gradient of the degree-two, order-one surface spherical harmonic; Mitrovica *et al.*, 2001). As an example, Tamisiea *et al.* (2002) demonstrated that rotational feedback has by far the largest influence on GIA predictions of the $\dot{C}_{2,1}$ and $\dot{S}_{2,1}$ Stoke's coefficients, and negligible impact on the zonal harmonics. Indeed, the potentially large imprint of rotational feedback on maps of the GIA-induced rates of change of geoid height (via the perturbation in the degree-two, order-one harmonic) has led to suggestions that this signal should be a target of analysis of data from satellite missions such as *GRACE* (Peltier, 2004). In this section we review recent work that suggests that the feedback signal has been significantly overestimated in predictions based on traditional GIA rotation theory.

Figure 4(a) shows a global prediction of the present-day rate of change of the geoid height due to GIA under the assumption of a nonrotating Earth model. (In the GIA literature, where sea-level changes are computed using an equilibrium, static theory, the term 'geoid' is used interchangeably with sea-surface height.) The prediction, based on a model with $\nu_{UM} = 10^{21}$ Pa s and $\nu_{LM} = 2 \times 10^{21}$ Pa s, is similar to many others that have appeared in the literature (e.g., Mitrovica and Peltier, 1991b), though we have truncated these particular calculations at degree and order 32. The map is characterized by upwarping over previously glaciated regions at rates that reach ~1 mm yr^{-1} over Laurentia and Antarctica. Outside of these zones, there is a broad downwarping at a rate of ~0.2 mm yr^{-1}. The latter trend is associated with the movement of water from the far-field oceans toward the subsiding solid surface at both the periphery of the ancient ice centers and offshore from continents (see Mitrovica and Milne (2002) for a detailed review of this physics).

In **Figure 4(b)** we show a prediction analogous to **Figure 4(a)**, with the exception that the signal due to rotational feedback has been included. The latter signal is computed using the feedback equations derived by Milne and Mitrovica (1996) and Mitrovica *et al.* (2001) (i.e., equations governing the response to the perturbed centrifugal potential associated with a given TPW) and the 'traditional' equations governing GIA-induced TPW derived by Wu and Peltier (1984). (We note that the feedback equations derived by Peltier (1998) include an error in sign; this error, and the associated predictions, were later corrected in Douglas and Peltier (2002) and Peltier (2004)). The large degree-two, order-one, quadrential geometry associated with rotational feedback is clearly evident in **Figure 4(b)** and the orientation of the rotational signature is as one would expect from **Figure 3**.

Mitrovica *et al.* (2005) have recently demonstrated that the traditional equations governing the rotational stability of a viscoelastic planet in response to surface loading, as derived by Wu and Peltier (1984), include an inaccurate assumption that acts to strongly destabilize the rotation pole. They have furthermore shown that this assumption leads to a significant overestimation of the size of the signal due to rotational feedback. The basic physics underlying the old rotation theory and a revised formulation derived by Mitrovica *et al.* (2005) is summarized, schematically, in **Figures 5(a)** and **5(b)**.

Any theory governing the rotational stability of the Earth in consequence of the ice age cycles requires an expression for the response of the model to loading associated with surface masses and the rotational driving potential, as well as an expression for the background oblateness of the model. Both the old and new rotation theories adopt the same Love number formulation for treating the response to loading, but they diverge in their treatment of the background oblateness. In particular, the old theory assumes that the background oblateness of the Earth can be accurately predicted using the same Love number formulation used to compute the response to the loading. As illustrated in **Figure 5(a)**, this assumption leads to an expression for the oblateness that is a function of the adopted thickness of the elastic lithosphere; the thicker the lithosphere (denoted by the blue outer shell on the top row of the figure), the less oblate the background form (as illustrated by the reduced flattening of frame A2 relative to frame A1). The background oblateness of the Earth, that is, the shape upon which ice age

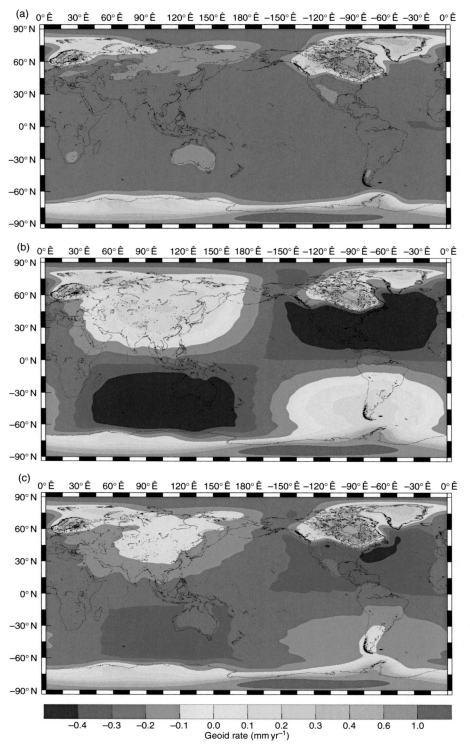

Figure 4 Maps of the predicted present-day rate of change of geoid height due to GIA computed using an Earth model with a lithospheric thickness of 120 km, an upper-mantle viscosity of 10^{21} Pa s and a lower-mantle viscosity of 2×10^{21} Pa s. The predictions are distinguished on the basis of the treatment of the rotational feedback due to TPW: (a) no feedback (i.e., a nonrotating Earth model); (b) feedback computed using the traditional GIA rotation theory (e.g., Wu and Peltier, 1984); and (c) feedback computed using the new GIA rotation theory (Mitrovica *et al.*, 2005).

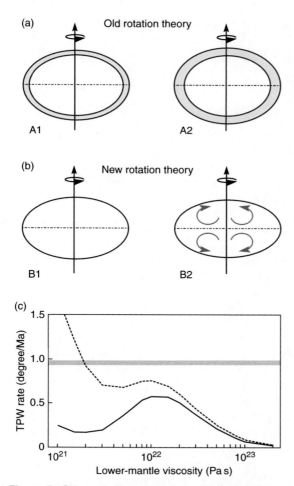

Figure 5 Schematic illustration of the difference in the underlying choice for the background oblateness of the Earth made within the (a) traditional (e.g., Wu and Peltier, 1984) and (b) new (Mitrovica *et al.*, 2005) GIA rotation theories (see text). (c) Predicted present-day true polar wander rate due to GIA as a function of the lower mantle viscosity of the Earth model. The dotted and solid lines are generated using the traditional and new rotation theories, respectively. All calculations adopt an elastic lithospheric thickness of 100 km, and an upper-mantle viscosity of 10^{21} Pa s. The shaded region represents an observational constraint on the TPW rate from Gross and Vondrak (1999).

forcings are superimposed, should not be a function of the Earth model adopted for the loading calculations. Rather, the observed oblateness is largely a sum of two parts: a hydrostatic form associated with the current rotation rate (B1), and an excess ellipticity due to the dynamic effects of internal convective flow (as denoted by the red swirls in frame B2) (Nakiboglu, 1982). The new rotation theory derived by Mitrovica *et al.* (2005) incorporates an improved

expression for the background oblateness that is tied to the observed form; since the rotation pole is stabilized by the rotational bulge, the old theory which underestimates the background oblateness also underestimates the stability of the pole.

As an example, **Figure 5c** shows a prediction of the present-day rate of TPW due to GIA as a function of ν_{LM} computed using the old (dotted line) and new (solid line) rotation theory (see also Mitrovica *et al.*, 2006). The new theory generates a significantly more stable pole, particularly for Earth models with $\nu_{LM} < 10^{22}$ Pa s, and removes the rather enigmatic nonmonotonic sensitivity to ν_{LM} evident in many previous GIA studies of TPW rate. Note that the rate of TPW computed for the model with $\nu_{LM} = 2 \times 10^{21}$ Pa s, as adopted in **Figure 4**, is reduced by 70% in the new theory relative to the old.

This stabilization is evident in geoid rate map in **Figure 4(c)**, where the rotational feedback is computed using the new TPW theory. In this case, the imprint of the feedback mechanism is significantly more muted.

As an example, consider a site located at colatitude 135° and east longitude 105°. The prediction for the nonrotating model (**Figure 4(a)**) yields a rate of -0.21 mm yr^{-1}. In contrast, the subsidence predicted on the basis of the old rotation theory is three times this value, -0.63 mm yr^{-1} (note the dark blue zone in the region encompassing Australia in **Figure 4(b)**). The new rotation theory predicts a subsidence rate of -0.36 mm yr^{-1}. Note that the perturbation relative to the nonrotating calculation generated using the old (0.42 mm yr^{-1}) and new (0.15 mm yr^{-1}) rotation theories is roughly proportional to the present-day rate of TPW in **Figure 5(c)** predicted, for $\nu_{LM} = 2 \times 10^{21}$ Pa s, based on these two theories. Clearly, the new, more accurate rotation theory, yields a more moderate feedback into the geoid (sea surface) rate field, and it may thus be significantly harder to detect using GRACE satellite-gravity measurements.

The revision in the TPW rate predictions based on the new rotation theory also has implications for the sea-level enigma mentioned above (Munk, 2002). In particular, a model with $\nu_{LM} = 2 \times 10^{21}$ Pa s, which reconciled the old constraint on the $\dot{J}_2$ datum (**Figure 1(a)**), also fits the observational constraint on the TPW rate (shown by the shaded region in **Figure 5(c)**) when the latter prediction is based on the old rotation theory (Wu and Peltier, 1984). However, this fit to the TPW rate disappears when the new theory (Mitrovica *et al.*, 2005) is invoked;

therefore, if a viscosity model of this class is correct then a signal from ongoing melting of ice reservoirs is required to fit the TPW observation.

We noted above that many previous efforts to simultaneously fit viscosity and ongoing melt rates using long-wavelength harmonics $\dot{J}_l$ and TPW (e.g., James and Ivins, 1997; Johnston and Lambeck, 1999) had neglected the constraint on integrated rotation rate provided by ancient eclipse data; these studies would also be modified by the new rotation theory which, at the least, would alter the inferred contributions from the Greenland and the Antarctic ice sheets to any ongoing sea-level rise – mass flux from the former is an effective driver of TPW, while melting from the latter, situated at the south pole, is not.

7.5 The Influence of Lateral Variations in Mantle Viscosity

The vast majority of published predictions of GIA, including those discussed above, have been based on calculations which adopt a Maxwell (linear) viscoelastic field theory, although there have been efforts to consider the importance of non-Maxwell linear rheologies (e.g., Sabadini *et al.*, 1986; Peltier *et al.*, 1986; Yuen *et al.*, 1986), as well as nonlinear creep laws (e.g., Giunchi and Spada, 2000). Moreover, these published predictions have assumed that the parameters governing the viscoelastic structure of the Earth model vary with depth alone. A set of recent studies have begun to explore the impact of lateral variations in mantle viscosity on GIA observables (e.g., Martinec, 2000; Wu and van der Wahl, 2003; Zhong *et al.*, 2003; Latychev *et al.*, 2005a), though to date only one study has focused on long-wavelength secular trends in the geopotential (Latychev *et al.*, 2005b).

The Latychev *et al.* (2005b) results were based on a finite-volume formulation of the equations governing the deformational and gravitational response of a Maxwell viscoelastic Earth model. The numerical model was benchmarked against a suite of predictions generated using spherically symmetric Earth models, including low-degree $\dot{J}_l$ predictions, and the agreement was excellent (Latychev *et al.*, 2005a). The input 3-D mantle viscosity field was derived from a model of mantle shear wave heterogeneity (Ritsema *et al.*, 1999) in a three-step process: (1) the shear wave model was converted to a model for density using a velocity-to-density scaling profile derived by Forte and Woodward (1997); (2) the

density field was mapped into variations in temperature, δT, using a depth-dependent value for the coefficient of thermal expansion (Chopelas and Boehler, 1992); (3) temperature was converted to viscosity using the following simple equation:

$$\nu(\theta, \phi, r) = \nu_0(r)e^{-\epsilon \delta T(\theta, \phi, r)} \qquad [3]$$

where ϵ is a free parameter (units of inverse degrees) used to vary the strength of the lateral viscosity variations, and ν_0 is a radially dependent scaling that ensures that the mean (logarithmic) value of the viscosity field at any depth is the same as the spherically symmetric model to which the 3-D results will be compared.

We will adopt, for the calculations in this section, a background radial model defined by a lithospheric thickness of 100 km, an upper-mantle viscosity of 5×10^{20} Pa s, and a lower-mantle viscosity of 5×10^{21} Pa s (henceforth model GIA2). As an illustration of the results, we investigate the impact on the GIA2 predictions of introducing heterogeneity in the lower mantle alone. Since we will largely be concerned with the longest-wavelength zonal harmonics, $\dot{J}_l$ for $l = 2 - 4$, lateral variations in upper-mantle viscosity or lithospheric thickness should have less impact on the predictions. Following Latychev *et al.* (2005b), we consider a suite of results in which ϵ is varied from 0.0 to 0.06. For each of these ϵ values we computed a histogram of the logarithmic perturbation in viscosity in each lower-mantle node relative to the background value of model GIA2 (5×10^{21} Pa s); the ± 1 standard deviation range for ϵ values of 0.015, 0.03, and 0.06, for example, indicated a spread of ~ 1.0, 1.5, and 3.0 orders of magnitude, respectively (Latychev *et al.*, 2005b). The peak-to-peak range in viscosity was much higher.

As an example of the viscosity field, we show, in **Figure 6**, the logarithmic perturbation in viscosity at a depth of 2000 km generated using $\epsilon = 0.055$. The model shows zones of high deep-mantle viscosity beneath the Arctic, North America, and the Antarctic. There are also two large regions of low viscosity that sample the megaplume structures beneath the Pacific and southern Africa (Ritsema *et al.*, 1999) which are thought to drive topographic uplift on the overlying lithosphere.

In **Figure 7** we reproduce (in the solid lines) the $\dot{J}_l$ predictions versus lower-mantle viscosity shown in **Figure 2** (based on the ICE-3G load history). The dashed lines map out the perturbations to

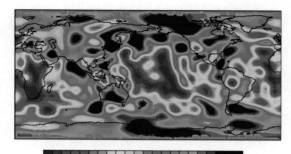

Figure 6 Map of the perturbation in the logarithm of viscosity relative to the mean value associated with model GIA2 (5×10^{21} Pa s) at depth of 2000 km in the mantle. The field was generated by adopting $\epsilon = 0.055$ in eqn [3] (see text for details).

these predictions generated by considering a suite of 3-D viscosity models defined by increasing ϵ values up to 0.06 (top abscissa scale). These calculations indicate that the $\dot{J}_2$ prediction is relatively insensitive to the introduction of 3-D viscosity structure; in contrast, the sensitivity to lateral viscosity variations is much higher for the $\dot{J}_4$ datum.

The peak perturbation in **Figure 7(c)** is 1.8×10^{-11} yr^{-1}, or a ~40% reduction relative to the GIA2 prediction of $\dot{J}_4$. This difference is significantly larger than the discrepancy associated with a change in ice geometry in **Figure 2(c)** (for GIA2) and it is sufficient to significantly bias inferences of deep-mantle viscosity based on the assumption of spherical symmetry. Indeed, our 3-D calculation with $\epsilon = 0.06$ (and an average lower-mantle viscosity of 5×10^{21} Pa s) yields the same $\dot{J}_4$ prediction as a 1-D model with $\nu_{LM} = 2.5 \times 10^{21}$ Pa s. The perturbation to $\dot{J}_4$ associated with a net twentieth century melting from the Greenland and the Antarctic ice sheets is ~3.0×10^{-11} yr^{-1} per mm yr^{-1} of equivalent ESL rise (Mitrovica and Peltier, 1993). Thus, the perturbation due to 3-D structure ($\epsilon = 0.06$) is the same as the signal associated with 0.6 mm yr^{-1} of ongoing melting from the polar ice complexes. This perturbation is also over an order of magnitude greater than the $\dot{J}_4$ signal associated with Meier's (1984) sources (10^{-12} yr^{-1}; Mitrovica and Peltier, 1993).

Finally, in **Figure 8**, we show a map of the difference between predictions of the present-day rate of change of the geoid height due to GIA computed using a 3-D Earth model ($\epsilon = 0.055$) and the spherically symmetric model GIA2. The inclusion of lateral variations in lower-mantle viscosity structure acts to reduce the geoid upwarping

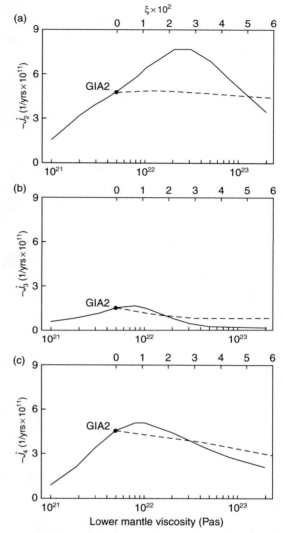

Figure 7 Solid lines – predictions of (a) $\dot{J}_2$, (b) $\dot{J}_3$, and (c) $\dot{J}_4$ due to GIA as a function of the lower-mantle viscosity (bottom abscissa scale) of the Earth model (reproduced from **Figure 2**). The dot labeled GIA2 is the prediction obtained with a lower-mantle viscosity of 5×10^{21} Pa s. The dashed line on each frame indicates the perturbation in the prediction GIA2 associated with increasing levels of lateral heterogeneity in the lower-mantle viscosity field, as reflected by the parameter ϵ (top abscissa scale; see text and eqn [3]). All calculations adopt an elastic lithospheric thickness of 100 km, and an upper-mantle viscosity of 5×10^{20} Pa s.

over Laurentia and the Antarctic by 0.35 mm yr^{-1} and 0.25 mm yr^{-1}, respectively, relative to the 1-D model. This reduction appears to be due to the much higher-than-average deep-mantle viscosity beneath these two regions associated with the 3-D

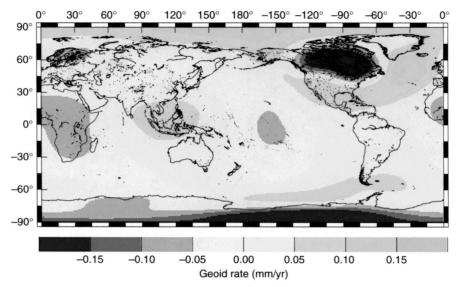

Figure 8   Perturbation in the prediction of the present-day rate of change of the geoid height due to GIA associated with the introduction of lateral variations in mantle viscosity within the lower mantle defined by $\epsilon = 0.055$ (as in **Figure 6**); that is, the prediction for the 3-D model minus the prediction based on the spherically symmetric model GIA2 (in each case, rotational feedback is not included).

model (**Figure 6**). Thus, these high-viscosity zones are acting to retard the relaxation in both high-latitude regions, leading to the reduction in the predicted amplitude of the even zonal harmonics $\dot{J}_2$ and $\dot{J}_4$ in **Figures 7(a)** and **7(c)**.

7.6   Summary

The impact of the Late Pleistocene ice age on present-day trends in the long-wavelength gravity field of the Earth has been the focus of continuous study since the first satellite-based estimates of $\dot{J}_2$ were published a quarter century ago. The geophysical applications of these studies have evolved with the growing recognition that a suite of processes may contribute significantly to the observed trends. In particular, early efforts to infer deep-mantle viscosity, under the assumption that the observed $\dot{J}_2$ datum was solely due to GIA, have been supplanted by simultaneous estimates of mantle structure and ongoing melting rates based on trends in a set of Stoke's coefficients augmented, in some cases, by other GIA observables (e.g., TPW, integrated changes in the length-of-day inferred from eclipse records, etc.)

As a companion to these efforts, theoretical treatments of the impact of GIA on the long-wavelength gravity field have also advanced beyond the earliest calculations based on spherically symmetric, nonrotating Earth models. In recent years, the feedback of glaciation-induced polar wander has been included in the predictions and these have had to be updated in consequence of a significant improvement in the theory governing load-induced perturbations in the rotation vector. Moreover, an extension to Earth models with complex, 3-D viscoelastic structure has also become possible. These new calculations indicate that lateral variations in deep-mantle viscosity can significantly perturb predictions of the $\dot{J}_l$ coefficients for $l > 2$.

The lessons learned from these studies, and the theoretical advances they motivated, will remain important as attention moves increasingly toward the constraints on the Earth's time-variable gravity field provided by the *GRACE* satellite mission. Indeed, regional studies of gravity trends over regions such as Alaska (Tamisiea *et al.*, 2005; Chen *et al.*, 2006a), Greenland and Antarctica (Velicogna and Wahr, 2005, 2006a, 2006b; Chen *et al.*, 2006b; Luthcke *et al.*, 2006), with the goal of estimating recent mass flux, have each been compelled to assess the possible range of GIA signal.

Acknowledgments

The work was supported by the Natural Sciences and Engineering Research Council of Canada, The Canadian Institute for Advanced Research, GEOIDE, and the National Aeronautics and Space Administration (grant NNG04GF09G).

References

Alexander JC (1983) Higher harmonic effects of the Earth's gravitational field from post-glacial rebound as observed from LAGEOS. *Geophysical Research Letters* 10: 1085–1087.

Argus DF (1996) Postglacial rebound from VLBI geodesy: On establishing vertical reference. *Geophysical Research Letters* 23: 973–977.

Benjamin D, Wahr J, Ray R, Egbert G, and Desai SD (2006) Constraints on Mantle anelasticity from geodetic observations, and implications for the J_2 anomaly. *Geophysical Journal International* 165: 3–16.

Cathles LM (1975) *The Viscosity of the Earth's Mantle*. Princeton, NJ: Princeton University Press.

Cazenave AA, Gegout P, Ferhat G, and Biancale R (1996) Temporal variations of the gravity field from LAGEOS 1 and LAGEOS 2 observations. In: Rapp R, Cazenave AA, and Nerem RS (eds.) *Global Gravity Field and Its Temporal Variations*, pp. 141–151. New York: Springer.

Chao BF (1995) Anthropogenic impact on global geodynamics due to reservoir water impoundment. *Geophysical Research Letters* 22: 3529–3532.

Chao BF and Gross RS (1987) Changes in the Earth's rotation and low-degree gravitational field induced by earthquakes. *Geophysical Journal of the Royal Astronomical Society* 91: 569–596.

Chen JL, Tapley BD, and Wilson CR (2006a) Alaskan mountain glacial melting observed by satellite gravimetry. *Earth and Planetary Science Letters* 248: 368–378 (doi:10.1016/j.epsl.2006.05.039).

Chen JL, Wilson CR, and Tapley BD (2006b) Satellite gravity measurements confirm accelerated melting of Greenland ice sheet. *Science* 313: 1957–1960 (doi:10.1126/science.1129007).

Cheng MK, Eanes RJ, Shum CK, Schutz BE, and Tapley BD (1989) Temporal variations in low degree zonal harmonics from Starlette orbit analysis. *Journal of Geophysical Research* 16: 393–396.

Cheng MK, Shum CK, and Tapley BD (1997) Determination of long-term changes in the Earth's gravity field from satellite laser ranging observations. *Journal of Geophysical Research* 102: 22377–22390.

Cheng M and Tapley BD (2004) Variations in the Earth's oblateness during the past 28 years. *Journal of Geophysical Research* 109: B09402.

Chopelas A and Boehler R (1992) Thermal expansivity in the lower mantle. *Geophysical Research Letters* 19: 1983–1986.

Cox CM, Boy A and Chao BF (2003) Time-variable gravity: Using satellite-laser-ranging as a tool for observing long-term changes in the Earth system. Noomen R, Klosko S, Noll C, and Pearlman M (eds.) *Proceedings of the 13th International Workshop on Laser Ranging*. NASA/CP-2003-212248, 9–19.

Cox CM and Chao BF (2002) Detection of a large-scale mass redistribution in the terrestrial system since 1998. *Science* 297: 831–832.

Devoti R, Luceri V, Sciarretta C, et al. (2001) The SLR secular gravity variations and their impact on the inference of mantle rheology and lithospheric thickness. *Geophysical Research Letters* 28: 855–858.

Douglas BC and Pelter WR (2002) The puzzle of global sea-level rise. *Physics Today* 55: 35–40.

Dumberry M and Bloxham J (2006) Azimuthal flows in the Earth's core and changes in length of day at millennial timescales. *Geophysical Journal International* 16: 32–46.

Dziewonski AM and Anderson DL (1981) Preliminary reference Earth model (PREM). *Physics of Earth and Planetary Interiors* 25: 297–356.

Fang M, Hager BH, and Herring TA (1996) Surface deformation caused by pressure changes in the fluid core. *Geophysical Research Letters* 23: 1493–1496.

Forte AM and Woodward RL (1997) Seismic-geodynamic constraints on three-dimensional structure, vertical flow, and heat transfer in the mantle. *Journal of Geophysical Research* 102: 17981–17994.

Gasperini P, Sabadini R, and Yuen DA (1986) Excitation of the Earth's rotational axis by recent glacial discharges. *Geophysical Research Letters* 13: 533–536.

Gross RS and Vondrak J (1999) Astrometric and space-geodetic observations of polar wander. *Geophysical Research Letters* 26: 2085–2088.

Giunchi C and Spada G (2000) Postglacial rebound in a non-Newtonian spherical Earth. *Geophysical Research Letters* 27: 2065–2068.

Han D and Wahr J (1989) Post-glacial rebound analysis for a rotating Earth. In: Cohen S and Vanicek P (eds.) *AGU Geophysical Monograph Series 49: Slow Deformations and Transmission of stress in the Earth*, pp. 1–6. Washington, DC: AGU.

Ivins ER, Sammis CG, and Yoder CF (1993) Deep mantle viscous structure with prior estimate and satellite constraint. *Journal of Geophysical Research* 98: 4579–4609.

James TS and Ivins ER (1995) Present-day Antarctic ice mass changes and crustal motion. *Journal of Geophysical Research* 22: 973–976.

James TS and Ivins ER (1997) Global geodetic signatures of the Antarctic ice sheet. *Journal of Geophysical Research* 102: 605–633.

James TS and Lambert A (1993) A comparison of VLBI data with the ICE-3G glacial rebound model. *Geophysical Research Letters* 20: 871–874.

Johansson JM, Davis JL, Scherneck H-G, et al. (2002) Continuous measurements of post-glacial adjustment in Fennoscandia, 1: Geodetic results. *Journal of Geophysical Research* 107: doi:10.1029/2001.

Johnston P and Lambeck K (1999) Postglacial rebound and sea level contributions to changes in the geoid and the Earth's rotation axis. *Geophysical Journal International* 136: 537–5889.

Lambert A, Courtier N, and James TS (2006) Long-term monitoring by absolute gravimetry: Tides to postglacial rebound. *Journal of Geodynamics* 41: 307–317.

Latychev K, Mitrovica JX, Tromp J, et al. (2005a) Glacial isostatic adjustment on 3-D Earth models: A new finite-element formulation. *Geophysical Journal International* 161: 421–444.

Latychev K, Mitrovica JX, Tamisiea ME, Tromp J, Christara C, and Moucha R (2005b) GIA-induced secular variations in the Earth's long wavelength gravity field: Influence of 3-D viscosity variations. *Earth and Planetary Science Letters* 240: 322–327.

Luthcke SB, Zwally HJ, Abdalati W, et al. (2006) Recent Greenland ice mass loss by drainage system from satellite gravity observations. *Science* 314: 1286–1289 (doi:10.1126/science.1130776).

Martinec Z (2000) Spectral-finite element approach to three-dimensional viscoelastic relaxation in a spherical earth. *Geophysical Journal International* 142: 117–141.

McConnell RK (1968) Viscosity of the mantle from relaxation time spectra of isostatic adjustment. *Journal of Geophysical Research* 73: 7089–7105.

Meier MF (1984) Contributions of small glaciers to global sea level. *Science* 226: 1418–1421.

Milne GA, Davis JL, Mitrovica JX, et al. (2001) Space-Geodetic constraints on glacial isostatic adjustment in fennoscandia. *Science* 291: 2381–2385.

Milne GA and Mitrovica JX (1996) Postglacial sea-level change on a rotating Earth: First results from a gravitationally self-consistent sea-level equation. *Geophysical Journal International* 126: F13–F20.

Milne GA and Mitrovica JX (1998) Postglacial sea-level change on a rotating Earth. *Geophysical Journal International* 133: 1–10.

Mitrovica JX, Davis JL, and Shapiro II (1993) Constraining proposed combinations of ice history and Earth rheology using VLBI determined baseline length rates in North America. *Geophysical Research Letters* 20: 2387–2390.

Mitrovica JX, Davis JL, and Shapiro II (1994) A spectral formalism for computing three-dimensional deformations due to surface loads – II. present-day glacial isostatic adjustment. *Journal of Geophysical Research* 99: 7075–7101.

Mitrovica JX and Forte AM (1997) Radial profile of mantle viscosity: Results from the joint inversion of convection and postglacial rebound observables. *Journal of Geophysical Research* 102: 2751–2769.

Mitrovica JX and Forte AM (2004) A new Inference of mantle viscosity based upon a joint inversion of convection and glacial isostatic adjustment data. *Earth and Planetary Science Letters* 225: 177–189.

Mitrovica JX and Milne GA (2002) On the origin of postglacial ocean syphoning. *Quaternary Science Reviews* 21: 2179–2190.

Mitrovica JX, Milne GA, and Davis JL (2001) Glacial isostatic adjustment of a rotating Earth. *Geophysical Journal International* 147: 562–578.

Mitrovica JX and Peltier WR (1989) Pleistocene deglaciation and the global gravity field. *Journal of Geophysical Research* 96: 13651–13657.

Mitrovica JX and Peltier WR (1991) A complete formalism for the inversion of post-glacial rebound data: Resolving power analysis. *Geophysical Journal International* 104: 267–288.

Mitrovica JX and Peltier WR (1991b) On post-glacial geoid subsidence over the equatorial oceans. *Journal of Geophysical Research* 96: 20053–20071.

Mitrovica JX and Peltier WR (1993) Present-day secular variations in the zonal harmonics of the Earth's geopotential. *Journal of Geophysical Research* 98: 4509–4526.

Mitrovica JX, Wahr J, Matsuyama I, and Paulson A (2005) The rotational stability of an ice age Earth. *Geophysical Journal International* 161: 491–506.

Mitrovica JX, Wahr J, Matsuyama I, and Paulson A (2006) Reanalysis of ancient eclipse, astronomic and geodetic data: A possible route to resolving the enigma of global sea-level rise. *Earth and Planetary Science Letters* 243: 390–399.

Munk W (2002) Twentieth century sea level: An enigma. *Proceedings of the National Academy of Sciences* 99: 6550–6555.

Nakada M and Lambeck K (1989) Late Pleistocene and Holocene sea-level change in the Australian region and mantle rheology. *Geophysical Journal International* 96: 497–517.

Nakiboglu SM and Lambeck K (1980) Deglaciation effects upon the rotation of the Earth. *Geophysical Journal of the Royal Astronomical Society* 62: 49–58.

Nerem RS and Klosko SM (1996) Secular variations of the zonal harmonics and polar motion as geophysical constraints. In: Rapp R, Cazenave AA, and Nerem RS (eds.) *Global Gravity Field and Its Temporal Variations*, pp. 152–163. New York: Springer.

Peltier WR (1983) Constraint on deep mantle viscosity from LAGEOS acceleration data. *Nature* 304: 434–436.

Peltier WR (1988) Global sea level and Earth rotation. *Science* 240: 895–901.

Peltier WR (1985) The LAGEOS constraint on deep mantle viscosity: Results from a new normal mode method for the inversion of viscoelastic relaxation spectra. *Journal of Geophysical Research* 90: 9411–9421.

Peltier WR (1995) VLBI baselines for the ICE-4G model of postglacial rebound. *Geophysical Research Letters* 22: 465–469.

Peltier WR (1998) Postglacial variations in the level of the sea: Implications for climate dynamics and solid-Earth geophysics. *Reviews of Geophysics* 36: 603–689.

Peltier WR (2004) Global glacial isostasy and the surface of the ice-age Earth: The ICE-5G (VM2) model and GRACE. *Annual Review of Earth and Planetary Sciences* 32: 111–149.

Peltier WR and Andrews JT (1976) Glacial-isostatic adjustment I – the forward problem. *Geophysical Journal of the Royal Astronomical Society* 46: 605–646.

Peltier WR, Drummond RA, and Tushingham AM (1986) Post-glacial rebound and transient lower mantle rheology. *Geophysical Journal of the Royal Astronomical Society* 87: 79–116.

Peltier WR, Forte AM, Mitrovica JX, and Dziewonski AM (1992) Earth's gravitational field: Seismic tomography resolves the enigma of the Laurentian anomaly. *Geophysical Research Letters* 19: 1555–1558.

Ritsema J, van Heijst HJ, and Woodhouse JH (1999) Complex shear wave velocity structure imaged beneath Africa and Iceland. *Science* 286: 1925–1928.

Rubincam D (1984) Postglacial rebound observed by LAGEOS and the effective viscosity of the lower mantle. *Journal of Geophysical Research* 89: 1077–1087.

Sabadini R, Yuen DA, and Gasperini P (1986) The effects of transient rheology on the interpretation of lower mantle rheology. *Geophysical Research Letters* 12: 361–365.

Sabadini R, DiDonato G, Vermeersen LLA, Devoti R, Luceri V, and Bianco G (2002) Ice mass loss in Antarctica and stiff lower mantle viscosity inferred from the long wavelength time dependent gravity field. *Geophysical Research Letters* 29: 10.1029/2001GL014016.

Sabadini R and Peltier WR (1981) Pleistocene deglaciation and the Earth's rotation: Implications for mantle viscosity. *Geophysical Journal of the Royal Astronomical Society* 66: 553–578.

Sabadini R, Yuen DA, and Gasperini P (1988) Mantle rheology and satellite signatures from present-day glacial forcings. *Journal of Geophysical Research* 12: 437–447.

Simons M and Hager BH (1997) Localization of the gravity field and the signature of glacial rebound. *Nature* 390: 500–504.

Stephenson FR and Morrison LV (1995) Long-term fluctuations in the Earth's rotation: 700 BC to AD 1990. *Proceedings of the Royal Society of London A* 351: 165–202.

Tamisiea M, Mitrovica JX, Tromp J, and Milne GA (2002) Present-day secular variations in the low-degree harmonics of the geopotential: Sensitivity analysis on spherically symmetric Earth models. *Journal of Geophysical Research* 107: (doi:10.1029/2001JB000696).

Tamisiea ME, Leuliette E, Davis JL, and Mitrovica JX (2005) Constraining hydrological and cryospheric mass flux in southeastern Alaska using space-based gravity measurements. *Geophysical Research Letters* 32: L20501 (doi:10.1029/2005GL023961).

Tamisiea ME, Mitrovica JX, and Davis JL (2007) Grace gravity data constrain ancient ice geometries and continental dynamics over Laurentia. *Science* 316: 881–883.

Tushingham AM and Peltier WR (1991) ICE-3G: A new global model of late Pleistocene deglaciation based upon geophysical predictions of post-glacial relative sea level change. *Journal of Geophysical Research* 96: 4497–4523.

Velicogna I and Wahr J (2005) Ice mass balance in Greenland from GRACE. *Journal of Geophysical Research* 32: L18505 (10.1029/2005GL023955).

Velicogna I and Wahr J (2006a) Measurements of time-variable gravity show mass loss in Antarctica. *Science* 311: 1754–1756 (doi:10.1126/science.1123785).

Velicogna I and Wahr J (2006b) Acceleration of Greenland ice mass loss in spring 2004. *Nature* 433: 329–331 (doi:10.1038/nature05168).

Vermeersen LLA, Sabadini R, Spada G, and Vlaar NJ (1994) Mountain building and Earth rotation. *Geophysical Journal International* 117: 610–624.

Vermeersen LLA, Sabadini R, Devoti R, *et al.* (1998) Mantle viscosity from joint inversions of Pleistocene deglaciation-induced changes in geopotential with a new SLR analysis and polar wander. *Geophysical Research Letters* 25: 4261–4264.

Walcott RI (1972) Late Quaternary vertical movements in Eastern North America: Quantitative evidence of glacio-isostatic rebound. *Reviews of Geophysics and Space Physics* 10: 849–884.

Wu P and Peltier WR (1983) Glacial isostatic adjustment and the free air gravity anomaly as a constraint on deep mantle viscosity. *Geophysical Journal of the Royal Astronomical Society* 74: 377–449.

Wu P and Peltier WR (1984) Pleistocene deglaciation and the Earth's rotation: A new analysis. *Geophysical Journal of the Royal Astronomical Society* 76: 753–791.

Wu P and van der Wal W (2003) Postglacial sealevels on a spherical, self-gravitating viscoelastic Earth: Effects of lateral viscosity variations in the upper mantle on the inference of viscosity contrasts in the lower mantle. *Earth and Planetary Science Letters* 211: 57–68.

Yoder C, Williams J, Dickey J, Schutz B, Eanes R, and Tapley B (1983) Secular variations of the Earth's gravitational harmonic J_2 coefficient from Lageos and nontidal acceleration of Earth rotation. *Nature* 303: 757–762.

Yoder CF and Ivins ER (1985) Changes in the Earth's gravity field from Pleistocene deglaciation and present-day glacial melting. *Eos Transactions American Geophysical Union* 66(18): 245.

Yuen DA and Sabadini R (1985) Viscosity stratification of the lower mantle as inferred from the J_2 observation. *Annales de Geophysciae* 3: 647–654.

Yuen DA, Sabadini R, Gasperini P, and Boschi EV (1986) On transient rheology and glacial isostasy. *Journal of Geophysical Research* 91: 11420–11438.

Yuen DA, Gasperini P, Sabadini R, and Boschi E (1987) Azimuthal dependence in the gravity field induced by recent and past cryospheric forcings. *Geophysical Research Letters* 8: 812–815.

Zhong S, Paulson A, and Wahr J (2003) Three-dimensional finite element modelling of Earth's viscoelastic deformation: Effects of lateral variations in lithospheric thickness. *Geophysical Journal International* 155: 679–695.

8 Time Variable Gravity from Satellites

J. M. Wahr, University of Colorado, Boulder, Co, USA

8.1 Introduction

The Earth's gravity field is a product of its mass distribution: mass both deep within the Earth and at and above its surface. That mass distribution is constantly changing. Tides in the ocean and solid Earth cause large mass variations at 12 h and 24 h periods. Atmospheric disturbances associated with synoptic storms, seasonal climatic variations, etc., lead to variations in the distribution of mass in the atmosphere, the ocean, and the water stored on land. Mantle convection causes mass variability throughout the mantle that has large amplitudes compared to those associated with climatic variability, but that generally occurs slowly relative to human timescales.

Because of these and other processes, the Earth's gravity field varies with time. Observations of that variability using either satellites or ground-based instrumentation can be used to study a wide variety of geophysical processes that involve changes in mass

(Dickey *et al.*, 1997). Solid Earth geophysics is not the prime beneficiary of time variable gravity measurements. Instead, most of the time-variable signal comes from the Earth's fluid envelope: the oceans, the atmosphere, the polar ice sheets and continental glaciers, and the storage of water and snow on land. Fluids (water and gasses) are much more mobile than rock.

Solid Earth deformation does have a significant indirect effect on ground-based gravity measurements. A gravimeter on the Earth's surface is sensitive to vertical motion of that surface. When the surface goes up, the gravimeter moves further from the center of the Earth and so it sees a smaller gravitational acceleration. For most solid Earth processes the signal from the vertical displacement of the meter is far larger than the actual gravity change caused by the displaced mass. Thus, a surface gravimeter can, in effect, be viewed as a vertical positioning instrument. A satellite, on the other hand, is not fixed to the surface, and so the gravity signals it detects are due entirely to the underlying mass distribution. Thus, satellite gravity provides direct constraints on that mass.

8.1.1   Nonuniqueness

One serious limitation when interpreting gravity observations is that the inversion of gravity for density is nonunique. There are always an infinite number of possible internal density distributions that can produce the same external gravity field. Even perfect knowledge of the external gravity field would not provide a unique density solution.

As a simple illustration of this nonuniqueness, consider the gravity field outside a sphere. The external gravitational acceleration is $g = MG/r^2$, where M is the total mass of the sphere, G is the gravitational acceleration, and r is the distance to the center of the sphere. This same expression holds whether the mass is uniformly distributed throughout the sphere, or is localized entirely at the outer surface, or has any other radially dependent distribution. By observing the external gravity field in this case, all that could be learned is the total mass of the sphere and the fact that the internal density is spherically symmetric. The details of how the density is distributed with radius would remain unknown. This nonuniqueness would

disappear if the gravity field everywhere inside the sphere were also known. But knowledge of the external field alone is not enough.

This nonuniqueness is a major limitation when interpreting the Earth's static gravity field. For example, **Figure 1** shows a map of the Earth's static geoid anomaly, as determined by Lemoine *et al.* (1998) from decades of satellite and surface observations. The geoid is the surface of constant potential that coincides with mean sea level over the ocean. The geoid anomaly is the elevation of the geoid above its mean ellipsoidal average. This is a common method of representing the Earth's gravity field, one that emphasizes the long wavelength characteristics of the field. There is a trade-off between amplitude and depth when using this map to constrain the Earth's time-averaged mass distribution. For example, from this map alone it is not possible to know whether the large red feature over Indonesia is caused by a large positive mass anomaly in the crust, or a much larger mass anomaly deeper in the mantle.

However, **Figure 1** clearly does contain information about the Earth's internal density. Not every density distribution can produce the same gravity field. The results provide a constraint on a weighted vertical average of the underlying mass anomalies. Static gravity observations are particularly useful

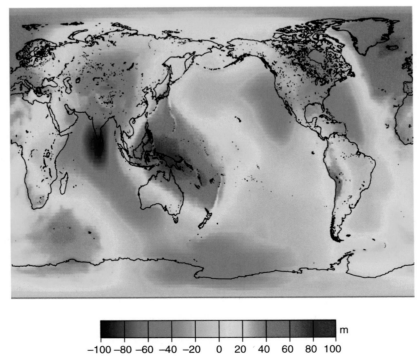

−100 −80 −60 −40 −20 0 20 40 60 80 100   m

Figure 1   The time-averaged geoid anomaly from EGM96 (Lemoine *et al.*, 1998).

when combined with independent information or assumptions about the depth of the density anomaly, or its amplitude, or its spatial pattern.

8.1.2 Time-Variable Gravity

Nonuniqueness is much less of an issue for time-varying gravity. Time-varying signals, if they vary rapidly enough, can usually be assumed to come from mass variability at the Earth's surface rather than from deep within the Earth. For example, **Figure 2(a)** shows the amplitude of the annual cycle in the geoid as observed from the Gravity Recovery and Climate Experiment (GRACE; see below). It is almost certain that this signal is coming from some combination of the atmosphere, the oceans, and the water/snow/ice stored on or just below the land surface. Few solid Earth processes are likely to vary this rapidly, let alone to show an annual cycle. The only exceptions are the body tide, which can be modeled and removed to an accuracy far better than the accuracy of the GRACE gravity observations; and the deformational response of the solid Earth to the surface mass load. That deformation signal, which is typically only a few percent of the signal from the load itself, can be linearly related to the load signal through scale-dependent, well-modeled, proportionality factors (load Love numbers; see below).

Thus, the seasonal mass anomaly can be assumed to be concentrated within a few kilometers of the surface. The inversion for mass anomalies still depends, in principle, on the exact depth of the load. But since the few kilometers uncertainty in vertical position is much smaller than the horizontal scales of the signals shown in **Figure 2(a)**, the corresponding uncertainty in the amplitude of the inferred mass anomaly is negligible. It is still not possible to tell, without additional information, whether a mass anomaly in a continental region, for example, is in the atmosphere, or in the water and snow on the surface, or in the water-stored underground. But at least the total amplitude of the mass anomaly can be determined.

The difficulty with time-variable gravity is that the amplitudes are small. A comparison of **Figures 1** and **2(a)**, for example, shows that the annually varying geoid is over 1000 times smaller than the lateral variation in the static field. Most of the Earth's mass, after all, is tied up in it's rocky interior, and remains relatively immobile on human timescales.

Advances in ground-based instrumentation over the last few decades have made it possible to begin to

(a)

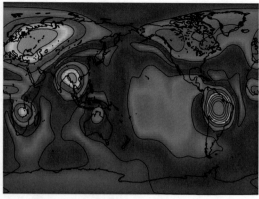

mm of geoid

0.0 0.5 1.0 1.5 2.0 2.5 3.0 3.5 4.0 4.5 9.0

(b)

cm of water

0 1 2 3 4 5 6 7 8 9 25

Figure 2 (a) The amplitude of the annual cycle in the geoid, fit to 4 years (spring 2002 to spring 2006) of monthly gravity field solutions from GRACE. A 750-km Gaussian smoothing function has been applied to the results. (b) The corresponding amplitude of the annual cycle in surface mass, also smoothed with a 750-km Gaussian, in units of centimeters of water thickness. The solid lines represent contour intervals of 1 mm in the top panel and 4 cm in the bottom panel.

observe time-variable gravity at local scales. Modern, high-precision gravimeters can detect surface displacements caused by solid Earth processes, as well as local gravitational changes caused by variations in the overlying atmosphere and underlying water storage.

But the recovery of large-scale time-varying signals requires satellite measurements. Until the launch of Challenging Microsatellite Payload (CHAMP) in 2001 and, especially, GRACE in 2002, satellite time-variable gravity solutions were based entirely on satellite laser ranging (SLR) observations. The most useful SLR measurements have involved LAGEOS (launched by NASA in 1976) and LAGEOS II (launched jointly by NASA and the Italian Space Agency in 1993). Both satellites are orbiting at 6000 km altitude. They are passive spheres, with outer surfaces covered with corner cube reflectors. A powerful laser on Earth sends a laser pulse up to the satellite, where the light is reflected back to the laser. The round-trip travel time is measured, and so the distance between the laser and the satellite is determined. By monitoring these distances from lasers around the Earth's surface, the satellite's orbital motion is computed. Since the orbital motion is determined by the Earth's gravity field, this allows for global gravity field solutions at regular time intervals. Differences between solutions for different time periods provide estimates of time-variable gravity.

8.1.3 Changes in the Earth's Oblateness

The first satellite identification of a nontidal time-varying signal was the recovery of a secular change in the Earth's oblateness. The oblateness is a global-scale component, and is the easiest laterally varying component to detect with a satellite. There are two reasons for this. Let $N(\theta, \phi)$ be the height of the geoid above the Earth's mean spherical surface at latitude θ and eastward longitude ϕ. It is usual to expand N as a sum of Legendre functions (see, e.g., Chao and Gross (1987)):

$$N(\theta, \phi) = a \sum_{l=2}^{\infty} \sum_{m=0}^{l} \tilde{P}_{lm}(\cos\theta)(C_{lm}\cos(m\phi) + S_{lm}\sin(m\phi)) \qquad [1]$$

where a is the radius of the Earth, the $\tilde{P}_{lm}$ are normalized associated Legendre functions, and the C_{lm} and S_{lm} are dimensionless (Stokes) coefficients. Global gravity field solutions are typically provided in the form of a set of Stokes coefficients. The indices l and m in [1] are the degree and order, respectively, of the Legendre function. The horizontal scale of any term in [1] is inversely proportional to the value of l. The half-wavelength of a (l, m) harmonic serves as an approximate representation of this scale, and is roughly $(20\,000/l)$ km. Note that the sum over l in [1] begins at $l = 2$. The $l = 0$ term vanishes because N is defined as the departure from the mean spherical surface; and the $l = 1$ terms vanish by requiring the

geoid to be centered about the Earth's center of mass. Thus, the $l = 2$ terms are the longest wavelength terms in the series expansion [1]. The Earth's oblateness is proportional to C_{20}.

Satellite determinations of gravity are sensitive to the gravity field at the altitude of the satellite, not at the Earth's surface. The gravitational potential from any (l, m) term in [1] decreases with increasing radius, r, as $(a/r)^{(l+1)}$. Thus, terms with the smallest values of l (i.e., the longest wavelengths) are the least attenuated up at the satellite altitude, and so tend to be the easiest to determine. This tends to favor the recovery of $l = 2$ Stokes coefficients, relative to coefficients with $l > 2$.

At the same time, terms with $m = 0$ are better determined than terms with $m > 0$. This is because an $m = 0$ term does not depend on longitude. For example, **Figure 3** shows the patterns of $(l, m) = (2, 0)$ and $= (2, 2)$ terms. Suppose we track a satellite orbiting in the $(2, 0)$ pattern shown in panel (a). As

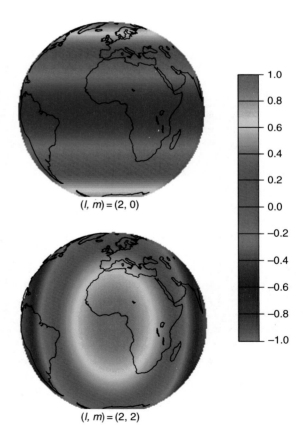

$(l, m) = (2, 0)$

$(l, m) = (2, 2)$

Figure 3 The spatial patterns of the Legendre functions $\tilde{P}_{20}$ and $\tilde{P}_{22}$.

the satellite makes its first orbit, traveling from near the north pole down to near the south pole and back again, it passes through the gravity pattern of red/green/blue shown in the figure, and its orbit gets perturbed accordingly. By the time it begins its second orbit, the Earth has rotated about the polar axis, but because there is no longitude dependence the satellite passes through the same red/green/blue pattern on its second orbit, and so that orbit gets perturbed in the same direction. This happens for every orbit, so the perturbation gradually builds up to large values and is easily seen in the ranging observations. On the other hand, for the (2, 2) pattern in panel (b), every time the satellite begins a new orbit the underlying pattern is different because the Earth's rotation has carried that pattern to the east. Thus, the orbital perturbations do not tend to add constructively and are harder to see.

Early SLR solutions showed a secular increase in C_{20} (Yoder *et al.*, 1983; Rubincam, 1984) which is consistent with a steady migration of mass from low latitudes toward high latitudes. The signal was first interpreted as due to postglacial rebound (PGR), the Earth's ongoing response to the removal of the ice loads at the end of the last ice age. The areas that lie beneath the ice loads centered over Hudson Bay and over the region around the North and Baltic Seas, are still depressed from the weight of those ancient ice sheets, and they are still gradually uplifting as material deep within the mantle flows in from lower latitudes. In fact, since its first detection, the observed secular change in C_{20} has been used in PGR models to help constrain the Earth's viscosity profile.

More recent SLR solutions give C_{20} trends that are in general agreement with those early estimates (e.g., Cox and Chao, 2002; Cheng and Tapley, 2004), though the actual rate tends to be sensitive to the time span of the data and the analysis method used (e.g., Benjamin *et al.*, 2006). A representative C_{20} time series is shown in **Figure 4** (data provided by Chris Cox (personal communication, 2005)). There is large seasonal variability, due presumably to a combination of atmospheric pressure variations and variations in the distribution of water in the oceans and on land (e.g., Chao and Au, 1991; Dong, *et al.*, 1996; Cheng and Tapley, 1999; Nerem *et al.*, 2000). A trend is also clearly evident in the results, and is more pronounced after the data have been low-pass filtered (the red line in **Figure 4**). But there is also evidence of interannual variability. In particular, notice the anomalous wiggle during 1998–2002 (Cox and Chao, 2002). This feature has been variously explained as the result of climatically

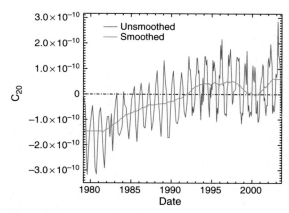

Figure 4 The blue line shows a time series of C_{20}, determined from more than 20 years of SLR measurements (data provided by Chris Cox, personal communication, 2005). The red line is a smoothed version of those values, obtained by first fitting and removing seasonal terms and then applying a 23-month moving average to the residuals.

driven oscillations in the ocean (Cox and Chao, 2002; Dickey *et al.*, 2002), in the storage of water, snow, and ice on land (Dickey *et al.*, 2002), and as partly the consequence of the effects of anelasticity on the 18.6-year solid Earth tide (Benjamin *et al.*, 2006). Whatever its origin, its presence illustrates why solutions for the secular trend depend on the time span.

In addition, it has become increasingly evident in recent years that there could be other processes that involve enough mass transfer between low and high latitudes to have a significant impact on the C_{20} trend, and so to confuse the PGR interpretation. The most important of these processes are likely to be changes in ice of the Greenland and Antarctic ice sheets. For example, a rate of Antarctic ice mass loss equivalent to 0.6 mm yr^{-1} of global sea-level rise averaged over the last 30 years, would cause a C_{20} rate of increase that is about equal in magnitude to the SLR value, though with the opposite sign (e.g., Trupin, 1993). If the ice mass trend was even a sizable fraction of this amount, it would have a significant impact on the C_{20} PGR constraint.

These uncertainties arise because knowledge of the single harmonic, C_{20}, is not sufficient to determine the spatial location of the signal. SLR has provided time-variable solutions for a handful of other harmonics (Cheng *et al.*, 1997; Cheng and Tapley, 1999; Nerem *et al.*, 2000; Moore *et al.*, 2005). But there are not nearly enough of these harmonics to give the spatial resolution necessary to confidently address these issues. The basic limitation comes from the high altitude of LAGEOS (6000 km) and the other SLR satellites. Shorter-scale harmonics in [1] are sufficiently attenuated at those high altitudes that their time dependence cannot be easily detected. The solution to this problem is to use a satellite in a lower-altitude orbit. That is, the motivation for CHAMP (Reigber *et al.*, 2002) and, especially, for GRACE (Tapley *et al.*, 2004a, 2004b).

8.2   GRACE

The GRACE mission design makes it particularly useful for time-variable gravity studies. Launched jointly by NASA and the German space agency (DLR) in March 2002, GRACE consists of two identical satellites in identical orbits, one following the other by about 220 km. The satellites use microwaves to continually monitor their separation distance to an accuracy of better than 1 μm – about 1/100th the thickness of a human hair. This distance changes with time as the satellites fly through spatial gradients in the gravity field, and so by monitoring those changes the gravity field can be determined. The satellite altitude is less than 500 km, which makes GRACE considerably more sensitive than SLR to short wavelength terms in the gravity field. The disadvantage of having such a low altitude is that GRACE experiences greater atmospheric drag, which can cause large and unpredictable changes in the intersatellite distance. To reduce this problem, each GRACE satellite has an on-board accelerometer to measure nongravitational accelerations. Those measurements are transmitted to the ground where they are used to correct the satellite-to-satellite distance measurements. Each spacecraft also has an on-board GPS receiver, used to determine the orbital motion of each spacecraft in the global GPS reference frame and to improve the gravity field solutions at global-scale wavelengths.

8.2.1   Gravity Solutions

GRACE transmits raw science instrument and satellite housekeeping data to the ground, where they are transformed into physically meaningful quantities: for example, satellite-to-satellite distances, nongravitational accelerations, and spacecraft attitudes. These quantities, called level-1 data, are made publically available and can be used to construct gravity field solutions. Since few users have the capability of constructing their own gravity solutions from these data, the GRACE

project does that as well, and makes those solutions, referred to as level-2 data, available on the web.

The level-2 gravity products consist of complete sets of harmonic (Stokes) coefficients [1] out to some maximum degree and order (typically $l_{max} = 60$), averaged over monthly intervals. Larger sets of coefficients averaged over longer time intervals are also provided to represent the static field. Harmonic coefficients can be used to generate geoid, gravity, or mass solutions at individual locations, or averaged over specific regions, as described below. Level-2 products are generated at several project-related processing centers (i.e., the Center for Space Research at the University of Texas, GeoForschungsZentrum in Potsdam, Germany, and the Jet Propulsion Laboratory), and each of these products is made available to users.

Harmonic solutions are traditional in satellite geodesy. Harmonics help with the problem of upward and downward continuing the gravity field between the surface and the satellite altitude, during the solution process. Specifically, the gravitational potential caused by any (l, m) term in [1] has a particularly simple radial dependence, decreasing with increasing radius, r, as $(a/r)^{(l+1)}$.

Nevertheless, users sometimes generate nonharmonic solutions directly from the level-1 data. Various methods have been derived for doing this, most of which involve partitioning the time-variable surface mass field into small regions, and using the level-1 data to directly determine the mass in each of those regions. For example, the first and second time derivatives of the satellite-to-satellite distance are the along-track differences in velocity and acceleration of the two satellites. These can be used to determine the along-track gradients of the gravitational potential and acceleration, respectively. These gradients can then be fit to upward-continued mass signals from specific regions, to determine the amplitudes of those mass signals (see, e.g., Jekeli (1999), Visser *et al.* (2003), Han *et al.* (2005a, 2006a), and Schmidt *et al.* (2006a).

Another approach involves the construction of 'mascon' solutions (e.g., Rowlands *et al.*, 2005; Watkins and Yuan, 2006; Yuan and Watkins, 2006). Mascons, in this context, are mass anomalies spread uniformly over either regular- (usually rectangular) or irregular-shaped blocks (Luthcke *et al.*, 2006) at the Earth's surface. Each such mass anomaly has an overall scale factor, which is determined from the level-1 data.

These alternative methods are usually designed to estimate regional mass anomalies, rather than to generate results everywhere over the globe. They thus sometimes ingest only those level-1 data that are acquired when the satellites are over the region of interest. This tends to reduce a problem common to global harmonic solutions, in which errors, either in the satellite measurements or in the geophysical background models, that affect the satellites in one region, end up leaking into the gravity field solutions in distant regions.

8.2.2 Using the Harmonic Solutions to Solve for Mass

Most users do not have the resources to process level-1 data, and rely instead on the standard level-2 gravity field products: the harmonic solutions. For most applications the gravity field itself is not of direct interest. Instead, it is usually the mass distribution causing the gravity field that is the desired quantity. Here, we describe how that mass distribution can be inferred from the harmonic gravity solutions. We focus specifically on the time-variable components of the gravity and mass fields. The methods described here are described in more detail in section 8.2.1 of Wahr *et al.* (1998) (see also Chao and Gross (1987)).

The time-variable component of the gravity field is obtained by removing the long-term mean of the Stokes coefficients from each monthly value. The mean can be obtained from one of the static fields available as level-2 products. Or, perhaps more usefully, it can be estimated by simply constructing the average of all the monthly fields used in the analysis. The reason for removing the mean field is that it is dominated by the static density distribution inside the solid Earth. It thus has no bearing on attempts to learn about, say, the distribution of water stored on land or in the ocean. Removing the static field, though, means that all contributions from the mean stored water are also removed. Thus, only the time-variable component of the water storage can be recovered.

The time-variable gravity field is then used to solve for the time-variable mass field. This solution is nonunique, as described in Section 8.1. Let ΔC_{lm} and ΔS_{lm} be the time-variable components of the (l, m) Stokes coefficients for some month. Let $\Delta\rho(r, \theta, \Phi)$ be the density redistribution that causes this time-dependent change in gravity. Then

$$\begin{Bmatrix} \Delta C_{lm} \\ \Delta S_{lm} \end{Bmatrix} = \frac{3}{4\pi a \rho_{\text{ave}}(2l+1)} \int \Delta\rho(r, \theta, \phi) \left(\frac{r}{a}\right)^{l+2}$$
$$\times \tilde{P}_{lm}(\cos\theta) \begin{Bmatrix} \cos(m\phi) \\ \sin(m\phi) \end{Bmatrix} \sin\theta \, \mathrm{d}\theta \, \mathrm{d}\phi \, \mathrm{d}r \quad [2]$$

where ρ_{ave} is the average density of the Earth ($=5517 \, \text{kg m}^{-3}$).

Suppose the density is expanded as a sum of Legendre functions:

$$\Delta\rho(r, \theta, \phi) = \sum_{l=0}^{\infty} \sum_{m=0}^{l} \bar{P}_{lm}(\cos\theta)$$
$$\times \left(\Delta\rho_{lm}^c(r)\cos(m\phi) + \Delta\rho_{lm}^s(r)\sin(m\phi)\right) \quad [3]$$

Using [3] in [2], and employing orthogonality relations for Legendre functions, [2] reduces to

$$\begin{Bmatrix} \Delta C_{lm} \\ \Delta S_{lm} \end{Bmatrix} = \frac{3}{a\rho_{\text{ave}}(2l+1)} \int \begin{Bmatrix} \Delta\rho_{lm}^c(r) \\ \Delta\rho_{lm}^s(r) \end{Bmatrix} \left(\frac{r}{a}\right)^{l+2} \mathrm{d}r \quad [4]$$

This result, [4], can be used to place constraints on $\Delta\rho(r, \theta, \phi)$ from measurements of ΔC_{lm} and ΔS_{lm}. The nonuniqueness is evident here in the fact that ΔC_{lm} and ΔS_{lm} provide information only on the radial integral of the density coefficients. There is no way of determining how the density depends on depth within the Earth.

Suppose, though, we have reason to believe that the observed ΔC_{lm} and ΔS_{lm} are caused by mass variability concentrated within a thin layer of thickness H near the Earth's surface; a layer containing those regions of the atmosphere, oceans, ice sheets, and land water storage that are subject to significant mass fluctuations. H, in this case, would be mostly determined by the thickness of the atmosphere, and is of the order of 10 km. If H is thin compared to the horizontal resolution of the observations, then the amplitude of the density anomaly can be uniquely determined, as follows.

Suppose the observed gravity field is accurate enough to resolve gravity anomalies down to scales of R km. That means the ΔC_{lm} and ΔS_{lm}'s contain useful information for values of l up to $l_{\max} = 20000/R$. At present, GRACE has a typical resolution of ~750 km, though resolutions as small as ~300 km can be obtained by employing postprocessing methods (Swenson and Wahr, 2006a). Thus, at present $l_{\max} \sim 65$. Suppose H is thin enough that

$$(l_{\max} + 2)H/a << 1 \quad [5]$$

Then, $(r/a)^{l+2} \approx 1$ for all usable values of l, and so [2] reduces to

$$\begin{Bmatrix} \Delta C_{lm}^{\text{surf mass}} \\ \Delta S_{lm}^{\text{surf mass}} \end{Bmatrix} = \frac{3}{4\pi a \rho_{\text{ave}}(2l+1)} \int \Delta\sigma(\theta, \phi)$$
$$\times \bar{P}_{lm}(\cos\theta) \begin{Bmatrix} \cos(m\phi) \\ \sin(m\phi) \end{Bmatrix} \sin\theta \, \mathrm{d}\theta \, \mathrm{d}\phi \quad [6]$$

where $\Delta\sigma$ is the change in surface density (i.e., mass/area), defined as the radial integral of $\Delta\rho$ through the surface layer:

$$\Delta\sigma(\theta, \phi) = \int_{\text{thin layer}} \Delta\rho(r, \theta, \phi)\, dr \qquad [7]$$

The assumption that the density anomaly is concentrated within this thin layer is incorrect. Any change in mass load at the surface will induce deformation within the solid Earth, leading to a density anomaly at depth as well. The gravity signal caused by these solid Earth mass anomalies is typically a few percent of the gravity anomaly caused by the surface mass, and fortunately can be easily represented in terms of load Love numbers, k_l (see, e.g., Farrell (1972), Chao (1994), eqn [6]). Specifically, if $\Delta C_{lm}^{\text{solid Earth}}$ and $\Delta S_{lm}^{\text{solid Earth}}$ represent the contributions to the gravity field from the load-induced deformation in the solid Earth, then

$$\left\{ \begin{array}{c} \Delta C_{lm}^{\text{solid Earth}} \\ \Delta S_{lm}^{\text{solid Earth}} \end{array} \right\} = k_l \left\{ \begin{array}{c} \Delta C_{lm}^{\text{surf mass}} \\ \Delta S_{lm}^{\text{surf mass}} \end{array} \right\} \qquad [8]$$

Thus, the total dependence of the Stokes coefficients on the surface mass density is

$$\left\{ \begin{array}{c} \Delta C_{lm} \\ \Delta S_{lm} \end{array} \right\} = \frac{3}{4\pi a \rho_{\text{ave}}} \frac{1+k_l}{(2l+1)} \int \Delta\sigma(\theta, \phi)$$
$$\times \tilde{P}_{lm}(\cos\theta) \left\{ \begin{array}{c} \cos(m\phi) \\ \sin(m\phi) \end{array} \right\} \sin\theta\, d\theta\, d\phi \qquad [9]$$

By expanding $\Delta\sigma(\theta, \phi)$ as a sum of Legendre Coefficients, similar to the expansion shown in [3] for $\Delta\rho$, and using the orthogonality of the Legendre functions to obtain a result similar to [4], we find

$$\Delta\sigma(\theta, \phi) = \frac{a\rho_{\text{ave}}}{3} \sum_{l=0}^{\infty} \sum_{m=0}^{l} \frac{2l+1}{1+k_l} \tilde{P}_{lm}(\cos\theta)(\Delta C_{lm}\cos(m\phi)$$
$$+ \Delta S_{lm}\sin(m\phi)) \qquad [10]$$

The results above assume that the surface layer is thin enough that [5] is valid. If we assume that $l_{\max} = 65$, and that the layer includes the atmosphere so that $H \sim 10$ km, then [5] is violated at about the 10% level. This is a large enough inaccuracy that it might be important, for some applications, to include the radial distribution of atmospheric density fluctuations. Methods of doing this are described in Swenson and Wahr (2002a). Mass variations in the oceans and in the water stored on land occur almost entirely within 1 km of the surface, and usually much closer to the surface than that. For a 1 km thick layer and $l_{\max} = 65$, [5] is

accurate to $\sim$1%, which is easily good enough for oceanographic and hydrological applications.

8.2.3 Love Numbers

The use of [10] to recover surface mass requires knowledge of the load Love numbers k_l. As a guide, one set of results for those Love numbers is given in **Table 1** (D. Han, personal communication, 1998) for a few values of l up to 200. These results are computed as described by Han and Wahr (1995), using Earth structural parameters from the Preliminary Reference Earth Model (PREM) of Dziewonski and Anderson (1981). Results for other values of $l < 200$ can be obtained by linear interpolation of the **Table 1** results. Linearly interpolating the **Table 1** results, instead of using exact results, introduces errors of less than 0.05% for all $l < 200$.

These results for k_l do not include anelastic effects. These effects increase with increasing period but are apt to be negligible for our applications. For example, Wahr and Bergen (1986) concluded that at an annual period the anelastic effect on the $l = 2$ body tide Love number, k_2^{body}, would probably be less than 2%, corresponding to an effect on $(1 + k_2^{\text{body}})$ of less than

Table 1 Elastic Love Numbers k_l computed by Dazhong Han as described by Han and Wahr (1995), for Earth Model PREM

l	k_l
0	+0.000
1	+0.027
2	−0.303
3	−0.194
4	−0.132
5	−0.104
6	−0.089
7	−0.081
8	−0.076
9	−0.072
10	−0.069
12	−0.064
15	−0.058
20	−0.051
30	−0.040
40	−0.033
50	−0.027
70	−0.020
100	−0.014
150	−0.010
200	−0.007

The $l = 1$ value assumes the origin of the coordinate system is the center of figure of the solid Earth's surface (see text). (Dziewonski and Anderson, 1981).

1%. Even allowing for larger effects at longer periods, and perhaps a somewhat greater effect for load Love numbers than for body tide Love numbers (since load Love numbers are more sensitive to upper mantle structure where the anelastic effects could be larger), we tentatively conclude that anelasticity would not perturb the results for $(1 + k_l)$ by more than a few percent.

The Love numbers in [10] with $l = 0$ and $l = 1$ require discussion. The $l = 0$ term is proportional to the total mass of the Earth where 'the Earth' includes not only the solid Earth, but also its fluid envelope (the oceans, atmosphere, etc.). This total mass does not change with time, and so ΔC_{00} from GRACE can be assumed to vanish. Suppose, though, the objective is to use [10] to find the surface mass contribution from just one component of the surface mass: say, the ocean, for example. The total mass of the ocean need not be constant, due to exchange of water with the atmosphere or the land surface. So the oceanic contributions to $\Delta \hat{C}_{00}$ need not vanish. But this nonzero $\Delta \hat{C}_{00}$ will not induce an $l = 0$ response in the solid Earth: that is, the load does not cause a change in the total solid Earth mass. Thus, $k_0 = 0$.

The $l = 1$ terms are proportional to the position of the Earth's center of mass relative to the center of the coordinate system and so depend on how the coordinate system is chosen. One possibility is to choose a system where the origin always coincides with the Earth's instantaneous center of mass. In that case all $l = 1$ terms in the geoid are zero by definition, and so the GRACE results for $\Delta C_{lm} = \Delta S_{lm} = 0$ for all $l = 1$. This is the coordinate system used for the geoid representation shown in [1]. Again, the $l = 1$ coefficients for an individual component of the total surface mass need not vanish. Redistribution of mass in the ocean, for example, can change the center of mass of the ocean. But that will induce a change in the center of mass of the solid Earth, so that the center of mass of the ocean + solid Earth remains fixed. So, for this choice of coordinate system, $k_{l=1} = -1$.

Another possibility is to define the coordinate system so that its origin coincides with the center of figure of the Earth's solid outer surface. That is the most sensible way of defining the origin when recovering the Earth's time-variable mass distribution, since hydrological, oceanographic, and atmospheric models are invariably constructed in a system fixed to the Earth's surface. In that case the $l = 1$ GRACE results for $\Delta C_{lm} = \Delta S_{lm}$ need not vanish, and the Love number $k_{l=1}$ is defined so that the $l = 1$ terms in [10] describe the offset between 'the center

of mass of the surface mass + deformed solid Earth' and 'the center of figure of the deformed solid Earth surface'. It is shown by Trupin et al. (1992, equation (10)) that for this coordinate system $k_{l=1} = -(h_{l=1} + 2\ell_{l=1})/3$, where $h_{l=1}$ and $\ell_{l=1}$ are the $l = 1$ displacement Love numbers when the origin is the center of mass of the deformed solid Earth. For this choice of origin, the numerical value of $k_{l=1} = -(h_{l=1} + 2\ell_{l=1})/3$ is given in **Table 1**.

8.2.4 Spatial Averaging

Equation [10] is the starting point for using GRACE estimates of ΔC_{lm} and ΔS_{lm} to recover changes in surface mass density. Because the errors in the GRACE results become large for large l (i.e., short scales), and because terms with large l values can make important contributions to the sum in [10] (note the $2l + 1$ factor in the numerator of [10]), the use of [10] as written can lead to highly inaccurate results.

To obtain accurate results it is necessary to somehow reduce the large-l contributions to the sum [10]. This involves the insertion of some additional multiplicative factor into [10], that is small for large values of l. Any such modification means that the sum will no longer be an exact representation of the surface mass at (θ, ϕ). Since most applications require the surface mass in the spatial domain, it is useful to choose a multiplicative factor in such a way that the sum still has some meaningful connection to the spatially dependent surface mass. Any multiplicative factor applied in the spectral (l, m) domain is equivalent to convolving with some corresponding weighting function in the spatial domain. The problem is to choose a factor that reduces the errors, but that keeps the weighting function localized. The issues are similar to those encountered when designing filters for time-series analysis, where the generic problem is to construct a filter that removes noise but that still provides a meaningful estimate of the true signal in the time domain.

Various methods have been used for improving the GRACE mass estimates in this way, though most of them are similar to one another (Wahr et al., 1998; Swenson and Wahr, 2002b; Swenson et al., 2003; Seo and Wilson, 2005; Chen et al., 2006a; Han et al., 2005b). These methods fall into one of two categories: smoothing the surface mass results, or averaging over specific regions.

8.2.4.1 Smoothing

The simplest way of modifying [10] to obtain accurate results is to introduce degree-dependent weighting factors W_l into the sum, so that

$$\overline{\Delta\sigma}(\theta, \phi) = \frac{a\rho_{ave}}{3}\sum_{l,m}\frac{2l+1}{1+k_l}W_l\tilde{P}_{lm}(\cos\theta)$$

$$\times [\Delta C_{lm}\cos(m\phi) + \Delta S_{lm}\sin(m\phi)] \quad [11]$$

$\overline{\Delta\sigma}$ then represents a smoothed version of the surface mass anomaly, given by

$$\overline{\Delta\sigma}(\theta, \phi) = \int \sin\theta'\,d\theta'\,d\phi'\,\Delta\sigma(\theta', \phi')W(\alpha) \quad [12]$$

where α is the angle between (θ, ϕ) and (θ', ϕ'), and $W(\alpha)$ is a smoothing function corresponding to the choice of the W_l's:

$$W(\alpha) = \frac{1}{4\pi}\sum_l\sqrt{2l+1}\,W_l\tilde{P}_{l0}(\alpha) \quad [13]$$

One obvious way of smoothing is simply to truncate the sum over l so that the inaccurate coefficients at large-l are not included. This is equivalent to choosing $W_l = 1$ for values of l less than some l_{max}, and $W_l = 0$ for $l \geq l_{max}$. This approach can, indeed, give accurate results for the sum if l_{max} is chosen to be small enough. The disadvantage of using this step-function weighting is that the equivalent convolution function, $W(\alpha)$, 'rings' in the spatial domain (see panels (a) and (b) in **Figure 5**). The results for $\overline{\Delta\sigma}$ in this case are an average not only of the true values of $\Delta\sigma$ at points close to (θ, ϕ), but also of $\Delta\sigma$ values at points all around the globe, and where the smoothing function has an oscillating sign.

This ringing can be avoided by choosing W_l to decrease smoothly with l. A convenient choice of smoothing coefficients (see, e.g., Wahr *et al.* (1998)) are the Gaussian values developed by Jekeli (1981) to improve estimates of the Earth's gravity

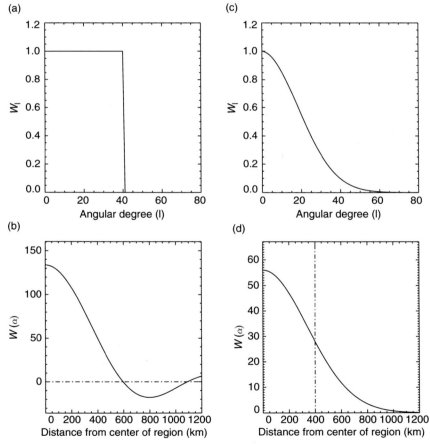

(a)

(b)

(c)

(d)

Figure 5 (a) the spectral smoothing coefficients equivalent to truncation at degree $l = 40$, and (b) the corresponding smoothing function in the spatial domain. (c) The spectral smoothing coefficients for Gaussian smoothing with a 400-km radius, and (d) the corresponding smoothing function in the spatial domain.

field. These coefficients can be found using the recursion relations

$$W_0 = 1$$

$$W_1 = \frac{1 + e^{-2b}}{1 - e^{-2b}} - \frac{1}{b} \qquad [14]$$

$$W_{l+1} = -\frac{2l + 1}{b} W_l + W_{l-1}$$

These coefficients correspond to a smoothing function

$$W(\alpha) = \frac{b \exp[-b(1 - \cos \alpha)]}{1 - e^{-2b}} \qquad [15]$$

where

$$b = \frac{\ln(2)}{(1 - \cos(r/a))} \qquad [16]$$

and r is the distance on the Earth's surface at which W has decreased to $1/2$ its value at $\alpha = 0$ (the distance on the Earth's surface $= a\alpha$). We will refer to r as the

smoothing radius. As an example, panels (c) and (d) in **Figure 5** show $W(\alpha)$ and W_l for $r = 400$ km. Note that the convolution function, $W(\alpha)$, decreases smoothly to zero at large angular distances, and does not oscillate. In practice, there will always be some oscillation, since no satellite gravity field model will ever provide Stokes coefficients out to infinite degree. But as long as the W_l are small out at the value of the maximum degree in the gravity model, the ringing is minimal.

The annual amplitudes shown in **Figure 2(b)** are obtained by applying a Gaussian smoothing function with a 750-km radius, to monthly GRACE mass solutions between the spring of 2002 and the spring of 2006. For comparison, the top panel of **Figure 6** shows results for a single month (after the temporal mean has been subtracted) for a 400 km radius. Note the notably increased noise for the shorter averaging radius. This occurs because the high-degree terms in [10] are not attenuated as effectively for shorter

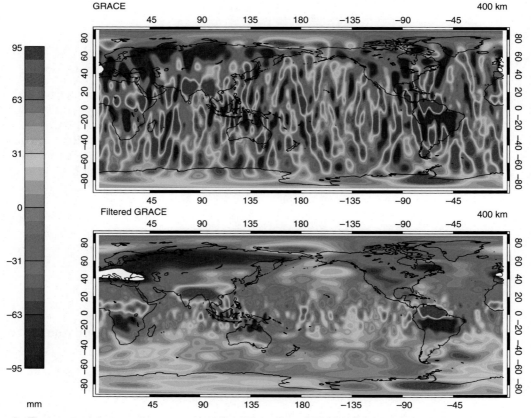

Figure 6 The top panel shows surface mass anomalies deduced from GRACE for a single month, after the temporal mean has been removed. The units are millimeters of water thickness. The bottom panel shows the same thing, but after postprocessing the Stokes coefficients to reduce noise as described by Swenson and Wahr (2006a). Figure provided by Sean Swenson.

smoothing radii. The disadvantage of using longer smoothing radii is that the results in the spatial domain are less able to pick up short-scale structure in the mass anomalies.

The results shown in the top panel of **Figure 6** suggest that 400-km resolution is beyond the current capabilities of GRACE. Note that the noise seems to be oriented in north–south stripes. This is a familiar characteristic of GRACE gravity solutions; and is not found, for example, in SLR gravity fields. It occurs because the GRACE satellites measure gravity gradients along-track, and since the GRACE inclination is $89°$, the tracks are oriented north–south. Thus, there is little east–west sensitivity and so any errors in the measurements or in the processing tend to be put into east–west gradients. Postprocessing methods can be used to remove those stripes. The bottom panel of **Figure 6** shows results for the same 400-km smoothing radius as the top panel, but after applying the postprocessing method described in Swenson and Wahr (2006a). Simulations show that this method reduces stripes with only minimal impact on real signal. Note that the stripes in the bottom panel are, indeed, greatly reduced, and that features that look like true signal are now clearly evident.

8.2.4.2 Regional averaging

Many applications require estimates of mass variability for specific regions; for example, estimating changes in mass of the Antarctic ice sheet, or changes in water storage in the Mississippi River Basin. These sorts of problems are better addressed by constructing specific averaging functions optimized for those regions, than by employing the sort of generic smoothing functions described above.

For example, an exact regional average would take the form

$$\Delta\sigma_{\text{region}} = \frac{1}{\Omega_{\text{region}}} \int \Delta\sigma(\theta, \phi)\vartheta(\theta, \phi)\sin\theta \, d\theta \, d\phi \quad [17]$$

where Ω_{region} is the angular area of the region of interest, and where

$$\vartheta(\theta, \phi) = \begin{cases} 0 \text{ outside the basin} \\ 1 \text{ inside the basin} \end{cases} \quad [18]$$

The result [17] can be expressed as a sum of Stokes coefficients:

$$\Delta\sigma_{\text{region}} = \frac{a\,\rho_{\text{ave}}}{3\,\Omega_{\text{region}}} \sum_{l=0}^{\infty} \sum_{m=0}^{l} \frac{(2l+1)}{(1+k_l)} \times \left(\vartheta_{lm}^c \Delta C_{lm} + \vartheta_{lm}^s \Delta S_{lm}\right) \quad [19]$$

where ϑ_{lm}^c and ϑ_{lm}^s are the harmonic coefficients of $\vartheta(\theta, \phi)$. Since the averaging function, $\vartheta(\theta, \phi)$ in this case, changes abruptly from 1 to 0 along the edge of the region, it has power at short spatial scales. Thus, ϑ_{lm}^c and ϑ_{lm}^s can be relatively large at high degrees, and so this estimate of $\Delta\sigma_{\text{region}}$ can be inaccurate.

The way around this problem is to smooth the averaging function, so that it is close to 1 inside the region and close to 0 outside, and varies smoothly between 0 and 1 along the edges. We replace [17] with

$$\overline{\Delta\sigma}_{\text{region}} = \frac{1}{\Omega_{\text{region}}} \int \Delta\sigma(\theta, \phi)\overline{W}(\theta, \phi)\sin\theta \, d\theta \, d\phi \quad [20]$$

where the averaging function

$$\overline{W}(\theta, \phi) = \frac{1}{4\pi} \sum_{lm} \tilde{P}_{lm}(\cos\theta) \times \{W_{lm}^c \cos(m\phi) + W_{lm}^s \sin(m\phi)\} \quad [21]$$

is chosen to closely approximate $\vartheta(\theta, \phi)$, but to vary smoothly enough that its expansion coefficients W_{lm}^c and W_{lm}^c are small for large values of l. In that case, the spectral equivalent to [20]

$$\overline{\Delta\sigma}_{\text{region}} = \sum_{l,m} \frac{a\,\rho_{\text{ave}}}{3\Omega_{\text{region}}} \frac{(2l+1)}{(1+k_l)} \times (W_{lm}^c \Delta C_{lm} + W_{lm}^s \Delta S_{lm}) \quad [22]$$

will be both reasonably representative of the true regional average, and reasonably accurate. Methods of optimizing the choice of $\overline{W}(\theta, \phi)$, based on estimates of the true signal characteristics, are described by Swenson and Wahr (2002b) and Swenson et al. (2003) (see also Seo and Wilson (2005)). In general, the larger the region, the more accurate the results. Examples of optimal averaging functions for Antarctica and for the Mississippi Basin are shown in **Figure 7**. Note that in both cases the averaging function is smaller than 1 inside the region, and remains larger than 0 for some distance outside the region.

8.2.5 Estimating Errors and Accounting for Leakage

Errors in a surface mass estimate separate into two categories: those due to errors in the Stokes coefficients, and those caused by leakage from other signals. Errors in the Stokes coefficients can be caused by instrumental, data processing, or aliasing errors. Temporal aliasing errors in the GRACE

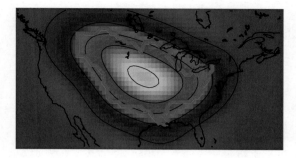

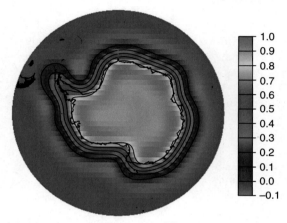

1.0
0.9
0.8
0.7
0.6
0.5
0.4
0.3
0.2
0.1
0.0
−0.1

Figure 7 Averaging functions, $\overline{W}$ (see eqn [21]), for the Mississippi River Basin and for Antarctica. The solid lines represent contour intervals of 0.2.

monthly gravity fields are caused by short-period (submonthly) variations in gravity. The satellite does not monitor the entire global field continually during a month, but samples the gravity field only along its orbital path. Infrequent sampling of a short-period signal can cause aliasing into the monthly averages. The best way to reduce these aliasing errors is to independently model and remove the effects of short-period gravity variations before constructing monthly averages. For GRACE, this means modeling and removing the effects of solid Earth and ocean tides, of atmospheric mass variability over land (using global, gridded atmospheric fields available from the European Centre for Medium-Range Weather Forecasts: ECMWF), and of short period variations in ocean bottom pressure (using an ocean general circulation model). Errors in any of those models cause aliasing errors in the monthly gravity field solutions (Knudsen and Andersen, 2002; Song and Zlotnicki, 2004; Han *et al.*, 2004, 2005c; Thompson *et al.*, 2004; Schrama, 2004; Ray and Luthcke, 2006).

To see how errors in the Stokes coefficients from any source (i.e., instrumental, processing, aliasing)

map into errors in a mass estimate, let δC_{lm} and δS_{lm} be the root-mean-square (rms) errors in the Stokes coefficients. The smoothed estimates [11] and the regional averages [22] are both of the form

$$\sigma = \sum_{l,m}\left[F_l^m \Delta C_{lm} + G_l^m \Delta S_{lm}\right] \qquad [23]$$

Suppose the errors in the different Stokes coefficients are uncorrelated with one another. Then the corresponding rms error in σ would be

$$\delta\sigma = \sqrt{\sum_{l,m}\left(F_{lm}^2 \delta C_{lm}^2 + G_{lm}^2 \delta S_{lm}^2\right)} \qquad [24]$$

The errors in different Stokes coefficients are unlikely to be uncorrelated. For GRACE, those correlations are responsible for the stripes evident in the top panel of **Figure 6**. Knowledge of the full error covariance matrix can improve the estimate of $\delta\sigma$. But even without the full covariance, [24] provides a reasonable first approximation for $\delta\sigma$, if δC_{lm} and δS_{lm} can be estimated.

To understand leakage errors, consider an application where the goal is to use the Stokes coefficients to assess a regional water-storage model. For example, suppose a surface mass average of the form [22] is constructed and interpreted as an estimate of water-storage variability in some chosen river basin. Leakage errors are the contributions to [22] caused by gravity signals from outside the basin.

These leakage errors can come from time variable mass anomalies either vertically above or below the river basin, or from mass anomalies off to the side of the basin. Signals above or below would come from the overlying atmosphere or the underlying solid Earth, and cannot be separated from the river basin signal, no matter how complete and accurate the gravity field estimation. This is a consequence of the nonuniqueness of gravity-based inversions for density, as described above. The only recourse is to independently model and remove the atmospheric and solid Earth signals. Any inaccuracy in those models is thus a source of errors for the hydrology estimates (Velicogna *et al.*, 2001).

Leakage from mass anomalies off to the side, in neighboring river basins for example, can be minimized using a weighting function that is as localized as possible to the river basin of interest. As described above, though, an averaging function should usually be smoother than the basin function to provide an accurate estimate.

For some applications, this horizontal leakage is not an issue. For example, suppose the objective is to compare the satellite estimates of σ for the Mississippi River Basin, with the output of a hydrology model. The leakage into the satellite estimate will come mainly from the river basins that border the Mississippi Basin. If the same averaging function is applied to the model output, then both will be subject to the same leakage. The model–satellite comparison will then actually be a comparison over a somewhat broader region than just the Mississippi Basin, but they will both be affected by leakage in the same way.

But for many applications the goal is to estimate mass variability within a specific region with no contamination from regions outside. In that case, leakage is an inescapable source of error. The only way to estimate the likely impact of that error is to apply the averaging function to simulated data. This sort of problem commonly arises in time-series analysis. Our averaging process is basically a low pass filter. A high pass filter not only removes high frequencies, but also reduces the low-frequency signal; that is, each filter has a characteristic gain function. The effects of the gain function must be determined and removed from the filtered data, in order to estimate the true low-frequency signal in the time domain.

The examples shown in **Figure 7**, that is, the Mississippi Basin and Antarctica, illustrate two types of situations. For the Mississippi, the averaging function will downweight the true Mississippi signal, since the averaging function is smaller than 1 over the entire basin. In effect, the averaging function replaces some of the signal located inside the Mississippi Basin, with signals located outside in neighboring basins. The amount of leakage thus depends on whether the external hydrology signal does or does not look like the internal signal. It basically depends on a comparison between the correlation length of the hydrology signal (which tends to be controlled by the scale length of the precipitation) and the resolution of the averaging kernel (which is usually chosen based on the resolution of the gravity field). For a reasonably homogeneous region like this portion of the interior United States, the signal just outside the basin is similar enough to the signal just inside, that the leakage from the averaging kernel shown in **Figure 7** is not severe. Nonetheless, the leakage is nonzero, and should be estimated using hydrology model output.

For Antarctica, the extension of the averaging function over the ocean means that some Antarctic signal is being replaced by ocean signal. There is likely to be no correlation at all between the Antarctic and ocean signals. For example, suppose the object is to determine the linear trend in Antarctic mass over some multiyear period. It is probable that there would be little or no multiyear trend over the ocean. So the averaging process under-represents the contribution from the trend in Antarctic mass, and replaces it with a negligible trend from the ocean. This can lead to serious underestimates of the Antarctic mass trend. The situation is similar for any region where the signal of interest is much larger than the signal in surrounding areas. Again, the only way to assess and correct for this effect is to apply the averaging function to simulated data for the Antarctic ice sheet and the surrounding ocean. Velicogna and Wahr (2006a), for example, found that the Antarctic averaging kernel shown in **Figure 7** underestimates the true Antarctic signal by about 35–40%. This correction, which Velicogna and Wahr refer to as scaling, is equivalent to correcting for the gain of the spatial filter represented by W_{lm}^c and W_{lm}^s in [22].

In principle, the sum in [22] should include all l in the range $0 \leq l \leq \infty$. In practice, the sum for GRACE is limited to $2 \leq l \leq l_{max}$, where l_{max} can be no larger than the maximum degree of the GRACE fields. The truncation to $l \leq l_{max}$ causes ringing: sensitivity to mass variability well outside the region of interest; though this sensitivity is weak if l_{max} is large. The restriction to $l \geq 2$ arises because GRACE does not recover $l = 0$, 1 coefficients. The $l = 0$ coefficient is proportional to the Earth's total mass. Since that mass remains constant, $\Delta C_{00} = 0$ is a reasonable assumption. But the omission of $l = 1$ terms in [22] has the potential of degrading estimates of $\overline{\Delta\sigma}_{region}$. These terms are proportional to the displacement of the geocenter (the offset between the Earth's center of mass and the center of figure of the surface), and are particularly affected by the seasonal transfer of water between the continents and the ocean. Their omission from [22] means, in effect, that the averaging function has a small-amplitude tail that extends around the globe, causing distant signals to leak into $\overline{\Delta\sigma}_{region}$. This leakage can be estimated either by using independent estimates of geocenter motion from other techniques (i.e., SLR or GPS), or by using hydrological and oceanographic models.

8.3 Applications

Time-variable satellite gravity measurements can be used to address a wide variety of problems, from across a broad spectrum of the Earth sciences. Any geophysical process that causes a significant redistribution of mass over scales of hundreds of kilometers is a possible target.

8.3.1 Hydrology

The largest-amplitude and most-varied time-dependent signals are related to water-storage variability on land. **Figure 2(b)**, for example, shows that the annually varying signals on land are much larger than those in the ocean. When water is placed on land, a sizable fraction often stays there for some time, either infiltrating into the soil or remaining on the surface as water or snow. But when a parcel of water is placed on the ocean, its natural tendency is to flow away. Note that the largest features evident in **Figure 2(b)** are easily recognized: for example, heavy-rainfall regions near the equator, the strong monsoon in Southeast Asia, the seasonal snow cycle in Eurasia and northern North America. A higher-resolution (300 km Gaussian smoothing) example is shown in **Figure 8**. Features clearly evident include the rain forest in Central America, the heavy mountain

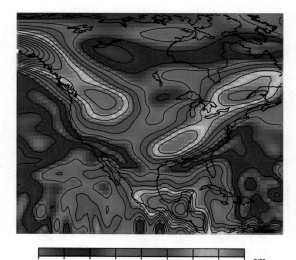

Figure 8 The amplitude of the annually varying mass signal, in units of centimeters of water thickness, recovered from GRACE during spring 2002 through spring 2006. The results have been postprocessed to reduce noise (Swenson and Wahr, 2006a), and smoothed with a 300-km Gaussian. The solid lines represent contour intervals of 1 cm.

0 1 2 3 4 5 6 7 15 cm

snows that stretch from southern Alaska down through the Central Rocky Mountains, the desert region of the Southwest United States, the region of high precipitation running from the lower Mississippi Basin up through Kentucky, and the high precipitation region along the upper Saint Lawrence River.

Time-variable gravity measurements are sensitive to the total water storage integrated through the entire water column (see [7]). This includes water and snow on the surface, and water in both the soil and subsoil layers. The measurements cannot distinguish between these stores, but can recover only the sum. This hydrological product is unique, both in its sensitivity to subsoil water storage and in its ability to recover results at large spatial scales. Other types of satellite-based instruments, either already on orbit or still in the planning stage, can detect water stored within the upper few centimeters of the soil, or can monitor surface water. But time-variable gravity missions provide the only available means of monitoring deeper water storage from space. Ground-based observations from such things as soil moisture probes and the monitoring of well levels, can provide information on subsurface storage at individual points. But probably no region in the world has a dense enough observational network to provide total water storage at scales of a few hundred km with the accuracy of GRACE.

8.3.1.1 Comparing with land surface models

Water-storage estimates obtained from time-variable gravity are of potential value both as stand-alone quantities and when used in combination with other data types. As an end product they can be compared with the total water storage predicted by land surface models, to help assess and improve those models (Ramillien *et al.*, 2005; Andersen and Hinderer, 2005; Andersen *et al.*, 2005; Niu and Yang, 2006; Nakaegawa, 2006; Swenson and Milly, 2006; Neumeyer *et al.*, 2006; Seo *et al.*, 2006; Schmidt *et al.*, 2006b; Hinderer *et al.*, 2006; Frappart *et al.*, 2006); and with soil moisture, ground water, and/or snow mass measurements to help validate and understand those measurements (Swenson *et al.*, 2006; Frappart *et al.*, 2006; Yeh *et al.*, 2006). For example, **Figure 9** (Sean Swenson, personal communication) shows comparisons between GRACE water-storage estimates and those predicted by the GLDAS/Noah water-storage model (Rodell *et al.*, 2004a), for three river basins. The GRACE error bars are defined so that if the disagreement for any month is larger than the error bars, we can be 68.3% confident that it is the model

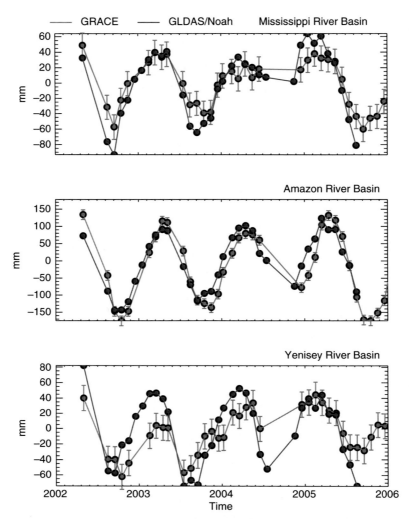

Figure 9 Water-storage results for three river basins, obtained using specially constructed averaging kernels for those basins. GRACE results. with their 68.3% confidence limits, are shown in red (postprocessed to reduce noise, as described by Swenson and Wahr (2006a)). Results from the GLDAS/Noah land surface model (Rodell, *et al.*, 2004a) are in blue. Figure provided by Sean Swenson.

that is in error (Wahr *et al.*, 2006). The agreement is excellent for the Mississippi, which is reassuring given the high density of observations used to improve the atmospheric forcing fields in that region. For the Amazon the phase of the model tends to slightly lead the phase of GRACE, and for the Yenisey (in northern Siberia) the phase disagreement is more pronounced, with the model losing mass perhaps a couple months too early in the early springs of 2003 and 2004. Comparisons like these can provide an indication of where model improvements are necessary. Eventually, gravity-based water-storage estimates could even be assimilated directly into the hydrology models.

8.3.1.2 Anthropogenic effects and sea-level contributions

Another application of these water-storage estimates as a stand-alone product is the general issue of hydrological contributions to sea-level change: what regions are important contributors, and at what timescales? The results shown in **Figure 9**, for example, can be loosely interpreted as the contribution to global sea-level change from those river basins (after scaling by the ratio of the land area to the area of the ocean, and reversing the sign). Though the connection is not that simple, of course, since the water that leaves a river basin does not necessarily go directly into the ocean.

The variability evident in **Figure 9** is mostly seasonal. Of more relevance to the issue of rising sea level would be regions that display linear trends. Trends can be an indication of anthropogenic influence. Groundwater is particularly susceptible to anthropogenic changes, both negative and positive; for example, aquifer pumping to obtain water for agricultural and urban use, and groundwater infiltration from irrigation. Because few large-scale land surface models include groundwater storage, and fewer still include anthropogenic effects, contributions such as these cannot be extracted from models. Time-variable satellite gravity measurements offer a means of monitoring this variability (Boy and Chao, 2002; Rodell and Famiglietti, 2002).

8.3.1.3 Precipitation (P) minus evapotranspiration (ET)

P and ET have an important impact on climate, because their difference largely determines the exchange of mass and latent heat between the atmosphere and underlying Earth. Estimates of $P - ET$ can be obtained from atmospheric models using moisture flux convergence parameters (e.g., Trenberth, 1997). Alternatively, for a land surface (hydrology) model P and ET are typically computed using a water and energy balance approach (Roads et al., 2003). These models are the best available tools for making long-range predictions of both natural and anthropogenic climate variability. However, because of the difficulty of obtaining relevant measurements using traditional methods, it has proved to be difficult to assess these model components, particularly at the synoptic scales that characterize the most energetic atmospheric disturbances. At seasonal and longer time periods, it is often assumed that storage changes are negligible, and that therefore $P - ET$ should balance the discharge. A model's ability to achieve this balance is sometimes used to assess the accuracy of its $P - ET$ estimates (Gutowski et al., 1997; Roads, 2002). But water-storage changes certainly do exist (see **Figures 2(b)**, **8**, and **9**, for example), and at seasonal periods are typically of the same order as the discharge.

Time variable gravity offers a new opportunity for determining $P - ET$ (see Rodell et al. (2004b) and Swenson and Wahr (2006b)). The water budget equation is

$$dS/dt = P - ET - R \qquad [25]$$

where S is total water storage and R is discharge. Time-variable gravity measurements can be used to estimate S in a river basin. If the river that drains that basin is gauged, then the discharge can be measured and so $P - ET$ can be determined. As an example, the bottom panel of **Figure 10** (provided by Sean Swenson) compares $P - ET$ estimates for the Ob River in Siberia from GRACE and river discharge, with atmospheric model estimates from ECMWF and NCEP. Clearly, these models do a good job at reproducing $P - ET$ in this basin.

As a variation of this application, suppose atmospheric models are believed to accurately predict $P - ET$ within some river basin. Time-variable gravity estimates of S can then be used in [25], along with the $P - ET$ results, to estimate the river discharge (Syed et al., 2005). This offers a means of determining discharge for rivers that are not adequately gauged.

8.3.2 Cryosphere

One of the most important likely consequences of rising global temperatures is increased global sea levels caused by accelerated mass loss of the Antarctic and Greenland ice sheets. There is enough frozen water in those ice sheets to raise the world's oceans by 70 m if they melted completely. Even a relatively small change in ice mass could thus have a significant impact on sea level. There have been recent, significant improvements in ice-sheet monitoring, using a variety of techniques, including radar- and laser-altimeter measurements of changes in ice-sheet elevations, radar-based measurements of the velocities and thinning rates of outlet glaciers, and ground-based mass balance studies that compare accumulation with discharge and melting (e.g., Church et al., 2001; Rignot and Thomas, 2002; Davis et al., 2005; Zwally et al., 2005; Rignot and Kanagaratnam, 2006). The conclusions of different studies are not always in good agreement. Improved monitoring of ice-sheet variability would help in understanding the present mass imbalance of the ice sheets, and could significantly improve predictions of future change.

Time-variable gravity provides a method of monitoring changes in ice-sheet mass that is not only independent of other methods, but that is arguably the most promising method for estimating the mass imbalance of an entire ice sheet. There have already been several GRACE estimates for Antarctica and Greenland (Velicogna and Wahr, 2005, 2006a, 2006b; Chen et al., 2006b, 2006c; Luthcke et al., 2006). Satellite gravity has two distinct advantages over other

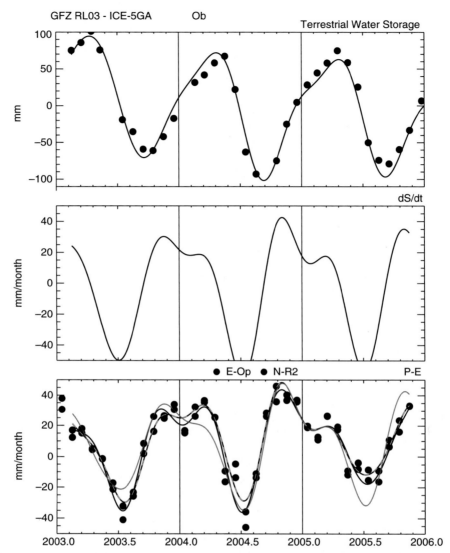

Figure 10   Top: Red dots are GRACE water-storage estimates (*S*) for the Ob River; the black line is a smoothed version. Middle: the time derivative (d*S*/d*t*) of the smoothed water storage. Bottom: Red line is the sum of d*S*/d*t* and the measured Ob discharge (*R*), and so is an estimate of *P – ET*. Green and orange dots are estimates of *P – ET*, using moisture convergence parameters from the ECMWF (green) and NCEP (orange) models. Solid black and hatched black lines are smoothed version of ECMWF and NCEP, respectively. Figure provided by Sean Swenson.

techniques. First, gravity measurements provide a direct estimate of mass, which is obviously the most relevant quantity for understanding mass imbalance. Other methods do not determine mass loss directly, but rely on independent assumptions to relate measured quantities to mass. Second, gravity signals at the altitude of a satellite are determined by mass variations averaged over a broad region of the underlying surface, not just at the point directly beneath the satellite. Thus,

satellite gravity inherently averages over large regions. Other methods tend to sample an ice sheet at relatively small, often nonoverlapping footprints, so that their estimates of total mass imbalance are subject to interpolation and extrapolation errors.

Time-variable gravity has its weaknesses, of course, For one thing, it cannot provide small-scale resolution, and so has trouble isolating the exact location of a mass anomaly. For another, time-variable gravity estimates are particularly sensitive to PGR errors. Both Antarctica and Greenland experienced significant melting at the end of the last ice age, and the underlying Earth is still rebounding. This rebound affects altimeter estimates of ice-sheet thickness change: if the crust rises (or falls), the ice sheet's surface will rise (fall) along with it, and so the altimeter data will imply the ice sheet is getting thicker (thinner). It affects satellite gravity estimates because it produces a gravity signal that is inseparable from the gravity signal caused by the ongoing ice change. Because rock in the upper mantle is 3–4 times as dense as ice, PGR's relative impact on gravity is 3–4 times as large as its impact on altimeter estimates. If the Earth's surface uplifts by 1 cm, the altimeter sees the ice-sheet surface rise by 1 cm. But a gravity measurement sees a gravity signal that is the equivalent of the signal from 3–4 cm of ice. Thus, although PGR models are generally used to remove the PGR signals from both altimeter and gravity measurements, any residual errors in those models cause more problems for gravity than for altimetry. Ultimately, the best approach will be to combine time-variable gravity and altimeter estimates, as well as GPS observations of vertical crustal motion where available, to reduce the PGR errors in both techniques (Wahr *et al.*, 2000; Velicogna and Wahr, 2002).

8.3.3 Solid Earth

Although the Earth's mean gravity field is caused almost entirely by mass within the solid Earth, changes in the distribution of that mass generally occur too slowly, or produce gravity signals that are too small or too localized, to be practical targets of time variable satellite gravity studies. The most notable exception is PGR. The PGR signal over Canada is already clearly visible in the 4 years of GRACE data presently available, and has proved to be useful in helping to constrain the Earth's viscosity profile (Tamisiea *et al.*, 2007; Paulson *et al.*, 2007). **Figure 11(a)**, for example, shows the best-fitting linear trend in surface mass, smoothed with a 400-km Gaussian. **Figure 11(b)** shows the expected PGR signal over that same region, computed using the ICE-5G ice deglaciation model and VM2 viscosity profile (Peltier 2004). There is clearly excellent agreement with the GRACE observations over Hudson Bay, as evidenced in **Figure 11(c)** which shows that after removing the ICE-5G results from GRACE, the GRACE Hudson Bay anomaly almost completely disappears. The remaining negative anomaly over southern Alaska has been interpreted as the effects of shrinking glaciers (Tamisiea *et al.*, 2005; Chen *et al.*, 2006d).

PGR signals in Scandinavia, Antarctica, and Greenland are, as expected, proving harder to recover using GRACE, due to the problems of separating those signals from other sources of gravity trends: present-day ice mass variability within Antarctica and Greenland, and long-period hydrological and oceanographic signals in Scandinavia and northern Europe. A longer data span will improve the recovery in both Scandinavia and Canada, by averaging out more of the competing hydrological and

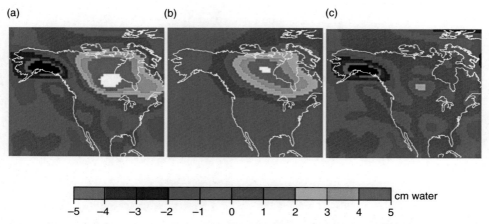

Figure 11 (a) The best-fitting linear trend for the GRACE fields between spring 2002 and spring 2004, after postprocessing the fields as described by Swenson and Wahr (2006a). Results have been smoothed with a 400-km Gaussian smoothing function. Units are in cm yr^{-1} of water thickness. (b) Predictions based on the ICE-5G PGR ice deglaciation model (Peltier, 2004). (c) The difference between (a) and (b).

oceanographic signals in those regions. For Antarctica and Greenland, the PGR signals can conceivably be recovered by combining time-variable gravity with ice-sheet altimetry and GPS observations, as mentioned above (Wahr *et al.*, 2000; Velicogna and Wahr, 2002).

Other solid Earth applications are possible, though most likely in the form of isolated events. A good example is the 2004 Sumatran Earthquake, an event that was (of course) unexpected, but with an associated signal that is clearly evident in GRACE data (see, e.g., Han *et al.* (2006b)). This was an unusually energetic earthquake. Nevertheless, its presence in the GRACE data raises the possibility of using time-variable satellite gravity to look not only at co-seismic events, but also to search for postseismic signals. The recovery of such signals would depend not only on the accuracy of the measurements, but also on how well the contamination from hydrological and oceanographic gravity signals can be reduced.

There is, in addition, an indirect way in which time-variable satellite gravity measurements can contribute to solid Earth studies. Global Positioning System (GPS) observations are widely used to monitor tectonic displacements of the Earth's surface. But the Earth's surface can deform in response to surface loading, as well. Load deformation is a source of noise for tectonic applications. It can be especially troublesome for campaign-style GPS observing programs, in which a site might be occupied for a few days, and then not re-occupied for perhaps several years. In that situation, seasonal and other short-period loading can alias into apparent long-term variability.

Time-variable satellite gravity observations can be used to model and remove the large-scale component of the load deformation, and so to reduce this source of noise. The surface mass variability recovered from those observations can be convolved with solid Earth Green's functions to estimate the loading. Preliminary studies using GRACE data are described by Davis *et al.* (2004), King *et al.* (2006), and van Dam *et al.* (2007).

8.3.4 Oceanography

The time-variable mass signal in the ocean is small compared to that from land, as can be seen for the annual cycle from **Figure 2(b)**. Bottom pressure variability is not dominated by an annual signal to the same extent as land water storage. But even when all temporal variations are included, ocean mass

fluctuations are still relatively small. **Figure 12**, for example, shows that the rms surface mass variability over the oceans, smoothed with a 750-km Gaussian, is typically only 2–3 cm or less. Presumably, the large red regions along coastlines mostly reflect the effects of the much larger land water signals leaking into the ocean estimates. This illustrates the danger of using time-variable gravity to study the ocean near the coast. This leakage can be reduced by decreasing the smoothing radius. But a smaller radius leads to more inaccurate estimates, which makes it harder to recover the relatively small ocean signal.

Still, ocean mass signals are clearly evident in GRACE data. For example, one of the direct oceanographic applications of these mass estimates is to combine them with sea-surface height measurements from altimetry, to separately estimate steric and nonsteric contributions to sea-surface variability. A satellite radar altimeter monitors sea-surface heights along its ground track. Suppose the altimeter detects a sea-surface rise in some region. The altimeter data cannot determine whether the rise was due to increased water mass in the region, or whether it was due to the water becoming warmer (and/or less salty) and expanding. Change in volume (i.e., 'steric' changes) does not cause a change in gravity. Thus, satellite gravity measurements detect only the nonsteric contributions. The steric contributions are then the difference between the altimeter and time-variable gravity results.

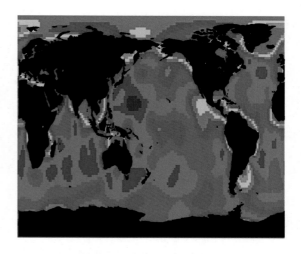

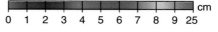

Figure 12 The rms about the mean of ocean surface mass variability deduced from GRACE for spring 2002 through spring 2006, smoothed with a 750-km Gaussian.

Figure 13 (results provided by Don Chambers, personal communication) shows how well this technique works on a global and seasonal scale (see also Chambers *et al.* (2004), Chambers (2006a, 2006b), and Garcia *et al.* (2006)). The red curve is the total ocean mass deduced from GRACE. The blue curve shows an altimetric estimate of sea-surface height, corrected for steric effects using temperature and salinity profiles collected from in situ data. The altimetric and steric signals are the long-term seasonal averages of data extending well back before the launch of GRACE. Thus, the blue curve does not show actual results for 2002.5–2004.5, but shows only the best-fitting seasonal cycle, as fit to a decade or more of prior data. Even so, the agreement is excellent.

If the time-variable steric signal can be estimated for some region by differencing altimeter and mass results, it is possible to recover changes in the heat content of that region, as

$$\Delta H = \frac{\rho c_p}{\alpha}\left(\Delta\eta - \frac{1}{\rho_0}\Delta\sigma\right) \qquad [26]$$

where ΔH is the change in ocean heat storage, ρ is the ocean density, c_p and α are the heat capacity and thermal expansion coefficient of sea water, $\Delta\sigma$ is the change in mass estimated from time-variable gravity, and $\Delta\eta$ is the change in sea surface height measured by the altimeter (Jayne *et al.*, 2003). This result, which extends the methodology of Chambers *et al.* (1997) to include mass variability, assumes that α is independent of depth, and that the effects of salinity variations are either negligible or can be independently modeled and removed.

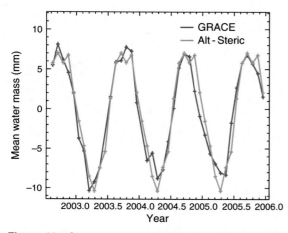

Figure 13 Compares seasonal estimates of total ocean mass from GRACE, with estimates deduced from a combination of satellite altimetry and *in situ* temperature and salinity measurements. Results provided by Don Chambers.

The exchange of heat between the ocean and atmosphere is one of the most significant examples of energy transfer within the Earth's climate system. Because of the large heat capacity of water, the ocean can store enormous amounts of energy. Therefore, it can act not only as a moderator of climate extremes, but also as an energy source for severe storms. Knowledge of the ocean's time-varying heat storage is of considerable importance for things such as climate change prediction, long-range weather forecasting, and hurricane strength prediction. Despite its great importance in climate, the ocean's time-varying heat content is greatly under-sampled because of the sparse coverage of in-situ observations. Therefore, accurate satellite mapping of the ocean's time-varying heat storage would be attractive for its global and repeating coverage.

Time-variable mass estimates can be used for other types of oceanographic applications, as well. Surface mass anomalies, $\Delta\sigma$, are proportional to variations in ocean bottom pressure

$$\Delta P_{\text{bott}}(\theta,\ \phi) = g\Delta\sigma(\theta,\ \phi) \qquad [27]$$

where $\Delta\sigma$ [7] is integrated from the bottom of the ocean to the top of the atmosphere. Thus, time-variable gravity over the ocean provides estimates of seafloor pressure variability at the spatial and temporal resolution of the gravity measurements. GRACE, for example, can provide monthly seafloor pressure maps at scales of several hundred kilometers and greater. These can be used to assess and improve oceanographic models (Condi and Wunsch, 2004; Bingham and Hughes, 2006; Zlotnicki *et al.*, 2007), and to compare with measurements from bottom pressure recorders (Kanzow *et al.*, 2005; Morison *et al.*, 2007) to separate the effects of regional and local signals.

The bottom pressure estimates can also be combined with the geostrophic assumption (which assumes a balance between pressure and Coriolis forces) to determine changes in deep ocean velocities at the temporal and spatial resolutions of the gravity field observation. For GRACE, this means that the results are averaged over scales of several hundred kilometers or more. This large spatial scale can make it difficult to apply the geostrophic assumption at the seafloor in the presence of short-scale topography. However, simulations (Wahr *et al.*, 2002) have shown that pressure variability at the seafloor is about the same as at 2 km depth, or at even shallower depths in many cases. Thus, the inferred currents can be interpreted to a high degree of accuracy in terms of the variability of currents at 2 km depth.

8.4 Summary

Although SLR has been providing time-variable gravity measurements for several decades, it is the much higher spatial resolution now available from GRACE that permits the kinds of applications described in this chapter. The figures shown here are computed using the GRACE gravity fields available at the time of this writing. The fields will continue to improve as processing methods mature and background geophysical models get better. Any such future improvements will be retroactively applied to all the fields, through reprocessing of the entire data set. In addition, as the GRACE time series lengthens, it will become easier to separate different geophysical signals. Only with a long time series, for example, will it be possible to clearly distinguish between multiyear variability and true secular signals.

GRACE, of course, has a finite lifetime; it was designed for a 5-year mission but may last on the order of a decade. Plans for a next-generation mission are presently being formulated and assessed (Watkins et al., 2000). The use of laser tracking for better monitoring the intersatellite distance, and the introduction of a drag-free propulsion system to reduce atmospheric drag at lower altitudes, could lead to order-of-magnitude improvements in measurement accuracy. This would increase the spatial resolution even further, down to perhaps $\sim$100 km, and would enable a whole new class of applications.

References

Andersen OB and Hinderer J (2005) Global inter-annual gravity changes from GRACE: Early results. *Geophysical Research Letters* 32(1): L01402.

Andersen OB, Seneviratne SI, Hinderer J, *et al.* (2005) GRACE-derived terrestrial water storage depletion associated with the 2003 European heat wave. *Geophysical Research Letters* 32(18): L18405.

Benjamin D, Wahr J, Ray RD, Egbert GD, and Desai SD (2006) Constraints on mantle anelasticity from geodetic observations, and implications for the J_2 anomaly. *Geophysical Journal International* 165: 3–16.

Bingham RJ and Hughes CW (2006) Observing seasonal bottom pressure variability in the North Pacific with GRACE. *Geophysical Research Letters* 33(8): L08607.

Boy JP and Chao BF (2002) Time-variable gravity signal during the water impoundment of China's Three-Gorges Reservoir. *Geophysical Research Letters* 29(24): 2200.

Chambers DP (2006a) Evaluation of new GRACE time-variable gravity data over the ocean. *Geophysical Research Letters* 33(17): L17603.

Chambers DP (2006b) Observing seasonal steric sea level variations with GRACE and satellite altimetry. *Journal of Geophysical Research* 111: 3010 (doi: 10.1029/2005JC002914).

Chambers DP, Tapley BD, and Stewart RH (1997) Long-period ocean heat storage rates and basin-scale heat fluxes from TOPEX. *Journal of Geophysical Research - Oceans* 102: 10525–10533.

Chambers DP, Wahr J, and Nerem RS (2004) Preliminary observations of global ocean mass variations with GRACE. *Geophysical Research Letters* 31: L13310 (doi:10.1029/2004GL020461).

Chao BF (1994) The geoid and Earth rotation. In: Vanicek P and Christou N (eds.) *Geoid and Its Geophysical Interpretations*, pp. 285–298. Boca Raton, FL: CRC Press.

Chao BF and Au AY (1991) Temporal variation of the Earth's low-degree zonal gravitational field caused by atmospheric mass distribution, 1980–1988. *Journal of Geophysical Research* 96(B4): 6569–6575.

Chao BF and Gross RS (1987) Changes in the Earth's rotation and low-degree gravitational field induced by earthquakes. *Geophysical Journal of the Royal Astronomical Society* 91: 569–596.

Chen JL, Tapley BD, and Wilson CR (2006d) Alaskan mountain glacial melting observed by satellite gravimetry. *Earth and Planetary Science Letters* 248(1-2): 368–378.

Chen JL, Wilson CR, Blankenship DD, *et al.* (2006b) Antarctic mass rates from GRACE. *Geophysical Research Letters* 33(11): L11502.

Chen JL, Wilson CR, and Seo KW (2006a) Optimized smoothing of gravity recovery and climate experiment (GRACE) time-variable gravity observations. *Journal of Geophysical Research* 111: B06408.

Chen JL, Wilson CR, and Tapley BD (2006c) Satellite gravity measurements confirm accelerated melting of Greenland ice sheet. *Science* 313: 1958–1960.

Cheng M, Eanes R, Shum C, Schutz B, and Tapley B (1989) Temporal variations in low degree zonal harmonics from Starlette orbit analysis. *Geophysical Research Letters* 16: 393–396.

Cheng M and Tapley BD (2004) Variations in the Earth's oblateness during the past 28 years. *Journal of Geophysical Research* 109: B09402.

Cheng MK, Shum CK, and Tapley BD (1997) Determination of long-term changes in the Earth's gravity field from satellite laser ranging observations. *Journal of Geophysical Research* 102(B10): 22377–22390.

Cheng MK and Tapley BD (1999) Seasonal variations in low degree zonal harmonics of the Earth's gravity field from satellite laser ranging observations. *Journal of Geophysical Research* 104(B2): 2667–2681.

Church JA, Gregory JM, Huybrechts P, *et al.* (2001) Changes in sea level. In: Houghton JT, Ding Y, Griggs DJ, *et al.* (eds.) *The Intergovernmental Panel on Climate Change, IPCC Third Assessment Report, Climate Change 2001: The Scientific Basis*, pp. 639–694. Cambridge, Ny: Cambridge University Press.

Condi F and Wunsch C (2004) Measuring gravity field variability, the geoid, ocean bottom pressure fluctuations, and their dynamical implications. *Journal of Geophysical Research* 109(C2): C02013.

Cox CM and Chao BF (2002) Detection of a large-scale mass redistribution in the terrestrial system since 1998. *Science* 297: 831–832.

Davis JL, Elosequi P, Mitrovica JX, *et al.* (2004) Climate-driven deformation of the solid Earth from GRACE and GPS. *Geophysical Research Letters* 31(24): L24605.

Davis CH, Li Y, McConnell JR, Frey MM, and Hanna E (2005) Snowfall-driven growth in East Antarctic ice sheet mitigates recent sea-level rise. *Science* 308: 5730.

Dickey JO, Bentley CR, Bilham R, *et al.* (1997) *Satellite gravity and the geosphere, National Research Council Report*, 112 pp. Washington, DC: National Academy Press.

Dickey JO, Marcus SL, de Viron O, and Fukumori I (2002) Recent Earth oblateness variations: Unraveling climate and postglacial rebound effects. *Science* 298: 1975–1977.

Dong D, Gross RS, and Dickey JO (1996) Seasonal variations of the Earth's gravitational field: An analysis of atmospheric and oceanic tidal excitation. *Geophysical Research Letters* 23: 725–728.

Dziewonski A and Anderson DL (1981) Preliminary reference Earth model. *Physics of the Earth and Planetary Interiors* 25: 297–356.

Farrell WE (1972) Deformation of the Earth by surface loading. *Reviews of Geophysics* 10: 761–797.

Frappart F, Ramillien G, Biancamaria S, *et al.* (2006) Evolution of high-latitude snow mass derived from the GRACE gravimetry mission (2002–2004). *Geophysical Research Letters* 33(2): L02501.

Garcia D, Chao BF, Del Rio J, *et al.* (2006) On the steric and mass-induced contributions to the annual sea level variations in the Mediterranean Sea. *Journal of Geophysical Research* 111(C9): C09030.

Gutowski WJ, Chen Y, and Otles Z (1997) Atmospheric water vapor transport in NCEP-NCAR reanalyses: Comparison with river discharge in the central United States. *Bulletin of the American Meteorological Society* 78: 1957–1969.

Han D and Wahr J (1995) The viscoelastic relaxation of a realistically stratified Earth, and a further analysis of postglacial rebound. *Geophysical Journal International* 120: 287–311.

Han SC, Jekeli C, and Shum CK (2004) Time-variable aliasing effects of ocean tides, atmosphere, and continental water mass on monthly mean GRACE gravity field. *Journal of Geophysical Research* 109(B4): B04403.

Han SC, Shum CK, Bevis M, and Kuo CY (2006b) Crustal dilatation observed by GRACE after the 2004 Sumatra–Andaman earthquake. *Science* 313: 658–662.

Han SC, Shum CK, Jekeli C, and Alsdorf D (2005a) Improved estimation of terrestrial water storage changes from GRACE. *Geophysical Research Letters* 32(7): L11502, L07302.

Han SC, Shum CK, Jekeli C, *et al.* (2005b) Non-isotropic filtering of GRACE temporal gravity for geophysical signal enhancement. *Geophysical Journal International* 163: 18–25.

Han SC, Shum CK, and Jekeli C (2006a) Precise estimation of *in situ* geopotential differences from GRACE low-low satellite-to-satellite tracking and accelerometer data. *Journal of Geophysical Research* 111: B04411.

Han SC, Shum CK, and Matsumoto K (2005c) GRACE observations of M-2 and S-2 ocean tides underneath the Filchner–Ronne and Larsen ice shelves, Antarctica. *Geophysical Research Letters* 32(20): L20311.

Hinderer J, Andersen O, Lemoine F, *et al.* (2006) Seasonal changes in the European gravity field from GRACE: A comparison with superconducting gravimeters and hydrology model predictions. *Journal of Geodynamics* 41(1-3): 59–68.

Jayne SR, Wahr JM, and Bryan FO (2003) Observing ocean heat content using satellite gravity and altimetry. *Journal of Geophysical Research-Oceans* 108: 3031 (doi:10.1029/2002JC001619).

Jekeli C (1981) *Alternative methods to smooth the Earth's gravity field. Report 327,* Department of Geodetic Sciences and Surveying. : University, Columbus, OH.

Jekeli C (1999) The determination of gravitational potential differences from satellite-to-satellite tracking. *Celestial Mechanics and Dynamical Astronomy* 75: 85–100.

Kanzow T, Flechtner F, Chave A, *et al.* (2005) Seasonal variation of ocean bottom pressure derived from gravity

recovery and climate experiment (GRACE): Local validation and global patterns. *Journal of Geophysical Research* 110(C9): C09001.

King M, Moore P, Clarke P, *et al.* (2006) Choice of optimal averaging radii for temporal GRACE gravity solutions, a comparison with GPS and satellite altimetry. *Geophysical Journal International* 166(1): 1–11.

Knudsen P and Andersen O (2002) Correcting GRACE gravity fields for ocean tide effects. *Geophysical Research Letters* 29(8): 1178.

Lemoine FG, Kenyon SC, Factor JK, *et al.* (1998) The development of the joint NASA GSFC and NIMA geopotential model EGM96 Goddard Space Flight Center. NASA/TP-1998-20681.

Luthcke SB, Zwally HJ, Abdalati W, *et al.* (2006) Recent Greenland ice mass loss by drainage system from satellite gravity observations. *Science* 24, doi:10.1126/Science. 1130776.

Moore P, Zhang Q, and Alothman A (2005) Annual and semiannual variations of the Earth's gravitational field from satellite laser ranging and CHAMP. *Journal of Geophysical Research* 110(B6): B06401.

Morison J, Wahr J, Kwok R, and Peralta-Ferriz C (2007) Recent trends in Arctic Ocean mass distribution revealed by GRACE. *Geophysical Research Letters* 34, L07602, doi: 10.1029/2006GL029016.

Nakaegawa T (2006) Detectability assessment of interannual variations in terrestrial water storage from satellite gravimetry using an offline land surface model simulation. *Hydrological Processes* 20(6): 1347–1364.

Nerem RS, Eanes RJ, Thompson PF, and Chen JL (2000) Observations of annual variations of the Earth's gravitational field using satellite laser ranging and geophysical models. *Geophysical Research Letters* 27(12): 1783–1786.

Neumeyer J, Barthelmes F, Dierks O, *et al.* (2006) Combination of temporal gravity variations resulting from superconducting gravimeter (SG) recordings, GRACE satellite observations and global hydrology models. *Journal of Geodesy* 79(10-11): 573–585.

Niu GY and Yang ZL (2006) Assessing a land surface model's improvements with GRACE estimates. *Geophysical Research Letters* 33(7): L07401.

Paulson A, Zhong S and Wahr J (2007) Inference of mantle viscosity from GRACE and relative sea level data. *Geophysical Journal International* (in press).

Peltier WR (2004) global glacial isostasy and the surface of the Ice-Age Earth: The ICE-5G(VM2) model and GRACE. *Annual Review of Earth and Planetary Sciences* 32: 111–149.

Ramillien G, Cazenave A, and Brunau O (2004) Global time variations of hydrological signals from GRACE satellite gravimetry. *Geophysical Journal International* 158(3): 813–826.

Ramillien G, Frappart F, Cazenave A, *et al.* (2005) Time variations of land water storage from an inversion of 2 years of GRACE geoids. *Earth and Planetary Science Letters* 235(1–2): 283–301.

Ray R and Luthcke S (2006) Tide model errors and GRACE gravimetry. *Geophysical Journal International* 167: 1055–1059.

Reigber C, Luhr H, and Schwintzer P (2002) CHAMP mission status. *Advances in Space Research* 30(2): 129–134.

Rignot E and Thomas R (2002) Mass balance of polar ice sheets. *Science* 297: 1502.

Rignot E and Kanagaratnam P (2006) Changes in the velocity structure of the Greenland ice sheet. *Science* 311: 5763.

Roads J (2002) Closing the water budget. *GEWEX NEWS* 12(1): 1–6.

Roads J, Lawford R, Bainto E, *et al.* (2003) GCIP water and energy budget synthesis (WEBS). *Journal of Geophysical Research* 108(16): 8609 (doi:10.1029/2002JD002583).

Rodell M and Famiglietti JS (2002) The potential for satellite-based monitoring of groundwater storage changes using GRACE: The High Plains aquifer, Central US. *Journal of Hydrology* 263(1-4): 245–256.

Rodell M, Famiglietti JS, Chen J, *et al.* (2004b) Basin scale estimates of evapotranspiration using GRACE and other observations. *Geophysical Research Letters* 31: 20504.

Rodell M, Houser PR, Jambor U, *et al.* (2004a) The Global Land Data Assimilation System. *Bulletin of the American Meteorological Society* 85: 381–394.

Rowlands DD, Luthcke SB, Klosko SM, *et al.* (2005) Resolving mass flux at high spatial and temporal resolution using GRACE intersatellite measurements. *Geophysical Research Letters* 32(4): L04310.

Rubincam DP (1984) Postglacial rebound observed by Lageos and the effective viscosity of the lower mantle. *Journal of Geophysical Research* 89: 1077–1088.

Schmidt M, Han SC, Kusche J, *et al.* (2006a) Regional high-resolution spatiotemporal gravity modeling from GRACE data using spherical wavelets. *Geophysical Research Letters* 33(8): L08403.

Schmidt R, Schwintzer P, Flechtner F, *et al.* (2006b) GRACE observations of changes in continental water storage. *Global and Planetary Change* 50(1-2): 112–126.

Schrama EJO (2004) Impact of limitations in geophysical background models on follow-on gravity missions. *Earth Moon and Planets* 94: 143–163.

Seo KW and Wilson CR (2005) Simulated estimation of hydrological loads from GRACE. *Journal of Geodesy* 78: 442–456 (doi 10.1007/s00190-004-0410-5).

Seo KW, Wilson CR, Famiglietti JS, *et al.* (2006) Terrestrial water mass load changes from gravity recovery and climate experiment (GRACE). *Water Resources Research* 42(5): W05417.

Song YT and Zlotnicki V (2004) Ocean bottom pressure waves predicted in the tropical Pacific. *Geophysical Research Letters* 31(5): L05306.

Swenson S and Milly PCM (2006) Climate-model biases in seasonality of continental water storage recovered by satellite gravimetry. *Water Resources Research* 42: W03201 (doi:10.1029/2005WR004628).

Swenson S and Wahr J (2002a) Estimated effects of the vertical structure of atmospheric mass on the time-variable geoid. *Journal of Geophysical Research* 107(B9): 2194 (doi:10.1029/2001JB000515).

Swenson S and Wahr J (2002b) Methods for inferring regional surface-mass anomalies from GRACE measurements of time-variable gravity. *Journal of Geophysical Research* 107(B9): 2193.

Swenson S and Wahr J (2006a) Post-processing removal of correlated errors in GRACE data. *Geophysical Research Letters* 33: L08402 (doi:10.1029/2005GL025285).

Swenson S and Wahr J (2006b) Estimating large-scale precipitation minus evapotranspiration from GRACE satellite gravity Measurements. *Journal of Hydrometeorology* 7: 252–270.

Swenson S, Wahr J, and Milly PCD (2003) Estimated accuracies of regional water storage variations inferred from the gravity recovery and climate experiment (GRACE). *Water Resource Research* 39(8): 1223 (doi:10.1029/2002WR001808).

Swenson S, Yeh PJF, Wahr J, and Famiglietti J (2006) A comparison of terrestrial water storage variations from GRACE with *in situ* measurements from Illinois. *Geophysical Research Letters* 33: L16401.

Syed TH, Famiglietti JS, Chen J, *et al.* (2005) Total basin discharge for the Amazon and Mississippi river basins from GRACE and a land–atmosphere water balance. *Geophysical Research Letters* 32(24): L24404.

Tamisiea ME, Mitrovica JX, and Davis JL (2007) GRACE gravity data constrain ancient ice geometries and continental dynamics over Laurentia (in press).

Tamisiea ME, Leuliette EW, Davis JL, and Mitrovica JX (2005) Constraining hydrological and cryospheric mass flux in Southeastern Alaska using space-based gravity measurements. *Geophysical Research Letters* 32: L20501.

Tapley BD, Bettadpur S, Ries JC, *et al.* (2004a) GRACE measurements of mass variability in the Earth system. *Science* 305(5683): 503–505.

Tapley BD, Bettadpur S, Watkins M, *et al.* (2004b) The gravity recovery and climate experiment: Mission overview and early results. *Geophysical Research Letters* 31(9): L09607.

Thompson PF, Bettadpur SV, and Tapley BD (2004) Impact of short period, non-tidal, temporal mass variability on GRACE gravity estimates. *Geophysical Research Letters* 31(6): L06619.

Trenberth KE (1997) Using atmospheric budgets as a constraint on surface fluxes. *Journal of Climate* 10: 2796–2809.

Trupin A (1993) Effects of polar ice on the Earth's rotation and gravitational potential. *Geophysical Journal International* 113: 273–283.

Trupin AS, Meier MF, and Wahr JM (1992) The effect of melting glaciers on the Earth's rotation and gravitational field: 1965–1984. *Geophysical Journal International* 108: 1–15.

van Dam T,Wahr J and Lavallee D (2007) A comparison of annual vertical crustal displacements from GPS and Gravity Recovery and Climate Experiment (GRACE) over Europe. *Journal of Geophysical Research* 112: B03404, doi: 10.1029/2006JB004335.

Velicogna I and Wahr J (2002) A method for separating Antarctic postglacial rebound and ice mass balance using future ICESat geoscience laser altimeter system, gravity recovery and climate experiment, and GPS satellite data. *Journal of Geophysical Research* 107(B10): 2263 (doi:10.1029/2001JB000708).

Velicogna I and Wahr J (2005) Greenland Mass balance from GRACE. *Geophysical Research Letters* 32: L18505 (doi:10.1029/2005GL023458).

Velicogna I and Wahr J (2006a) Measurements of time variable gravity shows mass loss in Antarctica. *Science* 311: 1754–1756.

Velicogna I and Wahr J (2006b) Significant acceleration of Greenland ice mass loss in spring, 2004. *Nature,* 443, (doi: 10.1038/nature05168).

Velicogna I, Wahr J, and van den Dool H (2001) Can surface pressure be used to remove atmospheric contributions from GRACE data with sufficient accuracy to recover hydrological signals?. *Journal of Geophysical Research* 106: 16415–16434.

Visser P, Sneeuw N, and Gerlach C (2003) Energy integral method for gravity field determination from satellite orbit coordinates. *Journal of Geodesy* 77: 207–216.

Wahr JM and Bergen Z (1986) The effects of mantle anelasticity on nutations, Earth tides, and tidal variations in rotation rate. *Geophysical Journal of the Royal Astronomical Society* 87: 633–668.

Wahr J, Molenaar M, and Bryan F (1998) Time variability of the Earth's gravity field: Hydrological and oceanic effects and their possible detection using GRACE. *Journal of Geophysical Research* 103(B12): 30205–30229.

Wahr J, Molenaar J, and Bryan F (1998) Time variability of the Earth's gravity field: Hydrological and oceanic effects and their possible detection using GRACE. *Journal of Geophysical Research* 103(B12): 30205–30229.

Wahr J, Wingham D, and Bentley CR (2000) A method of combining GLAS and GRACE satellite data to constrain Antarctic mass balance. *Journal of Geophysical Research* 105: 16279–16294.

Wahr J, Jayne SR, and Bryan FO (2002) A method of inferring deep ocean currents from satellite measurements of time variable gravity. *Journal of Geophysical Research –Oceans* 107: 3218 (doi:10.1029/2001JC001274).

Wahr J, Swenson S, and Velicogna I (2006) The Accuracy of GRACE mass estimates. *Geophysical Research Letters* 33: L06401 (doi:10.1029/2005GL025305).

Watkins MM, Folkner WM, Chao B, and Tapley BD (2000) EX-5: A Laser Interferometer Follow-On to the GRACE Misssion, Gravity, Geoid, and Geodynamics: IAG International Symposium GGG2000, Banff, Alberta, Canada, v. 123, Sideris M (eds.) New York: Springer-Verlag.

Watkins MM and Yuan DN (2006) Beyond Harmonics: Recent Mascon Solutions from GRACE. *Geophysical Research Abstracts*, 8,09474, 2006; SRef-ID: 1607–7962/gra/EGU06-A-09474.

Yeh PJ-F, Swenson S, Famiglietti J, and Rodell M (2006) Remote sensing of groundwater storage changes in Illinois using GRACE. *Water Resource Research* 42: W12203.

Yoder CF, Williams JG, Dickey JO, Schutz BE, Eanes RJ, and Tapley BD (1983) Secular variations of Earth's gravitational harmoic *J sub2* coefficient from Lageos and the non-tidal acceleration of Earth rotation. *Nature* 303: 757–762.

Yuan D and Watkins MM (2006) Recent Mascon Results from GRACE, Hotine–Marussi Symposium of Theoretical and Computational Geodesy: Challenge and Role of Modern Geodesy, Wuhan University.

Zwally J, Giovinetto MB, Li J, *et al.* (2005) Mass changes of the Greenland and Antarctic ice sheets and shelves and contributions to sea-level rise: 1992–2002. *Journal of Glaciology* 51: 175.

Zlotnicki V, Wahr J, Fukumori I, and Song Y-T (2007) The Antarctic circumpolar Current transport variability during 2002–2005 from GRACE. *Journal of Physical Oceanography* 37(2): 230–244.

9 Earth Rotation Variations – Long Period

R. S. Gross, Jet Propulsion Laboratory, California Institute of Technology, Pasadena, CA, USA

Nomenclature

a radius of a sphere having the same volume as the Earth (m)

h angular momentum vector due to motion relative to the rotating, body-fixed terrestrial reference frame ($kg\,m^2\,s^{-1}$)

h_x x-component of the angular momentum due to motion relative to the rotating, body-fixed terrestrial reference frame ($kg\,m^2\,s^{-1}$)

h_y y-component of the angular momentum due to motion relative to the rotating, body-fixed terrestrial reference frame ($kg\,m^2\,s^{-1}$)

h_z z-component of the angular momentum due to motion relative to the rotating, body-fixed terrestrial reference frame ($kg\,m^2\,s^{-1}$)

i $\sqrt{-1}$

k_2 degree-2 body tide Love number of the Earth (dimensionless)

k'_2 degree-2 load Love number of the Earth (dimensionless)

k_r parameter accounting for the effects of rotational deformation on the length-of-day (dimensionless)

m	complex-valued position of the rotation pole within the terrestrial reference frame (arcseconds)	**u**	velocity vector (m s^{-1})
		u	eastward velocity (m s^{-1})
m	amplitude of the free wobble of a rigid axisymmetric body (arcseconds)	*v*	northward velocity (m s^{-1})
		$\hat{\mathbf{x}}$	unit vector in direction of the *x*-coordinate axis of the terrestrial reference frame
m_x	*x*-component of the position of the rotation pole within the terrestrial reference frame (arcseconds)	$\hat{\mathbf{y}}$	unit vector in direction of the *y*-coordinate axis of the terrestrial reference frame
m_y	*y*-component of the position of the rotation pole within the terrestrial reference frame (arcseconds)	$\hat{\mathbf{z}}$	unit vector in direction of the *z*-coordinate axis of the terrestrial reference frame
m_z	time rate-of-change of (UT1–TAI) (seconds/day)	*A*	least principal moment of inertia of the Earth (kg m^2) area of ring laser gyroscope (m^2) amplitude of periodic motion (s or arcseconds)
n	complex-valued correction to the nutations in longitude and obliquity (arcseconds); also unit vector normal to a plane	
		A	transformation matrix that transforms coordinates from the terrestrial to the intermediate reference frames (dimensionless)
$\dot{n}$	tidal acceleration of the Moon (arcseconds/century^2)	
n_o	parameter related to the change in the mean moment of inertia of the Earth caused by the purely radial component of the rotational potential (dimensionless)	A'	average of the least and intermediate principal moments of inertia of the Earth (kg m^2)
		A_c	least principal moment of inertia of the core (kg m^2)
		A_m	least principal moment of inertia of the crust and mantle (kg m^2)
p	complex-valued position of the Celestial Intermediate Pole within the terrestrial reference frame (arcseconds)	A'_m	average of the least and intermediate principal moments of inertia of the crust and mantle (kg m^2)
p_x	*x*-component of the position of the Celestial Intermediate Pole within the terrestrial reference frame (arcseconds)	A_p	amplitude of prograde component of periodic polar motion (arcseconds)
p_y	*y*-component of the position of the Celestial Intermediate Pole within the terrestrial reference frame (arcseconds)	A_r	amplitude of retrograde component of periodic polar motion (arcseconds)
		B	intermediate principal moment of inertia of the Earth (kg m^2)
p_z	time rate-of-change of (UT1–TAI) (seconds/day)	*C*	greatest principal moment of inertia of the Earth (kg m^2)
r	position vector (m)	C_c	greatest principal moment of inertia of the core (kg m^2)
r	position in radial direction (m)	
r_c	vector of the coordinates of an observing station given in the celestial reference frame (m)	C_m	greatest principal moment of inertia of the crust and mantle (kg m^2)
r_i	vector of the coordinates of an observing station given in the intermediate reference frame (m)	*D*	coefficient relating changes in the inertia tensor of the Earth to changes in the wobble of the crust and mantle (kg m^2)
r_t	vector of the coordinates of an observing station given in the terrestrial reference frame (m)	*D*	Delaunay argument for the mean elongation of the Moon from the Sun (deg)
t	time (s)	$\bar{D}$	coefficient relating changes in the inertia tensor of the Earth to changes in the length-of-day (kg m^2)
t_o	reference time (s)	

E	coefficient relating changes in the relative angular momentum of the core to changes in the wobble of the crust and mantle $(\text{kg m}^2\,\text{s}^{-1})$	Y	y-component of wobble coordinate transformation matrix (dimensionless)
E'	coefficient relating changes in the relative angular momentum of the core to changes in the wobble of the crust and mantle $(\text{kg m}^2\,\text{s}^{-1})$	α	phase of periodic motion (deg)
		α_p	phase of prograde component of periodic polar motion (deg)
		α_r	phase of retrograde component of periodic polar motion (deg)
$\bar{E}$	coefficient relating changes in the relative angular momentum of the core to changes in the length-of-day $(\text{kg m}^2\,\text{s}^{-1})$	α_3	factor modifying the degree-2 load Love number of the Earth because of core decoupling (dimensionless)
F	Delaunay argument for $L-\Omega$, where L is the mean longitude of the Moon (deg)	γ	Greenwich mean sidereal time reckoned from the lower culmination of the vernal equinox (GMST$+\pi$) (deg)
G	Newtonian gravitational constant $(\text{m}^3\,\text{kg}^{-1}\,\text{s}^{-2})$	δD	coefficient relating changes in the inertia tensor of the Earth to changes in the wobble of the crust and mantle (kg m^2)
H	hour angle of the true equinox of date (s)		
I	inertia tensor (kg m^2)	δD_{12}	coefficient relating changes in the inertia tensor of the Earth to changes in the wobble of the crust and mantle (kg m^2)
l	Delaunay argument for the mean anomaly of the Moon (deg)		
l'	Delaunay argument for the mean anomaly of the Sun (deg)	δD_{13}	coefficient relating changes in the inertia tensor of the Earth to changes in the Earth's rotation (kg m^2)
I_o	initial inertia tensor (kg m^2)		
L	angular momentum vector $(\text{kg m}^2\,\text{s}^{-1})$	δD_{23}	coefficient relating changes in the inertia tensor of the Earth to changes in the Earth's rotation (kg m^2)
M	mass of the Earth (kg)		
M_{atm}	mass of the atmosphere (kg)	δf	Sagnac frequency, (rad s^{-1})
M_c	mass of the core (kg)	δh_x	x-component of the change in relative angular momentum due to motion in the core caused by a rigid rotation of the crust and mantle $(\text{kg m}^2\,\text{s}^{-1})$
M_m	mass of the crust and mantle (kg)		
M_{ocn}	mass of the oceans (kg)		
N	nutation coordinate transformation matrix (dimensionless)		
P	precession coordinate transformation matrix (dimensionless)	δh_y	y-component of the change in relative angular momentum due to motion in the core caused by a rigid rotation of the crust and mantle $(\text{kg m}^2\,\text{s}^{-1})$
P	optical path length of ring laser gyroscope (m)		
Q	quality factor of periodic motion (dimensionless)	δh_z	z-component of the change in relative angular momentum due to motion in the core caused by a rigid rotation of the crust and mantle $(\text{kg m}^2\,\text{s}^{-1})$
S	spin coordinate transformation matrix (dimensionless)		
T	period of periodic motion (s)		
TAI	Temps Atomique International (International Atomic Time) (s)	δI_{xz}	change in the (x,z)-element of the inertia tensor of the Earth due to changes in the rotation of the Earth (kg m^2)
UT0	Universal Time (s)		
UT1	Universal Time corrected for the effects of polar motion (s)		
V	volume (m^3)	δI_{yz}	change in the (y,z)-element of the inertia tensor of the Earth due to changes in the rotation of the Earth (kg m^2)
W	antisymmetric rotation matrix (rad s^{-1})		
X	x-component of wobble coordinate transformation matrix (dimensionless)		

δI_{zz} — change in the (z,z)-element of the inertia tensor of the Earth due to changes in the rotation of the Earth (kg m^2)

$\delta\varepsilon$ — correction to the nutation in obliquity (arcseconds)

$\delta\psi$ — correction to the nutation in longitude (arcseconds)

ε_a — ellipticity of the surface of the Earth (dimensionless)

ε_c — ellipticity of the surface of the core (dimensionless)

ε_o — mean obliquity of the ecliptic (degrees)

λ — East longitude (deg); also wavelength of laser beam used in ring laser gyroscope (m)

ρ — density (kg m^{-3})

σ — frequency (rad s^{-1})

σ_c — frequency given in the celestial reference frame (rad s^{-1})

σ_t — frequency given in the terrestrial reference frame (rad s^{-1})

σ_r — frequency of the free wobble of a rigid triaxial Earth (rad s^{-1})

σ_{ra} — frequency of the free wobble of a rigid axisymmetric Earth (rad s^{-1})

σ_{cw} — frequency of the free wobble of a deformable Earth with fluid core (rad s^{-1})

σ_o — observed complex-valued frequency of the Chandler wobble (rad s^{-1})

$\boldsymbol{\tau}$ — torque vector $(\text{kg m}^2\,\text{s}^{-2})$

τ — amplitude decay time constant (s)

ϕ — North latitude (deg)

$\phi_{r,x}$ — x-component of the excitation function of a rigid triaxial Earth (arcseconds)

$\phi_{r,y}$ — y-component of the excitation function of a rigid triaxial Earth (arcseconds)

$\phi_{r,z}$ — z-component of the excitation function of a rigid triaxial Earth (seconds/day)

$\phi_{ra,x}$ — x-component of the excitation function of a rigid axisymmetric Earth (arcseconds)

$\phi_{ra,y}$ — y-component of the excitation function of a rigid axisymmetric Earth (arcseconds)

χ — complex-valued excitation function of a deformable Earth with fluid core (arcseconds)

χ_x — x-component of the excitation function of a deformable Earth with fluid core (arcseconds)

χ_y — y-component of the excitation function of a deformable Earth with fluid core (arcseconds)

χ_z — z-component of the excitation function of a deformable Earth with fluid core (arcseconds)

$\boldsymbol{\omega}$ — angular velocity vector (rad s^{-1})

$\boldsymbol{\omega}_o$ — initial angular velocity vector (rad s^{-1})

ω_x — x-component of the angular velocity (rad s^{-1})

ω_y — y-component of the angular velocity (rad s^{-1})

ω_z — z-component of the angular velocity (rad s^{-1})

Λ_o — nominal length-of-day (s)

$\Delta\phi$ — variation of latitude (arcseconds)

$\Delta k'_{an}$ — modification of the degree-2 load Love number of the Earth due to mantle anelasticity (dimensionless)

$\Delta k_{ocn,s}$ — oceanic Love number for the spin of the Earth (dimensionless)

$\Delta k_{ocn,w}$ — oceanic Love number for the wobble of the Earth (dimensionless)

$\Delta \mathbf{I}$ — perturbation to the initial inertia tensor (kg m^2)

ΔI_{xx} — (x,x)-element of the perturbation to the initial inertia tensor (kg m^2)

ΔI_{xy} — (x,y)-element of the perturbation to the initial inertia tensor (kg m^2)

ΔI_{xz} — (x,z)-element of the perturbation to the initial inertia tensor (kg m^2)

ΔI_{yy} — (y,y)-element of the perturbation to the initial inertia tensor (kg m^2)

ΔI_{yz} — (y,z)-element of the perturbation to the initial inertia tensor (kg m^2)

ΔI_{zz} — (z,z)-element of the perturbation to the initial inertia tensor (kg m^2)

$\Delta\omega_z$ — z-component of the perturbation to the initial angular velocity (rad s^{-1})

$\Delta\Lambda$ — change in the length-of-day (s)

$\Delta\boldsymbol{\omega}$ — perturbation to the initial angular velocity vector (rad s^{-1})

Ω — mean angular velocity of Earth (rad s^{-1})

Ω — Delaunay argument for the mean longitude of the ascending node of the Moon (deg)

9.1 Introduction

The Earth is a dynamic system – it has a fluid, mobile atmosphere and oceans, a continually changing global distribution of ice, snow, and water, a fluid core that is undergoing some type of hydromagnetic motion, a mantle both thermally convecting and rebounding from the glacial loading of the last ice age, and mobile tectonic plates. In addition, external forces due to the gravitational attraction of the Sun, Moon, and planets also act upon the Earth. These internal dynamical processes and external gravitational forces exert torques on the solid Earth, or displace its mass, thereby causing the Earth's rotation to change.

Changes in the rotation of the solid Earth are studied by applying the principle of conservation of angular momentum to the Earth system. Under this principle, the rotation of the solid Earth changes as a result of (1) applied external torques, (2) internal mass redistribution, and (3) the transfer of angular momentum between the solid Earth and the fluid regions with which it is in contact; concomitant torques are due to hydrodynamic or magneto-hydrodynamic stresses acting at the fluid/solid Earth interfaces.

Here, changes in the Earth's rotation that occur on timescales greater than a day are discussed. Using the principle of conservation of angular momentum, the equations governing small variations in both the rate of rotation and in the position of the rotation vector with respect to the Earth's crust are first derived. These equations are then rewritten in terms of the Earth rotation parameters that are actually reported by Earth rotation measurement services. The techniques that are used to monitor the Earth's rotation by the measurement services are then reviewed, a description of the variations that are observed by these techniques is given, and possible causes of the observed variations are discussed.

9.2 Theory of Earth Rotation Variations at Long Periods

9.2.1 Instantaneous Rotation Vector

In a rotating reference frame that has been attached in some manner to the solid body of the Earth, the Eulerian equation of motion that relates changes in the angular momentum $\mathbf{L}(t)$ of the Earth to the external torques $\boldsymbol{\tau}(t)$ acting on it is (Munk and MacDonald, 1960; Lambeck, 1980, 1988; Moritz and Mueller, 1988; Eubanks, 1993)

$$\frac{\partial \mathbf{L}(t)}{\partial t} + \boldsymbol{\omega}(t) \times \mathbf{L}(t) = \boldsymbol{\tau}(t) \qquad [1]$$

where, strictly speaking, $\boldsymbol{\omega}(t)$ is the angular velocity of the rotating frame with respect to inertial space. But since the rotating frame has been attached to the solid body of the Earth, it is also interpreted as being the angular velocity of the Earth with respect to inertial space. In general, the angular momentum $\mathbf{L}(t)$ can be written as the sum of two terms: (1) that part $\mathbf{h}(t)$ due to motion relative to the rotating reference frame, and (2) that part due to the changing inertia tensor $I(t)$ of the Earth which is changing because the distribution of the Earth's mass is changing:

$$\mathbf{L}(t) = \mathbf{h}(t) + I(t) \cdot \boldsymbol{\omega}(t) \qquad [2]$$

Combining eqns [1] and [2] yields the Liouville equation

$$\begin{aligned} \frac{\partial}{\partial t} &[\mathbf{h}(t) + I(t) \cdot \boldsymbol{\omega}(t)] \\ &+ \boldsymbol{\omega}(t) \times [\mathbf{h}(t) + I(t) \cdot \boldsymbol{\omega}(t)] = \boldsymbol{\tau}(t) \qquad [3] \end{aligned}$$

The external torques acting on the Earth due to the gravitational attraction of the Sun, Moon, and planets cause the Earth to nutate and precess. Since the nutations and precession of the Earth are discussed in Chapter 10 of this volume, the external torques $\boldsymbol{\tau}(t)$ in eqn [3] will be set to zero. Note, however, that tidal effects on the Earth's rotation, which are also caused by the gravitational attraction of the Sun, Moon, and planets, are discussed here in Sections 9.4.1.3 and 9.4.2.3

The Earth's rotation deviates only slightly from a state of uniform rotation, the deviation being a few parts in 10^8 in speed, corresponding to changes of a few milliseconds (ms) in the length of the day, and about a part in 10^6 in the position of the rotation axis with respect to the crust of the Earth, corresponding to a variation of several hundred milliarcseconds (mas) in polar motion. Such small deviations in rotation can be studied by linearizing eqn [3]. Let the Earth initially be uniformly rotating at the constant rate Ω about the z-coordinate axis of the body-fixed reference frame and orient the frame within the

Earth in such a manner that the inertia tensor of the Earth is diagonal in this frame:

$$\boldsymbol{\omega}_o = \Omega \hat{z} \qquad [4]$$

$$I_o = \begin{pmatrix} A & 0 & 0 \\ 0 & B & 0 \\ 0 & 0 & C \end{pmatrix} \qquad [5]$$

where the hat denotes a vector of unit length, Ω is the mean angular velocity of the Earth, and A, B, and C are the mean principal moments of inertia of the Earth ordered such that $A < B < C$. In this initial state, in which all time-dependent quantities vanish, the Earth is rotating at a constant rate about its figure axis, there are no mass displacements, and there is no relative angular momentum. So, for example, the atmosphere, oceans, and core are at rest with respect to the solid Earth and merely corotate with it.

Now let this initial state be perturbed by the appearance of mass displacements and relative angular momentum. In general, since the crust and mantle of the Earth can deform, they can undergo motion relative to the rotating reference frame and hence can contribute to the relative angular momentum. However, let the body-fixed reference frame in the perturbed state be oriented in such a manner that the relative angular momentum due to motion of the crust and mantle vanishes. In this frame, which is known as the Tisserand mean-mantle frame of the Earth (Tisserand, 1891), the motion of the atmosphere, oceans, and core have relative angular momentum, but the motion of the crust and mantle does not.

In the Tisserand mean-mantle frame, the perturbed instantaneous rotation vector and inertia tensor of the Earth can be written without loss of generality as

$$\boldsymbol{\omega}(t) = \boldsymbol{\omega}_o + \Delta\boldsymbol{\omega}(t)$$
$$= \Omega\hat{z} + \Omega[m_x(t)\,\hat{x} + m_y(t)\,\hat{y} + m_z(t)\,\hat{z}] \qquad [6]$$

$$I(t) = I_o + \Delta I(t)$$
$$= \begin{pmatrix} A & 0 & 0 \\ 0 & B & 0 \\ 0 & 0 & C \end{pmatrix} + \begin{pmatrix} \Delta I_{xx}(t) & \Delta I_{xy}(t) & \Delta I_{xz}(t) \\ \Delta I_{xy}(t) & \Delta I_{yy}(t) & \Delta I_{yz}(t) \\ \Delta I_{xz}(t) & \Delta I_{yz}(t) & \Delta I_{zz}(t) \end{pmatrix} \qquad [7]$$

where the terms with the subscript 'o' denote the initial values given by eqns [4] and [5], $\Omega m_i(t)$ are the elements of the time-dependent perturbation $\Delta\boldsymbol{\omega}(t)$ to the rotation vector, and the $\Delta I_{ij}(t)$ are the elements of the time-dependent perturbation $\Delta I(t)$ to the inertia tensor.

The equation that relates small changes in the Earth's rotation to the mass displacements and relative angular momenta that are causing the rotation to change can be derived by substituting eqns [6] and [7] into eqn [3] and then linearizing the resulting expression by assuming that $h_i(t) \ll \Omega C$, $m_i(t) \ll 1$, and $\Delta I_{ij}(t) \ll C$. By keeping terms to first order in these small quantities, the equatorial and axial components of eqn [3] can be written as

$$\frac{1}{\sigma_r}\frac{\partial m_x(t)}{\partial t} + \left[\frac{B(C-B)}{A(C-A)}\right]^{1/2} m_y(t)$$
$$= -\left(\frac{B}{A}\right)^{1/2}\left[\frac{1}{\Omega}\frac{\partial\phi_{r,x}(t)}{\partial t} - \phi_{r,y}(t)\right] \qquad [8]$$

$$\frac{1}{\sigma_r}\frac{\partial m_y(t)}{\partial t} - \left[\frac{A(C-A)}{B(C-B)}\right]^{1/2} m_x(t)$$
$$= -\left(\frac{A}{B}\right)^{1/2}\left[\frac{1}{\Omega}\frac{\partial\phi_{r,y}(t)}{\partial t} + \phi_{r,x}(t)\right] \qquad [9]$$

$$\frac{1}{\Omega}\frac{\partial m_z(t)}{\partial t} = -\frac{1}{\Omega}\frac{\partial\phi_{r,z}(t)}{\partial t} \qquad [10]$$

where the external torques have been set to zero,

$$\sigma_r^2 = \left(\frac{C-A}{A}\right)\left(\frac{C-B}{B}\right)\Omega^2 \qquad [11]$$

and the $\phi_{r,i}(t)$, known as excitation functions, are

$$\phi_{r,x}(t) = \frac{h_x(t) + \Omega\,\Delta I_{xz}(t)}{\Omega\,\sqrt{(C-A)(C-B)}} \qquad [12]$$

$$\phi_{r,y}(t) = \frac{h_y(t) + \Omega\,\Delta I_{yz}(t)}{\Omega\,\sqrt{(C-A)(C-B)}} \qquad [13]$$

$$\phi_{r,z}(t) = \frac{1}{C\Omega}[h_z(t) + \Omega\,\Delta I_{zz}(t)] \qquad [14]$$

Equations [8] and [9] are coupled, first-order differential equations that describe the motion of the rotation pole in the rotating, body-fixed reference frame as it responds to the applied excitation. In the absence of excitation, the solution of these equations, which describes the natural or free motion of the rotation pole, can be written as

$$m_x(t) = m\cos(\sigma_r t + \alpha) \qquad [15]$$

$$m_y(t) = \left[\frac{A(C-A)}{B(C-B)}\right]^{1/2} m\sin(\sigma_r t + \alpha) \qquad [16]$$

where m is the amplitude of the motion along the x-axis and α is the phase of the motion. The natural motion described by eqns [15] and [16] is prograde undamped elliptical motion of frequency σ_r. Using

Table 1 Geodetic parameters of the Earth

Parameter	Value	Source
G	$6.67259 \times 10^{-11}\,\mathrm{m^3\,kg^{-1}\,s^{-2}}$	(a)
M_{atm}	$5.1441 \times 10^{18}\,\mathrm{kg}$	(b)
M_{ocn}	$1.4 \times 10^{21}\,\mathrm{kg}$	(c)
Whole Earth (observed)		
Ω	$7.292115 \times 10^{-5}\,\mathrm{rad\,s^{-1}}$	(a)
M	$5.9737 \times 10^{24}\,\mathrm{kg}$	(a)
C	$8.0365 \times 10^{37}\,\mathrm{kg\,m^2}$	(a)
B	$8.0103 \times 10^{37}\,\mathrm{kg\,m^2}$	(a)
A	$8.0101 \times 10^{37}\,\mathrm{kg\,m^2}$	(a)
$C-A$	$2.6398 \times 10^{35}\,\mathrm{kg\,m^2}$	(a)
$C-B$	$2.6221 \times 10^{35}\,\mathrm{kg\,m^2}$	(a)
$B-A$	$1.765 \times 10^{33}\,\mathrm{kg\,m^2}$	(a)
Whole Earth (modeled)		
a	$6371.0\,\mathrm{km}$	(d)
n_o	0.15505	(e)
k_2	0.298	(f)
$\Delta k_{ocn,w}$	0.047715	(g)
$\Delta k_{ocn,s}$	0.043228	(g)
k_r	0.997191	(g)
k_2'	-0.305	(f)
$\Delta k_{an}'$	$-0.011 + i0.003$	(f)
α_3	0.792	(h)
Crust and mantle (PREM)		
ε_a	3.334×10^{-3}	(d)
M_m	$4.0337 \times 10^{24}\,\mathrm{kg}$	(d)
C_m	$7.1236 \times 10^{37}\,\mathrm{kg\,m^2}$	(d)
A_m	$7.0999 \times 10^{37}\,\mathrm{kg\,m^2}$	(d)
Core (PREM)		
ε_c	2.546×10^{-3}	(d)
M_c	$1.9395 \times 10^{24}\,\mathrm{kg}$	(d)
C_c	$9.1401 \times 10^{36}\,\mathrm{kg\,m^2}$	(d)
A_c	$9.1168 \times 10^{36}\,\mathrm{kg\,m^2}$	(d)

PREM, preliminary reference Earth model (Dziewonski and Anderson, 1981).
Sources: (a) Groten (2004), (b) Trenberth and Guillemot (1994), (c) Yoder (1995), (d) Mathews *et al.* (1991), (e) Dahlen (1976), (f) Wahr (2005), (g) this paper, (h) Wahr (1983).

the values in **Table 1** for A, B, and C of the whole Earth, the period of the natural frequency, given by $2\pi/\sigma_r$, is found to be 304.46 sidereal days.

Euler (1765) first predicted that the Earth should freely wobble as it rotates, and that the period of this free wobble, assuming that the Earth is rigid, would be about 10 months. However, it was not until 1891 that the free wobble of the Earth was first detected in astronomical observations by Seth Carlo Chandler, Jr. (Chandler, 1891), albeit at a period of 14 months. The free wobble of the Earth is now known as the Chandler wobble in his honor.

Equations [8]–[14] describe changes in the rotation of a rigid body of arbitrary shape that is subject to small perturbing excitation. By recognizing that $(B-A)/A = 2.2 \times 10^{-5} \ll 1$ for the Earth (see **Table 1**), so that dynamically the Earth is nearly

axisymmetric, eqns [8] and [9] can be simplified by replacing A and B in them with the average $A' = (A+B)/2$ of the equatorial principal moments of inertia of the Earth:

$$\frac{1}{\sigma_{ra}}\frac{\partial m_x(t)}{\partial t} + m_y(t) = \phi_{ra,y}(t) - \frac{1}{\Omega}\frac{\partial \phi_{ra,x}(t)}{\partial t} \quad [17]$$

$$\frac{1}{\sigma_{ra}}\frac{\partial m_y(t)}{\partial t} - m_x(t) = -\phi_{ra,x}(t) - \frac{1}{\Omega}\frac{\partial \phi_{ra,y}(t)}{\partial t} \quad [18]$$

where the excitation functions $\phi_{ra,i}(t)$ of a rigid axisymmetric body are

$$\phi_{ra,x}(t) = \frac{b_x(t) + \Omega\,\Delta I_{xz}(t)}{(C-A')\,\Omega} \quad [19]$$

$$\phi_{ra,y}(t) = \frac{b_y(t) + \Omega\,\Delta I_{yz}(t)}{(C-A')\,\Omega} \quad [20]$$

In the absence of excitation, the free motion of a rigid axisymmetric body is prograde undamped circular motion of natural frequency

$$\sigma_{\mathrm{ra}} = \left(\frac{C - A'}{A'}\right)\Omega \qquad [21]$$

Note that eqns [10] and [14], which describe changes in the rate of rotation of a rigid body, are the same whether the body is dynamically axisymmetric or triaxial.

Equations [10], [14], and [17]–[21] describe changes in the rotation of a rigid axisymmetric body that is subject to small perturbing excitation. But the Earth is not rigid – it has an atmosphere and oceans, a fluid core, and a solid crust and mantle that can deform in response not only to the applied excitation but also to changes in rotation that are caused by the excitation. In general, changes in rotation can be expected to cause changes in both the Earth's inertia tensor and in relative angular momentum. However, by definition of the Tisserand mean-mantle frame, there are no changes in relative angular momentum caused by motion of the crust and mantle. Furthermore, if it is assumed that the oceans stay in equilibrium as the rotation of the solid Earth changes so that no oceanic currents are generated by the changes in rotation, then there are also no changes in relative angular momentum due to motion of the oceans. And effects of the atmosphere can be ignored here because of its relatively small mass (see **Table 1**). Thus, only the core will contribute to changes in relative angular momentum caused by changes in rotation.

Smith and Dahlen (1981) used the results of Hough (1895) to show that the change $\delta h_i(\sigma)$ in relative angular momentum due to core motion caused by a rigid rotation of an axisymmetric crust and mantle is

$$\begin{pmatrix} \delta h_x(\sigma) \\ \delta h_y(\sigma) \\ \delta h_z(\sigma) \end{pmatrix} = \begin{pmatrix} E & iE' & 0 \\ -iE' & E & 0 \\ 0 & 0 & \bar{E} \end{pmatrix} \begin{pmatrix} m_x(\sigma) \\ m_y(\sigma) \\ m_z(\sigma) \end{pmatrix} \qquad [22]$$

where to first order in the ellipticity ε_c of the surface of the core and at frequencies $\sigma \ll \Omega$

$$E = \left(\sigma^2/\Omega\right) A_c \qquad [23]$$

$$E' = -\sigma\left(1 - \varepsilon_c\right)A_c \qquad [24]$$

$$\bar{E} = -\Omega C_c \qquad [25]$$

where A_c and C_c are the equatorial and axial principal moments of inertia of the core and eqn [25] for $\bar{E}$ has

been inferred here by realizing that the core cannot respond to axial changes in the rotation of the mantle if the core–mantle boundary is axisymmetric and if there is no coupling between the core and the mantle (Merriam, 1980; Wahr et al., 1981; Yoder et al., 1981). Note that because of the assumption of dynamical axisymmetry, the equatorial components of eqn [22] are uncoupled from the axial, so that no spin–wobble coupling is introduced by the response of the core to changes in rotation.

Dahlen (1976) studied the passive influence of the oceans on the Earth's rotation, including the changes δI_{ij} in the Earth's inertia tensor that are caused by changes in the rotation of the Earth. In the absence of oceans, and assuming that the Earth responds to the centripetal potential associated with changes in rotation in exactly the same manner that a nonrotating Earth would respond to a static potential of the same amplitude and type, Dahlen (1976) found

$$\begin{pmatrix} \delta I_{xz} \\ \delta I_{yz} \\ \delta I_{zz} \end{pmatrix} = \frac{a^5\Omega^2}{3G} \begin{pmatrix} k_2 & 0 & 0 \\ 0 & k_2 & 0 \\ 0 & 0 & n_0 + \dfrac{4}{3}k_2 \end{pmatrix} \begin{pmatrix} m_x \\ m_y \\ m_z \end{pmatrix} \qquad [26]$$

where k_2 is the second-degree body tide Love number of the whole Earth (not of just the mantle (see Smith and Dahlen, 1981, p. 239), although for a different opinion see Dickman (2005)), n_0 arises from the change in the mean moment of inertia of the Earth caused by the term in the centripetal potential that gives rise to a purely radial deformation of the Earth, G is the Newtonian gravitational constant, and a is the mean radius of the Earth (that is, the radius of a sphere having the same volume as that of the Earth).

When equilibrium oceans are present, Dahlen (1976) found that eqn [26] for the changes in the inertia tensor caused by changes in rotation is modified to

$$\begin{pmatrix} \delta I_{xz} \\ \delta I_{yz} \\ \delta I_{zz} \end{pmatrix} = \begin{pmatrix} D + \delta D & \delta D_{12} & \delta D_{13} \\ \delta D_{12} & D - \delta D & \delta D_{23} \\ \delta D_{13} & \delta D_{23} & \bar{D} \end{pmatrix} \begin{pmatrix} m_x \\ m_y \\ m_z \end{pmatrix} \qquad [27]$$

where

$$D = \left(k_2 + \Delta k_{\mathrm{ocn,w}}\right)\frac{a^5\,\Omega^2}{3G} \qquad [28]$$

$$\bar{D} = \left[n_0 + \frac{4}{3}\left(k_2 + \Delta k_{\mathrm{ocn,s}}\right)\right]\frac{a^5\,\Omega^2}{3G} \qquad [29]$$

where the influence of equilibrium oceans has been written in terms of an 'oceanic Love number' Δk_{ocn}

which modifies the second-degree body tide Love number k_2. Dahlen (1976) found that because of the nonuniform distribution of the oceans, the oceanic Love number is different for each component. However, the average of the equatorial components has been taken here to define a mean oceanic Love number $\Delta k_{ocn,w}$ for the wobble. From eqn [27], it is seen that the nonuniform distribution of the oceans has also coupled the equatorial components to the axial via the off-diagonal elements δD_{13} and δD_{23}. However, the numerical results of Dahlen (1976) for Earth model 1066A (Gilbert and Dziewonski, 1975) show that this coupling is very weak, with $\delta D_{13}/D = 2.17 \times 10^{-3}$ and $-\delta D_{23}/D = 0.55 \times 10^{-3}$. The coupling between the equatorial components is also very weak, with $-\delta D_{12}/D = 3.15 \times 10^{-3}$.

Using eqns [22]–[25] to account for the relative angular momentum of the core caused by changes in rotation, eqns [27]–[29] to account for both the rotational deformation of the Earth and the passive response of equilibrium oceans to changes in rotation, and keeping terms to first order in small quantities, the linearized Liouville equation becomes

$$\frac{1}{\sigma_{cw}}\frac{\partial m_x(t)}{\partial t} + m_y(t) = \chi_y(t) - \frac{1}{\Omega}\frac{\partial \chi_x(t)}{\partial t} \quad [30]$$

$$\frac{1}{\sigma_{cw}}\frac{\partial m_y(t)}{\partial t} - m_x(t) = -\chi_x(t) - \frac{1}{\Omega}\frac{\partial \chi_y(t)}{\partial t} \quad [31]$$

$$\frac{1}{\Omega}\frac{\partial m_z(t)}{\partial t} = -\frac{1}{\Omega}\frac{\partial \chi_z(t)}{\partial t} \quad [32]$$

where the theoretical frequency of the Chandler wobble is

$$\sigma_{cw} = \left(\frac{C - A' - D}{A'_m + \varepsilon_c A_c + D}\right)\Omega \quad [33]$$

where $A'_m = A' - A_c$ is the equatorial principal moment of inertia of the crust and mantle, and the excitation functions $\chi_i(t)$ are

$$\chi_x(t) = \frac{h_x(t) + \Omega(1 + k'_2)\Delta I_{xz}(t)}{(C - A' - D)\Omega} \quad [34]$$

$$\chi_y(t) = \frac{h_y(t) + \Omega(1 + k'_2)\Delta I_{yz}(t)}{(C - A' - D)\Omega} \quad [35]$$

$$\chi_z(t) = k_r\frac{h_z(t) + \Omega(1 + \alpha_3 k'_2)\Delta I_{zz}(t)}{C_m\Omega} \quad [36]$$

where C_m is the axial principal moment of inertia of the crust and mantle and k_r is a factor, whose value is

near unity (see **Table 1**), that accounts for the effects of rotational deformation on the axial component:

$$k_r = \left\{1 + \left[n_o + \frac{4}{3}(k_2 + \Delta k_{ocn,s})\right]\frac{a^5\Omega^2}{3G}\frac{1}{C_m}\right\}^{-1} \quad [37]$$

The deformation of the Earth associated with surficial excitation processes that load the solid Earth has been taken into account in eqns [34]–[36] by including the second-degree load Love number k'_2 where, because of core decoupling, the load Love number in the axial component is modified by a factor of α_3 (Merriam, 1980; Wahr, 1983; Nam and Dickman, 1990; Dickman 2003). Expressions for the excitation functions for processes that do not load the solid Earth can be recovered from eqns [34]–[36] by setting the load Love number k'_2 to zero.

Equations [30]–[36] describe changes in the rotation of an elastic axisymmetric body having a fluid core and equilibrium oceans that is subject to small perturbing excitation. Equation [33] for the theoretical Chandler frequency of such a body was first derived by Smith and Dahlen (1981). Applying this result to the Earth, they found that elasticity of the solid Earth lengthens the period of the Chandler wobble from the rigid Earth value by 143.0 sidereal days, deformation of the oceans lengthens it a further 29.8 sidereal days, and the presence of a fluid core decreases it by 50.5 sidereal days. Using the values in **Table 1** for the Earth, the theoretical period of the Chandler wobble is found to be 426.8 sidereal days, or about 7.4 sidereal days shorter than the observed period of 434.2 ± 1.1 (1σ) sidereal days (Wilson and Vicente, 1990). This discrepancy between the theoretical and observed periods of the Chandler wobble is probably mainly due to the effects of mantle anelasticity, since departures of the oceans from equilibrium as large as 1% increase the Chandler period by only 0.3 sidereal days (Smith and Dahlen, 1981).

Mantle anelasticity modifies the body tide Love number k_2 and hence, via D, the frequency of the Chandler wobble and the equatorial excitation functions. It also modifies the load Love number k'_2. In the absence of accurate models of mantle anelasticity at the frequencies of interest here, namely, at frequencies $\sigma < \Omega$, a hybrid approach is taken to include its effects. The observed complex-valued frequency σ_0 of the Chandler wobble is substituted for its theoretical value in eqns [30] and [31]. It is also substituted for its theoretical value in the excitation functions after rewriting them in terms of the theoretical value by using eqn [33] to eliminate D. The results are

$$\frac{1}{\sigma_o} \frac{\partial m_x(t)}{\partial t} + m_y(t) = \chi_y(t) - \frac{1}{\Omega} \frac{\partial \chi_x(t)}{\partial t} \quad [38]$$

$$\frac{1}{\sigma_o} \frac{\partial m_y(t)}{\partial t} - m_x(t) = -\chi_x(t) - \frac{1}{\Omega} \frac{\partial \chi_y(t)}{\partial t} \quad [39]$$

$$m_z(t) = -\chi_z(t) \quad [40]$$

where the excitation functions become

$$\chi_x(t) = \frac{b_x(t) + \Omega\left[1 + (k_2' + \Delta k_{an}')\right] \Delta I_{xz}(t)}{[C - A' + A_m' + \varepsilon_c A_c]\sigma_o} \quad [41]$$

$$\chi_y(t) = \frac{b_y(t) + \Omega\left[1 + (k_2' + \Delta k_{an}')\right] \Delta I_{yz}(t)}{[C - A' + A_m' + \varepsilon_c A_c]\sigma_o} \quad [42]$$

$$\chi_z(t) = k_r \frac{b_z(t) + \Omega\left[1 + \alpha_3(k_2' + \Delta k_{an}')\right] \Delta I_{zz}(t)}{C_m \Omega} \quad [43]$$

where $\Delta k_{an}'$ accounts for the effects of mantle anelasticity on the load Love number. Eqns [38]–[43] are the final expressions for the changes $m_i(t)$ in the rotation of the Earth caused by small excitation $\chi_i(t)$. Numerically, using 434.2 sidereal days (Wilson and Vicente, 1990) for the observed period of the Chandler wobble and the values in **Table 1** for the other constants, the real parts of the excitation functions can be written as

$$\chi_x(t) = \frac{1.608 \left[b_x(t) + 0.684\,\Omega\,\Delta I_{xz}(t)\right]}{(C - A')\,\Omega} \quad [44]$$

$$\chi_y(t) = \frac{1.608 \left[b_y(t) + 0.684\,\Omega\,\Delta I_{yz}(t)\right]}{(C - A')\,\Omega} \quad [45]$$

$$\chi_z(t) = \frac{0.997}{C_m \Omega}\left[b_z(t) + 0.750\,\Omega\,\Delta I_{zz}(t)\right] \quad [46]$$

These results agree with those of Wahr (1982, 1983, 2005) to within 2%, with most of the disagreement being due to differences in the values of the numerical constants.

The approach used here to derive linearized Liouville equations that can be used to study small changes in the Earth's rotation follows the approach of Smith and Dahlen (1981) and Wahr (1982, 1983, 2005). Other approaches have been given by Barnes *et al.* (1983), Eubanks (1993; also see Aoyama and Naito, 2000), and Dickman (1993, 2003). Dickman (2003) compares these different approaches and discusses the implications of nonzero coupling between the core and mantle; Wahr (2005) discusses the implications of mantle anelasticity. The influence

of triaxiality on oceanless elastic bodies with a fluid core, with application to the rotation of Mars, has been studied by Yoder and Standish (1997) and Van Hoolst and Dehant (2002).

9.2.2 Celestial Intermediate Pole

Small changes in the Earth's rotation caused by small changes in relative angular momentum or small changes in the Earth's inertia tensor can be studied using eqns [38]–[43], where the $\Omega m_i(t)$ are the elements of the change $\Delta\boldsymbol{\omega}(t)$ to the Earth's rotation vector, so that $m_x(t)$, $m_y(t)$, and $1 + m_z(t)$ are the direction cosines of the rotation vector with respect to the coordinate axes of the rotating, body-fixed terrestrial reference frame. Alternatively, $m_x(t)$ and $m_y(t)$ can be interpreted as being the angular offsets of the rotation vector from the $\hat{\mathbf{z}}$-axis of the rotating reference frame in the $\hat{\mathbf{x}}$ and $\hat{\mathbf{y}}$ directions. That is, $m_x(t)$ and $m_y(t)$ specify the location of the rotation pole within the rotating, body-fixed terrestrial reference frame, where the rotation pole is that point defined by the intersection of the rotation axis with the surface of the Earth near the North Pole. But Earth rotation measurement services do not report the location of the rotation pole within the rotating, body-fixed terrestrial reference frame. Instead, they report the location of the celestial intermediate pole (CIP).

Just three time-dependent angles, the Euler angles, are required to directly transform the coordinates of some station from the terrestrial frame to the celestial frame. These angles are time dependent, of course, because the Earth, and hence the body-fixed terrestrial reference frame, is rotating. But, by tradition, an intermediate reference frame is used with the result that five angles are required to completely transform station coordinates from the terrestrial to the celestial reference frames (e.g., Sovers *et al.*, 1998):

$$\mathbf{r}_c(t) = \mathsf{PNSXY}\,\mathbf{r}_t(t) \quad [47]$$

where $\mathbf{r}_t(t)$ are, in general, the time-dependent coordinates of the station in the rotating, body-fixed terrestrial frame, $\mathbf{r}_c(t)$ are the coordinates of the station in the celestial frame, and P, N, S, X, and Y are the classical transformation matrices with P accounting for the precession of the Earth, N accounting for nutation, S accounting for spin, and X and Y accounting for the x- and y-components of polar motion. By first applying X and Y, the terrestrial coordinates of the station are transformed to an intermediate frame whose reference pole is the CIP; S represents a spin through a large angle about the $\hat{\mathbf{z}}$-axis of the intermediate frame; P and N finally transform the intermediate frame to the celestial frame. This approach of using an intermediate frame and five angles (two polar motion parameters, two nutation parameters, and a spin parameter) to transform station coordinates between the terrestrial and celestial reference frames has been traditionally followed in order to separate polar motion from precession–nutation. This separation is done in such a manner that the precessional and nutational motion of the Earth is long period when observed in the celestial reference frame, and polar motion is long period when observed in the terrestrial reference frame.

Earth rotation measurement services report the parameters that are needed to carry out the transformation given by eqn [47], namely, the polar motion parameters $p_x(t)$ and $p_y(t)$ that are required in X and Y and that give the location of the CIP in the rotating, body-fixed terrestrial reference frame, the nutation parameters $\delta\psi(t)$ and $\delta\varepsilon(t)$ that are required in N and that are corrections in longitude and obliquity to the adopted nutation model that are needed to give the location of the CIP in the celestial reference frame, and a spin parameter $\mathrm{UT}1(t)$ that is required in S and that represents the angle through which the Earth has rotated. The precession transformation matrix P depends on the lunisolar and planetary precession constants.

The relationship between the polar motion parameters $p_x(t)$ and $p_y(t)$ that are reported by Earth rotation measurement services and the elements $\Omega m_x(t)$ and $\Omega m_y(t)$ of the Earth's rotation vector that are needed in eqns [38] and [39] can be derived by considering the properties of transformation matrices (Goldstein, 1950). The transformation of station coordinates between two frames having a common origin implies a rotation. Applying the transformation matrix to the position vector of some station to get its coordinates in a new frame is equivalent to a rotation of the coordinate axes. If the initial reference frame is the terrestrial reference frame of the Earth and the final frame is the celestial reference frame, and because the terrestrial reference frame has been fixed to the body of the Earth, then the equivalent rotation of the coordinate axes is simply the rotation of the Earth. Because an intermediate frame has been used to separate polar motion from precession–nutation, in order to derive the relationship between the polar motion parameters $p_x(t)$ and $p_y(t)$ and the elements $\Omega m_x(t)$ and $\Omega m_y(t)$ of the Earth's rotation vector, it is sufficient to consider the transformation matrix $\mathsf{A}^\mathsf{T}=\mathsf{S}\,\mathsf{X}\,\mathsf{Y}$ that

transforms station coordinates between the terrestrial and intermediate reference frames:

$$\mathbf{r}_i(t) = \mathsf{A}^\mathsf{T}(t)\,\mathbf{r}_t(t) \qquad [48]$$

where the superscript 'T' denotes the transpose and $\mathbf{r}_i(t)$ are the coordinates of the station in the intermediate frame. The elements ω_i of the rotation vector that is associated with this transformation matrix, that is, of the rotation vector of the Earth, are the three independent elements of the antisymmetric matrix $\mathsf{W}(t)$ (Kinoshita et al., 1979; Gross, 1992):

$$\mathsf{W}(t)=\frac{\partial\mathsf{A}(t)}{\partial t}\mathsf{A}^\mathsf{T}(t)=\begin{pmatrix}0 & \omega_z & -\omega_y\\ -\omega_z & 0 & \omega_x\\ \omega_y & -\omega_x & 0\end{pmatrix} \qquad [49]$$

From Sovers et al. (1998), the classical X, Y, and S transformation matrices are

$$\mathsf{X}(t)=\begin{pmatrix}\cos p_x(t) & 0 & -\sin p_x(t)\\ 0 & 1 & 0\\ \sin p_x(t) & 0 & \cos p_x(t)\end{pmatrix} \qquad [50]$$

$$\mathsf{Y}(t)=\begin{pmatrix}1 & 0 & 0\\ 0 & \cos p_y(t) & \sin p_y(t)\\ 0 & -\sin p_y(t) & \cos p_y(t)\end{pmatrix} \qquad [51]$$

$$\mathsf{S}(t)=\begin{pmatrix}\cos H(t) & -\sin H(t) & 0\\ \sin H(t) & \cos H(t) & 0\\ 0 & 0 & 1\end{pmatrix} \qquad [52]$$

where by tradition the positive direction of $p_y(t)$ is taken to be toward 90° W longitude, and H is the hour angle of the true equinox of date which is related to UT1 and the hour angle of the mean equinox of date by the equation of the equinoxes. By forming $\mathsf{A}^\mathsf{T}=\mathsf{S}\,\mathsf{X}\,\mathsf{Y}$, using eqn [49], and keeping terms to first order in small quantities, the desired elements of the rotation vector of the terrestrial frame with respect to the intermediate frame is obtained (Brzezinski, 1992; Gross, 1992; Brzezinski and Capitaine, 1993):

$$\omega_x(t)=\Omega\,p_x(t)-\frac{\partial p_y(t)}{\partial t} \qquad [53]$$

$$\omega_y(t)=-\Omega\,p_y(t)-\frac{\partial p_x(t)}{\partial t} \qquad [54]$$

$$\omega_z(t)=[1+p_z(t)]\,\Omega \qquad [55]$$

where the time rate-of-change of H has been set equal to $(1 + p_z)\Omega$ where $p_z(t) = m_z(t)$ represents small departures from uniform spin at the mean sidereal rotation rate Ω of the Earth.

In complex notation, with $\mathbf{m}(t) = m_x(t) + im_y(t)$ and $\mathbf{p}(t) = p_x(t) - ip_y(t)$, where the negative sign accounts for $p_y(t)$ being positive toward 90° W longitude, eqns [53] and [54] can be written as

$$\mathbf{m}(t) = \mathbf{p}(t) - \frac{i}{\Omega} \frac{\partial \mathbf{p}(t)}{\partial t} \qquad [56]$$

For frequencies of motion $\sigma \ll \Omega$, the second term on the right-hand side of eqn [56] becomes much smaller than the first and the motion of the rotation pole becomes the same as the motion of the CIP. But for frequencies of motion $|\sigma| \approx \Omega$, the motions of the rotation and celestial intermediate poles are very different. This difference becomes important when studying rapid motions such as those caused by the diurnal and semidiurnal ocean tides (see Section 9.4.2.3). For example, the amplitude of the prograde diurnal tide-induced motion of the rotation pole is about twice as large as that of the CIP.

Using eqns [53] and [54] for the relationship between the reported polar motion parameters $p_x(t)$ and $p_y(t)$ and the elements $\omega_x(t)$ and $\omega_y(t)$ of the Earth's rotation vector, eqns [38] and [39] can be written in complex notation as (Brzezinski, 1992; Gross, 1992)

$$\mathbf{p}(t) + \frac{i}{\sigma_o} \frac{\partial \mathbf{p}(t)}{\partial t} = \boldsymbol{\chi}(\boldsymbol{\tau}) \qquad [57]$$

where $\boldsymbol{\chi}(t) = \chi_x(t) + i\chi_y(t)$. The axial component, eqn [40], is usually written in terms of changes $\Delta\Lambda(t)$ of the length of the day as

$$\frac{\Delta\Lambda(t)}{\Lambda_O} = \chi_z(t) \qquad [58]$$

where Λ_O is the nominal length-of-day (LOD) of 86 400 s. Equations [57] and [58] are the final expressions, written in terms of the parameters actually reported by Earth rotation measurement services, for the changes in the rotation of the Earth caused by small excitation $\chi_i(t)$, where the excitation functions are given by eqns [41]–[43].

Since five angles are traditionally used to transform station coordinates between the terrestrial and celestial reference frames when only three are required, the five traditional Earth orientation parameters (EOPs) are not independent of each other. Because the frequency σ_c of some motion as observed in the celestial reference frame is related to the

frequency σ_t of that same motion as observed in the terrestrial reference frame by

$$\sigma_c = \sigma_t + \Omega \qquad [59]$$

then motion having a retrograde nearly diurnal frequency in the terrestrial reference frame ($\sigma_t \approx -\Omega$) will be of low frequency (long period) when observed in the celestial reference frame. That is, retrograde, nearly diurnal polar motions are equivalent to nutations. In particular, the two polar motion parameters are related to the two nutation parameters by (Brzezinski, 1992; Brzezinski and Capitaine, 1993)

$$\mathbf{p}(t) = -\mathbf{n}(t)\, e^{-i\Omega t} \qquad [60]$$

where Greenwich mean sidereal time (GMST) has been approximated by Ωt in the exponent and $\mathbf{n}(t) = \delta\psi(t)\,\sin\varepsilon_o + i\delta\varepsilon(t)$ with ε_o being the mean obliquity of the ecliptic.

A degeneracy also exists between the EOPs and different realizations of the terrestrial reference frame. Since by eqn [49] the Earth's rotation vector can be determined from the transformation matrix that transforms station coordinates between the terrestrial and celestial reference frames, if the realization of the terrestrial reference frame changes, then the elements of the rotation vector will change. In particular, a positive change in the x-component of polar motion is equivalent to a left-handed (clockwise) rotation of the terrestrial reference frame about the $\hat{\mathbf{y}}$-axis; a positive change in the y-component of polar motion, remembering that $p_y(t)$ is defined to be positive toward 90° W longitude, is equivalent to a left-handed rotation of the terrestrial reference frame about the $\hat{\mathbf{x}}$-axis; and a positive change in UT1 is equivalent to a right-handed (counter-clockwise) rotation of the terrestrial reference frame about the $\hat{\mathbf{z}}$-axis.

The polar motion parameters $p_x(t)$ and $p_y(t)$ give the location in the terrestrial frame of the reference pole of the intermediate frame, whatever that intermediate frame may be. The 1980 International Astronomical Union (IAU) theory of nutation adopted the celestial ephemeris pole (CEP) as the reference pole of the intermediate frame (Seidelmann, 1982), defining it to be a pole that exhibits no nearly diurnal motions in either the body-fixed terrestrial frame or in the celestial frame. The CEP was chosen (Seidelmann, 1982) to be the B-axis of Wahr (1981), which is the axis of figure for the Tisserand mean outer surface of the Earth, where the averaging procedure is such that the resulting *B*-axis does not move in response to body tides. Since observing stations are attached to the outer surface of the Earth, their measurements are sensitive to the motion of the Earth's surface in space. Wahr (1981) thus generalized the concept of the Tisserand mean-mantle frame to that of the Tisserand mean-surface frame, with his *B*-axis being the reference axis of the Tisserand mean-surface frame that moves in space with the mean motion of the observing stations.

In 2000, the IAU adopted the CIP as the reference pole of the intermediate frame (McCarthy and Petit, 2004, chapter 5). The definition of the CIP extends that of the CEP by clarifying the definition of polar motion and precession–nutation. The CEP was defined in such a manner that it exhibits no nearly diurnal motion in either the terrestrial or celestial reference frames. That is, precession– nutation was considered to be motion of the CEP as viewed in the celestial reference frame with the frequency of motion ranging between −0.5 and +0.5 cycles per sidereal day (cpsd), and polar motion was considered to be motion of the CEP as viewed in the terrestrial reference frame with the frequency of motion in that frame ranging between −0.5 and +0.5 cpsd (Capitaine, 2000). Since frequencies of motion in the two frames are related by eqn [59], in the celestial reference frame polar motion was motion of the CEP with frequencies ranging between +0.5 and +1.5 cpsd. Thus, motion of the CEP in the celestial frame was defined for frequencies between −0.5 and +1.5 cpsd, with the division between polar motion and precession–nutation being at a frequency of +0.5 cpsd. Motion of the CEP outside this celestial frequency band was undefined. Similarly, motion of the CEP in the terrestrial reference frame was defined for frequencies between −1.5 and +0.5 cpsd with the division between polar motion and precession–nutation being at a frequency of −0.5 cpsd. Motion of the CEP outside this terrestrial frequency band was also undefined (see **Figure 1**).

Since 1980, when the CEP was adopted as the reference pole of the intermediate frame, models of polar motion with frequencies outside the terrestrial frequency band within which the motion of the CEP was defined became available. These polar motions were due to the effects of diurnal and semidiurnal ocean tides. Models of nutations having frequencies outside the celestial frequency band within which the CEP was defined also became available, as did space-geodetic measurements having subdaily temporal resolution.

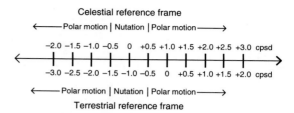

Figure 1 Schematic illustration of the relationship between the frequency of some motion as viewed in the celestial (top half of figure) and terrestrial (bottom half of figure) reference frames (see eqn [59]). By convention, nutation is motion of the CIP having frequencies in the range [−0.5, +0.5] cpsd as viewed in the celestial frame. As viewed in the terrestrial frame, this same nutational motion of the Earth has frequencies in the range [−1.5, −0.5] cpsd. Motion at all other frequencies is considered to be polar motion.

With these improvements in models and measurements came the need to extend the definition of the CEP to all possible frequencies of motion, not just those between −0.5 and +1.5 cpsd in the celestial frame, or −1.5 to +0.5 cpsd in the terrestrial frame. At the time that the definition of the intermediate pole was extended, its name was changed to the CIP. When defining the CIP, the concept used in defining the CEP of a pole having no nearly diurnal motions in either the terrestrial or celestial reference frames had to be abandoned because ocean tides can cause polar motions having frequencies near +1 cpsd as viewed in the terrestrial reference frame.

The CIP is still chosen to be the axis of figure for the Tisserand mean outer surface of the Earth, as it was for the CEP. And the CIP is defined in such a manner that precession–nutation is still considered to be motion of the CIP as viewed in the celestial reference frame with the frequency of motion ranging between −0.5 and +0.5 cpsd (Capitaine, 2000). But now polar motion is considered to be motion of the CIP in the celestial frame at all other frequencies, or motion of the CIP in the terrestrial frame at all frequencies except those between −1.5 and −0.5 cpsd (see **Figure 1**). This has the effect of including in nutation those ocean tidal terms having retrograde nearly diurnal frequencies as viewed in the terrestrial reference frame, and of including in polar motion those nutation terms having frequencies less than −0.5 cpsd or greater than +0.5 cpsd as viewed in the celestial reference frame (see **Table 2**).

Table 2   Coefficients of those nutation terms that are included in polar motion by definition of the CIP

Degree of potential	Fundamental argument						Period (solar days)	$p_x(t)$, (μas)		$p_y(t)$, (μas)	
	γ	l	l'	F	D	Ω		sin	cos	sin	cos
2	1	1	0	0	0	0	0.96244	0.76	−0.43	0.43	0.76
2	1	0	0	0	0	−1	0.99712	1.93	−1.11	1.11	1.93
2	1	0	0	0	0	0	0.99727	14.27	−8.19	8.19	14.27
2	1	0	0	−2	2	−2	1.00275	−4.76	2.73	−2.73	−4.76
2	1	−1	0	0	0	0	1.03472	0.84	−0.48	0.48	0.84
2	1	0	0	−2	0	−2	1.07581	−11.36	6.52	−6.52	−11.36
2	1	0	0	−2	0	−1	1.07598	−2.14	1.23	−1.23	−2.14
2	1	1	0	−2	−2	−2	1.22346	−0.44	0.25	−0.25	−0.44
2	1	−1	0	−2	0	−2	1.11951	−2.31	1.32	−1.32	−2.31
2	1	−1	0	−2	0	−1	1.11970	−0.44	0.25	−0.25	−0.44
3	0	1	0	1	0	1	13.719	1.28	0.16	−0.16	1.28
3	0	0	0	1	0	0	27.212	2.62	0.32	−0.32	2.62
3	0	0	0	1	0	1	27.322	16.64	2.04	−2.04	16.64
3	0	0	0	1	0	2	27.432	−0.87	−0.11	0.11	−0.87
3	0	1	0	1	−2	1	193.560	2.10	0.27	−0.27	2.10
3	0	0	0	1	−1	1	365.242	1.31	0.20	−0.20	1.31
3	0	1	1	−1	0	−1	411.807	1.05	0.27	−0.27	1.05
3	0	1	1	−1	0	0	438.360	−0.63	0.12	−0.12	−0.63
3	0	−1	0	1	0	0	2190.35	−2.78	−0.31	0.31	−2.78
3	0	−1	0	1	0	1	3231.50	−16.16	−1.83	1.83	−16.16
3	0	−1	0	1	0	2	6159.14	0.78	0.09	−0.09	0.78
3	0	1	0	−1	0	−2	−6159.14	−0.68	−0.09	0.09	−0.68
3	0	1	0	−1	0	−1	−3231.50	12.32	1.59	−1.59	12.32
3	0	1	0	−1	0	0	−2190.35	1.86	0.24	−0.24	1.86
3	0	−1	0	−1	2	−1	−193.560	0.81	0.10	−0.10	0.81
3	0	0	0	−1	0	−2	−27.432	−0.82	−0.10	0.10	−0.82
3	0	0	0	−1	0	−1	−27.322	15.75	1.93	−1.93	15.75
3	0	0	0	−1	0	0	−27.212	2.48	0.30	−0.30	2.48
3	0	−1	0	−1	0	−1	−13.719	1.39	0.17	−0.17	1.39

Rate of secular polar motion (μas/yr) due to the zero-frequency tide

4	0	0	0	0	0	0			−3.80		−4.31

Terms with amplitudes less than 0.5 microarcsecond (μas) are not tabulated. γ is GMST reckoned from the lower culmination of the vernal equinox (GMST + π). l, l', F, D, and Ω are the Delaunay arguments, expressions for which are given in Simon et al. (1994). The period, given in solar days, is the approximate period of the term as viewed in the terrestrial reference frame. Terms having positive (negative) periods indicate prograde (retrograde) circular motion. Summing prograde and retrograde circular motions having the same period yields elliptical motion. The nearly diurnal terms, like the nearly diurnal and nearly semidiurnal ocean tidal terms, are not included in the polar motion parameters reported by Earth rotation measurement services. However, the secular rate and the long-period terms, like the long-period ocean tidal terms, are included in the reported polar motion parameters. Also see McCarthy and Petit (2004, table 5.1).
Reproduced from Mathews PM and Bretagnon P (2003) Polar motions equivalent to high frequency nutations for a nonrigid Earth with anelastic mantle. *Astronomy and Astrophysics* 400: 1113–1128, with permission from Springer.

9.3 Earth Rotation Measurement Techniques

Changes in the Earth's rate of rotation become apparent when comparing time kept by the rotating Earth, known as Universal Time (UT), to uniform timescales based either upon atomic clocks or upon the motion of the Sun and other celestial bodies. Prior to the development of atomic clocks, the most accurate measurements of changes in the Earth's rate of rotation were obtained by timing the occultations of stars

by the Moon. With the advent of atomic clocks in 1955, a uniform atomic timescale became available that could be used as a reference when measuring the transit times of stars as they pass through the local meridian. Changes in the Earth's rate of rotation could then be determined more accurately from optical astrometric measurements of star transits than they could from measurements of lunar occultations. And prior to the development of space-geodetic techniques, optical astrometric measurements of changes in the apparent latitudes of observing stations yielded the most accurate estimates of polar motion. The space-geodetic techniques of very long baseline interferometry (VLBI), global navigation satellite systems (GNSSs) like the global positioning system (GPS), and satellite and lunar laser ranging (LLR) are now the most accurate techniques available for measuring changes in both the Earth's rate of rotation and in polar motion.

9.3.1 Lunar Occultation

The most recent re-reduction of lunar occultation measurements for Universal Time and Length-of-day changes is that of Jordi *et al.* (1994) who analyzed about 53 000 observations of lunar occultations spanning 1830.0–1955.5. They used a reference frame defined by the FK5 star catalog, the LE200 lunar ephemeris, and corrections for the limb profile of the Moon. The Universal Time series they obtained consists of values and 1σ uncertainties for the difference between Terrestrial Time (TT) and Universal Time (TT–UT1) spanning 1830.0–1955.5 at 4-month intervals. Terrestrial Time is a dynamical timescale that can be related to International Atomic Time (TAI) by adding 32.184 s to TAI (Seidelmann *et al.*, 1992).

Jordi *et al.* (1994) extended their (TT–UT1) series to 1992 by using values of (UT1–TAI) obtained from the Bureau International de l'Heure (BIH) and the International Earth Rotation and Reference Systems Service (IERS). They then derived a LOD series spanning 1830–1987 at 4-month intervals by finite differencing and smoothing the extended UT1 series. Gross (2001) combined the lunar occultation measurements of Jordi *et al.* (1994) with optical astrometric and space-geodetic measurements to produce a smoothed LOD series spanning 1832.5–1997.5 at yearly intervals. Other UT1 and LOD series based upon lunar occultation measurements are those of Morrison (1979), Stephenson and Morrison (1984), McCarthy and Babcock (1986), and Liao and Greiner-Mai (1999).

9.3.2 Optical Astrometric

The International Latitude Service (ILS) was established by the International Association of Geodesy (IAG) in 1895 for the purpose of monitoring the wobbling motion of the Earth that had been detected by Seth Carlo Chandler, Jr., in 1891. As the Earth wobbles, the apparent latitude of an astronomical observing station will vary. To measure this variation of latitude and infer the underlying polar motion that

is causing it, the ILS established six observing stations that were well distributed in longitude and that were all located at nearly the same latitude of 39° 8′ N. A seventh station, Kitab, was added in 1930 to replace the station at Tschardjui that ceased operations in 1919 due to a nearby river changing its course and adversely affecting the seeing conditions at Tschardjui. Locating all the ILS stations at nearly the same latitude allowed common star pairs to be observed by the same Horrebow–Talcott method (Munk and MacDonald, 1960, chapter 7), thereby allowing the polar motion to be determined from the latitude observations free of first-order errors in the reference star catalog.

The use of different star catalogs, standards, and data reduction procedures during the history of the ILS observing program can introduce discontinuities in the polar motion series derived from the ILS observations. In order to produce a homogeneous polar motion series unaffected by these sources of error, Yumi and Yokoyama (1980) re-reduced 772 395 latitude observations taken at the seven ILS observing stations using the Melchior and Dejaiffe (1969) star catalog and the 1964 IAU System of Astronomical Constants. The resulting polar motion series, known as the homogeneous ILS series, spans October 1899 to December 1978 at monthly intervals.

During the 20th century, numerous other optical astrometric measurements of latitude and longitude were taken at other stations and by other methods besides those of the ILS. At the BIH, Li (1985) and Li and Feissel (1986) re-reduced 240 140 optical astrometric measurements of longitude and 259 159 measurements of latitude taken at 136 observing stations. In order to produce an Earth orientation series independent of the ILS series, no measurements taken at the ILS stations were used. The resulting UT1 and polar motion series, which is known as the BIH series, spans January 5.0, 1962, to December 31.0, 1981, at 5-day intervals.

The High Precision Parallax Collecting Satellite (HIPPARCOS) star catalog, being constructed from observations taken above the Earth's atmosphere, is substantially more accurate than earlier ground-based catalogs such as the Melchior and Dejaiffe (1969) catalog used by Yumi and Yokoyama (1980). The improved accuracy of the HIPPARCOS star catalog motivated Vondrák (1991, 1999) and Vondrák *et al.* (1992, 1995, 1997, 1998) to once again re-reduce optical astrometric measurements for EOPs. All available optical astrometric measurements, numbering 4 315 628 from 48 instruments including those taken at the ILS stations, were collected and corrected for instrumental errors and such systematic effects as plate tectonic motion, ocean loading, and tidal variations. The corrected measurements were then used to estimate nutation, polar motion, and Universal Time. The resulting Earth orientation series, known as the HIPPARCOS series, consists of values and uncertainties for polar motion and nutation spanning 1899.7–1992.0 at quasi-5-day intervals, with UT1 estimates starting in 1956 shortly after atomic clocks first became available.

9.3.3 Space Geodetic

An integral part of geodesy has always been the definition and realization of a terrestrial, body-fixed reference frame, a celestial, space-fixed reference frame, and the determination of the EOPs (precession, nutation, spin, and polar motion) that link these two reference frames together. But with the advent of space geodesy – with the placement of laser retroreflectors on the Moon by Apollo astronauts and Soviet landers, the launch of the LAser GEOdynamics Satellite (LAGEOS), the development of VLBI, and the development of GNSSs like the GPS – a quantum leap has been taken in our ability to realize the terrestrial and celestial reference frames and to determine the EOPs.

The only space-geodetic measurement technique capable of independently determining all of the EOPs is multibaseline VLBI (*see* Chapter 11). All of the other techniques need to either apply external constraints to the determined EOPs or can determine only subsets of the EOPs, only linear combinations of the EOPs, or only their time rates-of-change.

9.3.3.1   Very long baseline interferometry

Radio interferometry is routinely used to make highly accurate measurements of UT1 and polar motion with observing sessions lasting from about an hour to a day. The VLBI technique measures the difference in the arrival time of a radio signal at two or more radio telescopes that are simultaneously observing the same distant extragalactic radio source (Lambeck, 1988, chapter 8; Robertson, 1991; Sovers *et al.*, 1998). This technique is therefore sensitive to processes that change the relative position of the radio telescopes with respect to the source, such as a change in the orientation of the Earth in space or a change in the position of the telescopes due to, for example, tidal displacements or tectonic motions. If

just two telescopes are observing the same source, then only two components of the Earth's orientation can be determined. A rotation of the Earth about an axis parallel to the baseline connecting the two radio telescopes does not change the relative position of the telescopes with respect to the source, and hence this component of the Earth's orientation is not determinable from VLBI observations taken on that single baseline. Multibaseline VLBI observations with satisfactory geometry can determine all of the components of the Earth's orientation including their time rates-of-change.

The International VLBI Service for Geodesy and Astrometry (IVS; Schlüter *et al.*, 2002), a service of both the IAG and the IAU, was established on 11 February 1999 to support research in geodesy, geophysics, and astrometry. As part of its activities, it coordinates the acquisition and reduction of VLBI observations for the purpose, in part, of monitoring changes in the Earth's rotation and defining and maintaining the international terrestrial and celestial reference frames. VLBI data products, including EOPs determined from both single and multibaseline observations, are available through the IVS website (see **Table 3**).

9.3.3.2   Global navigation satellite system

GNSSs consist of two major elements: (1) a space-based element consisting of a constellation of transmitting satellites, and (2) a ground-based element consisting of a network of receivers. In the GPS of the United States, the satellites, including spares, are at altitudes of 20 200 km in orbits with periods of 11 h 58 min located in six orbital planes, each inclined at 55° to the Earth's equator with four or more satellites in each plane. In the GLObal NAvigation Satellite System (GLONASS) of Russia, the satellites are at altitudes of 19 100 km in circular orbits with periods of 11 h 15 min located in three orbital planes, each inclined at 64.8° to the Earth's equator with eight satellites in each plane. In the future Galileo system of Europe, which is expected to be fully operational in 2008, the satellites will be at altitudes of 23 222 km in circular orbits with periods of 14 h 22 min located in three orbital planes, each inclined at 56° to the Earth's equator with nine satellites and one spare in each plane.

In GNSSs, the navigation signals are broadcast by the satellites at more than one frequency, thereby enabling first-order corrections to be made for ionospheric refraction effects. The ground-based multichannel receivers detect the navigation signals being broadcast by those satellites that are above the

Table 3 Sources of Earth orientation data

Data type	URL
Very long baseline interferometry	
IVS	http://ivscc.gsfc.nasa.gov/
Global navigation satellite system	
IGS	http://igscb.jpl.nasa.gov/
Satellite and lunar laser ranging	
ILRS	http://ilrs.gsfc.nasa.gov/
DORIS	
IDS	http://ids.cls.fr/
Intertechnique combinations	
IERS	http://www.iers.org/
JPL	ftp://euler.jpl.nasa.gov/keof/

URL, Uniform Resource Locator, IVS, International VLBI Service for Geodesy and Astrometry; VLBI, very long baseline interferometry; IGS, International GNSS Service; GNSS, global navigation satellite system; ILRS, International Laser Ranging Service; DORIS, Doppler orbitography and radio positioning integtrated by satellite; IDS International DORIS Service; IERS, International Earth Rotation and Reference Systems Service; JPL, Jet Propulsion Laboratory.

horizon (up to the number of channels in the receiver). In principle, trilateration can then be used to determine the position of each receiver, and by extension the orientation of the network of receivers as a whole. In practice, in order to achieve higher accuracy, more sophisticated analysis techniques are employed to determine the EOPs and other quantities such as orbital parameters of the satellites, positions of the stations, and atmospheric parameters such as the zenith path delay (Bock and Leppard, 1990; Blewitt, 1993; Beutler *et al.*, 1996; Hofmann-Wellenhof *et al.*, 1997; Leick, 2003).

Only polar motion and its time rate-of-change can be independently determined from GNSS measurements. UT1 cannot be separated from the orbital elements of the satellites and hence cannot be determined from GNSS data. The time rate-of-change of UT1, which is related to the length of the day, can be determined from GNSS measurements. But because of the corrupting influence of orbit error, VLBI measurements are usually used to constrain the GNSS-derived LOD estimates.

The International GNSS Service (IGS; Beutler *et al.*, 1999), a service of the IAG, was established on 1 January 1994 under its former name of the International GPS Service to support Earth science research. As part of its activities, it coordinates the acquisition and reduction of GNSS observations for the purpose, in part, of maintaining the ITRF and monitoring changes in the Earth's rotation and geocenter. GNSS data products, including EOPs, are available through the IGS website (see **Table 3**).

9.3.3.3 Satellite and lunar laser ranging

In the technique of satellite laser ranging (SLR), the round trip time-of-flight of laser light pulses are accurately measured as they are emitted from a laser system located at some ground-based observing station, travel through the Earth's atmosphere to some artificial satellite orbiting the Earth, are reflected by retroreflectors carried onboard that satellite, and return to the same observing station from which they were emitted (Lambeck, 1988, chapter 6). This time-of-flight range measurement is converted into a distance measurement by using the speed of light and correcting for a variety of known or modeled effects such as atmospheric path delay and satellite center-of-mass offset. Although a number of satellites carry retroreflectors for tracking and navigation purposes, the LAGEOS I and II satellites were specifically designed and launched to study geodetic properties of the Earth including its rotation and are the satellites most commonly used to determine EOPs. Including range measurements to the Etalon I and II satellites has been found to strengthen the solution for the EOPs, so these satellites are now often included when determining EOPs.

The EOPs are recovered from the basic range measurements in the course of determining the satellite's orbit. The basic range measurement is sensitive to any geophysical process that changes the distance between the satellite and the observing station, such as displacements of the satellite due to perturbations of the Earth's gravitational field, motions of the observing station due to tidal displacements or plate tectonics, or a change in the orientation of the Earth (which changes the location of the observing station with respect to the satellite). These and other geophysical processes must be modeled when fitting the satellite's orbit to the range measurements as obtained at a number of globally distributed tracking stations. Adjustments to the *a priori* models used for these effects can then be obtained during the orbit determination procedure, thereby enabling, for example, the determination of station positions and EOPs (Smith *et al.*, 1985, 1990, 1994; Tapley *et al.*, 1985, 1993). However, because variations in UT1 cannot be separated from variations in the orbital node of the satellite, which are caused by the effects of unmodeled forces acting on the satellite, it is not possible to independently determine UT1 from SLR measurements. But independent estimates of the time rate-of-change of UT1, or equivalently, of LOD, can be determined from SLR measurements, as can polar motion and its time rate-of-change.

The technique of LLR is similar to that of SLR except that the laser retroreflector is located on the Moon instead of on an artificial satellite (Mulholland, 1980; Lambeck, 1988, chapter 7; Williams *et al.*, 1993; Dickey *et al.*, 1994a; Shelus, 2001). LLR is technically more challenging than SLR because of the need to detect the much weaker signal that is returned from the Moon. Larger, more powerful laser systems with more sophisticated signal detection systems need to be employed in LLR; consequently, there are far fewer stations that range to the Moon than to artificial satellites. In fact, there are currently only two stations that regularly range to the Moon: the McDonald Observatory in Texas and the Observatoire de la Côte d'Azur in the south of France.

The EOPs are typically determined from LLR data by analyzing the residuals at each station after the lunar orbit and other parameters such as station and reflector locations have been fit to the range measurements (Stolz *et al.*, 1976; Langley *et al.*, 1981a; Dickey *et al.*, 1985). From this single-station technique, two linear combinations of UT1 and the polar motion parameters $p_x(t)$ and $p_y(t)$ can be determined, namely, UT0 and the variation of latitude $\Delta\phi_i(t)$ at that station:

$$\Delta\phi_i(t) = p_x(t)\cos\lambda_i - p_y(t)\sin\lambda_i \quad [61]$$

$$\begin{aligned} \text{UT0}(t) - \text{TAI}(t) = \text{UT1}(t) - \text{TAI}(t) + [\,p_x(t)\sin\lambda_i \\ + p_y(t)\cos\lambda_i]\tan\phi_i \quad [62] \end{aligned}$$

where ϕ_i and λ_i are the nominal latitude and longitude of station i. A rotation of the Earth about

an axis connecting the station with the origin of the terrestrial reference frame does not change the distance between the station and the Moon, and hence this component of the Earth's orientation cannot be determined from single-station LLR observations.

The International Laser Ranging Service (ILRS; Pearlman *et al.*, 2002, 2005) is a service of the IAG that was established on 22 September 1998 to support research in geodesy, geophysics, and lunar science. As part of its activities, the ILRS coordinates the acquisition and reduction of SLR and LLR observations for the purpose, in part, of maintaining the ITRF and monitoring the Earth's rotation and geocenter motion. SLR and LLR data products, including EOPs, are available through the ILRS website (see **Table 3**).

9.3.3.4 Doppler orbitography and radio positioning integrated by satellite

Like GNSSs, the French Doppler orbitography and radio positioning integrated by satellite (DORIS) system also consists of space-based and ground-based elements (Tavernier *et al.*, 2005; Willis *et al.*, 2006). But in the DORIS system, the transmitting beacons are located on the ground and the receivers are located on the satellites. Currently, there are 56 globally distributed beacons broadcasting to receivers onboard five satellites (SPOT-2, SPOT-4, SPOT-5, Jason-1, and the Environmental Satellite (ENVISAT)). The ocean topography experiment (TOPEX)/positioning ocean solid earth ice dynamics orbiting/orbital navigator (POSEIDON) (TOPEX/POSEIDON) and decommissioned SPOT-3 satellites also carried DORIS receivers and the future Jason-2 satellite will carry a DORIS receiver.

In the DORIS system, the Doppler shift of two transmitted frequencies is accurately measured. The use of two frequencies allows corrections to be made for ionospheric effects. Processing these Doppler measurements, usually as range-rate (Tapley *et al.*, 2004), allows the orbit of the satellite to be determined along with other quantities such as station positions and EOPs. As with other satellite techniques, UT1 cannot be determined from DORIS measurements, but its time rate-of-change can be determined, as can polar motion and its rate-of-change (Willis *et al.*, 2006).

The International DORIS Service (IDS; Tavernier *et al.*, 2005) is a service of the IAG that was established on 1 July 2003 to support research in geodesy and geophysics. As part of its activities, the

IDS coordinates the acquisition and reduction of DORIS observations for the purpose, in part, of maintaining the ITRF and monitoring the Earth's rotation and geocenter motion. DORIS data products, including EOPs, are available through the IDS website (see **Table 3**).

9.3.4 Ring Laser Gyroscope

Ring laser gyroscopes are a promising emerging technology for determining the Earth's rotation. In a ring laser gyroscope, two laser beams propagate in opposite directions around a ring. Since the ring laser gyroscope is rotating with the Earth, the effective path length of the beam that is corotating with the Earth is slightly longer than the path that is counter-rotating with it. Because the effective path lengths of the two beams differ, their frequencies differ, so they interfere with each other to produce a beat pattern (Stedman, 1997). The beat frequency $\delta f(t)$, which can be measured and is known as the Sagnac frequency, is proportional to the instantaneous angular velocity $\boldsymbol{\omega}(t)$ of the Earth:

$$\delta f(t) = \frac{4A}{\lambda P} \mathbf{n} \cdot \boldsymbol{\omega}(t) \qquad [63]$$

where A is the area of the ring, P is the optical path length, λ is the wavelength of the laser beam in the absence of rotation, and $\mathbf{n}$ is the unit vector normal to the plane of the ring. By making the instrument rigid and the laser stable so that A, P, and λ are constants, the component of the Earth's instantaneous rotation vector that is parallel to the normal of the plane of the ring can be determined from measurements of the Sagnac frequency. All three components of the Earth's rotation vector are determinable from three mutually orthogonal colocated ring laser gyroscopes, or from a globally distributed network of gyroscopes.

Ring laser gyroscopes measure the absolute rotation of the Earth in the sense that, in principle, just a single measurement of the Sagnac frequency is required to determine $\mathbf{n} \cdot \boldsymbol{\omega}$. All of the other techniques discussed above are relative sensors because they infer the Earth's rotation from the change in the orientation of the Earth that takes place between at least two measurements that are separated in time. Thus, ring laser gyroscopes are fundamentally different from the space-geodetic techniques that are currently used to regularly monitor the Earth's rotation. Measurements taken by ring laser gyroscopes can therefore be expected to contribute to our understanding of the Earth's changing rotation, particularly

of changes that occur on subdaily to daily timescales (Schreiber *et al.*, 2004).

9.3.5 Intertechnique Combinations

EOPs can be determined from measurements taken by each of the techniques discussed above. But each technique has its own unique strengths and weaknesses in this regard. Not only is each technique sensitive to a different subset and/or linear combination of the EOPs, but the averaging time for their determination is different, as is the interval between observations, the precision with which they can be determined, and the duration of the resulting EOP series. By combining the individual series determined by each technique, a series of the Earth's orientation can be obtained that is based upon independent measurements and that spans the greatest possible time interval. Such a combined Earth orientation series is useful for a number of purposes, including a variety of scientific studies and as an *a priori* series for use in data reduction procedures. However, care must be taken in generating such a combined series in order to account for differences in the underlying reference frames within which each individual series is determined, which can lead to differences in the bias and rate of the Earth orientation series, as well as to properly assign weights to the observations prior to combination.

The IERS, a service of both the IAG and the IAU, was established on 1 January 1988 under its former name of the International Earth Rotation Service to serve the astronomical, geophysical, and geodetic communities by, in part, providing the International Celestial Reference Frame (ICRF), the ITRF, and the EOPs that are needed to transform station coordinates between these two frames (Feissel and Gambis, 1993; Vondrák and Richter, 2004). As part of its activities, the IERS combines and predicts EOPs determined by the space-geodetic techniques using a weighted-average approach (Luzum *et al.*, 2001; Gambis, 2004; Johnson *et al.*, 2005). The EOPs produced by the IERS are available through its website (see **Table 3**).

Since the 1980s, a Kalman filter has been used at the Jet Propulsion Laboratory (JPL) to combine and predict EOPs in support of interplanetary spacecraft tracking and navigation (Freedman *et al.*, 1994a; Gross *et al.*, 1998). A Kalman filter has many properties that make it an attractive method of combining Earth orientation series. It allows the full accuracy of the measurements to be used, whether they are degenerate or full rank, are irregularly or regularly spaced in time, or are corrupted by systematic or other errors that can be described by stochastic models. And by using a stochastic model for the process, a Kalman filter can objectively account for the growth in the uncertainty of the EOPs between measurements. The combined and predicted Earth orientation series produced at JPL using a Kalman filter can be obtained by anonymous ftp from the address given in **Table 3**.

9.4 Observed and Modeled Earth Rotation Variations

The Earth's rotation changes on all observable time-scales, from subdaily to decadal and longer. The wide range of timescales on which the Earth's rotation changes reflects the wide variety of processes that are causing it to change, including external tidal forces, surficial fluid processes involving the atmosphere, oceans, and hydrosphere, and internal processes acting both within the solid Earth itself and between the fluid core and solid Earth. The changes in the Earth's rotation that are observed and the models that have been developed to explain them are reviewed here.

9.4.1 UT1 and LOD Variations

LOD observations (**Figure 2**) show that it consists mainly of (1) a linear trend of rate +1.8 ms/cy, (2) decadal variations having an amplitude of a few milliseconds, (3) tidal variations having an amplitude of about 1 ms, (4) seasonal variations having an amplitude of about 0.5 ms, and (5) smaller amplitude variations occurring on all measurable timescales. Here, the length-of-day variations that are observed and the models that have been developed to explain them are reviewed (*see* Chapter 10).

The length of the day is the rotational period of the Earth. Changes $\Delta\Lambda(t)$ in the length of the day are related to changes in Universal Time and to changes $\Delta\omega_z(t) = \Omega m_z(t)$ in the axial component of the Earth's angular velocity by

$$\frac{\Delta\Lambda(t)}{\Lambda_0} = -m_z(t) = -\frac{\partial(\text{UT1}-\text{TAI})}{\partial t} \qquad [64]$$

By eqns [43] and [58], the LOD will change as a result of changes in the axial component $h_z(t)$ of relative angular momentum and of changes in the axial component $\Omega \Delta I_{zz}(t)$ of angular momentum due to changes in the mass distribution of the Earth. Modeling the observed changes in the LOD therefore requires computing both types of changes in

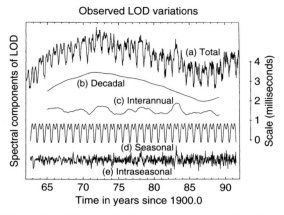

Observed LOD variations

Figure 2   (a) Observed LOD variations during 1962–91 from the COMB91 combined optical astrometric and space-geodetic Earth orientation series and its decomposition into variations on (b) decadal, (c) interannual, (d) seasonal, and (e) intraseasonal timescales. Tidal variations are not shown. From Dickey JO, Marcus SL, Hide R, Eubanks TM, and Boggs DH (1994b) Angular momentum exchange among the soild Earth, atmosphere and oceans: A case study of the 1982–83. El Niño event. *Journal of Geophysical Research* 99(B12): 23921–23937.

angular momentum for the different components of the Earth system. By definition, the angular momentum vector $\mathbf{L}(t)$ of some component of the Earth system such as the atmosphere or oceans is given by

$$\mathbf{L}(t) = \int_{V(t)} \rho(\mathbf{r}, t)\, \mathbf{r} \times [\boldsymbol{\omega} \times \mathbf{r} + \mathbf{u}(\mathbf{r}, t)]\, \mathrm{d}V \qquad [65]$$

where the integral is taken over the, in general, time-dependent volume $V(t)$ of that component of the Earth system under consideration, the first term in the square brackets represents changes in the angular momentum due to changes in mass distribution, the second term represents changes due to relative motion, and $\mathbf{r}$ is the position vector of some mass element of density $\rho(\mathbf{r},t)$ that is moving with Eulerian velocity $\mathbf{u}(\mathbf{r},t)$ with respect to a terrestrial reference frame that (1) is fixed to the solid Earth, (2) is oriented such that its x- and y-coordinate axes are along the Greenwich and 90° E meridians, respectively, with both axes lying in the equatorial plane, (3) has an origin located at the center of the Earth, and (4) is rotating with angular velocity $\boldsymbol{\omega}$ with respect to an inertial reference frame. In general, $\boldsymbol{\omega}$ is not constant in time but exhibits variations in both magnitude (related to changes in the LOD) and direction (related to polar motion). However, these changes are small, and for the purpose of deriving expressions for the angular momentum of different components of the Earth system it can be assumed that the terrestrial reference frame is uniformly rotating at the rate Ω about the z-coordinate axis: $\boldsymbol{\omega} = \Omega\,\hat{\mathbf{z}}$. In this case, the axial component of the mass term of the angular momentum can be written as

$$\Omega\,\Delta I_{zz}(t) = \Omega \int_{V(t)} \rho(\mathbf{r},\,t)\, r^2 \cos^2\phi\, dV \qquad [66]$$

where ϕ is north latitude. Similarly, the axial component of the motion term of the angular momentum can be written as

$$h_z(t) = \int_{V(t)} \rho(\mathbf{r},\,t)\, r \cos\phi\, u(\mathbf{r},\,t)\, dV \qquad [67]$$

where $u(\mathbf{r}, t)$ is the eastward component of the velocity.

9.4.1.1   Secular trend, tidal dissipation, and glacial isostatic adjustment

Tidal dissipation causes the Earth's angular velocity and hence rotational angular momentum to decrease (*see* Chapter 7). Since the angular momentum of the Earth–Moon system is conserved, the orbital angular momentum of the Moon must increase to balance the decrease in the Earth's rotational angular momentum. The increase in the orbital angular momentum of the Moon is accomplished by an increase in the radius of the Moon's orbit and a decrease in the Moon's orbital angular velocity. But early observations of the Moon's position showed that it was apparently accelerating, not decelerating, in its orbit. This apparent acceleration of the Moon was a result of assuming that the Earth is rotating with a constant, rather than decreasing, angular velocity when predicting the Moon's position. If the Earth's angular velocity is actually decreasing but is assumed to be constant when predicting the position of the Moon, then the observed position of the Moon will appear to be ahead of its predicted position, that is, the Moon will appear to be accelerating in its orbit. That the Moon is apparently accelerating in its orbit was first noted by Halley (1695). But it was not until 1939 that Spencer Jones (1939) was able to conclusively demonstrate that the angular velocity of the Earth is actually decreasing and that the apparent acceleration of the Moon in its orbit was an artifact of assuming that the angular velocity of the Earth was constant.

Halley (1695) also seems to have been the first to appreciate the importance of ancient and medieval records of lunar and solar eclipses for determining the apparent acceleration of the Moon and the corresponding decrease in the angular velocity of the Earth over the past few thousand years. The change in the Earth's rate of rotation can be deduced from the discrepancy between when and where eclipses should have been observed if the angular velocity of the Earth were constant to when and where they were actually observed as recorded in Babylonian clay tablets and Chinese, European, and Arabic books and manuscripts (Stephenson, 1997).

When using eclipse observations to deduce the secular change in the length of the day over the past few thousand years, the positions of the Sun and Moon must be accurately known. Of primary importance in this regard is the value for the tidal acceleration $\dot{n}$ of the Moon since it controls the long-term behavior of the Moon's motion. The tidal acceleration of the Moon can be determined from observations of the timings of transits of Mercury (e.g., Spencer Jones, 1939; Morrison and Ward, 1975) as well as from satellite and lunar laser ranging measurements. Tidal forces distort the figure of the Earth and hence its gravitational field which in turn perturbs the orbits of artificial satellites. SLR measurements can detect these tidal perturbations in the satellites' orbits and can therefore be used to construct tide models and hence determine the tidal acceleration of the Moon. Using this approach, Christodoulidis *et al.* (1988) report a value of -25.27 ± 0.61 arcseconds per century^2 ($''/\text{cy}^2$) for the tidal acceleration $\dot{n}$ of the Moon due to dissipation by solid Earth and ocean tides. Other SLR-derived values for $\dot{n}$ have been reported by Cheng *et al.* (1990, 1992), Marsh *et al.* (1990, 1991), Dickman (1994), Lerch *et al.* (1994), and Ray (1994).

Like the orbits of artificial satellites, the orbit of the Moon is also perturbed by tidal forces. Since LLR measurements can detect tidal perturbations in the Moon's orbit, they can be used to determine the tidal acceleration of the Moon. In addition to being sensitive to orbital perturbations caused by tides on the Earth, LLR measurements, unlike SLR measurements, are also sensitive to orbital perturbations caused by tides on the Moon. Using LLR measurements, Williams *et al.* (2001) report a value of $-25.73 \pm 0.5''/\text{cy}^2$ for the tidal acceleration of the Moon, which by Kepler's law corresponds to an increase of $3.79 \pm 0.07\,\text{cm yr}^{-1}$ in the semimajor axis of the Moon's orbit, and which includes a contribution of $+0.29\ ''/\text{cy}^2$ from dissipation within the Moon itself. Only about half of the discrepancy between the SLR- and LLR-derived values for $\dot{n}$ due to dissipation by just tides on the Earth is currently understood (Williams *et al.*, 2001). Other

LLR-derived values for $\dot{n}$ have been reported by Newhall *et al.* (1988), Dickey *et al.* (1994a), and Chapront *et al.*, (2002).

By *a priori* adopting a value for the tidal acceleration $\dot{n}$ of the Moon, lunar and solar eclipse observations can be used to determine the secular increase in the length of the day over the past few thousand years. The most recent re-reductions of lunar and solar eclipse observations for LOD changes are those of Stephenson and Morrison (1995) and Morrison and Stephenson (2001). Besides using eclipse observations spanning 700 BC to AD 1600, they also used lunar occultation observations spanning 1600–1955.5 and optical astrometric and space-geodetic measurements spanning 1955.5–1990. Adopting a value of $-26.0''/\text{cy}^2$ for $\dot{n}$, Morrison and Stephenson (2001) found that the LOD has increased at a rate of $+1.80 \pm 0.1$ ms/cy on average during the past 2700 years (see **Figure 3**). In addition to a secular trend, Stephenson and Morrison (1995) and Morrison and Stephenson (2001) also found evidence for a fluctuation in the LOD that has a peak-to-peak amplitude of about 8 ms and a period of about 1500 years (**Figure 3**).

By conservation of angular momentum, a tidal acceleration of the Moon of $-26.0''/\text{cy}^2$ should be accompanied by a $+2.3$ ms/cy increase in the length of the day (Stephenson and Morrison, 1995). Since the observed increase in the length of the day is only $+1.8$ ms/cy (Morrison and Stephenson, 2001), some other mechanism or combination of mechanisms

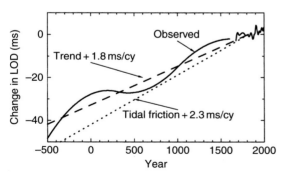

Figure 3 Secular change in the LOD during the past 2500 years estimated from lunar and solar eclipse, lunar occultation, optical astrometric, and space-geodetic observations. The difference between the observed secular trend and that caused by tidal friction is due to the effects of glacial isostatic adjustment and other processes such as ice sheet mass change and the accompanying nonsteric change in sea level. From Morrison LV and Stephenson FR (2001) Historical eclipses and the variability of the Earth's rotation. *Journal of Geodynamics* 32: 247–265.

must be acting to change the length of the day by -0.5 ms/cy. By eqns [43] and [58], changes in both the axial component of relative angular momentum and in the polar moment of inertia of the Earth can cause the LOD to change. A secular trend in the general circulation of fluids like the atmosphere and oceans, and hence in atmospheric and oceanic angular momentum, is unlikely to be sustained over the course of a few thousand years. In fact, using results from a 100-year run of the Hadley Centre general circulation model of the atmosphere, de Viron *et al.* (2004) found that the modeled secular trend in atmospheric angular momentum (AAM) during 1870–1997 causes a secular trend in LOD of only $+0.08$ ms/cy.

One of the most important mechanisms acting to cause a secular trend in the LOD on timescales of a few thousand years is glacial isostatic adjustment (GIA). The isostatic adjustment of the solid Earth in response to the decreasing load on it following the last deglaciation causes the figure of the Earth to change, and hence the LOD to change. Since the solid Earth is rebounding in the regions at high latitude where the ice load was formerly located, the figure of the Earth is becoming less oblate, the Earth's rotation is accelerating, and the LOD is decreasing. Models of GIA show that its effect on the LOD is very sensitive to the assumed value of lower-mantle viscosity (e.g., Wu and Peltier, 1984; Peltier and Jiang, 1996; Vermeersen *et al.*, 1997; Mitrovica and Milne, 1998; Johnston and Lambeck, 1999; Tamisea *et al.*, 2002; Sabadini and Vermeersen, 2004). But by deriving a model for the radial viscosity profile of the Earth that fits both postglacial decay times and free-air gravity anomalies associated with mantle convection, Mitrovica and Forte (1997) found that GIA should cause a secular trend in the LOD amounting to -0.5 ms/cy, a value in remarkable agreement with that needed to explain the difference between the observed secular trend in the length of the day and that caused by tidal dissipation.

However, GIA is not the only mechanism that will cause a secular trend in the length of the day. The present-day change in glacier and ice sheet mass and the accompanying change in nonsteric sea level will also cause a secular trend in LOD (e.g., Peltier, 1988; Trupin *et al.*, 1992; Mitrovica and Peltier, 1993; Trupin, 1993; James and Ivins, 1995, 1997; Nakada and Okuno, 2003; Tosi *et al.*, 2005). But the effect of this mechanism on LOD is very sensitive to the still-unknown present-day mass change of glaciers and ice sheets, particularly of the Antarctic ice sheet. By adopting various scenarios for the mass change of Antarctica, models predict that its mass change alone should cause a secular trend in the LOD ranging anywhere from −0.72 to +0.31 ms/cy (James and Ivins, 1997). Other mechanisms that may cause a secular trend in LOD include tectonic processes taking place under nonisostatic conditions (Vermeersen and Vlaar, 1993; Vermeersen et al., 1994; Sabadini and Vermeersen, 2004), plate subduction (Alfonsi and Spada, 1998), earthquakes (Chao and Gross, 1987), and deformation of the mantle caused by pressure variations acting at the core–mantle boundary that are associated with motion of the fluid core (Fang et al., 1996; Dumberry and Bloxham, 2004; Greff-Lefftz et al., 2004).

The fluctuation in the LOD of 1500-year period found by Stephenson and Morrison (1995) and Morrison and Stephenson (2001) is currently of unknown origin. However, given its large amplitude, which is too large to be caused by atmospheric and oceanic processes but which is comparable in size to the amplitude of the decadal variations, it is probably caused (Dumberry and Bloxham, 2006) by the same core–mantle interactions, such as gravitational coupling (Rubincam, 2003), that are known to cause decadal variations in the LOD.

9.4.1.2 Decadal variations and core–mantle interactions

While lunar and solar eclipse observations are valuable for studying the secular trend in the LOD, they are too sparse and inaccurate to reveal decadal variations. Instead, lunar occultation observations, which are available since the early 1600s (Martin, 1969; Morrison et al., 1981), are used to study decadal variations in the length of the day. **Figure 4** compares three different LOD series derived from lunar occultation observations. Gross (2001; black curve) derived a LOD series spanning 1832.5–1997.5 at yearly intervals by analyzing UT1 measurements obtained from lunar occultation observations by Jordi et al. (1994), from the HIPPARCOS optical astrometric series of Vondrák (1991, 1999) and Vondrák et al. (1992, 1995, 1997, 1998), and from the COMB97 combined optical astrometric and space-geodetic series of Gross (2000a). McCarthy and Babcock (1986; red curve) derived a LOD series spanning 1657.0–1984.5 at half-yearly intervals by analyzing UT1 measurements obtained from lunar occultation observations by Martin (1969) and Morrison (1979), from the optical astrometric series of McCarthy (1976), and from a series obtained from the BIH. Stephenson and Morrison (1984; green curve) obtained a LOD series spanning 1630–1980 at 5-year intervals before 1780 and at yearly intervals afterward by analyzing the lunar occultation and solar eclipse observations cataloged by Morrison (1978) and

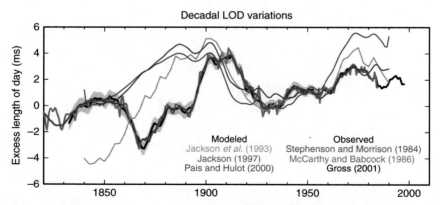

Figure 4 Plots of observed and modeled LOD variations on decadal timescales. The observed LOD series are those of Gross (2001; black curve with gray shading representing ± 1σ standard error), McCarthy and Babcock (1986; red curve), and Stephenson and Morrison (1984; green curve). Note that after about 1955 the uncertainties of the Gross (2001) LOD values are less than the width of the black line. The modeled CAM series are those of Jackson et al. (1993; teal curve), the *uvm-s* model of Jackson (1997; light blue curve), and the PH-inversion model of Pais and Hulot (2000; purple curve). A secular trend of +1.8 ms/cy has been added to the modeled CAM series to match the observed secular trend (see Section 9.4.1.1). An arbitrary bias has also been added to the modeled series in order to facilitate their comparison with the observed series. Adapted from Gross RS (2001) A combined length-of-day series spanning 1832–1997: LUNAR97. *Physics of the Earth and Planetary Interiors* 123: 65–76.

Morrison *et al.* (1981), combining them with an LOD series derived from optical astrometric measurements of UT1 obtained from the BIH. As can be seen from **Figure 4**, during their common time span, these three different LOD series are consistent with each other to within the 1σ standard error of the Gross (2001) series. Other LOD series that have been derived from lunar occultation observations are those of Morrison (1979), Jordi *et al.* (1994), and Liao and Greiner-Mai (1999).

The peak-to-peak variation in the LOD seen in **Figure 4** is about 7 ms, a variation far too large to be caused by atmospheric and oceanic processes (Gross *et al.*, 2005). That atmospheric processes cannot cause such large decadal-scale LOD variations can be easily demonstrated by considering the change in the LOD that would be caused if the motion of the atmosphere were to stop entirely. Because of the pole-to-equator temperature gradient, the atmosphere superrotates with respect to the solid Earth at an average rate of about $7\,\mathrm{m\,s^{-1}}$. If the superrotation of the atmosphere were to stop so that the atmosphere just passively corotates with the solid Earth, then by conservation of angular momentum the LOD would decrease by about 3 ms (Hide *et al.*, 1980). Since stopping the atmospheric motion entirely does not cause a LOD change as large as that observed in **Figure 4**, some other mechanism or combination of mechanisms must be acting to cause the large decadal-scale LOD variations that are observed.

The most important mechanism acting to cause decadal variations in the length of the day is core–mantle coupling. While it has been recognized for quite some time that the core is the only viable source of the large decadal LOD variations that are observed (e.g., Munk and MacDonald, 1960; Lambeck, 1980), it was not until 1988 that Jault *et al.* (1988) were able to model the core angular momentum (CAM) and show that it causes decadal LOD variations that agree reasonably well with those observed.

The flow of the fluid at the top of the core can be inferred from surface observations of the secular variation of the magnetic field by assuming that (1) the mantle is a perfect insulator so that the magnetic field can be expressed as the gradient of a potential, which facilitates the downward continuation of the magnetic field to the top of the core; (2) the core is a perfect conductor so that the magnetic field is 'frozen' into the core fluid and is thus advected by the horizontal flow at the top of the core; (3) the flow at the top of the core is tangentially geostrophic so that it is governed by the balance between the horizontal components of just the pressure gradient and the coriolis force at the top of the core (Bloxham and Jackson, 1991; Whaler and Davis, 1997); and (4) the flow is large scale. The last two assumptions are required in order to reduce the inherent nonuniqueness of core surface flow determinations. Other assumptions about the core surface flow fields that have been made are that the flow is purely toroidal so that it has no radial component (Whaler, 1980; Bloxham, 1990), that the flow is steady in time (Gubbins, 1982; Voorhies and Backus, 1985), that the flow is steady within a drifting reference frame (Davis and Whaler, 1996; Holme and Whaler, 2001), or that the flow includes a helical component (Amit and Olson, 2006).

While surface magnetic field observations can be used to infer the flow at the top of the core, the flow everywhere within the core must be known in order to compute the angular momentum of the core and hence the effect of core motion on the length of the day. Jault *et al.* (1988) realized from dynamical considerations that for geostrophic flows on decadal timescales, the axisymmetric, or zonal, component of the core flow can be described by relative motion of nested cylinders that are coaxial with the rotation axis. This model for the motion of the core at depth allows the axial angular momentum of the core, and hence the effect of core motion on the LOD, to be computed from the flow fields at the top of the core that are inferred from surface magnetic field observations (Jault *et al.*, 1988; Jackson *et al.*, 1993; Jackson, 1997; Hide *et al.*, 2000; Pais and Hulot, 2000; Holme and Whaler, 2001; Pais *et al.*, 2004; Amit and Olson, 2006). Models of the CAM that have been computed in this manner are available through the IERS Special Bureau for the Core (SBC) website (see **Table 4**). **Figure 4** compares the observed decadal LOD variations with the modeled results obtained by Jackson *et al.* (1993), Jackson (1997), and Pais and Hulot (2000). As can be seen, while the agreement is not perfect, CAM models produce decadal LOD variations that have about the same amplitude and phase as those observed.

In the coaxial nested cylinder model of the zonal core flow, the rotation of adjacent cylinders are coupled because of the magnetic field lines that thread through them. If for some reason the rotation of the cylinders is disturbed, then the magnetic field provides a restoring force that will cause the cylinders to oscillate. As first proposed by Braginsky (1970), it is the exchange with the mantle of the angular momentum associated with these torsional

Table 4   Sources of angular momentum models

Model type	URL
Global geophysical fluids	
IERS GGFC	http://www.ecgs.lu/ggfc/
Atmospheric	
IERS SBA	http://www.aer.com/scienceResearch/diag/sb.html
Oceanic	
IERS SBO	http://euler.jpl.nasa.gov/sbo/
Hydrologic	
IERS SBH	http://www.csr.utexas.edu/research/ggfc/
Mantle	
IERS SBM	http://bowie.gsfc.nasa.gov/ggfc/mantle.htm
Core	
IERS SBC	http://www.astro.oma.be/SBC/main.html
Tides	
IERS SBT	http://bowie.gsfc.nasa.gov/ggfc/tides/

URL, Uniform Resource Locator; IERS, International Earth Rotation and Reference Systems Service; GGFC, Global Geophysical Fluids Center; SBA, Special Bureau for the Atmosphere; SBO, Special Bureau for the Oceans; SBH, Special Bureau for Hydrology; SBM, Special Bureau for the Mantle; SBC, Special Bureau for the Core; SBT, Special Bureau for Tides.

oscillations of the fluid core that causes the LOD to change (Zatman and Bloxham, 1997, 1998, 1999; Buffett, 1998; Hide *et al.*, 2000; Pais and Hulot, 2000; Buffett and Mound, 2005; Mound and Buffett, 2005).

While the exchange of CAM with the solid Earth can clearly cause decadal LOD variations of approximately the right amplitude and phase, the mechanism or mechanisms by which the angular momentum is exchanged between the core and solid Earth is less certain. Possible core–mantle coupling mechanisms are viscous torques, topographic torques, electromagnetic torques, and gravitational torques (Jault, 2003; Ponsar *et al.*, 2003). Viscous coupling is caused by the drag of the core flow on the core–mantle boundary, with the strength of the coupling depending on the viscosity of the core fluid. Given current estimates of core viscosity, it is generally agreed that viscous torques are too weak to be effective in coupling the core to the mantle (Rochester, 1984).

If the core–mantle boundary is not smooth but exhibits undulations or 'bumps', then the flow of the core fluid can exert a torque on the mantle due to the fluid pressure acting on the boundary topography (Hide, 1969, 1977, 1989, 1993, 1995a; Jault and Le Mouël, 1989, 1990, 1991; Hide *et al.*, 1993; Jault *et al.*, 1996; Buffett, 1998; Kuang and Chao, 2001; Mound and Buffett, 2005; Asari *et al.*, 2006). The strength of this topographic coupling, a mechanism first suggested by Hide (1969), depends on the

amplitude of the topography at the core–mantle boundary. Because of uncertainties in the size of this topography and a controversy about how the topographic torque should be computed (Bloxham and Kuang, 1995; Hide, 1995b, 1998; Kuang and Bloxham, 1997; Jault and Le Mouël, 1999), there is as yet no consensus on the importance of topographic coupling as a mechanism for exchanging angular momentum between the core and mantle.

Electromagnetic torques arise from the interaction between the magnetic field within the core and the flow of electric currents in the weakly conducting mantle that are induced by both time variations of the magnetic field and by diffusion of electric currents from the core into the mantle (Bullard *et al.*, 1950; Rochester, 1960, 1962; Roden, 1963; Roberts, 1972; Stix and Roberts, 1984; Jault and Le Mouël, 1991; Love and Bloxham, 1994; Stewart *et al.*, 1995; Holme, 1998a, 1998b, 2000; Wicht and Jault, 1999, 2000; Mound and Buffett, 2005). The strength of this electromagnetic torque, a mechanism first suggested by Bullard *et al.* (1950), depends on both the conductivity of the mantle and on the strength of the magnetic field crossing the core–mantle boundary. If the conductivity of the mantle, or of a narrow layer at the base of the mantle, is sufficiently large, then electromagnetic torques can produce decadal LOD variations as large as those observed. But because of uncertainties in the conductivity at the base of the mantle, the importance of electromagnetic coupling, like that of topographic coupling, as a mechanism for exchanging angular momentum between the core and mantle remains unclear.

Gravitational attraction between density heterogeneities in the fluid core and mantle can exert a torque on the mantle, leading to changes in the length of the day (Jault and Le Mouël, 1989; Buffett, 1996a). The strength of the gravitational torque depends upon the size of the mass anomalies in the core and mantle, which are poorly known. As a result, there have been few quantitative estimates of the magnitude of the gravitational torque. However, Buffett (1996a, 1996b) has suggested that the inner core may be gravitationally locked to the mantle. If so, then any rotational disturbance of the inner core, possibly caused by electromagnetic torques acting on the inner core, will be transmitted to the mantle, causing LOD changes. While Buffett (1998) and Mound and Buffett (2005) consider this last mechanism to be the most viable mechanism for exchanging angular momentum between the core and mantle, Zatman (2003) finds that it is inconsistent with a model of inner core rotation rate determined from the core flow within the tangent cylinder.

9.4.1.3 Tidal variations and solid Earth, oceanic, and atmospheric tides

Tidal forces due to the gravitational attraction of the Sun, Moon, and planets deform the solid and fluid parts of the Earth, causing the Earth's inertia tensor to change and hence the Earth's rotation to change (*see* Chapter 6). Jeffreys (1928) was the first to predict that the periodic displacement of the solid and fluid masses of the Earth associated with the tides should cause periodic changes in the Earth's rate of rotation at the tidal frequencies (*see* Chapter 8). While Markowitz (1955, 1959) reported observing such periodic variations at the fortnightly and monthly tidal frequencies from observations taken at two photographic zenith tubes, these observations were later thrown into doubt by the error analysis of Fliegel and Hawkins (1967). By combining observations from about 55 instruments, Guinot (1970) was able to detect variations in the Earth's rate of rotation at the fortnightly and monthly tidal frequencies that had amplitudes significantly greater than the level of observation noise. Today, the high accuracy of the space-geodetic measurement systems allow long-period tidal effects on UT1 and LOD to be unambiguously observed (e.g., Hefty and Capitaine, 1990; Nam and Dickman, 1990; McCarthy and

Luzum, 1993; Robertson *et al.*, 1994; Schastok *et al.*, 1994; Chao *et al.*, 1995a; Dickman and Nam, 1995; Schuh and Schmitz-Hübsch, 2000; Bellanger *et al.*, 2002).

In a seminal paper, Yoder *et al.* (1981) derived a model for the long-period tidal variations in the rotation of the Earth assuming that the crust and mantle of the Earth is elastic, that the core is decoupled from the mantle, and that the ocean tides are in equilibrium. **Table 5** gives their results for the long-period tidal variations in UT1 and LOD, where the LOD results have been derived here from their UT1 results by using eqn [64]. On intraseasonal timescales, the largest effects are found to be at the fortnightly and monthly tidal periods. The large tidal variations at annual and semiannual periods are obscured in Earth rotation observations by meteorological effects (see Section 9.4.1.4), while those having periods of 9.3 and 18.6 years are obscured by the decadal variations in the Earth's rotation (see Section 9.4.1.2).

Effects of mantle anelasticity on the tidal variations in the Earth's rate of rotation have been discussed by Merriam (1984, 1985), Wahr and Bergen (1986), and Defraigne and Smits (1999). Dissipation associated with mantle anelasticity causes the deformational and hence rotational response of the Earth to lag behind the forcing tidal potential. As a result, not only does mantle anelasticity modify the in-phase rotational response of the Earth to the tidal potential, but out-of-phase terms are introduced as well. Anelastic effects are found to modify the elastic rotational response of the Earth by a few percent.

Defraigne and Smits (1999) also considered the effects of nonhydrostatic structure within the Earth on the tidal variations in the Earth's rotation. A nonhydrostatic initial state of the Earth was determined by computing the buoyancy-driven flow in the mantle due to the seismically observed mass anomalies there, while accounting for the associated flow-induced boundary deformation, potential readjustment, and mass readjustment in the outer and inner cores. Their results indicate that nonhydrostatic structure within the Earth modifies the rotational response of the Earth to the zonal tide generating potential by less than 0.1%.

Dynamic effects of long-period ocean tides on the Earth's rotation using ocean tide models based upon Laplace's tidal equations have been computed by Brosche *et al.* (1989), Seiler (1990, 1991), Wünsch and Busshoff (1992), Dickman (1993), Gross (1993), and Seiler and Wünsch (1995). But the accuracy of ocean tide models greatly improved when

Table 5 Modeled variations in UT1 and LOD caused by elastic solid body and equilibrium ocean tides

Argument					Period (days)	ΔUT1 (μs) sin	ΔΛ(t) (μs) cos
l	l'	F	D	Ω			
1	0	2	2	2	5.64	−2.35	2.62
2	0	2	0	1	6.85	−4.04	3.71
2	0	2	0	2	6.86	−9.87	9.04
0	0	2	2	1	7.09	−5.08	4.50
0	0	2	2	2	7.10	−12.31	10.90
1	0	2	0	0	9.11	−3.85	2.66
1	0	2	0	1	9.12	−41.08	28.30
1	0	2	0	2	9.13	−99.26	68.29
3	0	0	0	0	9.18	−1.79	1.22
−1	0	2	2	1	9.54	−8.18	5.38
−1	0	2	2	2	9.56	−19.74	12.98
1	0	0	2	0	9.61	−7.61	4.98
2	0	2	−2	2	12.81	2.16	−1.06
0	1	2	0	2	13.17	2.54	−1.21
0	0	2	0	0	13.61	−29.89	13.80
0	0	2	0	1	13.63	−320.82	147.86
0	0	2	0	2	13.66	−775.69	356.77
2	0	0	0	−1	13.75	2.16	−0.99
2	0	0	0	0	13.78	−33.84	15.43
2	0	0	0	1	13.81	1.79	−0.81
0	−1	2	0	2	14.19	−2.44	1.08
0	0	0	2	−1	14.73	4.70	−2.00
0	0	0	2	0	14.77	−73.41	31.24
0	0	0	2	1	14.80	−5.26	2.24
0	−1	0	2	0	15.39	−5.08	2.07
1	0	2	−2	1	23.86	4.98	−1.31
1	0	2	−2	2	23.94	10.06	−2.64
1	1	0	0	0	25.62	3.95	−0.97
−1	0	2	0	0	26.88	4.70	−1.10
−1	0	2	0	1	26.98	17.67	−4.11
−1	0	2	0	2	27.09	43.52	−10.09
1	0	0	0	−1	27.44	53.39	−12.22
1	0	0	0	0	27.56	−826.07	188.37
1	0	0	0	1	27.67	54.43	−12.36
0	0	0	1	0	29.53	4.70	−1.00
1	−1	0	0	0	29.80	−5.55	1.17
−1	0	0	2	0	31.66	11.75	−2.33
−1	0	0	2	1	31.81	−182.36	36.02
−1	0	0	2	2	31.96	13.16	−2.59
1	0	0	−2	−1	32.61	1.79	−0.34
−1	−1	0	2	0	34.85	−8.55	1.54
0	2	2	−2	2	91.31	−5.73	0.39
0	1	2	−2	1	119.61	3.29	−0.17
0	1	2	−2	2	121.75	−188.47	9.73
0	0	2	−2	0	173.31	25.10	−0.91
0	0	2	−2	1	177.84	117.03	−4.13
0	0	2	−2	2	182.62	−4824.74	166.00
0	2	0	0	0	182.63	−19.36	0.67
2	0	0	−2	−1	199.84	4.89	−0.15
2	0	0	−2	0	205.89	−54.71	1.67
2	0	0	−2	1	212.32	3.67	−0.11
0	−1	2	−2	1	346.60	−4.51	0.08
0	0	2	−2	−1	346.64	9.21	−0.17
1	0	0	−2	0	365.22	82.81	−1.42
0	1	0	0	0	365.26	−1535.87	26.42
0	1	0	0	1	386.00	−13.82	0.22
1	0	0	−2	−1	411.78	3.48	−0.05
1	0	−2	0	0	−1095.17	−13.72	−0.08
1	−1	0	0	−1	1305.47	42.11	−0.20
−1	0	0	0	−1	3232.85	−4.04	0.01
0	0	0	0	2	−3399.18	789.98	1.46
0	0	0	0	1	−6798.38	−161726.81	−149.47

The tabulated coefficients are derived using a constant value of $k/C = 0.94$ as recommended by Yoder et al. (1981). Terms with UT1 amplitudes less than 2 μs are not tabulated l, l', F, D, and Ω are the Delaunay arguments, expressions for which are given in Simon et al. (1994). The period, given in solar days, is the approximate period of the term as viewed in the terrestrial reference frame.
Source: Yoder CF, Williams JG, and Parke ME (1981) Tidal Variations of Earth rotation. *Journal of Geophysical Research* 86: 881–891.

Table 6   Modeled variations in UTI and LOD caused by long-period dynamic ocean tides

Tide	\multicolumn{5}{c}{Argument}	Period (days)	ΔUT1 mass (μs)		ΔUT1 motion (μs)		$\Delta\Lambda$(t) mass (μs)		$\Delta\Lambda$(t) motion (μs)	
	l	l'	F	D	Ω		sin	cos	sin	cos	cos	sin	cos	sin
M_f	0	0	2	0	2	13.66	−102.8	33.0	−1.4	15.4	47.28	15.18	0.64	7.08
M_m	1	0	0	0	0	27.56	−119.1	8.8	5.7	10.7	27.15	2.01	−1.30	2.44

The tabulated coefficients are from the assimilated (A) model of Kantha *et al.* (1998). *l*, *l'* *F*, *D*, and Ω are the Delaunay arguments, expressions for which are given in Simon *et al.* (1994). The period, given in solar days, is the approximate period of the term as viewed in the terrestrial reference frame.
Source: Kantha LH, Stewart JS, and Desai SD (1998) Long-period lunar fortnightly and monthly ocean tides. *Journal of Geophysical Research* 103(C6): 12639–12647.

TOPEX/POSEIDON (T/P) sea surface height measurements became available. Dynamic effects of long-period ocean tides on the Earth's rotation using tide models based upon T/P sea surface height measurements have been computed by Kantha *et al.* (1998) and Desai and Wahr (1999). **Table 6** gives the results obtained by Kantha *et al.* (1998) from their ocean tide model that assimilates T/P-derived tides. As expected, dynamic tide effects are seen to be larger at the fortnightly tidal frequency than they are at the monthly frequency, with the amplitude of the out-of-phase mass and motion terms (the cosine coefficients for UT1, the sine coefficients for LOD) each being larger for the fortnightly tide than they are for the monthly tide.

Ocean tides in the diurnal and semidiurnal tidal bands also affect the Earth's rate of rotation. While Yoder *et al.* (1981) were the first to predict these effects using theoretical ocean tide models based upon Laplace's tidal equations, they were not actually observed until Dong and Herring (1990) detected them in VLBI measurements. Subdaily variations in UT1 and LOD at the diurnal and semidiurnal tidal frequencies have now been unambiguously observed in measurements taken by VLBI (Brosche *et al.*, 1991; Herring and Dong, 1991, 1994; Wünsch and Busshoff, 1992; Herring, 1993; Sovers *et al.*, 1993; Gipson, 1996; Haas and Wünsch, 2006), SLR (Watkins and Eanes, 1994), and GPS (Lichten *et al.*, 1992; Malla *et al.*, 1993; Freedman *et al.*, 1994b; Grejner-Brzezinska and Goad, 1996; Hefty *et al.*, 2000; Rothacher *et al.*, 2001; Steigenberger *et al.*, 2006).

Following Yoder *et al.* (1981), predictions of subdaily tidal variations in the Earth's rate of rotation using theoretical ocean tide models have been made by Brosche (1982), Baader *et al.* (1983), Brosche *et al.* (1989), Seiler (1990, 1991), Wünsch and Busshoff (1992), Dickman (1993), Gross (1993), and Seiler and Wünsch (1995). But as with long-period ocean tide

models, the accuracy of diurnal and semidiurnal tide models greatly improved when T/P sea surface height measurements became available. Dynamic effects of diurnal and semidiurnal ocean tides on the Earth's rotation computed from tide models incorporating T/P sea surface height measurements have been given by Ray *et al.* (1994), Chao *et al.* (1995b, 1996), and Chao and Ray (1997). McCarthy and Petit (2004, chapter 8) give extensive tables of coefficients for the effect of diurnal and semidiurnal ocean tides on UT1 and LOD. Updated models for these effects are available through the IERS Special Bureau for Tides (SBT) website (see **Table 4**).

Comparisons of observations with models show the dominant role that ocean tides play in causing subdaily UT1 and LOD variations (**Figure 5**(a)), with as much as 90% of the observed UT1 variance being explained by diurnal and semidiurnal ocean tides (Chao *et al.*, 1996). Apart from errors in observations and models, the small difference that remains (e.g., Schuh and Schmitz-Hübsch, 2000) may be due to nontidal atmospheric and oceanic effects.

The diurnally varying solar heating of the atmosphere excites diurnal and semidiurnal tidal waves in the atmosphere that travel westward with the Sun (Chapman and Lindzen, 1970; Haurwitz and Cowley, 1973; Volland, 1988, 1997; Dai and Wang, 1999). These radiational tides are much larger than the gravitational tides in the atmosphere, with the amplitude of the surface pressure variations due to the radiational tides being about 20 times larger than the amplitude due to the gravitational tides. While gravitational tides in the atmosphere have no discernible effect on the Earth's rotation, the radiational tides do have an effect (Zharov and Gambis, 1996; Brzezinski *et al.*, 2002a; de Viron *et al.*, 2005). Using the National Centers for Environmental Prediction (NCEP)/National Center for Atmospheric Research (NCAR) reanalysis wind

Subdaily Earth orientation variations

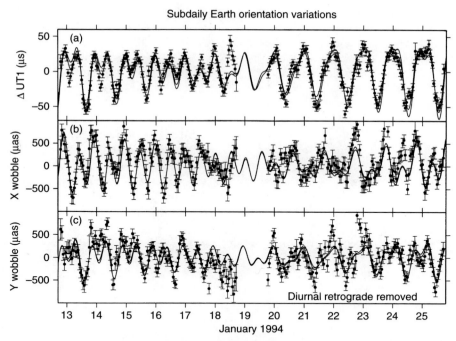

Figure 5 Plots of observed and modeled (a) UTI, (b) x-component of polar motion, and (c) y-component of polar motion during the Cont94 measurement campaign of 12–26 January 1994. The dots with 1σ error bars are the hourly VLBI observations. Polar motion variations in the retrograde nearly diurnal frequency band have been removed from the observed series and are not included in the modeled series. The solid lines are the predicted effects from the diurnal and semidiurnal T/P ocean tide models B and C of Chao *et al.* (1996). Each tide model explains about 90% of the observed UTI variance and about 60% of the observed polar motion variance. From Chao BF, Ray RD, Gipson JM, Egbert GD, and Ma C (1996) Diurnal/semidiurnal polar motion excited by oceanic tidal angular momentum. *Journal of Geophysical Research* 101(B9): 20151–20163.

and pressure fields, Brzezinski *et al.* (2002a) predict that the UT1 variations caused by the diurnal S_1 radiational tide should have an amplitude of about 0.5 µs and that the LOD variations should have an amplitude of about 3.3 µs. Since the NCEP/NCAR reanalysis winds and pressure are given every 6 h, the semidiurnal tidal frequency band is incompletely sampled, making it difficult to estimate the effect on UT1 and LOD of the semidiurnal S_2 radiational tide. In addition, since the oceans will respond dynamically to the tidal variations in the atmospheric wind and pressure fields, the oceans will also contribute to the excitation of UT1 and LOD by the radiational tides. In fact, the effect of radiational tides on UT1 and LOD is typically included in tables of the effects of diurnal and semidiurnal ocean tides on the Earth's rate of rotation (see, e.g., McCarthy and Petit, 2004, tables 8.3a and 8.3b).

9.4.1.4 Seasonal variations
Seasonal variations in the length of the day were first detected by Stoyko (1937). Numerous studies have since shown that the observed annual and semiannual

variations in the LOD are primarily caused by annual and semiannual changes in the angular momentum of the zonal winds (e.g., Munk and MacDonald, 1960; Lambeck, 1980, 1988; Hide and Dickey, 1991; Eubanks, 1993; Rosen, 1993; Höpfner, 1998, 2000; Aoyama and Naito, 2000). In fact, to within the uncertainty of the earlier LOD measurements, seasonal variations in the LOD could be accounted for solely by seasonal variations in the zonal winds (Rosen and Salstein, 1985, 1991; Naito and Kikuchi, 1990, 1991; Dickey *et al.*, 1993a; Rosen, 1993). But as measurement accuracies improved, discrepancies between the observed and modeled variations became noticeable.

With the advent of models of the general circulation of the oceans it became possible to evaluate the effect of the oceans on seasonal LOD variations (Brosche and Sündermann, 1985; Brosche *et al.*, 1990, 1997; Frische and Sündermann, 1990; Dickey *et al.*, 1993b; Segschneider and Sündermann, 1997; Marcus *et al.*, 1998; Johnson *et al.*, 1999; Ponte and Stammer, 2000; Ponte *et al.*, 2001, 2002; Gross, 2004; Gross *et al.*, 2004; Kouba and Vondrák, 2005; Yan *et al.*, 2006). For example, **Table 7** gives the results obtained by Gross *et al.* (2004). They found that during 1992–2000 the amplitude of the observed annual LOD variation was 369.0 ± 6.4 μs with that caused by the effect of zonal winds integrated to a height of 10 hectopascals (hPa) being 414.8 ± 5.0 μs, by atmospheric surface pressure variations being 37.4 ± 0.8 μs, by oceanic currents being 7.6 ± 0.3 μs, and by ocean-bottom pressure variations being 7.9 ± 0.3 μs. The sum of the effects of atmospheric winds up to 10 hPa, surface pressure, oceanic currents, and bottom pressure had an annual amplitude of 363.0 ± 5.2 μs, the same as that observed to within measurement uncertainty, and a phase difference of only 4.5°. But the effects of the winds above 10 hPa have not been taken into account yet.

Although only 1% of the atmospheric mass is located in the region of the atmosphere above 10 hPa, the strength of the zonal winds there is great enough that they have a noticeable effect on seasonal LOD variations (Rosen and Salstein, 1985, 1991; Dickey *et al.*, 1993a; Rosen, 1993; Höpfner, 2001). Gross *et al.* (2004) found that during 1992–2000 the annual LOD variation caused by winds above 10 hPa had an amplitude of 20.5 ± 0.3 μs, larger than the sum of the effects of oceanic currents and bottom pressure. Since the annual stratospheric winds are out of phase with the annual tropospheric winds, including the effect of the winds above 10 hPa brings the amplitude of the modeled annual LOD variations down to 343.5 ± 5.2 μs. Thus, when the effects of stratospheric winds are included, the annual LOD budget is not closed. **Table 7** also gives the results obtained by Gross *et al.* (2004) for the semiannual and terannual (3 cpy) LOD variations. Like the annual LOD budget, the semiannual budget is also not closed. But to within measurement uncertainty, LOD variations at the terannual frequency are completely accounted for by the effects of the atmosphere and oceans.

Apart from errors in observations and models, the residual that remains after modeled atmospheric and oceanic effects have been removed from the observations may be caused by hydrologic processes (Chao and O'Conner, 1988; Chen *et al.*, 2000; Zhong *et al.*, 2003; Chen, 2005). For example, Chen (2005) finds that during 1993.0–2004.3, modeled hydrologic processes cause an annual change in LOD of amplitude 17.2 μs which when added to oceanic effects is reduced to 12.5 μs. However, when the effects of balancing mass within the atmosphere, ocean, and

hydrology system are included, this is further reduced to 1.4 μs, which is not large enough to close the annual LOD budget. His results for the semiannual LOD variations are similar. Thus, while models of atmospheric, oceanic, and hydrologic angular momentum have greatly improved, the seasonal LOD budget is still not closed, except perhaps at the terannual frequency.

Since meteorological processes are the predominant cause of seasonal LOD variations, and since these processes can change from year to year, there is no reason to expect that the seasonal LOD variations should be the same from year to year, either in amplitude or in phase. In fact, Feissel and Gavoret (1990) showed that the amplitude of the annual LOD oscillation was about twice as large as normal during the 1982–83 El Niño/Southern Oscillation (ENSO) event, with the amplitude of the semiannual LOD oscillation being about half as large as normal. Gross *et al.* (1996a, 2002) extended this study, showing that during 1962–2000 the amplitudes of the seasonal LOD and wind-driven AAM variations at both annual and semiannual frequencies have not been constant but have fluctuated by as much as 50%. They also showed that the changing amplitudes of the annual and semiannual LOD and AAM variations are significantly correlated with the Southern Oscillation index (SOI), an index defined to be the normalized difference in surface pressure between Darwin and Tahiti. Since the SOI is also correlated with LOD and AAM variations occurring on interannual timescales (see Section 9.4.1.5), the significant correlation they observed between changes in the amplitude of the seasonal cycle and the SOI is evidence of a linkage between the seasonal cycle and interannual LOD and AAM variations, a linkage that can only occur through nonlinear interactions.

9.4.1.5 Interannual variations and the ENSO

Like seasonal variations in the length of the day, variations on interannual timescales are also predominantly caused by changes in the angular momentum of the zonal winds (e.g., Hide and Dickey, 1991; Eubanks, 1993; Rosen, 1993). The most prominent feature of the climate system on these timescales is the ENSO phenomenon. ENSO is a global-scale oscillation of the coupled atmosphere–ocean system characterized by fluctuations in atmospheric surface pressure and ocean temperatures in the tropical Pacific (e.g., Philander, 1990).

Table 7 Observed and modeled nontidal LOD variations at seasonal frequencies during 1992–2000

Excitation process	Annual		Semiannual		Terannual	
	Amplitude (μs)	Phase (degrees)	Amplitude (μs)	Phase (degrees)	Amplitude (μs)	Phase (degrees)
Observed	369.0 ± 6.4	31.6 ± 1.0	294.3 ± 6.3	−116.5 ± 1.2	52.4 ± 6.4	20.9 ± 7.0
Atmospheric						
Winds (grounds to 10 hPa)	414.8 ± 5.0	34.7 ± 0.7	244.3 ± 5.0	−110.0 ± 1.2	54.1 ± 5.0	30.4 ± 5.3
Winds (10–0.3 hPa)	20.5 ± 0.3	−161.0 ± 0.9	29.4 ± 0.3	−122.7 ± .0.6	3.5 ± 0.3	−165.7 ± 5.1
All winds (ground to 0.3 hPa)	395.1 ± 5.0	35.5 ± 0.7	273.1 ± 5.0	−111.3 ± 1.0	50.7 ± 5.0	31.4 ± 5.6
Surface pressure (IB)	37.4 ± 0.8	−154.6 ± 1.3	9.2 ± 0.8	113.3 ± 5.1	2.6 ± 0.8	−48.8 ± 18.1
All winds and surface pressure	358.3 ± 5.1	36.5 ± 0.8	266.7 ± 5.1	−112.7 ± 1.1	51.2 ± 5.1	28.6 ± 5.7
Oceanic						
Currents	7.6 ± 0.3	−165.5 ± 2.3	0.9 ± 0.3	115.5 ± 18.7	1.6 ± 0.3	35.9 ± 11.1
Bottom pressure	7.9 ± 0.3	−148.4 ± 2.2	3.1 ± .0.3	176.7 ± 5.5	1.0 ± 0.3	−67.3 ± 16.4
Currents and bottom pressure	15.3 ± 0.5	−156.8 ± 2.0	3.6 ± 0.5	163.4 ± 8.6	1.7 ± 0.5	−0.6 ± 18.4
Atmospheric and oceanic						
All winds and currents	388.0 ± 5.0	35.9 ± 0.7	272.5 ± 5.0	−111.5 ± 1.1	52.3 ± 5.0	31.6 ± 5.5
Surface and bottom pressure	45.2 ± 0.9	−153.5 ± 1.2	10.9 ± 0.9	127.9 ± 4.9	3.6 ± 0.9	−54.1 ± 15.1
Total of all atmos. and oceanic						
Without winds above 10 hPa	363.0 ± 5.2	36.1 ± 0.8	238.1 ± 5.2	−112.4 ± 1.3	56.1 ± 5.2	26.9 ± 5.3
With winds above 10 hPa	343.5 ± 5.2	37.1 ± 0.9	267.1 ± 5.2	−113.5 ± 1.1	52.7 ± 5.2	27.7 ± 5.7

IB, inverted barometer. The amplitude A and phase α of the observed and modeled seasonal LOD variations are defined by $\Delta\Lambda(t) = A \cos[\sigma(t - t_o) - \alpha]$ where σ is the annual, semiannual, or terannual frequency and the reference date t_o is January 1.0, 1990.

Source: Gross RS, Fukumori I, Menemenlis D, and Gegout P (2004) Atmospheric and oceanic excitation of length-of-day variations during 1980–2000. *Journal of Geophysical Research* 109:B01406 (doi:10.1029/2003JB002432).

During an ENSO event, in which the SOI decreases, the tropical easterlies collapse causing the AAM to increase. By the conservation of angular momentum, as the AAM increases, the solid Earth's angular momentum decreases and the LOD increases. Numerous studies (e.g., Stefanick, 1982; Chao, 1984, 1988, 1989; Rosen et al., 1984; Eubanks et al., 1986; Salstein and Rosen, 1986; Feissel and Gavoret, 1990; Gambis, 1992; Dickey et al., 1992a, 1993a, 1994b, 1999; Jordi et al., 1994; Abarca del Rio et al., 2000; Zhou et al., 2001; Zheng et al., 2003) have shown that observed LOD variations on interannual timescales, as well as inter-annual variations in the angular momentum of the zonal winds, are (negatively) correlated with the SOI, reflecting the impact on the LOD of changes in the zonal winds associated with ENSO. For example, Chao (1984) reported finding a correlation between interann-ual LOD variations and the SOI during 1957–83 with a maximum negative value for the correlation coefficient of –0.56 obtained when the SOI leads the interannual LOD by 1 month; Eubanks et al. (1986) reported a maximum negative correlation coefficient during 1962–84 of about –0.5 when the SOI leads the inter-annual LOD by 3 months; Chao (1988) reported a maximum negative value for the correlation coefficient during 1972–86 of –0.68 for a 2-month lead time; and Dickey et al. (1993a) reported a maximum negative correlation coefficient during 1964–89 of –0.67 for a 1-month lead time.

In a detailed study of the 1982–83 ENSO event, Dickey et al. (1994b) showed that up to 92% of the observed interannual LOD variance could be explained by atmospheric wind and pressure fluctua-tions. They suggested that variations in oceanic angular momentum could explain the remaining sig-nal. But studies (Johnson et al., 1999; Ponte et al., 2002; Gross et al., 2004) of the effects of oceanic processes show that they are only marginally effective in caus-ing interannual LOD variations. For example, Gross et al. (2004) found that during 1980–2000 atmospheric winds were the dominant mechanism causing the LOD to change on interannual timescales, explaining 85.8% of the observed variance, and having a corre-lation coefficient of 0.93 with the observations. The effect of atmospheric surface pressure changes explained only 2.6% of the observed variance and was not significantly correlated with the observa-tions. However, including the effect of surface pressure changes with that of the winds increased the observed variance explained from 85.8% to 87.3%. Oceanic currents and bottom pressure changes were found to have only a marginal effect

on interannual LOD variations, each explaining less than 1% of the observed variance, and neither being significantly correlated with the observations. However, including their effects with those of atmo-spheric winds and surface pressure changes increased the observed variance explained from 87.3% to 87.9%, and increased the correlation coefficient with the observations from 0.93 to 0.94.

The interannual LOD signal that remains after atmospheric and oceanic effects are removed may be caused by hydrologic processes (Chen et al., 2000; Chen, 2005). For example, Chen (2005) finds a significant correlation between the sum of the mass-balanced oceanic and hydrologic angular momenta with the observed interannual LOD variations from which atmospheric effects have been removed, although the modeled variations are of smaller amplitude than those observed. Like seasonal varia-tions, better atmospheric, oceanic, and hydrologic models are needed to close the LOD budget on interannual timescales.

There is also a persistent oscillation in LOD over the last century with a period of about 6 years and an amplitude of about 0.12 ms that is not caused by atmospheric fluctuations (Vondrák, 1977; Djurovic and Paquet, 1996; Abarca del Rio et al., 2000). Mound and Buffett (2003) suggest that it is caused by the exchange of angular momentum between the mantle and core arising from gravitational coupling between the mantle and inner core.

9.4.1.6 Intraseasonal variations and the Madden–Julian oscillation

Like the seasonal and interannual variations in the length of the day, variations on intraseasonal timescales are also predominantly caused by changes in the angular momentum of the zonal winds (e.g., Hide and Dickey, 1991; Eubanks, 1993; Rosen, 1993). The Madden–Julian oscillation (Madden and Julian, 1971, 1972, 1994) with a period of 30–60 days is the most prominent feature in the atmosphere on these timescales and a number of studies have shown that fluctuations in the zonal winds associated with this oscillation cause the LOD to change (Feissel and Gambis, 1980; Langley et al., 1981b; Anderson and Rosen, 1983; Feissel and Nitschelm, 1985; Madden, 1987; Dickey et al., 1991; Hendon, 1995; Marcus et al., 2001).

Studies of the effects of oceanic processes show that they are only marginally effective in causing intraseasonal LOD variations (Ponte, 1997; Johnson et al., 1999; Ponte and Stammer, 2000; Ponte and Ali, 2002; Gross et al., 2004; Kouba and Vondrák, 2005).

For example, Gross *et al.* (2004) found that during 1992–2000 atmospheric winds were the dominant mechanism causing the LOD to change on intraseasonal timescales, explaining 85.9% of the observed intraseasonal variance and having a correlation coefficient of 0.93 with the observations. Atmospheric surface pressure, oceanic currents, and ocean-bottom pressure were found to have only a minor effect on intraseasonal LOD changes, each explaining only about 3–4% of the observed variance. However, including the effect of surface pressure changes with that of the winds increased the variance explained from 85.9% to 90.2% and increased the correlation coefficient with the observations from 0.93 to 0.95. Additionally, including the effects of changes in oceanic currents and bottom pressure further increased the variance explained from 90.2% to 92.2% and further increased the correlation coefficient from 0.95 to 0.96. Thus, although the impact of the oceans is relatively minor, closer agreement with the observations in the intraseasonal frequency band is obtained when the effects of oceanic processes are added to that of atmospheric.

Hydrologic effects on intraseasonal LOD variations are thought to be relatively insignificant (Chen *et al.*, 2000; Chen, 2005), although the monthly sampling interval of current hydrologic models makes it difficult to study such rapid variations.

9.4.2 Polar Motion

Observations of polar motion (**Figure 6**) show that it consists mainly of (1) a forced annual wobble having a nearly constant amplitude of about 100 mas, (2) the free Chandler wobble having a period of about 433 days and a variable amplitude ranging from about 100 to 200 mas, (3) quasiperiodic variations on decadal timescales having amplitudes of about 30 mas known as the Markowitz wobble, (4) a linear trend having a rate of about 3.5 mas yr^{-1} and a direction towards 79° W longitude, and (5) smaller amplitude variations occurring on all measurable timescales. Here, the polar motion variations that are observed and the models that have been developed to explain them are reviewed.

The motion of the pole is usually described by giving the *x*- and *y*-components of its location in the terrestrial reference frame. But since polar motion is inherently a two-dimensional quantity, periodic motion of the pole can also be described by giving the amplitude *A* and phase α of its prograde and retrograde components defined by

$$\mathbf{p}(t) = p_x(t) - i\, p_y(t)$$
$$= A_p\, e^{i\alpha_p} e^{i\sigma(t-t_o)} + A_r e^{i\alpha_r} e^{-i\sigma(t-t_o)} \qquad [68]$$

where the subscript p denotes prograde, the subscript r denotes retrograde, σ is the strictly positive frequency of motion, and t_o is the reference date. Prograde motion of the pole is circular motion in a counter-clockwise direction; retrograde motion is circular motion in a clockwise direction. In general, the sum of prograde and retrograde circular motion is elliptical motion, with linear motion resulting when the amplitudes of the prograde and retrograde components are the same.

By eqns [41]–[42] and [57], the observed polar motion variations are excited by changes in the equatorial components $h_x(t)$ and $h_y(t)$ of relative angular momentum and by changes in the equatorial components $\Omega\,\Delta I_{xz}(t)$ and $\Omega\,\Delta I_{yz}(t)$ of angular momentum due to changes in the mass distribution of the Earth. Like modeling the observed LOD changes, modeling the observed polar motion excitation requires computing both types of changes in angular momentum for the different components of the Earth system. From eqn [65], and again assuming for this purpose that the terrestrial reference frame is uniformly rotating at the rate Ω about the z-coordinate axis, the equatorial components of the mass term of the angular momentum can be written as

$$\Omega\,\Delta I_{xz}(t) = -\Omega \int_V \rho(\mathbf{r},\, t)\, r^2 \sin\phi \cos\phi \cos\lambda\, \mathrm{d}V \qquad [69]$$

$$\Omega\,\Delta I_{yz}(t) = -\Omega \int_V \rho(\mathbf{r},\, t)\, r^2 \sin\phi \cos\phi \sin\lambda\, \mathrm{d}V \qquad [70]$$

where ϕ is north latitude and λ is east longitude. Similarly, the equatorial components of the motion term of the angular momentum can be written as

$$h_x(t) = \int_V \rho(\mathbf{r},\, t)\, [r \sin\lambda\, \nu(\mathbf{r},\, t) \\ - r \sin\phi \cos\lambda\, u(\mathbf{r},\, t)]\, \mathrm{d}V \qquad [71]$$

$$h_y(t) = \int_V \rho(\mathbf{r},\, t)\, [- r \cos\lambda\, \nu(\mathbf{r},\, t) \\ - r \sin\phi \sin\lambda\, u(\mathbf{r},\, t)]\, \mathrm{d}V \qquad [72]$$

where $u(\mathbf{r},t)$ is the eastward component of the velocity and $\nu(\mathbf{r},t)$ is the northward component.

9.4.2.1 True polar wander and GIA

Determining an unbiased estimate of the linear trend in the observed path of the pole over the past 100 years is complicated by the presence of

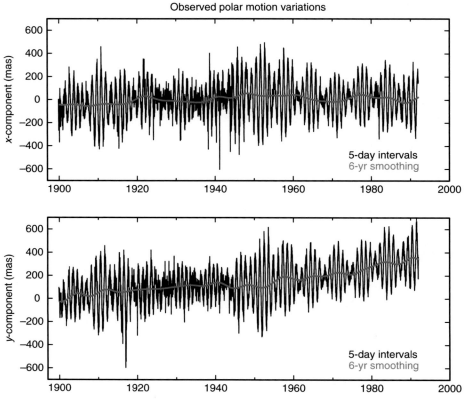

Figure 6   Observed polar motion variations from the HIPPARCOS optical astrometric series. The low-frequency variations shown in red were obtained by applying to the 5-day HIPPARCOS series a low-pass boxcar filter having a 6-year cutoff period. The beating between the 12-month annual wobble and the 14-month Chandler wobble is readily apparent.

the annual, Chandler, and Markowitz wobbles. Various approaches have been taken to estimate the trend in the presence of these large-amplitude periodic and quasiperiodic variations (see **Table 8** for the resulting trend estimates). The annual wobble has been removed by both a least-squares fit (Wilson and Gabay, 1981; McCarthy and Luzum, 1996; Schuh *et al.*, 2000, 2001), and by a seasonal adjustment of the polar motion series (Wilson and Vicente, 1980). The Chandler wobble has been removed by both a least-squares fit for periodic terms (Dickman, 1981; McCarthy and Luzum, 1996; Schuh *et al.*, 2000, 2001), and by deconvolution (Wilson and Vicente, 1980; Wilson and Gabay, 1981; Vicente and Wilson, 2002). Smoothing has also been used to remove the annual and Chandler wobbles (Okamoto and Kikuchi, 1983; Höpfner, 2004). The decadal-scale variations have been removed by modeling them as being strictly periodic at a single frequency of $1/31$ cpy and then least-squares fitting a sinusoid at this single frequency (Dickman, 1981; McCarthy and Luzum, 1996).

Rather than modeling the decadal variations as being strictly periodic at a single frequency, Gross and Vondrák (1999) devised a method to account for their quasiperiodic nature when estimating the linear trend. The annual and Chandler wobbles were first removed from the observations by applying to the polar motion series a low-pass boxcar filter having a 6-year cutoff period. A spectrum of the resulting low-pass-filtered polar motion series was then computed and the linear trend estimated by a simultaneous weighted least-squares fit for a mean, trend, and periodic terms at the frequencies of all the peaks evident in the spectrum. The estimates for the linear trend that they obtained by applying this technique to the ILS, HIPPARCOS, and SPACE96 polar motion series are given in **Table 8**. **Figure 7** shows the observed low-pass-filtered polar motion series they used (solid lines), the modeled series they obtained by the simultaneous weighted least-squares fit for a mean, trend, and all periodic terms evident in the spectrum (dashed lines that coincide with the solid lines and are therefore difficult to see),

Table 8   Observed linear trend in the path of the pole

Rate (mas yr^{-1})	Direction (degrees W)	Data span	Source
HIPPARCOS polar motion			
3.39	78.5	1899.7–1992.0	(a)
3.51 ± 0.01	**79.2 ± 0.20**	**1900.0–1992.0**	**(b)**
3.31 ± 0.05	76.1 ± 0.80	1899.7–1992.0	(c)
HIPPARCOS polar motion excitation			
2.84	73.03	1899–1992	(d)
ILS polar motion			
3.4	78	1900–1977	(e)
3.521 ± 0.094	80.1 ± 1.6	1899.8–1979.0	(f)
3.52	79.4	1899.8–1979.0	(g)
3.456	80.56	1899.0–1979.0	(h)
3.81 ± 0.07	75.5 ± 1.0	1899.8–1979.0	(b)
ILS polar motion excitation			
3.4	66	1900–77	(e)
3.3	65	1901–70	(i)
3.49	79.5	1899–1979	(d)
Latitude observations			
3.62	89	1900–78	(j[*])
3.51	79	1900–78	(j[†])
3.24	84.9	1899.8–1979.0	(k[†])
2.97	77.7	1899.8–1979.0	(k[‡])
Space-geodetic polar motion			
3.39 ± 0.53	85.4 ± 4.0	1976–94	(l)
4.123 ± 0.002	73.9 ± 0.03	1976.7–97.1	(b)
Combined astrometric and space-geodetic polar motion			
3.29	78.2	1900.0–84.0	(m)
3.33 ± 0.08	75.0 ± 1.1	1899.8–1994.1	(l)
3.901 ± 0.022	65.17 ± 0.22	1891.0–1999.0	(n)
Combined astrometric and space-geodetic excitation			
3.35	76.3	1900–99	(d)
3.54	69.92	1900–99	(d)

The recommended estimate is given in bold.
Sources: (a) Vondrák *et al.* (1998); (b) Gross and Vondrák (1999); (c) Schuh *et al.* (2000, 2001); (d) Vicente and Wilson (2002); (e) Wilson and Vicente (1980); (f) Dickman (1981); (g) Chao (1983); (h) Okamoto and Kikuchi (1983); (i) Wilson and Gabay (1981); (j[*]) Zhao and Dong (1988) based on measurements taken at nine latitude observing stations; (j[†]) Zhao and Dong (1988) based on measurements taken at the 5 ILS latitude observing stations located at Mizusawa, Kitab, Carloforte, Gaithersburg, and Ukiah; (k[†]) Vondrák (1994) based on measurements taken at the 5 ILS latitude observing stations located at Mizusawa, Kitab, Carloforte, Gaithersburg, and Ukiah; (k[‡]) Vondrák (1994) based on measurements taken at the 4 ILS latitude observing stations located at Mizusawa, Kitab, Carloforte, and Gaithersburg; (l) McCarthy and Luzum (1996); (m) Vondrák (1985); (n) Höpfner (2004).

and the resulting linear trend estimates (dotted lines). As can be seen, their model is an excellent fit to the observations and hence their estimates for the linear trend should be unbiased by the presence of the quasiperiodic decadal-scale polar motion variations. For this reason, and because the HIPPARCOS series spans a greater length of time than the other series that they studied, the recommended estimate for the linear trend in the pole path is the estimate that they obtained for the HIPPARCOS series, namely, a trend of rate 3.51 ± 0.01 mas yr^{-1} toward 79.2 ± 0.20° W longitude.

The estimates given in **Table 8** for the observed linear trend in the path of the pole are with respect to a terrestrial reference frame that has been attached to the mean lithosphere in such a manner that within it the tectonic plates exhibit no net rotation. Argus and Gross (2004) argue that ideally the motion of the pole should be given with respect to a reference frame that is attached to the mean solid Earth. They argue that it would therefore be better to use a reference frame that is attached to hotspots rather than the mean lithosphere because hotspots move slower than the

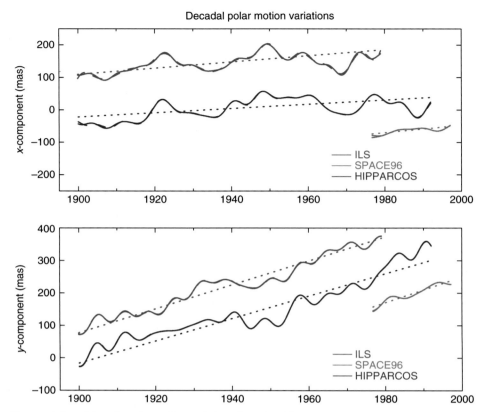

Figure 7   Observed decadal-scale polar motion variations from the ILS optical astrometric series (solid green curve), the HIPPARCOS optical astrometric series (solid blue curve), and the SPACE96 combined space-geodetic series (solid red curve). The decadal-scale variations were obtained by applying to the original series a low-pass boxcar filter having a 6-year cutoff period. The dotted lines show the linear trends in the pole path that were estimated from these series by Gross and Vondrák (1999). The dashed lines that coincide with the solid lines and that are therefore hidden from view show the model of the decadal variations that they used when estimating the trend. For clarity of display the curves have been offset from each other by an arbitrary amount. From Gross RS and Vondrák J (1999) Astrometric and space-geodetic observations of polar wander. *Geophysical Research Letters* 26(14): 2085–2088.

mean lithosphere with respect to the mean solid Earth and hence better represent the mean solid Earth. Transforming the HIPPARCOS trend results of Gross and Vondrák (1999) to a hotspot reference frame, they find that during the past century the linear trend in the path of the pole relative to hotspots was 4.03 mas yr^{-1} towards 68.4° W longitude, or about 15% faster and in a more eastward direction than the trend relative to the mean lithosphere.

One of the most important mechanisms acting to cause a linear trend in the path of the pole on timescales of a few thousand years is GIA. The isostatic adjustment of the solid Earth as it responds to the decreasing load on it following the last deglaciation causes the figure of the Earth to change, and hence the pole to drift. Models of GIA show that its effect on the pole path is sensitive to the assumed value of

lower-mantle viscosity, to the assumed thickness and rheology of the lithosphere, to the treatment of the density discontinuity at 670 km depth, and to the assumed compressibility of the Earth model (e.g., Wu and Peltier, 1984; Peltier and Jiang, 1996; Vermeersen and Sabadini, 1996, 1999; Vermeersen *et al.*, 1996, 1997; Mitrovica and Milne, 1998; Johnston and Lambeck, 1999; Nakada, 2000, 2002; Sabadini and Vermeersen, 2002; Tamisea *et al.*, 2002; Sabadini and Vermeersen, 2004; Mitrovica *et al.*, 2005).

However, GIA is not the only mechanism that will cause a trend in the pole path. The present-day change in glacier and ice sheet mass and the accompanying change in nonsteric sea level will also cause a linear trend in polar motion (e.g., Gasperini *et al.*, 1986; Peltier, 1988; Trupin *et al.*, 1992; Trupin, 1993;

James and Ivins, 1995, 1997; Nakada and Okuno, 2003). But the effect of this mechanism is very sensitive to the still-unknown present-day mass change of glaciers and ice sheets, particularly of the Antarctic ice sheet. By adopting various scenarios for the mass change of Antarctica, models predict that its mass change alone should cause a linear trend in the pole path ranging anywhere in rate from 0.31 to 4.46 mas yr^{-1} and in direction from $101°$ to $281°$ W longitude (James and Ivins, 1997). Other mechanisms that may cause a linear trend in the path of the pole include tectonic processes taking place under nonisostatic conditions (Vermeersen and Vlaar, 1993; Vermeersen et al., 1994; Sabadini and Vermeersen, 2004), plate subduction (Ricard et al., 1992, 1993; Spada et al., 1992; Richards et al., 1997; Alfonsi and Spada, 1998), mantle convection (Steinberger and O'Connell, 1997; Richards et al., 1999), upwelling mantle plumes (Steinberger and O'Connell, 2002; Greff-Lefftz, 2004), and earthquakes (Chao and Gross, 1987).

9.4.2.2 Decadal variations, the Markowitz wobble, and core–mantle interactions

By analyzing ILS latitude measurements taken during 1900–59, Markowitz (1960, 1961) found that at long periods the motion of the pole with respect to the Earth's crust and mantle includes a periodic component superimposed on a linear drift. He found that the periodic component, now known as the Markowitz wobble in his honor, had a period of 24 years and an amplitude of 22 mas and that the linear drift had a rate of 3.2 mas yr^{-1} and a direction towards $60°$ W longitude, remarkably close to the recent determination of 3.51 mas yr^{-1} toward $79.2°$ W longitude (Gross and Vondrák, 1999) that is recommended here.

In more recent re-reductions of the ILS and other optical astrometric measurements, which extend more than three decades past those analyzed by Markowitz (1960, 1961), rather than appearing as a strictly periodic phenomenon of well-defined frequency, the Markowitz wobble now appears as a quasiperiodic variation on decadal timescales having an amplitude of about 30 mas (see **Figure 7**). However, since optical astrometric measurements are known to be corrupted by systematic errors, there has always been some doubt about the reality of the decadal variations seen in the ILS and HIPPARCOS series. Since the highly accurate space-geodetic measurements are less susceptible to systematic error than are optical astrometric

measurements, any decadal variations seen in the space-geodetic measurements can be considered to be reliable. **Figure 7** compares the decadal variations seen in the SPACE96 combined space-geodetic series with those seen in the ILS and HIPPARCOS optical astrometric series. While decadal variations are seen to be present in the SPACE96 series, and hence can be considered to be real, they have a smaller amplitude and a different phase than those seen in either the ILS or HIPPARCOS series. Thus, space-geodetic measurements both confirm the reality of decadal polar motion variations and demonstrate that the decadal variations seen in the ILS and HIPPARCOS series are unreliable and should be used only to place an upper bound on the size of the decadal variations.

The cause of the decadal-scale polar motion variations is currently unknown. Gross et al. (2005) found that redistribution of mass within the atmosphere and oceans cannot be the main excitation source of decadal polar motion variations during 1949–2002 since it amounts to only 20% (x-component) and 38% (y-component) of that observed, and with the modeled excitation being $180°$ out of phase with that observed. However, the ocean model used in their study was not forced by mass changes associated with precipitation, evaporation, or runoff from rivers including that from glaciers and ice sheets, and so had a constant total mass. Thus, their study did not address the question of the excitation of decadal polar motion by processes that change the total mass of the oceans, such as a nonsteric sea level height change associated with glacier and ice sheet mass change. Wilson (1993) noted that an oscillation in global sea level on decadal timescales would excite decadal polar motion variations with a polarization similar to that observed, implying that mass change of the oceans is responsible for exciting the observed decadal polar motions. In fact, using a climate model, Celaya et al. (1999) showed that changes in Antarctic snow pack are capable of inducing decadal polar motion variations of nearly the same amplitude as that observed. But realistic estimates of mass change in glaciers and the Antarctic and other ice sheets, along with estimates of the accompanying nonsteric change in sea level, are required to further evaluate this possible source of decadal polar motion excitation.

Since core–mantle processes are known to cause decadal variations in the length of the day, they may also excite decadal variations in polar motion. But electromagnetic coupling between the core and mantle appears to be 2–3 orders of magnitude too weak

(Greff-Lefftz and Legros, 1995), and topographic coupling appears to be too weak by a factor of 3–10 (Greff-Lefftz and Legros, 1995; Hide *et al.*, 1996; Hulot *et al.*, 1996; Asari *et al.*, 2006). In addition, the modeled decadal polar motion variations resulting from these studies show little agreement in phase with the observed variations.

Following the study of Jochmann (1989), Greiner-Mai *et al.* (2000, 2003) and Greiner-Mai and Barthelmes (2001) have suggested that irregular motion of a tilted oblate inner core may excite decadal polar motion variations, although they did not account for the dynamic effects of the fluid core in their model. Dumberry and Bloxham (2002), who did account for the dynamic effects of the fluid core, suggested that a tilt of the inner core with respect to the mantle of only 0.07°, perhaps caused by an electromagnetic torque acting on the inner core, would generate gravitational and pressure torques on the mantle strong enough to excite decadal polar motion variations having amplitudes as large as those observed and, like the observed variations, having a preferred direction of motion. But when Mound (2005) examined this mechanism, he concluded that while torsional oscillations may excite both decadal LOD and polar motion variations, when the oscillations are constrained to match the observed LOD amplitude, the resulting electromagnetic torque on the inner core is too weak to excite decadal polar motions to their observed level. Despite these conflicting conclusions, invoking the inner core when modeling core–mantle processes may ultimately provide the long-sought explanation for the cause of the Markowitz wobble.

9.4.2.3 Tidal wobbles and oceanic and atmospheric tides

Tidally induced deformations of the solid Earth caused by the second-degree zonal tide raising potential cause long-period changes in the Earth's rate of rotation (see Section 9.4.1.3). But since this potential is symmetric about the polar axis, tidal deformations of the axisymmetric solid Earth cannot excite polar motion. However, due to the nonaxisymmetric shape of the coastlines, the second-degree zonal tide raising potential acting on the oceans can generate polar motion via the exchange of nonaxial oceanic tidal angular momentum with the solid Earth. Yoder *et al.* (1981) discussed the possible existence of long-period tidal variations in polar motion and tables of such variations predicted from theoretical ocean tide models have been given by Seiler

(1990, 1991), Dickman (1993), Gross (1993), Brosche and Wünsch (1994), and Seiler and Wünsch (1995).

Long-period tidal variations in polar motion were first observed by Chao (1994) and have been discussed by Gross *et al.* (1996b, 1997). **Table 9** gives the results of Gross *et al.* (1997) for the observed variations. As discussed by them, the observations at the two fortnightly tidal frequencies M_f and $M_{f'}$, and perhaps at the other tidal frequencies as well, are suspect, because at such close frequencies the oceans, and hence the rotational response of the Earth, should have the same relative response to the tidal potential. This implies that the results at the M_f and $M_{f'}$ tidal frequencies should have the same phase and an amplitude ratio, 2.4, the same as that of the tidal potential at these frequencies. But they do not. Better observations of the effect of long-period ocean tides on polar motion are clearly needed, as are better models for these effects because, as also discussed by Gross *et al.* (1997), predictions from the available theoretical ocean tide models do not agree with each other or with the observations. And predictions from the T/P tide model of Desai and Wahr (1995) as quoted by Gross *et al.* (1997) have not converged even though data through T/P cycle 130 were used.

Ocean tides in the diurnal and semidiurnal tidal bands also cause polar motion variations. While Yoder *et al.* (1981) discussed the possible existence of such polar motion variations, they were not actually observed until Dong and Herring (1990) detected them in VLBI measurements. Subdaily variations in polar motion at the diurnal and semidiurnal tidal frequencies have now been observed in measurements taken by VLBI (Herring and Dong, 1994; Herring, 1993; Sovers *et al.*, 1993; Gipson, 1996; Haas and Wünsch, 2006), SLR (Watkins and Eanes, 1994), and GPS (Hefty *et al.*, 2000; Rothacher *et al.*, 2001; Steigenberger *et al.*, 2006). Predictions of subdaily tidal variations in polar motion using theoretical ocean tide models have been given by Seiler (1990, 1991), Dickman (1993), Gross (1993), Brosche and Wünsch (1994), and Seiler and Wünsch (1995). Predictions of subdaily tidal variations in polar motion using tide models based upon T/P sea surface height measurements have been given by Chao *et al.* (1996) and Chao and Ray (1997). McCarthy and Petit (2004, chapter 8) give extensive tables of coefficients for the effect of diurnal and semidiurnal ocean tides on polar motion. Updated models for these effects are available through the IERS Special Bureau for Tides website (see **Table 4**).

Table 9 Observed variations in polar motion and polar motion excitation caused by long-period ocean tides

Tide	Argument					Period (days)	Polar motion				Polar motion excitation			
							Prograde		Retrograde		Prograde		Retrograde	
	l	l'	F	D	Ω		amp (µas)	phase (degrees)	amp (µas)	phase (degrees)	amp (µas)	phase (degrees)	amp (µas)	phase (degrees)
M_t'	1	0	2	0	1	9.12	14.65	174.3	12.47	140.8	680.82	−5.8	604.66	140.7
M_t	1	0	2	0	2	9.13	4.73	−161.1	6.84	−36.4	219.41	18.7	330.95	−36.6
M_f'	0	0	2	0	1	13.63	44.42	−149.4	41.65	69.2	1366.47	30.4	1364.30	69.0
M_f	0	0	2	0	2	13.66	55.16	−64.1	79.01	31.5	1693.09	115.8	2583.21	31.3
M_m	1	0	0	0	0	27.56	43.97	−85.0	58.62	−55.2	646.96	94.8	979.84	−55.3

l, l', F, D, and Ω are the Delaunay arguments, expressions for which are given in Simon et al. (1994). The period, given in solar days, is the approximate period of the term as viewed in the terrestrial reference frame. The amplitude (amp) and phase of the prograde and retrograde components of polar motion are defined by eqn [68]. The amplitude and phase of the prograde and retrograde components of polar motion excitation are similarly defined but with $\chi(t) = \chi_x(t) + i\,\chi_y(t)$.

Source: Gross RS, Chao BF, and Desai S (1997) Effect of long-period ocean tides on the Earth's polar motion. Progress in Oceanography 40: 385–397.

Comparisons of observations with models show the major role that ocean tides play in causing subdaily polar motion variations (**Figures 5(b)** and **5(c)**), with as much as 60% of the observed polar motion variance being explained by diurnal and semidiurnal ocean tides (Chao *et al.*, 1996). Apart from errors in observations and models, the difference that remains (e.g., Schuh and Schmitz-Hübsch, 2000) may be due to nontidal atmospheric and oceanic effects.

The effects on polar motion of the radiational tides in the atmosphere have been studied by Zharov and Gambis (1996), Brzezinski *et al.* (2002a), and de Viron *et al.* (2005). However, since the oceans will respond dynamically to subdaily tidal variations in the atmospheric wind and pressure fields, the oceans will also contribute to the excitation of polar motion by the radiational tides. Brzezinski *et al.* (2004) have predicted the effects on polar motion of the subdaily radiational tides in the atmosphere and oceans using a barotropic ocean model forced by the 6-h NCEP/NCAR reanalysis wind and pressure fields (also see Gross, 2005a). **Table 10** gives their results at prograde nearly diurnal frequencies. They find the largest effect to be at the prograde S_1 tidal frequency with an amplitude of 9 μas that is primarily excited by ocean-bottom pressure variations. Their predictions at retrograde, nearly diurnal frequencies affect the nutations and are therefore not reproduced here. Since the NCEP/NCAR reanalysis winds and pressure are given every 6 h, the semidiurnal tidal frequency band is incompletely sampled, making it difficult to estimate atmospheric and oceanic effects on polar motion at these frequencies. More complete models for the effects of the radiational tides at nearly semidiurnal frequencies must await the availability of atmospheric fields sampled more often than every 6 h.

9.4.2.4 Chandler wobble and its excitation

Any irregularly shaped solid body rotating about some axis that is not aligned with its figure axis will freely wobble as it rotates (Euler, 1765). The Eulerian free wobble of the Earth is known as the Chandler wobble in honor of Seth Carlo Chandler, Jr., who first observed it (Chandler, 1891). Unlike the forced wobbles of the Earth, such as the annual wobble, whose periods are the same as the periods of the forcing mechanisms, the period of the free Chandler wobble is a function of the internal structure and rheology of the Earth (see Section 9.2.1), and its decay time constant, or quality factor Q, is a function of the dissipation mechanisms acting to dampen it. The observed values for the period and Q of the Chandler wobble can therefore be used to better understand the internal structure of the Earth and the dissipation mechanisms, such as mantle anelasticity, that dampen the Chandler wobble causing its amplitude to decay in the absence of excitation.

Determining an unbiased estimate for the period and Q of the Chandler wobble is complicated by the relatively short duration of the observational record and by incomplete and inaccurate models of the mechanisms acting to excite it. In the absence of any knowledge of its excitation, statistical models of the excitation can be adopted (e.g., Jeffreys, 1972; Wilson and Haubrich, 1976; Ooe, 1978; Wilson and Vicente, 1980, 1990). Since atmospheric, oceanic, and hydrologic processes are thought to be major sources of Chandler excitation, its period and Q have also been estimated by using AAM data (Furuya and Chao, 1996; Kuehne *et al.*, 1996), atmospheric and oceanic angular momentum data (Gross, 2005b), and atmospheric, oceanic, and hydrologic angular momentum data

Table 10   Modeled variations in polar motion caused by the diurnal radiational tide in the atmosphere and oceans

| Tide | Fundamental argument | | | | | | Period (solar days) | $p_x(t)$, μas | | $p_y(t)$, μas | |
	γ	I	I'	F	D	Ω		sin	cos	sin	cos
P_1	1	0	0	−2	2	−2	1.00275	0.1 ± 0.4	0.0 ± 0.4	−0.0 ± 0.4	0.1 ± 0.4
S_1	1	0	−1	0	0	0	1.00000	8.3 ± 0.4	−3.4 ± 0.4	3.4 ± 0.4	8.3 ± 0.4
K_1	1	0	0	0	0	0	0.99727	−0.1 ± 0.5	0.5 ± 0.5	−0.5 ± 0.5	−0.1 ± 0.5

γ is GMST reckoned from the lower culmination of the vernal equinox (GMST + π). I, I', F, D, and Ω are the Delaunay arguments, expressions for which are given in Simon *et al.* (1994). The period, given in solar days, is the approximate period of the term as viewed in the terrestrial reference frame. Since the diurnal radiational tide is seasonally modulated, it also affects polar motion at the P_1 and K_1 tidal frequencies.
Source: Brzezinski A, Ponte RM, and Ali AH (2004) Nontidal oceanic excitation of nutation and diurnal/semidiurnal polar motion revisited. *Journal of Geophysical Research* 109: B11407 (doi:10.1029/2004JB003054).

Table 11 Estimated period and Q of Chandler wobble

Period (solar days)	Q	Data span (years)	Source
Statistical excitation			
433.2 ± 2.2	63 (36, 192)	67.6	(a)
434.0 ± 2.6	100 (50, 400)	70	(b)
434.8 ± 2.0	96 (50, 300)	76	(c)
433.3 ± 3.1	170 (47, 1000)	78	(d)
433.0 ± 1.1	**179 (74, 789)**	**86**	**(e)**
433.1 ± 1.7		93	(f)
Atmospheric excitation			
439.5 ± 2.1	72 (30, 500)	8.6	(g)
433.7 ± 1.8	49 (35, 100)	10.8	(h)
430.8	41	10	(i)
Atmospheric and oceanic excitation			
429.4	107	10	(i)
431.9	83	51	(i)

The recommended estimate is given in bold. The 1σ confidence interval for the Q estimates is given in parentheses.
Sources: (a) Jeffreys (1972); (b) Wilson and Haubrich (1976); (c) Ooe (1978); (d) Wilson and Vicente (1980); (e) Wilson and Vicente (1990); (f) Vicente and Wilson (1997); (g) Kuehne *et al.* (1996); (h) Furuya and Chao (1996); (i) Gross (2005b).

(Wilson and Chen, 2005). **Table 11** gives the resulting estimates for the period and *Q* of the Chandler wobble.

Gross (2005b) discusses the sensitivity of the estimated period and *Q* of the Chandler wobble to the length of the data sets that he analyzed, concluding that data sets spanning at least 31 years are needed to obtain stable estimates. He also notes the need for Monte Carlo simulations to determine corrections to the bias of the estimated *Q* values. Since Gross (2005b) did not do this, but since it was done by Wilson and Vicente (1990) who also used data spanning 86 years to estimate the period and *Q* of the Chandler wobble, the recommended estimate is that determined by them, namely, a period of 433.0 ± 1.1 (1σ) solar days and a *Q* of 179 with a 1σ range of 74–789.

With these recommended values for the period *T* and quality factor *Q* of the Chandler wobble, in the absence of excitation it would freely decay to the minimum rotational energy state of rotation about the figure axis with an *e*-folding amplitude decay time constant $\tau = 2QT/2\pi$ of about 68 years. But a damping time of 68 years is short on a geological timescale, and since the amplitude of the Chandler wobble has at times been observed to actually increase, some mechanism or mechanisms must be acting to excite it. Since its discovery, many possible excitation mechanisms have been studied, including core–mantle interactions (Gire and Le Mouël, 1986; Hinderer *et al.*, 1987; Jault and Le Mouël, 1993), earthquakes (Souriau and Cazenave, 1985; Gross,

1986), continental water storage (Chao *et al.*, 1987; Hinnov and Wilson, 1987; Kuehne and Wilson, 1991), atmospheric wind and surface pressure variations (Wilson and Haubrich, 1976; Wahr, 1983; Furuya *et al.*, 1996, 1997; Aoyama and Naito, 2001; Aoyama *et al.*, 2003; Aoyama, 2005; Stuck *et al.*, 2005), and oceanic current and bottom pressure variations (Ponte and Stammer, 1999; Gross, 2000b; Brzezinski and Nastula, 2002; Brzezinski *et al.*, 2002b; Gross *et al.*, 2003; Liao *et al.*, 2003; Seitz *et al.*, 2004, 2005; Liao, 2005; Ponte, 2005; Seitz and Schmidt, 2005; Thomas *et al.*, 2005).

While there is growing agreement that the Chandler wobble is excited by a combination of atmospheric, oceanic, and hydrologic processes, the relative contribution of each process to its excitation is still being debated. For example, Gross *et al.* (2003) studied the excitation of the Chandler wobble during 1980–2000, finding that atmospheric winds and surface pressure and oceanic currents and bottom pressure combined are significantly coherent with and have enough power to excite the Chandler wobble. They found that during this time interval the observed power in the Chandler band is 1.90 mas², with the sum of all atmospheric and oceanic excitation processes having more than enough power, at 2.22 mas², to excite the Chandler wobble. Ocean-bottom pressure variations were found to be the single most effective process exciting the Chandler wobble, with atmospheric surface pressure variations having about two-thirds as much power as ocean-bottom pressure variations, and with the power of winds and currents combined being less than one-third the power of the combined effects of surface and bottom pressure. The conclusion that oceanic processes are more effective than atmospheric in exciting the Chandler wobble has also been reached by Gross (2000b), Brzezinski and Nastula (2002), Brzezinski et al. (2002b), and Liao et al. (2003). Studies of the excitation of the Chandler wobble using coupled atmosphere–ocean climate models also confirm that atmospheric and oceanic processes have enough power to maintain the Chandler wobble and indicate that the relative contribution of individual atmospheric and oceanic processes changes with time (Celaya et al., 1999; Leuliette and Wahr, 2002; Ponte et al., 2002). However, other studies have concluded that atmospheric processes alone have enough power to excite the Chandler wobble (Furuya et al., 1996, 1997; Aoyama and Naito, 2001; Aoyama et al., 2003; Aoyama, 2005). Resolution of the debate about the relative importance of atmospheric, oceanic, and hydrologic processes to exciting the Chandler wobble awaits the availability of more accurate models of these processes and, because the power needed to maintain the Chandler wobble depends upon its damping (Gross, 2000b), of more accurate estimates of its period and Q.

9.4.2.5 Seasonal wobbles

While continuing to analyze variation of latitude measurements, Chandler (1892) soon discovered that the wobbling motion of the Earth includes an annual component as well as a 14-month component. The annual wobble is a forced wobble of the Earth that is caused largely by the annual appearance of a high atmospheric pressure system over Siberia every winter (e.g., Munk and MacDonald, 1960; Lambeck, 1980, 1988; Eubanks, 1993).

Chao and Au (1991) have shown that during 1980–88 the amplitude of the prograde annual polar motion excitation can be accounted for by atmospheric wind and pressure fluctuations, with equatorial winds contributing about 25% and pressure fluctuations contributing about 75% to the total atmospheric excitation. But even though the amplitude of the observed prograde annual polar motion excitation is consistent with atmospheric wind and pressure fluctuations, there is a rather large phase discrepancy of about 30°. Furthermore, no agreement was found by Chao and Au (1991) between the observed retrograde annual polar motion excitation and that due to atmospheric

processes, with atmospheric excitation being about twice as large as the observed excitation. Discrepancies as large as a factor of 2 in amplitude were also found by Chao and Au (1991) between observed semiannual polar motion excitation and that due to atmospheric processes (also see Wilson and Haubrich, 1976; Merriam, 1982; Barnes et al., 1983; Wahr, 1983; King and Agnew, 1991; Chao, 1993; Aoyama and Naito, 2000; Nastula and Kolaczek, 2002; Kolaczek et al., 2003; Stuck et al., 2005).

Since the agreement with atmospheric excitation is poor except for the prograde annual amplitude, other processes must be contributing to the excitation of seasonal wobbles. Recently, near-global general circulation models of the oceans have been used to investigate the contribution that nontidal oceanic processes make to exciting the seasonal wobbles of the Earth (Furuya and Hamano, 1998; Ponte et al., 1998, 2001; Johnson et al., 1999; Ponte and Stammer, 1999; Nastula et al., 2000, 2003; Wünsch, 2000; Gross et al., 2003; Chen et al., 2004; Gross, 2004; Seitz et al., 2004; Johnson, 2005; Seitz and Schmidt, 2005; Zhou et al., 2005; Zhong et al., 2006). These studies have shown that adding nontidal oceanic excitation to atmospheric improves the agreement with the observed excitation. For example, **Table 12** gives the results obtained by Gross et al. (2003) for the atmospheric and oceanic excitation of polar motion at the the annual and semiannual frequencies. They found that atmospheric processes were more effective than oceanic in exciting the annual and semiannual wobbles. Atmospheric surface pressure variations were found to be the single most important mechanism exciting the annual and semiannual wobbles, with the sum of surface and ocean-bottom pressure variations being about 2–3 times as effective as the sum of winds and currents.

A rather large residual remains after the effects of the atmosphere and oceans are removed from the observed seasonal polar motion excitation. This residual is probably at least partly due to errors in the atmospheric and oceanic models, but could also be due to the neglect of other excitation processes such as hydrologic processes (Hinnov and Wilson, 1987; Chao and O'Conner, 1988; Kuehne and Wilson, 1991; Chen et al., 2000; Wünsch, 2002; Chen and Wilson, 2005; Nastula and Kolaczek, 2005). Wünsch (2002) has summarized the contribution of soil moisture and snow load to exciting the annual and semiannual wobbles. Although the available soil moisture models exhibit large differences, Wünsch (2002) concludes that soil moisture and snow load effects are important

Table 12　Observed and modeled nontidal polar motion excitation at annual and semiannual frequencies

	Annual				Semiannual			
	Prograde		Retrograde		Prograde		Retrograde	
Excitation process	Amplitude (mas)	Phase (degrees)	Amplitude (mas)	Phase (degrees)	Amplitude (mas)	Phase (degrees)	Amplitude (mas)	Phase (degrees)
Observed excitation	14.52 ± 0.33	−62.77 ± 1.30	7.53 ± 0.33	−120.09 ± 2.50	5.67 ± 0.33	107.56 ± 3.32	5.80 ± 0.33	123.66 ± 3.24
Atmospheric								
Winds	2.97 ± 0.12	−34.92 ± 2.30	2.04 ± 0.12	12.06 ± 3.36	0.36 ± 0.12	71.76 ± 19.0	0.55 ± 0.12	−134.43 ± 12.4
Surface pressure (IB)	15.12 ± 0.17	−101.92 ± 0.66	15.05 ± 0.17	−105.30 ± 0.66	2.60 ± 0.17	47.45 ± 3.81	4.75 ± 0.17	103.77 ± 2.08
Winds & surface pressure	16.51 ± 0.23	−92.38 ± 0.81	14.23 ± 0.23	−98.00 ± 0.94	2.93 ± 0.23	50.36 ± 4.58	4.49 ± 0.23	109.77 ± 2.99
Oceanic								
Currents	2.31 ± 0.09	39.72 ± 2.13	2.11 ± 0.09	50.70 ± 2.33	1.32 ± 0.09	176.16 ± 3.71	1.41 ± 0.09	−143.46 ± 3.49
Bottom pressure	3.45 ± 0.11	63.18 ± 1.87	3.42 ± 0.11	110.24 ± 1.89	0.77 ± 0.11	133.89 ± 8.41	1.48 ± 0.11	−136.46 ± 4.35
Currents & bottom pressure	5.64 ± 0.16	53.81 ± 1.61	4.84 ± 0.16	88.22 ± 1.87	1.96 ± 0.16	160.89 ± 4.62	2.89 ± 0.16	−139.87 ± 3.14
Atmospheric & oceanic								
Winds & currents	4.22 ± 0.15	−3.09 ± 2.08	3.91 ± 0.15	31.71 ± 2.24	1.28 ± 0.15	160.38 ± 6.83	1.96 ± 0.15	−140.92 ± 4.48
Surface & bottom pressure	11.82 ± 0.20	−97.61 ± 0.99	12.43 ± 0.20	−114.51 ± 0.94	2.75 ± 0.20	63.61 ± 4.26	4.22 ± 0.20	121.55 ± 2.78
Total of all modeled excitation	12.23 ± 0.28	−77.50 ± 1.32	9.43 ± 0.28	−101.18 ± 1.71	2.90 ± 0.28	89.70 ± 5.57	4.41 ± 0.28	147.63 ± 3.66

IB, inverted barometer. The amplitude and phase of the prograde and retrograde components of polar motion excitation are defined by eqn [68] but with $\chi(t) = \chi_x(t) + i\,\chi_y(t)$. The reference date t_0 for the phase is January 1.0 1990.

Source: Gross RS, Fukumori I, and Menemenlis D (2003) Atmospheric and oceanic excitation of the Earth's wobbles during 1980–2000. *Journal of Geophysical Research* 108(B8): 2370 (doi: 10.1029/2002JB002143).

contributors to exciting the annual and semiannual wobbles. Climate models have also been used to study atmospheric, oceanic, and hydrologic excitation of seasonal polar motion (Celaya *et al.*, 1999; Ponte *et al.*, 2002; Zhong *et al.*, 2003).

9.4.2.6   Nonseasonal wobbles

Like the seasonal wobbles, the wobbling motion of the Earth on interannual timescales is a forced response of the Earth to its excitation mechanisms. Abarca del Rio and Cazenave (1994) compared the observed excitation of the Earth's wobbles to atmospheric excitation during 1980–91, finding similar fluctuations in both components on timescales between 1 and 3 years, and in the *y*-component on timescales between 1.2 and 8 years, but only when the atmospheric excitation is computed assuming that the oceans fully transmit the imposed atmospheric pressure variations to the floor of the oceans (rigid ocean approximation). Chao and Zhou (1999) studied the correlation of the observed polar motion excitation functions on interannual timescales during 1964–94 with the SOI and the North Atlantic Oscillation index (NAOI). Although little agreement was found with the SOI, significant agreement was found with the NAOI, especially for the *x*-component of the observed excitation function, indicating a possible meteorological origin of the interannual wobbles (also see Zhou *et al.*, 1998). Johnson *et al.* (1999) compared the observed polar motion excitation functions on interannual timescales during 1988–98 with atmospheric and oceanic excitation functions, finding only weak agreement between the observed and modeled excitation functions. For periods between 1 and 6 years excluding the annual cycle, oceanic processes were found by Gross *et al.* (2003) to be much more important than atmospheric in exciting interannual polar motions, explaining 33% of the observed variance compared to 6% for atmospheric processes, and having a correlation coefficient of 0.59 with the observations compared to 0.28 for atmospheric processes. Ocean-bottom pressure variations were found to be the single most important excitation mechanism, explaining 30% of the observed variance and having a correlation coefficient of 0.59 with the observations. Although the other processes were not nearly as effective as ocean-bottom pressure variations in exciting interannual polar motions, when all the processes were combined they were found to explain 40% of the observed variance and to have a correlation coefficient of 0.64 with the observations (also see Johnson, 2005; Chen and Wilson, 2005).

Like the seasonal and interannual wobbles, the wobbling motion of the Earth on intraseasonal timescales is a forced response of the Earth to its excitation mechanisms. Eubanks *et al.* (1988) were the first to study the Earth's wobbles on timescales between 2 weeks and several months, concluding that these rapid polar motions during 1983.75–86.75 are at least partially driven by atmospheric surface pressure changes and suggesting that the remaining excitation may be caused by dynamic ocean-bottom pressure variations. Subsequent studies confirmed the importance of atmospheric processes in exciting rapid polar motions (Salstein and Rosen, 1989; Nastula *et al.*, 1990; Gross and Lindqwister, 1992; Nastula, 1992, 1995, 1997; Kuehne *et al.*, 1993; Salstein, 1993; Kosek *et al.*, 1995; Nastula and Salstein, 1999; Stieglitz and Dickman, 1999; Kolaczek *et al.*, 2000a, 2000b; Schuh and Schmitz-Hübsch, 2000; di Leonardo and Dickman, 2004), although the existence of significant discrepancies indicates that nonatmospheric processes must also play an important role.

The contribution of oceanic processes to exciting rapid polar motions has been studied using both barotropic (Ponte, 1997; Nastula and Ponte, 1999) and baroclinic (Ponte *et al.*, 1998, 2001; Johnson *et al.*, 1999; Nastula *et al.*, 2000; Gross *et al.*, 2003; Chen *et al.*, 2004; Nastula and Kolaczek, 2005; Zhou *et al.*, 2005; Lambert *et al.*, 2006) models of the oceans. Such studies have shown that while better agreement with the observations is obtained when oceanic excitation is added to that of the atmosphere, significant discrepancies still remain. For example, Gross *et al.* (2003) found that on intraseasonal timescales, with periods between 5 days and 1 year excluding the seasonal cycles, atmospheric processes were found to be more effective than oceanic in exciting polar motion, explaining 45% of the observed variance compared to 19% for oceanic processes, and having a correlation coefficient of 0.67 with the observations compared to 0.44 for oceanic processes. Of these processes, atmospheric surface pressure variations were found to be the single most effective process exciting intraseasonal polar motions, with winds being the next most important process; ocean-bottom pressure was about half as effective as atmospheric surface pressure, and currents were about half as effective as winds. Atmospheric winds and surface pressure and oceanic currents and bottom pressure combined explained 65% of the observed variance and had a correlation coefficient of 0.81 with the observations.

On the shortest intraseasonal timescales, Ponte and Ali (2002) used the NCEP/NCAR reanalysis atmospheric model and a barotropic ocean model forced by the NCEP/NCAR reanalysis surface pressure and wind stress fields to study atmospheric and oceanic excitation of polar motion during July 1996 through June 2000 on timescales of 2–20 days. They confirmed the importance of oceanic processes in exciting rapid polar motions, finding that about 60% of the observed polar motion excitation variance between periods of 6–20 days is explained by atmospheric excitation, increasing to about 80% when oceanic excitation is added to that of the atmosphere (also see Kouba, 2005).

Spectra of subdaily atmospheric and oceanic excitation functions (Brzezinski *et al.*, 2002a; Gross, 2005a) reveal the presence of atmospheric normal modes: (1) a peak at about -0.83 cycles per day (cpd) due to the ψ_1^1 atmospheric normal mode; (2) a peak of smaller amplitude at about $+1.8$ cpd due to the ξ_2^1 atmospheric normal mode; and (3) slight enhancement in power at about -0.12 cpd due to the ψ_3^1 atmospheric normal mode. Because of their small amplitudes, the effects of these normal modes on polar motion are difficult to detect, although the effect of the ψ_3^1 mode was detected by Eubanks *et al.* (1988).

Acknowledgments

The work described in this paper was performed at the Jet Propulsion Laboratory, California Institute of Technology, under contract with the National Aeronautics and Space Administration. Support for this work was provided by the Solid Earth and Natural Hazards program of NASA's Science Mission Directorate.

References

Abarca del Rio R and Cazenave A (1994) Interannual variations in the Earth's polar motion for 1963–1991: Comparison with atmospheric angular momentum over 1980–1991. *Geophysical Research Letters* 21(22): 2361–2364.
Abarca del Rio R, Gambis D, and Salstein DA (2000) Interannual signals in length of day and atmospheric angular momentum. *Annales Geophysicae* 18: 347–364.
Alfonsi L and Spada G (1998) Effect of subductions and trends in seismically induced Earth rotational variations. *Journal of Geophysical Research* 103(B4): 7351–7362.
Amit H and Olson P (2006) Time-average and time-dependent parts of core flow. *Physics of the Earth and Planetary Interiors* 155: 120–139.

Anderson JD and Rosen RD (1983) The latitude-height structure of 40–50 day variations in atmospheric angular momentum. *Journal of Atmospheric Science* 40: 1584–1591.
Aoyama Y (2005) Quasi-14 month wind fluctuation and excitation of the Chandler wobble. In: Plag H-P, Chao BF, Gross RS, and van Dam T (eds.) *Forcing of Polar Motion in the Chandler Frequency Band: A Contribution to Understanding Interannual Climate Change*, Cahiers du Centre Européen de Géodynamique et de Séismologie, vol. 24, pp. 135–141. Luxembourg: Cahiers du centre Européen de Géodynamique et de Séismologie.
Aoyama Y and Naito I (2000) Wind contributions to the Earth's angular momentum budgets in seasonal variation. *Journal of Geophysical Research* 105(D10): 12417–12431.
Aoyama Y and Naito I (2001) Atmospheric excitation of the Chandler wobble, 1983–1998. *Journal of Geophysical Research* 106(B5): 8941–8954.
Aoyama Y, Naito I, Iwabuchi T, and Yamazaki N (2003) Atmospheric quasi-14 month fluctuation and excitation of the Chandler wobble. *Earth Planets Space* 55: e25–e28.
Argus DF and Gross RS (2004) An estimate of motion between the spin axis and the hotspots over the past century. *Geophysical Research Letters* 31: L06614 (doi:10.1029/2004GL019657).
Asari S, Shimizu H, and Utada H (2006) Variability of the topographic core–mantle torque calculated from core surface flow models. *Physics of Earth and Planetary Interiors* 154: 85–111.
Baader H-R, Brosche P, and Hovel W (1983) Ocean tides and periodic variations of the Earth's rotation. *Journal of Geophysics* 52: 140–142.
Barnes RTH, Hide R, White AA, and Wilson CA (1983) Atmospheric angular momentum fluctuations, length-of-day changes and polar motion. *Proceedings of the Royal Society of London A* 387: 31–73.
Bellanger E, Blanter EM, Le Mouël J-L, and Shnirman MG (2002) Estimation of the 13.63-day lunar tide effect on length of day. *Journal of Geophysical Research* 107(B5): 2102 (doi:10.1029/ 2000JB000076).
Beutler G, Hein GW, Melbourne WG, and Seeber G (eds.) (1996) GPS Trends in Precise Terrestrial, Airborne, and Spaceborne Applications. *Proceedings of the IAG Symposium no. 115*, 351p. New York: Springer.
Beutler G, Rothacher M, Schaer S, Springer TA, Kouba J, and Neilan RE (1999) The International GPS Service (IGS): An interdisciplinary service in support of the Earth sciences. *Advances in Space Research* 23(4): 631–653.
Blewitt G (1993) Advances in Global Positioning System technology for geodynamics investigations: 1978–1992. In: Smith DE and Turcotte DL (eds.) *Contributions of Space Geodesy to Geodynamics: Technology*, American Geophysical Union Geodynamics Series, vol. 25, pp. 195–213. Washington, DC: American Geophysical Union.
Bloxham J (1990) On the consequences of strong stable stratification at the top of the Earth's outer core. *Geophysical Research Letters* 17(12): 2081–2084.
Bloxham J and Jackson A (1991) Fluid flow near the surface of the Earth's core. *Reviews of Geophysics* 29(1): 97–120.
Bloxham J and Kuang W (1995) Comment on "The topographic torque on a bounding surface of a rotating gravitating fluid and the excitation by core motions of decadal fluctuations in the Earth's rotation". *Geophysical Research Letters* 22(24): 3561–3562.
Bock Y and Leppard N (eds.) (1990) Global Positioning System: An Overview. *Proceedings of the IAG Symposium no. 102*, 459p. New York: Springer.
Braginsky SI (1970) Torsional magnetohydrodynamic vibrations in the Earth's core and variations in day length. *Geomagnetizm I Aeronomiya* (English translation) 10: 1–8.

Brosche P (1982) Oceanic tides and the rotation of the Earth. In: Fricke W and Teleki G (eds.) *Sun and Planetary System*, pp. 179–184. Dordrecht, Holland: Reidel.

Brosche P and Sündermann J (1985) The Antarctic circumpolar current and its influence on the Earth's rotation. *Deutsche Hydrographische Zeitschrift* 38: 1–6.

Brosche P and Wünsch J (1994) On the "rotational angular momentum" of the oceans and the corresponding polar motion. *Astronomische Nachrichten* 315: 181–188.

Brosche P, Seiler U, Sündermann J, and Wünsch J (1989) Periodic changes in Earth's rotation due to oceanic tides. *Astronomy and Astrophysics* 220: 318–320.

Brosche P, Wünsch J, Frische A, Sündermann J, Maier-Reimer E, and Mikolajewicz U (1990) The seasonal variation of the angular momentum of the oceans. *Naturwissenschaften* 77: 185–186.

Brosche P, Wünsch J, Campbell J, and Schuh H (1991) Ocean tide effects in Universal Time detected by VLBI. *Astronomy and Astrophysics* 245: 676–682.

Brosche P, Wünsch J, Maier-Reimer E, Segschneider J, and Sündermann J (1997) The axial angular momentum of the general circulation of the oceans. *Astronomische Nachrichten* 318: 193–199.

Brzezinski A (1992) Polar motion excitation by variations of the effective angular momentum function: Considerations concerning deconvolution problem. *Manuscripta Geodaetica* 17: 3–20.

Brzezinski A and Capitaine N (1993) The use of the precise observations of the celestial ephemeris pole in the analysis of geophysical excitation of Earth rotation. *Journal of Geophysical Research* 98(B4): 6667–6675.

Brzezinski A and Nastula J (2002) Oceanic excitation of the Chandler wobble. *Advances in Space Research* 30: 195–200.

Brzezinski A, Bizouard Ch, and Petrov S (2002a) Influence of the atmosphere on Earth rotation: What new can be learned from the recent atmospheric angular momentum estimates? *Surveys in Geophysics* 23: 33–69.

Brzezinski A, Nastula J, and Ponte RM (2002b) Oceanic excitation of the Chandler wobble using a 50-year time series of ocean angular momentum. In: Adám J and Schwarz K-P (eds.) *Vistas for Geodesy in the New Millennium*, IAG Symposia, vol. 125, pp. 434–439. New York: Springer.

Brzezinski A, Ponte RM, and Ali AH (2004) Nontidal oceanic excitation of nutation and diurnal/semidiurnal polar motion revisited. *Journal of Geophysical Research* 109: B11407 (doi:10.1029/2004JB003054).

Buffett BA (1996a) Gravitational oscillations in the length of day. *Geophysical Research Letters* 23(17): 2279–2282.

Buffett BA (1996b) A mechanism for decade fluctuations in the length of day. *Geophysical Research Letters* 23(25): 3803–3806.

Buffett BA (1998) Free oscillations in the length of day: Inferences on physical properties near the core–mantle boundary. In: Gurnis M, Wysession ME, Knittle E, and Buffett BA (eds.) *The Core–Mantle Boundary Region*, American Geophysical Union Geodynamics Series, vol. 28, pp. 153–165. Washington, DC: American Geophysical Union.

Buffett BA and Mound JE (2005) A Green's function for the excitation of torsional oscillations in the Earth's core. *Journal of Geophysical Research* 110: B08104 (doi:10.1029/2004JB003495).

Bullard EC, Freeman C, Gellman H, and Nixon J (1950) The westward drift of the Earth's magnetic field. *Philosophical Transactions of the Royal Society London A* 243: 61–92.

Capitaine N (2000) Definition of the celestial ephemeris pole and the celestial ephemeris origin. In: Johnston KJ, McCarthy DD, Luzum BJ, and Kaplan GH (eds.) *Towards Models and Constants for Sub-Microarcsecond Astrometry*, Proc. IAU Colloquium 180, US Naval Obs, pp. 153–163. Washington, DC: US Naval Observatory.

Celaya MA, Wahr JM, and Bryan FO (1999) Climate-driven polar motion. *Journal of Geophysical Research* 104(B6): 12813–12829.

Chandler SC (1891) On the variation of latitude, II. *The Astronomical Journal* 11: 65–70.

Chandler SC (1892) On the variation of latitude, VII. *The Astronomical Journal* 12: 97–101.

Chao BF (1983) Autoregressive harmonic analysis of the Earth's polar motion using homogeneous International Latitude Service data. *Journal of Geophysical Research* 88(B12): 10299–10307.

Chao BF (1984) Interannual length-of-day variations with relation to the Southern Oscillation/El Niño. *Geophysical Research Letters* 11(5): 541–544.

Chao BF (1988) Correlation of interannual length-of-day variation with El Niño/Southern Oscillation, 1972–1986. *Journal of Geophysical Research* 93(B7): 7709–7715.

Chao BF (1989) Length-of-day variations caused by El Niño/Southern Oscillation and Quasi-Biennial Oscillation. *Science* 243: 923–925.

Chao BF (1993) Excitation of Earth's polar motion by atmospheric angular momentum variations, 1980–1990. *Geophysical Research Letters* 20(2): 253–256.

Chao BF (1994) Zonal tidal signals in the Earth's polar motion. *Eos Transactions of the American Geophysical Union* 75(44): 158.

Chao BF and Au AY (1991) Atmospheric excitation of the Earth's annual wobble: 1980–1988. *Journal of Geophysical Research* 96(B4): 6577–6582.

Chao BF and Gross RS (1987) Changes in the Earth's rotation and low-degree gravitational field introduced by earthquakes. *Geophysical Journal of the Royal Astronomical Society* 91: 569–596.

Chao BF and O'Connor WP (1988) Global surface-water-induced seasonal variations in the Earth's rotation and gravitational field. *Geophysical Journal of the Royal Astronomical Society* 94: 263–270.

Chao BF and Ray RD (1997) Oceanic tidal angular momentum and Earth's rotation variations. *Progress in Oceanography* 40: 399–421.

Chao BF and Zhou Y-H (1999) Meteorological excitation of interannual polar motion by the North Atlantic Oscillation. *Journal of Geodynamics* 27: 61–73.

Chao BF, O'Connor WP, Chang ATC, Hall DK, and Foster JL (1987) Snow load effect on the Earth's rotation and gravitational field, 1979–1985. *Journal of Geophysical Research* 92(B9): 9415–9422.

Chao BF, Merriam JB, and Tamura Y (1995a) Geophysical analysis of zonal tidal signals in length of day. *Geophysical Journal International* 122: 765–775.

Chao BF, Ray RD, and Egbert GD (1995b) Diurnal/semidiurnal oceanic tidal angular momentum: Topex/Poseidon models in comparison with Earth's rotation rate. *Geophysical Research Letters* 22(15): 1993–1996.

Chao BF, Ray RD, Gipson JM, Egbert GD, and Ma C (1996) Diurnal/semidiurnal polar motion excited by oceanic tidal angular momentum. *Journal of Geophysical Research* 101(B9): 20151–20163.

Chapman S and Lindzen RS (1970) *Atmospheric Tides*. Dordrecht, The Netherlands: D. Reidel.

Chapront J, Chapront-Touzé M, and Francou G (2002) A new determination of lunar orbital parameters, precession constant and tidal acceleration from LLR measurements. *Astronomy and Astrophysics* 387: 700–709.

Chen J (2005) Global mass balance and the length-of-day variations. *Journal of Geophysical Research* 110: B08404 (doi:10.1029/2004JB003474).

Chen JL and Wilson CR (2005) Hydrological excitations of polar motion, 1993–2002. *Geophysical Journal International* 160: 833–839.

Chen JL, Wilson CR, Chao BF, Shum CK, and Tapley BD (2000) Hydrological and oceanic excitations to polar motion and length-of-day variation. *Geophysical Journal International* 141: 149–156.

Chen JL, Wilson CR, Hu X-G, Zhou Y-H, and Tapley BD (2004) Oceanic effects on polar motion determined from an ocean model and satellite altimetry: 1993–2001. *Journal of Geophysical Research* 109: B02411 (doi:10.1029/2003JB002664).

Cheng MK, Eanes RJ, and Tapley BD (1992) Tidal deceleration of the Moon's mean motion. *Geophysical Journal International* 108: 401–409.

Cheng MK, Shum CK, Eanes RJ, Schutz BE, and Tapley BD (1990) Long-period perturbations in Starlette orbit and tide solution. *Journal of Geophysical Research* 95(B6): 8723–8736.

Christodoulidis DC, Smith DE, Williamson RG, and Klosko SM (1988) Observed tidal braking in the Earth/Moon/Sun system. *Journal of Geophysical Research* 93(B6): 6216–6236.

Dahlen FA (1976) The passive influence of the oceans upon the rotation of the Earth. *Geophysical Journal of the Royal Astronomical Society* 46: 363–406.

Dai A and Wang J (1999) Diurnal and semidiurnal tides in global surface pressure fields. *Journal of Atmospheric Science* 56: 3874–3891.

Davis RG and Whaler KA (1996) Determination of a steady velocity in a rotating frame of reference at the surface of the Earth's core. *Geophysical Journal International* 126: 92–100.

Defraigne P and Smits I (1999) Length of day variations due to zonal tides for an inelastic Earth in non-hydrostatic equilibrium. *Geophysical Journal International* 139: 563–572.

Desai SD and Wahr JM (1995) Empirical ocean tide models estimated from Topex/Poseidon altimetry. *Journal of Geophysical Research* 100(C12): 25205–25228.

Desai SD and Wahr JM (1999) Monthly and fortnightly tidal variations of the Earth's rotation rate predicted by a Topex/Poseidon empirical ocean tide model. *Geophysical Research Letters* 26(8): 1035–1038.

de Viron O, Salstein D, Bizouard Ch, and Fernandez L (2004) Low-frequency excitation of length of day and polar motion by the atmosphere. *Journal of Geophysical Research* 109: B03408 (doi:10.1029/2003JB002817).

de Viron O, Schwarzbaum G, Lott F, and Dehant V (2005) Diurnal and subdiurnal effects of the atmosphere on the Earth rotation and geocenter motion. *Journal of Geophysical Research* 110: B11404 (doi:10.1029/2005JB003761).

Dickey JO, Bender PL, Faller JE, et al. (1994a) Lunar laser ranging: A continuing legacy of the Apollo program. *Science* 265: 482–490.

Dickey JO, Gegout P, and Marcus SL (1999) Earth-atmosphere angular momentum exchange and ENSOs: The rotational signature of the 1997–98 event. *Geophysical Research Letters* 26(16): 2477–2480.

Dickey JO, Ghil M, and Marcus SL (1991) Extratropical aspects of the 40–50 day oscillation in length-of-day and atmospheric angular momentum. *Journal of Geophysical Research* 96(D12): 22643–22658.

Dickey JO, Marcus SL, Eubanks TM, and Hide R (1993a) Climate studies via space geodesy: Relationships between ENSO and interannual length-of-day variations. In: McBean GA and Hantel M (eds.) *Interactions Between Global Climate Subsystems: The Legacy of Hann*, American Geophysical Union Geophysical Monograph Series, vol. 75, pp. 141–155. Washington, DC: American Geophysical Union.

Dickey JO, Marcus SL, and Hide R (1992) Global propagation of interannual fluctuations in atmospheric angular momentum. *Nature* 357: 482–488.

Dickey JO, Marcus SL, Hide R, Eubanks TM, and Boggs DH (1994b) Angular momentum exchange among the solid Earth, atmosphere and oceans: A case study of the 1982–83 El Niño event. *Journal of Geophysical Research* 99(B12): 23921–23937.

Dickey JO, Marcus SL, Johns CM, Hide R, and Thompson SR (1993b) The oceanic contribution to the Earth's seasonal angular momentum budget. *Geophysical Research Letters* 20: 2953–2956.

Dickey JO, Newhall XX, and Williams JG (1985) Earth orientation from lunar laser ranging and an error analysis of polar motion services. *Journal of Geophysical Research* 90: 9353–9362.

Dickman SR (1981) Investigation of controversial polar motion features using homogeneous International Latitude Service data. *Journal of Geophysical Research* 86(B6): 4904–4912.

Dickman SR (1993) Dynamic ocean-tide effects on Earth's rotation. *Geophysical Journal International* 112: 448–470.

Dickman SR (1994) Ocean tidal effects on Earth's rotation and on the lunar orbit. In: Schutz BE, Anderson A, Froidevaux C, and Parke M (eds.) *Gravimetry and Space Techniques Applied to Geodynamics and Ocean Dynamics*, American Geophysical Union Geophysical Monograph Series, vol. 82, pp. 87–94. Washington, DC: American Geophysical Union.

Dickman SR (2003) Evaluation of "effective angular momentum function" formulations with respect to core–mantle coupling. *Journal of Geophysical Research* 108(B3): 2150 (doi:10.1029/2001JB001603).

Dickman SR (2005) Rotationally consistent Love numbers. *Geophysical Journal International* 161: 31–40.

Dickman SR and Nam YS (1995) Revised predictions of long-period ocean tidal effects on Earth's rotation rate. *Journal of Geophysical Research* 100(B5): 8233–8243.

di Leonardo SM and Dickman SR (2004) Isolation of atmospheric effects on rapid polar motion through Wiener filtering. *Geophysical Journal International* 159: 863–873.

Djurovic D and Pâquet P (1996) The common oscillations of solar activity, the geomagnetic field, and the Earth's rotation. *Solar Physics* 167: 427–439.

Dong D and Herring T (1990) Observed variations of UT1 and polar motion in the diurnal and semidiurnal bands. *EOS Transactions of the American Geophysical Union* 71: 482.

Dumberry M and Bloxham J (2002) Inner core tilt and polar motion. *Geophysical Journal International* 151: 377–392.

Dumberry M and Bloxham J (2004) Variations in the Earth's gravity field caused by torsional oscillations in the core. *Geophysical Journal International* 159: 417–434.

Dumberry M and Bloxham J (2006) Azimuthal flows in the Earth's core and changes in length of day at millennial time-scales. *Geophysical Journal International* 165: 32–46.

Dziewonski AM and Anderson DL (1981) Preliminary reference Earth model. *Physics of Earth and Planetary Interiors* 25: 297–356.

Eubanks TM (1993) Variations in the orientation of the Earth. In: Smith DE and Turcotte DL (eds.) *Contributions of Space Geodesy to Geodynamics: Earth Dynamics*, American Geophysical Union Geodynamics Series, vol. 24, pp. 1–54. Washington, DC: American Geophysical Union.

Eubanks TM, Steppe JA, and Dickey JO (1986) The El Niño, the Southern Oscillation and the Earth's rotation. In: Cazenave A (ed.) *Earth Rotation: Solved and Unsolved Problems*, pp. 163–186. Hingham, MA: D. Reidel.

Eubanks TM, Steppe JA, Dickey JO, Rosen RD, and Salstein DA (1988) Causes of rapid motions of the Earth's pole. *Nature* 334: 115–119.

Euler L (1765) *Theoria Motus Corporum Solidorum*. Rostock, Germany: Litteris et impensis A.F. Röse.

Fang M, Hager BH, and Herring TA (1996) Surface deformation caused by pressure changes in the fluid core. *Geophysical Research Letters* 23(12): 1493–1496.

Feissel M and Gambis D (1980) La mise en evidence de variations rapides de la durée du jour. *Comptes Rendus De L Academie Des Sciences Paris, Series B* 291: 271–273.

Feissel M and Gambis D (1993) The International Earth Rotation Service: Current results for research on Earth rotation and reference frames. *Advances in Space Research* 13(11): 143–150.

Feissel M and Gavoret J (1990) ENSO-related signals in Earth rotation, 1962–87. In: McCarthy DD and Carter WE (eds.) *Variations in Earth Rotation*, AGU Geophysical Monograph Series, vol. 59, pp. 133–137. Washington, DC: American Geophysical Union.

Feissel M and Nitschelm C (1985) Time dependent aspects of the atmospheric driven fluctuations in the duration of the day. *Annales Geophysicae* 3: 180–186.

Fliegel HF and Hawkins TP (1967) Analysis of variations in the rotation of the Earth. *Astronomical Journal* 72(4): 544–550.

Freedman AP, Ibañez-Meier R, Herring TA, Lichten SM, and Dickey JO (1994b) Subdaily Earth rotation during the Epoch '92 campaign. *Geophysical Research Letters* 21(9): 769–772.

Freedman AP, Steppe JA, Dickey JO, Eubanks TM, and Sung L-Y (1994a) The short-term prediction of universal time and length of day using atmospheric angular momentum. *Journal of Geophysical Research* 99(B4): 6981–6996.

Frische A and Sündermann J (1990) The seasonal angular momentum of the thermohaline ocean circulation. In: Brosche P and Sündermann J (eds.) *Earth's Rotation From Eons to Days*, pp. 108–126. New York: Springer.

Furuya M and Chao BF (1996) Estimation of period and Q of the Chandler wobble. *Geophysical Journal International* 127: 693–702.

Furuya M and Hamano Y (1998) Effect of the Pacific Ocean on the Earth's seasonal wobble inferred from National Center for Environmental Prediction ocean analysis data. *Journal of Geophysical Research* 103(B5): 10131–10140.

Furuya M, Hamano Y, and Naito I (1996) Quasi-periodic wind signal as a possible excitation of Chandler wobble. *Journal of Geophysical Research* 101(B11): 25537–25546.

Furuya M, Hamano Y, and Naito I (1997) Importance of wind for the excitation of Chandler wobble as inferred from wobble domain analysis. *Journal of Physics of the Earth* 45: 177–188.

Gambis D (1992) Wavelet transform analysis of the length of the day and the El-Niño/Southern Oscillation variations at intra-seasonal and interannual time scales. *Annales Geophysicae* 10: 429–437.

Gambis D (2004) Monitoring Earth orientation using space-geodetic techniques: State-of-the-art and prospective. *Journal of Geodesy* 78: 295–303.

Gasperini P, Sabadini R, and Yuen DA (1986) Excitation of the Earth's rotational axis by recent glacial discharges. *Geophysical Research Letters* 13(6): 533–536.

Gilbert F and Dziewonski AM (1975) An application of normal mode theory to the retrieval of structural parameters and source mechanisms from seismic spectra. *Philosophical Transactions of the Royal Society London* A 278: 187–269.

Gipson JM (1996) Very long baseline interferometry determination of neglected tidal terms in high-frequency Earth orientation variation. *Journal of Geophysical Research* 101(B12): 28051–28064.

Gire C and Le Mouël J-L (1986) Flow in the fluid core and Earth's rotation. In: Cazenave A (ed.) *Earth Rotation: Solved and Unsolved Problems*, pp. 241–258. Dordrecht, Holland: D. Reidel.

Goldstein H (1950) *Classical Mechanics*. Reading, MA: Addison-Wesley.

Greff-Lefftz M (2004) Upwelling mantle plumes, superswells, and true polar wander. *Geophysical Journal International* 159: 1125–1137.

Greff-Lefftz M and Legros H (1995) Core–mantle coupling and polar motion. *Physics of Earth and Planetary Interiors* 91: 273–283.

Greff-Lefftz M, Pais MA, and Le Mouël J-L (2004) Surface gravitational field and topography changes induced by the Earth's fluid core motions. *Journal of Geodesy* 78: 386–392.

Greiner-Mai H and Barthelmes F (2001) Relative wobble of the Earth's inner core derived from polar motion and associated gravity variations. *Geophysical Journal International* 144: 27–36.

Greiner-Mai H, Jochmann H, and Barthelmes F (2000) Influence of possible inner-core motions on the polar motion and the gravity field. *Physics of Earth and Planetary Interiors* 117: 81–93.

Greiner-Mai H, Jochmann H, Barthelmes F, and Ballani L (2003) Possible influences of core processes on the Earth's rotation and the gravity field. *Journal of Geodynamics* 36: 343–358.

Grejner-Brzezinska DA and Goad CC (1996) Subdaily Earth rotation determined from GPS. *Geophysical Research Letters* 23(19): 2701–2704.

Gross RS (1986) The influence of earthquakes on the Chandler wobble during 1977–1983. *Geophysical Journal of the Royal Astronomical Society* 85: 161–177.

Gross RS (1992) Correspondence between theory and observations of polar motion. *Geophysical Journal International* 109: 162–170.

Gross RS (1993) The effect of ocean tides on the Earth's rotation as predicted by the results of an ocean tide model. *Geophysical Research Letters* 20(4): 293–296.

Gross RS (2000a) Combinations of Earth orientation measurements: SPACE97, COMB97, and POLE97. *Journal of Geodesy* 73: 627–637.

Gross RS (2000b) The excitation of the Chandler wobble. *Geophysical Research Letters* 27(15): 2329–2332.

Gross RS (2001) A combined length-of-day series spanning 1832–1997: LUNAR97. *Physics of Earth and Planetary Interiors* 123: 65–76.

Gross RS (2004) Angular momentum in the Earth system. In: Sanso F (ed.) *V Hotine-Marussi Symposium on Mathematical Geodesy*, IAG Symposia, vol. 127, pp. 274–284. New York: Springer.

Gross RS (2005a) Oceanic excitation of polar motion: A review. In: Plag H-P, Chao BF, Gross RS, and van Dam T (eds.) *Forcing of Polar Motion in the Chandler Frequency Band: A Contribution to Understanding Interannual Climate Change,* Cahiers du Centre Européen de Géodynamique et de Séismologie, vol. 24, pp. 89–102. Luxembourg: Cahiers du Centre Européen de Géodynamique et de Séismologie.

Gross RS (2005b) The observed period and Q of the Chandler wobble. In: Plag H-P, Chao BF, Gross RS, and van Dam T (eds.) *Forcing of Polar Motion in the Chandler Frequency Band: A Contribution to Understanding Interannual Climate Change,* Cahiers du Centre Européen de Géodynamique et de Séismologie, vol. 24, pp. 31–37. Luxembourg: Cahiers du Centre Européen de Géodynamique et de Séismologie.

Gross RS and Lindqwister UJ (1992) Atmospheric excitation of polar motion during the GIG '91 measurement campaign. *Geophysical Research Letters* 19(9): 849–852.

Gross RS and Vondrák J (1999) Astrometric and space-geodetic observations of polar wander. *Geophysical Research Letters* 26(14): 2085–2088.

Gross RS, Marcus SL, Eubanks TM, Dickey JO, and Keppenne CL (1996a) Detection of an ENSO signal in seasonal length-of-day variations. *Geophysical Research Letters* 23(23): 3373–3376.

Gross RS, Chao BF, and Desai S (1997) Effect of long-period ocean tides on the Earth's polar motion. *Progress in Oceanography* 40: 385–397.

Gross RS, Eubanks TM, Steppe JA, Freedman AP, Dickey JO, and Runge TF (1998) A Kalman-filter-based approach to combining independent Earth-orientation series. *Journal of Geodesy* 72: 215–235.

Gross RS, Fukumori I, and Menemenlis D (2003) Atmospheric and oceanic excitation of the Earth's wobbles during 1980–2000. *Journal of Geophysical Research* 108(B8): 2370 (doi:10.1029/2002JB002143).

Gross RS, Fukumori I, Menemenlis D, and Gegout P (2004) Atmospheric and oceanic excitation of length-of-day variations during 1980–2000. *Journal of Geophysical Research* 109: B01406 (doi:10.1029/2003JB002432).

Gross RS, Fukumori I, and Menemenlis D (2005) Atmospheric and oceanic excitation of decadal-scale Earth orientation variations. *Journal of Geophysical Research* 110: B09405 (doi:10.1029/2004JB003565).

Gross RS, Hamdan KH, and Boggs DH (1996b) Evidence for excitation of polar motion by fortnightly ocean tides. *Geophysical Research Letters* 23(14): 1809–1812.

Gross RS, Marcus SL, and Dickey JO (2002) Modulation of the seasonal cycle in length-of-day and atmospheric angular momentum. In: Adám J and Schwarz K-P (eds.) *Vistas for Geodesy in the New Millennium*, IAG Symposia, vol. 125, pp. 457–462. New York: Springer.

Groten E (2004) Fundamental parameters and current (2004) best estimates of the parameters of common relevance to astronomy, geodesy, and geodynamics. *Journal of Geodesy* 77: 724–731.

Gubbins D (1982) Finding core motions from magnetic observations. *Philosophical Transactions of the Royal Society London A* 306: 247–254.

Guinot B (1970) Short-period terms in Universal Time. *Astronomy and Astrophysics* 8: 26–28.

Haas R and Wünsch J (2006) Sub-diurnal earth rotation variations from the VLBI CONT02 campaign. *Journal of Geodynamics* 41: 94–99.

Halley E (1695) Some account of the ancient state of the city of Palmyra; with short remarks on the inscriptions found there. *Philosophical Transactions of the Royal Society London* 19: 160–175.

Haurwitz B and Cowley AD (1973) The diurnal and semidiurnal barometric oscillations, global distribution and annual variation. *Pure and Applied Geophysics* 102: 193–222.

Hefty J and Capitaine N (1990) The fortnightly and monthly zonal tides in the Earth's rotation from 1962 to 1988. *Geophysical Journal International* 103: 219–231.

Hefty J, Rothacher M, Springer T, Weber R, and Beutler G (2000) Analysis of the first year of Earth rotation parameters with a sub-daily resolution gained at the CODE processing center of the IGS. *Journal of Geodesy* 74: 479–487.

Hendon HH (1995) Length of day changes associated with the Madden–Julian oscillation. *Journal of Atmospheric Science* 52(13): 2373–2383.

Herring TA (1993) Diurnal and semidiurnal variations in Earth rotation. *Advances in Space Research* 13(11): 281–290.

Herring TA and Dong D (1991) Current and future accuracy of Earth rotation measurements. In: Carter WE (ed.) *Proceedings of the AGU Chapman Conference on Geodetic VLBI: Monitoring Global Change*, NOAA Technical Report NOS 137 NGS 49, pp. 306–324. Washington, DC: NOAA.

Herring TA and Dong D (1994) Measurement of diurnal and semidiurnal rotational variations and tidal parameters of Earth. *Journal of Geophysical Research* 99(B9): 18051–18071.

Hide R (1969) Interaction between the Earth's liquid core and solid mantle. *Nature* 222: 1055–1056.

Hide R (1977) Towards a theory of irregular variations in the length of the day and core–mantle coupling. *Philosophical Transactions of the Royal Society London A* 284: 547–554.

Hide R (1989) Fluctuations in the Earth's rotation and the topography of the core–mantle interface. *Philosophical Transactions of the Royal Society London A* 328: 351–363.

Hide R (1993) Angular momentum transfer between the Earth's core and mantle. In: Le Moüel J-L, Smylie DE, and Herring T (eds.) *Dynamics of the Earth's Deep Interior and Earth Rotation*, American Geophysical Union Geophysical Monograph Series, vol. 72, pp. 109–112. Washington, DC: American Geophysical Union.

Hide R (1995a) The topographic torque on a bounding surface of a rotating gravitating fluid and the excitation by core motions of decadal fluctuations in the Earth's rotation. *Geophysical Research Letters* 22(8): 961–964.

Hide R (1995b) Reply to the comment by Bloxham J. and Kuang W. on a paper entitled "The topographic torque on a bounding surface of a rotating gravitating fluid and the excitation by core motions of decadal fluctuations in the Earth's rotation". *Geophysical Research Letters* 22(24): 3563–3565.

Hide R (1998) A note on topographic core–mantle coupling. *Physics of Earth and Planetary Interiors* 109: 91–92.

Hide R and Dickey JO (1991) Earth's variable rotation. *Science* 253: 629–637.

Hide R, Birch NT, Morrison LV, Shea DJ, and White AA (1980) Atmospheric angular momentum fluctuations and changes in the length of the day. *Nature* 286: 114–117.

Hide R, Clayton RW, Hager BH, Spieth MA, and Voorhies CV (1993) Topographic core–mantle coupling and fluctuations in the Earth's rotation. In: Aki K and Dmowska R (eds.) *Relating Geophysical Structures and Processes: The Jeffreys Volume*, American Geophysical Union Geophysical Monograph Series, vol. 76, pp. 107–120. Washington, DC: American Geophysical Union.

Hide R, Boggs DH, Dickey JO, Dong D, Gross RS, and Jackson A (1996) Topographic core–mantle coupling and polar motion on decadal time scales. *Geophysical Journal International* 125: 599–607.

Hide R, Boggs DH, and Dickey JO (2000) Angular momentum fluctuations within the Earth's liquid core and torsional oscillations of the core–mantle system. *Geophysical Journal International* 143: 777–786.

Hinderer J, Legros H, Gire C, and Le Moüel J-L (1987) Geomagnetic secular variation, core motions and implications for the Earth's wobbles. *Physics of Earth and Planetary Interiors* 49: 121–132.

Hinnov LA and Wilson CR (1987) An estimate of the water storage contribution to the excitation of polar motion. *Geophysical Journal of the Royal Astronomical Society* 88: 437–459.

Hofmann-Wellenhof B, Lichtenegger H, and Collins J (1997) *Global Positioning System: Theory and Practice.* New York: Springer.

Holme R (1998a) Electromagnetic core–mantle coupling—I. Explaining decadal changes in the length of day. *Geophysical Journal International* 132: 167–180.

Holme R (1998b) Electromagnetic core–mantle coupling II: Probing deep mantle conductance. In: Gurnis M, Wysession ME, Knittle E, and Buffett BA (eds.) *The Core-Mantle Boundary Region*, American Geophysical Union Geodynamics Series, vol. 28, pp. 139–151. Washington, DC: American Geophysical Union.

Holme R (2000) Electromagnetic core–mantle coupling III. Laterally varying mantle conductance. *Physics of Earth and Planetary Interiors* 117: 329–344.

Holme R and Whaler KA (2001) Steady core flow in an azi-muthally drifting reference frame. *Geophysical Journal International* 145: 560–569.

Höpfner J (1998) Seasonal variations in length of day and atmospheric angular momentum. *Geophysical Journal International* 135: 407–437.

Höpfner J (2000) Seasonal length-of-day changes and atmospheric angular momentum oscillations in their temporal variability. *Journal of Geodesy* 74: 335–358.

Höpfner J (2001) Atmospheric, oceanic, and hydrological contributions to seasonal variations in length of day. *Journal of Geodesy* 75: 137–150.

Höpfner J (2004) Low-frequency variations, Chandler and annual wobbles of polar motion as observed over one century. *Surveys in Geophysics* 25: 1–54.

Hough SS (1895) The oscillations of a rotating ellipsoidal shell containing fluid. *Philosophical Transactions of the Royal Society London A* 186: 469–506.

Hulot G, Le Huy M, and Le Mouël J-L (1996) Influence of core flows on the decade variations of the polar motion. *Geophysical and Astrophysical Fluid Dynamics* 82: 35–67.

Jackson A (1997) Time-dependency of tangentially geostrophic core surface motions. *Physics of Earth and Planetary Interiors* 103: 293–311.

Jackson A, Bloxham J, and Gubbins D (1993) Time-dependent flow at the core surface and conservation of angular momentum in the coupled core–mantle system. In: Le Mouël J-L, Smylie DE, and Herring T (eds.) *Dynamics of the Earth's Deep Interior and Earth Rotation*, American Geophysical Union Geophysical Monograph Series, vol. 72, pp. 97–107. Washington, DC: American Geophysical Union.

James TS and Ivins ER (1995) Present-day Antarctic ice mass changes and crustal motion. *Geophysical Research Letters* 22(8): 973–976.

James TS and Ivins ER (1997) Global geodetic signatures of the Antarctic ice sheet. *Journal of Geophysical Research* 102(B1): 605–633.

Jault D (2003) Electromagnetic and topographic coupling and LOD variations. In: Jones CA, Soward AM, and Zhang K (eds.) *Earth's core and lower mantle*, pp. 56–76. London: Taylor & Francis.

Jault D and Le Mouël J-L (1989) The topographic torque associated with tangentially geostrophic motion at the core surface and inferences on the flow inside the core. *Geophysical and Astrophysical Fluid Dynamics* 48: 273–296.

Jault D and Le Mouël J-L (1990) core–mantle boundary shape: Constraints inferred from the pressure torque acting between the core and the mantle. *Geophysical Journal International* 101: 233–241.

Jault D and Le Mouël J-L (1991) Exchange of angular momentum between the core and the mantle. *Journal of Geomagnetism and Geoelectricity* 43: 111–129.

Jault D and Le Mouël J-L (1993) Circulation in the liquid core and coupling with the mantle. *Advances in Space Research* 13(11): 221–233.

Jault D and Le Mouël J-L (1999) Comment on "On the dynamics of topographical core-mantle coupling" by Weijia Kuang and Jeremy Bloxham. *Physics of Earth and Planetary Interiors* 114: 211–215.

Jault D, Gire C, and Le Mouël J-L (1988) Westward drift, core motions and exchanges of angular momentum between core and mantle. *Nature* 333: 353–356.

Jault D, Hulot G, and Le Mouël J-L (1996) Mechanical core–mantle coupling and dynamo modelling. *Physics of Earth and Planetary Interiors* 98: 187–191.

Jeffreys H (1928) Possible tidal effects on accurate time keeping. *Monthly Notices of the Royal Astronomical Society Geophysical Supplement* 2: 56–58.

Jeffreys H (1972) The variation of latitude. In: Melchior P and Yumi S (eds.) *Rotation of the Earth*, Int. Astron. Union Symp. No. 48, pp. 39–42. Dordrecht, Holland: D. Reidel.

Jochmann H (1989) Motion of the Earth's inner core and related variations of polar motion and the rotational velocity. *Astronomische Nachrichten* 310: 435–442.

Johnson TJ (2005) The interannual spectrum of the atmosphere and oceans. In: Plag H-P, Chao BF, Gross RS, and van Dam T (eds.) *Forcing of Polar Motion in the Chandler Frequency Band: A Contribution to Understanding Interannual Climate Change,* Cahiers du Centre Européen de Géodynamique et de Séismologie, vol. 24, pp. 69–75. Luxembourg: Cahiers du Centre Européen de Géodynamique et de Séismologie.

Johnson TJ, Wilson CR, and Chao BF (1999) Oceanic angular momentum variability estimated from the Parallel Ocean Climate Model, 1988–1998. *Journal of Geophysical Research* 104(B11): 25183–25195.

Johnson TJ, Luzum BJ, and Ray JR (2005) Improved near-term Earth rotation predictions using atmospheric angular momentum analysis and forecasts. *Journal of Geodynamics* 39: 209–221.

Johnston P and Lambeck K (1999) Postglacial rebound and sea level contributions to changes in the geoid and the Earth's rotation axis. *Geophysical Journal International* 136: 537–558.

Jordi C, Morrison LV, Rosen RD, Salstein DA, and Rosselló G (1994) Fluctuations in the Earth's rotation since 1830 from high-resolution astronomical data. *Geophysical Journal International* 117: 811–818.

Kantha LH, Stewart JS, and Desai SD (1998) Long-period lunar fortnightly and monthly ocean tides. *Journal of Geophysical Research* 103(C6): 12639–12647.

King NE and Agnew DC (1991) How large is the retrograde annual wobble? *Geophysical Research Letters* 18(9): 1735–1738.

Kinoshita H, Nakajima K, Kubo Y, Nakagawa I, Sasao T, and Yokoyama K (1979) Note on nutation in ephemerides. *Publications of the International Latitude Observatory of Mizusawa* 12(2): 71–108.

Kolaczek B, Kosek W, and Schuh H (2000a) Short-period oscillations of Earth rotation. In: Dick S, McCarthy D, and Luzum B (eds.) *Polar Motion: Historical and Scientific Problems, IAU Colloq. 178,* Astron. Soc. Pacific Conf. Ser, vol. 208, pp. 533–544. San Francisco: Astronomical Society of Pacific.

Kolaczek B, Nuzhdina M, Nastula J, and Kosek W (2000b) El Niño impact on atmospheric polar motion excitation. *Journal of Geophysical Research* 105(B2): 3081–3087.

Kolaczek B, Nastula J, and Salstein D (2003) El Nino-related variations in atmosphere-polar motion interactions. *Journal of Geodynamics* 36: 397–406.

Kosek W, Nastula J, and Kolaczek B (1995) Variability of polar motion oscillations with periods from 20 to 150 days in 1979–1991. *Bulletin Géodesique* 69: 308–319.

Kouba J (2005) Comparison of polar motion with oceanic and atmospheric angular momentum time series for 2-day to Chandler periods. *Journal of Geodesy* 79: 33–42.

Kouba J and Vondrák J (2005) Comparison of length of day with oceanic and atmospheric angular momentum series. *Journal of Geodesy* 79: 256–268.

Kuang W and Bloxham J (1997) On the dynamics of topographical core–mantle coupling. *Physics of Earth and Planetary Interiors* 99: 289–294.

Kuang W and Chao B (2001) Topographic core–mantle coupling in geodynamo modeling. *Geophysical Research Letters* 28(9): 1871–1874.

Kuehne J and Wilson CR (1991) Terrestrial water storage and polar motion. *Journal of Geophysical Research* 96(B3): 4337–4345.

Kuehne J, Johnson S, and Wilson CR (1993) Atmospheric excitation of nonseasonal polar motion. *Journal of Geophysical Research* 98(B11): 19973–19978.

Kuehne J, Wilson CR, and Johnson S (1996) Estimates of the Chandler wobble frequency and Q. *Journal of Geophysical Research* 101(B6): 13573–13579.

Lambeck K (1980) *The Earth's Variable Rotation: Geophysical Causes and Consequences*. New York: Cambridge University Press.

Lambeck K (1988) *Geophysical Geodesy: The Slow Deformations of the Earth*. New York: Oxford University Press.

Lambert SB, Bizouard C, and Dehant V (2006) Rapid variations in polar motion during the 2005–2006 winter season. *Geophysical Research Letters* 33: L13303 (doi:10.1029/2006GL026422).

Langley RB, King RW, and Shapiro II (1981a) Earth rotation from lunar laser ranging. *Journal of Geophysical Research* 86: 11913–11918.

Langley RB, King RW, Shapiro II, Rosen RD, and Salstein DA (1981b) Atmospheric angular momentum and the length of the day: A common fluctuation with a period near 50 days. *Nature* 294: 730–733.

Leick A (2003) *GPS Satellite Surveying*. New York: Wiley.

Lerch FJ, Nerem RS, Putney BH, *et al.* (1994) A geopotential model from satellite tracking, altimeter, and surface gravity data: GEM-T3. *Journal of Geophysical Research* 99(B2): 2815–2839.

Leuliette EW and Wahr JM (2002) Climate excitation of polar motion. In: Adám J and Schwarz K-P (eds.) *Vistas for Geodesy in the New Millennium*, IAG Symposia, vol. 125, pp. 428–433. New York: Springer.

Li Z (1985) Earth rotation from optical astrometry, 1962.0–1982.0. In: *Bureau International de l'Heure Annual Report for 1984*, pp. D31–D63. Paris: Observation de Paris.

Li Z and Feissel M (1986) Determination of the Earth rotation parameters from optical astrometry observations, 1962.0–1982.0. *Bulletin Géodesique* 60: 15–28.

Liao D-C (2005) A brief review of atmospheric and oceanic excitation of the Chandler wobble. In: Plag H-P, Chao BF, Gross RS, and van Dam T (eds.) *Forcing of Polar Motion in the Chandler Frequency Band: A Contribution to Understanding Interannual Climate Change*, Cahiers du Centre Européen de Géodynamique et de Séismologie, vol. 24, pp. 155–162. Luxembourg: Cahiers du Centre Européen de Géodynamique et de Séismologie.

Liao DC and Greiner-Mai H (1999) A new ΔLOD series in monthly intervals (1892.0–1997.0) and its comparison with other geophysical results. *Journal of Geodesy* 73: 466–477.

Liao D, Liao X, and Zhou Y (2003) Oceanic and atmospheric excitation of the Chandler wobble. *Geophysical Journal International* 152: 215–227.

Lichten SM, Marcus SL, and Dickey JO (1992) Sub-daily resolution of Earth rotation variations with global positioning system measurements. *Geophysical Research Letters* 19(6): 537–540.

Love JJ and Bloxham J (1994) Electromagnetic coupling and the toroidal magnetic field at the core–mantle boundary. *Geophysical Journal International* 117: 235–256.

Luzum BJ, Ray JR, Carter MS, and Josties FJ (2001) Recent improvements to IERS Bulletin A combination and prediction. *GPS Solutions* 4(3): 34–40.

Malla RP, Wu SC, and Lichten SM (1993) Geocenter location and variations in Earth orientation using global positioning system measurements. *Journal of Geophysical Research* 98(B3): 4611–4617.

Madden RA (1987) Relationships between changes in the length of day and the 40- to 50-day oscillations in the tropics. *Journal of Geophysical Research* 92: 8391–8399.

Madden RA and Julian PR (1971) Detection of a 40–50 day oscillation in the zonal wind in the tropical Pacific. *Journal of Atmospheric Science* 28: 702–708.

Madden RA and Julian PR (1972) Description of global-scale circulation cells in the tropics with a 40–50 day period. *Journal of Atmospheric Science* 29: 1109–1123.

Madden RA and Julian PR (1994) Observations of the 40–50-day tropical oscillation — A review. *Monthly Weather Review* 122: 814–837.

Marcus SL, Chao Y, Dickey JO, and Gegout P (1998) Detection and modeling of nontidal oceanic effects on Earth's rotation rate. *Science* 281: 1656–1659.

Marcus SL, Dickey JO, and de Viron O (2001) Links between intraseasonal (extended MJO) and ENSO timescales: Insights via geodetic and atmospheric analysis. *Geophysical Research Letters* 28(18): 3465–3468.

Markowitz W (1955) The annual variation in the rotation of the Earth, 1951–54. *Astronomical Journal* 60: 171.

Markowitz W (1959) Variations in rotation of the Earth, results obtained with the dual-rate Moon camera and photographic zenith tubes. *Astronomical Journal* 64: 106–113.

Markowitz W (1960) Latitude and longitude, and the secular motion of the pole. In: Runcorn SK (ed.) *Methods and Techniques in Geophysics*, pp. 325–361. New York: Interscience Publishers.

Markowitz W (1961) International determination of the total motion of the pole. *Bulletin Géodesique* 59: 29–41.

Marsh JG, Lerch FJ, Putney BH, *et al.* (1990) The GEM-T2 gravitational model. *Journal of Geophysical Research* 95(B13): 22043–22071.

Marsh JG, Lerch FJ, Putney BH, *et al.* (1991) Correction to "The GEM-T2 gravitational model". *Journal of Geophysical Research* 96(B10): 16651.

Martin CF (1969) *A Study of the Rate of Rotation of the Earth from Occultations of Stars by the Moon 1627–1860*. PhD Thesis, 144p. Yale University, New Haven.

Mathews PM and Bretagnon P (2003) Polar motions equivalent to high frequency nutations for a nonrigid Earth with anelastic mantle. *Astronomy and Astrophysics* 400: 1113–1128.

Mathews PM, Buffett BA, Herring TA, and Shapiro II (1991) Forced nutations of the Earth: Influence of inner core dynamics 2. Numerical results and comparisons. *Journal of Geophysical Research* 96: 8243–8257.

McCarthy DD (1976) The determination of Universal Time at the U.S. Naval Observatory. *U.S. Naval Obs. Circ.*, No. 154.

McCarthy DD and Babcock AK (1986) The length of day since 1656. *Physics of Earth and Planetary Interiors* 44: 281–292.

McCarthy DD and Luzum BJ (1993) An analysis of tidal variations in the length of day. *Geophysical Journal International* 114: 341–346.

McCarthy DD and Luzum BJ (1996) Path of the mean rotational pole from 1899 to 1994. *Geophysical Journal International* 125: 623–629.

McCarthy DD and Petit G (eds.) (2004) *IERS Conventions (2003)*. IERS Tech. Note no. 32, 127p. Frankfurt, Germany: Bundesamts für Kartographie und Geodäsie.

Melchior P and Dejaiffe R (1969) Calcul des déclinaisons et mouvements propres des étoiles du Service International des Latitudes à partir des catalgues méridiens. *Annales Observatoire Royal de Belgique 3e Serie* 10: 63–339.

Merriam JB (1980) Zonal tides and changes in the length of day. *Geophysical Journal of the Royal Astronomical Society* 62: 551–561.

Merriam JB (1982) Meteorological excitation of the annual polar motion. *Geophysical Journal of the Royal Astronomical Society* 70: 41–56.

Merriam JB (1984) Tidal terms in Universal Time: Effects of zonal winds and mantle Q. *Journal of Geophysical Research* 89(B12): 10109–10114.

Merriam JB (1985) LAGEOS and UT measurements of long-period Earth tides and mantle Q. *Journal of Geophysical Research* 90(B11): 9423–9430.

Mitrovica JX and Forte AM (1997) Radial profile of mantle viscosity: Results from the joint inversion of convection and postglacial rebound observables. *Journal of Geophysical Research* 102(B2): 2751–2769.

Mitrovica JX and Milne GA (1998) Glaciation-induced perturbations in the Earth's rotation: A new appraisal. *Journal of Geophysical Research* 103(B1): 985–1005.

Mitrovica JX and Peltier WR (1993) Present-day secular variations in the zonal harmonics of Earth's geopotential. *Journal of Geophysical Research* 98(B3): 4509–4526.

Mitrovica JX, Wahr J, Matsuyama I, and Paulson A (2005) The rotational stability of an ice-age Earth. *Geophysical Journal International* 161: 491–506.

Moritz H and Mueller II (1988) *Earth Rotation: Theory and Observation*. New York: Ungar.

Morrison LV (1978) Catalogue of observations of occultations of stars by the Moon for the years 1943 to 1971. *Royal Greenwich Observatory Bulletin* No.183.

Morrison LV (1979) Re-determination of the decade fluctuations in the rotation of the Earth in the period 1861–1978. *Geophysical Journal of the Royal Astronomical Society* 58: 349–360.

Morrison LV and Stephenson FR (2001) Historical eclipses and the variability of the Earth's rotation. *Journal of Geodynamics* 32: 247–265.

Morrison LV and Ward CG (1975) An analysis of the transits of Mercury: 1677–1973. *Monthly Notices of the Royal Astronomical Society* 173: 183–206.

Morrison LV, Lukac MR, and Stephenson FR (1981) Catalogue of observations of occultations of stars by the Moon for the years 1623–1942 and solar eclipses for the years 1621–1806. *Royal Greenwich Observatory Bulletin* No.186.

Mound J (2005) Electromagnetic torques in the core and resonant excitation of decadal polar motion. *Geophysical Journal International* 160: 721–728.

Mound JE and Buffett BA (2003) Interannual oscillations in length of day: Implications for the structure of the mantle and core. *Journal of Geophysical Research* 108(B7): 2334 (doi:10.1029/2002JB002054).

Mound JE and Buffett BA (2005) Mechanisms of core–mantle angular momentum exchange and the observed spectral properties of torsional oscillations. *Journal of Geophysical Research* 110: B08103 (doi:10.1029/2004JB003555).

Mulholland JD (1980) Scientific advances from ten years of lunar laser ranging. *Reviews of Geophysics and Space Physics* 18: 549–564.

Munk WH and MacDonald GJF (1960) *The Rotation of the Earth: A Geophysical Discussion*. New York: Cambridge University Press.

Naito I and Kikuchi N (1990) A seasonal budget of the Earth's axial angular momentum. *Geophysical Research Letters* 17: 631–634.

Naito I and Kikuchi N (1991) Reply to Rosen and Salstein's comment. *Geophysical Research Letters* 18: 1927–1928.

Nakada M (2000) Effect of the viscoelastic lithosphere on polar wander speed caused by the Late Pleistocene glacial cycles. *Geophysical Journal International* 143: 230–238.

Nakada M (2002) Polar wander caused by the Quaternary glacial cycles and fluid Love number. *Earth and Planetary Science Letters* 200: 159–166.

Nakada M and Okuno J (2003) Perturbations of the Earth's rotation and their implications for the present-day mass balance of both polar ice caps. *Geophysical Journal International* 152: 124–138.

Nam YS and Dickman SR (1990) Effects of dynamic long-period ocean tides on changes in Earth's rotation rate. *Journal of Geophysical Research* 95(B5): 6751–6757.

Nastula J (1992) Short periodic variations in the Earth's rotation in the period 1984–1990. *Annales Geophysicae* 10: 441–448.

Nastula J (1995) Short periodic variations of polar motion and hemispheric atmospheric angular momentum excitation functions in the period 1984–1992. *Annales Geophysicae* 13: 217–225.

Nastula J (1997) The regional atmospheric contributions to the polar motion and EAAM excitation functions. In: Segawa J, Fujimoto H, and Okubo S (eds.) *Gravity, Geoid, and Marine Geodesy*, IAG Symposia, vol. 117, pp. 281–288. New York: Springer.

Nastula J and Kolaczek B (2002) Seasonal oscillations in regional and global atmospheric excitation of polar motion. *Advances in Space Research* 30(2): 381–386.

Nastula J and Kolaczek B (2005) Analysis of hydrological excitation of polar motion. In: Plag H-P, Chao BF, Gross RS, and van Dam T (eds.) *Forcing of Polar Motion in the Chandler Frequency Band: A Contribution to Understanding Interannual Climate Change*, Cahiers du Centre Européen de Géodynamique et de Séismologie, vol. 24, pp. 149–154. Luxembourg: Cahiers du Centre Européen de Géodynamique et de Séismologie.

Nastula J and Ponte RM (1999) Further evidence for oceanic excitation of polar motion. *Geophysical Journal International* 139: 123–130.

Nastula J and Salstein D (1999) Regional atmospheric angular momentum contributions to polar motion excitation. *Journal of Geophysical Research* 104: 7347–7358.

Nastula J, Gambis D, and Feissel M (1990) Correlated high-frequency variations in polar motion and length of the day in early 1988. *Annales Geophysical* 8: 565–570.

Nastula J, Ponte RM, and Salstein DA (2000) Regional signals in atmospheric and oceanic excitation of polar motion. In: Dick S, McCarthy D, and Luzum B (eds.) *Polar Motion: Historical and Scientific Problems, IAU Colloq. 178*, Astron. Soc. Pacific Conf. Ser, vol. 208, pp. 463–472. San Francisco: Astronomical Society of Pacific.

Nastula J, Salstein DA, and Ponte RM (2003) Empirical patterns of variability in atmospheric and oceanic excitation of polar motion. *Journal of Geodynamics* 36: 383–396.

Newhall XX, Williams JG, and Dickey JO (1988) Earth rotation from lunar laser ranging. In: Babcock AK and Wilkins GA (eds.) *The Earth's Rotation and Reference Frames for Geodesy and Geodynamics*, pp. 159–164. Dordrecht, Holland: D. Reidel.

Okamoto I and Kikuchi N (1983) Low frequency variations of homogeneous ILS polar motion data. *Publications of the International Latitude Observatory of Mizusawa* 16: 35–40.

Ooe M (1978) An optimal complex ARMA model of the Chandler wobble. *Geophysical Journal of the Royal Astronomical Society* 53: 445–457.

Pais A and Hulot G (2000) Length of day decade variations, torsional oscillations, and inner core superrotation: Evidence from recovered core surface zonal flows. *Physics of Earth and Planetary Interiors* 118: 291–316.

Pais MA, Oliveira O, and Nogueira F (2004) Nonuniqueness of inverted core-mantle boundary flows and deviations from tangential geostrophy. *Journal of Geophysical Research* 109: B8105 (doi:10.1029/2004JB003012).

Pearlman MR, Degnan JJ, and Bosworth JM (2002) The International Laser Ranging Service. *Advances in Space Research* 30(2): 135–143.

Pearlman M, Noll C, Dunn P, *et al.* (2005) The International Laser Ranging Service and its support for IGGOS. *Journal of Geodynamics* 40: 470–478.

Peltier WR (1988) Global sea level and Earth rotation. *Science* 240: 895–901.

Peltier WR and Jiang X (1996) Glacial isostatic adjustment and Earth rotation: Refined constraints on the viscosity of the deepest mantle. *Journal of Geophysical Research* 101(B2): 3269–3290.

Philander SG (1990) *El Niño, La Niña, and the Southern Oscillation*. San Diego, CA: Academic Press.

Ponsar S, Dehant V, Holme R, Jault D, Pais A, and Van Hoolst T (2003) The core and fluctuations in the Earth's rotation. In: Dehant V, Creager KC, Karato S, and Zatman S (eds.) *Earth's Core: Dynamics, Structure, Rotation*, American Geophysical Union Geodynamics Series, vol. 31, pp. 251–261. Washington, DC: American Geophysical Union.

Ponte RM (1997) Oceanic excitation of daily to seasonal signals in Earth rotation: Results from a constant-density numerical model. *Geophysical Journal International* 130: 469–474.

Ponte RM (2005) What do we know about low frequency signals in ocean angular momentum? In: Plag H-P, Chao BF, Gross RS, and van Dam T (eds.) *Forcing of Polar Motion in the Chandler Frequency Band: A Contribution to Understanding Interannual Climate Change, Cahiers du Center Européen de Géodynamique et de Séismologie*, vol. 24, pp. 77–82. Luxembourg: Cahiers du Center Européen de Géodynamique et de Séismologie.

Ponte RM and Ali AH (2002) Rapid ocean signals in polar motion and length of day. *Geophysical Research Letters* 29(15): 1711 (10.1029/2002GL015312).

Ponte RM and Stammer D (1999) Role of ocean currents and bottom pressure variability on seasonal polar motion. *Journal of Geophysical Research* 104: 23393–23409.

Ponte RM and Stammer D (2000) Global and regional axial ocean angular momentum signals and length-of-day variations (1985–1996). *Journal of Geophysical Research* 105: 17161–17171.

Ponte RM, Stammer D, and Marshall J (1998) Oceanic signals in observed motions of the Earth's pole of rotation. *Nature* 391: 476–479.

Ponte RM, Stammer D, and Wunsch C (2001) Improving ocean angular momentum estimates using a model constrained by data. *Geophysical Research Letters* 28: 1775–1778.

Ponte RM, Rajamony J, and Gregory JM (2002) Ocean angular momentum signals in a climate model and implications for Earth rotation. *Climate Dynamics* 19: 181–190.

Ray RD (1994) Tidal energy dissipation: Observations from astronomy, geodesy, and oceanography. In: Majumdar SK, Miller EW, Forbes GS, Schmalz RF, and Panah AA (eds.) *The Oceans: Physical-Chemical Dynamics and Human Impact*, pp. 171–185. Pittsburgh: Pennsylvania Academic of Science.

Ray RD, Steinberg DJ, Chao BF, and Cartwright DE (1994) Diurnal and semidiurnal variations in the Earth's rotation rate induced by oceanic tides. *Science* 264: 830–832.

Ricard Y, Sabadini R, and Spada G (1992) Isostatic deformations and polar wander induced by redistribution of mass within the Earth. *Journal of Geophysical Research* 97(B10): 14223–14236.

Ricard Y, Spada G, and Sabadini R (1993) Polar wandering of a dynamic Earth. *Geophysical Journal International* 113: 284–298.

Richards MA, Ricard Y, Lithgow-Bertelloni C, Spada G, and Sabadini R (1997) An explanantion for Earth's long-term rotational stability. *Science* 275: 372–375.

Richards MA, Bunge H-P, Ricard Y, and Baumgardner JR (1999) Polar wandering in mantle convection models. *Geophysical Research Letters* 26(12): 1777–1780.

Roberts PH (1972) Electromagnetic core–mantle coupling. *Journal of Geomagnetism and Geoelectricity* 24: 231–259.

Robertson DS (1991) Geophysical applications of very-long-baseline interferometry. *Reviews of Modern Physics* 63(4): 899–918.

Robertson DS, Ray JR, and Carter WE (1994) Tidal variations in UT1 observed with very long baseline interferometry. *Journal of Geophysical Research* 99(B1): 621–636.

Rochester MG (1960) Geomagnetic westward drift and irregularities in the Earth's rotation. *Philosophical Transactions of the Royal Society of London A* 252: 531–555.

Rochester MG (1962) Geomagnetic core–mantle coupling. *Journal of Geophysical Research* 67: 4833–4836.

Rochester MG (1984) Causes of fluctuations in the rotation of the Earth. *Philosophical Transactions of the Royal Society of London A* 313: 95–105.

Roden RB (1963) Electromagnetic core–mantle coupling. *Geophysical Journal of the Royal Astronomical Society* 7: 361–374.

Rosen RD (1993) The axial momentum balance of Earth and its fluid envelope. *Surveys in Geophysics* 14: 1–29.

Rosen RD and Salstein DA (1985) Contribution of stratospheric winds to annual and semi-annual fluctuations in atmospheric angular momentum and the length of day. *Journal of Geophysical Research* 90: 8033–8041.

Rosen RD and Salstein DA (1991) Comment on "A seasonal budget of the Earth's axial angular momentum" by Naito and Kikucho. *Geophysical Research Letters* 18: 1925–1926.

Rosen RD, Salstein DA, Eubanks TM, Dickey JO, and Steppe JA (1984) An El Niño signal in atmospheric angular momentum and Earth rotation. *Science* 225: 411–414.

Rothacher M, Beutler G, Weber R, and Hefty J (2001) High-frequency variations in Earth rotation from global positioning system data. *Journal of Geophysical Research* 106(B7): 13711–13738.

Rubincam DP (2003) Gravitational core-mantle coupling and the acceleration of the Earth. *Journal of Geophysical Research* 108(B7): 2338 (doi:10.1029/2002JB002132).

Sabadini R and Vermeersen BLA (2002) Long-term rotation instabilities of the Earth: A Reanalysis. In: Mitrovica JX and Vermeersen BLA (eds.) *Ice Sheets, Sea Level and the Dynamic Earth*, American Geophysical Union Geodynamics Series, vol. 29, pp. 51–67. Washington, DC: American Geophysical Union.

Sabadini R and Vermeersen B (2004) *Global Dynamics of the Earth: Applications of Normal Mode Relaxation Theory to Solid-Earth Geophysics*. Dordrecht, The Netherlands: Kluwer.

Salstein DA (1993) Monitoring atmospheric winds and pressures for Earth orientation studies. *Advances in Space Research* 13(11): 175–184.

Salstein DA and Rosen RD (1986) Earth rotation as a proxy for interannual variability in atmospheric circulation, 1860-present. *Journal of Climate and Applied Meteorology* 25: 1870–1877.

Salstein DA and Rosen RD (1989) Regional contributions to the atmospheric excitation of rapid polar motions. *Journal of Geophysical Research* 94: 9971–9978.

Schastok J, Soffel M, and Ruder H (1994) A contribution to the study of fortnightly and monthly zonal tides in UT1. *Astronomy and Astrophysics* 283: 650–654.

Schreiber KU, Velikoseltsev A, Rothacher M, Klügel T, Stedman GE, and Wiltshire DL (2004) Direct measurment of diurnal polar motion by ring laser gyroscopes. *Journal of Geophysical Research* 109: B06405 (doi:10.1029/2003JB002803).

Schlüter W, Himwich E, Nothnagel A, Vandenberg N, and Whitney A (2002) IVS and its important role in the maintenance of the global reference systems. *Advances in Space Research* 30(2): 145–150.

Schuh H and Schmitz-Hübsch H (2000) Short period variations in Earth rotation as seen by VLBI. *Surveys in Geophysics* 21: 499–520.

Schuh H, Richter B, and Nagel S (2000) Analaysis of long time series of polar motion. In: Dick S, McCarthy D, and Luzum B (eds.) *Polar Motion: Historical and Scientific Problems, IAU Colloq. 178*, Astron. Soc. Pacific Conf. Ser, vol. 208, pp. 321–331. San Francisco: Astronomical Society of Pacific.

Schuh H, Nagel S, and Seitz T (2001) Linear drift and periodic variations observed in long time series of polar motion. *Journal of Geodesy* 74: 701–710.

Segschneider J and Sündermann J (1997) Response of a global circulation model to real-time forcing and implications to Earth's rotation. *Journal of Physical Oceanography* 27: 2370–2380.

Seidelmann PK (1982) 1980 IAU theory of nutation: The final report of the IAU working group on nutations. *Celestial Mechanics* 27: 79–106.

Seidelmann PK, Guinot B, and Doggett LE (1992) Time. In: Seidelmann PK (ed.) *Explanatory Supplement to the Astronomical Almanac*, pp. 39–93. Mill Valley, CA: University Science Books.

Seiler U (1990) Variations of the angular momentum budget for tides of the present ocean. In: Brosche P and Sündermann J (eds.) *Earth's Rotation from Eons to Days*, pp. 81–94. New York: Springer.

Seiler U (1991) Periodic changes of the angular momentum budget due to the tides of the world ocean. *Journal of Geophysical Research* 96(B6): 10287–10300.

Seiler U and Wünsch J (1995) A refined model for the influence of ocean tides on UT1 and polar motion. *Astronomische Nachrichten* 316: 419–423.

Seitz F and Schmidt M (2005) Atmospheric and oceanic contributions to Chandler wobble excitation determined by wavelet filtering. *Journal of Geophysical Research* 110: B11406 (doi:10.1029/2005JB003826).

Seitz F, Stuck J, and Thomas M (2004) Consistent atmospheric and oceanic excitation of the Earth's free polar motion. *Geophysical Journal International* 157: 25–35.

Seitz F, Stuck J, and Thomas M (2005) White noise Chandler wobble excitation. In: Plag H-P, Chao BF, Gross RS, and van Dam T (eds.) *Forcing of Polar Motion in the Chandler Frequency Band: A Contribution to Understanding Interannual Climate Change,* Cahiers du Centre Européen de Géodynamique et de Séismologie, vol. 24, pp. 15–21. Luxembourg: Cahiers du Centre Européen de Géodynamique et de Séismologie.

Shelus PJ (2001) Lunar laser ranging: Glorious past and a bright future. *Surveys in Geophysics* 22: 517–535.

Simon JL, Bretagnon P, Chapront J, Chapront-Touzé M, Francou G, and Laskar J (1994) Numerical expressions for precession formulae and mean elements for the Moon and the planets. *Astronomy and Astrophysics* 282: 663–683.

Smith DE, Christodoulidis DC, Kolenkiewicz R, *et al.* (1985) A global geodetic reference frame from LAGEOS ranging (SL5.1AP). *Journal of Geophysical Research* 90: 9221–9233.

Smith DE, Kolenkiewicz R, Dunn PJ, *et al.* (1990) Tectonic motion and deformation from satellite laser ranging to LAGEOS. *Journal of Geophysical Research* 95: 22013–22041.

Smith DE, Kolenkiewicz R, Nerem RS, *et al.* (1994) Contemporary global horizontal crustal motion. *Geophysical Journal International* 119: 511–520.

Smith ML and Dahlen FA (1981) The period and Q of the Chandler wobble. *Geophysical Journal of the Royal Astronomical Society* 64: 223–281.

Souriau A and Cazenave A (1985) Re-evaluation of the seismic excitation of the Chandler wobble from recent data. *Earth and Planetary Science Letters* 75: 410–416.

Sovers OJ, Jacobs CS, and Gross RS (1993) Measuring rapid ocean tidal Earth orientation variations with very long baseline interferometry. *Journal of Geophysical Research* 98(B11): 19959–19971.

Sovers OJ, Fanselow JL, and Jacobs CS (1998) Astrometry and geodesy with radio interferometry: Experiments, models, results. *Reviews of Modern Physics* 70(4): 1393–1454.

Spada G, Ricard Y, and Sabadini R (1992) Excitation of true polar wander by subduction. *Nature* 360: 452–454.

Spencer Jones H (1939) The rotation of the Earth and the secular acceleration of the Sun, Moon, and planets. *Monthly Notices of the Royal Astronomical Society* 99: 541–558.

Stedman GE (1997) Ring-laser tests of fundamental physics and geophysics. *Reports on Progress in Physics* 60: 615–688.

Stefanick M (1982) Interannual atmospheric angular momentum variability 1963–1973 and the southern oscillation. *Journal of Geophysical Research* 87: 428–432.

Steigenberger P, Rothacher M, Dietrich R, Fritsche M, Rülke A, and Vey S (2006) Reprocessing of a global GPS network. *Journal of Geophysical Research* 111: B05402 (doi:10.1029/2005JB003747).

Steinberger BM and O'Connell RJ (1997) Changes of the Earth's rotation axis inferred from advection of mantle density heterogeneities. *Nature* 387: 169–173.

Steinberger B and O'Connell RJ (2002) The convective mantle flow signal in rates of true polar wander. In: Mitrovica JX and Vermeersen BLA (eds.) *Ice Sheets, Sea Level and the Dynamic Earth*, American Geophysical Union Geodynamics Series, vol. 29, pp. 233–256. Washington, DC: American Geophysical Union.

Stephenson FR (1997) *Historical Eclipses and Earth's Rotation*. New York: Cambridge University Press.

Stephenson FR and Morrison LV (1984) Long-term changes in the rotation of the Earth: 700 B.C. to A.D. 1980. *Philosophical Transactions of the Royal Society London* A 313: 47–70.

Stephenson FR and Morrison LV (1995) Long-term fluctuations in the Earth's rotation: 700 BC to AD 1990. *Philosophical Transactions of the Royal Society London* A 351: 165–202.

Stewart DN, Busse FH, Whaler KA, and Gubbins D (1995) Geomagnetism, Earth rotation and the electrical conductivity of the lower mantle. *Physics of Earth and Planetary Interiors* 92: 199–214.

Stieglitz TC and Dickman SR (1999) Refined correlations between atmospheric and rapid polar motion excitation. *Geophysical Journal International* 139: 115–122.

Stix M and Roberts PH (1984) Time-dependent electromagnetic core-mantle coupling. *Physics of Earth and Planetary Interiors* 36: 49–60.

Stolz A, Bender PL, Faller JE, *et al.* (1976) Earth rotation measured by lunar laser ranging. *Science* 193: 997–999.

Stoyko N (1937) Sur la périodicité dans l'irrégularité de la rotation de la terre. *Comptes Rendus De L Academie Des Sciences* 205: 79–81.

Stuck J, Seitz F, and Thomas M (2005) Atmospheric forcing mechanisms of polar motion. In: Plag H-P, Chao BF, Gross RS, and van Dam T (eds.) *Forcing of Polar Motion in the Chandler Frequency Band: A Contribution to Understanding Interannual Climate Change,* Cahiers du Centre Européen de Géodynamique et de Séismologie, vol. 24, pp. 127–133. Luxembourg: Cahiers du Centre Européen de Géodynamique et de Séismologie.

Tamisea ME, Mitrovica JX, Tromp J, and Milne GA (2002) Present-day secular variations in the low-degree harmonics of the geopotential: Sensitivity analysis on spherically symmetric Earth models. *Journal of Geophysical Research* 107(B12): 2378 (doi:10.1029/2001JB000696).

Tapley BD, Schutz BE, and Eanes RJ (1985) Station coordinates, baselines, and Earth rotation from LAGEOS laser ranging: 1976-1984. *Journal of Geophysical Research* 90: 9235–9248.

Tapley BD, Schutz BE, Eanes RJ, Ries JC, and Watkins MM (1993) Lageos laser ranging contributions to geodynamics, geodesy, and orbital dynamics. In: Smith DE and Turcotte DL (eds.) *Contributions of Space Geodesy to Geodynamics: Earth Dynamics*, American Geophysical Union Geodynamics Series, vol. 24, pp. 147–173. Washington, DC: American Geophysical Union.

Tapley BD, Schutz BE, and Born GH (2004) *Statistical Orbit Determination*. Burlington, MA: Elsevier.

Tavernier G, Fagard H, Feissel-Vernier M, *et al.* (2005) The International DORIS Service (IDS). *Advances in Space Research* 36(3): 333–341.

Thomas M, Dobslaw H, Stuck J, and Seitz F (2005) The ocean's contribution to polar motion excitation – As many solutions as numerical models? In: Plag H-P, Chao BF, Gross RS, and van Dam T (eds.) *Forcing of Polar Motion in the Chandler Frequency Band: A Contribution to Understanding Interannual Climate Change*, Cahiers du Centre Européen de Géodynamique et de Séismologie, vol. 24, pp. 143–148 Luxembourg: Cahiers du Centre Européen de Géodynamique et de Séismologie.

Tisserand F (1891) *Traité de Mécanique Céleste*, vol. II. Paris: Gauthier-Villars.

Trenberth KE and Guillemot CJ (1994) The total mass of the atmosphere. *Journal of Geophysical Research* 99(D11): 23079–23088.

Tosi N, Sabadini R, Marotta AM, and Vermeersen LLA (2005) Simultaneous inversion for the Earth's mantle viscosity and ice mass imbalance in Antarctica and Greenland. *Journal of Geophysical Research* 110: B07402 (doi:10.1029/2004JB003236).

Trupin AS (1993) Effects of polar ice on the Earth's rotation and gravitational potential. *Geophysical Journal International* 113: 273–283.

Trupin AS, Meier MF, and Wahr JM (1992) Effect of melting glaciers on the Earth's rotation and gravitational field: 1965–1984. *Geophysical Journal International* 108: 1–15.

Van Hoolst T and Dehant V (2002) Influence of triaxiality and second-order terms in flattenings on the rotation of terrestrial planets I. Formalism and rotational normal modes. *Physics of Earth and Planetary Interiors* 134: 17–33.

Vermeersen LLA and Sabadini R (1996) Significance of the fundamental mantle rotational relaxation mode in polar wander simulations. *Geophysical Journal International* 127: F5–F9.

Vermeersen LLA and Sabadini R (1999) Polar wander, sea-level variations and ice age cycles. *Surveys in Geophysics* 20: 415–440.

Vermeersen LLA and Vlaar NJ (1993) Changes in the Earth's rotation by tectonic movements. *Geophysical Research Letters* 20(2): 81–84.

Vermeersen LLA, Sabadini R, Spada G, and Vlaar NJ (1994) Mountain building and Earth rotation. *Geophysical Journal International* 117: 610–624.

Vermeersen LLA, Sabadini R, and Spada G (1996) Compressible rotational deformation. *Geophysical Journal International* 126: 735–761.

Vermeersen LLA, Fournier A, and Sabadini R (1997) Changes in rotation induced by Pleistocene ice masses with stratified analytical Earth models. *Journal of Geophysical Research* 102(B12): 27689–27702.

Vicente RO and Wilson CR (1997) On the variability of the Chandler frequency. *Journal of Geophysical Research* 102(B9): 20439–20445.

Vicente RO and Wilson CR (2002) On long-period polar motion. *Journal of Geodesy* 76: 199–208.

Volland H (1988) *Atmospheric Tidal and Planetary Waves*. Dordrecht, The Netherlands: Kluwer.

Volland H (1997) Atmospheric tides. In: Wilhelm H, Zürn W, and Wenzel H-G (eds.) *Tidal Phenomena*, Springer-Verlag Lecture Notes in Earth Sciences, vol. 66, pp. 221–246. Berlin: Springer.

Vondrák J (1977) The rotation of the Earth between 1955.5 and 1976.5. *Studia Geophysica Et Geodaetica* 21: 107–117.

Vondrák J (1985) Long-period behaviour of polar motion between 1900.0 and 1984.0. *Annales Geophysicae* 3: 351–356.

Vondrák J (1991) Calculation of the new series of the Earth orientation parameters in the HIPPARCOS reference frame. *Bulletin of the Astronomical Institutes of Czechoslovakia* 42: 283–294.

Vondrák J (1994) Secular polar motion, crustal movements, and International Latitude Service observations. *Studia Geophysica Et Geodaetica* 38: 256–265.

Vondrák J (1999) Earth rotation parameters 1899.7–1992.0 after reanalysis within the Hipparcos frame. *Surveys in Geophysics* 20(2): 169–195.

Vondrák J and Richter B (2004) International Earth Rotation and Reference Systems Service (IERS). *Journal of Geodesy* 77: 585–586.

Vondrák J, Feissel M, and Essaïfi N (1992) Expected accuracy of the 1900–1990 Earth orientation parameters in the Hipparcos reference frame. *Astronomy and Astrophysics* 262: 329–340.

Vondrák J, Ron C, Pesek I, and Cepek A (1995) New global solution of Earth orientation parameters from optical astrometry in 1900–1990. *Astronomy and Astrophysics* 297: 899–906.

Vondrák J, Ron C, and Pesek I (1997) Earth rotation in the Hipparcos reference frame. *Celestial Mechanics & Dynamical Astronomy* 66: 115–122.

Vondrák J, Pesek I, Ron C, and Cepek A (1998) *Earth Orientation Parameters 1899.7–1992.0 in the ICRS Based on the HIPPARCOS reference frame*, Publication No. 87. Ondrejov, Czech Republic: Astronomical Institute of the Academy of Sciences of the Czech Republic.

Voorhies CV and Backus GE (1985) Steady flows at the top of the core from geomagnetic field models: The steady motions theorem. *Geophysical and Astrophysical Fluid Dynamics* 32: 163–173.

Wahr JM (1981) The forced nutations of an elliptical, rotating, elastic and oceanless earth. *Geophysical Journal of the Royal Astronomical Society* 64: 705–727.

Wahr JM (1982) The effects of the atmosphere and oceans on the Earth's wobble — I. Theory. *Geophysical Journal of the Royal Astronomical Society* 70: 349–372.

Wahr JM (1983) The effects of the atmosphere and oceans on the Earth's wobble and on the seasonal variations in the length of day — II. Results. *Geophysical Journal of the Royal Astronomical Society* 74: 451–487.

Wahr J (2005) Polar motion models: Angular momentum approach. In: Plag H- P, Chao BF, Gross RS, and Van Dam T (eds.) *Forcing of Polar Motion in the Chandler Frequency Band: A Contribution to Understanding Interannual Climate Change*, Cahiers du Centre Européen de Géodynamique et de Séismologie, vol. 24, pp. 89–102. Luxembourg: Cahiers du Centre Européen de Géodynamique et de Séismologie.

Wahr J and Bergen Z (1986) The effects of mantle anelasticity on nutations, Earth tides, and tidal variations in rotation rate. *Geophysical Journal of the Royal Astronomical Society* 87: 633–668.

Wahr JM, Sasao T, and Smith ML (1981) Effect of the fluid core on changes in the length of day due to long period tides. *Geophysical Journal of the Royal Astronomical Society* 64: 635–650.

Watkins MM and Eanes RJ (1994) Diurnal and semidiurnal variations in Earth orientation determined from LAGEOS laser ranging. *Journal of Geophysical Research* 99(B9): 18073–18079.

Whaler KA (1980) Does the whole of the Earth's core convect? *Nature* 287: 528–530.

Whaler KA and Davis RG (1997) Probing the Earth's core with geomagnetism. In: Crossley DJ (ed.) *Earth's Deep Interior*, pp. 114–166. Amsterdam: Gordon and Breach.

Wicht J and Jault D (1999) Constraining electromagnetic core-mantle coupling. *Physics of Earth and Planetary Interiors* 111: 161–177.

Wicht J and Jault D (2000) Electromagnetic core-mantle coupling for laterally varying mantle conductivity. *Journal of Geophysical Research* 105(B10): 23569–23578.

Williams JG, Newhall XX, and Dickey JO (1993) Lunar laser ranging: Geophysical results and reference frames. In: Smith DE and Turcotte DL (eds.) *Contributions of Space Geodesy to Geodynamics: Earth Dynamics*, American Geophysical Union Geodynamics Series, vol. 24, pp. 83–88. Washington, DC: American Geophysical Union.

Williams JG, Boggs DH, Yoder CF, Ratcliff JT, and Dickey JO (2001) Lunar rotational dissipation in solid body and molten core. *Journal of Geophysical Research* 106(E11): 27933–27968.

Willis P, Jayles C, and Bar-Sever Y (2006) DORIS: From orbit determination for altimeter missions to geodesy. *Comptes Rendus Geoscience* 338: 968–979 (doi:10.1016/j.crte.2005.11.013).

Wilson CR (1993) Contributions of water mass redistribution to polar motion excitation. In: Smith DE and Turcotte DL (eds.) *Contributions of Space Geodesy to Geodynamics: Earth Dynamics*, American Geophysical Union Geodynamics Series, vol. 24, pp. 77–82. Washington, DC: American Geophysical Union.

Wilson CR and Chen J (2005) Estimating the period and Q of the Chandler wobble. In: Plag H-P, Chao BF, Gross RS, and van Dam T (eds.) *Forcing of Polar Motion in the Chandler Frequency Band: A Contribution to Understanding Interannual Climate Change*, Cahiers du Centre Européen de Géodynamique et de Séismologie, vol. 24, pp. 23–29. Luxembourg: Cahiers du Centre Européen de Géodynamique et de Séismologie.

Wilson CR and Gabay S (1981) Excitation of the Earth's polar motion: A reassessment with new data. *Geophysical Research Letters* 8: 745–748.

Wilson CR and Haubrich RA (1976) Meteorological excitation of the Earth's wobble. *Geophysical Journal of the Royal Astronomical Society* 46: 707–743.

Wilson CR and Vicente RO (1980) An analysis of the homogeneous ILS polar motion series. *Geophysical Journal of the Royal Astronomical Society* 62: 605–616.

Wilson CR and Vicente RO (1990) Maximum likelihood estimates of polar motion parameters. In: McCarthy DD and Carter WE (eds.) *Variations in Earth Rotation*, American Geophysical Union Geophysical Monograph Series, vol. 59, pp. 151–155. Washington, DC: American Geophysical Union.

Wu P and Peltier WR (1984) Pleistocene deglaciation and the Earth's rotation: a new analysis. *Geophysical Journal of the Royal Astronomical Society* 76: 753–791.

Wünsch J (2000) Oceanic influence on the annual polar motion. *Journal of Geodynamics* 30: 389–399.

Wünsch J (2002) Oceanic and soil moisture contributions to seasonal polar motion. *Journal of Geodynamics* 33: 269–280.

Wünsch J and Busshoff J (1992) Improved observations of periodic UT1 variations caused by ocean tides. *Astronomy and Astrophysics* 266: 588–591.

Yan H, Zhong M, Zhu Y, Liu L, and Cao X (2006) Nontidal oceanic contribution to length-of-day changes estimated from two ocean models during 1992–2001. *Journal of Geophysical Research* 111: B02410 (doi:10.1029/2004JB003538).

Yoder CF (1995) Astrometric and geodetic properties of Earth and the solar system. In: Ahrens TJ (ed.) *Global Earth Physics: A Handbook of Physical Constants*, American Geophysical Union Reference Shelf 1, pp. 1–31. Washington, DC: American Geophysical Union.

Yoder CF and Standish EM (1997) Martian precession and rotation from Viking lander range data. *Journal of Geophysical Research* 102(E2): 4065–4080.

Yoder CF, Williams JG, and Parke ME (1981) Tidal variations of Earth rotation. *Journal of Geophysical Research* 86: 881–891.

Yumi S and Yokoyama K (1980) *Results of the International Latitude Service in a Homogeneous System, 1899.9-1979.0.* Mizusawa, Japan: Publication of the Central Bureau of the International Polar Motion Service and the International Latitude Observatory of Mizusawa.

Zatman S (2003) Decadal oscillations of the Earth's core, angular momentum exchange, and inner core rotation. In: Dehant V, Creager KC, Karato S, and Zatman S (eds.) *Earth's Core: Dynamics, Structure, Rotation*, American Geophysical Union Geodynamics Series, vol. 31, pp. 233–240. Washington, DC: American Geophysical Union.

Zatman S and Bloxham J (1997) Torsional oscillations and the magnetic field within the Earth's core. *Nature* 388: 760–763.

Zatman S and Bloxham J (1998) A one-dimensional map of B_s from torsional oscillations of the Earth's core. In: Gurnis M, Wysession ME, Knittle E, and Buffett BA (eds.) *The Core-Mantle Boundary Region*, American Geophysical Union Geodynamics Series, vol. 28, pp. 183–196. Washington, DC: American Geophysical Union.

Zatman S and Bloxham J (1999) On the dynamical implications of models of B_s in the Earth's core. *Geophysical Journal International* 138: 679–686.

Zhao M and Dong D (1988) A new research for the secular polar motion in this century. In: Babcock AK and Wilkins GA (eds.) *The Earth's Rotation and Reference Frames for Geodesy and Geodynamics*, Int. Astron. Union Symp. No. 128, pp. 385–392. Dordrecht, Holland: D. Reidel.

Zharov VE and Gambis D (1996) Atmospheric tides and rotation of the Earth. *Journal of Geodesy* 70: 321–326.

Zheng D, Ding X, Zhou Y, and Chen Y (2003) Earth rotation and ENSO events: combined excitation of interannual LOD variations by multiscale atmospheric oscillations. *Global and Planetary Change* 36: 89–97.

Zhong M, Naito I, and Kitoh A (2003) Atmospheric, hydrological, and ocean current contributions to Earth's annual wobble and length-of-day signals based on output from a climate model. *Journal of Geophysical Research* 108(B1): 2057 (doi:10.1029/2001JB000457).

Zhong M, Yan H, Wu X, Duan J, and Zhu Y (2006) Non-tidal oceanic contribution to polar wobble estimated from two oceanic assimilation data sets. *Journal of Geodynamics* 41: 147–154.

Zhou YH, Chen JL, Liao XH, and Wilson CR (2005) Oceanic excitations on polar motion: A cross comparison among models. *Geophysical Journal International* 162: 390–398.

Zhou Y, Zheng D, Zhao M, and Chao BF (1998) Interannual polar motion with relation to the North Atlantic Oscillation. *Global and Planetary Change* 18: 79–84.

Zhou YH, Zheng DW, and Liao XH (2001) Wavelet analysis of interannual LOD, AAM, and ENSO: 1997-98 El Niño and 1998-99 La Niña signals. *Journal of Geodesy* 75: 164–168.

10 Earth Rotation Variations

V. Dehant, Royal Observatory of Belgium, Brussels, Belgium

P. M. Mathews, University of Madras, Chennai, India

Nomenclature

Notation	Definition	Note
$\boldsymbol{\Omega}_0$	mean angular velocity vector	
Ω_0	mean Earth angular velocity	$= 2\pi$ radian/sidereal day
		$= 7.292115\,10^{-5}\,\mathrm{rad\,s^{-1}}$
		$= 1$ cpsd
a	mean radius of the Earth	$= 6371$ km
a_e	equatorial radius of the Earth	$= 6378$ km
θ	co-latitude	
λ	longitude	
T_d	period of the relative orbital motion of the Earth and the Sun in solar days	$T_d = 365.2422$ days

H_d	dynamical ellipticity of the Earth (astronomy community)		$H_d = \dfrac{C-A}{C}$
e	dynamical ellipticity of the Earth (geodesy community)		$e = \dfrac{C-A}{A}, \quad H_d = \dfrac{e}{(1+e)}$
e'	triaxiallity coefficient of the Earth		$e' = \dfrac{B-A}{A+B} = (B-A)/(2\bar{A})$
A, B	equatorial moments of inertia		$A \approx B$
$\bar{A}$	mean equatorial moments of inertia		$\bar{A} = \dfrac{A+B}{2}$
C	polar moments of inertia		$C > B > A$
ϵ	obliquity		
ϵ_0	mean obliquity at J2000		
$\psi_A(t)$	precession in longitude of the figure axis	more preciseily of the CIP	
$\dot{\psi}_A$	precession rate in longitude of the figure axis	more preciseily of the CIP	
$\Delta\psi$	nutation in longitude of the figure axis	more preciseily of the CIP	
$\epsilon_A(t)$	mean obliquity of date		
$\dot{\epsilon}_A$	precession rate in obliquity of the figure axis	more preciseily of the CIP	
$\Delta\epsilon$	nutation in obliquity of the figure axis	more preciseily of the CIP	
X, Y	coordinates of the CIP in the GCRF		$X + iY = \Delta\psi \sin\epsilon + i\Delta\epsilon$
$\tilde{\eta}$	nutation of the CIP, nutation of the pole of the TRF in space		$\tilde{\eta} = \Delta\psi \sin\epsilon + i\Delta\epsilon = X + iY$
			$\tilde{\eta} = X + iY = \tilde{\eta}(\omega_n)\, e^{i(\omega_n t + \chi_n)}$
ω_n	frequency of nutation in the CRF		
ν	frequency of nutation in the CRF in cpsd		$\omega_n = \nu\Omega_0$
χ_n	phase of component of frequency ω_n of $\tilde{\eta}$		
$\tilde{\eta}(\omega_n)$	amplitude of component of frequency ω_n of $\tilde{\eta}$		
$\tilde{\eta}_R(\omega_n)$	rigid Earth nutation amplitude of component of frequency ω_n		
$\Xi(t)$ or $\Xi_\omega(t)$	nutation argument		$\Xi(t) = n_1\ell + n_2\ell' + n_3 F + n_4 D$
			$+ n_5\Omega + \sum_{i=1}^{9} n_i\lambda_i$
ℓ	mean anomaly of the Moon		
ℓ'	mean anomaly of the Sun		
F	one of the fundamental nutation arguments		$F = L - \Omega$
L	mean longitude of the Moon		
D	mean elongation of the Moon from the Sun		
Ω	mean longitude of the Moon ascending node		
λ_i	mean longitude of the planets		
p_A	precession		
A, A', A''	amplitudes of the nutation in longitude		$\Delta\psi(t) = \sum_\omega [(A_\omega + A'_\omega t)\sin\Xi_\omega(t)$
			$+ A''_\omega \cos\Xi_\omega(t)]$
B, B', B''	amplitudes of the nutation in obliquity		$\Delta\epsilon(t) = \sum_\omega [(B_\omega + B'_\omega t)\cos\Xi_\omega(t)$
			$+ B''_\omega \sin\Xi_\omega(t)]$
X_s, X_c	sine and cosine components of X		$X = X_s \sin\Xi + X_c \cos\Xi$
Y_s, Y_c	sine and cosine components of Y		$Y = Y_c \cos\Xi + Y_s \sin\Xi$
$\eta^{p,ip}$	amplitude of the prograde in phase component		$\eta^{p,ip} = \dfrac{1}{2}\left(\dfrac{\omega_n}{|\omega_n|}X_s - Y_c\right)$
$\eta^{p,op}$	amplitude of the prograde out of phase component		$\eta^{p,op} = \dfrac{1}{2}\left(X_c + \dfrac{\omega_n}{|\omega_n|}Y_s\right)$
$\eta^{r,ip}$	amplitude of the retrograde in phase component		$\eta^{r,ip} = -\dfrac{1}{2}\left(\dfrac{\omega_n}{|\omega_n|}X_s + Y_c\right)$
$\eta^{r,op}$	amplitude of the retrograde out of phase component		$\eta^{r,op} = \dfrac{1}{2}\left(X_c - \dfrac{\omega_n}{|\omega_n|}Y_s\right)$
Ω	Earth angular velocity vector wobble		$\Omega = (\Omega_x, \Omega_y, \Omega_z)$
m	wobble		$\Omega = \Omega_0 + \Omega_0 m$

m_z	wobble axial component	$m_z = \dfrac{\Omega_z}{\Omega_0} - 1 = \dfrac{-\Delta\mathrm{LOD}}{\mathrm{LOD}}$
m_x, m_y	wobble equatorial components	$m_x = \dfrac{\Omega_x}{\Omega_0}$ and $m_y = \dfrac{\Omega_y}{\Omega_0}$
$\tilde{m}$	complex sum of the equatorial components	$\tilde{m} = m_x + i m_y$ $\tilde{m} = \tilde{m}(\omega_w) e^{i(\omega_w t + \chi_w)}$ $\Omega_0 \tilde{m} = i\dot{\tilde{\eta}} e^{-i\Omega_0(t+t_0)}$
$\omega_w = \omega$	frequency of the equatorial wobble	$\omega_w = \omega_n - \Omega_0$
σ	frequency of the equatorial wobble in cpsd	$\sigma = \dfrac{\omega_w}{\Omega_0} = \nu - 1$
$\Theta_w(t) = \omega t + \chi_w$	argument of the equatorial wobble	
χ_w	phase of the equatorial wobble	
$\tilde{m}(\omega_w)$	amplitude of the component of frequency ω_w of $\tilde{m}$	$\tilde{m}(\omega_w) = -\dfrac{\omega_n}{\Omega_0}\tilde{\eta}(\omega_n) = \dfrac{\omega_n}{\Omega_0}\tilde{p}(\omega_p)$
H	angular momentum vector	
Γ	torque vector	
(H_X, H_Y, H_Z)	components of **H** in space	
$(\Gamma_X, \Gamma_Y, \Gamma_Z)$	components of Γ in space	
X_H, Y_H	normalized components of **H** in space	$X_H = \dfrac{H_X}{C\Omega_0}, \; Y_H = \dfrac{H_Y}{C\Omega_0}$
$\Delta\psi_H$	nutation in longitude of the angular momentum axis	
$\Delta\epsilon_H$	nutation in obliquity of the angular momentum axis	
$\psi_{A,H}(t)$	precession in longitude of the angular momentum axis	
$\dot{\psi}_{A,H}$	precession rate in longitude of the angular momentum axis	
(H_x, H_y, H_z)	components of **H** in TRF	
$(\Gamma_x, \Gamma_y, \Gamma_z)$	components of Γ in TRF	
$\tilde{\Gamma}$	complex sum of the first two above components	$\tilde{\Gamma} = \Gamma_x + i\Gamma_y$
r	position vector of a point within the Earth relative to the geocenter	$r = (x, y, z)$
$\mathbf{r}_B$	position vector of a celestial body B relative to the geocenter	$\mathbf{r}_B = (x_B, y_B, z_B)$ in TRF $\mathbf{r}_B = (X_B, X_B, X_B)$ in CRF
r_B	distance from the geocenter to B	
$\tilde{r}_B$	complex sum of the two first two components of $\mathbf{r}_B$	$\tilde{r}_B = x_B + i y_B$
$W_B(r)$	gravitational potential of a celestial body B	
G	constant of gravitation	
M_B	mass of a celestial body B	
M_E	mass of the Earth	
θ_B	co-latitude of a celestial body B	
λ_B	longitude of a celestial body B	
$\rho(\mathbf{r})$	density of matter at **r**	
ϕ_1, ϕ_2	time-dependent coefficients of xz and yz in degree 2 tesseral tide generating potential	
$\tilde{\phi}$	complex sum of the two above components	$\tilde{\phi} = \phi_1 + i\phi_2$ $\tilde{\Gamma} = -iAe\Omega_0^2\tilde{\phi} = \left(\dfrac{15}{8\pi}\right)^{1/2} iAeF_{21}$
$F_{nm}(r_B, \theta_B, \lambda_B)$	coefficient of the tide generating potential	
$F_{nm}^*(r_B, \theta_B, \lambda_B)$	coefficient of the tide generating potential	
H_ω^{nm}	spectral component amplitude in terms of 'height' of the tide generating potential	
$\Theta_\omega(t)$	argument of the tidal component	$\Theta_\omega(t) = \omega t + \chi_\omega$
ω	tidal frequency	
χ_ω	tidal phase	
τ	Doodson argument called lunar time	$\tau = 180° + \mathrm{GMST} - s + \lambda$
s	Doodson argument, mean tropic longitude of the Moon	

h	Doodson argument, mean tropic longitude of the Sun	
p	Doodson argument, mean tropic longitude of the Moon's perigee	
N'	Doodson argument, mean tropic longitude of the Moon's node	
p_s	Doodson argument, mean tropic longitude of the Sun's perigee	
GMST	Greenwich hour angle of the mean equinox	$GMST = GMST_0 + \Omega_0 t + \cdots$ $= 280.4606° + \Omega_0 t + \cdots$
GST	Greenwich hour angle of the true equinox	
C_{nm}, S_{nm}	geopotential coefficients	$C_{n0} = -J_2$
K	Hamiltonian for the computation of rigid Earth nutation	
c	semi-major axis of the orbit of B around the Earth; also distance from geocenter of B on a circular motion in the orbital plane	B could be apparent Sun
p	mean angular velocity of B around the Earth in uniform motion along a circle of radius c centered at the geocenter	for the simple case considered in the chapter $= 2\pi$ radians per revolution $p^2 = GM_B/c^3$
n_S	mean angular velocity of the apparent Sun	$= 2\pi$ radians per year
F	normalization factor of Γ in the CRF	$F = (3GM_B/2c^3)\, Ae \sin \epsilon$
V	vector of which the equatorial part executes an elliptical motion	
V_1, V_2	components of **V**	
A_s	amplitude of the sine component of (V_1, V_2)	$V_1 = A_s \sin(\omega t + \alpha)$
B_c	amplitude of the cosine component of (V_1, V_2)	$V_2 = B_c \cos(\omega t + \alpha)$
ω	frequency of the motion	
α	phase at time $t = 0$	
V^+, V^-	amplitudes of the circular motions	$V^+ = (A_s - B_c)/2$ $V^- = -(A_s + B_c)/2$
V^p	amplitudes of the circular progade motions	$V^p = \frac{1}{2}\left(\frac{\omega}{\lvert\omega\rvert}A_s - B_c\right)$
V^r	amplitudes of the circular retrograde motions	$V^r = -\frac{1}{2}\left(\frac{\omega}{\lvert\omega\rvert}A_s + B_c\right)$
$T_n(\omega_n)$	nutation transfer function	$T_n(\omega_n) \equiv \dfrac{\tilde{\eta}(\omega_n)}{\tilde{\eta}_R(\omega_n)}$
$T_w(\omega_w)$	wobble transfer function	$T_w(\omega_w) \equiv \dfrac{\tilde{m}(\omega_w)}{\tilde{m}_R(\omega_w)}$ $T_n(\omega_n) = T_w(\omega_w)$
c_{ij}	components of the incremental inertia tensor	
$\tilde{c}$	complex sum of the c_{ij} components	$\tilde{c} = c_{13} + ic_{23}$
k	Love number for mass redistribution potential	
k_f	fluid Love number	$k_f = 3(C - A)G/(\Omega_0^2 a^5)$
$\boldsymbol{\Omega}_f$	incremental core angular velocity vector	
m_{fx}, m_{fy}	incremental core wobble equatorial components	
$\tilde{m}_f$	complex sum of the m_{fx}, m_{fy} components	$\tilde{m}_f = m_{fx} + im_{fy}$
$\mathbf{H}_f$	angular momentum vector of the core	
$\tilde{c}_f$	complex sum of two of the three components of the core incremental inertia tensor	$\tilde{c}_f = c_{f13} + ic_{f23}$
κ	compliance for the Earth incremental moment of inertia; concerns the centrifugal and external potential	$\kappa = \dfrac{k}{k_f}e$

ξ	compliance for the Earth incremental moment of inertia; concerns the core centrifugal potential	
γ	compliance for the core incremental moment of inertia; concerns the centrifugal and external potential	
β	compliance for the core incremental moment of inertia; concerns the core centrifugal potential	
σ_E	CW frequency for a rigid Earth also calle Euler frequency	$\sigma_E = e\Omega_0$
$\sigma_{CW}^{elastic}$	CW frequency for an elastic Earth	$\sigma_{CW}^{elastic} = \Omega_0(e - \kappa)$
σ_{CW}	CW frequency for Earth with a fluid core	$\sigma_{CW} = \Omega_0 \dfrac{A}{A_m}(e - \kappa)$
m_{sx}, m_{sy}	incremental inner core wobble equatorial components	
$\tilde{m}_s$	complex sum of the m_{sx}, m_{sy} components	$\tilde{m}_s = m_{sx} + i m_{sy}$
$\tilde{\Theta}_s$	tilt of the solid inner core	$\tilde{\Theta}_s = \tilde{m}_s$
K_{CMB}	coupling constant at the core–mantle boundary	
K_{ICB}	coupling constant at the inner core boundary	
$(x_p, -y_p)$	coordinates of the CIP in the TRF	
$\tilde{p}$	complex sum of the above components	$\tilde{p} = x_p - i y_p$
		$\tilde{p} = \tilde{p}(\omega_p) e^{i(\omega_p t + \chi_p)}$
		$\tilde{p} = -(X + iY)e^{-i\Omega_0(t - t_0)}$
		$\tilde{p} = -\tilde{\eta} e^{-i\Omega_0(t - t_0)}$
ω_p	frequency of polar motion in TRF	$\omega_p = \omega_n - \Omega_0$
χ_p	phase of the component of frequency ω_p of $\tilde{p}$	
$\tilde{p}(\omega_p)$	amplitude of the component of frequency ω_p of $\tilde{p}$	$\tilde{p}(\omega_p) = -\tilde{\eta}(\omega_n)$
$\rho_0(r)$	mean density at radius r	
$\mu_0(r)$	rigidity modulus at radius r	$\mu_0(r) = 0$ in the fluid part
$\kappa_0(r)$	incompressibility modulus at radius r	
$\lambda_0(r), \mu_0(r)$	Lamé parameters	$\lambda_0(r) = \kappa_0(r) - \dfrac{2}{3}\mu_0(r)$
$\phi_0(r)$	mean initial self-gravitational potential at radius r	
$\psi_0(r)$	centrifugal potential	
$\Phi_0(r)$	mean initial gravity potential at radius r, used for hydrostatic equilibrium	$\Phi_0(r) = \phi_0(r) + \psi_0(r)$
$\tilde{g}$	gradiant of the gravity potential	$\tilde{g} = \dfrac{d\phi_0}{dr}$
g	gravity	$g = 2\pi G\rho_0 r + \dfrac{r}{2}\dfrac{d^2\phi_0}{dr^2}$
σ	frequency expressed in cpsd	
$D_{mn}^{\ell}(\theta, \lambda)$	Generalized spherical harmonics (GSH)	
ℓ	GSH degree	ℓ in $[, +\infty[$
m	GSH order	m in $[-\ell, +\ell]$
n	GSH third integer	n in $[-\ell, +\ell]$
$\hat{e}_-, \hat{e}_0, \hat{e}_+$	basis vector used in GSH approach	
Φ^{ext}	external gravitational potential	
Φ_1^E	mass redistribution potential	components $\Phi_{1\ell}^{Em}(r)$
$g_{1\ell}^{Em}(r)$	related to derivative of the spherical harmonic (ℓ, m)-components of the mass redistribution potential $\Phi_{1\ell}^{Em}(r)$ and the spherical harmonic (ℓ, m)-components of radial displacement U_ℓ^m	$g_{1\ell}^{Em}(r) = \dfrac{d\Phi_{1\ell}^{Em}(r)}{dr} + 4\pi G\rho_0 U_\ell^m(r)$
$\mathbf{s}$	displacement vector	
$S_\ell^{m-}, S_\ell^{m0}, S_\ell^{m+}$	components of $\mathbf{s}$	
$U_\ell^m(r)$	radial component of $\mathbf{s}$	$U_\ell^m(r) = S_\ell^{m0}(r)$
$V_\ell^m(r)$	spheroidal tangential component of $\mathbf{s}$	$V_\ell^m(r) = S_\ell^{m+}(r) + S_\ell^{m-}(r)$
$W_\ell^m(r)$	toroidal tangential component of $\mathbf{s}$	$W_\ell^m(r) = S_\ell^{m+}(r) - S_\ell^{m-}(r)$

X_ℓ^m	dilatiation	$X_\ell^m = \dfrac{dU_\ell^m}{dr} + \dfrac{1}{r}(L_0^\ell V_\ell^m + 2U_\ell^m)$
$T(r, t)$	incremental stress tensor	components $T_{\ell i}^{mj}(r)$
$P_\ell^m(r)$	Pressure	$P_\ell^m(r) = T_{\ell 0}^{m0}(r)$
$Q_\ell^m(r)$	Spheroidal tangential stress	$Q_\ell^m(r) = T_{\ell 0}^{m+}(r) + T_{\ell 0}^{m-}(r)$
$R_\ell^m(r)$	Toroidal tangential stress	$R_\ell^m(r) = T_{\ell 0}^{m+}(r) - T_{\ell 0}^{m-}(r)$
$\eta_{1\ell}^{Um}, \eta_{2\ell}^{Um}, \eta_{3\ell}^{Um}$	Intermediate notation used in the scalar equations in the	
$\eta_{1\ell}^{Vm}, \eta_{2\ell}^{Vm}, \eta_{3\ell}^{Vm}$	frequency domain and using GSH	
$\eta_{1\ell}^{Wm}, \eta_{2\ell}^{Wm}, \eta_{3\ell}^{Wm}$		

10.1 Introduction

10.1.1 Manifestations of Variations in Earth Rotation and Orientation

Variations in the orientation in space of an Earth-fixed reference frame (Earth orientation variations) are driven by variations in Earth rotation, i.e., in the angular velocity vector of Earth rotation. They are manifested as variations in direction of Earth-related axes in space (precession and nutation, see **Figure 1**) as well as relative to a terrestrial reference frame (wobble, polar motion), and also as variations in the angular speed of rotation (or spin rate, for brevity) which translate into variations in the length of day (LOD variations).

The axes of interest are the figure axis, rotation axis, and the angular momentum axis. The term "figure axis" is used herein for the axis of maximum moment of inertia of the "static" Earth, i.e., of the Earth with time dependent deformations caused by tidal and other forces disregarded. (The axis of maximum moment of inertia of the dynamically deforming Earth has indeed been called the figure axis in some of the literature, but it is of little

interest.) The rotation axis is simply the direction of the angular velocity vector (also called the rotation vector). All the axes may be thought of as unit vectors in the directions of the respective axes. Nutations of all three axes are distinct, yet closely related.

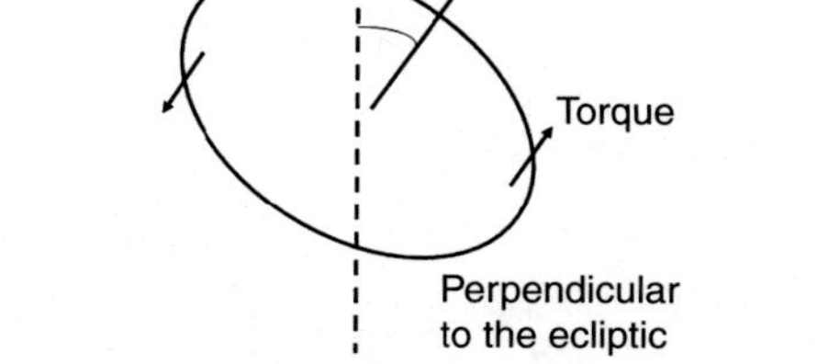

With every axis is associated a pole which is the intersection of the axis with a geocentric sphere of unit radius (in arbitrary units). The motion of the axis is faithfully reflected in the motion of the associated pole, and also in the motion of the plane perpendicular to the axis (the equator of that axis); so the nutation, for instance, may and often is, referred to as the nutation of the associated pole or equator.

Precession of any axis is the secular (smoothly varying) part of the motion of the axis in space, i.e. relative to the directions of "fixed" stars; it is the 'mean' motion of the axis. (Closest to the ideal of fixed stars are the quasars, the most distant objects in space.) Nutation is the oscillatory part of the motion of the axis, with a large multiplicity of periods; it causes the instantaneous position of the axis to move around its mean position (See **Figures 1–3**).

The term "wobble" is used herein in a very broad sense, for any periodic or quasi-periodic motion of the instantaneous rotation axis with respect to the figure of the Earth, irrespective of the frequency or the physical

Figure 1 Definition of precession and nutation.

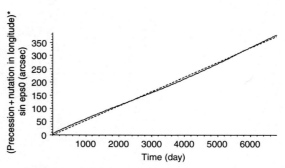

Figure 2 Representation of the changes in longitude of the axis undergoing precession and nutations.

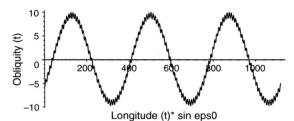

Figure 3 Representation of the changes in obliquity of the axis undergoing nutations.

origin of the motion. The essence of wobble is the temporally varying offset (angular separation) of the direction of the rotation axis from that of the figure axis. Wobble is characterized by the variation of the equatorial components of the rotation axis in a terrestrial frame of reference. The Earth is made up broadly of three regions, the mantle and the two core regions within it-the fluid outer core and the solid inner core. The term "solid Earth", as generally used, includes all three regions, while excluding the fluid layers at the surface, namely, the oceans and the atmosphere. The three regions of the solid Earth have different rotation vectors, though their motions are mutually coupled. Therefore each region has its own wobble. The term 'wobble', when used without specific mention of the region, is intended to be that of the mantle.

The term "polar motion" has been in use to refer generally to any motion of the Earth's pole of rotation relative to the Earth, and in particular, to the wobble as well as for secular polar motion. The latter is a steady drift of the mean pole of rotation (with the periodic wobbles averaged out) relative to the Earth; it may be viewed as the present-day trend of the polar wander which has been taking place over geological time-scales. In recent usage, polar motion has taken on a technical meaning, to be explained later on (see Sections 10.2.5.1 and 10.12.1), as the motion of a conceptual pole called the celestial intermediate pole (CIP) with frequencies outside a specified range.

A different aspect of the variations in Earth rotation is the variation in its spin rate, or its equivalent, the LOD variation. The mean rate of Earth rotation is one cycle per sidereal day (cpsd), equivalent to an angular speed $\Omega_0 = 2\pi$ radian per sidereal day $= 7.292115 \times 10^{-5}$ rad s^{-1}. (The sidereal day is the period of rotation of the Earth relative to the most distant 'fixed' stars. It is $T_d/(T_d + 1)$ times the solar day of 24 h, where $T_d = 365.2422$. is the period of the relative orbital motion of the Earth and the Sun in solar days.) Deviations from this uniform rotation give rise to LOD variations.

10.1.2 Free Rotational Modes

The variations in Earth rotation/orientation are made up of both free motions and forced motions of the various axes, apart from variations in the spin rate.

The free motions, which are not driven by any external torque, conserve the Earth's total angular momentum. The angular momentum is a product of the moment of inertia (which, in general, is a tensor) and the angular velocity, and its conservation requires that if one changes with time, the other must undergo compensating changes. This happens whenever the rotation (angular velocity) vector of the mantle or other region of the Earth is not along a principal axes, causing the inertia tensor (in a space fixed frame) to vary as the rotation proceeds; this, in turn, causes a compensatory variation of the rotation vector. The free nutation/wobble normal modes are the manifestations of this phenomenon. The three-layer Earth has four such modes: the Chandler wobble (CW) with a period of about 430 days; the free core nutation (FCN) with a period of about −430 days in the celestial frame, along with its associated wobble (the Nearly Diurnal Free Wobble or NDFW) which has nearly diurnal frequency in the terrestrial frame as its name implies; the less prominent free inner-core nutation (FICN, also called the prograde free core nutation or PFCN), which has a rather long period – apparently around 1000 days – in the celestial frame; and the inner-core wobble (ICW) with a still longer period in the terrestrial frame (see Appendix 1 for further explanations). They are excited by various geophysical processes. They have been found, observationally, to have variable amplitudes, and are not amenable to theoretical prediction. Since the frequencies of the FCN and FICN/PFCN modes lie in the middle of the low-frequency band wherein the most important nutation components lie, the resonances associated with these modes play an important role in the nutational responses of the Earth. Though the Chandler wobble frequency lies well outside this band when viewed in the celestial frame, its influence on nutation is not completely negligible unlike that of the ICW.

10.1.3 Forced Motions

The forced nutation, wobble, and precessional motions of the axes are excited almost wholly by the gravitational pulls of the Sun, Moon, and planets on the nonspherical (nearly ellipsoidal) Earth, and are different aspects of the gyroscopic response of the rotating Earth to the gravitational torque. The deformations produced by the direct action of the spatially varying gravitational force (as distinct from the gravitational torque) on the solid Earth, as well as by the loading effect of ocean tides and the gravitational attraction of the solid Earth by the redistributed ocean mass, play an important role in rotation variations; so does the deforming action of the centrifugal perturbation caused by the rotation variations themselves. Torques exerted by atmospheric pressure tides (arising primarily from the varying thermal input from solar radiation) also make minor contributions to variations in the directions of the axes. The forced motions of the axes due to the gravitational torques on the solid Earth are accurately predictable, given a sufficiently accurate knowledge of the relevant Earth parameters and of the orbital motions of the various bodies.

The nutational motions consist of numerous periodic components with a discrete spectrum of frequencies determined by the spectrum of the gravitational potential, at the Earth, of the solar system bodies. The frequencies of spectral components with nonnegligible amplitudes extend from large negative values to large positive values for the nutations of any of the axes mentioned in paragraph 1.2 above. (By 'large' we mean several times the frequency of the diurnal Earth rotation.) Positive frequencies correspond to prograde motions, which are in the same sense as that of the diurnal Earth rotation, while negative frequencies are for retrograde motions (in the opposite sense).

Figure 4 shows the observations of length-of-day variations, polar motion, and nutations.

The dominant causes of polar motion (other than the tidally excited wobbles) are the atmosphere and the ocean, at timescales of a few days to a few years in the terrestrial frame (*see* Chapter 9). These geophysical fluids are also primarily responsible for the LOD variations at timescales of a few days to a few years in the terrestrial frame, while the interaction of magnetic field in the fluid core with the mantle is invoked to explain the variations at decadal timescales. There is a discrete spectrum in the LOD variations, arising primarily from the changes in the axial moment of inertia resulting from the deforming action of the tidal potential, and to a very much

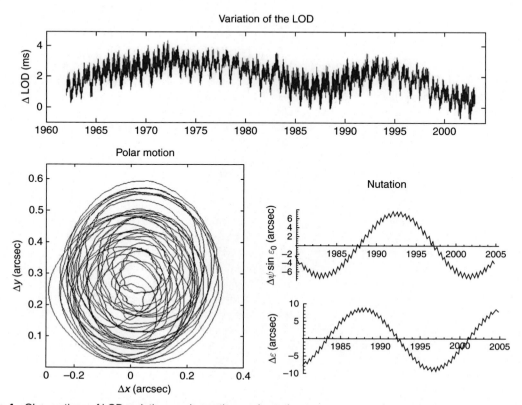

Figure 4 Observations of LOD variations, polar motion, and nutation.

smaller extent by tidal torques: the axial component of the tidal torque depends on the difference between the two principal moments of inertia in the equatorial plane, which is less than 1% of the difference between the polar and equatorial moments of inertia.

10.1.4   Earth's Response to Gavitational Torques

The extent of the Earth's response to gravitational torques is determined by the values of a number of Earth, parameters which represent various aspects of the structure and properties of the Earth's interior: the matter distribution within the Earth, the existence and mutual couplings of the fluid outer-core and solid inner-core regions, the couplings of these regions to the mantle, and rheological and other properties of the Earth's interior. By far the most important of the Earth parameters is the dynamical ellipticity, which is the fractional difference between the moments of the Earth's inertia about the polar and equatorial axes; it is a reflection of the Earth's 'equatorial bulge'. The response to the external forcing is modified by the tidal and other excitations of the fluid layers at the surface, namely the oceans and the atmosphere.

10.1.5   Low-Frequency Nutations

The spectrum of the nutational motion is discrete, and is determined by the spectra of the orbital motions of the solar system bodies. The major spectral components of the nutation have frequencies within the low-frequency band (between -0.5 and 0.5 cpsd); many of these components have amplitudes that are larger than those of the spectral components in high-frequency bands (frequencies exceeding 0.5 cpsd in magnitude) by five orders of magnitude. The low-frequency forced nutations arise from the action of external gravitational potentials on the axially symmetric part of the matter distribution within the Earth. The distribution is, in fact, both axially symmetric and ellipsoidal to a very high degree of approximation. If the small deviations from such a structure are ignored, the principal moments of inertia are A, A, and C, with $C > A$, and the dominant role in Earth rotation variations (other than those in the LOD) is played by the dynamic ellipticity defined either by $H_d = (C - A)/C$ or by $e = (C - A)/A \approx 1/304.5$, the former parameter being favored by astronomers and the latter by geophysicists. There exists a residual (nonellipsoidal) axially symmetric part of

the density distribution; it too gives rise to low-frequency nutations, but with amplitudes much lower than those of the dominant nutations due to e. These are closely followed by the semidiurnal nutations (frequencies between 1.5 and 2.5 cpsd) arising from the triaxiality of the Earth (a small difference between the two equatorial moments of inertia: $B > A$ with $(B - A) \approx 2e(C - A)$). The high-frequency nutations, for which the nonaxially symmetric part of the density distribution is responsible, have even smaller amplitudes as already stated.

In view of the above, most theoretical formulations of Earth rotation variations are, justifiably, based on axially symmetric ellipsoidal Earth models characterized by the dynamic ellipticity e, and are therefore concerned with low-frequency nutations only.

10.1.6   Gravitational Perturbations and Displacement Fields

Different elements of matter in the Earth are subjected to differing gravitational accelerations by the Moon (or Sun or other body), because they are at different distances from the external body. Therefore, they undergo position-dependent displacements from their unperturbed positions (i.e., positions in the absence of the gravitational forces). In the case of a deformable Earth model, the displacement field $\mathbf{s}(\mathbf{r})$ over the whole Earth can be analyzed into a rotational part wherein the relative distances between different points remain undisturbed, and a deformational part. One approach to the theory of the gravitational perturbations of the Earth seeks to solve the equation of motion governing the field $\mathbf{s}(\mathbf{r})$, given a detailed model of the matter density and elastic constants as functions of position in the Earth's interior, to determine thereby both the deformations (the so-called solid Earth tides) and the rotational motions (nutations and precession) produced by the perturbing potential. Treatments employing the displacement field approach use only axially symmetric ellipsoidal Earth models. Spherically symmetric models are constructed from information provided by seismic data; the axially symmetric ellipsoidal structure is then derived from such models by assuming hydrostatic equilibrium: a balance between the Earth's gravitational potential and the centrifugal potential of Earth rotation at the mean rate of rotation. Among the most widely used models is PREM of Dziewonski and Anderson

(1981), modified to the ellipsoidal structure based on hydrostatic equilibrium theory.

10.1.7 Gravitational Torque on the Earth

A simpler approach focuses directly on the rotation variations by employing the equation of motion governing rotational motion, namely the angular momentum balance equation (torque equation). The gravitational torque on the Earth is dependent solely on the matter distribution within the Earth and on the positions of the gravitating bodies relative to the terrestrial reference frame. Information on the motions of the solar system bodies is derived from the ephemerides. As for the matter distribution, one could rely on detailed models of the spherically symmetric density profiles within the Earth constructed from seismic studies which are converted to axially symmetric ellipsoidal models using a theory which determines the flattening of the constant density surfaces that is caused by the centrifugal effects of the Earth's mean rotation at a constant rate around the polar axis. The torque on an idealized model with the above symmetry involves only one Earth parameter, namely, the dynamical ellipticity which represents the fractional difference between the moments of inertia about polar and equatorial axes. This parameter is sufficient to account for all but a very small part of the nutational motion. But when a fuller account is needed, one has to consider other features of the Earth's structure which are represented by the higher-degree geopotential coefficients, on which other parts of the torque depend. These coefficients are estimated from space geodetic observations of the orbits of artificial Earth satellites.

10.1.8 Earth Response to the Torque

When considering the rotation variations of a wholly solid model Earth, its angular momentum can be expressed as a product of its inertia tensor and angular velocity vector. The torque equation in the rotating terrestrial frame then becomes an equation for the angular velocity components; the wobble and spin rate variations are related directly to these components and may be obtained by solving this equation. But when the existence of a fluid core region and the freedom it has to rotate with a different angular velocity from that of the mantle are taken into account, the torque equation for the whole Earth involves the angular velocities of both the mantle and the fluid core, and so one needs to consider also the

torque equation for the core region alone. Couplings of the core to the mantle enter into the latter equation, and influence the rotation variations of the mantle which are observable. Examples of such couplings are the so-called inertial coupling due to the fluid pressure on the ellipsoidally shaped core-mantle boundary (CMB), and the electromagnetic coupling and possibly also viscous coupling at the CMB. The differential wobble motions of the mantle and the fluid core, on the one hand, and the presence of magnetic fields crossing over from the fluid core to the mantle side, on the other, cause induction of currents in the conducting layers at the bottom of the mantle; and the interaction of the magnetic field with the induced currents all over the conducting mantle layer is responsible for the electromagnetic coupling. Simultaneous solution of the equations for the whole Earth and for the core yields the wobbles of both the mantle and the core. Inclusion of the inner core into the theory brings in a pair of additional equations, one governing its angular velocity variations and the other, its orientation relative to the mantle.

10.1.9 Role of Deformations in Earth Rotation

The inertia tensor of each of the regions (or layers) is not just that of the unperturbed Earth: it includes the perturbations associated with deformations produced directly by the gravitational potentials of the external bodies as well as by the centrifugal potentials due to the variations in the angular velocity vectors of all three layers. The perturbations of each of the inertia tensors is made up, therefore, of parts which are proportional to the strength of the perturbing potential and to the wobbles of the three different regions. The proportionality constants are deformability parameters (also called compliances) which are intrinsic properties of the Earth model; they may be computed by integration of the equations governing the deformations of the Earth, given a detailed model of the matter density and elastic constants as functions of position in the Earth's interior. While the compliance parameters are real for an elastic Earth, the anelastic behavior of the mantle results in small complex increments to their elastic-Earth values. Loading of the crust by the ocean tides raised by the gravitational attraction of the external bodies causes further deformation which again increments the inertia tensors and thereby influences the rotation variations. The increments due to ocean tidal effects, including the effect of ocean tidal currents, are non-trivial, and have been calculated by different methods. One method that has been recently employed represents the deformational effect of ocean tides in terms of effective increments to the compliance parameters, which are not only complex but also frequency dependent.

10.1.10   Interplay of Various Phenomena

In order to get results of high accuracy for Earth rotation variations, it is also necessary to take into account the mutual influences of a variety of phenomena. The tidal deformations of the Earth, including the contribution from the centrifugal perturbation caused by Earth rotation variations, as well as from ocean loading, affect the rotations of all three regions through the changes in their inertia tensors; the deformations also perturb the gravitational potential of the Earth itself, with a reciprocal effect on the deformations themselves; they affect the ocean tides through movements of the ocean floor and through the effect of the perturbation of the Earth's gravitational potential on the ocean surface; and the consequent modification of the ocean tidal loading

and tidal currents influence both the deformations and the rotation variations. In brief, there is an interplay of the various phenomena which is a factor in determining the effects of each one of them, and it is necessary to take account of this interplay while seeking to obtain accurate results for any of these effects. A chart summarizing all the interaction between the phenomena involved is presented in **Figure 5**.

10.1.11   Precision of Observations: Challenge to Theory

The precision of recent nutation estimates has surpassed 0.2 milliarcseconds (mas), and that of estimates of individual spectral components is of the order of 10 microarcseconds (μas) for many of the more prominent components except for those with periods exceeding one year, for which the uncertainties are higher (1 mas $= 4.848 \times 10^{-9}$ rad, 1 μas $= 4.848 \times 10^{-12}$ rad). To match this level of precision, all effects which make contributions at the μas level to nutations have to be identified and taken into account. One such effect which had been ignored till very recently is torque produced by the action

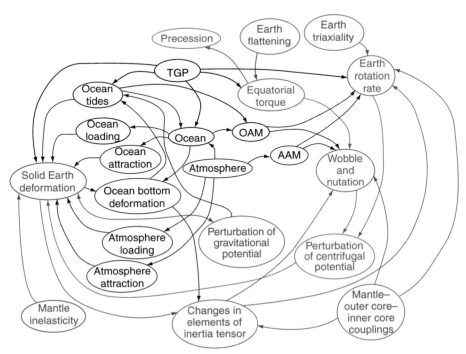

Figure 5   Chart representing all the interactions and physical processes involved in Earth orientation variations and in Earth deformations.

of the gravitational potential of solar system bodies on the perturbations in the matter distribution within the Earth due to the action of the potential itself. Recent work has shown that the small contributions to nutation and precession arise from such second-order torque terms are not negligible in this context. General relativistic effects (geodesic precession and nutation) and effects of atmospheric pressure are also significant.

10.2 Earth Orientation/Rotation Variables: Reference Frames

As a preliminary observation, we note that the direction of any of the coordinate axes or other axes like the Earth's rotation axis or the angular momentum axis may be conveniently represented by its point of intersection with a geocentric unit sphere called the celestial sphere; the point is called the pole of the axis. Variations of the coordinates of the pole reflect the motions of the axis. The definition of the variables representing the motions of a chosen axis (or its pole) calls for the use of a suitable reference frame. For motions in space (precession and nutation), a celestial reference frame has to be used, while wobble and polar motion are to be referred to a terrestrial frame. The celestial and terrestrial frames considered here are all geocentric: the coordinate axes originate from the geocenter.

There are two types of celestial reference frames in use.

10.2.1 Ecliptic Celestial Reference Frame

It has the ecliptic as the principal plane (the plane containing the first two Cartesian coordinate axes) to which the third axis is perpendicular. The ecliptic plane may be thought of as the plane of the Earth's orbit around the Sun (or more correctly, as the plane of the orbit of the center of mass of the Earth–Moon system). The circle along which this plane intersects the celestial sphere provides a useful representation of the ecliptic plane, and may also be referred to as the ecliptic; and the intersection point of the normal to the ecliptic with the celestial sphere is called the pole of the ecliptic. Similarly, the intersection of the equatorial plane (perpendicular to the axis of interest) with the celestial sphere will be called the equator, and the intersection of the axis itself with the celestial sphere is known as the pole of the equator. The intersections of the equator and the ecliptic

are the equinoxes. The equinox at which the Sun crosses over from south of the equator to the north in its annual journey along the ecliptic (as perceived from the Earth) is the vernal (spring) equinox and the other is the autumn equinox. The first axis of the ecliptic reference frame is chosen to be along the nodal line (line of intersection) of the ecliptic and equatorial planes and is taken in the direction of the vernal equinox.

10.2.1.1 Precession

The angle between the equatorial and ecliptic planes (or between their polar axes) is the obliquity ϵ. A steady rotational motion of the equatorial plane or its polar axis about the first axis of the ecliptic frame constitute the precession in obliquity which is extremely small and is often ignored; the additional oscillatory (multiply periodic) rotation about the same axis is the nutation in obliquity, $\Delta\epsilon$. The precession in longitude is a steady rotational motion of the Earth-related axis around the normal to the ecliptic plane, keeping an angle ϵ_0 to this normal, where $\epsilon_0 = \epsilon - \Delta\epsilon$. Precession takes the pole of this axis over an arc of a circle of radius $\sin\epsilon_0$ centered on the normal. The angular size of this arc as seen from the center of the circle is denoted by $\psi_A(t)$; its linear dimension is evidently $\psi_A(t)\sin\epsilon_0$. The nutation in longitude is an additional oscillatory rotation $\Delta\psi$ about the same normal (over the same arc). The direction of the precessional motion of the Earth's pole is opposite to that of the Earth–Moon system around the normal to the ecliptic, and this is defined to be the direction of increasing longitude.

At the present rate of precession, which is very close to 50 arcsec year, the pole of the Earth's axis would take about 25 770 years to complete a full circuit around the pole of the ecliptic.

In defining an ecliptic-based frame, one has to contend with a complication: the ecliptic itself has a very slow rotation in space around a vector lying in the ecliptic. One may choose a reference frame based on the 'ecliptic of epoch', that is, the ecliptic at a fixed epoch (instant of time). The standard choice of epoch is J2000, which is at 12 h Universal Time on 1 January 2000). Alternatively, one may choose a frame based on the ecliptic plane of the time of interest (ecliptic of date). The nutation and precession referred to the fixed ecliptic represent the actual motion of the nutating axis in space. The precession in this sense is only in longitude; it is often called the luni-solar precession because it is caused almost entirely by torques exerted by the Moon and Sun.

But the classical definition of nutation and precession is with reference to the moving ecliptic; the precession in longitude thus defined includes a small part arising from the motion of the ecliptic, and is called the general precession. There is also a small precession in obliquity arising from the ecliptic motion.

10.2.1.2 Nutation

The classical nutation variables $\Delta\psi(t)$ and $\Delta\epsilon(t)$ representing the oscillatory part of the motions in longitude and obliquity were defined, strictly speaking, relative to the moving ecliptic. The purely periodic components of both $\Delta\psi$ and $\Delta\epsilon$ remain unaltered even if the fixed ecliptic is used, though the so-called Poisson terms do change (see Bretagnon et al. (1997)). Poisson terms are products of t or powers of t multiplied by sine or cosine factors.

$\Delta\psi$ goes up to about ±19 arcsec and $\Delta\epsilon$ up to about ±10 arcsec.

10.2.2 Geocentric Celestial Reference Frame

This is a frame having the mean equatorial plane of J2000 as its principal plane and having no rotation in space. The qualification 'mean' in this context means that the nutational part of the displacement of the figure axis/equator is excluded while fixing the position of the equator at this epoch. The first axis (X-axis) of the frame is in the direction of the vernal equinox. This is the direction of increasing longitude referred to the ecliptic of J2000, and the Y-axis of the Geocentric Celestial Reference Frame (GCRF) is in the direction of increasing obliquity.

In the GCRF, the displacement of the pole of axis of interest due to the precession and nutation is represented by the coordinates $X(t)$, $Y(t)$. For times that are not too far from J2000, one may take

$$X(t) = [\psi_A(t) + \Delta\psi(t)]\sin\epsilon_0$$
$$Y(t) = \epsilon_A(t) + \Delta\epsilon(t) \qquad [1]$$

where ϵ_0 is the mean obliquity at J2000. But when the time difference from J2000 gets large, corrections have to be applied to these lowest order expressions. The rate of precession in obliquity, $\dot\epsilon_A(t)$, is only about 0.25 milliarcseconds per year (mas/yr), which is extremely small compared to $\dot\psi_A(t)$ which is about 50400 mas yr^{-1}.

10.2.3 International Terrestrial Reference Fame

A terrestrial frame, by definition, is fixed to the Earth. The International Terrestrial Reference Frame (ITRF) is defined through coordinates assigned to a number of sites taking account of various effects which cause time-dependent displacements of the sites (due to tectonic plate motions, solid Earth tides, etc.), so that its motion reflects the rotation of the Earth as a whole. It has the plane of the true equator, very close to be perpendicular to the instantaneous direction of the figure axis, as its principal plane, and its first axis (x-axis) on the Greenwich meridian. When we refer to a terrestrial frame, it is to be understood, by default, to be the ITRF unless otherwise specified.

10.2.4 Principal Axis Frame

This is a terrestrial frame which differs from the ITRF only in having the two equatorial coordinate axes along the two principal axes of inertia in the equatorial plane. Its x-axis lies about 15° West of the Greenwich meridian. In setting up the torque equations governing the rotational motion of the Earth, it is convenient and customary to make use of this reference frame.

The wobble variables are defined, whether in the ITRF or in the principal axis frame, as

$$m_x = \Omega_x/\Omega_0 \text{ and } m_y = \Omega_y/\Omega_0 \qquad [2]$$

where Ω_x and Ω_y are the first two components of the angular velocity vector $\mathbf{\Omega}$ of the Earth (or more correctly, of the mantle), and Ω_0 is the mean angular velocity. Wobble is an inescapable accompaniment of nutation and vice versa. Wobble is intimately related to the nutation of the figure axis, as we shall see, and plays an essential role in various approaches to the theoretical treatment of nutation.

10.2.5 The CIP and CIO

The most direct and most precise information about the varying orientation of the terrestrial reference frame in space is obtained from Very Long Baseline Interferometry. This technique uses observations of the time delays between the arrival times of radio signals from the most distant (and hence the most 'space fixed') sources of radio waves in space at different members of an array of radio antennas spread over the globe to obtain precise estimates of the orientation of the terrestrial frame relative to a celestial frame defined by such sources. The observations are made at intervals of several days (typically 5 or 7), and have been carried out for over two decades; they make it possible to estimate, with very high precision, the variations in orientation, that is, nutations of the figure axis and rotation rate variations, having periods exceeding a few days. Similarly, it is the low-frequency variations in direction of the rotation axis in the terrestrial reference frame that are susceptible to estimation with high precision. The conventions adopted by the International Astronomical Union (IAU) conform to this reality.

10.2.5.1 Nutation and polar motion in the Conventions: CIP

A pole called the Celestial Intermediate Pole (CIP), intermediate between the poles of the terrestrial and celestial reference frames, is introduced, such that its nutation, that is, its motion relative to the space fixed frame (GCRS), has only low-frequency spectral components (between −0.5 and 0.5 cpsd, see **Figure 6**). It is this motion of the CIP that is defined in the conventions as nutation. This low-frequency band in space maps into the 'retrograde diurnal' band in the terrestrial frame, with frequencies between −1.5 and −0.5 cpsd, because of the prograde rotation of this frame in space at the rate of 1 cpsd. Motions of the CIP in the terrestrial frame with frequencies outside this band is defined, according to the conventions, as polar motion. The low-frequency part of this band, which includes the Chandler wobble and the tidally induced wobbles, dominates the polar motion. However, the amplitude of the polar motion is quite different from that of the wobble to which it is related, as will be shown in Section 10.7.

10.2.5.2 LOD variations in the conventions: CIO

Variation in the spin rate around the z-axis, represented by m_z, and the consequent variation Δ LOD in the length of day, are related to the z-component Ω_z of $\mathbf{\Omega}$:

$$m_z = (\Omega_z/\Omega_0) - 1$$
$$\Delta\text{LOD} = -(\text{LOD})m_z \quad [3]$$

In the scheme of analysis of very long baseline interferometry (VLBI) data, variations in the angular velocity component in the direction of the CIP are estimated rather than those in the z-component; these observations are performed at the milliarcsecond (mas) level. Computation of the LOD variations from the spin rate variations estimated in this manner is in accord with current conventions. The spin rate, in this context, is the rate of change of the angle of rotation around the CIP axis, measured from a point on the equator of the CIP which has no rotation about this axis. This point is called the conventional international origin (CIO).

10.3 Equations of Rotational Motion

The most direct approach to the theoretical treatment of Earth rotation variations is based on the torque equations, which are statements of angular momentum conservation.

10.3.1 Equation of Motion in a Celestial Frame

In an inertial (space fixed) reference frame, the torque equation is simply

$$\frac{d\mathbf{H}}{dt} = \mathbf{\Gamma} \quad [4]$$

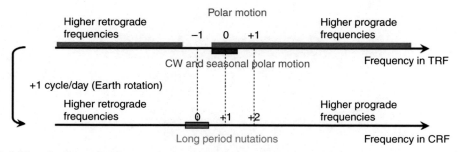

Figure 6 Definition of polar motion frequencies and precession-nutation frequency.

So we have, trivially,

$$\mathbf{H} = \int \Gamma \, dt \qquad [5]$$

This result is independent of the structure of the body which is subjected to the torque.

The components H_X, H_Y are very small compared to H_Z which is closely approximated by $C\Omega_0$; so $|\mathbf{H}| \approx C\Omega_0$. The X- and Y-components of the 'unit vector' in the direction of $\mathbf{H}$, which characterize the precession and nutation 'of the angular momentum axis', are then

$$X_H = \frac{H_X}{C\Omega_0} = \frac{\int \Gamma_X \, dt}{C\Omega_0}$$

$$Y_H = \frac{H_Y}{C\Omega_0} = \frac{\int \Gamma_Y \, dt}{C\Omega_0} \qquad [6]$$

where the subscript H identifies quantities pertaining to the angular momentum axis. The precession and nutation in longitude, $\psi_{A,H}$ and $\Delta\psi_H$, and the nutation $\Delta\epsilon_H$, are related to X_H and Y_H through equations of the same form as [1]:

$$X_H = (\psi_{A,H} + \Delta\psi_H)\sin\epsilon, \qquad Y_H = \Delta\epsilon_H \qquad [7]$$

Explicit evaluation of the nutation quantities has to wait till the torque acting on the Earth and its time dependence are determined.

10.3.2   Equation of Motion in a Terrestrial Frame

Consider now the description of the Earth's rotational motion as seen from an Earth fixed reference frame. The frame itself is rotating in space, and we denote its angular velocity vector by $\mathbf{\Omega}$. The torque equation governing the variation of the angular momentum vector $\mathbf{H}$ in this reference frame is

$$\frac{d\mathbf{H}}{dt} + \mathbf{\Omega} \times \mathbf{H} = \Gamma \qquad [8]$$

where the vectors $\mathbf{H}$ and Γ are now referred to the terrestrial frame. On writing the equation out in terms of the Cartesian components, we get

$$\dot{H}_x + \Omega_y H_z - \Omega_z H_y = \Gamma_x$$
$$\dot{H}_y + \Omega_z H_x - \Omega_x H_z = \Gamma_y \qquad [9]$$
$$\dot{H}_z + \Omega_x H_y - \Omega_y H_x = \Gamma_z$$

where $\dot{H}_x = dH_x/dt$, and so on.

The above equations, unlike the corresponding ones in an inertial frame, involve not just $\mathbf{H}$ but $\mathbf{\Omega}$ too. Therefore it is necessary, in order to solve these equations, to make use of the relationship of the former to the latter. If the Earth were wholly solid, the relation would be $\mathbf{H} = [\mathbf{C}] \cdot \mathbf{\Omega}$, where $[\mathbf{C}]$ is the Earth's inertia tensor. This is no longer true if core regions are present. While the mantle rotates with angular velocity $\mathbf{\Omega}$, the other regions rotate, in general, with angular velocities differing from $\mathbf{\Omega}$; these differences would result in additional contributions to $\mathbf{H}$ from the core regions. Furthermore, the inertia tensors of the different regions will have to include contributions from tidal deformations of the respective regions when the deformability of the Earth is to be taken into account. Thus, it is necessary to use as inputs various aspects of the Earth's structure and properties before the torque equation in the terrestrial frame can be formulated; and additional equations which govern the variation of the angular velocities of the core regions will have to be considered simultaneously. We consider first the simplest case of a rigid Earth.

10.3.2.1   Euler's equations for a rigid body
For any rigid body, eqns [9] reduce to the Euler equations when coordinate axes parallel to the principal axes of the body are chosen. With such a choice,

$$H_x = A\Omega_x, \quad H_y = B\Omega_y, \quad H_z = C\Omega_z \qquad [10]$$

where A, B, C stand for the principal moments of inertia in increasing order of magnitude, as usual. Substituting into eqns [9], we obtain

$$A\dot{\Omega}_x + (C-B)\Omega_y\Omega_z = \Gamma_x$$
$$B\dot{\Omega}_y + (A-C)\Omega_z\Omega_x = \Gamma_y \qquad [11]$$
$$C\dot{\Omega}_z + (B-A)\Omega_x\Omega_y = \Gamma_z$$

These are the celebrated Euler equations.

For the rotation variations of the Earth, the fractional variations in the angular velocity, namely m_x, m_y, m_z, are far smaller than unity. (For the real Earth, they are actually of the order of 10^{-8} or smaller for the forced wobbles, and about 10^{-6} for the free Chandler wobble.) So one can approximate the

Euler equations to the first order in these small quantities:

$$A\Omega_0 \frac{dm_x}{dt} + \Omega_0^2(C-B)m_y = \Gamma_x$$

$$A\Omega_0 \frac{dm_y}{dt} + \Omega_0^2(A-C)m_x = \Gamma_y \qquad [12]$$

$$C\Omega_0 \frac{dm_z}{dt} = \Gamma_z$$

10.3.2.2 Axially symmetric ellipsoidal case: Wobble motion

We specialize now to the Earth considered as an axially symmetric, ellipsoidal, rigid body with no core.

The $(B-A)$ term in the third of the Euler equations [12] then drops out, and $(C-B)$ in the first equation becomes $(C-A)$; and Γ_z vanishes too, because of axial symmetry. It follows that H_z remains constant, equal to $C\Omega_0$, that is, $m_z = 0$. After these simplifications, the addition of i times the second of the above equations to the first one yields a single equation involving the complex quantities

$$\tilde{m} = m_x + im_y, \quad \tilde{\Gamma} = \Gamma_x + i\Gamma_y \qquad [13]$$

The combined (complex) equation is

$$\frac{d\tilde{m}}{dt} - ie\Omega_0\tilde{m} = \frac{\tilde{\Gamma}}{A\Omega_0} \qquad [14]$$

where e the dynamical ellipticity in the sense in which this terms is commonly used in the geophysics literature:

$$e = (C-A)/A \qquad [15]$$

Solution of the above equation is a trivial matter:

$$\tilde{m}(t) = e^{ie\Omega_0 t}\left[\tilde{m}(0) + \int \frac{\tilde{\Gamma}(t)}{A\Omega_0}e^{-ie\Omega_0 t}\,dt\right] \qquad [16]$$

10.3.2.2.(i) Free wobble The first term, $\tilde{m}(0)e^{ie\Omega_0 t}$, does not involve the torque. It is a free motion of the angular velocity vector around the figure axis with frequency $e\Omega_0$ or e cpsd, as may be seen readily by decomposing $\tilde{m}$ into its real and imaginary parts: $m_x = \tilde{m}(0)\cos e\Omega_0 t$, $m_y = \tilde{m}(0)\sin e\Omega_0 t$, assuming $\tilde{m}(0)$ to be real. It represents, in fact, the Eulerian free wobble, with e cpsd as its eigenfrequency. In the case of more realistic nonrigid Earth models, the free mode becomes the Chandler wobble, with a frequency differing from $e\Omega_0$.

10.3.2.2.(ii) Forced wobble The second term in [16] is the forced wobble, which can be evaluated once the time dependence of $\tilde{\Gamma}$ is known.

Let us consider a single spectral component of the torque, with frequency $\sigma\Omega_0$, that is, σ cpsd:

$$\tilde{\Gamma}(t) = \tilde{\Gamma}(\sigma)e^{i\sigma\Omega_0 t} \qquad [17]$$

Then eqn [14] simplifies to

$$i(\sigma - e)\tilde{m}(\sigma) = \tilde{\Gamma}(\sigma)/(A\Omega_0^2) \qquad [18]$$

so that

$$\tilde{m}(\sigma) = -i\frac{\tilde{\Gamma}(\sigma)}{A\Omega_0^2(\sigma - e)} \qquad [19]$$

The factor $1/(\sigma - e)$ represents the 'resonance' in the forced wobble that is associated with the Eulerian free wobble having the eigenfrequency e cpsd.

In later sections where we consider both rigid and nonrigid Earth models, we identify the rigid case by adding a subscript R (standing for 'rigid'). Thus $\tilde{m}$, m_x, etc., of this section would be denoted by $\tilde{m}_R$, m_{Rx}, etc.

10.4 The Tidal Potential and Torque

Before going on to the treatment of more realistic Earth models, let us examine the gravitational effects of a celestial body B of mass M_B on the Earth.

10.4.1 Potential, Acceleration, Torque

If the position vector of B relative to the geocenter is $\mathbf{r}_B$, then its gravitational potential at $\mathbf{r}$ in the Earth is

$$W_B(\mathbf{r}) = -\frac{GM_B}{|\mathbf{r}_B - \mathbf{r}|}$$

$$= -\frac{GM_B}{r}\left(1 - \frac{2\mathbf{r}\cdot\mathbf{r}_B}{r_B^2} + \frac{r^2}{r_B^2}\right)^{-1/2} \qquad [20]$$

Since $r = |\mathbf{r}| \le a_e \approx 6378$ km, and the nearest solar system body is the Moon at a geocentric distance of about $400\,000$ km, $r/r_B < 0.016$ and therefore an expansion in powers of (r/r_B) is useful:

$$W_B(\mathbf{r}) = -\frac{GM_B}{r_B}\left(1 + \frac{\mathbf{r}\cdot\mathbf{r}_B}{r_B^2} - \frac{r^2}{2r_B^2} + \frac{3(\mathbf{r}\cdot\mathbf{r}_B)^2}{2r_B^4} + \cdots\right) \qquad [21]$$

The acceleration of a mass element at $\mathbf{r}$ is then

$$-\nabla W_B(\mathbf{r}) = \frac{GM_B}{r_B}\left(\frac{\mathbf{r}_B}{2r_B^2} + \frac{3(\mathbf{r}\cdot\mathbf{r}_B)\mathbf{r}_B - r_B^2\mathbf{r}}{r_B^4} + \cdots\right) \qquad [22]$$

The first term represents a uniform acceleration toward the external body. In case the body is the Sun or the Moon, it is this centripetal acceleration which keeps the Earth and the other body going in an orbit around each other. The second term in the acceleration is position dependent, and causes different mass elements in the Earth to move relative to one another. The result is a deformation of the solid Earth which is manifested as the 'solid Earth tide', and a redistribution of water mass in the oceans, the 'ocean tide'. The nonuniform acceleration is referred to as the tidal acceleration. The term in the potential W_B which gave rise to this tidal acceleration is designated as the 'tidal potential' of degree 2. Succeeding terms in the expansion of W_B, which are not explicitly shown, constitute tidal potentials of degrees >2. The torque on the whole Earth, which is responsible for the forced variations in Earth rotation, is

$$\boldsymbol{\Gamma} = \int -\rho(\mathbf{r})\mathbf{r} \times \nabla W_B(\mathbf{r})\mathrm{d}^3\mathbf{r}$$
$$= \frac{GM_B}{r_B} \int \rho(\mathbf{r})\left(\frac{\mathbf{r} \times \mathbf{r}_B}{r_B^2} + \frac{3(\mathbf{r}\cdot\mathbf{r}_B)(\mathbf{r}\times\mathbf{r}_B)}{r_B^4} + \cdots\right) \quad [23]$$

where the integration is over the volume of the Earth. The first term in [23] involves $\int \rho(\mathbf{r})\mathbf{r}\mathrm{d}^3r$, which is simply the Earth's mass times the position vector of the center of mass; this vector is null because the center of mass is the origin (geocenter) itself. The torque is therefore given by the integral of the second term in the above expression, apart from the much smaller higher degree terms represented by the set of dots at the end.

10.4.2   Torque Components in Terrestrial Frame

We evaluate the components of the torque integral in a terrestrial reference frame by expressing $\mathbf{r}$ and $\mathbf{r}_B$ in terms of its components with respect to that frame. We choose the principal axis frame, which is defined as one in which the inertia tensor is diagonal. The x-axis of this frame is the principal axis of the least moment of inertia A; the longitude α of this axis in the ITRF is about $-14°.93$.

Since the elements C_{ij} of the tensor are defined by

$$C_{ij} = \int \rho(\mathbf{r})\left(r^2 - x_ix_j\right)\mathrm{d}^3\mathbf{r} \quad [24]$$

we have, in the principal axis reference frame,

$$\begin{pmatrix} C_{11} \\ C_{22} \\ C_{33} \end{pmatrix} = \int \rho(\mathbf{r})\begin{pmatrix} (y^2+z^2) \\ (z^2+x^2) \\ (x^2+y^2) \end{pmatrix}\mathrm{d}^3r = \begin{pmatrix} A \\ B \\ C \end{pmatrix} \quad [25]$$

$$\int \rho(\mathbf{r})xy\mathrm{d}^3r = \int \rho(\mathbf{r})yz\mathrm{d}^3r = \int \rho(\mathbf{r})zx\mathrm{d}^3r = 0 \quad [26]$$

If the time-dependent variations in the density function $\rho(\mathbf{r})$ and hence in the inertia tensor due to tidal deformations and other causes were also taken into account, the result of the integrations in eqn [25] would get modified to $A+c_{11}, B+c_{22}, C+c_{33}$ and those in [26] to $-c_{12}, -c_{23}, -c_{31}$, respectively, where the c_{ij} are the contributions from the deformations. They would lead to second-order contributions to the torque which are very small; they will be ignored in the following. But they are not entirely negligible at the levels of accuracy required at present, and we shall consider second-order terms in Section 10.10 on nonrigid Earth numerical model.

Now, we can express the factor $(\mathbf{r}\cdot\mathbf{r}_B)(\mathbf{r}\times\mathbf{r}_B)$ in the torque integral [23] in terms of the components of $\mathbf{r}$ and $\mathbf{r}_B$ and then evaluate the integral using [25] and [26]. Thus, we obtain

$$\begin{pmatrix} \Gamma_x \\ \Gamma_y \\ \Gamma_z \end{pmatrix} = \frac{3GM_B}{r_B^5}\begin{pmatrix} (C-B)y_Bz_B \\ (A-C)x_Bz_B \\ (B-A)x_By_B \end{pmatrix} \quad [27]$$

Replacing the coordinates of the celestial body in terms of the polar coordinates $(r_B, \theta_B, \lambda_B)$, where θ_B is its colatitude, λ_B is its longitude in the terrestrial frame that we are using, we rewrite the equatorial components of the torque as

$$\Gamma_x = \frac{3GM_B}{r_B^3}\bar{A}(e-e')\cos\theta_B\sin\theta_B\sin\lambda_B$$
$$\Gamma_y = -\frac{3GM_B}{r_B^3}\bar{A}(e+e')\cos\theta_B\sin\theta_B\cos\lambda_B \quad [28]$$

wherein $\bar{A}=(A+B)/2$, and

$$e = \frac{C-\bar{A}}{\bar{A}}, \quad e' = \frac{B-A}{2\bar{A}} \quad [29]$$

The parameter e', proportional to $(B-A)$, is a manifestation of the triaxiality of the Earth, and e as defined here is the dynamical ellipticity of the triaxial Earth; the latter dominates the torque because e' is only about $e/300$.

The complex combination $\tilde{\Gamma}=\Gamma_x+i\Gamma_y$ is

$$\tilde{\Gamma} = -i\frac{3GM_B}{r_B^3}\bar{A}\cos\theta_B\sin\theta_B\left(e\,e^{i\lambda_B} + e'e^{-i\lambda_B}\right) \quad [30]$$

Now, as seen from the rotating Earth, all celestial objects move westward in sky, that is, in a retrograde sense (opposite to the Eastward rotation of the Earth itself, as a consequence of this rotation). Their angular velocities $d\lambda_B/dt$ are close to -1 cycle per sidereal day. In fact, the temporal variation of the torque due to the orbital motion of the Sun or the Moon has spectral components with frequencies on both sides of -1 cpsd even if the orbits are taken to be circular, as will be demonstrated in the next couple of sections. The fact that the orbits are elliptical, and that even these are perturbed by the influence of third bodies (e.g., perturbation of the lunar orbit around the Earth by the attraction of the Sun and the planets on the Moon) leads to a rich spectrum of frequencies bunched very close to and centered at -1 cpsd. This band of frequencies is referred to as the 'retrograde diurnal band'. An examination of eqns [28] shows now that the main part proportional to $(C - \bar{A})$ of the equatorial projection (Γ_x, Γ_y) of the torque vector rotates in the retrograde sense as a consequence of the negative rate of change of λ_B, and that this motion has the same retrograde diurnal spectrum; the triaxiality part of the vector, on the other hand, rotates in the prograde sense, and has a prograde diurnal spectrum.

Another point of interest is that the products $y_B z_B$ and $x_B z_B$ present in Γ_x and Γ_y in [27] come from the terms

$$-\frac{3GM_B}{r_B^5}(x_B z_B\, xz + y_B z_B\, yz) = -\Omega_0^2(\phi_1 xz + \phi_2 yz) \quad [31]$$

of the potential [21]. The dimensionless quantities ϕ_1, ϕ_2, and their complex combination $\tilde{\phi} = \phi_1 + i\phi_2$ were introduced by Sasao et al. (1980), and have been used extensively in further developments based on their formalism. A potential which is a linear combination of xz and yz is called a 'tesseral' potential of degree 2. Thus, the equatorial components of the gravitational torque due to the degree 2 terms in the potential are produced solely by the tesseral part of these terms; and the only Earth parameters on which this torque depends are the differences among the three principal moments of inertia, which, in turn, depend on the dynamical ellipticity e and the triaxiality (fractional difference between the two equatorial moments of inertia). Since higher degree terms in the expansion of the full potential are of much smaller magnitude as noted earlier, the degree 2 tesseral potential, which has its spectrum in the retrograde diurnal band, is responsible for dominant part of the wobbles and hence of the nutations

too, as we shall see. The residual small part of the wobble and nutation arise from the action of potentials of degree >2 on moments of the Earth's matter distribution that are of degree >2 (the moments of inertia being of degree 2).

10.4.3 Expansion of the Potential: Tidal Spectrum

The terms up to degree 2 in (r/r_B) in the expansion of the potential $W_B(\mathbf{r})$ of an external body B were shown in eqn [21]. A further expansion of each of the tidal terms (of degree $n \geq 2$) can be made into parts of different orders m from 0 to n which are distinguished by the nature of their dependence on the direction of the position $\mathbf{r}_B$ of the body B and that of the Earth point $\mathbf{r}$ relative to the terrestrial reference frame (ITRF). This dependence is through the surface spherical harmonic functions Y_n^m of the colatitude θ and longitude λ of $\mathbf{r}$ in the ITRF, as well as the Y_n^m of θ_B and λ_B of $\mathbf{r}_B$ in the same frame. (Note that λ and λ_B in this frame are equal to α plus the corresponding quantities relative to the principal axis frame of the last section.)

The spherical harmonics are defined by

$$Y_n^m(\theta, \lambda) = N_{nm} P_n^m(\cos\theta)e^{im\lambda} \quad [32]$$

and

$$Y_n^{-m} = (-1)^m Y_n^{m*} \quad [33]$$

with asterisk standing for complex conjugation and with

$$N_{nm} = (-1)^m \left(\frac{2n+1}{4\pi}\right)^{1/2}\left(\frac{(n-m)!}{(n+m)!}\right)^{1/2} \quad [34]$$

for $n = 0, 1, 2, \ldots$, $m = 0, 1, \ldots, n$. The P_n^m are the associated Legendre functions; they are derived from the Legendre polynomials P_n which are defined through the expansion

$$(1 - 2cw + w^2)^{-1/2} = \sum_{n=0}^{\infty} P_n(c)w^n \quad [35]$$

$$P_n^m(c) = (1-c^2)^{m/2}\frac{d^m P_n(c)}{dc^m} \quad [36]$$

$P_n(c)$ is a polynomial of degree n in c, and has unit value at $c = 1$. The spherical harmonics Y_n^m for all n and $m = -n, -n+1, \ldots, n-1, n$, form an orthonormal set:

$$\int Y_q^{p*}(\theta, \lambda)Y_n^m(\theta, \lambda)\sin\theta\, d\theta\, d\lambda = \delta_{qn}\delta_{pm} \quad [37]$$

The integration here is over the unit sphere and δ is the Kronecker delta function: δ_{qn} is 1 if $q = n$ and is 0 otherwise. The expansion of the potential [20] in terms of the Y_n^m is, after dropping terms of degrees $n = 1, 2$ which do not produce any tidal deformations or torques,

$$W_B(\mathbf{r}) = -\sum_{n=2}^{\infty} r^n \sum_{m=0}^{n} \left[F_{nm}^*(r_B, \theta_B, \lambda_B) Y_n^m(\theta, \lambda) \right.$$
$$\left. + F_{nm}(r_B, \theta_B, \lambda_B) Y_n^{m*}(\theta, \lambda) \right] \qquad [38]$$

with

$$F_{nm}(r_B, \theta_B, \lambda_B) = \frac{GM_B}{r_B^{n+1}} \frac{(2 - \delta_{m0}) 2\pi}{2n + 1} Y_n^m(\theta_B, \lambda_B) \qquad [39]$$

and the complex conjugate of it provides the expression for $F_{nm}^*(r_B, \theta_B, \lambda_B)$. It may be observed that the part of W_B belonging to any specific degree n and order m contains the phase factors $e^{im(\lambda - \lambda_B)}$ and $e^{im(\lambda - \lambda_B)}$. Each of the phases is 0 on the meridian $\lambda = \lambda_B$; and as the celestial bodies move westward in the sky (i.e., in the retrograde sense) at a rate close to once per day, as seen from the terrestrial frame, the above meridian moves with it; other constant phase meridians move in tandem. So, for any $m \neq 0$, this term constitutes a retrograde tidal wave sweeping over the volume of the Earth, with m cycles over the globe (because of the factor m multiplying λ).

Representation of the time dependence of the potential through its spectral expansion is of great value. The most widely used form is that of the expansion by Cartwright and Tayler (1971), though a number of tidal tables of much higher accuracy are now available (Hartmann and Soffel, 1994; Hartmann and Wenzel, 1995a, 1995b; Hartman et al., 1999, Roosbeek and Dehant, 1998; Roosbeek, 1999):

$$F_{nm}^*(r_B, \theta_B, \lambda_B) = (g_e/a_e^n) \sum_{\omega} H_{\omega}^{nm} \zeta_{nm} e^{i\Theta_{\omega}(t)} \qquad [40]$$

wherein the amplitude of the spectral component is given in units of length as the equivalent 'height' H_{ω}^{nm}. The factor ζ_{nm} is defined to be unity when $n - m$ is even and $-i$ when $n - m$ is odd. With these values for ζ_{nm}, the dependence of the tidal component on the longitude λ and time appears through the factor $\cos(\lambda + \Theta_{\omega}(t))$ for $(n - m)$ even and $\sin(\lambda + \Theta_{\omega}(t))$ for $(n - m)$ odd, when the real part is taken in [38].

$\Theta_{\omega}(t)$ is the argument of the tidal component; the frequency ω is non-negative, and is equal to $d\Theta_{\omega}/dt$.

$$\Theta_{\omega}(t) = \omega t + \chi_{\omega} \qquad [41]$$

where χ_{ω} is a constant. $\Theta_{\omega}(t)$ does have a dependence on n and m, which is not indicated explicitly. Now, the mean motion of B in the sky (i.e., the mean value of $d\lambda_B/dt$) relative to the terrestrial frame is close to -1 cpsd; and $e^{i\Theta_{\omega}(t)}$, for given n and m, originates from $e^{-im\lambda_B}$ (modulated by other factors depending on θ_B and r_B), as is evident from the expression for F_{nm}^*. Hence, the frequency spectrum of $e^{i\Theta_{\omega}(t)}$ must lie in a band around m cpsd. Since the angle of a spectral component of the wave of order m is now $(\lambda + \Theta_{\omega}(t))$, it constitutes a retrograde wave with frequency $-\omega$ lying within the retrograde band around $-m$ cpsd.

The roles of the various solar system bodies in the argument $\Theta_{\omega}(t)$ are displayed through its expansion in terms of Doodson's fundamental arguments of the tides:

$$\Theta_{\omega} = n_1(\tau - \lambda) + n_2 s + n_3 h + n_4 p + n_5 N' + n_6 p_s + \cdots \qquad [42]$$

where $n_1 = m$ and the set of dots stands for terms that are multiples of the mean tropic longitudes of the planets from Mercury to Saturn. Of the fundamental arguments that are explicitly shown, $s, h, p, -N, p_s$ are the mean tropic longitudes of the Moon, Sun, Moon's perigee, Moon's node, and the Sun's perigee, respectively. The periods relating to these arguments range from 27.32 days for s to about 209 years for p_s. The Doodson argument τ is $180° + GMST - s + \lambda$ when expressed in degrees. GMST is the Greenwich Mean Sidereal Time which is the time measured by the angle traversed along the mean equator of date by the Greenwich meridian at the mean sidereal rotation rate Ω_0:

$$GMST = GMST_0 + \Omega_0 t + \dots \qquad [43]$$

$$GMST_0 = 280°.4606 \dots \qquad [44]$$

where t is measured from J2000. The period associated with $\tau - \lambda + s$ is that of the mean sidereal rotation, namely, 1 sidereal day. It is the term $n_1 \tau = m\tau$ in Θ_{ω} that causes the frequencies in potentials of order m to lie within a band centered at m cpsd.

10.4.4 Degree 2 Potential and Torque: Spectra

The tidal torque on an ellipsoidal Earth arises from the degree 2 part of the potential W_B, as has been noted earlier. The degree 2 potential, in turn, is made up of zonal, tesseral, and sectorial parts which are of orders $m = 0, 1$, and 2, respectively.

The associated Legendre functions appearing in the three orders are

$$P_2(\cos\theta) = \frac{3}{2}\cos^2\theta$$
$$P_2^1(\cos\theta) = 3\sin\theta\cos\theta \qquad [45]$$
$$P_2^2(\cos\theta) = 3\sin^2\theta$$

These are factors in the spherical harmonics $Y_2^m(\theta,\lambda)$, and appear with the coefficients $F_{2m}^*(\theta_B,\lambda_B)$ in the tidal potential. Their time dependence enters through $e^{i\Theta\omega(t)}$ with $n_1 = m = 0, 1, 2$, respectively, as we see from eqns [40] and [42]. Therefore, the spectra of the zonal, tesseral, and sectorial potential lie in the low frequency, prograde diurnal, and prograde semidiurnal bands, respectively, as will be clear from the discussion at the end of the last section.

10.4.4.1 Equatorial Components of Torque

We can now carry over $\tilde{\Gamma}$ of eqn [30], which was with respect to the principal axis frame, to the ITRF. In doing so, we have to make the replacement $\lambda_B \to (\lambda_B - \alpha)$, as the x-axis of the principal axis frame is at longitude α in the ITRF as already noted. Second, the transformation of the vector components between the two frames makes $\tilde{\Gamma}$ in the ITRF $e^{i\alpha}$ times $\tilde{\Gamma}$ of the other frame though Γ_z remains unchanged. In consequence of these two changes, eqn [30] goes over into

$$\tilde{\Gamma} = -i\frac{GM_B}{r_B^3}P_2^1(\cos\theta_B)\bar{A}\left(ee^{i\lambda_B} + e'e^{2i\alpha}e^{-i\lambda_B}\right)$$
$$= i\frac{GM_B}{r_B^3}\left(\frac{24\pi}{5}\right)^{1/2}\bar{A}\left(eY_2^1(\theta_B,\lambda_B)\right.$$
$$\left. + e'e^{2i\alpha}Y_2^{1*}(\theta_B,\lambda_B)\right) \qquad [46]$$

where the definitions [32] and [34] have been used in the second step. One sees immediately by comparison with eqn [39] that the ellipticity term in the last expression is F_{21}^* apart from constant factors, and that the triaxiality term is proportional to F_{21}. Then eqn [40] taken together with the ensuing discussion shows that the spectrum of these two terms lies in retrograde diurnal and prograde diurnal bands, respectively. We have observed earlier that it is the tesseral part which produces the equatorial torque on an ellipsoidal Earth; this torque drives the wobbles. The form employed in eqn [31] for the tesseral potential may be related readily to F_{21}. We note that

$$-\Omega_0^2(\phi_1 xz + \phi_2 yz) = -\frac{\Omega_0^2 r^2}{3}\mathrm{Re}\left(\tilde{\phi}^*(t)P_2^1(\cos\theta)e^{i\lambda}\right) \quad [47]$$

where $\tilde{\phi} = \phi_1 + i\phi_2$. Comparison with the $n = 2, m = 1$ term in [38] after writing $P_{21}e^{i\lambda}$ as $-(24\pi/5)^{1/2}Y_{21}$ enables us to identify the dimensionless quantity $\tilde{\phi}^*$ as

$$\tilde{\phi}^*(t) = -\left(\frac{15}{8\pi}\right)^{1/2}\frac{F_{21}^*}{\Omega_0^2} \qquad [48]$$

We turn out attention now to the equatorial torque $\tilde{\Gamma}$. One can express it in terms of F_{21} and F_{21}^* by using the definition of Y_2^1 from [32]. One finds, after some elementary algebra, that

$$\tilde{\Gamma} = \left(\frac{15}{8\pi}\right)^{1/2}i\bar{A}\left(eF_{21} + e'e^{2i\alpha}F_{21}^*\right) \qquad [49]$$

An alternative expression for $\tilde{\Gamma}$ in terms of $\tilde{\phi}$ may be obtained by using [48]:

$$\tilde{\Gamma} = -i\bar{A}\Omega_0^2\left(e\tilde{\phi} + e'e^{2i\alpha}\tilde{\phi}^*\right) \qquad [50]$$

This form, especially for the ellipticity part, is widely used in works following the approach of Sasao et al. (1980).

We write down now the spectral expansion of $\tilde{\phi}$ by making use of the relation [48] and the expansion of F_{21}^* from [40]:

$$\tilde{\phi} = -i\left(\frac{15}{8\pi}\right)^{1/2}\frac{g_e}{\alpha_e^2\Omega_0^2}\sum_\omega H_\omega^{21}e^{-i\Theta_\omega(t)} \qquad [51]$$

The expansion of the torque follows immediately. An important point to note is that the ellipticity part of the torque involves $\tilde{\phi}$ with the time dependence $e^{-\Theta\omega(t)} = e^{-i(\omega t + \chi\omega)(t)}$ for the spectral terms, so that the spectral frequencies are negative (since ω is positive); when combined with the fact that $m = 1$ for the tesseral potential, it follows that the spectrum is in the retrograde diurnal band.

In the work of Sasao et al. (1980), and in many other related works, it is customary to use the notation $\sigma\Omega_0$ for the (negative) spectral frequencies involved, and $\tilde{\phi}(\sigma)$ for the amplitude of the spectral component. With this notation, a typical component of $\tilde{\phi}$ is given by

$$\tilde{\phi}_\sigma(t) = -i\tilde{\phi}(\sigma)e^{i(\sigma\Omega_0 t + \chi_\sigma)} \qquad [52]$$

with

$$\sigma = -\omega/\Omega_0, \quad \chi_\sigma = -\chi_\omega \qquad [53]$$

with the amplitude $\tilde{\phi}(\sigma)$ related to the height H_ω^{21} of Cartwright and Tayler (1971) by

$$\tilde{\phi}(\sigma) = \left(\frac{15}{8\pi}\right)^{1/2}\frac{g_e}{a_e^2\Omega_0^2}H_\omega^{21} \qquad [54]$$

Finally, we obtain an explicit expression for the wobble amplitude $\tilde{m}_R(\sigma)$ of a rigid axially symmetric Earth by introducing the ellipticity part of the torque [50] into the solution [19] of the torque equation:

$$\tilde{m}_R(\sigma) = -\frac{e\tilde{\phi}(\sigma)}{\sigma - e} \qquad [55]$$

Tables of the components of the tidal potential list the values of H_ω for the various components identified by the respective sets of values of the multipliers $n_1, n_2, \ldots, n_6$ of the Doodson arguments. The amplitude $\tilde{\phi}(\sigma)$ of any tide of interest may be computed using [54] after picking out the H_ω of that tide from a tide table. The rigid Earth wobble amplitude may then be computed from [55], given the value of e.

10.4.4.2 Axial component of the torque

The z-component of the torque, given in equation [27], may be rewritten with the replacement $\lambda_B \rightarrow (\lambda_B - \alpha)$ as

$$\Gamma_z = -i\frac{3GM_B}{4r_B^3}(B-A)\sin^2\theta_B(e^{2i(\lambda_B-\alpha)} - e^{-2i(\lambda_B-\alpha)})$$

$$= -i\frac{GM_B}{r_B^3}(B-A)\left(\frac{6\pi}{5}\right)^{1/2}\left[e^{-2i\alpha}Y_2^2 - e^{2i\alpha}Y_2^{2*}\right] \qquad [56]$$

This translates into

$$\Gamma_z = -i\left(\frac{15}{8\pi}\right)^{1/2}\bar{A}e'\left(e^{-2i\alpha}F_{22} - e^{2i\alpha}F_{22}^*\right) \qquad [57]$$

The spectral expansion of this torque may be obtained as a special case of [40]. Since $m = \pm 2$ in the present case, the two terms within parentheses have their spectra in the retrograde and prograde semidiurnal bands, respectively. The amplitudes are very small because e'/e is only about $1/300$ as already noted.

10.4.5 Terms of General Degree and Order in the Torque

The torque generated by any higher degree term in the potential may be obtained by taking $(-\mathbf{r} \times \nabla)$ of the (nm) part of [38], then multiplying by $\rho(\mathbf{r})$, and integrating over the volume of the Earth. On using the fact that

$$\left[(\mathbf{r} \times \nabla)_x + i(\mathbf{r} \times \nabla)_y\right]Y_n^m = i[(n-m) \\ \times (n+m+1)]^{1/2}Y_n^{m+1} \qquad [58]$$

for any m, positive or negative, and that $Y_n^{m*} = (-1)^m Y_n^{-m}$, it turns out that the equatorial part of the torque is given by

$$\tilde{\Gamma}^{nm} = iM_E a_e^n \frac{2n+1}{16\pi N_{nm}}\left\{\left(C_{n,m-1} - iS_{n,m-1}\right)\frac{2F_{n,m}}{(2-\delta_{m,1})}\right.$$
$$\left. -(n+m)(n-m+1)\left(C_{n,m+1} + iS_{n,m+1}\right)F_{n,m}^*\right\}$$

where C_{nm} and S_{nm} are geopotential coefficients defined by

$$M_E a_e^n\left(C_{n,m} + iS_{nm}\right) = (2-\delta_{m,0})\frac{(n-m)!}{(n+m)!} \\ \times \int \rho(\mathbf{r})r^n P_n^m(\cos\theta)e^{im\lambda}d^3r \qquad [59]$$

Here M_E is the mass and a_e the equatorial radius of the Earth. It may be seen from the spectral representation [40] of F_{nm}^* and its complex conjugate that the spectra of the two terms of [59] lie, in general, in the two bands centered at $-m$ cpsd and m cpsd, respectively. (In the special case $m = 0$, it turns out that both parts of the torque involve $(C_{n1} + iS_{n1})$ and that the spectra of the two parts together make up the band of width 1 cpsd centered at $m = 0$.)

The coefficients C_{n0} of order 0, also denoted by $-\mathcal{J}_n$, represent axially symmetric structures. They appear only through the first term of [59] with $m = 1$. The torque resulting from the existence of C_{n0} may be seen to reduce to

$$i\frac{GM_B M_E}{r_B}\left(\frac{a_e}{r_B}\right)^n C_{n0}P_{n1}(\cos\theta_B)e^{i\lambda_B} \qquad [60]$$

(Note that all S_{n0} vanish.)

The axially symmetric ellipsoidal Earth is a special case with $n = 2$, for which

$$C_{20} = -\mathcal{J}_2 = -\left(\bar{A}/M_E a_e^2\right)e \qquad [61]$$

It is of some interest to note that the triaxiality enters through

$$C_{22} = \left(\bar{A}/2M_E a_e^2\right)e'\cos 2\alpha$$
$$S_{22} = \left(\bar{A}/2M_E a_e^2\right)e'\sin 2\alpha \qquad [62]$$

In this context, α stands for the longitude of the direction, in the ITRF, of the equatorial axis of the least moment of inertia A. (This angle was introduced in Section 10.4.2. It is to be kept in mind that α is used elsewhere to denote the right ascension.)

The z-component of the general torque may also be obtained by making use of the identity $(\mathbf{r} \times \nabla)_z Y_n^m = im Y_n^m$. One finds that

$$\Gamma_z^{nm} = -iM_E a_e^n \frac{2n+1}{4\pi}\frac{m}{N_{nm}(2-\delta_{m0})} \\ \times \left\{(C_{nm} - iS_{nm})F_{nm} - \left(C_{n,m} + iS_{n,m}\right)F_{n,m}^*\right\}$$

10.5 Torque in Celestial Frame: Nutation-Precession in a Simple Model

10.5.1 Torque Components in Celestial Frame

We considered in the last section the torque and rotation variations referred to the terrestrial reference frame. One might ask: What if we employed the components (X, Y, Z) of $\mathbf{r}$ and (X_B, Y_B, Z_B) of $\mathbf{r}_B$ in a space fixed equatorial reference frame (say, the GCRF), and wished to compute $(\Gamma_X, \Gamma_Y, \Gamma_Z)$ in that frame on the same lines in the foregoing section ? The problem that one encounters in attempting such a course is that, neither the Earth's z-axis (axis of maximum moment of inertia) nor the Earth's rotation vector stays aligned with the Z-axis of the celestial frame, as a result of precession–nutation and wobble. However, we may ignore the small time-dependent offsets among these axes for the limited purpose of a lowest-order calculation of the torque, and then, if we also ignore the small deviation of the Earth from axial symmetry, the axial symmetry persists about the Z-axis too. With these approximations, one may evaluate the integral [23] in the space fixed frame by expressing the integrand in terms of coordinates (X, Y, Z) and (X_B, Y_B, Z_B) and noting that the results [25] and [26] remain valid with these replacements. The components of $\mathbf{\Gamma}$ in the space fixed frame are then found to be

$$\Gamma_X = \frac{3GM_B}{r_B^5}(C-A)\,Y_B Z_B$$
$$\Gamma_Y = -\frac{3GM_B}{r_B^5}(C-A)\,Z_B X_B \tag{63}$$

with $\Gamma_Z = 0$.

The values of r_B, δ, α as functions of time may be taken, for any of the celestial bodies, from the appropriate ephemerides such as VSOP for the Sun and the planets (Bretagnon and Francou, 1988) and ELP for the Moon (Chapront-Touzé and Chapront, 1988). The time dependence of the torque components is then known. The nutation of the angular momentum axis may then be obtained by direct integration in view of eqns [6] and [7].

An interesting fact that is of importance is that though the orbital motions of both the Sun and the Moon, as seen from a nonrotating geocentric frame, are in the prograde sense relative to Earth's rotation, the torque exerted by each of the bodies contains both prograde and retrograde components relative to such a frame. A treatment based on a simplified

model of the orbital motion of the Sun will bring out clearly why this happens, besides providing a concrete example of the calculation of nutation and precession.

10.5.2 Simple Model: Sun in a Circular Orbit

We consider an axially symmetric ellipsoidal Earth subjected to the tidal potential of the Sun which will be supposed, for illustrative purposes, to be in uniform motion with angular velocity p along a circle of radius c centered at the geocenter and lying in the plane of the ecliptic. We take the space fixed reference frame to be equatorial celestial frame (GCRF), with its X-axis along in the direction of the vernal equinox. (This is the direction of increasing ecliptic longitude, that is, of positive $\Delta\psi$. The motion in obliquity of the figure axis is in the direction of the Y-axis.) The angle between the principal plane and the ecliptic is of course the obliquity ϵ, which we shall treat as a constant. Then the coordinates of the Sun in our reference system are

$$X_B = c\cos pt, \quad Y_B = c\cos\epsilon\sin pt$$
$$Z_B = c\sin\epsilon\sin pt \tag{64}$$

Note that (X_B, Y_B) describes the projection of the Sun's circular orbit from the ecliptic plane on to the equatorial plane; the projected orbit is, not surprisingly, an ellipse.

10.5.3 The Torque in the Simple Model

The time dependence of the torque [63] can now be displayed in explicit form:

$$\Gamma_X = F\cos\epsilon(1-\cos 2pt)$$
$$\Gamma_Y = -F\sin 2pt, \quad \Gamma_Z = 0 \tag{65}$$

where

$$F = \left(3GM_B/2c^3\right)Ae\sin\epsilon \tag{66}$$

We know from Kepler's laws that $GM_B = n_B^2 c^3$, where n_B is the 'mean motion', that is, the mean orbital angular velocity of the body, which is what we denoted by p above. (In the case of an elliptical orbit, c has to be the semimajor axis of the orbit in the Keplerian relation.)

Thus, the overall magnitude of the solar torque on the Earth is determined by p and the obliquity ϵ, besides the Earth parameter $(C-A)$. One may get an idea of the strength of the torque from the value of F. For the rotation variations induced by the torque, the nondimensional quantity $F/\Omega_0^2 A)$ is more relevant:

$$\frac{F}{\Omega_0^2 A} = \frac{3}{2}\frac{p^2}{\Omega_0^2} e \sin \epsilon \qquad [67]$$

We know that $p/\Omega_0 = 1/366.2422$, since the period of the relative orbital motion of the Sun and the Earth is one year of 365.2422 solar days or 366.2422 sidereal days. We also have, from the IERS Conventions 2000 (McCarthy, 2003), $\sin \epsilon = 0.397777$, and $\cos \epsilon = 0.917482$. Therefore, $F/(\Omega_0^2 A) = 4.448 \times 10^{-6} e$, it is a dimensionless measure of the amplitude of the torque. On using the value $e = 0.0032845$ from a recent estimate, one finds it to be 1.461×10^{-8}.

Returning to [65] and [67], we take note of three important features that are evident: the presence of a constant (time independent) term in the X-component of the torque, the semiannual frequency of the periodic part of both the equatorial components of the torque, and the fact that the periodic part of the variation of (Γ_X, Γ_Y) is an elliptical motion in the equatorial plane.

10.5.4   Nutation and Precession in the Model

With the components of Γ given in the celestial frame by [65], explicit computation of the motion of the angular momentum axis in space may be done trivially using eqns [5] according to which $X_H = \int \Gamma_X dt$ and $Y_H = \int \Gamma_Y dt$. We find thus that

$$X_H = \frac{3p^2}{2\Omega_0} H_d \sin \epsilon \cos \epsilon [t - (1/2p) \sin 2pt]$$
$$Y_H = \frac{3p^2}{2\Omega_0} H_d \sin \epsilon [(1/2p)\cos 2pt] \qquad [68]$$

where $H_d = (C - A)/C = e/(1 + e)$. The periodic parts of X_H and Y_H evidently describe an 'elliptical' nutation.

Now, we resolve the two-vectors with components X_H, Y_H into two parts as follows:

$$\begin{pmatrix} X \\ Y \end{pmatrix} = D \left[(1 + \cos \epsilon) \begin{pmatrix} -\sin 2pt \\ \cos 2pt \end{pmatrix} \right.$$
$$\left. + (1 - \cos \epsilon) \begin{pmatrix} \sin 2pt \\ \cos 2pt \end{pmatrix} \right] \qquad [69]$$

where $D = (3p/8\Omega_0) H_d \sin \epsilon$. The first vector on the right-hand side rotates in the prograde sense (the same as the rotation of the Earth, from above positive direction of the X axis toward that of the Y-axis), and

the second one's rotation is retrograde. Thus, the above equation constitutes a resolution of the elliptical motion of the angular momentum axis into prograde and retrograde circular motions. The factor $(1 - \cos \epsilon)$ in the amplitude of the retrograde component makes it clear that its presence (despite the motion of the Sun being in the prograde sense in the ecliptic plane) is because of the nonzero obliquity of the ecliptic relative to the equator. It is another matter that D becomes zero and therfore the torque as a whole vanishes if $\epsilon = 0$.

The complex combination of X_H and Y_H is

$$X_H + iY_H = i(3p/8\Omega_0)H_d \sin \epsilon [(1 + \cos \epsilon)e^{2ipt} + (1 - \cos \epsilon)e^{-2ipt}] \qquad [70]$$

The prograde part is characterized by a positive frequency $(2p)$ and the retrograde part by a negative frequency $(-2p)$. This is a very general property, as will be seen in the next section.

It is worth emphasizing here that the results [68] and [71] are for the angular momentum axis and not for the figure axis. These results are independent of whether the Earth is rigid or nonrigid and of whether or not it has a core, since there was no need to input any information about the structure and properties of the Earth (other than that of axial symmetry) in order to obtain the solution for the angular momentum vector. In contrast, the motions of the figure axis and the rotation axis are strongly influenced by the Earth's properties. As a consequence, studies on these motions are the ones that serve to provide glimpses into the properties of the Earth's interior.

We return now to eqns [64] and rewrite their periodic parts in terms of the classical nutation variables $\Delta\psi_H$ and $\Delta\epsilon_H$ using eqns [1]:

$$\Delta\psi_H = -(3p/4\Omega_0)H_d \cos \epsilon \sin 2pt$$
$$\Delta\epsilon_H = (3p/4\Omega_0)H_d \sin \epsilon \cos 2pt \qquad [71]$$

The coefficients of $\sin 2pt$ and $\cos 2pt$ in these expressions are known as the 'coefficients of nutation' in longitude and obliquity.

For this semiannual nutation of angular frequency $2p = 2n_s = 4\pi$ rad yr^{-1}, one finds the coefficients in longitude and latitude to be -1269 and 550 mas, respectively (this corresponds to a quasi-circular motion at the Earth's surface of about 16 m as seen from space). All these numbers are very close to those obtained from accurate treatments of the problem. The largest of the nutation terms is of lunar origin. It has a period of about 18.6 years, which arises from the precession of the lunar orbit around the ecliptic. The

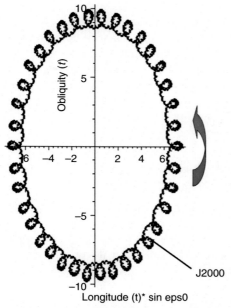

Figure 7 Representation of the changes in the position of the axis undergoing nutations alone.

coefficients in longitude and latitude of this nutation are about $-17\,000$ mas and 9200 mas, respectively. A representation of the nutational motion is provided in **Figure 7**.

As for the term in X_H which is linear in t, it represents precession in longitude. The precession rate is

$$\dot{\psi}_{A,H} = \frac{1}{\sin\epsilon}\frac{\mathrm{d}X_{Hsec}}{\mathrm{d}t} = (3p^2/2\Omega_0)H_\mathrm{d}\cos\epsilon \quad [72]$$

On using the values given earlier for the various parameters, one finds that the precession rate due to solar gravitational torque as given by [72] is 7.7291×10^{-5} rad yr^{-1} or equivalently, $15\,943$ mas yr^{-1}. Lunar attraction too gives rise to precessional motion, a little more than double that due to the Sun; the greater role of the Moon despite its mass being very much smaller than that of the Sun is a consequence of its being very much closer to the Earth. The total precession rate is about $50\,000$ mas yr^{-1}.

10.6 Elliptical Motions: Prograde and Retro-Grade Circular Components

We have observed in the above example that the periodic part of the motion of the vector (X_H, Y_H) describes an elliptical path. Periodic motions of the figure axis or the rotation axis in space are also elliptical. For our purposes, the resolution of the elliptical nutations of the axes and of torque vectors responsible for the nutations into pairs of counter-rotating circular components is important, because the Earth's nutational responses to the two circular components can be far from similar. In the following, we consider specifically the elliptical motions in the celestial frame, but it may be kept in mind that the resolution into circular components can be carried out equally well for motions, such as the wobble, in the terrestrial frame (and indeed, for any elliptical motion in general).

10.6.1 The Spectrum and the Fundamental Arguments of Nutation

The example of the last section is a gross over-simplification of the orbital motion of the Sun as seen from the GCRS. The actual orbit is elliptical and not circular, and even the elliptical orbit is perturbed by the Moon's pull on the Earth and other influences; consequently, the orbital motion has a spectrum of frequencies, which will be reflected in the spectrum of the solar torque on the Earth. The contributions from the orbital motions of the Moon and the Sun (perturbed by the planetary influences) make up all but a small part of the spectrum; the remainder consists of the relatively minor but nonignorable contributions from the planets. An argument $\Xi(t)$ of the sine and cosine terms in a typical spectral component of the torque bears therefore the imprint of more than one of these bodies, and is expressed as a linear combination, with integer coefficients, of a number of 'fundamental arguments', each of which relates to the motion of one or the other of the solar system bodies. The fundamental arguments relating to the lunar and solar motions are the Delaunay arguments. They are as follows:

- $\ell =$ mean anomaly of the Moon,
- $\ell' =$ mean anomaly of the Sun,
- $F = L - \Omega$, L being the mean longitude of the Moon,
- $D =$ mean elongation of the Moon from the Sun, and
- $\Omega =$ mean longitude of the ascending node of the Moon.

Additional fundamental arguments relating to the planets are $\lambda_1, \lambda_2, \ldots, \lambda_8$ which are the mean longitudes with respect to the fixed equinox and ecliptic of J2000, in a solar system barycentric reference frame, of Mercury, Venus, Earth, Mars, Jupiter,

Saturn, Uranus, and Neptune, respectively. Each of these planetary arguments is taken to be linear in time. One more argument which appears in planetary terms in nutation is the precession angle p_a, in which a t^2 term is also retained.

The argument of a general term is then

$$\Xi(t) = n_1\ell + n_2\ell' + n_3F + n_4D + n_5\Omega + \sum_{i=1}^{9} n_i\lambda_i \quad [73]$$

where $\lambda_9 = p_A$. The frequency of this spectral term is $d\Xi/dt$.

A listing of the fundamental arguments as functions of time may be found in the IERS Conventions 2000 (McCarthy, 2003). The periods corresponding to the frequencies of the lunar and solar arguments ℓ, ℓ', F, D, and Ω are 27.5545, 365.2596, 27.2122, 29.5306, and −6798.384 solar days, respectively. The last of these periods, which is about 18.6 years, is the period of the precession of the orbital plane of the Moon around the normal to the ecliptic. The period corresponding to L is 27.3216 solar days; it is the period of revolution of the Moon in its orbit. All these arguments are very nearly but not exactly linear in t, meaning that the periods are very slowly varying with time. The periods used to identify them are the values pertaining to J2000. The periods of the planets, which relate to the arguments λ_i, $(i = 1, 2, \ldots, 8)$ range from about 88 days for Mercury to about 60 000 days for Neptune; and the period relating to p_a is about 25 770 years. It is clear then that the frequencies of the fundamental arguments are much smaller than once per day; all the $\Xi(t)$ of interest to the nutations of an axially symmetric ellipsoidal Earth have the same property. Thus, they are all in the low-frequency band, that is, between −0.5 and 0.5 cpsd.

Bretagnon et al. (1997) observed that ℓ' and Ω are so close to $2\ell - 2D - \lambda_3$ and $\lambda_3 + D - F$, respectively, that observations over periods of the order of thousands of years would be needed to distinguish them. Therefore, in developing their nutation series, they dropped ℓ and Ω from the fundamental set, replacing them by the above combinations plus small corrections that are linear in t; they also dropped the argument p_a which has a very long period. Furthermore, the highly accurate nutation series that they have developed in terms of the truncated set of 11 fundamental arguments are not for the classical $\Delta\psi$ and $\Delta\epsilon$ which are referred to the moving ecliptic, but are for the periodic and Poisson terms in two of the Euler angles (which they denote by ψ and ω) of the transformation from the fixed

ecliptic of J2000 to the terrestrial equatorial frame of date. An unfortunate consequence of these differences is that direct comparison of their SMART97 nutation series with other series is not possible. One has to keep in mind, in addition, differences in signs between the different sets of variables: for example, $\omega_0 = -\epsilon_0$ and $\Delta\omega$ corresponds to $-\Delta\epsilon$ with the effects of ecliptic motion ignored.

10.6.2   Resolution of Elliptical Motions into Pairs of Circular Motions

With the coordinate frame chosen to be the GCRS in which the X- and Y-axis are in the directions of increasing longitude and latitude, respectively, the components Γ_1 and Γ_2 of the torque acting on an ellipsoidal Earth have the forms

$$\begin{pmatrix} \Gamma_1 \\ \Gamma_2 \end{pmatrix} = \begin{pmatrix} P_s \sin \Xi(t) \\ Q_c \cos \Xi(t) \end{pmatrix} \quad [74]$$

with

$$\Xi(t) = (\omega_n t + \alpha_\omega) \quad [75]$$

where ω_n is the nutation frequency and α_ω is the initial phase. The curve traced by the tip of this two-dimensional vector is circular if $Q_c = \pm P_s$, and elliptical otherwise. It should be kept in mind that Γ_1, Γ_2, P_s, Q_c (and other similar quantities introduced below) refer to a particular spectral component, though we have avoided adding an index ω to keep the notation simpler.

The Earth's response to the gravitational tidal forcing is not strictly in phase with the forcing, for a variety of reasons: anelasticity of the mantle causes the tidal deformation of the solid Earth to lag behind, the ocean tide raised by the potential, and hence the deformation produced by ocean loading, are very much out of phase with the forcing, and so on; and these effects affect the rotational Earth's rotational response to the gravitational torque. Consequently, the Cartesian components $X(t)$, $Y(t)$ of the nutation, for example, will not be in phase with the torque; $X(t)$ and $Y(t)$ will each have both cosine and sine terms:

$$\begin{pmatrix} X(t) \\ Y(t) \end{pmatrix} = \begin{pmatrix} X_s \sin \Xi + X_c \cos \Xi \\ Y_c \cos \Xi + Y_s \sin \Xi \end{pmatrix} \quad [76]$$

X_s and Y_c belong to the part of the nutation that is 'in phase' with the torque, while X_c and Y_s pertain to the 'out-of-phase' part.

In most of the existing literature, the expansions into cosine and sine terms of various frequencies are written for $\Delta\psi$ and $\Delta\epsilon$ rather than for X and Y, and the coefficients in those expansions are called the 'coefficients of nutation'; thus $X_s/\sin\epsilon$ and Y_c are the in phase coefficients of nutation, and $X_c/\sin\epsilon$ and Y_s are the out-of-phase ones.

We can now resolve the two-vectors with components (X, Y) into parts represented by the following two-vectors each of which executes a prograde or retrograde circular motion:

$$\begin{pmatrix} \sin\Xi \\ -\cos\Xi \end{pmatrix}, \quad \begin{pmatrix} -\sin\Xi \\ -\cos\Xi \end{pmatrix}, \quad \begin{pmatrix} \cos\Xi \\ \sin\Xi \end{pmatrix}, \quad \begin{pmatrix} \cos\Xi \\ -\sin\Xi \end{pmatrix} \quad [77]$$

A little reflection shows that if the $\Xi(t)$ is increasing with time, that is, if the frequency ω_n is positive, the first and third of the above two-vectors rotate in the sense from the positive direction of the first axis of the equatorial reference frame toward that of the second axis. This is the same sense in which the Earth rotates, and is therefore said to be 'prograde'. The second and fourth two-vectors rotate in the opposite or 'retrograde' sense. The roles are reversed if ω_n is negative.

Expanding in terms of these prograde and retrograde vectors, we now have

$$\begin{pmatrix} X \\ Y \end{pmatrix} = \eta^{+,\text{ip}} \begin{pmatrix} \sin\Xi \\ -\cos\Xi \end{pmatrix} + \eta^{-,\text{ip}} \begin{pmatrix} -\sin\Xi \\ -\cos\Xi \end{pmatrix}$$
$$+ \eta^{+,\text{op}} \begin{pmatrix} \cos\Xi \\ \sin\Xi \end{pmatrix} + \eta^{-,\text{op}} \begin{pmatrix} \cos\Xi \\ -\sin\Xi \end{pmatrix} \quad [78]$$

wherein the parts which come from the in-phase and out-of-phase parts are identified by the superscripts 'ip' and 'op', respectively. The relations connecting the 'amplitudes' $X^{\pm,\text{ip}}$ of the in-phase parts of the nutation and $X^{\pm,\text{op}}$ of the out-of-phase parts to the 'coefficients of nutation' present in eqn [76] may be immediately written down. We have

$$\eta^{+,\text{ip}} = \frac{1}{2}(X_s - Y_c), \quad \eta^{-,\text{ip}} = -\frac{1}{2}(X_s + Y_c)$$
$$\eta^{+,\text{op}} = \frac{1}{2}(X_c + Y_s), \quad \eta^{-,\text{op}} = \frac{1}{2}(X_c - Y_s) \quad [79]$$

Recall now that if $\omega_n > 0$, the prograde components of the motion are those with the coefficients $\eta^{+,\text{ip}}$ and $\eta^{+,\text{op}}$ and the retrograde ones have the coefficients $\eta^{-,\text{ip}}$ and $\eta^{+,\text{op}}$, while the connections are reversed if $\omega_n < 0$. So, if we identify the prograde and retrograde amplitudes by superscripts p and r, respectively, in the place of $\pm$, we see that for $\omega_n = \pm|\omega_n|$,

$\eta^{p,\text{ip}} = \eta^{\pm,\text{ip}}$ and $\eta^{p,\text{op}} = \eta^{\pm,\text{op}}$ while $\eta^{r,\text{ip}} = \eta^{\mp,\text{ip}}$ and $\eta^{r,\text{op}} = \eta^{\mp,\text{op}}$. These relations may be written, with the use of eqns [79], as the following expressions for the prograde and retrograde amplitudes in terms of the coefficients of nutation:

$$\eta^{p,\text{ip}} = \frac{1}{2}\left(\frac{\omega_n}{|\omega_n|}X_s - Y_c\right)$$
$$\eta^{r,\text{ip}} = -\frac{1}{2}\left(\frac{\omega_n}{|\omega_n|}X_s + Y_c\right)$$
$$\eta^{p,\text{op}} = \frac{1}{2}\left(X_c + \frac{\omega_n}{|\omega_n|}Y_s\right) \quad [80]$$
$$\eta^{r,\text{op}} = \frac{1}{2}\left(X_c - \frac{\omega_n}{|\omega_n|}Y_s\right)$$

Conversely,

$$X_s = \frac{\omega_n}{|\omega_n|}\left(\eta^{p,\text{ip}} - \eta^{r,\text{ip}}\right), \quad Y_c = -\left(\eta^{p,\text{ip}} + \eta^{r,\text{ip}}\right)$$
$$X_c = \left(\eta^{p,\text{op}} + \eta^{r,\text{op}}\right), \quad Y_s = \frac{\omega_n}{|\omega_n|}\left(\eta^{p,\text{op}} - \eta^{r,\text{op}}\right) \quad [81]$$

In the literature, the above relations are practically always written in terms of $\Delta\psi$ and $\Delta\epsilon$ rather than in terms of X and Y. The transition from the relations obtained above is accomplished by the replacements

$$X_s \to A\sin\epsilon, \quad Y_c \to B$$
$$X_c \to A''\sin\epsilon, \quad Y_s \to B'' \quad [82]$$

In these relations, ϵ is treated as a constant equal to ϵ_0 (i.e., neglecting the small increments to ϵ_0 from nutation in obliquity).

Complex combinations like $\tilde{\Gamma} = \Gamma_1 + i\Gamma_2$ and $X + iY$ are of considerable interest in the context of nutations. (We have already employed such combinations to advantage in Section 10.3.2.2) It may be verified, by applying the developments presented above, that $X + iY$ may be written as

$$X + iY = \tilde{\eta}^{\text{P}}e^{i(\omega_n/|\omega_n|)\Xi} + \tilde{\eta}^{\text{r}}e^{-i(\omega_n/|\omega_n|)\Xi} \quad [83]$$

with

$$\tilde{\eta}^{\text{P}} = \eta^{p,\text{ip}} + i\eta^{p,\text{op}}, \quad \tilde{\eta}^{r} = \eta^{r,\text{ip}} + i\eta^{r,\text{op}} \quad [84]$$

$\tilde{\eta}^{p}$ and $\tilde{\eta}^{r}$ are the complex amplitudes of the prograde and retrograde components of the nutation with the in phase coefficients X_s, Y_c and out-of-phase coefficients X_c, Y_s as in [76].

It is worth keeping in mind for future reference that whichever term has a positive value for the frequency (i.e., for the coefficient of '$i t$' in the exponent) represents a prograde circular motion, and that any term with a negative value for this coefficient represents retrograde motion. In the first term of eqn [83] the coefficient of 'it' in the exponent is $(\omega/|\omega|)\omega = |\omega|$ and in the second term it is $-(\omega/|\omega|)\omega = -|\omega|$; that is why these two terms are prograde and retrograde, respectively, as indicated by the superscripts p and r.

10.6.3 Nutation Series

Up to this point, we have been focusing on one spectral component of the nutational motion. To represent the total motion, we have to add to the quantities appearing in the expressions on the right-hand side of eqn [76] an index n labeling the spectral terms of the various frequencies $\omega_n = d\Xi_n/dt$ present in the nutation, and then sum over all the terms. The resulting series provide the spectral representation of $X(t)$ and $Y(t)$. The series for $\Delta\psi$ is $1/\sin\epsilon$ times that of $X(t)$, while the series for $\Delta\epsilon$ is the same as that of $Y(t)$. However, the series that are thus obtained are not quite complete. A reason is that the spectral components of the torque are not strictly harmonic as would appear from the expressions in [74]; the amplitudes have a slow variation with time, which may be taken to be linear. Moreover, the fundamental arguments of nutation (and hence the argument $\Xi(t)$ which appears in the torque terms) are not strictly linear in time, which is to say that the nutation frequencies also are varying, though very slowly. The combined effect of the slow variations of the amplitudes and the arguments is that the amplitudes of the spectral terms in the nutation series are not constants. The time dependence of the amplitudes is adequately taken into account in the classical nutation series for $\Delta\psi$ and $\Delta\epsilon$ through a linear dependence on time. With this enhancement, the series referred to in the last paragraph take the following form:

$$\Delta\psi(t) = \sum_n [(A_n + A'_n t)\sin\Xi_n(t) + A''_n \cos\Xi_n(t)]$$
$$\Delta\epsilon(t) = \sum_n [(B_n + B'_n t)\cos\Xi_n(t) + B''_n \sin\Xi_n(t)] \quad [85]$$

Nutation series have the same general form as above irrespective of whether they are for the figure axis, rotation axis, or the angular momentum axis; but the coefficients of nutation differ from one to the other.

Nutations have periods extending from thousands of years down to fractions of a day. The largest of them are the 18.6 years, 9.3 years, annual, semiannual, and 13.66 day nutations. The in-phase coefficients A_n and B_n of the nutations of the figure axis are shown in **Tables 1** and **2** for these five periods. The values are from recent computations, and are given both for a

Table 1 Coefficients in longitude and obliquity of prominent nutations (in mas)

Period[a]	Rigid Earth		Nonrigid Earth	
	Longitude	Obliquity	Longitude	Obliquity
−18.6 years	−17281	9228	−17208	9205
−9.3 years	209	−90	207	−90
1 year	126	0	148	7
0.5 year	−1277	553	−1317	573
13.66 days	−222	95	−228	98

[a]The periods shown are the historically assigned values, a few of which are negative.

Table 2 Amplitudes of prograde and retrograde circular nutations (in mas)

Period[a]	Rigid Earth		Nonrigid Earth	
	Retrograde	Prograde	Retrograde	Prograde
18.6 years	−8051	−1177	−8025	−1180
9.3 years	87	4	86	4
1 year	−25	25	−33	26
0.5 year	−23	−531	−25	−548
13.66 days	−3	−92	−4	−94

[a]The periods shown are for the prograde nutations. The retrograde nutations have the corresponding negative values for their periods.

nonrigid Earth model and for a hypothetical rigid Earth. The entries in the table bring out the fact that the nutations of the figure axis are significantly affected by nonrigidity of the Earth, especially at some frequencies; in particular, the amplitude of the retrograde annual nutation of the nonrigid Earth is over 30% higher than that of the rigid Earth, while that of the prograde annual nutation is hardly affected by nonrigidity. The reason for the differing levels of sensitivity of different spectral components is that certain normal modes and associated resonances (especially the FCN) exist and affect the Earth's response to the torque in a frequency-dependent fashion in the case of the nonrigid Earth but not in the rigid case, as will be seen from the theory presented in later sections (see Sections 10.9 and 10.10).

10.7 Kinematical Relations between the Nutation of the Figure Axis and the Wobble

Consider the motion of the Earth's figure axis relative to the equatorial celestial reference frame of J2000 (see Section 10.2.2). Recall that the coordinates of the pole of this axis in this frame at an instant t are $X(t)$, $Y(t)$. The Z-coordinate of the pole is of no interest here.

The motion of the pole in the interval $(t, t + dt)$ changes its coordinates by $(dX, dY, 0)$. We transform this infinitesimal vector to the terrestrial reference frame now, in order to establish its relation to the components of the wobble. We seek to obtain the transformed components only to the first order in small quantities. For this limited purpose, we may take the transformation matrix to the zeroth order, meaning that we ignore the first-order offsets between the Z-axis of the celestial frame on the one hand, and the figure and rotation axes of the Earth on the other, and also ignore the variations in the spin rate. In this approximation, the equators of the celestial and terrestrial frames coincide, and the transformation from the former to the latter is simply a uniform rotation with angular velocity Ω_0 about the Z-axis of the celestial frame. So the components of the infinitesimal displacement vector of the figure axis pole in the terrestrial frame are

$$dx = dX \cos \Omega_0(t - t_0) + dY \sin \Omega_0(t - t_0)$$
$$dy = -dX \sin \Omega_0(t - t_0) + dY \cos \Omega_0(t - t_0) \qquad [86]$$

where t_0 is the instant of coincidence of the first axes of the two frames. The errors in these expressions are of the second order in the neglected small quantities.

The displacement (dx, dy) of the figure axis pole in the terrestrial frame is, however, determined directly by the angular velocity vector. The Earth's rotation with angular velocity $\mathbf{\Omega}$ has the effect that any mantle fixed vector $\mathbf{V}$ gets displaced by an amount $(\mathbf{\Omega} \times \mathbf{V})dt$ in an infinitesimal time interval dt. In particular, the displacement of the the z-axis during the interval $(t, t + dt)$ is $(\mathbf{\Omega} \times \hat{\mathbf{z}})dt$, with components $(\Omega_y, -\Omega_x, 0)dt = \Omega_0(m_y, -m_x, 0)dt$ relative to the terrestrial frame of the instant t. Since the z-axis is the figure axis, we have thus

$$dx = \Omega_0 m_y dt, \quad dy = -\Omega_0 m_x dt \qquad [87]$$

We obtain the relations that we seek on equating the expressions [86] and [87]:

$$\Omega_0 m_x = \dot{X} \sin \Omega_0(t - t_0) - \dot{Y} \cos \Omega_0(t - t_0)$$
$$\Omega_0 m_y = \dot{X} \cos \Omega_0(t - t_0) + \dot{Y} \sin \Omega_0(t - t_0) \qquad [88]$$

The familiar complex combination of this pair of equations now yields

$$\Omega_0 \bar{m} = i(\dot{X} + i\dot{Y})e^{-i\Omega_0(t - t_0)} = i\frac{d\bar{\eta}}{dt}e^{-i\Omega_0(t - t_0)} \qquad [89]$$

where

$$\bar{\eta} = X + iY \qquad [90]$$

It is important to note that no dynamical considerations have been invoked while deriving the result (89). Therefore it is valid for rigid and nonrigid Earth models alike. Since it is of purely kinematical origin, it is commonly referred to as the *kinematical relation*.

10.7.1   Kinematic Relations in the Frequency Domain

For a circular nutation of some frequency ω_n (using the subscript n here for nutation),

$$X = \bar{\eta}(\omega_n)\cos(\omega_n t + \chi_n)$$
$$Y = \bar{\eta}(\omega_n)\sin(\omega_n t + \chi_n) \qquad [91]$$
$$X + iY = \bar{\eta} = \bar{\eta}(\omega_n)e^{i(\omega_n t + \chi_n)}$$

where χ_n is a phase which depends on the frequency ω_n. On introducing this expression into the first of the equalities in [89], it becomes obvious that the form of $\bar{m}$ is given by

$$\bar{m} = \bar{m}(\omega_w)e^{i(\omega_w t + \chi_w)} \qquad [92]$$

with

$$\bar{m}(\omega_w) = -\frac{\omega_n}{\Omega_0}\bar{\eta}(\omega_n), \quad \omega_w = \omega_n - \Omega_0 \qquad [93]$$

The subscript w labels quantities pertaining to the wobble. These equations constitute the kinematic relations in the spectral domain; they correspond to the kinematical relation [89] in the time domain.

An additional relation, which is not usually written down explicitly, is obtained on considering the zero frequency part of $\dot{X}$ which is $\dot{\psi}_A \sin \epsilon$ where $\dot{\psi}_A$ is the rate of precession in longitude. It is evident from [89] that the corresponding spectral component of $\bar{m}$ has frequency $(-\Omega_0)$. It follows then that

$$\Omega_0 \bar{m}(-\Omega_0) = i\dot{\psi}_A \sin\epsilon - \dot{\epsilon}_A \qquad [94]$$

With ϵ approximated by ϵ_0, the above equation expresses the precession rate $\dot{\psi}_A$ in terms of the wobble amplitude at the frequency $-\Omega_0$ (-1 cpsd) as $-i(\Omega_0 / \sin \epsilon_0)\bar{m}(-\Omega_0)$.

Another relation that is of importance concerns polar motion. We recall from Section 10.2.5.1 that the motion of the CIP in the celestial frame involves only frequencies in the low-frequency bands. This means that for all other frequencies, the CIP coincides with the pole of the celestial frame. But these are the frequencies, after transforming to the terrestrial frame, that are contained in polar motion. It follows that polar motion is effectively the motion of the pole of the celestial frame relative to the terrestrial frame. But the reverse of this motion, that is, the motion of the pole of the terrestrial frame (which is the pole of the figure axis) in the celestial frame, is the nutation $\tilde{\eta}$ of the figure axis. Therefore, the spectral component of frequency ω_p of the polar motion as defined by the IAU Conventions may be obtained simply by transforming the negative of the corresponding spectral component of the nutation of the figure axis from eqn [91] to the terrestrial frame:

$$\tilde{p}_{\omega_p}(t) = -\tilde{\eta}(\omega_n)e^{i\omega_n t + \chi_n}e^{-i\Omega_0 t}$$

$$\omega_p = \omega_w = \omega_n - \Omega_0$$

[95]

On using the kinematic relation [93] between $\tilde{\eta}$ and $\tilde{m}$, we finally obtain the relation between the amplitudes of the polar motion and the corresponding wobble:

$$\tilde{p}(\omega_p) = -\tilde{\eta}(\omega_n) = \frac{\Omega_0}{\omega_n}\tilde{m}(\omega_w)$$

[96]

Clearly, a polar motion of the CIP, in the frequency domain in which it is defined, is not identical to the wobble of the same frequency.

10.7.2   Implications of the Kinematic Relations

Now we examine briefly the implications of the kinematic relations.

The first and foremost is that these relations provide a formal proof of the existence of a circular nutation associated with every circular wobble and vice versa; the frequency of the nutation is algebraically greater than that of the associated wobble by 1 cpsd. This important general relation is simply a consequence of the Earth's diurnal rotation in space with angular velocity Ω_0. This relationship was already seen in a qualitative way earlier: it was noted in Section 10.4.2 that the spectrum of the torque in the terrestrial frame is in the retrograde diurnal band, and it was shown explicitly in Section 10.5.2 that the spectrum in the celestial frame is in the low-frequency band.

The amplitude relation in [93] shows that the amplitude $\tilde{\eta}(\omega_n)$ of the nutation contains a factor $1/\omega_n$ which simulates a normal mode at $\omega_n = 0$. Unlike a proper normal mode, it has no connection to any property of the rotating Earth. It has been referred to in the literature as the 'tilt-over mode'

(TOM), following Smith (1974) and Wahr (1981). The resonance factor $1/\omega_n$ produces a great enhancement of the nutation amplitude compared to the associated wobble amplitude when the frequency is very small, their ratio $(-\Omega_0/\omega_n)$ being, apart from a minus sign, the period of the nutation in sidereal days. Thus, the amplitude of the retrograde 18.6-year nutation (period $= -6798$ solar days $= -6816$ sidereal days) is 6816 times as large as that of the corresponding retrograde diurnal wobble. There is no such resonant kinematic factor in other frequency bands, and this is a major reason for the overriding importance of the low frequency band in nutations. As far as other normal modes are concerned, each wobble normal mode and the resonance in the wobble amplitude that is associated with it have their counterparts in the nutation amplitude, of course with a shift in frequencies by 1 cpsd.

As an immediate application of the kinematic relations, we use the amplitude relation in [93] to arrive at the amplitude $\tilde{\eta}(\omega_n)$ of the nutation of the 'rigid Earth', that is produced by a torque $\tilde{\Gamma}(\sigma)$ having frequency σ cpsd in the terrestrial frame. Introducing the wobble amplitude [19] of the rigid Earth into this relation, we obtain

$$\tilde{\eta}_R(\omega_n) = i\frac{\tilde{\Gamma}(\omega_w)}{A(\omega_w - \Omega_0 e)(\Omega_0 + \omega_w)}$$

[97]

As for the precession rate, one finds from [19] at the particular frequency $\omega_w = -\Omega_0$, together with [94], that

$$\dot{\psi}_A = \frac{\tilde{\Gamma}(-\Omega_0)}{A\Omega_0^2(1 + e)}$$

[98]

10.7.3   Transfer Function

Consider two Earth models, one rigid and the other nonrigid, both of which are forced by the same torque $\tilde{\Gamma}(\omega_w)$. (In the simplest case when both are axially symmetric and ellipsoidal, this means that both must have the same value for the dynamical ellipticity e.) Let the wobbles of the rigid and nonrigid Earth models be $\tilde{m}_R(\omega_w)$ and $\tilde{m}(\omega_w)$, respectively. The former may be taken from $\tilde{m}$ of eqn [19] of Section 10.3.2.2 which is for the rigid Earth. Both $\tilde{m}_R$ and $\tilde{m}$ are proportional to $\tilde{\Gamma}(\omega_w)$, and therefore the ratio $\tilde{m}(\omega_w)/\tilde{m}_R(\omega_w)$ is independent of the torque. Moreover, as was observed in the last section, the kinematical relation [89] and its frequency domain version [93] are valid for any type of Earth model, rigid or nonrigid. Therefore, it follows from the latter equation that

$$T_n(\omega_n) \equiv \frac{\bar{\eta}(\omega_n)}{\bar{\eta}_R(\omega_n)} = \frac{\bar{m}(\omega_w)}{\bar{m}_R(\omega_w)} \equiv T_w(\omega_w) \quad [99]$$

These ratios, which are independent of the forcing torque as noted above, are the 'transfer functions' for nutation and wobble; the two are equal. It is useful to write these relations also with the alternative notation for frequencies:

$$T_n(\nu) \equiv \frac{\bar{\eta}(\nu)}{\bar{\eta}_R(\nu)} = \frac{\bar{m}(\sigma)}{\bar{m}_R(\sigma)} \equiv T_w(\sigma) \quad [100]$$

where ν and σ are the frequencies of the nutation and the wobble, respectively, expressed in cpsd:

$$\nu = \omega_n/\Omega_0, \quad \sigma = \omega_w/\Omega_0, \quad \nu = \sigma + 1 \quad [101]$$

the last equality being one of the kinematic relations.

The transfer functions are of great utility for the following reason. If one solves for wobble amplitude $\bar{m}(\sigma)$ of the Earth model of interest by solving the relevant equations of motion with inputs from the Earth structure pertaining to that model, one can compute the transfer function $T_w(\sigma)$ (which is independent of the amplitude of the forcing), by dividing the amplitude by $\bar{m}_R(\sigma)$ as given by [19] or [55]. Since $T_n(\nu) = T_w(\sigma)$, we can immediately obtain the nutation amplitudes of the chosen Earth model as

$$\bar{\eta}(\nu) = T_w(\sigma)\bar{\eta}_R(\nu) = -T_w(\sigma)\bar{m}_R(\sigma)/(1+\sigma) \quad [102]$$

where the last step makes use of the kinematic relations. Of course, $\bar{m}_R(\sigma)$ may be readily calculated using the tide tables, as explained in eqn [55].

Alternatively, $\bar{\eta}_R(\nu)$ in the last equation may be taken directly from tables of rigid Earth nutation amplitudes. Nutations of the rigid Earth have been studied extensively, and highly accurate results are available for the rigid Earth nutation amplitudes. The important point is that the computation of the torques exerted by the solar system bodies, starting with their time dependent positions as given by the ephemerides, need to be done only once, at the stage of working out the rigid Earth amplitudes. The use of the transfer functions enables one to avoid repeating this exercise for each nonrigid Earth model.

10.7.4 Relations Connecting the Nutations of Different Axes: Oppolzer Terms

The nutation of the rotation axis can be related to that of the figure axis with the aid of the kinematic relation [89], as we shall now show. This relation, like the kinematical relation, holds good whether the Earth is rigid or nonrigid.

By definition, the temporal variations of the vectors from the pole of the Z-axis of the celestial frame to the poles of the figure axis and the rotation axis, as seen from the celestial frame, constitute the nutations of these two axes. We denote the complex combinations of the (equatorial) components of the pole of the figure axis in the celestial frame by $(X + iY)$, and that of the rotation axis by $(X_R + iY_R)$. (Caution: In this section, R does not stand for 'rigid' as elsewhere, but for 'rotation'.) Now, it is evident that the difference $(X_R + iY_R) - (X + iY)$ gives the complex representation of the components (in the celestial frame) of the vector from the figure axis pole to the rotation pole. It is the variation of this vector, as seen in the 'terrestrial' frame, that constitutes the wobble, represented by $\bar{m}$; its components in the celestial frame will then be represented by $\bar{m}e^{i\Omega_0(t-t_0)}$, under the same approximations as in the second paragraph of Section 10.7. Consequently, we have

$$(X_R + iY_R) - (X + iY) = \bar{m}e^{i\Omega_0(t-t_0)} \quad [103]$$

We can now use the kinematic relation [89] to eliminate $\bar{m}$ from the above equation. The result is the relation that we seek

$$X_R + iY_R = (X + iY) + \frac{i}{\Omega_0}(\dot{X} + i\dot{Y}) \quad [104]$$

For a spectral component of the nutation with frequency ω_n, one writes

$$\begin{aligned} X + iY &= \bar{\eta}(\omega_n)e^{i(\omega_n t + \alpha)} \\ X_R + iY_R &= \bar{\eta}_R(\omega_n)e^{i(\omega_n t + \alpha)} \end{aligned} \quad [105]$$

and the foregoing relation becomes

$$\bar{\eta}_R(\omega_n) = \left(1 - \frac{\omega_n}{\Omega_0}\right)\bar{\eta}(\omega_n) \quad [106]$$

We see that the fractional difference between the nutation amplitudes of the rotation axis and the figure axis is the frequency of the nutation expressed in cpsd.

It is of interest to note that the precession of the figure axis in longitude, which is represented by $\dot{\psi}_A t$, leads to the precession part $\dot{\psi}_A(t + i/\Omega_0)$ in $(X_R + iY_R)$. Thus, one finds that the rotation pole has, expectedly, the same rate of precession in longitude as the pole of the figure axis; but the rotation pole is offset by a constant amount $(\dot{\psi}_A/\Omega_0)$ in obliquity.

One may ask whether the nutation of the angular momentum axis is also related in a simple manner to the nutations of the other two axes. The answer is in the negative, in general. The reason is that the relation of the angular momentum vector to the rotation vector depends very much on the structure of the Earth and so the relation between the nutations of the angular momentum and rotation axes differs from one Earth model to another. The particular case of a rigid Earth is of considerable interest, however, because classical treatments of the rotation variations of the rigid Earth have had, as their primary output, the nutations of the angular momentum axis; the nutations of the figure and rotation axes were then inferred. The relations used for this last step will now be derived, taking the Earth to be a rigid axially symmetric ellipsoid. The components of the angular momentum vector in a principal axis frame are then $H_x = A\Omega_0 m_x$, $H_y = A\Omega_0 m_y$, $H_z = C\Omega_0(1 + m_z)$, and the equatorial components of the unit vector along **H** are $(A/C)\, m_x$, $(A/C)m_y$ to the lowest order in m_x, m_y, m_z. These are the components of the vector offset of the pole of the angular momentum axis from that of the figure axis (the z-axis); they are simply (A/C) times the corresponding components (m_x, m_y) for the pole of the rotation axis. The scalar factor (A/C) is of course independent of whether the offset vectors of the angular momentum axis and the rotation axis are viewed from the terrestrial or the celestial frame. Therefore, we have

$$(X_H + iY_H) - (X + iY) = (A/C)[(X_R + iY_R) - (X + iY)]$$
[107]

This means that when considering the offset of the angular momentum axis, a relation of the form [103] holds with $\bar{m}$ replaced by $(A/C)\bar{m} = \bar{m}/(1 + e)$. Consequently we have, instead of [104], the relation

$$X_H + iY_H = (X + iY) + \frac{i}{(1 + e)\Omega_0}(\dot{X} + i\dot{Y})$$
[108]

The corresponding relation for the amplitudes of the respective spectral components of frequency ω_n is

$$\tilde{\eta}_H(\omega_n) = \left(1 - \frac{\omega_n}{(1 + e)\Omega_0}\right)\tilde{\eta}(\omega_n)$$
[109]

Therefore,

$$\tilde{\eta}(\omega_n) - \tilde{\eta}_H(\omega_n) = \frac{\omega_n}{(1 + e)\Omega_0 - \omega_n}\tilde{\eta}_H$$
[110]

This is called the 'Oppolzer term for the figure axis'. It enables one to compute the amplitude of the nutation of the figure axis, given that of the angular momentum axis. The 'Oppolzer term for the rotation axis' can be obtained by combining the above equation with [106]. The result is that

$$\tilde{\eta}_R(\omega_n) - \tilde{\eta}_H(\omega_n) = -\frac{e\omega_n}{(1 + e)\Omega_0 - \omega_n}\tilde{\eta}_H$$
[111]

It must be kept in mind that the expressions for the Oppolzer terms [110] and [111] are valid only for the rigid Earth model considered, unlike the kinematic relation and the relation [106] between $\tilde{\eta}_R$ and $\tilde{\eta}$ which are independent of the Earth model.

Traditionally, nutation tables for the rigid Earth have been presented as tables of the coefficients in the spectral expansions of $\Delta\psi_H(t)$ and $\Delta\epsilon_H(t)$ in terms of real simple harmonic (cosine and sine) functions of time, together with the corresponding coefficients for the Oppolzer terms. In the case of nonrigid Earth models, tables are presented, as a rule, for the nutations of the figure axis.

10.8   Rigid Earth Nutation

By 'rigid Earth' we mean a rigid Earth model having the same density function $\rho(\mathbf{r})$ as for the actual Earth. It has therefore the same principal moments of inertia, and other moments of higher degree of the matter disribution, as the actual Earth. These parameters are determined from the spatial structure of the Earth's own external gravitational field, inferred from its influence on the orbits of low-flying Earth satellites observed by satellite laser ranging (SLR). The determination of the rotation variations should then, in principle, be a simple matter, if the time dependence of the gravitational potentials of the Moon, Sun, and the planets at the Earth is known. This dependence is determined by the orbital motions of the Moon around the Earth and of the Earth and other solar system bodies around the Sun. The orbits deviate from simple Keplerian ellipses for a variety of reasons:

1. Perturbation of the lunar orbit around the Earth by the attraction of the Moon by the Sun, and of the Earth's orbit around the Sun by the Moon's gravitation (the so-called three-body effects);

2. additional perturbations of these orbits by the planets (resulting in the 'indirect planetary effect' on Earth rotation), and also the so-called planetary-tilt effect on the lunar orbit; and

3. the 'J_2-tilt effect' due to the perturbation of the Moon's orbit by the noncentral part of the Earth's gravitational potential that arises from the Earth's ellipticity.

All these orbit perturbations are reflected in the lunar and solar gravitational potentials at the Earth and hence in their torques on the Earth. Additional torques arise from the direct action of the planets on the ellipticity of the Earth, referred to as the 'direct planetary effect'.

The ELP2000 ephemerides provide the time-dependent position of the Moon in analytical form, and VSOP2000 presents similar information for the Sun and the planets. These ephemerides are recent versions of the ELP and VSOP series of ephemerides beginning with Chapront-Touzé and Chapront (1983) and Bretagnon and Francou's (1988), respectively. The ephemerides of the Sun and the Moon include the perturbations mentioned above.

Treatments of the rotation variations of the rigid Earth fall into two broad classes: those based on the Hamiltonian formulation of mechanics, and those making use of the torque equation governing rotational motion.

The Hamiltonian approach begins generally with a treatment of the 'main problem' where the influences of the Moon and Sun on the Earth's rotation are considered, taking into account the Sun's perturbation of the lunar orbit and the Moon's perturbation of the Earth–Sun orbital motion, and in recent treatments, also the planetary perturbations of both these orbits. The direct planetary torques on the Earth are treated separately, as are other second-order effects like the 'crossed-nutation coupling', a term which refers to the change in the torque on the Earth due to the the changes in the orientation of the Earth itself (due to precession-nutation) and the consequent increments to these motions.

The torque approach makes direct use of the ephemerides which already take account of all the orbit perturbations. In either approach, the Earth parameter which plays the dominant role in the gravitational torque on the Earth is the dynamical ellipticity, but triaxiality and the higher-degree geopotential coefficients which play lesser roles are also considered. In addition, second-order increments to the gravitational torque exerted by the Moon and Sun due to changes in Earth's orientation caused by nutation and precession are also taken into account in recent rigid Earth nutation theories.

10.8.1 Hamiltonian Approach

In the Hamiltonian approach, the starting point is the Hamiltonian K of the rotating Earth. (The standard notation for the Hamiltonian is H; it has been changed here because of the use of H for another quantity in the present context.) K is the sum of the Earth's rotational kinetic energy and the potential energy of its gravitational interaction with the Moon and Sun. (The relatively small interactions with the planets are dealt with separately.) Recent computations of the Earth's precession and nutation by Souchay *et al.* (1999) are the culmination of developments starting from the work of Kinoshita (1977) based on the Hamiltonian approach. The presentation below is a very brief overview of that work.

The Hamiltonian of a system is a function of a set of coordinate variables q_i and corresponding momentum variables p_i which, taken together, describe the state of the system at any given time; (q_i, p_i) for any given i is said to constitute a canonically conjugate pair. Given the expression $K(q, p)$ for the Hamiltonian, the equations of motion are given by

$$\dot{q}_i = \frac{\partial K}{\partial p_i}, \quad \dot{p}_i = -\frac{\partial K}{\partial q_i} \qquad [112]$$

for all i.

The Hamiltonian for the rotational motion of the Earth is

$$K = \sum_i \frac{L_x^2}{2A} + \frac{L_y^2}{2B} + \frac{L_z^2}{2C} + M_B W_E(r_B, \theta_B, \lambda_B) \qquad [113]$$

wherein the kinetic energy involves the components of the angular momentum vector in the terrestrial frame (which are denoted here by L_x, L_y, L_z in a departure from earlier notation), and the potential energy is expressed through the potential W_E of the Earth at the position of the body B. (To be specific, we take the body to be the Moon; the case of the Sun is entirely similar). Since θ_B and λ_B characterize the direction of the Moon in the terrestrial reference frame, they depend on the orientation of this frame in space, that is, in relation to an inertial reference frame. This orientation may be represented by the transformation which carries the latter frame to the former. It is customary, in Hamiltonian treatments of the rotation variations of the rigid Earth, to place the nutation and precession of its angular momentum axis in the foreground, and to infer from them subsequently the motions of the figure and rotation axes. Consequently, the inertial plane (which is the principal plane normal to the third axis of the inertial frame), the 'angular momentum plane' (or Andoyer plane) normal to the angular momentum vector, and the principal plane of the terrestrial frame (the equatorial plane), are all made use of in the transformation from the inertial frame to the terrestrial one. The transformation consists of (1) a rotation through an angle h about the third axis of the inertial frame to bring its first axis into a new position along the nodal line (intersection) of the inertial and angular momentum planes; (2) a rotation through an angle I about this new axis to carry the third axis of the inertial frame into the direction of the angular momentum vector; (3) a rotation about this vector through an angle g to take the first axis from the above-mentioned nodal line to the node of the angular momentum plane and the equatorial plane, (4) a rotation through angle $\mathcal{J}$ about this node to carry the third axis over from the angular momentum axis to the third axis of the terrestrial frame, and (5) a rotation through an angle l about the last-mentioned axis to bring the first axis into alignment with the first axis (at 0 longitude) of the terrestrial frame. The rotation angles h, g, l are taken as the coordinate variables q_i, ($i = 1, 2, 3$) describing Earth orientation, and the angular momentum components H, G, L along the axes about which these rotations are taken as the conjugate momentum variables p_i for the purpose of the Hamiltonian formulation. (H and L are evidently the components of the angular momentum vector along the Z-axis of the inertial frame and the z-axis of the terrestrial frame, respectively, and G is the magnitude of the angular momentum vector.) These three pairs of canonically conjugate variables are called the Andoyer variables (Andoyer, 1923). The two other quantities I and $\mathcal{J}$ are not independent of these: $\cos I = H/G$, $\cos \mathcal{J} = L/G$.

The nutation and precession of the angular momentum axis in longitude consist of the motions of this axis around the third axis of the inertial plane. This is precisely what variations Δh in the angle h describe. The nutation-precession in obliquity is represented by the variations ΔI in the inclination of the angular momentum axis to the inertial axis; this quantity may be expressed in terms of variations in the Andoyer variables H and G: $\Delta I = [(H/G)\Delta G - \Delta H]/(G \sin I)$. They may be determined, therefore, by solving the Hamiltonian equations for the Andoyer variables.

To set up the Hamiltonian equations, it is necessary to express the total energy given by [113] as a function of the Andoyer variables representing Earth rotation, on the one hand, and the position coordinates of the Moon with reference to the inertial frame, on the other. The orbital motion of the Moon makes its coordinates time dependent. The ephemerides (ELP2000 in the case of the Moon)

express the time dependence of each of the position coordinates as a spectral expansion in which the argument of each term is an integer linear combination of the Delaunay fundamental arguments and additional arguments referring to the motions of the planets (see Section 10.6.1); the planetary arguments enter because of the gravitational perturbations they produce on the lunar orbit. Expansions of the same kind are also provided for the functions of the coordinates which appear in the spherical harmonic expansion of the potential in the celestial frame. With the time dependence of the potential from the orbital motions of the solar system bodies thus made explicit, the Hamiltonian equations $\dot{h} = \partial K/\partial H$, $\dot{H} = -\partial K/\partial h$, and so on, yield the three pairs of time derivatives as functions of the six Andoyer variables and of the time variable appearing through the spectral expansions.

Kinoshita (1977) applied a method due to Hori (1966) to facilitate the solution of these equations by means of a transformation of the Andoyer variables and their functions, which preserves the form of Hamilton's equations while expressing them in terms of the new (transformed) variables. The transformation, which is carried out perturbatively, is designed to cause the new Hamiltonian to have only a secular variation with time, while all periodicities enter through the 'determining function' through which the transformation is accomplished. A detailed treatment of the transformation and the subsequent solution of the equations are beyond our scope. The interested reader may refer to Kinoshita (1977). For the inclusion of planetary influences, see Souchay and Kinoshita (1996, 1997). Various second-order effects are treated in later papers of Souchay, Kinoshita and collaborators (Kinoshita and Souchay, 1990, Souchay and Kinoshita, 1996, 1997, Folgueira *et al.*, 1998a, 1998b), and the complete results are presented by Souchay *et al.* (1999) as the REN-2000 nutation series. This is the rigid Earth series used by Mathews *et al.* (2002) for convolution with their transfer function in constructing the MHB2000 series on which the IAU2000A nutation series is based.

10.8.2 Torque Approach

This approach is based on the solution of the torque equation either in the celestial frame or in the terrestrial frame, introduced in Sections 10.3.1 and 10.3.2, respectively.

10.8.2.1 Axially symmetric Earth: Simplified treatment

The Earth's structure parameters (the principal moments of inertia and geopotential coefficients of higher orders) are constant in the terrestrial frame, and therefore the torque on the Earth is readily determined in this frame, as noted earlier. Transformation of the torque to the celestial frame is easy only if the Earth is taken to be axially symmetric and the effects of the Earth's nutation, precession, and wobble on the torque are ignored. We have seen in Section 10.4.5 that the structure parameters which appear in the torque on an axially symmetric Earth are the $\mathcal{J}_n$ (equal to $-C_{n0}$) and that the torque proportional to $\mathcal{J}_n$ are given by eqn [60] in the terrestrial frame. Under the approximations stated above, the corresponding torque in the celestial frame may be obtained simply by replacing θ_B and λ_B by $(\pi/2 - \delta_B)$ and α_B, respectively, where δ_B is the declination, and α_B the right ascension of the body B in the equatorial celestial reference frame. The reasoning is the same as in the special case $n = 2$ considered in Section 10.5.1 and will not be repeated here. The relations between the Cartesian coordinates used there and the polar coordinates δ and α are:

$$X_B = r_B \cos \delta_B \cos \alpha_B, \qquad Y_B = r_B \cos \delta_B \sin \alpha_B,$$
$$Z_B = r_B \sin \delta_B.$$

The time dependence of the torque may now be made explicit by using the ELP or VSOP ephemerides, as the case may be. The nutation of the angular momentum axis is then trivially obtained by integration of the torque. This approach has been used by Roosbeek and Dehant (1998); they identify the corrections that need to be applied to the nutation series obtained directly from the integration. Their final results constitute the RDAN nutation series. Elegant analytical derivations have been provided by Williams (1994) for the contributions to nutation and precession from various small effects.

10.8.2.2 Rigorous Treatment of the General Case

The transformation of the torque equations taking account of the full range of geopotential coefficients in the terrestrial frame (ITRF) to equivalent equations in the fixed Ecliptic Celestial Reference Frame of J2000 can be carried out rigorously, as was done by Bretagnon *et al.* (1997, 1998). (We denote this celestial frame by ECRF0, with 0 serving as a reminder that it is based on a fixed ecliptic.) The transformation is effected through a sequence of three rotations. The angles of the rotations, which these authors represent by ψ, ω, and φ, are the Euler angles of the

transformation. The first two rotations bring one over from the ecliptic plane to the equatorial plane while taking the first axis of ECRF0 over into the nodal line of the two planes; the rotation through φ is about the third axis of the ITRF, and takes the first axis from the nodal line to the first axis of the ITRF. If one desires that the final frame be the principal axis frame rather than the ITRF, one needs only to replace φ by $\bar{\varphi} = \varphi + \alpha$, where α is the longitude (in the ITRF) of the equatorial principal axis of the least moment of inertia A.

The variations of φ or $\bar{\varphi}$ are closely related to LOD variations. The variations of ψ and $(-\omega)$ as a result of the variable rotation of the Earth represent the nutation-precession in longitude and obliquity, respectively. The precession here is the so-called lunisolar precession reflecting the actual motions of the equator (or equivalently, of the figure axis) and differs from the general precession which is defined relative to the moving ecliptic and therefore includes an ecliptic motion term.

The torque equations in the principal axis frame are Euler's equations for the angular velocity components. The angular velocity components $\Omega_x, \Omega_y, \Omega_z$ in the principal axis frame, which Bretagnon *et al.* denote by (p, q, r), are related to $\dot{\psi}, \dot{\omega}, \dot{\varphi}$ through the (exact) kinematic relations

$$
\begin{aligned}
\Omega_x &= \dot{\psi} \sin \omega \sin \bar{\varphi} + \dot{\omega} \cos \bar{\varphi} \\
\Omega_y &= \dot{\psi} \sin \omega \cos \bar{\varphi} - \dot{\omega} \cos \bar{\varphi} \qquad [114] \\
\Omega_z &= \dot{\psi} \cos \omega + \dot{\varphi}
\end{aligned}
$$

On introducing these expressions for the angular velocity components into the Euler equations [11], one obtains a set of coupled second-order differential equations for the Euler angles, which involve small nonlinear terms arising from the small deviations of ω and $\dot{\varphi}$ from their constant mean values.

The torque components which appear on the right-hand side of the equations are in the principal axis frame. Their expressions may be obtained from the torque components referred to the ITRF, which are given in eqns [59] and [63]. For the complex combination of the equatorial components, all one needs to do is to multiply the former expression by $e^{i\alpha}$, where α is the longitude of the x-axis of the principal axis frame in the ITRF (see below eqns [62]). The axial component of the torque is the same in both the frames. Of course, the above expressions have to be summed over n and m to get the torque due to the full potential of the external body.

The time dependence of the torque produced by the Moon, Sun, or planet is obtainable by using the analytical expressions given in the relevant ephemerides for the position coordinates of that body. The coupled differential equations may then be solved to any desired accuracy (given the Earth's structure parameters to sufficient accuracy) by an iterative process which is needed because of the nonlinearties. Corrections for relativistic effects (Fukushima, 1991, Brumberg *et al.*, 1992, Soffel and Klioner, 1998, Klioner, 1998) have to be applied to the solutions thus obtained. The final nutation series constructed in this manner by Bretagnon *et al.* (1997, 1998) is named SMART97.

10.9 Axially Symmetric Ellipsoidal Nonrigid Earth: Torque Equations and Solutions

In the foregoing sections we considered the wobble and nutation in some simplified models, then established a number of basic results of general applicability, and briefly dealt with treatments of the nutations of the rigid Earth. We turn now to a study of more realistic models having one or more of the following features: deformability, existence of a fluid core and solid inner core, etc. The torque approach will be employed for these studies.

We observe at the outset that the basic equation of motion governing the rotation variations as seen from the surface of the Earth in relation to a terrestrial frame continues to be the torque equation [8] for the whole Earth; the angular velocity $\mathbf{\Omega}$ appearing in the equation is that of the mantle. An additional equation is needed to take account of the rotation variations of the fluid core which are coupled to, but not the same as, those of the mantle. Two more equations are needed when a solid inner core is present, as will be seen shortly.

10.9.1 Deformable wholly solid Earth

If the Earth were wholly solid, the only difference from the rigid case would be in the inertia tensor. The deformation from the direct action of the tidal potential, as well as indirect effects like the perturbations of the centrifugal resulting from Earth rotation variations, have the consequence that the inertia tensor [**C**] is no longer diagonal. So the angular momentum components become

$$\begin{pmatrix} H_x \\ H_y \\ H_z \end{pmatrix} = \begin{pmatrix} A+c_{11} & c_{12} & c_{13} \\ c_{21} & A+c_{22} & c_{23} \\ c_{31} & c_{32} & C+c_{33} \end{pmatrix} \begin{pmatrix} m_x\Omega_0 \\ m_y\Omega_0 \\ (1+m_z)\Omega_0 \end{pmatrix}$$

$$= \Omega_0 \begin{pmatrix} Am_x + c_{13} \\ Am_y + c_{23} \\ C(1+m_z) + c_{33} \end{pmatrix} \qquad [115]$$

with the neglect of second-order terms which involve the product of any of the c_{ij} with m_x or m_y or m_z. The components of $\mathbf{\Omega} \times \mathbf{H}$ may be obtained using the above. Its components, leaving out the second-order terms, are $Cm_y - (Am_y + c_{23})$, $-Cm_x + (Am_x + c_{13})$, 0). When these are substituted into the torque equation $d\mathbf{H}/dt + \mathbf{\Omega} \times \mathbf{H} = \mathbf{\Gamma}$, one obtains the following equations:

$$\Omega_0(A\dot{m}_x + \Omega_0\dot{c}_{13}) + (C-A)\Omega_0^2 m_y - \Omega_0^2 c_{23} = \Gamma_x$$
$$\Omega_0(A\dot{m}_y + \Omega_0\dot{c}_{23}) - (C-A)\Omega_0^2 m_x - \Omega_0^2 c_{23} = \Gamma_y \qquad [116]$$
$$C\Omega_0\dot{m}_z + \dot{c}_{33} = \Gamma_z$$

Now we take the complex combination of the first two equations, and denote $c_{13} + ic_{23}$ by $\tilde{c}_3$. Then, on using for the torque the expression [50] with $e' = 0$ (for axial symmetry), we obtain the following equation:

$$A\frac{d\tilde{m}}{dt} - ie\Omega_0 A\tilde{m} + \frac{d\tilde{c}_3}{dt} + i\Omega_0\tilde{c}_3 = -ie A\Omega_0\tilde{\phi} \qquad [117]$$

For a spectral component with frequency $\sigma\Omega_0$, we have then

$$A(\sigma - e)\tilde{m}(\sigma) + (\sigma + 1)\tilde{c}_3(\sigma) = -eA\tilde{\phi}(\sigma) \qquad [118]$$

In order to solve this equation for $\tilde{m}(\sigma)$, we need to know how $\tilde{c}_3$ is related to the wobble $\tilde{m}$ and to the external forcing $\tilde{\phi}$. One can show from deformation theory that the contribution to $\tilde{c}_3$ is $-A\kappa\tilde{\phi}$. Since the value of κ could be frequency dependent, as we shall see later, we take this relation to be in the frequency domain. The parameter κ is a measure of the deformability of the Earth under forcing by $\tilde{\phi}$, and is referred to as a compliance; it is closely related to the Love number k of the unperturbed Earth:

$$\kappa = \frac{\Omega_0^2 a^5}{3GA}k = \frac{ek}{k_f}, \quad k_f = \frac{3GAe}{\Omega_0^2 a^5} \qquad [119]$$

where a is the mean radius of the Earth and G is the constant of universal gravitation, $G = 6.67259 \ 10^{-11}$ $m^3 \ kg^{-1} \ s^{-2}$. (The Love number k is defined by the statement that the incremental gravitational produced at the surface of the Earth by the redistribution of matter within the Earth as a result of the action of the lunisolar potential is k times the lunisolar potential at the surface.) The so-called fluid Love number k_f plays the same role for a hypothetical wholly fluid Earth. Other Love numbers h and l relate the tide-induced vertical and horizontal displacements at the Earth's surface to the lunisolar potential. (In these cases, the displacement at the surface is h/g times the lunisolar potential.) There are also load Love numbers k', h', l' which pertain to the potential perturbation and the displacements caused by loading on the Earth's surface. The contribution from the incremental centrifugal potential (associated with the wobble $\tilde{m}$) to $\bar{c}_3$ is $A\kappa\tilde{m}$, for the simple reason that the difference between the centrifugal potentials associated with the perturbed and unperturbed angular velocities $\mathbf{\Omega}$ and $\mathbf{\Omega}_0$ is identical to the tesseral potential given by eqn [31], except for the replacement of ϕ_1, ϕ_2 by $-m_x$, $-m_y$. Thus,

$$\bar{c}_3 = A\kappa(\tilde{m} - \tilde{\phi}) \qquad [120]$$

On substituting this expression for $\bar{c}_3$ in [118], the equation becomes

$$[\sigma - e + (\sigma+1)\kappa]\tilde{m}(\sigma) = -[e - (\sigma+1)\kappa]\tilde{\phi}(\sigma) \qquad [121]$$

One sees, by setting $\tilde{\phi}(\sigma) = 0$, that the frequency $\sigma_E = e$ of the Eulerian free wobble mode of the rigid Earth is now replaced by $\sigma_1 = (e - \kappa)/(1 + \kappa)$. It turns out that κ is close to $e/3 \approx 0.001$, and so $\sigma_1 \approx (2/3)\sigma_E$, meaning that the period of the free wobble of the deformable Earth with no fluid regions is approximately 50% longer than that of the rigid Earth. As for the wobble response to forcing by $\tilde{\phi}(\sigma)$, the change from the rigid case consists in a replacement of e by the frequency-dependent effective value $e - (\sigma+1)\kappa$. It becomes clear on recalling that $(1+\sigma)$ lies within the low frequency band, that the modification becomes larger as σ moves away from -1, but is never very large, as $|\sigma+1|$ is at most of the order of 0.1 for spectral components of any significance.

10.9.2 Two-Layer Earth, with Mantle and Fluid Core

10.9.2.1 Equations of motion

When the rotation of the mantle is not about its symmetry axis, the fluid flow within the core consists of a rotation of the fluid as a whole about yet another axis, plus a residual flow needed to make the flow at the ellipsoidal core–mantle boundary (CMB) normal to the boundary. The angular velocity of the rotational flow is also not along the symmetry axis; we denote it by $\mathbf{\Omega}_f$. The differential angular velocity between the core and the mantle is then

$$\mathbf{\omega}_f = \mathbf{\Omega}_f - \mathbf{\Omega} \qquad [122]$$

with components $\Omega_0(m_{fx}, m_{fy}, m_{fz})$. (Superscripts or subscripts f are used to identify quantities pertaining to the fluid core. m_{fx} and m_{fy} represent the differential wobble between the core and the mantle.) The contribution $[\mathbf{C}]_f \cdot \mathbf{\omega}_f$ to the angular momentum of the Earth from this differential rotation has to be added now to $[\mathbf{C}] \cdot \mathbf{\Omega}$ to get the total $\mathbf{H}$. Consequently, the $[\mathbf{C}]_f \cdot (d\mathbf{\omega}_f/dt)$ and $\mathbf{\Omega} \times ([\mathbf{C}]_f \cdot \mathbf{\omega}_f)$ will appear on the left-hand side of the torque equation in addition to the terms in $[\mathbf{C}]$ that we had earlier. The components of these can be readily evaluated, noting that the components of $[\mathbf{C}]_f \cdot \mathbf{\omega}_f$ are $(A_f m_{fx}, A_f m_{fy}, C_f m_{fz})$ to the first order in small quantities. On using these we find the new equation of motion to be

$$A\Omega_0 \frac{d\tilde{m}}{dt} - ieA\Omega_0^2\tilde{m} + A_f\Omega_0\frac{d\tilde{m}_f}{dt}$$
$$+ iA_f\Omega_0^2\tilde{m}_f + \Omega_0\frac{d\bar{c}_3}{dt} + i\Omega_0^2\bar{c}_3 = \tilde{\Gamma} \qquad [123]$$

This is no longer an equation for $\tilde{m}$ alone: it involves $\tilde{m}_f$ also.

A second equation for $\tilde{m}$ and $\tilde{m}_f$ is provided by the torque equation for the fluid core alone. Unlike the equation for the whole Earth, wherein the internal interactions between its different parts were of no relevance, the equation for the core alone must include torques exerted by all bodies outside itself, namely, the mantle as well as the celestial bodies. The obvious interaction mechanism comes from the impact of the flowing fluid core on the CMB (which is nonspherical) and the resulting inertial reaction from the mantle on the core, when the axis of the global rotation of the core is not aligned with the symmetry axis. This reaction is reflected in the perturbations of the various dynamical quantities within the fluid, like the pressure, gravitational potential, and density. The combined effect of all this is an inertial torque on the core that is at the first order proportional to the differential wobble ω_f and the dynamical ellipticity e_f of the core (e_f is defined, analogously to e, as $(C_f - A_f)/A_f$). Mathematically, this result comes from a proof that the total torque $\mathbf{\Gamma}_f$ on the fluid core can be reduced to $(\mathbf{\Omega} + \mathbf{\omega}_f) \times \mathbf{H}_f$ to the first order in the surface flattening of the CMB. The torque arises from forces derived from the pressure P and the gravitational potential (of the Earth itself as well as of celestial bodies), and is represented by the integral

$$\mathbf{\Gamma}_f = -\int \mathbf{r} \times (\nabla P + \rho \nabla \phi_g)\, dV \qquad [124]$$

where the integration is over the instantaneous volume of the core. The proof is not very simple and will not be presented here (see, for instance, the annex in Mathews *et al.* (1991)). Given this result, the torque equation for the core, which is of exactly the same form as eqn [8] except for the replacement of $\mathbf{H}$ by $\mathbf{H}_f$ and $\mathbf{\Gamma}$ by $\mathbf{\Gamma}_f$, reduces to

$$\frac{d\mathbf{H}_f}{dt} - \boldsymbol{\omega}_f \times \mathbf{H}_f = 0 \qquad [125]$$

This is the equation in reduced form for the angular momentum of the core. An equivalent form was derived by Poincaré (1910) by taking the core fluid to be homogeneous and incompressible assuming for the flow velocity in the fluid a form that is linear in x, y, z and satisfies from the outset the condition of being normal to the core boundary. A derivation without making such simplifying assumptions was given in elegant form by Sasao *et al.* (1980), using ideas based on a much more complicated derivation by Molodenski (1961).

To obtain the equation of motion for $\tilde{m}_f$, we need to express the above equation in component form. The components of $\mathbf{H}_f$ may be obtained in the same manner as those of $\mathbf{H}$:

$$\mathbf{H}_f = \Omega_0 \begin{pmatrix} A_f(m_x + m_{fx}) + c_{13}^f \\ A_f(m_y + m_{fy}) + c_{23}^f \\ C_f(1 + m_z + m_{fz}) + c_{33}^f \end{pmatrix} \qquad [126]$$

Using these to express eqn [125] in terms of components, and then taking the complex combination of the first two components, we obtain

$$A_f \Omega_0 \dot{m} + A_f \Omega_0 \dot{m}_f + i(1 + e_f)A_f \Omega_0^2 \tilde{m}_f + \Omega_0 \dot{c}_{3f} = 0 \qquad [127]$$

The pair of coupled equations [123] and [127] govern the temporal variations of m and m_f. On going over to the frequency domain, the dynamical variable to be solved for are $\tilde{m}(\sigma)$ and $\tilde{m}_f(\sigma)$; and c_3 and c_{3f} go over into $c_3(\sigma)$ and $c_{3f}(\sigma)$. The last two quantities have to be written, as before, in terms of the forcing and the dynamical variables. We now have, in the frequency domain,

$$\tilde{c}_3 = A\left[\kappa(\tilde{m} - \tilde{\phi}) + \xi \tilde{m}_f\right]$$
$$\tilde{c}_{3f} = A_f\left[\gamma(\tilde{m} - \tilde{\phi}) + \beta \tilde{m}_f\right] \qquad [128]$$

The terms proportional to $\tilde{m}_f$ represent the deformations due to the centrifugal effect of the differential wobble of the core. There are now four compliances in all, representing the deformabilities of the Earth and of the core under the centrifugal forcing due to wobbles of the whole Earth or due to the differential wobbles of the core. All of them are of order 10^{-3} or smaller, though the precise values depend on the Earth model used:

$$\kappa \approx 1.0 \times 10^{-3}, \quad \xi \approx 2 \times 10^{-4}$$
$$\gamma \approx 2 \times 10^{-3}, \quad \beta \approx 6 \times 10^{-4} \qquad [129]$$

On substituting the above expressions into the spectral components of the eqns [123] and [127], we obtain the following coupled equations:

$$[\sigma - e + (\sigma + 1)\kappa]\tilde{m}(\sigma) + (\sigma + 1)(A_f/A + \xi)\tilde{m}_f(\sigma)$$
$$= -[e - (\sigma + 1)\kappa]\tilde{\phi}(\sigma) \qquad [130]$$

$$\sigma(1 + \gamma)\tilde{m}(\sigma) + (\sigma + 1 + \beta\sigma + e_f)\tilde{m}_f(\sigma) = \sigma\gamma\tilde{\phi}(\sigma) \qquad [131]$$

An important point to note is that for $\sigma = -1$, which corresponds to zero frequency relative to the celestial reference frame, the first of the above equations reduces to $(\sigma - e)\tilde{m}(\sigma) = -e\tilde{\phi}(\sigma)$, which is precisely the same as for the rigid Earth. This is an example of a general property which Poincaré named 'gyrostatic rigidity': at zero frequency in an inertial frame, the nutation of a nonrigid Earth and the associated wobble in the terrestrial frame are the same as for a rigid Earth with the same value of e.

10.9.2.2 Wobble normal modes

We are now in a position to determine the wobble normal modes and eigenfrequencies of the two-layer Earth by solving the homogeneous equations obtained by setting $\tilde{\phi}(\sigma) = 0$. To the lowest order in the ellipticies and compliance parameter, we find the eigenfrequencies to be

$$\sigma_1 = \sigma_{CW} = \frac{A}{A_m}(e - \kappa)$$
$$\sigma_2 = \sigma_{NDFW} = -\left(1 + \frac{A}{A_m}(e_f - \beta)\right) \qquad [132]$$

where A_m is the moment of inertia of the mantle: $A_m = A - A_f$. The labels CW and NDFW stand for 'Chandler wobble' and 'nearly diurnal free wobble', respectively. The Chandler wobble frequency of the two-layer Earth differs from that of the wholly solid Earth just by a factor (A/A_m), which is about 9/8. As for the NDFW, its frequency is $-1 - 1/430\,\text{cpsd}$, justifying the characterization as nearly diurnal.

(In nutation, this mode appears as the free core nutation (FCN), or more precisely, the retrograde free core nutation (RFCN), with the retrograde frequency $-(A/A_m)(e_f - \beta) \approx -1/430$ cpsd.) This mode has a very important role to play in the forced wobbles, because of its location in the middle of the retrograde diurnal band in which the forcing frequencies lie: its presence results in resonant enhancement of several of the prominent wobbles and the corresponding nutations. In particular, the amplitude of the retrograde annual nutation with frequency $\approx -1/366$ cpsd is enhanced by over 30% relative to that of the same nutation of the rigid Earth, because its frequency σ in the terrestrial frame is $\approx -1-1/366$ cpsd, very close to the NDFW eigenfrequency.

10.9.2.3 Solution of the wobble equations; Resonances; Nutation amplitudes

Solution of the pair of equation [130] and [131] with the forcing potential $\tilde{\phi}(\sigma)$ present yields the forced wobble $\tilde{m}(\sigma)$ of the mantle and the differential wobble $\tilde{m}_f(\sigma)$ of the fluid core. To the first order in the compliances and ellipticites, one finds that

$$\tilde{m}(\sigma) = -\frac{(\sigma + 1)[eA + \gamma\sigma A_f - (1 + \sigma)\kappa A]}{A_m(\sigma - \sigma_1)(\sigma - \sigma_2)}\tilde{\phi}(\sigma) \quad [133]$$

The two factors appearing in the denominator give rise, not surprisingly, to resonances associated with the eigenfrequencies σ_1 and σ_2. The expression for the amplitude $\tilde{m}_f(\sigma)$ of the differential wobble of the fluid core may be similarly written down.

The amplitude of the nutation associated with the wobble [133] may be obtained by applying the kinematical relation $\tilde{\eta}(\nu) = -\tilde{m}(\sigma)/\nu$, with $\nu = 1 + \sigma$. Given the values of the Earth parameters, numerical evaluation may be done using $\tilde{\phi}(\sigma)$ obtained from a tide table.

Alternatively, one may take the transfer function $T_w(\sigma) = T_n(\nu)$ obtained by dividing the above expression by the rigid Earth wobble amplitude $\tilde{m}_R(\sigma) = e\tilde{\phi}(\sigma)/(e - \sigma)$, and multiply it by rigid Earth nutation amplitude for the frequency $\nu = \sigma + 1$, for example, from REN2000 of Souchay et al. (1999), to obtain the nutation amplitude for the nonrigid Earth. In this case one does not need to use the amplitude of the tidal potential. This is the approach adopted by Mathews et al. (2002).

10.9.3 Coupling of the Core and the Mantle at the CMB

It was mentioned in the beginning of Section 10.9.2.1 that the pressure and gravitational forces acting on the fluid core result in an effective inertial coupling between the core and the mantle, which is proportional to the dynamical ellipticity of the core. No other couplings were included in the equations of that section. The efforts to refine nutation theory in order to achieve close agreement between theoretical predictions and the high precision observational data available now have led to the investigation of other coupling mechanisms that might be significant, especially electromagnetic and viscous couplings at the CMB.

The magnetic fields generated in the fluid core by the geodynamo mechanism extends into the mantle across the CMB. As differential motions between the solid and fluid sides of the CMB take place as a result of the differential wobble between the mantle and the fluid core, the field lines remain 'frozen' in the fluid because of the very high conductivity of the fluid and move with it, and the continuing portion of the lines on the mantle side tends to sweep through the mantle material. But if a highly conducting layer exists at the bottom of the mantle, the attempted motion of the field lines induces currents which, on the one hand, interact with the magnetic field and produce a Lorentz force on the matter, and on the other, produce a magnetic field which increments the main magnetic field $\mathbf{B}_0$ which existed in the absence of the wobbles. The Lorentz forces all through the conducting layer result in a torque on the mantle and an equal reaction by the mantle on fluid core. The strength of this torque which couples the two regions depends on the conductivity of the mantle layer, the frequency of the wobble causing the relative motion and on the strength and configuration of the main field over the surface of the CMB. If the field is sufficiently weak, the torque is $45°$ out of phase with the wobble and has a strength proportional to the mean squared of the radial component B_r of $\mathbf{B}_0$ over the boundary surface; otherwise the proportionality is only approximate, and the phase difference too depends on the overall strength and configuration of the field. In any case, the equatorial torque on the fluid core can be expressed as

$$\Gamma^{CMB} = -i\Omega_0^2 A_f K^{CMB} \tilde{m}_f \quad [134]$$

where K^{CMB} is a complex coupling constant depending on the physical quantities mentioned above. The viscosity of the fluid core is widely believed to be negligible; in case it is not, both electromagnetic and viscous effects can be encompassed under an expression of the same form as above, with K^{CMB} becoming dependent also on the coefficient of viscosity. The mathematical derivation of the expression for K^{CMB} in terms of the relevant physical parameters will not be presented here. The interested reader may refer to Buffett *et al.* (1996, 2002) and Mathews and Guo (2005).

A little reflection will show that the effect of these couplings can be incorporated into the theory of Section 10.9.2 simply by adding K^{CMB} to the coefficient of $\tilde{m}_f$ in eqn [131]. This replacement does not affect σ_{CW} of the Chandler wobble. But σ_{NDFW} does get modified; the change consists in the replacement of $(e_f - \beta)$ in [132] by $(e_f - \beta + K^{CMB})$, making it complex. Consequently, the NDFW resonance acquires a width.

10.9.4 Anelasticity and Ocean Tide Effects

It has been known, from seismological data on the oscillational normal modes of the Earth and on the propagation of seismic waves, that the response of the material of the Earth's mantle to stresses is not strictly elastic. A small part of the response is not instantaneous; it decays over a finite though short period of time. This part of the response is said to be 'anelastic'. When transformed to the frequency domain, the anelastic response manifests itself through a frequency dependence of the rheological parameters like the Young's modulus and rigidity modulus; in addition, these parameters become complex. Wahr and Bergen (1986) showed that anelasticity effects are in the range of tenths of mas on the amplitudes (both in phase and out of phase) of a number of spectral components of nutation. The anelasticity model that they employed would produce increments $\Delta^{AE}\kappa$, $\Delta^{AE}\gamma$, etc., proportional to

$$\left\{1 - \left(\frac{\omega_m}{\omega}\right)^{\alpha}\right\}\cot\frac{\alpha\pi}{2} + i\left(\frac{\omega_m}{\omega}\right)^{\alpha} \qquad [135]$$

to the various compliances. Here ω is the frequency of the tidal potential responsible for the nutation of interest and ω_m is a reference frequency in the band of frequencies of the seismic waves used for estimating the elastic rheological parameters; it is taken to be equivalent to a period of 1, 200, or 300 s in different models (Dziewonski and Anderson, 1981, Widmer *et al.*, 1991). The power law index α is typically

considered to be around 0.15 (between 0.1 and 0.2). Evaluation of the proportionality constant has to be done by integration of the deformation equations for the Earth model.

The computation of the effects of loading of the Earth's crust by ocean tides, as well as those of ocean tidal currents, were first done by Wahr and Sasao (1981) using the theoretical formalism of Sasao and Wahr (1981). They included the effect of ocean tides in eqns [123] and [127] by adding the ocean tidal contributions $\tilde{c}_3^O$ and $\tilde{c}_{3f}^O$ to $\tilde{c}_3$ and $\tilde{c}_{3f}$ respectively, where the added quantities reflect the changes in the inertia tensors of the whole Earth and of the fluid core due to the deformation produced by the tidal redistribution of ocean mass, together with the equivalent of the angular momentum content of the tidal currents in the oceans. Using estimates of these quantities obtained with the help of ocean tidal models, they found increments of up to about 1 mas to the amplitudes of the leading nutations. Mathews *et al.* (2002) expressed $\tilde{c}_3^O$ and $\tilde{c}_{3f}^O$ in terms of incremental compliances $\Delta^{OT}\kappa$ and $\Delta^{OT}\gamma$. These were computed using empirical formulae (separately for the mass redistribution part and for the tidal current part) representing their frequency dependences; the parameters in the formula were estimated with the help of available data, in particular, the ocean tide angular momentum data provided by Chao *et al.* (1996) for four prominent tidal components; the above reference may be consulted for details. The results that Mathews *et al.* (2002) obtained for the increments to the nutations from ocean tidal (starting from the results of Chao *et al.* (1996)) as well as anelasticity effects are close to those of earlier authors, for example, Wahr and Sasao (1981) and Wahr and Bergen (1986).

10.9.5 Inclusion of Solid Inner Core

In the presence of the solid inner core (SIC), the fluid layer is more appropriately referred to as the fluid outer core (FOC); and all the quantities labeled by the subscript 'f' now refer to the FOC. Quantities pertaining to the SIC will be identified by the subscript s. The SIC has a moment of inertia A_s that is only $1/1400$ of that of the whole Earth, and so it would seem that it could not influence the rotation of the Earth as a whole to any significant extent. (Note that the mantle moment of inertia is $A_m = A - A_f - A_s$ now.) Recent investigations, beginning with Mathews *et al.* (1991), De Vries and Wahr (1991), Dehant *et al.* (1993), and Legros *et al.* (1993), have revealed however that a new nearly diurnal free wobble appears as a result of the role of the inner core in the dynamics, and that the associated resonance has an appreciable effect on the amplitudes of a few nutations.

The inner core is floating within the fluid core, and its symmetry axis is free to go out of alignment with that of the mantle. Consequently, two new equations enter the picture. One is the torque equation governing the time variations of the differential wobble $\tilde{m}_s$ of the SIC, that is, of the complex combination of the equatorial components of $(\boldsymbol{\Omega}_s - \boldsymbol{\Omega})/\Omega_0$; the other is a kinematical equation relating $\tilde{m}_s$ to $\tilde{n}_s$, the complex version of equatorial components of the offset $(\mathbf{i}_s - \boldsymbol{\Omega}_0/\Omega_0)$ of the inner core symmetry axis $\mathbf{i}_s$ from that of the mantle. Thus, one has a set of four equations which couple the variables $\tilde{m}$, $\tilde{m}_f$, $\tilde{m}_s$, $\tilde{n}_s$ to one another.

The torque equation involves gravitational and other torques acting on the inner core. The gravitational torque is made up of the torque exerted by the celestial body and a torque arising from the misalignment between the ellipsoidal inner core and the rest of the Earth (which is also of ellipsoidal symmetry except for a slight realignment of the inner-core boundary because of the tilt of the inner core). Both parts are proportional to the ellipticity e_s of the SIC; the internal gravitational part is proportional also to the density contrast between the SIC and the fluid core just outside the inner core boundary (ICB), and involves a gravitational coupling constant α_g between the SIC and the rest of the Earth. An additional torque arises from electromagnetic (and possibly viscous) couplings at the ICB. It may be expressed as $\Gamma^{ICB} = -i\Omega_0^2 A_s K^{ICB} \tilde{m}_s$; the complex ICB coupling constant K^{ICB} appears in the torque equations for the both the regions coupled by it, namely, the FOC and the SIC.

The expression for $\mathbf{H}$ now gets an additional contribution from $\tilde{m}_s$, and the moments of inertia of all the regions bear the imprint of the misalignment $\tilde{n}_s$. Furthermore, $\tilde{c}_3$, $\tilde{c}_{3f}$, and the new $\tilde{c}_{3s}$, all involve terms proportional to $\tilde{m}_s$ besides those in $\tilde{m}$ and $\tilde{m}_f$; and a number of new compliances enter the picture.

10.9.5.1 Solution of the equations for the three-layer Earth

Solution of the four coupled linear equations with the tidal potential $\tilde{\phi}$ set equal to zero yields two new normal modes in addition to the earlier two. One has a frequency depending on e_s and the ratio of the

density contrast mentioned above to the mean density of the inner core; it is very close to -1 cpsd. The corresponding nutation, with a very small prograde frequency, is referred to as the prograde free core nutation (PFCN) or free inner core nutation (FICN). The resonance associated with this mode has a nontrivial effect on forced nutations at nearby frequencies. The frequency of the other wobble normal mode, called the inner core wobble (ICW) mode, is almost an order of magnitude smaller than that of the Chandler wobble, and its effect is quite ignorable. The forced wobbles $\tilde{m}$, $\tilde{m}_f$, $\tilde{m}_s$, as well as the offset $\tilde{n}_s$ of the inner-core axis for any excitation frequency σ are found by solving the set of four equations with $\tilde{\phi}(\sigma)$ taken to be nonzero. Division of $\tilde{m}(\sigma)$ by the rigid Earth amplitude $\tilde{m}_R(\sigma)$ from eqn [55] gives the transfer function $T_w(\sigma) = T_n(\nu)$, and multiplication of this quantity by the rigid Earth amplitude $\tilde{\eta}_R(\nu)$ of the nutation of frequency $\nu = \sigma + 1$ finally gives the nutation amplitude $\tilde{\eta}(\nu)$ for the nonrigid Earth. The extent to which the resonance associated with each normal mode affects the amplitudes of forced nutation for a given Earth model may be displayed through the resonance expansion of the transfer function for that model. For the three-layered Earth which has four eigenmodes, the expansion takes the form

$$T_w(\sigma) = R + R'(1 + \sigma) + \frac{R_{CW}}{\sigma - \sigma_{CW}} + \frac{R_{FCN}}{\sigma - \sigma_{FCN}} + \frac{R_{FICN}}{\sigma - \sigma_{FICN}} + \frac{R_{ICW}}{\sigma - \sigma_{ICW}} \quad [136]$$

The above form of the expansion presumes that the dynamical flattening used in the construction of the rigid Earth nutation series accurately represents the flattening of the real Earth. It was found by Mathews *et al.* (2002), on requiring the best fit of the results of nonrigid Earth nutation theory to observational data, that a slightly different value of the flattening parameter is called for. It is necessary then to modify the transfer function accordingly (see Mathews *et al.* (2002) for more detail).

10.9.6 Confronting Theory with Observations

The main aim of the nutation theory is to provide theoretical models from which the nutation amplitudes/coefficients can be computed accurately, making it possible to match the observationally estimated series by the computed nutation series at about the same level as the uncertainties in the observational estimates (a couple of tenths of mas) when observational inputs are used for the inherently unpredictable contributions to variations in orientation related to excitations of geophysical origin (atmospheric and nontidal ocean effects, the free wobbles and nutations, etc.). Such a theoretical model, combined with an accurate model for LOD variations, can then be used to make accurate predictions for the Earth's orientation in space at future epochs.

For highly accurate computations to be possible, one needs sufficiently accurate values of the relevant Earth parameters. The other essential input, namely, accurate values of the amplitudes of the components of the tidal potential, or alternatively, the values of the rigid Earth nutation amplitudes, is provided by existing theories. Given all these, the precession and low-frequency nutation due to the action of the degree 2 tesseral gravitational potential of the solar system bodies at the first order in the potential may be calculated to high accuracy using the theory outlined in Section 10.9 as applied to a three-layer Earth, taking into account the effects dealt with in Section 10.9.3 and subsequent sections. However, to get a complete accounting of the nutations, one needs to add some other contributions.

1. Contributions to nutations and precession from torques of the second order in the perturbing potential which arise from the action of each of the three parts (zonal, tesseral, and sectorial) of the potential on the increment to the density function $\rho(\mathbf{r})$ that is produced by the tidal deformation induced by the other two parts. These have been evaluated in part by Folgueira *et al.* (1998a, 1998b), Souchay and Folgueira (2000), and Mathews *et al.* (2002). A complete calculation done since by Lambert and Mathews (2006) shows that mutual cancellations limit the total effect to about 30 μas on the 18.6-year nutation (about a third of the earlier results) and that the effect on precession is practically nil.

2. Geodesic precession and nutation, arising from the difference between kinematical and dynamical quantities in general relativity. This relativistic contribution is about $-19.2\,\mathrm{mas\,yr^{-1}}$ to the precession, and about -30 and $+30\,\mu\mathrm{as}$, respectively, to the prograde and retrograde annual nutations; the effect on other nutations is much smaller.

3. The contribution from atmospheric pressure tides generated by solar thermal effects is of the order of

0.1 mas at the prograde annual period and possibly non-negligible at a couple of other periods.

4. Contributions of the order of 0.1 mas or smaller to a couple of low frequency nutations from the degree 3 tidal potential acting on $\mathcal{J}_3$, and about $25\,\mu\mathrm{as\,yr^{-1}}$ to precession from the degree 4 potential acting on $\mathcal{J}_4$ (see, for instance, Williams (1994) and Souchay *et al.* (1999)).

Other variations in Earth rotation with spectra outside the low frequency band in the terestrial frame are produced by potentials of degrees >2 and of orders $m \neq 1$; they are grouped under polar motions of the CIP and are not considered as nutations. For a treatment of these (see Mathews and Bretagnon (2003)).

Contributions from the effects (1–4) above form part of the observed nutation and precession. These contributions have to be subtracted out from the observational estimates of the precession rate and the coefficients/amplitudes of low frequency nutations, before comparing to the results of computations based on the theory of Section 10.9.

The values of the Earth parameters appearing in the theory are obtained by computations based on an Earth model. Only very rough estimates can be made of the electromagnetic coupling constants K^{CMB} and K^{ICB}, assuming some model for the structure and strengths of the core magnetic fields crossing the CMB and ICB. If, on using these parameter values, computational results on the precession rate and nutation amplitudes turn out to have unacceptably large residuals relative to their observational estimates, one is forced to conclude that the value(s) of one or more of the Earth parameters do not reflect the properties of the real Earth accurately enough, implying that the detailed Earth model used for computation of the parameter values needs to be suitably modified or refined. One has then the option of adjusting the values of those parameters to which the offending residuals are most sensitive. Such a course was adopted by Gwinn *et al.* (1986) and Herring *et al.* (1986), and it led to the inference that the ellipticity e_f of the core has to be about 5% higher than its value for the hydrostatic equilibrium Earth model. This deduction was based on the observation that the retrograde annual nutation, in which a large residual of about 2 mas was found, is strongly influenced by the FCN resonance as the FCN eigenfrequency is quite close to the retrograde annual, and that a rather small adjustment of e_f would change the eigenfrequency enough to eliminate the residual. The adjustment needed to e_f has since been modeled as a result of mantle convection which causes the flattening of the CMB to deviate from the hydrostatic equilibrium value (Defraigne *et al.*, 1996) – an effect which is not taken into account in Earth models constructed from radial seismic data taken together with the assumption of hydrostatic equilibrium. It had been noted even earlier by Wahr (1981), from observational data on the precession rate, that the ellipticity *e* of the whole Earth has to be higher than the hydrostatic equilibrium value by over 1%.

In recent versions of torque equations which take account of the inner core as well as the couplings at the CMB and the ICB between the fluid core and the adjoining solid regions, there are parameters like K^{CMB} and K^{ICB} for which no inputs are available from Earth models though the estimates of the mean squared magnetic field at the CMB inferred from satellite measurements of magnetic fields outside the Earth make an initial estimate for K^{CMB} possible. Confronting nutation theory with observational data provides a means of obtaining estimates with relatively small uncertainties for such parameters.

In constructing the nutation model MHB2000 on which the current IAU model IAU2000A is based, Mathews *et al.* (2002) employed a least-squares procedure to obtain the best fit of the results of their theory to estimates of nutation and precession (see Herring *et al.* (2002)) from a long VLBI data set which was then up-to-date. Before the fitting, the small contributions listed under items (1–4) above were subtracted from the observational estimates, for reasons explained earlier. The least-squares process involves repeated adjustments and optimization of the values of a few Earth parameters, namely, the ellipticities e, e_f, compliances κ, γ, the imaginary part of K^{CMB}, and the real and imaginary parts of K^{ICB}, while retaining values based on the PREM Earth model for all the other parameters. (The real part of K^{CMB} appears in combination with e_f in the equations of the theory, and the two could not be independently estimated. Instead, a theoretical relation between the real and imaginary part of K^{CMB} has been used, and e_f only has been estimated.) The theoretical values for the fit were obtained from the solution of the equations for the three-layer Earth as outlined in Section 10.9.5.1 including the various effects considered in the sections preceding it. The rigid Earth nutation amplitudes used to multiply the transfer function were taken by these authors from the REN2000 series of Souchay *et al.* (1999). In carrying out the numerical evaluation of the nutation amplitudes and the precession rate, and *a priori* set of values was employed for the Earth

parameters occurring in the equations. The values of the selected Earth parameters were optimized by a fitting process which seeks a least-squares minimization of the residuals between the observational estimates and the theoretically computed values. The process was iterative in view of the nonlinear dependence of the nutation amplitudes on the parameters being optimized. Within each step of the iterative process another iterative loop was necessary. The reason was that the strong frequency dependence of the ocean tide increments $\Delta^{OT} \kappa$, $\Delta^{OT} \gamma$, etc., to the compliance parameters that appear in the set of wobble equations was not known *a priori*; it had to be determined through an ocean response model involving free parameters as well as frequency-dependent body tide and load Love numbers which depend on the wobbles which, in turn, have to be found by solving the wobble equations – and these equations involve the incremental compliances just mentioned, which we are trying to determine. The ocean responses known observationally for a very small number of tidal frequencies served as the input in this inner loop. In this process, the mutual influences of nutation-wobble, solid Earth deformation, ocean tidal responses, all come into full play in determining each type of response.

The final results are a set of optimized values for the Earth parameters chosen for adjustment by the process of arriving at the "best fit" between the numbers from theory and observational data, on the one hand, and the MHB2000 nutation series on the other. The series is obtained from the nutation series that is output at the end of the iteration process (i.e., from the equations containing the optimized values for the adjusted parameters) by adding the small contributions which had been subtracted out earlier.

The best-fit values of e and e_f are higher than their hydrostatic equilibrium values; these are interpreted as effects of mantle convection, as mentioned earlier. The excess (nonhydrostatic) part of e_f corresponds to an extra difference of a little under 400 m between the equatorial and polar radii of the CMB. The estimates obtained for K^{CMB} and K^{ICB} call for higher root-mean-squared (rms) values for the magnetic fields at the CMB and ICB than what are suggested by geodynamo models in general, and also, in the case of the CMB, higher than the extrapolated values from satellite observations of magnetic fields along the satellite's orbit. No independent observations are available for the fields at the ICB. The higher estimate for the rms radial field strength at the CMB is, at least in part, a reflection of the short (spatial) wave length components that are not detectable at satellite heights. The possibility that the coupling constants might include the effect of viscous drag at the boundaries of the fluid core has been considered by Mathews and Guo (2005) and Deleplace and Cardin (2006).

10.10   Nutation-Precession from the Displacement Field

10.10.1   Displacement Field Approach

Perturbation by the tidal potential produces displacements of elements of matter from their equilibrium positions in the otherwise uniformly rotating Earth. The field of these displacements over the volume of the Earth may be analyzed into deformational and rotational parts; the first part is manifested as Earth tides and the second as rotation variations. The formulation and solution of the equations which govern the displacement field enable one to determine the precessional and nutational motions from the rotational part of the displacement field.

The approach of computing Earth orientation variations from the equations governing the field of displacements produced within the Earth by the tidal potential has been developed by Smith (1974), Wahr (1981), Dehant (1987, 1990), and Rogister (2001). Since this approach relies on detailed seismic models of the matter density and rheological parameters as functions of position within the Earth's interior, it can yield only accuracies commensurate with those of available Earth models.

10.10.1.1   *Reference system*

The reference frame we are using here has its origin at the geocentre (center of mass of the Earth) and is tied to the hydrostatic equilibrium Earth; it rotates with the uniform velocity $\mathbf{\Omega}_0 = \Omega_0 \hat{e}_z$. It is thus different from the reference frame used in the torque approach.

10.10.1.2   *Earth model parameters and unknowns used*

Consider first a spherical Earth model. In this case, the physical parameters depend only on the radius r. The following are the relevant quantities:

- $\rho_0(r)$: the mean density,
- $\mu_0(r)$: the rigidity modulus (it is 0 in the fluid part),
- $\kappa_0(r)$: the incompressibility modulus,
- $\lambda_0(r) = \kappa_0(r) - \frac{2}{3}\mu_0(r)$, and
- $\phi_0(r)$: gravity potential.

$\lambda_0(r)$ and $\mu_0(r)$ are called Lamé parameters; $\phi_0(r)$ is determined by the density function:

$$\phi_0(r) = -G \int_{V_{in}} \frac{\rho_0(r')}{|r-r'|} dV_{in} \qquad [137]$$

where V in is the initial equilibrium volume. As the equilibrium Earth in these theories is considered to be in uniform rotation around the z-axis of the above reference frame, it has an equatorial bulge in this frame that can be computed from the expression of the effect of the rotation at a long timescale (hydrostatic equilibrium). The equilibrium is between the elastic forces determined by the rheological parameters and the gradient of the geopotential or gravity potential V_{in} containing the centrifugal potential $-(1/2)[\Omega_0^2 r^2 - \mathbf{\Omega} \cdot \mathbf{r})^2]$ and of which the mean value on a sphere Φ_0 is made up of the mean gravitational ϕ_0 and the mean centrifugal potentials ψ_0 ($\Phi_0 = \phi_0 + \psi_0$). While any surface of constant density in the nonrotating Earth (which is assumed to be spherically symmeric) is a geocentric sphere of some radius r, the action of the centrifugal force in a rotating Earth causes this constant density surface to go over into an ellipsoidal surface with flattening $\epsilon(r)$ in the hydrostatic equilibrium state. Such a surface is also a surface on which the geopotential and the rheological parameters have constant values. The density and the other parameters are now functions of both r and θ. For the density, for instance, Clairaut's theory shows that, to the first order in the flattening,

$$\rho(\mathbf{r}) = \rho(r, \theta) = \rho_0(r) + \rho_2(r)P_2(\cos(\theta)) \qquad [138]$$

where

$$\rho_2(r) = \frac{2}{3}\epsilon(r_0)r_0 \frac{d\rho_0}{dr} \qquad [139]$$

and P_2 is the Legendre polynomial of degree 2. One can define similarly $\mu_2(r)$, $\lambda_2(r)$, and $\Phi_2(r)$. The above parameters are given as input to the deformation equations that will be integrated to obtain the nutations. The unknowns of our system are: (1) the components of the displacement vector $\mathbf{s}(\mathbf{r}, t)$, (2) the components of the stress tensor $\mathbf{T}(\mathbf{r}, t)$, and (3) the self-gravitational potential of the Earth ϕ_1^E that is produced by the mass redistribution resulting from the tidal deformation.

10.10.1.3   *Basic equations*

10.10.1.3.(i)   *Poisson equation*   The Poisson equation relates the mass redistribution potential to the displacement field, more specifically to the dilatation and the density gradient.

$$\nabla^2 \Phi_1^E = -4\pi G \nabla \cdot [(\rho_0 + \rho_2 P_2)\mathbf{s}] \qquad [140]$$

10.10.1.3.(ii) Stress–strain relationship The stress–strain relationship expresses the incremental stress tensor in terms of the deformation:

$$\mathbf{T} = (\lambda_0 + \lambda_2 P_2)(\nabla \cdot \mathbf{s})\mathbf{I} + (\mu_0 + \mu_2 P_2)\left(\nabla \mathbf{s} + (\nabla \mathbf{s})^{\mathrm{T}}\right)$$

$$[141]$$

where **I** is the unit matrix, and superscript 'T' indicates that we must take the transpose of the matrix.

10.10.1.3.(iii) Motion equation This is the equation relating the forces acting on a body and the deformations. It involves the internal stress, the self-gravity, the rotation of the reference frame, the external gravitational potential Φ^{ext}, and the mass redistribution potential:

$$\nabla \cdot \mathbf{T} = (\rho_0 + \rho_2 P_2)(\eta_1 + \eta_2 + \eta_3) \qquad [142]$$

where the η_i are defined by

$$\eta_1 = \frac{d^2\mathbf{s}}{dt^2} + \nabla \Phi_1^E + \nabla \Phi^{\mathrm{ext}}$$
$$- (\nabla \cdot \mathbf{s})\nabla \Phi_0 + \nabla(\mathbf{s} \cdot \nabla \Phi_0) \qquad [143]$$

$$\eta_2 = 2\Omega \wedge \frac{d\mathbf{s}}{dt} \qquad [144]$$

$$\eta_3 = -(\nabla \cdot \mathbf{s})\nabla(\Phi_2 P_2) + \nabla(\mathbf{s} \cdot \nabla(\Phi_2 P_2)) \qquad [145]$$

Equations [140]–[142] must be accompanied by appropriate boundary conditions.

10.10.1.4 GSH expansions: Radial functions

Solution of the equations of the problem is facilitated by reducing the equations to those for individual spectral components of the variables governed by the equations. For a spectral component of frequency σ cpsd, the time dependence appears through the factor $e^{i\sigma\Omega_0 t}$, and therefore the time derivative $\partial/\partial t$ appearing in eqns [143] and [144] may be replaced by $i\sigma\Omega_0$. The fundamental eqns [140]–[142] then become partial differential equations in the spatial variables only.

As a second step, one projects these equations on to a basis related to the generalized spherical harmonics (GSH) D_{mn}^ℓ which are functions of the colatitude θ and the longitude ϕ and which depend on three integer numbers: ℓ in $[0, +\infty]$ called the degree; m in $[-\ell, +\ell]$, called the order; and n, determined by the tenson quantity considered in $[-\ell, +\ell]$ (n is in $[-1, +1]$ for a vector, $[-2, +2]$ for a second-order tensor, etc.) (see Appendix 2). Then the coefficients of any particular GSH in the variables $\mathbf{s}$, Φ_1^E, etc., and hence

in the various terms of the above-mentioned partial differential equations, are functions of r only; and these equations reduce to coupled ordinary differential equations for the radial functions. For the displacement field $\mathbf{s}$, for instance, one writes the expansion

$$\mathbf{s}(r, \theta, \lambda) = s^- \hat{e}_- + s^0 \hat{e}_0 + s^+ \hat{e}_+$$
$$= \sum_{\ell=0}^{\infty} \sum_{m=-\ell}^{\ell} \begin{pmatrix} S_\ell^{m-}(r)D_{m-}^\ell(\theta, \lambda) \\ S_\ell^{m0}(r)D_{m0}^\ell(\theta, \lambda) \\ S_\ell^{m+}(r)D_{m+}^\ell(\theta, \lambda) \end{pmatrix} \qquad [146]$$

The unit vectors $\hat{e}_-, \hat{e}_0, \hat{e}_+$ constitute the canonical basis related to the classical $(\hat{e}_r, \hat{e}_\theta, \hat{e}_\lambda)$ by

$$\hat{e}_- = \frac{1}{\sqrt{2}}(\hat{e}_\theta - i\hat{e}_\lambda)$$
$$\hat{e}_0 = \hat{e}_r \qquad [147]$$
$$\hat{e}_+ = -\frac{1}{\sqrt{2}}(\hat{e}_\theta + i\hat{e}_\lambda)$$

For ℓ and m fixed, one defines the radial, spheroidal, and toroidal scalars of the displacement by

$$U_\ell^m(r) = S_\ell^{m0}(r) \qquad [148]$$
$$V_\ell^m(r) = S_\ell^{m+}(r) + S_\ell^{m-}(r) \qquad [149]$$
$$W_\ell^m(r) = S_\ell^{m+}(r) - S_\ell^{m-}(r) \qquad [150]$$

where the variation with θ and λ of the radial spheroidal fields, transverse spheroidal fields, and toroidal fields are the same as that of $\hat{\mathbf{r}}Y_\ell^m$, $r\nabla Y_\ell^m$, and $\mathbf{r} \times \nabla Y_\ell^m$, respectively. In the basis $(\hat{e}_r, \hat{e}_\theta, \hat{e}_\lambda)$

$$\mathbf{T} \cdot \hat{e}_r = \sum_{\ell=0}^{\infty} \sum_{m=-\ell}^{\ell} \begin{pmatrix} T_{\ell 0}^{m-} D_{m-}^\ell \\ T_{\ell 0}^{m0} D_{m0}^\ell \\ T_{\ell 0}^{m+} D_{m+}^\ell \end{pmatrix} \qquad [151]$$

and one defines the following scalars:

$$P_\ell^m(r) = T_{\ell 0}^{m0}(r) \qquad [152]$$
$$Q_\ell^m(r) = T_{\ell 0}^{m+}(r) + T_{\ell 0}^{m-}(r) \qquad [153]$$
$$R_\ell^m(r) = T_{\ell 0}^{m+}(r) - T_{\ell 0}^{m-}(r) \qquad [154]$$

The last two other unknowns are Φ_1^E and g_1^E defined by

$$\Phi_1^E(r, \theta, \lambda) = \sum_{\ell=0}^{\infty} \sum_{m=-\ell}^{\ell} \Phi_{1\ell}^{Em}(r)D_{m0}^\ell(\theta, \lambda) \qquad [155]$$

and

$$g_{1\ell}^{Em}(r) = \frac{d\Phi_{1\ell}^{Em}(r)}{dr} + 4\pi G\rho_0 U_\ell^m(r) \qquad [156]$$

From now on we shall omit to write the dependence on r, for simplifying the writing.

10.10.1.5   *Equations for radial functions*

10.10.1.5.(i)   Poisson equation   The projection of the Poisson equation on the GSH basis provides one with an equation of the second order in d/dr. After introducing the new variables, one finally obtains two equations of the first order in d/dr:

$$\frac{d\Phi_{1\ell}^{Em}}{dr} = g_{1\ell}^{Em} - 4\pi G\rho_0 U_\ell^m \qquad [157]$$

$$\frac{dg_{1\ell}^{Em}}{dr} = \frac{\ell(\ell+1)}{r^2}\Phi_{1\ell}^{Em} - \frac{2}{r}g_{1\ell}^{Em} - \frac{L_0^\ell}{r}4\pi G\rho_0 V_\ell^m$$

$$-4\pi G \sum_{\ell'=|\ell-2|}^{|\ell+2|} \begin{vmatrix} \ell & 2 & \ell' \\ 1 & 0 & 1 \\ m & 0 & m \end{vmatrix} \frac{L_0^\ell}{r}\rho_2 \left\{ \begin{matrix} V_{\ell'}^m \\ W_{\ell'}^m \end{matrix} \right\}$$

$$-4\pi G \sum_{\ell'=|\ell-2|}^{|\ell+2|} \begin{vmatrix} \ell & 2 & \ell' \\ 0 & 0 & 0 \\ m & 0 & m \end{vmatrix} \left[\frac{d\rho_2}{dr}U_{\ell'}^m + \frac{2}{r}U_{\ell'}^m \right.$$

$$\left. + \rho_2\left(\frac{1}{\beta_0}P_{\ell'}^m - \frac{\lambda_0}{\beta_0 r}\left(L_0^\ell V_{\ell'}^m + 2U_{\ell'}^m\right) \right) \right] \qquad [158]$$

The determinant-like symbols that appear in the equations are defined from the Wigner coefficients by

$$\begin{vmatrix} \ell & \ell_1 & \ell_2 \\ n & n_1 & n_2 \\ m & m_1 & m_2 \end{vmatrix} = (-1)^{m+n}(2\ell+1)$$

$$\times \begin{pmatrix} \ell & \ell_1 & \ell_2 \\ -n & n_1 & n_2 \end{pmatrix}\begin{pmatrix} \ell & \ell_1 & \ell_2 \\ -m & m_1 & m_2 \end{pmatrix} \qquad [159]$$

where the 3-j Wigner symbols are defined by

$$\begin{pmatrix} \ell & \ell_1 & \ell_2 \\ n & n_1 & n_2 \end{pmatrix} = (-1)^{\ell-\ell_1-n_2}$$

$$\times \sqrt{\frac{(\ell+\ell_1-\ell_2)!(\ell-\ell_1+\ell_2)!(-\ell+\ell_1+\ell_2)!}{(\ell+\ell_1+\ell_2+1)!}}$$

$$\times \sqrt{(\ell+n)!(\ell-n)!(\ell_1+n_1)!}$$

$$\times \sqrt{(\ell_1-n_1)!(\ell_2-n_2)!(\ell_2+n_2)!}$$

$$\times \sum_k \frac{(-1)^k}{k!(\ell+\ell_1-\ell_2-k)!(\ell-n-k)!(\ell_1+n_1-k)!}$$

$$\times \frac{1}{(\ell_2-\ell_1+n+k)!(\ell_2-\ell-n_1+k)!} \qquad [160]$$

The sum over k extends to all the integers that do not imply negative arguments in the factorials. The L_i^ℓ are defined by

$$L_i^\ell = \sqrt{\frac{(\ell+i)(\ell+1-i)}{2}} \qquad [161]$$

with

$$L_0^\ell = \sqrt{\frac{\ell(\ell+1)}{2}} \qquad [162]$$

$$L_2^\ell = \sqrt{\frac{(\ell+2)(\ell-1)}{2}} \qquad [163]$$

The expressions $\left\{ \begin{matrix} \text{term 1} \\ \text{term 2} \end{matrix} \right\}$ that one finds in eqns [157] and [158] (and later in eqns [165]–[179]) mean that one needs to take the upper term if ℓ' has the same parity as ℓ and the lower term if not. One notes also that $\beta_0 = \lambda_0 + 2\mu_0$ and $\beta_2 = \lambda_2 + 2\mu_2$:

$$X_\ell^m = \frac{dU_\ell^m}{dr} + \frac{1}{r}\left(L_0^\ell V_\ell^m + 2U_\ell^m\right) \qquad [164]$$

10.10.1.5.(ii)   Stress–strain relationship   The projection of the stress–strain relationship on the GSH basis provides three differential equations of the first order in d/dr:

$$\frac{dU_\ell^m}{dr} = \frac{1}{\beta_0}P_\ell^m - \frac{\lambda_0}{\beta_0 r}\left(L_0^\ell V_\ell^m + 2U_\ell^m\right) - \sum_{\ell'=|\ell-2|}^{|\ell+2|}\begin{vmatrix} \ell & 2 & \ell' \\ 0 & 0 & 0 \\ m & 0 & m \end{vmatrix}$$

$$\times \left[\frac{\beta_2}{\beta_0}P_{\ell'}^m - 2\frac{\lambda_0\mu_2 - \lambda_2\mu_0}{\beta_0^2 r}\left(L_0^\ell V_{\ell'}^m + 2U_{\ell'}^m\right) \right] \qquad [165]$$

$$\frac{dV_\ell^m}{dr} = \frac{1}{\mu_0}Q_\ell^m - \frac{1}{r}\left(V_\ell^m + 2L_0^\ell U_\ell^m\right)$$

$$- \frac{\mu_2}{\mu_0}\sum_{\ell'=|\ell-2|}^{|\ell+2|}\begin{vmatrix} \ell & 2 & \ell' \\ 1 & 0 & 1 \\ m & 0 & m \end{vmatrix}\left\{ \begin{matrix} Q_{\ell'}^m \\ R_{\ell'}^m \end{matrix} \right\} \qquad [166]$$

$$\frac{dW_\ell^m}{dr} = \frac{1}{\mu_0}R_\ell^m + \frac{1}{r}W_\ell^m - \frac{\mu_2}{\mu_0}\sum_{\ell'=|\ell-2|}^{|\ell+2|}\begin{vmatrix} \ell & 2 & \ell' \\ 1 & 0 & 1 \\ m & 0 & m \end{vmatrix}\left\{ \begin{matrix} R_{\ell'}^m \\ Q_{\ell'}^m \end{matrix} \right\} \qquad [167]$$

10.10.1.5.(iii)   Motion equation   Finally, the projection of the motion equation on the GSH basis provides three differential equations of the first order in d/dr:

$$\frac{dP_\ell^m}{dr} = -\frac{2}{r}P_\ell^m - \frac{L_0^\ell}{r}Q_\ell^m + \frac{2}{r}\left[\lambda_0 X_\ell^m + \frac{\mu_0}{r}\left(L_0^\ell V_\ell^m + 2U_\ell^m\right)\right]$$
$$+ \rho_0\left(\eta_{1\ell}^{Um} + \eta_{2\ell}^{Um} + \eta_{3\ell}^{Um}\right)$$
$$+ \sum_{\ell'=|\ell-2|}^{|\ell+2|}\begin{vmatrix} \ell & 2 & \ell' \\ 0 & 0 & 0 \\ m & 0 & m \end{vmatrix}\left[\rho_2\left(\eta_{1\ell'}^{Um} + \eta_{2\ell'}^{Um} + \eta_{3\ell'}^{Um}\right)\right.$$
$$\left.+ \frac{2}{r}\left(\lambda_2 X_{\ell'}^m + \frac{\mu_2}{r}\left(L_0^{\ell'} V_{\ell'}^m + 2U_{\ell'}^m\right)\right)\right] \quad [168]$$

$$\frac{dQ_\ell^m}{dr} = -\frac{3}{r}Q_\ell^m - \frac{2(L_0^\ell)^2}{r^2}\mu_0 V_\ell^m + \frac{2L_0^\ell}{r}\left[\lambda_0 X_\ell^m\right.$$
$$\left.+ \frac{\mu_0}{r}\left(L_0^\ell V_\ell^m + 2U_\ell^m\right)\right] + \rho_0\left(\eta_{1\ell}^{Vm} + \eta_{2\ell}^{Vm} + \eta_{3\ell}^{Vm}\right)$$
$$+ \sum_{\ell'=|\ell-2|}^{|\ell+2|}\begin{vmatrix} \ell & 2 & \ell' \\ 1 & 0 & 1 \\ m & 0 & m \end{vmatrix}\rho_2\left\{\begin{array}{c} \eta_{1\ell'}^{Vm} + \eta_{2\ell'}^{Vm} + \eta_{3\ell'}^{Vm} \\ \eta_{1\ell'}^{Wm} + \eta_{2\ell'}^{Wm} + \eta_{3\ell'}^{Wm} \end{array}\right\}$$
$$+ \frac{2L_0^\ell}{r}\sum_{\ell'=|\ell-2|}^{|\ell+2|}\begin{vmatrix} \ell & 2 & \ell' \\ 0 & 0 & 0 \\ m & 0 & m \end{vmatrix}$$
$$\times\left[\lambda_2 X_{\ell'}^m + \frac{\mu_2}{r}\left(L_0^{\ell'} V_{\ell'}^m + 2U_{\ell'}^m\right)\right]$$
$$+ \frac{2L_0^\ell}{r^2}\mu_0\sum_{\ell'=|\ell-2|}^{|\ell+2|}\begin{vmatrix} \ell & 2 & \ell' \\ 2 & 0 & 2 \\ m & 0 & m \end{vmatrix}L_2^{\ell'}\left\{\begin{array}{c} V_{\ell'}^m \\ W_{\ell'}^m \end{array}\right\} \quad [169]$$

$$\frac{dR_\ell^m}{dr} = -\frac{3}{r}R_\ell^m + \frac{2(L_0^\ell)^2}{r^2}\mu_0 W_\ell^m + \rho_0\left(\eta_{1\ell}^{Wm} + \eta_{2\ell}^{Wm} + \eta_{3\ell}^{Wm}\right)$$
$$+ \sum_{\ell'=|\ell-2|}^{|\ell+2|}\begin{vmatrix} \ell & 2 & \ell' \\ 1 & 0 & 1 \\ m & 0 & m \end{vmatrix}\rho_2\left\{\begin{array}{c} \eta_{1\ell'}^{Wm} + \eta_{2\ell'}^{Wm} + \eta_{3\ell'}^{Wm} \\ \eta_{1\ell'}^{Vm} + \eta_{2\ell'}^{Vm} + \eta_{3\ell'}^{Vm} \end{array}\right\}$$
$$+ \frac{2L_0^\ell}{r^2}\mu_0\sum_{\ell'=|\ell-2|}^{|\ell+2|}\begin{vmatrix} \ell & 2 & \ell' \\ 2 & 0 & 2 \\ m & 0 & m \end{vmatrix}L_2^{\ell'}\left\{\begin{array}{c} W_{\ell'}^m \\ V_{\ell'}^m \end{array}\right\} \quad [170]$$

and $\bar{g} = d\Phi_0/dr$ and $g = 2\pi G\rho_0 r + (r/2)(d^2\Phi_0/dr^2)$. The radial, spheroidal, and toroidal scalars $\eta_{i\ell'}^{Um}$, $\eta_{i\ell'}^{Vm}$, and $\eta_{i\ell'}^{Wm}$ are defined by

$$\eta_{1\ell}^{Um} = -\Omega^2\sigma^2 U_\ell^m + g_{1\ell}^{Em} - \frac{2(g+\bar{g})}{r}U_\ell^m - \frac{L_0^\ell}{r}\bar{g}V_\ell^m \quad [171]$$

$$\eta_{1\ell}^{Vm} = -\Omega^2\sigma^2 V_\ell^m - \frac{L_0^\ell}{r}\left(\Phi_{1\ell}^{Em} + \bar{g}U_\ell^m\right) \quad [172]$$

$$\eta_{1\ell}^{Wm} = -\Omega^2\sigma^2 W_\ell^m \quad [173]$$

$$\eta_{2\ell}^{Um} = 2\sigma\Omega^2\sum_{\ell'=|\ell-1|}^{|\ell+1|}\begin{vmatrix} \ell & 2 & \ell' \\ 0 & 1 & -1 \\ m & 0 & m \end{vmatrix}\left\{\begin{array}{c} V_{\ell'}^m \\ -W_{\ell'}^m \end{array}\right\} \quad [174]$$

$$\eta_{2\ell}^{Vm} = -2\sigma\Omega^2\sum_{\ell'=|\ell-1|}^{|\ell+1|}\left[\begin{vmatrix} \ell & 2 & \ell' \\ -1 & -1 & 0 \\ m & 0 & m \end{vmatrix}\left\{\begin{array}{c} 2U_{\ell'}^m \\ 0 \end{array}\right\}\right.$$
$$\left.+ \begin{vmatrix} \ell & 2 & \ell' \\ -1 & 0 & -1 \\ m & 0 & m \end{vmatrix}\left\{\begin{array}{c} V_{\ell'}^m \\ -W_{\ell'}^m \end{array}\right\}\right] \quad [175]$$

$$\eta_{2\ell}^{Wm} = 2\sigma\Omega^2\sum_{\ell'=|\ell-1|}^{|\ell+1|}\left[\begin{vmatrix} \ell & 2 & \ell' \\ -1 & -1 & 0 \\ m & 0 & m \end{vmatrix}\left\{\begin{array}{c} 0 \\ 2U_{\ell'}^m \end{array}\right\}\right.$$
$$\left.+ \begin{vmatrix} \ell & 2 & \ell' \\ -1 & 0 & -1 \\ m & 0 & m \end{vmatrix}\left\{\begin{array}{c} -W_{\ell'}^m \\ V_{\ell'}^m \end{array}\right\}\right] \quad [176]$$

$$\eta_{3\ell}^{Um} = \sum_{\ell'=|\ell-2|}^{|\ell+2|}\begin{vmatrix} \ell & 2 & \ell' \\ 0 & 0 & 0 \\ m & 0 & m \end{vmatrix}\left[\frac{d\Phi_2}{dr}\left(\frac{dU_{\ell'}^m}{dr} - X_{\ell'}^m\right) + \frac{d^2\Phi_2}{dr^2}\right]$$
$$+ \sqrt{3}\sum_{\ell'=|\ell-2|}^{|\ell+2|}\begin{vmatrix} \ell & 2 & \ell' \\ 0 & 1 & -1 \\ m & 0 & m \end{vmatrix}\left(\frac{1}{r}\frac{d\Phi_2}{dr} - \frac{\Phi_2}{r^2} - \frac{\Phi_2}{r}\frac{d}{dr}\right)$$
$$\times\left\{\begin{array}{c} V_{\ell'}^m \\ -W_{\ell'}^m \end{array}\right\} \quad [177]$$

$$\eta_{3\ell}^{Vm} = -\frac{2L_0^\ell}{r}\frac{d\Phi_2}{dr}\sum_{\ell'=|\ell-2|}^{|\ell+2|}\begin{vmatrix}\ell & 2 & \ell' \\ 0 & 0 & 0 \\ m & 0 & m\end{vmatrix}U_{\ell'}^m$$

$$-\sqrt{3}\frac{\Phi_2}{r}\sum_{\ell'=|\ell-2|}^{|\ell+2|}\begin{vmatrix}\ell & 2 & \ell' \\ 0 & 1 & -1 \\ m & 0 & m\end{vmatrix}\begin{Bmatrix}V_{\ell'}^m \\ \\ -W_{\ell'}^m\end{Bmatrix}$$

$$+2\sqrt{3}\frac{\Phi_2}{r}\sum_{\ell'=|\ell-2|}^{|\ell+2|}\begin{vmatrix}\ell & 2 & \ell' \\ 1 & 1 & 0 \\ m & 0 & m\end{vmatrix}\begin{Bmatrix}X_{\ell'}^m \\ \\ 0\end{Bmatrix} \quad [178]$$

$$\eta_{3\ell}^{Wm} = 2\sqrt{3}\frac{\Phi_2}{r}\sum_{\ell'=|\ell-2|}^{|\ell+2|}\begin{vmatrix}\ell & 2 & \ell' \\ 1 & 1 & 0 \\ m & 0 & m\end{vmatrix}\begin{Bmatrix}0 \\ \\ X_{\ell'}^m\end{Bmatrix} \quad [179]$$

where the superscripts U, V, and W indicate that the quantities $\eta_{il}^{U, V, W, m}$ appear in the equation for dU_ℓ^m/dr, dV_ℓ^m/dr, and dW_ℓ^m/dr, respectively.

10.10.1.6 Solutions of the radial equations

The fundamental vectorial equations governing the gravitational, displacement, and stress fields have now been transformed into the set of radial differential equations [157]–[164]. Let us now define

$$\boldsymbol{\sigma}_\ell^m = U_\ell^m D_\ell^{m0}\hat{e}_0 + \frac{1}{2}V_\ell^m\left(D_\ell^{m+}\hat{e}_+ + D_\ell^{m-}\hat{e}_-\right) \quad [180]$$

and

$$\boldsymbol{\tau}_\ell^m = \frac{1}{2}W_\ell^m\left(D_\ell^{m+}\hat{e}_+ - D_\ell^{m-}\hat{e}_-\right) \quad [181]$$

$\boldsymbol{\sigma}_\ell^m$ is the spheroidal part of the displacement field and $\boldsymbol{\tau}_\ell^m$, is the toroidal part. Both are vector fields, and

$$\mathbf{s} = \sum_{\ell=0}^{+\infty}\sum_{m=-\ell}^{\ell}\left(\boldsymbol{\sigma}_\ell^m + \boldsymbol{\tau}_\ell^m\right) \quad [182]$$

It is clear that the radial differential equations obtained above do not couple field components of different orders m. On the other hand, the spheroidal (including radial) displacements of degree ℓ are

coupled with the toroidal displacements of degrees $\ell-1$, $\ell+1$, $\ell-3$, $\ell+3,\dots$, and with spheroidal displacements of degrees $\ell-2$, $\ell+2$, $\ell-4$, $\ell+4$, $\dots$. Let us now arrange the set of radial functions contained in the spheroidal vector fields (including $\boldsymbol{\sigma}_\ell^m$) as a column vector $\bar{\sigma}_\ell^m$ and those in the toroidal fields (including $\boldsymbol{\tau}_\ell^m$) as a column vector $\bar{\tau}_\ell^m$. In view of the nature of the couplings of the different parts of the field as just explained, the differential equations for the $\bar{\sigma}_\ell^m$ and $\bar{\tau}_\ell^m$ have the structure of the following matrix-differential equation, wherein ε is a generic symbol standing for the matrix blocks made up of the first order parts of the matrix elements that arise from the ellipticity of the Earth's structure, and A_0 is for the zeroth order parts which remain nonvanishing when ellipticity is neglected. (The dimensions of the matrix blocks A_0 and the values of the elements of the blocks depend on their positions in the matrix below, and simarly for the blocks ε. All the blocks are rectangular.)

$$\begin{pmatrix}\ddots & & & & & & & & & \\ & \ddots & & & & & & & & \\ \varepsilon & \varepsilon & A_0+\varepsilon & \varepsilon & \varepsilon & 0 & 0 & 0 & 0 \\ 0 & \varepsilon & \varepsilon & A_0+\varepsilon & \varepsilon & \varepsilon & 0 & 0 & 0 \\ 0 & 0 & \varepsilon & \varepsilon & A_0+\varepsilon & \varepsilon & \varepsilon & 0 & 0 \\ 0 & 0 & 0 & \varepsilon & \varepsilon & A_0+\varepsilon & \varepsilon & \varepsilon & 0 \\ 0 & 0 & 0 & 0 & \varepsilon & \varepsilon & A_0+\varepsilon & \varepsilon & \varepsilon \\ & & & & & & & \ddots & \\ & & & & & & & & \ddots\end{pmatrix}$$

$$\times \begin{pmatrix}\vdots \\ \vdots \\ \bar{\sigma}_{\ell-2}^m \\ \bar{\tau}_{\ell-1}^m \\ \bar{\sigma}_\ell^m \\ \bar{\tau}_{\ell+1}^m \\ \bar{\sigma}_{\ell+2}^m \\ \vdots \\ \vdots\end{pmatrix} = \begin{pmatrix}0 \\ 0 \\ 0 \\ 0 \\ f(\phi^{\text{ext}}) \\ 0 \\ 0 \\ 0 \\ 0\end{pmatrix} \quad [183]$$

where the forcing potential ϕ^{ext} is taken to be of degree ℓ and order m, and $\bar{\sigma}_\ell^m$ and $\bar{\tau}_\ell^m$ stand for the following column vectors:

$$\bar{\sigma}_\ell^m = \begin{pmatrix} U_\ell^m \\ P_\ell^m \\ \Phi_{1\ell}^{Em} \\ g_{1\ell}^{Em} \\ V_\ell^m \\ Q_\ell^m \end{pmatrix} \qquad [184]$$

and

$$\bar{\tau}_\ell^m = \begin{pmatrix} W_\ell^m \\ R_\ell^m \end{pmatrix} \qquad [185]$$

The system of equations to be solved is infinite since ℓ may take any integer $\geq |m|$.

For m fixed, one has the associated deformation vectors:

$$\mathbf{s}_\tau = \boldsymbol{\tau}_{|m|}^m + \boldsymbol{\sigma}_{|m|+1}^m + \boldsymbol{\tau}_{|m|+2}^m + \boldsymbol{\sigma}_{|m|+3}^m + \cdots \qquad [186]$$

or

$$\mathbf{s}_\sigma = \boldsymbol{\sigma}_{|m|}^m + \boldsymbol{\tau}_{|m|+1}^m + \boldsymbol{\sigma}_{|m|+2}^m + \boldsymbol{\tau}_{|m|+3}^m + \cdots \qquad [187]$$

since $\ell \geq |m|$. In order to solve the system, it is necessary to truncate at the first order in the small quantities, considering that all $\bar{\sigma}_{\ell'}$ and $\bar{\tau}_{\ell'}^m$ with $\ell' < \ell - 1$ or $\ell' > \ell + 1$ are small as well. Then the infinite system of equations describing the Earth response to a forcing of degree ℓ and order m reduces to the following super-truncated system:

$$\begin{pmatrix} A_0 + \varepsilon & \varepsilon & \varepsilon \\ \varepsilon & A_0 + \varepsilon & \varepsilon \\ \varepsilon & \varepsilon & A_0 + \varepsilon \end{pmatrix} \begin{pmatrix} \bar{\tau}_{\ell-1}^m \\ \bar{\sigma}_\ell^m \\ \bar{\tau}_{\ell+1}^m \end{pmatrix} = \begin{pmatrix} 0 \\ f(\phi^{\text{ext}}) \\ 0 \end{pmatrix} \qquad [188]$$

consisting of just 10 equations for the 10 following variables: U_ℓ^m, P_ℓ^m, $\Phi_{1\ell}^{Em}$, $g_{1\ell}^{Em}$, V_ℓ^m, Q_ℓ^m, $W_{\ell-1}^m$, $R_{\ell-1}^m$, $W_{\ell+1}^m$, and $R_{\ell+1}^m$.

For the actual solution of these equations, one starts then from five independent solutions that do not diverge at $r = 0$. One integrates from the center up to the surface of the Earth using the rheological properties τ_0, μ_0, and the density function ρ_0 given as input and using the boundary conditions at the inner-core boundary and at the CMB (such as continuity of the potential, continuity of the radial displacement,

and continuity of the stress). One arrives at the surface with five independent solutions and one applies the five surface boundary conditions. This ensures the uniqueness of the solution and is done for one fixed frequency.

If one wants to search for the Earth's normal modes, one has to set $\phi^{\text{ext}} = 0$ in the truncated equations, integrate similarly inside the Earth (ignoring the right-hand side of equation [188]), and impose the boundary conditions on the solutions of the resulting system of equations. At the surface, the boundary conditions to be valid correspond to algebraic relations that have a solution only when the determinant of the matrix of the system equals to zero, which is a relation involving the frequency. The remaining steps are thus to search by successive iterations for the frequency that allows the system to be solved.

10.11 Atmospheric Tides and Nontidal Effects from Surficial Fluids

The effects of fluid masses at the Earth's surface on nutation must be considered as they produce observable effects. They consist of ocean tidal effects, atmospheric effects, and nontidal ocean effects. The first of these has been dealt with in Section 10.9.4.

Atmospheric effects on nutation are mainly on the prograde annual component corresponding to the one-solar day period in a terrestrial frame; the diurnal cycle in the temperature of the atmosphere is responsible for the effect. Modulation of the diurnal frequency by the annual and semiannual seasonal variations result in effects at additional frequencies which translate into semi-annual and ter-annual prograde frequencies and the retrograde annual and zero frequencies in the celestial frame. These effects are not easy to compute and in any case, the amplitudes of the atmospheric signals keep changing with time. Further consideration of the atmospheric effects on nutation should be made in the time domain.

Atmospheric effects on the ocean also induce changes in the nutation. These too are difficult to compute. In practice, the annual prograde nutation contribution from both direct atmospheric effect and induced atmospheric ocean effect on nutation are fitted on the observation. de Viron et al. (2005) have shown that the effects computed from general circulation models using the angular momentum approach astonishingly fit the 'observed' value.

10.12 New Conventions for Earth Rotation Variations

10.12.1 CIP and Relation of Its Motions to the Nutation and Wobble

It has been the practice in the analysis of observational data relating to the orientation of the Earth in space to choose a conceptual "intermediate" pole around which the diurnal rotation of the Earth is visualized as taking place at a rate which fluctuates around the mean rate Ω_0, and to analyze the variable rotation of the terrestrial frame in space into three factors: a motion of the pole of the terrestrial frame relative to the chosen conceptual pole, the diurnal rotation around the latter pole, and a motion of this pole relative to the pole of the celestial frame. The definition of the conceptual pole is done by specifying the spectra of frequencies assigned to its motions relative to the pole of each of the two reference frames. In the "classical" definition, the intermediate pole, called the Celestial Ephemeris Pole (CEP), was defined to have only low frequency spectral components ($\ll \Omega_0$ in magnitude) in its motions relative to the poles to both the celestial and the terrestrial reference frames. Recognizing that this specification does not allow for the full range of frequencies actually present in the motions of the poles of the celestial and terrestrial frames relative to each other, specification of a new intermediate pole, called the Celestial Intermediate Pole (CIP), was made by the IAU in 2000 and by the IUGG in 2003. The CIP is defined to have no motion in space with frequencies outside the range $-1/2$ to $1/2$ cpsd, and to have no motions relative to the ITRF with retrograde frequencies between $-3/2$ and $-1/2$ cpsd (which correspond to frequencies between $-1/2$ and $1/2$ cpsd in space) (see Capitaine, 2000, and Capitaine *et al.*, 2000). The CIP definition has been adopted by the IAU and the IUGG in 2000 and 2003 respectively. The nutational and precessional motions of the ITRF pole with periods greater than 2 days in magnitude are then the same as those of the CIP in space. The CIP differs from the classical CEP in having all frequencies outside the retrograde diurnal range in the spectrum of frequencies of its motion relative to the pole of the terrestrial reference frame; the banned part of the spectrum of the motions of the CIP in the terrestrial frame translates to the low frequency spectrum that has been already assigned to its motions in the celestial frame.

The position of the CIP in the ICRF is represented by two coordinates (X, Y) which include a constant offset from the pole of the ICRF besides the lunisolar precessions and nutations in longitude and obliquity. Reduction of observational data requires the use of the transformation between the ICRF and the ITRF, which involves a succession of elementary rotations (represented by matrices) which may be chosen in different ways. The expression is in terms of the rotation which carries the pole of the ICRF into the intermediate pole (CIP), a rotation around this pole, and the rotation which brings this pole into coincidence with the pole of the ITRF. This transformation classically involves the GST, Greenwich sidereal time (also called the Greenwich apparent sidereal time, GAST) of date, that is, the angle on the true equator of date between the true equinox of date and the Greenwich Meridian, which is the origin of longitude in the terrestrial frame.

The classical transformation in terms of the GST, prior to the adoption of the nonrotating origin, considers a rotation around an intermediate pole involving the Greenwich meridian motion in space with respect to the true equinox of date fixed is space. It therefore does not represent exactly the rotation of the Earth and its length-of-day variations as it is contaminated by the motion of the equinox induced by precession and nutation. This is the reason why the nonrotating origin (NRO) has been introduced.

10.12.2 Nonrotating Origin

Time based on Earth rotation, as classically defined, is the Greenwich Apparent Sidereal Time (GAST, or simply GST). GST is the arc length measured along the equator of date from the equinox of date to the Greenwich meridian which marks the axis Ox which serves as the origin of longitude in the ITRF. If the axis of rotation (the pole of the equator) is kept in a fixed direction in space (and if the ecliptic remained stationary), GST thus defined would give the angle of Earth rotation exactly. However, the equatorial plane is moving in space because of nutation and precession, and this gives rise to small motions of the equinox along the equator which mimic small rotations around the pole of the equator, though they are not caused by axial rotation around the pole of the equator. GST is therefore no longer an exact measure of the angle of axial rotation. To obtain a rigorous measure of time based on Earth rotation, it is necessary, therefore, to define two origins (one very close to the equinox, one close to the Greenwich meridian) on the instantaneous equator, which has no motion along this equator or, differently stated, which has no rotation around the pole of the equator. Such an origin is referred to as the nonrotating origin (NRO) (Guinot, 1979). These origins are called the Celestial Intermediate Origin (CIO) and the Terrestrial Intermediate Origin (TIO).

According to the conventions adopted by the IAU in 2000 and the IUGG in 2003, the axial rotation of the Earth is to be measured around the CIP; so the equator of the CIP is the relevant equator for the NRO which is to replace the equinox in the definition of time. To distinguish the newly defined NRO from other origins which had earlier been referred to by the same name, the nomenclature 'celestial intermediate origin' (CIO) and 'terrestrial intermediate origin' (TIO) have been adopted for the NROs on the CIP equator, by IAU2000 resolutions (see Capitaine *et al.*, (1986) and Capitaine (1986, 1990)).

If $\hat{n}$ is the unit vector in the direction of the CIP, and if $\boldsymbol{\Omega}$ is the angular velocity of the Earth, the defining condition on the NRO, namely that it should have no rotation component in the equator of the CIP, is given by (see Capitaine *et al.* (1986) and Capitaine (1986, 1990))

$$\hat{n} \cdot \boldsymbol{\Omega} = 0 \qquad [189]$$

An equivalent way of stating the condition is in terms of the velocity of the CIO: that the velocity, say $\dot{\mathbf{x}}$, should be parallel to the instantaneous direction of the CIP:

$$\dot{\mathbf{x}} = k\hat{n} \qquad [190]$$

On taking the scalar product with $\hat{n}$, one finds that $k = \mathbf{x} \cdot \hat{n}$, the last step being a consequence of the fact that $\mathbf{x} \cdot \hat{n}$ and its time derivative vanish since $\mathbf{x}$ is on the equator of $\hat{n}$. Hence, the defining condition takes the form introduced by Kaplan (2005):

$$\dot{\mathbf{x}} = -(\mathbf{x} \cdot \hat{n})\hat{n} \qquad [191]$$

Considering that $\dot{\mathbf{x}} = \boldsymbol{\Omega} \wedge \mathbf{x}$, and that consequently $\dot{\mathbf{x}} \cdot \boldsymbol{\Omega} = 0$, one immediately see that [189]–[191] are perfectly equivalent.

10.12.3 Definition of Universal Time

Until the implementation of the IAU2000 resolutions, the Universal Time UT1 was defined from GMST through the relation

$$\mathrm{GMST}(T_u, \ \mathrm{UT1}) = \mathrm{GMST}_{0h\mathrm{UT1}}(T_u) + r \ \mathrm{UT1}$$

which relates the increase of GMST during day T_u to the 'time of day' (UT1), with $\mathrm{GMST}_{0h\mathrm{UT1}}(T_u)$ given by a formula which is not strictly linear in T_u (see IERS Conventions 1996 (McCarthy, 1996) and Capitaine *et al.* (1986, 2000, 2003)).

Universal Time UT1 was intended to be a measure of the true accumulated rotation of the Earth after a specified initial epoch. The stellar angle θ is defined to be just that. GMST does not represent the true magnitude of the accumulated rotation. The new definition of UT1 is in terms of θ: $\theta = \theta_0 + k(\mathrm{UT1} - \mathrm{UT1}_0)$, where k is a scale factor that accounts for the fact that the rate of change of θ differs slightly from that of the old UT1. In this case, the new day of UT1 remains close to the mean solar day. The value of θ_0 is chosen to ensure continuity of the new UT1 with the old one at the epoch of switchover from the old to the new (2003 January 1.0, Julian Day 2452640.5).

10.12.4 Transformation between ICRF and ITRF

Reduction of observational data requires the use of the transformation between the ICRF and the ITRF, which involves a succession of elementary rotations (represented by matrices) which may be chosen in different ways. They may, for instance, be rotations through the classical Euler angles. An alternative expression is in terms of the rotation which carries the pole of the CRS into the intermediate pole, a rotation around this pole, and the rotation which brings this pole into coincidence with the pole of the ITRF. The matrix transformation which allows us to pass from the terrestrial reference frame to the celestial reference frame or vice versa was classically related to precession, nutation, LOD, and polar motion by

$$[\mathrm{TRF}] = W \, R_3(\mathrm{GST})\mathrm{PN} \, [\mathrm{CRF}] \qquad [192]$$

where W represents the polar motion contribution, R_3 indicates a rotation around the third axis, and PN contains the precession–nutation contributions. Here R_1, R_2, and R_3 are rotations around the x-axis, the y-axis, and the z-axis, respectively (see Mueller (1981)). GST is the Greenwich Sidereal Time of date (also called the Greenwich Apparent Sidereal Time, GAST), that is, the angle on the true equator of date between the true equinox of date and the Greenwich Meridian, which is the origin of longitude in the terrestrial frame (direction of Ox). The transformation [192] is the classical one involving an intermediate frame associated with the true equator and equinox of date. In the NRO approach, one has

$$[\mathrm{TRF}] = M R_3(\theta) M'[\mathrm{CRF}] \qquad [193]$$

where M and M' are expressed in terms of the CIP coordinates in space and in the terrestrial frame, respectively. The transformation [193] is with reference to an intermediate frame based on the concept of the CIP and NRO (adopted by the IAU in 2000 and the IUGG in 2003). As explained above, the CIP position can be obtained from the position of a conceptual pole, the motion of which corresponds to the precessional and nutational motions (with periods greater than 2 days) of the ITRF pole. The position in the ICRF is represented by two coordinates (X, Y) and includes a constant offset from the pole of the ICRF besides the lunisolar precession and nutations in longitude and obliquity. The position in the ITRF is represented by $(x_p, -y_p)$. One has

$$[\mathrm{CRF}] = R_1(-Y)R_2(X)R_3\left(\frac{XY}{2} + s\right)R_3(-\theta)$$
$$\times R_3\left(-s' + \frac{x_p y_p}{2}\right)R_2(x_p)R_1(y_p)[\mathrm{TRF}] \qquad [194]$$

or equivalently,

$$[\mathrm{TRF}] = R_1(-y_p)R_2(-x_p)R_3\left(s' - \frac{x_p y_p}{2}\right)R_3(\theta)$$
$$\times R_3\left(-\frac{XY}{2} - s\right)R_2(-X)R_1(Y)[\mathrm{CRF}] \qquad [195]$$

with

$$R_3(s)R_3(-\theta)R_3(-s') = R_3(-\mathrm{GST}) \qquad [196]$$

In the decomposition of the transformation between the celestial and terrestrial reference frames (CRF and TRF) into polar motion, axial rotation about the CIP, and nutation-precession, θ appears thus in combination with s and s' defined by Capitaine *et al.* (1986).

Appendix 1: Rotational Normal Mode Definitions

The existence of more than one rotational normal mode for the Earth is due to the presence of the core regions and the consequent possibility of differential rotations between different regions.

In each of the rotational modes, the rotation axes of the three regions, as well as the symmetry axis of the SIC, are offset from the symmetry axis of the mantle by different angles. (The magnitude of the angular offset of the rotation axis of each region from the mantle symmetry axis is the amplitude of the wobble of that region.) All four axes rotate with a common frequency (the frequency of the normal mode) around the mantle symmetry axis. The rotational normal modes not only contribute directly to Earth rotation variations, but also give rise to resonances in the forced nutations and wobbles when the frequency of forcing is close to one or the other of the normal mode eigenfrequencies. The four rotational modes are the following:

• The Chandler wobble (CW), which is a low frequency mode with a period of about 432 days in a terrestrial reference frame. If observed, it is characterized by an offset of the direction of the axis of rotation of the mantle from the figure axis of the ellipsoidal Earth, accompanied by a motion of the rotation axis around the figure axis in the prograde sense, that is, in the same sense as that of the diurnal rotation of the Earth. In this mode, the rotation axes of the three regions are nearly coincident. This is the only rotational eigenmode which could exist if the Earth were wholly solid. (The period would of course be different in that case.) The counterpart of the Chandler Wobble for a rigid Earth is the Eulerian wobble with a period close to 300 days, which is determined solely by the Earth's dynamical ellipticity $(C - A)/A$, that is, the fractional difference between the moments of inertia about the polar and equatorial axes. In fact, the period, when expressed in cpsd, is just $(1/e)$.

• The free core nutation (FCN) is a free mode related to the existence of a flattened fluid core inside the Earth. If observed, it is associated with an offset of the rotation axis of the fluid core from the mantle symmetry axis (i.e., the amplitude of the wobble of the FOC), far larger than that of the rotation axis of the mantle itself. The frequency of this mode is nearly diurnal as seen in a terrestrial frame; for this reason, the motion as seen from the terrestrial frame is referred to as the nearly diurnal free wobble (NDFW). The motion of the rotation axes is accompanied by a circular motion of the mantle symmetry axis in space, which is the FCN; it is retrograde, that is, in the opposite sense to that of the diurnal Earth rotation, and has a period of about 430 days. Ellipticity of the core mantle boundary is essential for the existence of the FCN mode. It causes the equatorial bulge of the fluid mass to impinge on the mantle at the CMB in the course of the differential rotation of the core and the mantle about noncoincident axes; the pressure of the fluid on the boundary, and the inertial reaction on the fluid core itself, sustain the differential rotation between the core and the mantle in the FCN mode.

- The free inner core nutation (FICN) is a mode related to the existence of a flattened inner core inside the fluid core. If excited, the amplitude of the wobble of the SIC dominates over the wobble amplitudes of the other regions; the figure axis of the SIC too is offset by almost the same amount as the rotation axis of the SIC. The FICN has, like the FCN, a nearly diurnal retrograde period in a terrestrial reference frame. The motion of the axes in space is prograde in this mode, unlike in the case of the FCN. For this reason, the FICN is also referred to as the prograde free core nutation (PFCN); when it is necessary to emphasize the distinction between this mode and the FCN, the latter is referred to as the retrograde free core nutation (RFCN). The period of the FICN in space is about 1000 days. As in the case of the FCN, the fluid pressure acting on the ellipsoidal boundary (the ICB in this case) plays an important role here; but an even larger role is played by the gravitational coupling of the SIC to the mantle due to the offset between the symmetry axes of the two ellipsoidal regions.

- The inner core wobble (ICW) is a mode also related to the existence of a flattened inner core inside the fluid core. If excited, the offset of the symmetry axis of the SIC from that of the mantle exceeds the offsets of the three rotation axes by several orders of magnitude. It has a long period (longer than the CW) in a terrestrial frame, and is prograde. The ellipsoidal structure of the SIC and of the rest of the Earth, and the density contrast between the SIC and the fluid at the ICB, are essential elements in the generation of this normal mode.

Appendix 2: Generalized Spherical Harmonics

The generalized spherical harmonic (GSH) function D_{mn}^{ℓ} is a function of θ and φ depending on three integer numbers ℓ in $[0, +\infty]$, called degree, m in $[-\ell, +\ell]$, called order, and n in $[-\ell, +\ell]$; n is determined by the tensor quantity considered; $n = 0$ for scalar, n goes from -1 to $+1$ for vector quantities, and n goes from -2 to $+2$ for 2nd order tensors

(tensors whose Cartesian components have two indices). Its definition is the following:

$$D_{mn}^{\ell}(\theta, \varphi) = (-1)^{m+n} P_{\ell}^{mn}(\mu) e^{im\varphi} \qquad [197]$$

where $\mu = \cos(\theta)$ and P_{ℓ}^{mn} is the generalized Legendre function:

$$P_{\ell}^{mn}(\mu) = \frac{(-1)^{\ell-n}}{2^{\ell}(\ell-n)!} \sqrt{\frac{(\ell-n)!(\ell+m)!}{(\ell+n)!(\ell-m)!}}$$
$$\times \sqrt{\frac{1}{(1-\mu)^{m-n}(1+\mu)^{m+n}}}$$
$$\times \frac{d^{\ell-m}}{d\mu^{\ell-m}}\left((1-\mu)^{\ell-n}(1+\mu)^{\ell+n}\right) \qquad [198]$$

The product of two GSH is expressed using the so-called $\mathcal{J}$-square coefficients related to the Wigner symbols as seen in the text.

$$D_{m_1 n_1}^{\ell_1} D_{m_2 n_2}^{\ell_2} = \sum_{\ell=|\ell_2-\ell_1|}^{\ell=|\ell_2+\ell_1|} \begin{vmatrix} \ell & \ell_1 & \ell_2 \\ n & n_1 & n_2 \\ m & m_1 & m_2 \end{vmatrix} D_{mn}^{\ell} \qquad [199]$$

where $n = n_1 + n_2$ and $m = m_1 + m_2$. The GSH form a basis on which the components of a tensor, of a vector, or of a scalar can be decomposed. The canonical component $(\alpha_1, \alpha_2, \ldots, \alpha_{\mathcal{J}})$ of a tensor $\mathbf{T}$ of order $\mathcal{J}$ can be written as

$$t^{\alpha_1, \alpha_2, \ldots, \alpha_{\mathcal{J}}}(r, \theta, \varphi) = \sum_{\ell=0}^{\infty} \sum_{m=-\ell}^{\ell} t^{\alpha_1, \alpha_2, \ldots, \alpha_{\mathcal{J}}}(r) D_{mn}^{\ell}(\theta, \varphi)$$
$$[200]$$

where $n = \sum_{i=1}^{\mathcal{J}} \alpha_i$. The canonical components of a vector

$$\mathbf{v}(r, \theta, \varphi) = v^-(r, \theta, \varphi)\hat{e}_- + v^0(r, \theta, \varphi)\hat{e}_0 + v^+(r, \theta, \varphi)\hat{e}_+$$
$$= \begin{bmatrix} v^-(r, \theta, \varphi) \\ v^0(r, \theta, \varphi) \\ v^+(r, \theta, \varphi) \end{bmatrix} \qquad [201]$$

can be decomposed as

$$\mathbf{v}(r, \theta, \varphi) = \sum_{\ell=0}^{\infty} \sum_{m=-\ell}^{\ell} \begin{bmatrix} v_{\ell}^{m-}(r) D_{m-}^{\ell}(\theta, \varphi) \\ v_{\ell}^{m0}(r) D_{m0}^{\ell}(\theta, \varphi) \\ v_{\ell}^{m+}(r) D_{m+}^{\ell}(\theta, \varphi) \end{bmatrix} \qquad [202]$$

For a scalar function $f(r, \theta, \varphi)$, the decomposition in GSH is

$$f(r, \theta, \varphi) = \sum_{\ell=0}^{\infty} \sum_{m=-\ell}^{\ell} f_{\ell}^m(r) D_{m0}^{\ell}(\theta, \varphi) \qquad [203]$$

References

Andoyer H (1923) *Cours de Mécanique Céleste, vol. 1*. Paris: Gauthier-Villar.

Bretagnon P and Francou G (1988) Planetary theories in rectangular and spherical variables. VSOP87 solutions. *Astronomy and Astrophysics* 202: 309–315.

Bretagnon P, Rocher P, and Simon JL (1997) Theory of the rotation of the rigid Earth. *Astronomy and Astrophysics* 319: 305–317.

Bretagnon P, Francou G, Rocher P, and Simon J-L (1998) SMART97: A new solution for the rotation of the rigid Earth. *Astronomy and Astrophysics* 329: 329–338.

Brumberg VA, Bretagnon P, and Francou G (1992) Analytical algorithms of relativistic reduction of astronomical observations. In: Capitaine N (ed.) *Proc. Journées Systèmes de Références spatio-temporels*, Observatoire de Paris, France, pp. 141–148.

Buffett BA (1996a) Gravitational oscillations in the length of day. *Geophysical Research Letters* 23: 2279–2282.

Buffett BA (1996b) A mechanism for decade fluctuations in the length of day. *Geophysical Research Letters* 23: 3803–3806.

Buffett BA, Mathews PM, and Herring TA (2002) Modeling of nutation and precession: Effects of electromagnetic coupling. *Journal of Geophysical Research* 107(B4): 2070 (doi:10.1029/2000JB000056).

Capitaine N (1986) The Earth rotation parameters: Conceptual and conventional definitions. *Astronomy and Astrophysics* 162: 323–329.

Capitaine N, Souchay J, and Guinot B (1986) A non-rotating origin on the instantaneous equator - Definition, properties and use. *Celestial Mechanics* 39(3): 283–307.

Capitaine N (1990) The celestial pole coordinates. *Celestial Mechanics and Dynamical Astronomy* 48: 127–143.

Capitaine N (2000) Definition of the celestial ephemeris pole and the celestial ephemeris origin. In: Johnson K, McCarthy D, Luzum B, and Kaplan G (eds.) *Towards Models and Constants for Sub-microarcsecond Astrometry.*, IAU Colloquium 180, pp. 153–163. Washington: US: Naval Observatory.

Capitaine N, Guinot B, and McCarthy DD (2000) Definition of the celestial ephemeris origin and of UT1 in the International Celestial Reference Frame. *Astronomy and Astrophysics* 355: 398–405.

Capitaine N, Chapront J, Lambert S, and Wallace P (2003) Expressions for the Celestial Intermediate Pole and Celestial Ephemeris Origin consistent with IAU2000A precession-nutation model. *Astronomy and Astrophysics* 400: 1145–1154.

Cartwright DE and Taylor RJ (1971) New computations in the tide-generating potential. *Geophysical Journal of the Royal Astronomical Society* 23: 45–74.

Chao BF, Ray RD, Gipson JM, Egbert GD, and Ma C (1996) Diurnal/semidiurnal polar motion excited by oceanic tidal angular momentum. *Journal of Geophysical Research* 101(B9): 20151–20164.

Chapront-Touzé M and Chapront J (1983) The lunar ephemeris ELP-2000. *Astronomy and Astrophysics* 124(1): 50–62.

Chapront-Touzé M and Chapront J (1988) ELP2000-85: A semi-analytical lunar ephemeris adequate for historical times. *Astronomy and Astrophysics* 190: 342–352.

Defraigne P, Dehant V, and Wahr JM (1996) Internal loading of an homogeneous compressible Earth with phase boundaries. *Geophysical Journal International* 125: 173–192.

Dehant V (1987) Tidal parameters for an inelastic Earth. *Physics of the Earth and Planetary Interiors* 49: 97–116.

Dehant V (1990) On the nutations of a more realistic Earths model. *Geophysical Journal International* 100: 477–483.

Dehant V, Hinderer J, Legros H, and Lefftz M (1993) Analytical approach to the computation of the Earth, the outer core and the inner core rotational motions. *Physics of the Earth and Planetary Interiors* 76: 259–282.

Deleplace B and Cardin P (2006) Visco-magnetic torque at the core mantle boundary. *Geophysical Journal International* 167(2): 557–566.

de Viron O, Schwarzbaum G, Lott F, and Dehant V (2005) Diurnal and sub-diurnal effects of the atmosphere on the Earth rotation and geocenter motion. *Journal of Geophysical Research* 110(B11): B11404 (doi:10.1029/2005JB003761).

De Vries D and Wahr JM (1991) The effects of the solid inner core and nonhydrosatic strcture on the Earth's forced nutations and Earth tides. *Journal of Geophysical Research* 96(B5): 8275–8293.

Dziewonski AD and Anderson DL (1981) Preliminary reference earth model. *Physics of the Earth and Planetary Interiors* 25: 297–356.

Folgueira M, Souchay J, and Kinoshita H (1998a) Effects on the nutation of the non-zonal harmonics of third degree. *Celestial Mechanics* 69(4): 373–402.

Folgueira M, Souchay J, and Kinoshita H (1998b) Effects on the nutation of C4m and S4m harmonics. *Celestial Mechanics* 70(3): 147–157.

Fukushima T (1991) Geodesic nutation. *Astronomy and Astrophysics* 244: 11–12.

Guinot B (1979) Basic problems in the kinematics of the rotation of the Earth. In: McCarthy DD and Pilkington PD (eds.) *Time and the Earths Rotation*, pp. 7–18. Dordrecht: D. Reidel.

Gwinn CR, Herring TA, and Shapiro II (1986) Geodesy by radio interferometry: Studies of the forced nutations of the Earth, 2. Interpretation. *Journal of Geophysical Research* 91: 4755–4765.

Hartmann T and Soffel M (1994) The nutation of a rigid Earth model: Direct influences of the planets. *Astronomical Journal* 108: 1115–1120.

Hartmann T and Wenzel H-G (1995a) Catalogue HW95 of the tide generating potential. *Bulletin d'Informations Marées Terrestres* 123: 9278–9301.

Hartmann T and Wenzel H-G (1995b) The HW95 tidal potential catalogue. *Geophysical Research Letters* 22(24): 9278–9301.

Hartmann T, Soffel M, and Ron C (1999) The geophysical approach towards the nutation of a rigid Earth. *Astronomy and Astrophysics Supplement* 134: 271–286.

Herring TA, Gwinn CR, and Shapiro II (1986) Geodesy by radio interferometry: Studies of the forced nutations of the Earth, 1. Data analysis. *Journal of Geophysical Research* 91: 4745–4754.

Herring TA, Mathews PM, and Buffett B (2002) Modeling of nutation-precession of a non-rigid Earth with ocean and atmosphere. *Journal of Geophysical Research* 107(B4): 2069 (doi: 10.1029/2001JB000165).

Hori GI (1966) Theory of general perturbations with unspecified canonical variables. *Publications of the Astronomical Society of Japan* 18(4): 287–296.

Kaplan G (2005) The IAU resolutions on astronomical reference systems, time scales, and Earth rotation models, Publ. United States Naval Observatory, Circular 179, 104p.

Kinoshita H (1977) Theory of the rotation of the rigid Earth. *Celestial Mechanics* 15: 277–326.

Kinoshita H and Souchay J (1990) The theory of the nutation for the rigid Earth model at the second order. *Celestial Mechanics* 48: 187–265.

Klioner S (1998) Astronomical reference frames in the PPN formalism. In: Vordrak J and Capitaine N (eds.) *Journées Systèmes de Références*, pp. 32–37. Prague: Czech Republic.

Lambert S and Mathews PM (2005) Second order torque on the tidal redistribution and the Earth's rotation. *Astronomy and Astrophysics* 453(1): 363–369.

Legros H, Hinderer J, Lefftz M, and Dehant V (1993) The influence of the solid inner core on gravity changes and spatial nutations induced by luni-solar tides and surface loading. *Physics of the Earth and Planetary Interiors* 76: 283–315.

Mathews PM, Buffett BA, Herring TA, and Shapiro II (1991) Forced nutations of the Earth: Influence of inner core dynamics. I. Theory. *Journal of Geophysical Research* 96(B5): 8219–8242.

Mathews PM, Herring TA, and Buffett BA (2002) Modeling of nutation and precession: New nutation series for nonrigid Earth and insights into the Earths interior. *Journal of Geophysical Research* 107(B4): 2068 (doi:10.1029/2001JB000390).

Mathews PM and Bretagnon P (2003) Polar motions equivalent to high frequency nutations for a nonrigid Earth with anelastic mantle. *Astronomy and Astrophysics* 400: 1113–1128.

Mathews PM and Guo JY (2005) Viscoelectromagnetic coupling in precession-nutation theory. *Journal of Geophysical Research* 110: B02402 (doi:10.1029/2003JB002915).

McCarthy DD (1996) IERS Conventions 1996, IERS Technical Note 22.

McCarthy DD (2003) IERS Conventions 2000, IERS Technical Note 29.

Molodensky MS (1961) The theory of nutation and diurnal Earth tides. *Communication de l'Observatoire Royal de Belgique* 188: 25–66.

Mueller II (1981) Titre. In: Gaposchkin EM and Kolaczek B (eds.) *'Reference Coordinate Systems for Earth Dynamics'* pp. 1–22. Dordrecht: Reidel.

Poincaré H (1910) Sur la precession des corps deformables. *Bulletin Astronomique* 27: 321–356.

Rogister Y (2001) On the diurnal and nearly diurnal free modes of the Earth. *Geophysical Journal International* 144(2): 459–470.

Roosbeek F and Dehant V (1998) RDAN97: An analytical development of rigid Earth nutation series using the torque approach. *Celestial Mechanics and Dynamical Astronomy* 70: 215–253.

Roosbeek F (1999) Diurnal and Subdiurnal Terms in Rdan97 Series. *Celestial Mechanics and Dynamical Astronomy* 74(4): 243–252.

Sasao T, Okubo S, and Saito M (1980) A simple theory on the dynamical effects of a stratified fluid core upon nutational motion of the Earth. In: Duncombe RL (ed.) *Proceedings of the IAU Symposium: Nutation and the Earth's Rotation*, pp. 165–183. Dordrecht, The Netherlands: D. Reidel.

Sasao T and Wahr JM (1981) An excitation mechanism for the free core nutation. *Geophysical Journal of the Royal Astronomical Society* 64: 729–746.

Smith M (1974) The scalar equations of infinitesimal elastic-gravitational motion for a rotating, slightly elliptical Earth. *Geophysical Journal of the Royal Astronomical Society* 37: 491–526.

Soffel M and Klioner S (1998) The present status of Einstein relativistic celestial mechanics. In: Vondrak J and Capitaine N (eds.) *Journees Systemes de References*, pp. 27–31. Prague: Czech Republic.

Souchay J and Kinoshita H (1996) Corrections and new developments in rigid Earth nutation theory: I. Lunisolar influence including indirect planetary effects. *Astronomy and Astrophysics* 312: 1017–1030.

Souchay J and Kinoshita H (1997) Corrections and new developments in rigid Earth nutation theory: II. Influence of second-order geopotential and direct planetary effect. *Astronomy and Astrophysics* 318: 639–652.

Souchay J, Loysel B, Kinoshita H, and Folgueira M (1999) Corrections and new developments in rigid Earth nutation theory, III. Final tables REN2000 including crossed nutation and spin-orbit coupling effects. *Astronomy and Astrophysics Supplement Series* 135: 111–139.

Souchay J and Folgueira M (2000) The effect of zonal tides on the dynamical ellipticity of the Earth and its influence on the nutation. *Earth Moon Planets* 81: 201–216.

Wahr JM (1981) The forced nutations of an elliptical, rotating, elastic and oceanless Earth. *Geophysical Journal of the Royal Astronomical Society* 64: 705–727.

Wahr JM and Sasao T (1981) A diurnal resonance in the ocean tide and in the Earth's load response due to the resonant free "core nutation". *Geophysical Journal of the Royal Astronomical Society* 64: 747–765.

Wahr JM and Bergen Z (1986) The effects of mantle anelasticity on nutations, Earth tides and tidal variations in rotation rate. *Geophysical Journal of the Royal Astronomical Society* 87: 633–668.

Widmer R, Masters G, and Gilbert F (1991) Spherically symmetric attenuation within the Earth from normal mode data. *Geophysical Journal International* 104: 541–553.

Williams JG (1994) Contributions to the Earth's obliquity rate, precession, and nutation. *Astronomical Journal* 108: 711–724.

11 GPS and Space-Based Geodetic Methods

G. Blewitt, University of Nevada, Reno, NV, USA

11.1 The Development of Space Geodetic Methods

11.1.1 Introduction

Geodesy is the science of accurately measuring and understanding three fundamental properties of the Earth: (1) its gravity field, (2) its geometrical shape, and (3) its orientation in space (Torge, 2001). In recent decades the growing emphasis has been on the time variation of these 'three pillars of geodesy' (Beutler *et al.*, 2004), which has become possible owing to the accuracy of new space-based geodetic methods, and also owing to a truly global reference system that only space geodesy can realize (Altamimi *et al.*, 2001, 2002) As each of these three properties are connected by physical laws and are forced by natural processes of scientific interest (Lambeck, 1988), space geodesy has become a highly interdisciplinary field,

intersecting with a vast array of geophysical disciplines, including tectonics, Earth structure, seismology, oceanography, hydrology, atmospheric physics, meteorology, climate change, and more. This richness of diversity has provided the impetus to develop space geodesy as a precise geophysical tool that can probe the Earth and its interacting spheres in ways never before possible (Smith and Turcotte, 1993).

Borrowing from the fields of navigation and radio astronomy and classical surveying, space geodetic methods were introduced in the early 1970s with the development of lunar laser ranging (LLR), satellite laser ranging (SLR), very long baseline interferometry (VLBI), soon to be followed by the Global Positioning System (GPS) (Smith and Turcotte, 1993). The near future promises other new space geodetic systems similar to GPS, which can be more generally called Global Navigation Satellite Systems (GNSS). In recent years the GPS has become commonplace, serving a diversity of applications from car navigation to surveying. Originally designed for few meter-level positioning for military purposes, GPS is now routinely used in many areas of geophysics (Dixon, 1991; Bilham, 1991; Hager *et al.*, 1991; Segall and Davis, 1997), for example, to monitor the movement of the Earth's surface between points on different continents with millimeter-level precision, essentially making it possible to observe plate tectonics as it happens.

The stringent requirements of geophysics are part of the reason as to why GPS has become as precise as it is today (Blewitt, 1993). As will be described here, novel techniques have been developed by researchers working in the field of geophysics and geodesy, resulting in an improvement of GPS precision by four orders of magnitude over the original design specifications. Owing to this high level of precision and the relative ease of acquiring GPS data, GPS has revolutionized geophysics, as well as many other areas of human endeavor.

Whereas perhaps the general public may be more familiar with the georeferencing applications of GPS, say, to locate a vehicle on a map, this chapter introduces space geodetic methods with a special focus on GPS as a high-precision geodetic technique, and introduces the basic principles of geophysical research applications that this capability enables. As an example of the exacting nature of modern GPS geodesy, **Figure 1** shows a geodetic GPS station of commonplace (but leading-edge) design now in the western United States, installed for purposes of measuring tectonic deformation across the boundary between the North American and Pacific Plates. This station was installed in 1996 at Slide Mountain, Nevada as part of the BARGEN network (Bennett *et al.*, 1998, 2003; Wernicke *et al.*, 2000). To mitigate the problem of very local, shallow surface motions (Wyatt, 1982) this station has a deep-brace Wyatt-type monument design, by which the antenna is held fixed to the Earth's crust by four welded braces that are anchored ~10 m below the surface (and are decoupled by padded boreholes from the surface). Tests have shown that such monuments exhibit less environmentally caused displacement than those installed to a (previously more common) depth of ~2 m (Langbein *et al.*, 1995). Time series of daily coordinate estimates from such sites indicate repeatability at the level of 1 mm horizontal, and 3 mm vertical, with a velocity uncertainty of 0.2 mm yr^{-1} (Davis *et al.*, 2003). This particular site detected ~10 mm of transient motion for 5 months during late 2003, concurrent with unusually deep seismicity below Lake Tahoe that was likely caused by intrusion of magma into the lower crust (Smith *et al.*, 2004).

11.1.2 The Limitations of Classical Surveying Methods

It is useful to consider the historical context of terrestrial surveying at the dawn of modern space geodesy around 1970 (Bomford, 1971). Classical geodetic work of the highest (~mm) precision was demonstrated during the 1970s for purposes of measuring horizontal crustal strain over regional scales (e.g., Savage, 1983). However, the limitations of classical geodesy discussed below implied that it was essentially impossible to advance geodetic research on the global scale.

Classical surveying methods were not truly three dimensional (3-D). This is because geodetic networks were distinctly separated into horizontal networks and height networks, with poor connections between them. Horizontal networks relied on the measurement of angles (triangulation) and distances (trilateration) between physical points (or 'benchmarks') marked on top of triangulation pillars, and vertical networks mainly depended on spirit leveling between height benchmarks. In principle, height networks could be loosely tied to horizontal networks by collocation of measurement techniques at a subset of benchmarks, together with geometrical observations of vertical angles. Practically this was difficult to achieve, because of the differing requirements on the respective networks. Horizontal benchmarks

Figure 1 Permanent IGS station at Slide Mountain, Nevada, USA.

could be separated further apart on hill tops and peaks, but height benchmarks were more efficiently surveyed along valleys wherever possible. Moreover, the measurement of angles is not particularly precise, and subject to significant systematic error, such as atmospheric refraction.

Fundamentally, however, the height measured with respect to the gravity field (by spirit leveling) is not the same quantity as the geometrical height, which is given relative to some conventional ellipsoid (that in some average sense represents sea level). Thus, the horizontal and height coordinate systems (often called a '2 + 1' system) could never be made entirely consistent.

A troublesome aspect of terrestrial surveying methods was that observations were necessarily made between benchmarks that were relatively close to each other, typically between nearest neighbors in a network. Because of this, terrestrial methods suffered from increase in errors as the distance increased across a network. Random errors would add together across a network, growing as a random walk process, proportional to the square root of distance.

Even worse, systematic errors in terrestrial methods (such as errors correlated with elevation, temperature, latitude, etc.) can grow approximately linearly with distance. For example, wave propagation for classical surveying occurs entirely within the troposphere, and thus errors due to refraction increase with the distance between stations. In contrast, no matter how far apart the stations, wave propagation for space geodetic techniques occurs almost entirely in the approximate vacuum of space, and is only subject to refraction within ~10 km optical thickness of the troposphere (and within the ionosphere in the case of microwave techniques, although ionsopheric refraction can be precisely calibrated by dual-frequency measurements). Furthermore, by modeling the changing slant depth through the troposphere (depending on the source position in the sky), tropospheric delay can be accurately estimated as part of the positioning solution.

There were other significant problems with terrestrial surveying that limited its application to geophysical geodesy. One was the requirement of interstation visibility, not only with respect to intervening terrain, but also with respect to the weather at the time of observation. Furthermore, the precision and accuracy of terrestrial surveying depended a lot on the skill and experience of the surveyors making the measurements, and the procedures developed to

mitigate systematic error while in the field (i.e., errors that could not readily be corrected after the fact).

Finally, the spatial extent of classical terrestrial surveying was limited by the extent of continents. In practice different countries often adopted different conventions to define the coordinates of their national networks. As a consequence, each nation typically had a different reference system. More importantly from a scientific viewpoint, connecting continental networks across the ocean was not feasible without the use of satellites. So in the classical geodetic era, it was possible to characterize the approximate shape of the Earth; however, the study of the change of the Earth's shape in time was for all practical purposes out of the question.

11.1.3 The Impact of Space Geodesy

Space geodetic techniques have since solved all the aforementioned problems of terrestrial surveying. Therefore, the impact of space geodetic techniques can be summarized as follows (as will be explained in detail later):

- They allow for true 3-D positioning.
- They allow for relative positioning that does not degrade significantly with distance.
- They do not require interstation visibility, and can tolerate a broader range of weather conditions.
- The precision and accuracy is far superior and position estimates are more reproducible and repeatable than for terrestrial surveying, where for space geodesy the quality is determined more by the quality of the instruments and data processing software than by the skill of the operator.
- They allow for global networks that can define a global reference frame, thus the position coordinates of stations in different continents can be expressed in the same system.

From a geophysical point of view, the advantages of space geodetic techniques can be summarized as follows:

- The high precision of space geodesy (now at the ~1 mm level), particularly over very long distances, allows for the study of Earth processes that could never be observed with classical techniques.
- The Earth's surface can be surveyed in one consistent reference frame, so geophysical processes can be studied in a consistent way over distance scales ranging ten orders of magnitude from $10^0 - 10^{10}$ m (Altamimi *et al.*, 2002). Global surveying

allows for the determination of the largest-scale processes on Earth, such as plate tectonics and surface mass loading.

- Geophysical processes can be studied in a consistent way over timescales ranging 10 orders of magnitude from 10^{-1} to 10^9 s. Space geodetic methods allow for continuous acquisition of data using permanent stations with communications to a data processing center. This allows for geophysical processes to be monitored continuously, which is especially important for the monitoring of natural hazards, but is also important for the characterization of errors, and for the enhancement of precision in the determination of motion. Sample rates from GPS can be as high as 50 Hz. Motion is fundamentally determined by space geodesy as a time series of positions relative to a global reference frame. Precise timing of the sampled positions in a global timescale ($<<0.1$ μs Universal Coordinated Time (UTC)) is an added bonus for some applications, such as seismology, and SLR.
- Space geodetic surveys are more cost efficient than classical methods; thus, more points can be surveyed over a larger area than was previously possible.

The benefits that space geodesy could bring to geophysics is precisely the reason why space geodetic methods were developed. For example, NASA's interest in directly observing the extremely slow motions (centimeters per year) caused by plate tectonics was an important driver in the development of SLR, geodetic VLBI, and geodetic GPS (Smith and Turcotte, 1993). SLR was initially a NASA mission dedicated to geodesy. VLBI and GPS were originally developed for other purposes (astronomy and navigation, respectively), though with some research and development (motivated by the potential geophysical reward) were adapted into high-precision geodetic techniques for geophysical research.

The following is just a few examples of geophysical applications of space geodesy:

- Plate tectonics, by tracking the relative rotations of clusters of space geodetic stations on different plates.
- Interseismic strain accumulation, by tracking the relative velocity between networks of stations in and around plate boundaries.
- Earthquake rupture parameters, by inverting measurements of co-seismic displacements of stations located within a few rupture lengths of the fault.

- Postseismic processes and rheology of the Earth's topmost layers, by inverting the decay signature (exponential, logarithmic, etc.) of station positions in the days to decades following an earthquake.
- Magmatic processes, by measuring time variation in the position of stations located on volcanoes or other regions of magmatic activity, such as hot spots.
- Rheology of the Earth's mantle and ice-sheet history, by measuring the vertical and horizontal velocities of stations in the area of postglacial rebound (glacial isostatic adjustment).
- Mass redistribution within the Earth's fluid envelope, by measuring time variation in Earth's shape, the velocity of the solid-Earth center of mass, Earth's gravity field, and Earth's rotation in space.
- Global change in sea level, by measuring vertical movement of the solid Earth at tide gauges, by measuring the position of space-borne altimeters in a global reference frame, and by inferring exchange of water between the oceans and continents from mass redistribution monitoring.
- Hydrology of aquifers by monitoring aquifer deformation inferred from time variation in 3-D coordinates of a network of stations on the surface above the aquifer.
- Providing a global reference frame for consistent georeferencing and precision time tagging of nongeodetic measurements and sampling of the Earth, with applications in seismology, airborne and space-borne sensors, and general fieldwork.

What characterizes modern space geodesy is the broadness of its application to almost all branches of geophysics, and the pervasiveness of geodetic instrumentation and data used by geophysicists who are not necessarily experts in geodesy. GPS provides easy access to the global reference frame, which in turn fundamentally depends on the complementary benefits of all space geodetic techniques (Herring and Perlman, 1993). In this way, GPS provides access to the stability and accuracy inherent in SLR and VLBI without need for coordination on the part of the field scientist. Moreover, GPS geodesy has benefited tremendously from earlier developments in SLR and VLBI, particularly in terms of modeling the observations.

11.1.4 LLR Development

Geodesy was launched into the space age by LLR, a pivotal experiment in the history of geodesy. The basic concept of LLR is to measure the distance to the Moon from an Earth-based telescope by timing the flight of a laser pulse that emitted by the telescope, reflects off the Moon's surface, and is received back into the same telescope. LLR was enabled by the Apollo 11 mission in July 1969, when Buzz Aldrin deployed a laser retro-reflector array on the Moon's surface in the Sea of Tranquility (Dickey *et al.*, 1994). Later, Apollo 14 and 15, and a Soviet Lunokhod mission carrying French-built retroreflectors have expanded the number of sites on the Moon. Since initial deployment, several LLR observatories have recorded measurements around the globe, although most of the routine observations have been made at only two observatories: MacDonald Observatory in Texas, USA, and the CERGA station in France. Today the MacDonald Observatory uses a 0.726 m telescope with a frequency-doubled neodymium-YAG laser, producing 1500 mJ pulses of 200 ps width at 532 nm wavelength, at a rate of 10 Hz.

The retroreflectors on the lunar surface are corner cubes, which have the desirable property that they reflect light in precisely the opposite direction, independent of the angle of incidence. Laser pulses take between 2.3 and 2.6 s to complete the 385 000 km journey. The laser beam width expands from 7 mm on Earth to several kilometers at the Moon's surface (a few kilometers), and so in the best conditions only one photon of light will return to the telescope every few seconds. By timing the flight of these single photons, ranges to the Moon can now be measured with a precision approaching 1 cm.

The LLR experiment has produced the following important research findings fundamental to geophysics (Williams *et al.*, 2001, 2004), all of which represent the most stringent tests to date:

- The Moon is moving radially away from the Earth at 38 mm yr^{-1}, an effect attributed to tidal friction, which slows down Earth rotation, hence increasing the Moon's distance so as to conserve angular momentum of the Earth–Moon system.
- The Moon likely has a liquid core.
- The Newtonian gravitational constant G is stable to <10–12.
- Einstein's theory of general relativity correctly explains the Moon's orbit to within the accuracy of LLR measurements. For example, the equivalence principle is verified with a relative accuracy of 10–13, and geodetic precession is verified to within <0.2% of general relativistic expectations.

11.1.5 SLR Development

SLR was developed in parallel with LLR and is based on similar principles, with the exception that the retroreflectors (corner cubes) are placed on artificial satellites (Degnan, 1993). Experiments with SLR began in 1964 with NASA's launch of the Beacon-B satellite, tracked by Goddard Space Flight Center with a range accuracy of several meters. Following a succession of demonstration tests, operational SLR was introduced in 1975 with the launch of the first dedicated SLR satellite, Starlette, launched by the French Space Agency, soon followed in 1976 by NASA's Laser Geodynamics Satellite (LAGEOS-1) in a near-circular orbit of 6000 km radius. Since then, other SLR satellites now include LAGEOS-2, Stella, Etalon-1 and -2, and Ajisai. There are now approximately 10 dedicated satellites that can be used as operational SLR targets for a global network of more than 40 stations, most of them funded by NASA for purposes of investigating geodynamics, geodesy, and orbital dynamics (Tapley *et al.*, 1993).

SLR satellites are basically very dense reflecting spheres orbiting the Earth. For example, LAGEOS-2 launched in 1992 is a 0.6 m sphere of mass 411 kg. The basic principle of SLR is to time the round-trip flight of a laser pulse shot from the Earth to the satellite. Precise time tagging of the measurement is accomplished with the assistance of GPS. The round-trip time of flight measurements can be made with centimeter-level precision, allowing for the simultaneous estimation of the satellite orbits, gravity field parameters, tracking station coordinates, and Earth rotation parameters. The reason the satellites have been designed with a high mass to surface area ratio is to minimize accelerations due to nonconservative forces such as drag and solar radiation pressure. This produces a highly stable and predictable orbit, and hence a stable dynamic frame from which to observe Earth rotation and station motions.

SLR made early contributions to the confirmation of the theory of plate tectonics (Smith *et al.*, 1994) and toward measuring and understanding contemporary crustal deformation in plate-boundary zones (Wilson and Reinhart, 1993; Jackson *et al.*, 1994). To date, SLR remains the premier technique for determining the location of the center of mass of the Earth system, and its motion with respect to the Earth's surface (Watkins and Eanes, 1997; Ray, 1998; Chen *et al.*, 1999). As an optical technique that is relatively less sensitive to water vapor in the atmosphere, SLR has also played a key role in the realization of reference frame scale (Dunn *et al.*, 1999). The empirical realization of scale and origin is very important for the testing of dynamic Earth models within the rigorous framework of the International Terrestrial Reference System (ITRS) (McCarthy, 1996).

Today SLR is used in the following research (Pearlman *et al.*, 2002):

- Mass redistribution in the Earth's fluid envelope, allowing for the study of atmosphere–hydrosphere–cryosphere–solid-Earth interactions. SLR can sense the Earth's changing gravity field (Nerem *et al.*, 1993; Gegout and Cazenave, 1993; Bianco *et al.*, 1977; Cheng and Tapley, 1999, 2004), the location of the solid-Earth center of mass with respect to the center of mass of the entire Earth system (Chen *et al.*, 1999). Also SLR determination of Earth rotation in the frame of the stable satellite orbits reveals the exchange of angular momentum between the solid Earth and fluid components of the Earth system (Chao *et al.*, 1987). SLR stations can sense the deformation of the Earth's surface in response to loading of the oceans, atmosphere, and hydrosphere, and can infer mantle dynamics from response to the unloading of ice from past ice ages (Argus *et al.*, 1999).
- Long-term dynamics of the solid Earth, oceans, and ice fields (Sabadini *et al.*, 2002). SLR can sense surface elevations unambiguously with respect to the Earth center of mass, such as altimeter satellite height and hence ice-sheet and sea-surface height. Thus, SLR is fundamental to the terrestrial reference frame and the long-term monitoring of sea-level change.
- Mantle–core interaction through long-term variation in Earth rotation (Eubanks, 1993).
- General relativity, specifically the Lens–Thirring effect of frame dragging (Ciufolini and Pavlis, 2004).

SLR is a relatively expensive and cumbersome technique, and so has largely been superseded by the GPS technique for most geophysical applications. SLR is still necessary for maintaining the stability of the International Terrestrial Reference Frame (ITRF), in particular, to aligning the ITRF origin with the specifications of ITRS (Altamimi *et al.*, 2002). SLR is also necessary to determine long-term variation in the low-degree components of the Earth's gravity field. SLR is maintained by NASA to support high-precision orbit determination (such as for satellite altimetry), though GPS is also now being used for that purpose.

11.1.6 VLBI Development

VLBI, originally a technique designed for observing distant celestial radio sources with high angular resolution, was from the late 1970s developed for high-precision geodetic applications by applying the technique 'in reverse' (Rogers *et al.*, 1978). Much of this development of geodetic VLBI was performed by the NASA Crustal Dynamics Project initiated in 1979 (Bosworth *et al.*, 1993) with the idea to have an alternative technique to SLR to provide independent confirmation of scientific findings.

Conceptually geodetic VLBI uses radio waves from distant quasars at known positions on the celestial sphere, and measures the difference in the time of arrival of signals from those quasars at stations (radio observatories) on the Earth's surface. Such data provide information on how the geometry of a network of stations evolves in time. This time-variable geometry can be inverted to study geophysical processes such as Earth rotation and plate tectonics, and can be used to define a global terrestrial reference frame with high precision. Unique to VLBI is that it can provide an unambiguous, stable tie between the orientation of the terrestrial reference frame and the celestial reference frame, that is, Earth orientation. However, as a purely geometric technique, it is not directly sensitive to the Earth's center of mass and gravity field, although inferences by VLBI on gravity can be made through models that connect gravity to Earth's shape, such as tidal and loading models.

Comparisons between VLBI and SLR proved to be important for making improvements in both methods. As a radio technique, VLBI is more sensitive to errors in atmospheric refraction (Davis *et al.*, 1985; Truehaft and Lanyi, 1987; Niell, 1996) than the optical SLR technique; however, VLBI has the advantage that the sources are quasars that appear to be essentially fixed in the sky, thus providing the ultimate in celestial reference frame stability. VLBI is therefore the premier technique for determining parameters describing Earth rotation in inertial space, namely precession, nutation, and UT1 (the angle of rotation with respect to UTC) (Eubanks, 1993). VLBI ultimately has proved to be more precise than SLR in measuring distances between stations.

However, VLBI has never been adapted for tracking Earth-orbiting platforms, and is highly insensitive to the Earth's gravity field, and thus cannot independently realize the Earth's center of mass as the origin of the global reference frame. On the other hand, the stability of scale in VLBI is unsurpassed. For most geophysical applications, GPS has superseded VLBI, except for the important reference frame and Earth-orientation tasks described above. VLBI remains important for characterizing long-wavelength phenomena such as postglacial rebound, with the highest precision among all techniques today, and therefore is integral to the stability of global terrestrial reference frames.

To summarize, geodetic VLBI's main contributions to scientific research involve (Schlüter *et al.*, 2002):

- unambiguous Earth-orientation parameters, which can be used to study angular momentum exchange between the solid Earth and its fluid reservoirs, and provide a service to astronomy and space missions by connecting the terrestrial reference frame to the celestial reference frame (Eubanks, 1993);
- providing a stable scale for the global terrestrial reference frame (Boucher and Altamimi, 1993); and
- providing the highest-precision measurements of long-wavelength Earth deformations, thus providing stability to the global frame, and constraints on large-scale geodynamics such as postglacial rebound and plate tectonics (Argus *et al.*, 1999; Stein, 1993).

11.1.7 GPS Development

As of September 2006, the GPS consists of 29 active satellites that can be used to position a geodetic receiver with an accuracy of millimeters within the ITRF. To do this requires geodetic-class receivers (operating at two frequencies, and with antennas designed to suppress signal multipath), currently costing a few thousand US dollars, and geodetic research-class software (developed by various universities and government institutions around the world). Such software embody leading-edge models (of the solid Earth, atmosphere, and satellite dynamics), and data processing algorithms (signal processing and stochastic parameter estimation). Many of the models have been developed as a result of much research conducted by the international geodetic and geophysical community, often specifically to improve the accuracy of GPS. Today it is even possible for a nonexpert to collect GPS data and produce receiver positions with centimeter accuracy by using an Internet service for automatic data processing.

The geodetic development of the GPS has been driven by a number of related factors (Blewitt, 1993):

- The foundation for many of the research-class models was already in place owing to the similarities between GPS and VLBI (as radio techniques), and GPS and SLR (as satellite dynamic techniques), thus giving an early boost to GPS geodesy. Continued collaboration with the space geodetic community has resulted in standard models such as those embodied by the ITRS Conventions (McCarthy, 1996), which aim to improve the accuracy and compatibility of results from the various space geodetic techniques.
- GPS is relatively low cost and yet has comparable precision to VLBI and SLR. Whereas the GPS system itself is paid for by the US taxpayer, the use of the system is free to all as a public good. This has made GPS accessible to university researchers, and the resulting research has further improved GPS accuracy through better models.
- GPS stations are easy to deploy and provide a practical way to sample the deformation field of the Earth's surface more densely, thus allowing space geodesy to address broader diversity scientific questions. This has opened up interdisciplinary research within geophysics, leading to discoveries in unforeseen areas, and to further improvements in GPS accuracy through improved observation models.
- GPS was readily adopted because of the ease of access to the ITRF on an *ad hoc* basis, without need for special global coordination from the point of view of an individual investigator. Furthermore, the ITRF gives implicit access to the best possible accuracy and stability that can be achieved by SLR and VLBI (Herring and Pearlman, 1993).

Following closely the historical perspectives of Evans *et al.* (2002) and Blewitt (1993), GPS has its roots as a successor to military satellite positioning systems developed in the 1960s, though the first geophysical applications of GPS were not realized until the early 1980s. In the run-up to the space age in 1955, scientists at the Naval Research Laboratory first proposed the application of satellite observations to geodesy. By optical observation methods, the first geodetic satellites were quickly used to refine parameters of the Earth's gravity field. Optical methods were eventually made obsolete by the Doppler technique employed by the Navy Navigation Satellite System (TRANSIT). As the name implies, Doppler positioning was based on measuring the frequency of the satellite signal as the relative velocity changed between the satellite and the observer. By the early 1970s, Doppler positioning with 10 m accuracy became possible on the global scale, leading to the precise global reference frame 'World Geodetic System 1972' (WGS 72), further improved by WGS 84, which was internally accurate at the 10 cm level. Having a global network of known coordinates together with the success of radiometric tracking methods set the stage for the development of a prototype GPS system in the late 1970s.

The US Department of Defense launched its first prototype Block-I GPS satellite, NAVSTAR 1, in February 1978. By 1985, 10 more Block-1 satellites had been launched, allowing for the development and testing of prototype geodetic GPS data processing software that used dual-frequency carrier phase observables. In February 1989 the first full-scale operational GPS satellite known as Block II was deployed, and by January 1994, a nominally full constellation of 24 satellites was completed, ensuring that users could see satellites of a sufficient number (at least five) at anytime, anywhere in the world. Initial operational capability was officially declared in December 1993, and full operational capability was declared in April 1995. From July 1997, Block IIRs began to replace GPS satellites. The first modified version block IIR-M satellite was launched in 2005 (for the first time emitting the L2C signal, which allows civilian users to calibrate for ionospheric delay). The current constellation of 29 satellites includes extra satellites as 'active spares' to ensure seamless and rapid recovery from a satellite failure. The first Block IIF satellite is scheduled to launch in 2008, and may transmit a new civil signal at a third frequency.

The GPS system design built on the success of Doppler by enabling the measurement of a biased range ('pseudorange') to the satellite, which considerably improved positioning precision. Carrier phase tracking technology further improved the signal measurement precision to the few millimeter level. As a radio technique, VLBI technology was adapted in NASA's prototype GPS geodetic receivers. The SERIES receiver, developed by MacDoran (1979) at the Jet Propulsion Laboratory (JPL), pointed at one source at a time using a directional antenna (a technique no longer used). Many key principles and benefits of the modern GPS geodesy were based on the omnidirectional instrument, MITES, proposed by Counselman and Shapiro (1979). This was

developed by the Massachusetts Institute of Technology (MIT) group into the Macrometer instrument, which proved centimeter-level accuracy using the innovative double-difference method for eliminating clock bias, a method which has its origins in radio navigation of the Apollo mission (Counselman et al., 1972).

By the mid-1980s, commercial receivers such as the Texas Instrument TI4100 became available (Henson et al., 1985) and were quickly deployed by geophysicists in several pioneering experiments to measure the slow motions associated with plate tectonics (Dixon et al., 1985; Prescott et al., 1989; Freymueller and Kellogg, 1990); Such experiments spurred the development of analysis techniques to improve precision at the level required by geophysics (Tralli et al., 1988; Larson and Agnew, 1991; Larson et al., 1991). Important developments during these early years include ambiguity resolution over long distances (Blewitt, 1989; Dong and Bock, 1989), precise orbit determination (King et al., 1984; Beutler et al., 1985; Swift, 1985; Lichten and Border, 1987), and troposphere modeling (Lichten and Border, 1987; Davis et al., 1987; Tralli and Lichten, 1990).

The development of geodetic GPS during the 1980s was characterized by intensive hardware and software development with the goal of subcentimeter positioning accuracy, over increasingly long distances. A prototype digital receiver known as 'Rogue' was developed by the JPL (Thomas, 1988), which procuded high-precision pseudorange data that could be used to enhance data-processing algorithms, such as ambiguity resolution. Several high-precision geodetic software packages that were developed around this time are still in use and far exceed the capabilities of commercial packages. These included the BERNESE developed at the University of Berne (Beutler et al., 1985; Gurtner et al., 1985; Rothacher et al., 1990), GAMIT-GLOBK developed at MIT (Bock et al., 1986; Dong and Bock, 1989; Herring et al., 1990), and GIPSY-OASIS developed at JPL (Lichten and Border, 1987; Sovers and Border; Blewitt, 1989, 1990).

GPS became fully operational in 1994, with the completion of a full constellation of 24 satellites. Developments toward high precision in the 1990s include (1) truly global GPS solutions made possible by the completion of the Block II GPS constellation and, simultaneously, installation and operation of the global network in 1994 (shown in its current configuration in **Figure 2**) by the International GPS Service (IGS, since renamed the International

Figure 2 The global network of the International GPS Service. Courtesy of A. Moore.

GNSS Service) (Beutler *et al.*, 1994a); (2) global-scale ambiguity resolution (Blewitt and Lichten, 1992); (3) further refinement to tropospheric modeling and the inclusion of tropospheric gradient parameters (Davis *et al.*, 1993; McMillan, 1995; Niell, 1996; Chen and Herring, 1997; Bar-Sever *et al.*, 1998; Rothacher *et al.*, 1998); (4) adoption of baseband-digital GPS receivers with the low-multipath choke-ring antenna developed originally at JPL, which remains the IGS standard design today; (5) improved orbit models, particularly with regard to GPS satellite attitude, and the tuning of stochastic models for solar radiation pressure (Fliegel *et al.*, 1992; Beutler *et al.*, 1994b; Bar-Sever, 1996; Fliegel and Gallini, 1996; Kuang *et al.*, 1996); (6) improved reference system conventions (McCarthy, 1996); and (7) simultaneous solution for both orbits and station positions (fiducial-free global analysis) (Heflin *et al.*, 1992).

The focus of developments in the decade have included (1) building on earlier work by Schupler *et al.* (1994), antenna phase center variation modeling and calibrations for both stations and the GPS satellites themselves (Mader, 1999; Mader and Czopek, 2002; Schmid and Rothacher, 2003; Schmid *et al.*, 2005; Ge *et al.*, 2005); (2) densification of stations in the ITRF and the installation of huge regional networks of geodetic GPS stations, such as the ~1000 station Plate Boundary Observatory currently being installed in the western North America (Silver *et al.*, 1999); (3) improved analysis of large regional networks of stations through common-mode signal analysis (Wdowinski *et al.*, 1997) and faster data processing algorithms (Zumberge *et al.*, 1997; Blewitt,

2006); (4) the move toward real-time geodetic analysis with applications such as GPS seismology (Nikolaidis *et al.*, 2001; Larson *et al.*, 2003) and tsunami warning systems (Blewitt *et al.*, 2006), including signal processing algorithms to filter out sidereally repeating multipath (Bock *et al.*, 2000, 2004; Choi *et al.*, 2004) ; and (6) further improvements in orbit determination (Ziebart *et al.*, 2002), tropospheric modeling (Boehm *et al.*, 2006), and higher order ionospheric models (Kedar *et al.*, 2003).

With the significant improvements to modeling since the inception of the IGS in 1994, data reprocessing of global GPS data sets has begun in earnest. Early results indicate superior quality of the GPS data products, such as station coordinate time series, orbit and clock accuracy, and Earth-orientation parameters (Steigenberger *et al.*, 2006).

11.1.8 Comparing GPS with VLBI and SLR

GPS geodesy can be considered a blend of the two earlier space geodetic techniques: VLBI and SLR. The most obvious similarities are that (1) SLR and GPS are satellite systems, and so are sensitive to Earth's gravity field, and (2) VLBI and GPS are radio techniques and so the observables are subject to atmospheric refraction in a similar way. Due to these similarities, GPS geodesy has benefited from earlier work on both VLBI and SLR observation modeling, and from reference system conventions already established through a combination of astronomical observation, VLBI and SLR observation, and geodynamics modeling. Moreover, Earth models such as tidal deformation are required by all global geodetic techniques, so GPS geodesy was in a position to exploit what had already been learned. Today the improvements in modeling any of the techniques can often be exploited by another technique.

The similarities between certain aspects of the different techniques lead to an overlap of strengths and weaknesses, error sources, and sensitivity to geophysical parameters of interest. The strengths and weaknesses can be summarized as follows:

- The main strength of SLR is the stability of the orbits due to custom designed satellites. This leads to high sensitivity to the low-degree gravity field harmonics and their long-term changes in time. This includes the degree-1 term (characterizing geocenter motion, the motion of the solid Earth with respect to the center of mass of the entire Earth system), which is important for realizing the origin of the ITRF. As an optical technique, SLR is insensitive to moisture in the atmosphere, and so has relatively small systematic errors associated with signal propagation. This inherently leads to more robust estimation of station height. On the other hand, SLR has problems working during daylight hours, and in cloudy conditions. It is also an expensive and bulky technique, and so suffers from a lack of geographical coverage.

- The main strength of VLBI is the stability and permanency of the sources, which are quasars. This leads to two important qualities: (1) VLBI is insensitive to systematic error in orbit dynamic models, and can potentially be the most stable system for detecting the changes over the longest observed time periods, and (2) VLBI is strongly connected to an external, celestial reference frame, a vantage point from which Earth orientation and rotation can be properly determined. A major weakness of VLBI is (similar to SLR) its expensiveness and bulkiness. Moreover, some VLBI observatories are used for astronomical purposes, and so cannot be dedicated to continuous geodetic measurement. VLBI antennas are very large structures which have their own set of problems, including the challenge to relate the observations to a unique reference point, and the stability of the structure with respect to wind and gravitational stress, and aging.

The main advantage of GPS is its low cost and ease of deployment, and all weather capability. Thus GPS can provide much better geographical coverage, continuously. The flexibility of deployment allows for

ties to be made between the terrestrial reference frames of the various techniques through collocation at SLR and VLBI sites. The disadvantage of GPS is that it is subject to both the systematic error associated with orbit dynamics, and atmospheric moisture. Furthermore, the omnidirectional antennas of GPS lead to multipath errors. Thus geodetic GPS is essential for improved sampling of the Earth in time and space, but ultimately depends on SLR and VLBI to put such measurements into a reference frame that has long-term stability. This synergy lies at the heart of the emerging concept the Global Geodetic Observing System (GGOS), under the auspices of the International Association of Geodesy (Rummel *et al.*, 2005).

11.1.9 GPS Receivers in Space: Low Earth Orbit GPS

GPS has proved extremely important for positioning space-borne scientific instruments in low Earth orbit (LEO) (sometimes called LEO GPS). Evans *et al.* (2002) provide an overview of space-borne GPS, which is only briefly summarized here. The LANDSAT 4 satellite launched in 1982 was the first to carry a GPS receiver, called GPSPAC. This was followed by three more missions using GPSPAC, including LANDSAT 5 in 1984 and on DoD satellites in 1983 and 1984. As the GPS satellite constellation grew during the 1980s, so the precision improved, enabling decimeter-level accuracy for positioning space-borne platforms. Following Evans *et al.* (2002), the applications of space-borne GPS can be categorized as: (1) precise orbit determination of the host satellite for applications such as altimetry; (2) measurement of the Earth's gravity field, such as the missions Challenging Minisatellite Payload (CHAMP) and Gravity Recovery and Climate Experiment (GRACE); (3) ionospheric imaging; and (4) indirect enhancements to global geodesy and remote sensing. In addition to these categories, space-borne GPS is also being used to invert for the refractivity of the Earth's neutral atmosphere by occultation measurements, which can be used, for example, to infer stratospheric temperatures for studies of global climate change.

11.1.10 The Future of GNSS

The success of GPS has led to the development of similar future systems, such as the European Galileo system which has been scheduled to become operational by approximately 2010. In general, such systems are generically referred to as GNSS. The Russian system Global Navigation Satellite System (GLONASS) has a growing number of satellites in orbit and may reach a full constellation within the next several years, depending on whether the current rate of deployment is maintained. Other systems may also be developed, for example, China's plans for its Compass system, and also future GPS following-on systems by the US. The main reason for the development of alternative systems to GPS is to ensure access to GNSS signals that are not under the control of any single nation, with implications for the military in times of war and national emergencies, and for civilian institutions such as national aviation authorities that have stringent requirements on guaranteed access to a sufficient number of GNSS signals at all times.

Thus, the future of GNSS is essentially guaranteed. By analogy with the Internet, navigation and geospatial referencing has become such an embedded part of the world's infrastructure and economy that it is now difficult to imagine a future world where GNSS is not pervasive. As GPS has proved, a GNSS system does not necessarily have to be designed with high-precision geodesy in mind in order for it to be used successfully as a high-precision geophysical tool. However, it is likely that future GNSS systems will take more into account the high-precision applications in their design, and thus may be even better suited to geophysical applications than GPS currently is. Much can be done to mitigate errors, for example, in the calibration of the phase center variation in the satellite transmitting antenna, or the transmission of signals at several different frequencies.

Satellite geodesy in the future will therefore use multiple GNSS systems interoperably and simultaneously. This will lead to improved precision and robustness of solutions. It will also allow for new ways to probe and hopefully mitigate systematic errors associated with specific GNSS systems and satellites. The continued downward spiral in costs of GNSS receiver systems will undoubtedly result in the deployment of networks with much higher density (reduced station spacing), which will benefit geophysical studies. For example, it would allow for higher-resolution determination of strain accumulation due to crustal deformation in plate-boundary zones.

11.1.11 International GNSS Service

Infrastructure development and tremendous international cooperation characterized the 1990s. GPS operations moved away from the campaigns, back to the model of permanent stations, familiar to VLBI and SLR. As the prototype receivers developed by research groups in the 1980s had become commercialized, the cost of installing a GPS station in the 1990s had fallen to ~$25 000, in contrast to the millions of dollars required for VLBI/SLR. Thus the long-range goal of the federal funding agencies was realized: dozens of GPS stations could be installed for the price of one VLBI station.

With the cooperation of ~100 research institutions around the world under the umbrella of the International GPS (now GNSS) Service (IGS), a global GPS network (now at ~350 stations, **Figure 2**) with a full geodetic analysis system came into full operation in 1994 (Beutler et al., 1994a). This backbone, together with the regional stations located in areas of tectonic activity, such as Japan and California, form a global-scale instrument capable of resolving global plate-tectonic motions and regional phenomena such as earthquake displacement. As a result of this international cooperation, a culture of data sharing has developed, with data freely available for research purposes via the Internet from IGS Global Data Centers. The establishment of a standard GPS measurement format known as Receiver Independent Exchange (RINEX) has facilitated this extensive exchange of data through IGS (see **Table 1** for IGS data availability).

The mission of the IGS is to provide the highest-quality data and products as the standard for GNSS in support of Earth science research and multidisciplinary applications. So although the IGS does not specifically carry out geophysical investigations, it does provide an essential service without which such investigations would be very costly and difficult to carry out. The 1990s has seen the development of collaborations with specific geophysical goals. Groups such as WEGENER (Plag et al., 1998) and UNAVCO have provided an umbrella for geoscientists using GPS geodesy as a tool. Such groups depend on IGS for their success; conversely, IGS as a volunteer organization depends on such users to contribute to its operations and technical working groups.

The infrastructure has indeed become quite complex, yet cooperative, and often with an efficient division between geodetic operations and geodynamics investigations. As an example of how infrastructure is developing, solutions are being exchanged in a standard Software Independent Exchange (SINEX) format to enable the construction of combined network solutions and, therefore, combined global solutions for Earth surface kinematics.

Table 1   IGS raw data types and availability

	Latency	Updates	Sample interval
Ground observations			
GPS and GLONASS	1 day	Daily	30 s
data	1 hour	Hourly	30 s
	15 min	15 min	1 s[a]
GPS Broadcast	1 day	Daily	
Ephemerides	1 hour	Hourly	NA
	15 min	15 min	
GLONASS Broadcast	1 day	Daily	NA
Ephemerides			
Meterological	1 day	Daily	5 min
	1 hour	Hourly	5 min
Low earth orbiter			
observations			
GPS	4 days	Daily	10 s

[a]Selected subhourly stations have sampling intervals 1 s < *t* < 10 s.
Source: IGS Central Bureau, http://igscb.jpl.nasa.gov.

This standard has since also been adopted by the other space geodetic techniques. Combination solutions have the advantage that (1) the processing burden is distributed among many groups who can check each other's solutions; (2) noise and errors are reduced through increased redundancy and quality control procedures; (3) coverage and density are increased; and (4) regional geodynamics can be interpreted in a self-consistent global context. An emerging focus of this decade (2000s) is the development of such combination solutions, and on the inversion of these solutions to infer geophysical parameters.

As the premier service for high-precision geodesy, the quality of IGS products is continually improving with time (**Figure 3**) and represents the current state of the art (Dow *et al.*, 2005b; Moore, 2007). The levels of accuracy claimed by the IGS for its various products are reproduced in **Table 2**.

Analogous to the IGS, geodetic techniques are organized as scientific services within the International Association of Geodesy (IAG). The IAG services are as follows:

- International Earth Rotation and Reference System Service (IERS) (IERS, 2004).
- International GNSS Service, formerly the International GPS Service (IGS) (Dow *et al.*, 2005b).
- International VLBI Service (IVS) (Schlüter *et al.*, 2002).
- International Laser Ranging Service (ILRS) (Pearlman *et al.*, 2002).
- International DORIS Service (IDS) (Tavernier *et al.*, 2005).

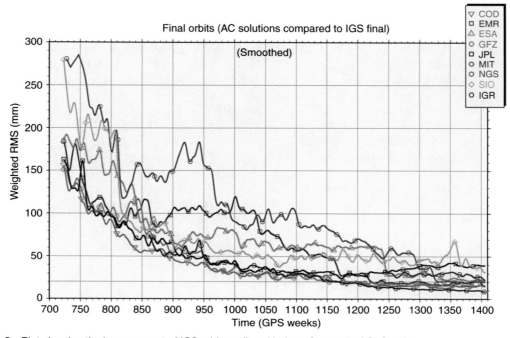

Figure 3   Plot showing the improvement of IGS orbit quality with time. Courtesy of G. Gendt.

Table 2 IGS (and broadcast) product availability and quality

		Accuracy[a]	Latency	Updates	Interval
GPS Satellite Ephemerides					
Broadcast[b]	Orbits	160 cm	Real time	NA	NA
	Satellite clocks	7 ns			
Ultrarapid	Orbits	10 cm	Real time	6 h	15 min
(predicted half)	Satellite clocks	5 ns			
Ultrarapid	Orbits	<5 cm	3 h	6 h	15 min
(observed half)	Satellite clocks	0.2 ns			
Rapid	Orbits	<5 cm			15 min
	All clocks	0.1 ns	17 h	Daily	5 min
Final	Orbits[c]	<5 cm			15 min
	All clocks[d]	0.1 ns	13 days	Weekly	5 min
GLONASS Satellite Ephemerides					
Final		15 cm	2 weeks	Weekly	15 min
IGS Station Coordinates[e]					
Positions	Horizontal	3 mm	12 days	Weekly	Weekly
	Vertical	6 mm			
Velocities	Horizontal	2 mm yr^{-1}	12 days	Weekly	~Years
	Vertical	3 mm yr^{-1}			
Earth Rotation Parameters[f]					
Ultrarapid	Pole position	0.3 mas			
(predicted half)	Pole rate	0.5 mas day^{-1}	Real time	6 h	6 h
	Length of day	0.06 ms			
Ultrarapid	Pole position	0.1 mas			
(observed half)	Pole rate	0.3 mas day^{-1}	3 h	6 h	6 h
	Length of day	0.03 ms			
Rapid	Pole position	<0.1 mas			
	Pole rate	<0.2 mas day^{-1}	17 h	Daily	Daily
	Length of day	0.3 ms			
Final	Pole position	0.05 mas			
	Pole rate	0.2 mas day^{-1}	13 Days	Weekly	Daily
	Length of day	0.2 ms			

[a]Generally, precision (based on scatter of solutions) is better than the accuracy (based on comparison with independent methods).
[b]Broadcast ephemerides only shown for comparison (but are also available from IGS).
[c]Orbit accuracy based on comparison with satellite laser ranging to satellites.
[d]Clock accuracy is expressed relative to the IGS timescale, which is linearly aligned to GPS time in 1 day segments.
[e]Station coordinate and velocity accuracy based on intercomparison statistics from ITRF.
[f]Earth rotation parameters based on intercomparison statistics by IERS. IGS uses VLBI results from IERS Bulletin A to calibrate for long-term LOD biases.
Source: IGS Central Bureau, http://igscb.jpl.nasa.gov, courtesy of A. Moore.

These scientific services, as well as gravity field services and an expected future altimetry service, are integral components of the future GGOS (Rummel *et al.*, 2005). Closer cooperation and understanding through GGOS is expected to bring significant improvements to the ITRF and to scientific uses of geodesy in general (Dow *et al.*, 2005a).

Figure 4 shows the current status of co-located space geodetic sites, which forms the foundation for ITRF and GGOS. Co-location is essential to exploit the synergy of the various techniques, and so

increasing the number and quality of co-located sites will be a high priority for GGOS.

11.2 GPS System and Basic Principles

11.2.1 Basic Principles

GPS positioning is based on the principle of 'trilateration', which is the method of determining position by measuring distances to points of known positions (not to be confused with triangulation,

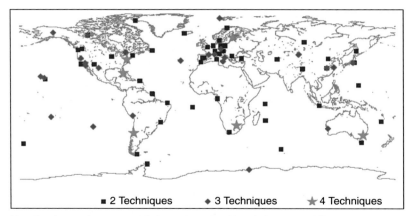

■ 2 Techniques   ◆ 3 Techniques   ★ 4 Techniques

Figure 4   Distribution of co-located space geodetic stations that have at least two different operational techniques of GPS, VLBI, SLR, and DORIS. Courtesy of Z. Altamimi.

which measures angles between known points). At a minimum, trilateration requires three ranges to three known points. In the case of GPS, the known points would be the positions of the satellites in view. The measured ranges would be the distances between the GPS satellites and a user's GPS receiver. (Note that GPS is a completely passive system from which users only receive signals). GPS receivers, on the other hand, cannot measure ranges directly, but rather 'pseudoranges'. A pseudorange is a measurement of the difference in time between the receiver's local clock and an atomic clock on board a satellite. The measurement is multiplied by the speed of light to convert it into units of range (meters):

$$\text{Pseudorange} = (\text{receiver time} - \text{satellite time}) \\ \times \text{ speed of light} \qquad [1]$$

The satellite effectively sends its clock time by an encoded microwave signal to a user's receiver. It does this by multiplying a sinusoidal carrier wave by a known sequence ('code') of $+1$ and -1, where the timing of the signal (both code and carrier wave) is controlled by the satellite clock. The receiver generates an identical replica code, and then performs a cross-correlation with the incoming signal to compute the required time shift to align the codes. This time shift multiplied by the speed of light gives the pseudorange measurement.

The reason the measurement is called a pseudorange is that the range is biased by error in the receiver's clock (typically a quartz oscillator). However, this bias at any given time is the same for all observed satellites, and so it can be estimated as one extra parameter in the positioning solution. There are also (much smaller)

errors in the satellites' atomic clocks, but GPS satellites handle this by transmitting another code that tells the receiver the error in its clock (which is routinely monitored and updated by the US Department of Defense).

Putting all this together, point positioning with GPS therefore requires pseudorange measurements to at least four satellites, where information on the satellite positions and clocks are also provided as part of the GPS signal. Three coordinates of the receiver's position can then be estimated simultaneously along with the receiver's clock offset. By this method, GPS positioning with few-meter accuracy can be achieved by a relatively low-cost receiver.

Hence GPS also allows the user to synchronize time to the globally accessible atomic standard provided by GPS. In fact, the GPS atomic clocks form part of the global clock ensemble that define UTC. Note that since GPS time began (6 January, 1980) there have accumulated a number of leap seconds (14 s as of 2006) between GPS time (a continuous timescale), and UTC (which jumps occasionally to maintain approximate alignment with the variable rotation of the Earth). Synchronization to GPS time (or UTC) can be achieved to <0.1 μs using a relatively low-cost receiver. This method is suitable for many time-tagging applications, such as in seismology, SLR, and even for GPS receivers themselves. That is, by using on-board point positioning software, GPS receivers can steer their own quartz oscillator clocks through a feedback mechanism such that observations are made within a certain tolerance of GPS time.

A fundamental principle to keep in mind is that GPS is a timing system. By use of precise timing information on radio waves transmitted from the GPS satellite, the user's receiver can measure the range to each satellite in view, and hence calculate its position. Positions can be calculated at every measurement epoch, which may be once per second when applied to car navigation (and in principle as frequently as 50 Hz). Kinematic parameters such as velocity and acceleration are secondary, in that they are calculated from the measured time series of positions.

11.2.2 GPS System Design and Consequences

The GPS system has three distinct segments:

1. The Space Segment, which includes the constellation of ~30 GPS satellites that transmit the signals from space down to the user, including signals that enable a user's receiver to measure the biased range (pseudorange) to each satellite in view, and signals that tell the receiver the current satellite positions, the current error in the satellite clock, and other information that can be used to compute the receiver's position.

2. The Control Segment (in the US Department of Defense) which is responsible for the monitoring and operation of the Space Segment, including the uploading of information that can predict the GPS satellite orbits and clock errors into the near future, which the Space Segment can then transmit down to the user.

3. The User Segment, which includes the user's GPS hardware (receivers and antennas) and GPS data-processing software for various applications, including surveying, navigation, and timing applications.

The satellite constellation is designed to have at least four satellites in view anywhere, anytime, to a user on the ground. For this purpose, there are nominally 24 GPS satellites distributed in six orbital planes. In addition, there is typically an active spare satellite in each orbital plane, bringing the total number of satellites closer to 30. The orientation of the satellites is always changing, such that the solar panels face the Sun, and the antennas face the centre of the Earth. Signals are transmitted and received by the satellite using microwaves. Signals are transmitted to the User Segment at frequencies L1 = 1575.42 MHz, and L2 = 1227.60 MHz in the direction of the Earth. This signal is encoded with the 'Navigation Message', which can be read by the user's GPS receiver. The Navigation Message includes orbit parameters (often called the 'Broadcast Ephemeris'), from which the receiver can compute satellite

coordinates (X,Y,Z). These are Cartesian coordinates in a geocentric system, known as WGS-84, which has its origin at the Earth centre of mass, Z axis pointing toward the North Pole, X pointing toward the Prime Meridian (which crosses Greenwich), and Y at right angles to X and Z to form a right-handed orthogonal coordinate system. The algorithm which transforms the orbit parameters into WGS-84 satellite coordinates at any specified time is called the 'Ephemeris Algorithm'. For geodetic purposes, precise orbit information is available over the Internet from civilian organizations such as the IGS in the Earth-fixed reference frame.

According to Kepler's laws of orbital motion, each orbit takes the approximate shape of an ellipse, with the Earth's centre of mass at the focus of the ellipse (**Figure 5**). For a GPS orbit, the eccentricity of the ellipse is so small (0.02) that it is almost circular. The semimajor axis (largest radius) of the ellipse is approximately 26 600 km, or approximately four Earth radii.

The six orbital planes rise over the equator at an inclination angle of 55°. The point at which they rise from the Southern to Northern Hemisphere across the equator is called the 'Right Ascension of the ascending node'. Since the orbital planes are evenly distributed, the angle between the six ascending nodes is 60°.

Each orbital plane nominally contains four satellites, which are generally not spaced evenly around the ellipse. Therefore, the angle of the satellite within its own orbital plane, the 'true anomaly', is only

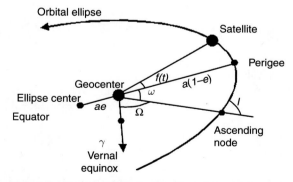

Figure 5 Diagram illustrating the Keplerian orbital elements: semimajor axis *a*, eccentricity *e*, inclination *I*, argument of perigee (closest approach) ω, Right Ascension of the ascending node Ω, and true anomaly *f* as a function of time *t*. The geocenter is the Earth center of mass; hence, satellite geodesy can realize the physical origin of the terrestrial reference system. This diagram is exaggerated, as GPS orbits are almost circular.

approximately spaced by 90°. The true anomaly is measured from the point of closest approach to the Earth (the perigee). Instead of specifying the satellite's anomaly at every relevant time, it is equivalent to specify the time that the satellite had passed perigee, and then compute the satellites future position based on the known laws of motion of the satellite around an ellipse. Finally, the argument of perigee specifies the angle between the equator and perigee. Since the orbit is nearly circular, this orbital parameter is not well defined, and alternative parameterization schemes are often used.

Taken together (the eccentricity, semimajor axis, inclination, Right Ascension of the ascending node, the time of perigee passing, and the argument of perigee), these six parameters define the satellite orbit (according to the Keplerian model). These parameters are known as Keplerian elements. Given the Keplerian elements and the current time, it is possible to calculate the coordinates of the satellite.

However, GPS satellites do not move in perfect ellipses, so additional parameters are necessary. Nevertheless, GPS does use Kepler's laws to its advantage, and the orbits are described in the Broadcast Ephemeris by parameters which are Keplerian in appearance. Additional parameters must be added to account for non-Keplerian behavior. Even this set of parameters has to be updated by the Control Segment every hour for them to remain sufficiently valid.

Several consequences of the orbit design can be deduced from the above orbital parameters, and Kepler's laws of motion. First of all, the satellite speed is $\sim 4\,\mathrm{km\,s^{-1}}$ relative to Earth's center. All the GPS satellites orbits are prograde, which means the satellites move in the direction of Earth's rotation. Therefore, the relative motion between the satellite and a user on the ground must be less than $4\,\mathrm{km\,s^{-1}}$. Typical values around $1\,\mathrm{km\,s^{-1}}$ can be expected for the relative speed along the line of sight (range rate).

The second consequence is the phenomena of 'repeating ground tracks' every day. The orbital period is approximately $\mathrm{T} = 11\,\mathrm{h}\,58\,\mathrm{min}$, therefore a GPS satellite completes two revolutions in 23 h 56 min. This is intentional, as it equals one sidereal day, the time it takes for the Earth to rotate 360°. Therefore, everyday (minus 4 min), the satellite appears over the same geographical location on the Earth's surface. The 'ground track' is the locus of points on the Earth's surface that is traced out by a line connecting the satellite to the centre of the Earth. The ground track is said to repeat. From the user's point of view, the same satellite appears in the same direction in the sky every day minus 4 min. Likewise, the 'sky tracks' repeat.

So from the point of view of a ground user, the entire satellite geometry repeats every sidereal day. Consequently, any errors correlated with satellite geometry will repeat from one day to the next. An example of an error tied to satellite geometry is 'multipath', which is due to the antenna also sensing signals from the satellite which reflect and refract from nearby objects. In fact, it can be verified that, because of multipath, observation residuals do have a pattern that repeats every sidereal day. Therefore such errors will not significantly affect the repeatability of coordinates estimated each day. However, the accuracy can be significantly worse than the apparent precision for this reason.

Another consequence of this is that the same subset of the 24 satellites will be observed everyday by someone at a fixed geographical location. Generally, not all 24 satellites will be seen by a user at a fixed location. This is one reason why there needs to be a global distribution of receivers around the globe to be sure that every satellite is tracked sufficiently well.

The inclination angle of 55° also has consequences for the user. Note that a satellite with an inclination angle of 90° would orbit directly over the poles. Any other inclination angle would result in the satellite never passing over the poles. From the user's point of view, the satellite's sky track would never cross over the position of the celestial pole in the sky. In fact, there would be a 'hole' in the sky around the celestial pole where the satellite could never pass. For a satellite constellation with an inclination angle of 55°, there would therefore be a circle of radius at least 35° around the celestial pole, through which the sky tracks would never cross. This has a big effect on the satellite geometry as viewed from different latitudes. An observer at the pole would never see a GPS satellite rise above 55° elevation. Most of the satellites would hover close to the horizon. Therefore, vertical positioning is slightly degraded near the poles. An observer at the equator would see some of the satellites passing overhead, but would tend to deviate away from points on the horizon directly to the north and south.

Due to a combination of Earth rotation, and the fact that the GPS satellites are moving faster than the Earth rotates, the satellites actually appear to move approximately north–south or south–north to an observer at the equator, with very little east–west motion. Therefore, the closer the observer is to the equator, the better determined becomes the north component of relative position as compared to the east component. An observer at mid-latitudes in the Northern Hemisphere would see satellites anywhere in the sky to the south, but there would be a large void toward the north. This has consequences for site selection, where a good view is desirable to the south, and the view to the north is less critical. For example, one might want to select a site in the Northern Hemisphere which is on a south-facing slope (and vice versa for an observer in the Southern Hemisphere).

11.2.3 Introducing High-Precision GPS

By measuring pseudoranges to at least four satellites with relatively low-cost equipment, GPS can readily provide users with a positioning accuracy of meters, and a timing accuracy of 0.1 μs. On the other hand, geodetic GPS positioning with an accuracy of a few millimeters requires a number of significant improvements to the technique described above, which will be emphasized in this section. For example, accurate positioning requires accurate knowledge of the GPS satellite positions and satellite clock offsets. For standard GPS positioning, this 'ephemeris' information is broadcast by the GPS satellites in the so-called 'Navigation Message'; however, it is not sufficiently accurate for geodetic applications.

In addition to the three GPS segments listed above, one could informally include the 'Service Segment' consisting of civilian networks that provide the User Segment with data and services to enhance positioning accuracy. This information can be transmitted to the user in a variety of ways, such as by the Internet, cell phone, and geostationary satellite. The part of this Service Segment that is relevant to geodetic positioning would be the IGS, an international collaboration of geodesists that provides high-accuracy data on satellite orbits and clocks. IGS also provides data from reference stations around the globe, at accurately known coordinates that account for plate tectonics and other geophysical movements such as earthquakes. Thus the IGS enables users to position their receivers anywhere on the globe with an accuracy of millimeters in a consistent terrestrial reference frame. But this only solves one of the many problems toward achieving geodetic precision.

In practice, high-precision geodesy requires a minimum of five satellites in view, because it is essential to estimate parameters to model tropospheric refraction. At an absolute minimum, one zenith delay is estimated,

which can be mapped to delay at any elevation angle using a 'mapping function' based on tropospheric models.

Geodetic applications require much more sophisticated GPS receivers that not only measure the pseudorange observable, but the so-called 'carrier phase' observable. The carrier phase observable is the difference between (1) the phase of the incoming carrier wave (upon which the codes are transmitted) and (2) the phase of a signal internally generated by the receiver which is synchronized with the receiver clock. When multiplied by the ~20 cm wavelength of the carrier wave, the result is a biased distance to the satellite. Indeed this is a type of pseudorange that is about 100 times more precise than the coded pseudoranges. The downside to the carrier phase observable is that in addition to the receiver clock bias, there is an additional bias of an unknown number of wavelengths. It is possible to resolve this bias exactly by so-called 'ambiguity resolution' techniques. Ambiguity resolution is essential to achieve the highest possible precision for geodetic applications. Hence in units of range, the observed carrier phase can be expressed:

$$\text{Carrier phase} = (\text{reference phase} - \text{signal phase} + \text{integer}) \times \text{carrier wavelength} \quad [2]$$

Note that the signal phase is generated by the satellite clock, and that the reference phase is generated by the receiver clock, hence eqn [2] is just a very precise form of eqn [1] for the pseudorange, except that it has an integer-wavelength ambiguity. (In fact this is why the sign of the phase difference was chosen by subtracting the incoming signal phase from the reference phase.) Therefore the observable models for eqns [1] and [2] are very similar, and relate to the theoretical difference between the reading of the receiver clock (time of reception) and the satellite clock (time of transmission), including clock biases.

This similarity of models has enabled the development of automatic signal processing algorithms to check the integrity of the data, such as the detection of data outliers and jumps in the integer ambiguity (so called 'cycle slips'), which occur when the receiver loses lock on the signal, for example, due to a temporary obstruction between the ground antenna and the satellite. In fact, the pseudorange data can be used together with the carrier phase data to correct for the initial integer ambiguity (Blewitt, 1989) and for subsequent cycle clips (Blewitt, 1990).

For geodetic positioning, both pseudoranges and carrier phases are measured at two different frequencies (L1 at 19.0 cm wavelength, and L2 at 24.4 cm), to provide a self-calibration of delay in the Earth's ionosphere. So in total there are four observations that are fundamental to high-precision GPS geodesy: two pseudoranges, and two carrier phases. This enables more algorithms to assure the integrity of the data, and allows for monitoring of the ionosphere itself.

Another requirement for geodetic positioning is the use of highly specialized stochastic multiparameter estimation software by modeling the carrier phase data, including modeling of the satellite-station geometry, Earth's atmosphere, solid-Earth tides, Earth rotation, antenna effects, circular polarization effects (phase wind-up), and relativistic effects (both special and general). In addition the software must be capable of detecting and correcting integer offsets in the carrier phase observables (cycle slips), and must be capable of resolving the integer ambiguity in the initial phase measurements.

In summary, therefore, geodetic GPS requires:

- geodetic-class GPS receivers capable of acquiring dual-frequency carrier phase data;
- geodetic-class satellite orbit and clock information, which is available from the IGS;
- simultaneous observations to a minimum of five satellites; and
- specialized postprocessing software (not on the receiver itself) that embodies high-accuracy observable models, carrier phase data processing algorithms, and simultaneous parameter estimation.

The quality of the IGS orbit and clock data depends on their latency, so generally there is a tradeoff between latency and accuracy. Currently, the ultrarapid IGS product is actually a prediction from 3–9 h ago. Even though there are atomic clocks on board the GPS satellites, the clock time is much more difficult to predict than the satellite orbits. In the case that sufficiently accurate clock data are not yet available, it is nevertheless possible to produce geodetic-class solutions for relative positions between ground stations. This is achieved either by (1) solving for satellite clock biases at every epoch as part of the positioning solution, or equivalently by (2) differencing data between ground stations to cancel out the clock bias. Furthermore, data can be differenced again ('double difference') between

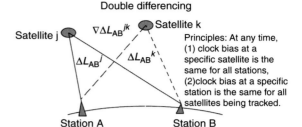

Figure 6 Diagram illustrating double differencing of GPS data. The idea is to difference away the satellite and station clock biases. Double differencing is equivalent to estimating the clock biases explicitly when processing undifferenced data.

satellites to cancel out the receiver clock bias rather than estimate it as a parameter (**Figure 6**).

In practice, the following different approaches to estimating positions all give results that are of geodetic quality (with errors measured in millimeters) and typically agree very well:

- Precise point positioning (PPP) of single stations using precise orbit and clock data.
- Relative positioning of networks by clock estimation, using precise orbit data.
- Relative positioning of networks by double-differenced data (**Figure 6**), using precise orbit data.

All of these three methods are in common use today for geophysical research purposes. In each case, dual-frequency pseudorange and carrier phase data types are used.

11.2.4 GPS Observable Modeling

This section describes how GPS observables are typically modeled by geodetic-quality software packages. First, however, a few more specific details on the GPS signals are required. The signals from a GPS satellite are fundamentally driven by an atomic clock precisely at frequency 10.23 MHz. Two sinusoidal carrier signals are generated from this signal by multiplying the frequency by 154 for the L1 channel (frequency = 1575.42 MHz; wavelength = 19.0 cm), and 120 for the L2 channel (frequency = 1227.60 MHz; wavelength = 24.4 cm). Information is encoded in the form of binary bits on the carrier signals by a process known as phase modulation. The binary digits 0 and 1 are actually represented by multiplying the electrical signals by either +1 or −1.

For purposes of observable modeling, here the observables (all in units of meters) will be called L1 and L2 for the two types of carrier phase, and P1 and P2 for the two types of pseudorange. The actual observable types are numerous due to different methods of correlating the signals; however, the fundamental observation equations can be written in the same generic way, with the exception that there is generally a bias associated with each observable types, including instrumental bias and, in the case of the carrier phase, an integer-wavelength bias. (For some older signal-squaring receivers, the bias is a half-integer wavelength.) Here it is simply assumed that such biases are not problematic, which is typically the case, and so the inter-observable biases are not explicitly modeled.

Taking eqns [1] and [2], the generic (pseudorange or carrier phase) GPS observation P_j^i at receiver j (subscript, on the ground) from satellite i (superscript, up in space) can be modeled:

$$P_j^i = c(T_j - \bar{T}^i) + B_j^i \qquad [3]$$

Since special and general relativity prove to be important in the model, care must be taken to define each term with respect to a reference frame. Thus, T_j is the time according to the receiver clock coincident with signal reception (used as the time-tag, recorded with the observation), $\bar{T}^i$ is the time according to the satellite clock coincident with signal transmission (which imprints its signature on the signal, hence the bar, which denotes time local to the satellite), B_j^i is a frame-invariant bias associated with this type of observation, and c is the frame-invariant speed of light in a vacuum. In addition, this observation is recorded at an epoch with time-tag T_j (the same for all satellites observed at that epoch).

The clock difference can be rewritten as the sum of four time differences:

$$P_j^i = c\{(T_j - t_j) + (t_j - t^i) + (t^i - \bar{t}^i) + (\bar{t}^i - \bar{T}^i)\} + B_j^i \qquad [4]$$

where t_j is the coordinate time at the receiver, t^i is the coordinate time at the satellite, and t^{-i} is the proper time at the satellite (the time kept by a perfect clock on board the satellite). 'Coordinate time' simply means the timescale that is actually used to compute the models. It is convenient to take coordinate time in the 'local Earth' frame (that of a perfect clock on the

geoid) (Ashby and Allan, 1984). Appropriate time-scales for this purpose include Terrestrial Dynamic Time (TDT) and International Atomic Time (TIA), but for the discussion here it is convenient to choose GPS time. The important thing to keep in mind (to cut through the confusion of all these conventions) is that all these timescales ideally run at the same rate as UTC, with the unit of time being the SI second (Kaplan, 1981), and so all these scales only differ by conventional constant offsets (and leap seconds).

The four time-difference terms found in eqn [4] can be written as follows. First, the difference in receiver clock time and coordinate time is simply the receiver clock bias, which we will model as an independent parameter τ_j at every epoch (at every value of T_j):

$$(T_j - t_j) = \tau_j \qquad [5]$$

The clock bias includes the sum of a clock error (with respect to proper time) plus a minor relativistic bias due to the geodetic location of the receiver clock.

The second term is the difference between coordinate time at the receiver and satellite, the so-called 'light-time equation' (here expressed as a range):

$$c(t_j - t^i) = r_j^i + \sum_{\text{prop}} \Delta r_{\text{prop}\,j}^{\,i}$$
$$= |\mathbf{r}_j(t_j) - \mathbf{r}^i(t^i)| + \Delta r_{\text{GR}\,j}^{\,i} + \Delta r_{\text{ion}\,j}^{\,i} + \Delta r_{\text{trop}\,j}^{\,i} + \Delta r_{\text{pev}\,j}^{\,i} + \Delta r_{\text{circ}\,j}^{\,i} + K \qquad [6]$$

where r_j^i is the Euclidean distance between the satellite and receiver, and $\Delta r_{\text{prop}\,j}^{\,i}$ represent various propagation delays, which are a function of station-satellite geometry, arising from space-time curvature (general relativity), ionosphere, troposphere, antenna phase center variations, circular polarization effects, and other propagation terms as necessary. In eqn [6], r_j is the geocentric receiver position at the time of reception, r^i is the geocentric satellite position at the time of transmission. The reference frame for the light-time equation is taken to be J2000, the conventional Earth-centered inertial (ECI) frame (so the axes do not co-rotate with the Earth), as this is most convenient for integrating the satellite equations of motion.

The general relativistic delay can be computed as:

$$\Delta r_{\text{GR}\,j}^{\,i} = \frac{2G^{M_\oplus}}{c^2} \ln \frac{r_j + r^i + r_i^j}{r_j + r^i - r_i^j} \qquad [7]$$

where $G^M \oplus$ is the Earth's gravitational constant, and in general, $r \equiv |\mathbf{r}|$. Antenna effects such as phase center variation $\Delta r_{\mathrm{pev}\,j}^{\,i}$ (Schupler *et al.*, 1994) and circular polarization $\Delta r_{\mathrm{circ}\,j}^{\,i}$ ('phase wind-up') (Wu *et al.*, 1993) are examples of important effects that have been researched and applied to improve positioning accuracy, but as non-geophysical effects they are beyond the scope of this text.

Ionospheric delay can be adequately modeled as being inversely proportional to the squared frequency f of the carrier wave:

$$\Delta r_{\mathrm{ion}\,j}^{\,i}(f) = \pm k \frac{\mathrm{TEC}_j^i}{f^2} \qquad [8]$$

where the positive sign is taken for pseudoranges, and the negative sign for carrier phase observations. The term TEC refers to 'total electron content', which is excited by solar radiation and so is highly variable through the day and is sensitive to geographic location. The constant k can be derived from the theory of electromagnetic wave propagation in plasmas. Delays at GPS frequencies can be as large as 100 m near the equator, peaking around 2 pm local time, and can be as small as centimeters at mid-latitudes between midnight and dawn. In contrast, higher-order terms are of the order of millimeters and are typically ignored, although including them in the model is one of many themes of current research (Kedar *et al.*, 2003). An appropriate linear combination of observations eliminates the frequency-squared term exactly, leaving all nondispersive terms in the model unchanged. In fact, the 'ionosphere-free' combination of carrier phases can be so defined (and similarly for the pseudoranges):

$$\mathrm{LC} = \frac{f_1^2 L1 - f_2^2 L2}{(f_1^2 f_2^2)}$$
$$\cong 2.546 L1 - 1.546 L2 \qquad [9]$$

Hence k and TEC are not explicitly needed to compute the ionosphere-free data. The coefficients above can be computed exactly by substituting $f_1 = 154$ and $f_2 = 120$, owing to the properties of the GPS signals described at the beginning of this section. As an aside, if the ionosphere is the geophysical scientific target of interest, then differencing the observations at two different frequencies results in a 'geometry-free' observation from which TEC can be estimated:

$$\mathrm{PI} = P1 - P2$$
$$= k \cdot \mathrm{TEC}\left(\frac{1}{f_1^2} - \frac{1}{f_2^2}\right) + \mathrm{bias} \qquad [10]$$

Using GPS stations located around the globe, this method is now routinely used to map ionospheric TEC. A side benefit of this method is the estimation of the interchannel bias between observables at L1 and L2 frequency, which can be monitored for long-term variability and used as input to ambiguity-resolution algorithms.

The tropospheric delay is almost entirely nondispersive (independent of frequency) at GPS L-band frequencies, and so must be handled in a different way. Whereas it is possible in principle to model tropospheric delay based on ground-based meteorological observations, in practice this has not proved to be sufficiently accurate. The key to successful tropospheric modeling is the estimation of the delay at zenith, by accurately modeling the relationship between zenith delay Z and delay at lower elevations ε, for example:

$$\Delta r_{\mathrm{trop}\,j}^{\,i} = \frac{Z_j}{\sin \varepsilon_j^i} \qquad [11]$$

where the inverse sign of elevation angle is the simplest example of a 'mapping function', which can be derived by assuming a horizontally layered troposphere over a flat Earth. This model breaks down rapidly for $\varepsilon < 20°$. More accurate modeling (Truehaft and Lanyi, 1987) requires modifying the mapping function to account for Earth curvature, and partitioning the delay into so-called dry and wet components which have different characteristic scale heights (~ 10 and ~ 2 km, respectively):

$$\Delta r_{\mathrm{trop}\,j}^{\,i} = \Delta r_{\mathrm{dry}\,j}^{\,i} + \Delta r_{\mathrm{wet}\,j}^{\,i}$$
$$= Z_{\mathrm{dry}} F_{\mathrm{dry}}(\varepsilon_j^i) + Z_{\mathrm{wet}} F_{\mathrm{wet}}(\varepsilon_j^i) \qquad [12]$$

Due to the inherent weakness in the determination of height with both GPS and VLBI, accurate modeling of mapping functions has always been and remains an active area of research. The wet delay is caused by the interaction of the electromagnetic (EM) wave with the static dipole of molecular water. The dry delay is due to the dynamic dipole induced by the EM wave on all component molecules in the atmosphere, including a (small) contribution from water (and so 'dry' is just a conventional, perhaps misleading term). Typical values for the dry and wet delay are 2.1, and 0.1 m, respectively, to within ~ 10 cm.

The dry component can be adequately modeled as a function of hydrostatic pressure at the altitude of the receiver. Nominal values can be computed in the absence of meteorological data by assuming a nominal surface pressure at sea level, and then subtracting a correction for altitude, assuming that pressure decays exponentially with altitude. The wet component is typically assumed to have a nominal value of zero, and Z_{wet} is then estimated from the GPS data along with the positioning solution. Note that in this case, the estimated value of Z_{wet} would absorb most (but not all) the obvious inadequacies of the nominal model for Z_{dry}. Whereas this is currently the standard method in high-precision GPS geodesy, the limitations of this approach is an active area of research (Tregoning and Herring, 2006).

The tropospheric delay model is important not only for solving for geodetic position, but also for the study of the troposphere itself. For this application, estimates of troposphere delay to solve for precipitable water vapor in the atmosphere, which can then be used as input for weather forecasting and climate modeling. For this application, surface meteorological data is essential to more accurately partition the dry and wet components of delay (Bevis et al., 1992).

Now returning to the light-time equation, even if we had perfect propagation models, the light-time equation needs to be solved iteratively rather than simply computed, because at first we do not have a nominal value for t^i, the coordinate time of signal transmission. The procedure is as follows.

- Starting with the observation time-tag T_j, use eqn [5] and a nominal value for the receiver clock bias τ_j (which may be zero, or a preliminary estimate) to compute the coordinate time of signal reception t_j. Note that the assumed clock bias affects the subsequent computation of geometric range, indicating the need for iterative estimation. (This problem can be more conveniently addressed by accounting for the range rate in the partial derivative with respect to the receiver clock parameter.)
- Given a modeled station position r_j at time t_j, and an interpolated table of modeled satellite positions r^i as a function of coordinate time t^i, iteratively compute the coordinate time of transmission using

$$ct^j[n+1] = ct^j[n]$$
$$+ \frac{ct_j - ct^j[n] - r^i_j[n] - \sum_{prob} \Delta r_{prop}{}^i_j[n]}{1 - \dot{r}^i[n] \cdot \hat{r}^i_j[n]/c} \quad [13]$$

where $\hat{r}^i_j = r^i_j/r^i_j$ are the direction cosines to the receiver from the satellite. It is to be understood that at the nth iteration, for example, the satellite

position r^i and velocity $\dot{r}^i$ are both interpolated to time $t^i[n]$. By virtue of this equation converging very quickly, it is sufficient to initialize the transmission time to the reception time $t^i[0] = t_j$.

Being in the ECI frame (J2000), the receiver position must account for Earth rotation and geophysical movements of the Earth's surface:

$$r_j(t_j) = \textbf{PNUXY} \left[x_{0j} + \sum_k \Delta x_{kj}(t_j) \right] \quad [14]$$

Here **PNUXY** is the multiple of 3×3 rotation matrices that account (respectively) for precession, nutation, rate of rotation, and polar motion (in two directions). The bracketed term represents the receiver position in the (co-rotating) conventional Earth-fixed terrestrial reference frame known as ITRF. Conventional station position x_{0j} is specified by station coordinates in ITRF at some conventional epoch, and $\Delta x_{kj}(t_j)$ represents the displacement from the epoch position due to geophysical process k, for example, accounting for the effects of plate tectonics, solid-Earth tides, etc. Equation [14] together with [6] form the fundamental basis of using GPS as a geophysical tool, and this will be explored later.

Returning now to the original observation eqn [4], the third term is the difference between coordinate time and proper time at the satellite. According to special relativity, GPS satellite clocks run slow relative to an observer on the Earth's surface due to relative motion. In contrast, general relativity predicts that the satellite clocks will appear to run faster when observed from the Earth's surface due to the photons gaining energy (gravitational blue shift) as they fall into a gravitational well. These two effects do not entirely cancel. With appropriate foresight in the design of GPS, the satellite clocks operate at a slightly lower frequency than 10.23 MHz to account for both special and general relativity, so that their frequency would be 10.23 MHz (on average) as viewed on the Earth's surface. The residual relativistic term can be computed from:

$$(t^i - \bar{t}^i) = 2r^i(t^i) \cdot \dot{r}^i(t^i)/c^2 \quad [15]$$

This result assumes an elliptical orbit about a point mass, and assumes that additional relativistic effects are negligible. The expected accuracy is at the level of 10^{-12}, comparable to the level of stability of the satellite atomic clocks.

The fourth term is the difference between proper time and clock time at the satellite, which is simply the negative error in the satellite clock:

$$\left(\bar{t}^i - \bar{T}^i\right) = -\tau^i \qquad [16]$$

Unlike errors in the station clock, errors in the satellite clock do not affect the model of geometric range (in the light-time equation) and so are not required in advance to compute that part of the model. However, they do affect the observable itself, and so not accounting for satellite clock error would result in an error in estimated receiver position. Despite the satellite clocks being atomic, they are not sufficiently stable to be predicted forward in time for geodetic applications (though this is the method for standard positioning with GPS). Therefore this term is estimated independently at every epoch as part of the positioning solution (or alternatively, observation equations can be differenced between pairs of observing receivers to eliminate this parameter). As a consequence, geodetic-quality precise point positioning of individual stations in real time presents a challenge that is the topic of current research.

Finally, each observation type has an associated bias. Typically receivers are designed so that these biases should be either calibrated or are stable in time. Like the case for the satellite clock error, these biases have no effect on the computed geometric range, and so do not need to be known in advance, however they are present in the observations themselves, and so can affect positioning accuracy unless they are absorbed by parameters in the least-squares solution. It turns out for the most part that biases between observable types can be ignored for purposes of positioning, because they can be absorbed into the station or satellite clock bias parameters as part of the least-squares positioning solution. For purposes of accurate timing, however, special considerations are required to calibrate such biases. Some of the interobservable biases are monitored by major GPS analysis centers and made routinely available if needed.

The most important bias to consider for geodetic applications is the carrier phase bias which has an integer ambiguity (Blewitt, 1989). The carrier phase bias is not predictable from models, and can vary by integer jumps occasionally (Blewitt, 1990). For an initial solution, the carrier phase biases can be nominally assumed to be zero (because they do not affect the light-time equation), and then estimated as real-valued parameters. Methods to resolve these integer ambiguities exactly, along with their discrete changes in time, will in the next section take us to the topic of data processing algorithms. Such automated algorithms are essential to achieve the highest positioning accuracies, and should be discussed in the context of understanding sources of error. Following that it will be explained how parameters of the model are estimated.

In summary, this has been a key section, in that the light eqn [6] represents the heart of GPS observable modeling, and a very specific component given by is the source of all geophysical applications that relate to precise positioning. What remains to be explained are the algorithms used to process the data, and the strategy to estimate biases in the assumed nominal values of the parameters, thereby realizing the full potential accuracy of the observation model.

11.2.5   Data Processing Software

Geodetic GPS data processing as implemented by research software packages can typically be generalized as a modular scheme (**Figure 7**). In this processing model, the input is raw data from GPS receivers, and the processing stops with the production of a set of station coordinates. Before discussing data processing in detail, it should be noted that the data processing does not stop with the initial production of station coordinates, but rather this is the first step toward time series analysis, velocity estimation, kinematic analysis, all leading to dynamic analysis and geophysical interpretation. It is convenient to separate the actual processing of GPS data shown above from the subsequent kinematic analysis, though for some geophysical applications (e.g., ocean tidal loading) this division is not correct, and geophysical parameters must be estimated directly in the solution.

Several software packages have been developed since the 1980s that are capable of delivering high-precision geodetic estimates over long baselines. These days, the processing of GPS data by these software packages is, to a large degree, automatic, or at least a 'black-box' approach is common. The black box can of course be tampered with for purposes of research into GPS methodology, but one big advantage of automation is reproducibility and consistency of results produced in an objective way.

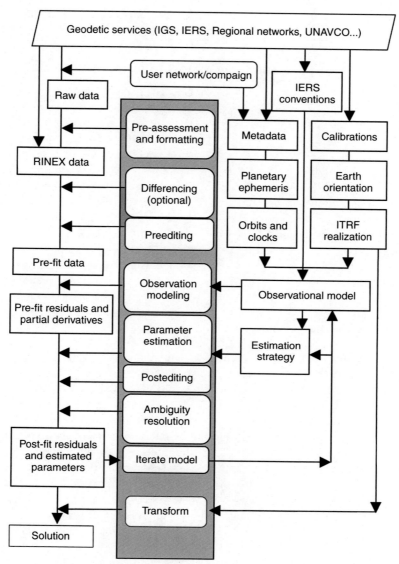

Figure 7   Generic modular scheme for geodetic GPS data processing.

Geodetic data processing software is a result of intensive geodetic research, mainly by universities and government research laboratories. Typical features of such software include:

- orbit integration with appropriate force models;
- accurate observation model (Earth model, media delay) with rigorous treatment of celestial and terrestrial reference systems;
- reliable data editing (cycle slips, outliers);
- estimation of all coordinates, orbits, tropospheric bias, receiver clock bias, polar motion, and Earth spin rate;

- ambiguity resolution algorithms applicable to long baselines; and
- estimation of reference frame transformation parameters and kinematic modeling of station positions to account for plate tectonics and co-seismic displacements.

The typical quality of geodetic results from processing 24 h of data can be summarized as follows:

- relative positioning at the level of few parts per billion of baseline length;
- geocentric (global) positioning to <6 mm in the ITRF;

- tropospheric delay estimated to <5 mm;
- GPS orbits determined to <5 cm;
- Earth pole position determined to <2 mm;
- clock synchronization (relative bias estimation) to <0.1 ns; and
- ionospheric TEC maps to <10 TEC units.

Two features of commercial software are often conspicuously absent from more advanced packages: (1) sometimes double differencing is not implemented, but instead, undifferenced data are processed, and clock biases are estimated; (2) network adjustment using baseline solutions is unnecessary, since advanced packages do a rigorous, one-step, simultaneous adjustment of station coordinates directly from all available GPS observations.

Some precise software packages incorporate a Kalman filter (or an equivalent formulism) (Bierman, 1977; Lichten and Border, 1987; Herring *et al.*, 1990). This allows for certain selected parameters to vary in time, according to a statistical ('stochastic') model. Typically this is used for the tropospheric bias, which can vary as a random walk in time (Tralli and Lichten, 1990). A filter can also be used to estimate clock biases, where 'white noise' estimation of clock bias approaches the theoretical equivalent of double differencing.

Although many more packages have been developed, there are three ultrahigh-precision software packages which are widely used around the world by researchers and are commonly referenced in the scientific literature:

- BERNESE software, by Astronomical Institute, University of Bern, Switzerland (Rothacher *et al.*, 1990);
- GAMIT-GLOBK software, by MIT, USA (King and Bock, 2005)
- GIPSY-OASIS II software, by JPL, California Institute of Technology, USA (Webb and Zumberge, 1993).

There are several other packages, but they tend to be limited to the institutions that wrote them. It should be noted that, unlike commercial software packages, use of the above software can require a considerable investment in time to understand the software and how best to use it under various circumstances. Expert training is essential.

11.3 Global and Regional Measurement of Geophysical Processes

11.3.1 Introduction

Geodesy is the science of the shape of the Earth, its gravity field, and orientation in space, and is therefore intrinsically connected to geophysics (Torge, 2001; Lambeck, 1988). Indeed, space geodetic techniques, such as GPS can be used to observe the Earth and hence probe geodynamical processes on a global scale (**Figure 8**). GPS contributes to geophysics through comparing the observed and modeled motion of the Earth's surface. Since the observed motion of the Earth's surface will represent the sum of the various effects, it is clear that geophysics must be modeled as a whole, even when investigating a specific problem. This creates a rich area of interdisciplinary research.

As the precision and coverage of GPS stations has improved over the last two decades, the depth and breadth of GPS geodesy's application to geodynamics has increased correspondingly. It has now matured to the point that it is viewed as an important and often primary tool for understanding the mechanics of Earth processes.

On the other hand, geophysical models are essential to GPS geodesy; as such, models are embedded in the reference systems we use to define high-accuracy positions. For example, if the reference system did not account for the tidal deformation of the solid Earth, the coordinates of some stations could vary as much as ~10 cm in the time frame of several hours. Therefore, reference systems to enable high-accuracy geodetic positioning have developed in parallel with progress in geodynamics, which in turn depends on geodetic positioning. Thus, this interdependent relationship between geodesy and geophysics is inextricable.

Table 3 shows examples of the various geophysical processes that affect space geodetic observables and thus are subject to investigation using space geodesy. Most of the applications assume the ability to track the position of geodetic stations with subcentimeter precision, but other possibilities include the determination of Earth's polar motion and rate of rotation, low-degree gravity field coefficients, and atmospheric delay in the troposphere and ionosphere. For example, global climate change could affect both the shape and gravity field through mass redistribution (e.g., melting polar ice caps), but also could affect large-scale tropospheric delay.

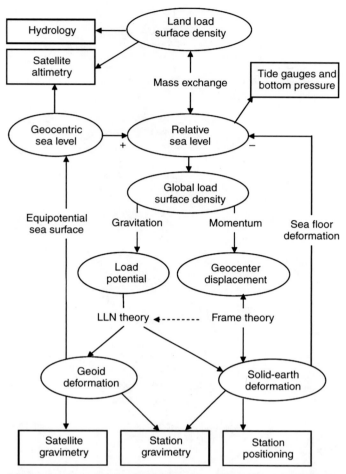

Figure 8 Schematic model of surface mass loading that incorporates self-consistency of the reference frame, loading dynamics, and passive ocean response. Closed-form inversion solutions have been demonstrated (Blewitt and Clarke, 2003; Gross *et al.*, 2004). Note that everything is a function of time, so 'continental water' in its most general sense would include the entire past history of ice sheets responsible for postglacial rebound. (Arrows indicate the direction toward the computation of measurement models, phenomena are in round boxes, measurements are in rectangles, and physical principles label the arrows).

In this section, the focus will be on providing examples of geodetic applications across the spatio-temporal spectrum, ranging from co-seismic rupture and seismic waves to plate rotations.

The subsequent section will then focus on how geodesy can be used to address large-scale loading problems.

11.3.2 Estimation of Station Velocity

All of the following examples use a time series of discrete station positions (whether they be relative station positions, or with respect to the reference frame origin). For many applications, it is convenient to first fit a 3-D station velocity to each time series. If

the velocity is intended to represent the secular motion of a station but the time series spans less than ~4.5 years, then it is important to simultaneously fit an empirical seasonal signal (Blewitt and Lavallée, 2002). The simplest seasonal model would fit amplitudes for an annual sine and cosine wave. In some locations the semi-annual signal may also be important, for example:

$$\mathbf{x}_i(t_j) = \mathbf{x}_{i0} + \dot{\mathbf{x}}_i(t_j - t_0) + \mathbf{a}_{1i}^C \cos(2\pi t_j) + \mathbf{a}_{1i}^S \sin(2\pi t_j)$$
$$+ \mathbf{a}_{2i}^C \cos(\pi t_j) + \mathbf{a}_{2i}^C \sin(\pi t_j) + \mathbf{v}_{ij}(x) \qquad [17]$$

where $\mathbf{x}_i(t_j)$ is the observed vector position of station i at epoch t_j (in Julian years), t_0 is an arbitrary user-specified time to define the modeled epoch position $\mathbf{x}_{i0}$, $\dot{\mathbf{x}}_i$ is the station velocity (independent of the

Table 3 Geophysical processes that affect geodetic observations as a function of spatial and temporal scale

Scale	Temporal				
Spatial	10^{-2}–10^3 s	10^0–10^1 h	10^0–10^2 day	10^0–10^2 years	10^2–10^6 years
10^0–10^1 km	Co-seismic rupture Volcanism	Creep events Volcanism	Afterslip Poro-elastic relaxation Dyke injection	Visco-elastic relaxation Inter-seismic strain	Earthquake cycle
10^1–10^2 km	M 6–7.5 seismic strain release Tropospheric moisture	Storm-surge loading Tsunami loading Tropospheric moisture	Rifting events Aquifer deformation Poro-elastic relaxation Lower crustal magmatism Lake loading Snow loading	Visco-elastic relaxation Block rotation Strain partition Mountain growth Glacial loading Sedimentary loading	Fault activation and evolution Mountain range building Denudation Regional topography Sedimentary loading
10^2–10^3 km	M 7.5–9 seismic strain release Traveling ionospheric disturbances Seismic waves	Coastal ocean loading	Atmospheric loading Regional hydrological loading	Mantle–crust coupling Ice-sheet loading	Plateau rise Mountain range building Glacial cycle Isostacy
10^3–10^4 km	M 9+ seismic strain release Seismic waves Free oscillations	Earth tides Tidal loading	Seasonal fluid transport Ocean bottom pressure	Core–mantle coupling Climate change Solar cycle	Plate rotations Mantle flow Continental evolution

choice of t_0), the harmonic vector amplitude $\mathbf{a}^C_{2i}$, for example, indicates the cosine amplitude of frequency 2 cycle yr^{-1} at station i, and $\mathbf{v}_{ij}$ represents the vector error. When using geocentric Cartesian coordinates, it is especially important to use a full 3×3 weight matrix in the inversion, because of the large difference ($\sim$factor of 3) in the magnitude of formal error in the vertical direction.

If the geophysical signals under investigation are seasonal in nature, then of course the harmonic amplitudes are interesting in their own right, and the velocity term may be considered the 'nuisance parameter'. It should be always kept in mind that the parameter estimates will absorb the sum of all relevant geophysical processes and errors that affect the specified data set. Seasonal systematic errors are particularly difficult to quantify. In the case of simultaneous geophysical processes, it is often the case that the larger-scale processes (e.g., global-scale plate tectonics) can be characterized first and used as

calibration or as boundary conditions for a smaller-scale study (e.g., plate-boundary deformation).

For some applications it may be sufficient to study the post-fit residual time series (i.e., estimates of $\mathbf{v}_{ij}$). However caution is warranted if the form of the signal under investigation is likely to correlate significantly with the velocity or harmonic amplitude parameters. If the exact form of the signal is known (e.g., a step function in the case of a co-seismic displacement field), then it is always better to augment the above model and estimate the extra parameters simultaneously with the above base set of parameters. On the other hand, if the exact form is not known but the signal is assumed to start with an event at a given time T, then a reasonable approach is to estimate the base parameters using only data prior to time T, then forming the residuals time series for all data using this model.

Finally, it should be noted that for some inversion problems it would be more rigorous to incorporate a stochastic model that accounts for temporal correlations in the position time series. There have been several attempts to infer the stochastic nature of errors in the time domain from spectral analysis (Mao et al., 1999; Williams, 2003). The consensus conclusion of such investigations is that GPS time series has the characteristics of flicker noise. The presence of random walk noise, which is quite damaging to the determination of station velocity, for example, is much less conclusive. The importance of these models has proven to lie largely in the realistic assignment of error bars on the estimated geophysical parameters, and not so much on the actual estimates themselves. Ultimately the accuracy of geophysical parameter estimates is better inferred by other external means, such as the smoothness of the inverted velocity field in regions where smoothness is expected from geological considerations.

As a general rule, estimation of station velocity can be achieved with precision $<1 \text{ mm yr}^{-1}$ using >2.5 years of continuous data. One measure of precision is to infer it from the smoothness of a velocity field across a network (Davis et al., 2003). In some sense, this approach gives a measure of 'accuracy', because the results are being compared to an assumed truth (the smoothness of the velocity field). Another method, which assesses the level of systematic error is to compare results using different software packages. For example, Hill and Blewitt (2006) compare velocities produced using the GAMIT and GIPSY software packages, where GAMIT processes double-difference data, and GIPSY processes undifferenced data. Using 4 years of data from a 30-station regional GPS network they found the RMS difference in GPS horizontal velocity is $<0.1 \text{ mm yr}^{-1}$ (after accounting for a 14-parameter reference frame transformation between the two solutions). The data processing by both packages was done in a black-box fashion, with minimal user intervention. This result indicates that errors in GPS station velocities are more than likely to be dominated by biases in common to both GIPSY and GAMIT, for example, multipath error, antenna phase center mismodeling, and nonsecular Earth deformations.

11.3.3 Plate-Tectonic Rotations

Once geodetic station velocities have been estimated (as outlined above), plate-tectonic rotations can be estimated using the following classical kinematic model (Larson et al., 1997):

$$\dot{\mathbf{x}}_j^p = \mathbf{\Omega}^p \times \mathbf{x}_j \qquad [18]$$

where $\mathbf{\Omega}^p$ is the angular velocity (sometimes called the 'Euler vector') of a plate called 'p' associated with station j. The magnitude $\Omega^p = |\mathbf{\Omega}^p|$ is the 'rate of rotation' of plate p (often expressed as degrees per million years, but computationally as radians per year), and the direction $\hat{\mathbf{\Omega}}^p = \mathbf{\Omega}^p/\Omega^p$ is called the 'Euler Pole' (often expressed as a spherical latitude and longitude, but computationally as Cartesian components, i.e., direction cosines) (Minster and Jordan, 1978). The Euler Pole can be visualized as the fixed point on the Earth's surface (not generally within the plate itself) about which the plate rotates. This rotation model essentially constrains the plate to move rigidly on the Earth's surface (no radial motion). The cross-product is taken between the angular velocity and station position in a geocentric reference frame; therefore the velocity is also expressed in the geocentric reference frame. The label p on $\dot{\mathbf{x}}_j^p$ simply identifies the assumed plate (not the reference frame). This notation becomes useful later when considering the relative motion at a plate boundary.

Figure 9 shows an example of an inversion of GPS velocities for rigid plate rotations from the REVEL model (Sella et al., 2002). In this figure, only the stations so indicated were used to invert for plate rotations, on the assumption that they are located on stable plate interiors. Stations that fall within deforming plate boundaries must be treated differently, as will be explained in the following subsection.

Several points are worth noting about the classical kinematic model of plate tectonics:

• The motions are instantaneous, in the sense that the time of observation is sufficiently short that the angular velocities are assumed to be constant in time. As the equation apparently works well for paleomagnetic data over a few million years (Minster and Jordan, 1978; DeMets et al., 1990, 1994), such an assumption is essentially perfect for geodetic observation periods of decades. Indeed, discrepancies between angular velocities from geodesy and paleomagnetic inversions can test whether plates might have significant angular accelerations.

• Plate-tectonic theory here assumes that plate motions are rigid, and that the motion is a rotation about a fixed point in common to all the Earth's surface. Thus the motions are purely horizontal on a spherical Earth.

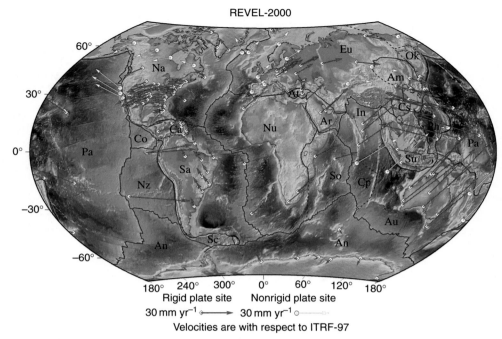

REVEL-2000

Rigid plate site   Nonrigid plate site
30 mm yr⁻¹ ⟶   30 mm yr⁻¹ ⚬⤑
Velocities are with respect to ITRF-97

Figure 9 The REVEL-2000 plate motion model derived from GPS velocities. From Sella G, Dixon T, and Mao A (2002) REVEL : A model for recent plate velocities from space geodesy. *Journal of Geophysical Research* 107(B4) (doi:10.1029/ 2000JB000033).

• The assumption of plate rigidity can be tested independently of the above model, for example, by observing changes of distance between stations supposedly on the same plate. Thus by using geodesy (together with independent evidence), the 'stable plate interior' can be defined empirically as the domain within a plate that, to within the errors of observations, is consistent with having zero deformation. The above equation is therefore more properly applied to such defined stable plate interiors. The relative motions between neighboring plate interiors therefore imposes boundary conditions on the deformational processes that are taking place in the plate-boundary region (Stein, 1993).

• Since the Earth is only approximately spherical, the above equation gives long systematic errors at the level of ∼0.2 mm yr⁻¹, including in the vertical direction (with respect to the WGS-84 reference ellipsoid). Because the errors have a very long wavelength, the induced artificial strain rates are negligible (<0.1 nstrain yr⁻¹).

• Even though vertical motions are predicted to be zero in the model, it is convenient to invert the above equation using Cartesian coordinates, and using the full weight matrix (inverse covariance) associated with the Cartesian components of velocity. In any case, the

resulting estimate of angular velocity will not be sensitive to errors in vertical velocity.

• If the true plate motions (for the part of plates exposed on the Earth's surface) are on average gravitationally horizontal (with respect to the geoid), then on average the motion must also be horizontal with respect to the reference ellipsoid (which is defined to align with the geoid on average). Such a reference ellipsoid is necessarily centered on the center of mass of the entire Earth system, CM. Therefore the fixed point of rotation can be taken to be CM, which is the ideal origin of ITRF. Due to the (verifiable) assumption that plate motions are constant, it is therefore important to use the long-term average CM rather than the instantaneous CM, which can move by millimeters relative to the mean Earth's surface (CF) over tidal and seasonal timescales (caused by redistribution of fluid mass).

• Any systematic error in the realization of CM will map into errors in the model for plate motions, and hence errors in estimates of plate angular velocities. Significantly, this will also affect model predictions of the relative velocities of stations across plate boundaries. Consider the velocity of station j which resides nominally on plate p (e.g., Pacific), in a reference frame co-rotating with plate n (e.g., North

America), then this can be expressed as the following relative velocity:

$$\dot{\mathbf{x}}_j^p = \mathbf{\Omega}^p \times \mathbf{x}_j$$
$$\dot{\mathbf{x}}_j^p - \dot{\mathbf{x}}_j^n = \mathbf{\Omega}^p \times \mathbf{x}_j - \mathbf{\Omega}^n \times \mathbf{x}_j \qquad [19]$$
$$\Delta\dot{\mathbf{x}}_j^{pn} = \Delta\mathbf{\Omega}^{pn} \times \mathbf{x}_j$$

Here $\Delta\mathbf{\Omega}^{pq}$ is the relative angular velocity between plates p and n, and $\Delta\dot{\mathbf{x}}_j^{pn}$ is the relative velocity between plates p and n at station j, in other words, the relative velocity of station j on plate p as viewed by an observer fixed to plate n. Note that if station j actually lies in the stable interior of plate p, then $\Delta\dot{\mathbf{x}}_j^{pn}$ represents the path integral of deformation plus rotation crossing the entire plate boundary going from the stable interior of plate n to station j. Hence systematic errors in $\Delta\mathbf{\Omega}^{pq}$ will negatively impact geophysical inferences on plate-boundary deformation. This proves that it is important to plate-tectonic applications of geodesy to realize the origin of the reference frame as the long-term center of mass of the entire Earth system. Thus from a physical standpoint, the SLR technique is essential (to realize the origin), even if it is not the primary tool for observing relative motions between plates. It is possible to realize an appropriate origin geometrically, by assuming that, after accounting for known geophysical processes that cause vertical motion (e.g., glacial isostatic adjustment), there should be no residual vertical motion in some average sense. There are various possible ways to define such an origin, and it remains a promising topic of research to understand which types of global reference frames (in terms of their realization of the velocity reference at the origin) are most appropriate for determining plate angular velocities.

• The angular-velocity parameters for any given plate are going to be best constrained by a network that maximally spans the rigid plate interior. This presents a problem if the plate is small. For very small plates (e.g., blocks in plate-boundary zones) the motion can be characterized by a horizontal translation to within the sensitivity of geodetic measurements. In this case there is a high correlation between the rate of rotation and the location of the Euler Pole normal to the direction plate motion, and so the concepts of rate of rotation and Euler Pole essentially lose their meaning. Nevertheless, what is important to geophysical processes is not the precision of the Euler Pole and rate of rotation, but rather the precision to which relative motion is known

across plate boundaries. Generally this will be constrained very well if the geodetic network spans those boundaries.

11.3.4 Plate-Boundary Strain Accumulation

Approximately 85% of the Earth's surface can be characterized by rigid plate tectonics. The remaining 15% can be characterized as plate-boundary zones, within which the Earth's crust deforms to accommodate the relative rotation between neighboring plates (Holt *et al.*, 2005). As these zones are responsible for generating destructive earthquakes, they are the subject of intense geodetic research. To accommodate crustal deformation, the model for rigid plate rotations can be modified to a continuum velocity field $\dot{\mathbf{x}}(\mathbf{x})$ as follows:

$$\dot{\mathbf{x}}(\mathbf{x}) = \mathbf{\Omega}(\mathbf{x}) \times \mathbf{x} \qquad [20]$$

where $\dot{\mathbf{x}}(\mathbf{x})$ as been parameterized in terms of a continuum angular velocity field $\mathbf{\Omega}(\mathbf{x})$, otherwise known as the 'rotational vector function' (Haines and Holt, 1993). The advantage of this reparameterization is that the angular velocity field is a constant within a stable plate interior, unlike the velocity field which appears as a rotation, depending on the defined reference frame. If a region can be defined *a priori* as being on a stable plate interior, then $\mathbf{\Omega}(\mathbf{x})$ can be constrained as a constant parameter in the model: $\mathbf{\Omega}(\mathbf{x}) = \mathbf{\Omega}^p$, and in these regions the formula reduces to the plate rotation model. Otherwise, spatial gradients of $\mathbf{\Omega}(\mathbf{x})$ correspond to deformation rates. Specifically, the three horizontal components of the deformational (symmetric, nonrotating) strain-rate tensor on a sphere can be written:

$$\dot{\varepsilon}_{\phi\phi} = \frac{\hat{\mathbf{\Theta}}}{\cos\theta} \cdot \frac{\partial\mathbf{\Omega}}{\partial\phi}$$
$$\dot{\varepsilon}_{\theta\theta} = -\hat{\mathbf{\Phi}} \cdot \frac{\partial\mathbf{\Omega}}{\partial\theta} \qquad [21]$$
$$\dot{\varepsilon}_{\phi\theta} = \frac{1}{2}\left(\hat{\mathbf{\Theta}} \cdot \frac{\partial\mathbf{\Omega}}{\partial\theta} - \frac{\hat{\mathbf{\Phi}}}{\cos\theta} \cdot \frac{\partial\mathbf{\Omega}}{\partial\phi}\right)$$

where $\hat{\mathbf{\Theta}}$ and $\hat{\mathbf{\Phi}}$; are unit vectors that point in the north and east directions, respectively. The contribution of vertical velocity to horizontal strain rates is neglected, because this is <2% for even rapid uplift rates of $10\,\text{mm}\,\text{yr}^{-1}$. Similarly, the vertical component of the rotation rate (the symmetric strain-rate tensor component) is:

$$w = \frac{1}{2}\left(\hat{\boldsymbol{\Theta}} \cdot \frac{\partial\Omega}{\partial\theta} + \frac{\hat{\boldsymbol{\Phi}}}{\cos\theta} \cdot \frac{\partial\Omega}{\partial\phi}\right) \qquad [22]$$

The Global Strain-Rate Map (GSRM) Project has implemented this approach to invert GPS station velocities for a global map of strain (Kreemer *et al.*, 2000, 2003; Holt et al., 2005). The GSRM website is housed and maintained at UNAVCO facility in Boulder, Colorado. The GSRM website has an introduction page where one can access information on the methodology used, the data and references, model results, and acknowledgments. A sample of a global strain-rate map is presented in **Figure 10**. Areas with no color (white) are constrained *a priori* to be stable (zero strain, corresponding to rigid plate rotation, as in the classical plate-tectonic model).

Haines (1982) showed that if the spatial distribution of strain rates is everywhere defined, then the full-velocity gradient tensor is uniquely defined. Given that geodetic stations only sample the continuum velocity field at a discrete set of locations, additional constraints are required to invert the equations. In early work the rotation vector function was expanded as polynomials (Holt *et al.*, 1991; Holt and Haines, 1993; Jackson *et al.*, 1992), but in all later work the bicubic Bessel interpolation on a curvilinear grid has been used for the Aegean (Jackson *et al.*, 1994), Asia (Holt *et al.*, 1995), Iran (Jackson *et al.*, 1995), Japan (Shen-Tu *et al.*, 1995), the Indian Ocean (Tinnon *et al.*, 1995), the western US (Shen-Tu *et al.*, 1999), New Zealand (Holt and Haines, 1995) and the Tonga subduction zone (Holt, 1995).

The art of designing appropriate constraint methods is a fertile area of research. In general, the constraints should be data driven where there are data, but should be averse to generating artifacts in sparsely sampled regions. Ideally the constraints should adapt the spatial resolution to the extreme nonhomogeneity that is the case for today's global network of continuous GPS stations. However, it should be kept in mind that no matter what the constraints, they will generally smooth the observed velocity field to some extent, and will generally generate anomalous spatially distributed artifacts around stations with velocity errors. Such is the nature of underdetermined inversion problems. The key to successful geophysical interpretation is to not over-analyze the results, and to use only information at wavelengths longer than a spatial resolution appropriate to the expected errors. One exception to this is the case where anomalous motion of a station is truly geophysical and related to an interesting localized

process. Strain-rate mapping can be used to help identify such candidates for further investigation.

One method of imposing constraints is using independent strain-rate inferences from earthquake moment tensors and geological fault slip data, through Kostrov's relation (Kostrov, 1974). Observed average seismic strain rates for any grid area can be obtained by summing moment tensors in the volume described by the product of the grid area and the assumed seismogenic thickness:

$$\dot{\varepsilon}_{ij} = \frac{1}{2\mu VT}\sum_{k=1}^{N} M_0 m_{ij} \qquad [23]$$

where N is the number of events in the grid area, μ is the shear modulus, V the cell volume, T is the time period of the earthquake record, M_0 is the seismic moment, and m_{ij} is the unit moment tensor. Similarly, average horizontal strain-rate components from Quaternary fault slip data can be obtained by a variant of Kostrov's summation (Kostrov, 1974) over N fault segments k within a grid area A:

$$\dot{\varepsilon}_{ij} = \frac{1}{2}\sum_{k=1}^{N} \frac{L_k \dot{u}_k}{A\sin\delta_k} m_{ij}^k \qquad [24]$$

where m_{ij}^k is the unit moment tensor defined by the fault orientation and unit slip vector, and the fault segment has length L_k, dip angle δ_k, and slip rate $\dot{u}_k$.

In this combined (geodetic + seismic + geological) scheme, an objective minimization function can then be defined that accommodates all three data types (e.g., Kreemer *et al.*, 2000). Typically geodetic data is given a strong weight in such schemes because, unlike the case for geodetic data, it is not clear to what extent a limited sample of earthquake moment tensors or Quaternary geological data represent strain rates today. Whereas this approach can be applied to produce a combined (geodetic plus seismic) solution for strain-rate mapping, an alternative approach is to produce an independent empirical geodetic solution from which to compare other geophysical data types.

A different approach to strain mapping is based on the concept of 2-D tomography (Spakman and Nyst, 2002). The idea is that the relative velocity between distant geodetic stations must equal the path integral of strain no matter what the path. Therefore faults can be assigned slip rates and block domains can be assigned rotations such that path integrals that cross these structures agree with the geodetic data. This approach requires the user to construct a variety of

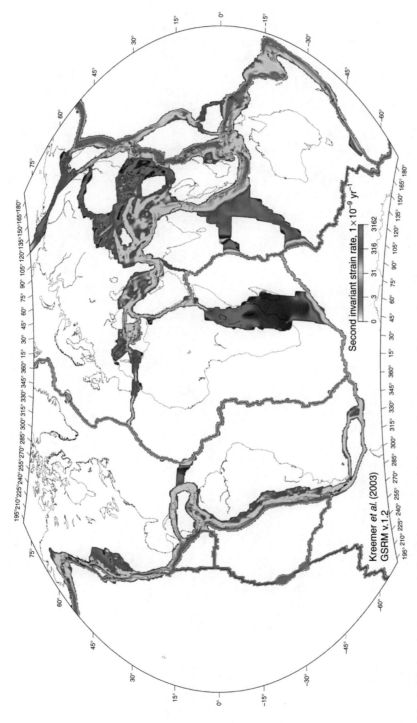

Figure 10 A global strain-rate map, derived from space geodetic data. From Kreemer C, Holt WE, Haines AJ (2003) An integrated global model of present-day plate motions and plate boundary deformation. *Geophysical Journal International* 154: 8-34.

path integrals that will ensure a well-conditioned solution. Although this introduces an additional level of nonuniqueness (due to the user's choices of path integral), the resulting strain-rate maps are insensitive to the choices made so long as their choice overdetermines the problem. Implicitly the rotation function approach also ensures that the path integral agrees in a least-squares sense with relative velocities between stations, and so both methods lead to similar solutions, assuming the *a priori* constraints (from non-geodetic evidence) are approximately equivalent.

The applications of strain-rate mapping in plate-boundary zones are numerous, ranging from under-standing mantle-scale processes to identifying areas of enhanced seismic hazard. The general pattern of the style of strain can point to the larger-scale picture of the driving dynamics. Deformations can be understood as the sum of dilatational strain (increase in surface area) and shear strain (distortion of shape). Regions of strain that are predominantly represented by shear relate to strike-slip faulting, which typically accommo-dates strain across transform boundaries, such as the North America–Pacific Plate boundary. Positive dila-tational strain is associated with zones of extension, which can be driven by a combination of gravitational collapse and boundary conditions on a region, as is the case of the Great Basin in western North America. Gravitational collapse is a predominant factor in the Himalaya and the broader zone of deformation in southeast Asia. Negative dilatational strain is of course associated with convergent plate boundaries. Whereas the above largely related to mantle-scale processes, on the smaller-scale combinations of all styles of strains can arise from inhomogeneities in the crust. For exam-ple, kinks in a strike-slip fault can create either a compressional fold or pull-apart basins.

Clearly all these processes can be and have been studied with nongeodetic techniques and geodesy should be considered as just one tool that can be brought to bear. Broadly speaking, what geodesy brings to the table are the following two basic advantages:

- Geodesy can provide a seamless, consistent map of strain rates spanning a broad range of distance scales, ranging from seismogenic thickness (~15 km) to the global scale (~10 000 km). As such,
  - geodesy can provide a spatial framework within which other types of geophysical evi-dence can be better interpreted;
  - geodesy can indicate to what extent strain can be attributed to broader mantle-scale processes versus more localized crustal-scale structures;

- a geodetic map of strain rate can provide boundary conditions for a study area within which more detailed fieldwork can be pre-formed and understood in the broader context.
- Geodesy can provide a seamless, consistent characterization of changes in strain rates over the timescales of seconds to decades. As such:
  - geodesy clearly represents what is happening today. Differences with other techniques may point to temporal evolution in recent geologi-cal time.
  - geodesy is an appropriate tool to study all phases of the earthquake cycle (the topic of the next section), ranging from co-seismic rup-ture, through postseismic relaxation, to steady-state interseismic strain accumulation.

11.3.5   The Earthquake Cycle

As pointed out in the previous section, geodesy can be used to investigate motions of the Earth's surface on timescales of seconds to decades, and so is an appropriate tool to study all phases of the earthquake cycle (Hammond, 2005). **Figure 11** schematically illustrates the expected characteristics of geodetic position time series as a function of time and distance from a fault through the earthquake cycle. In this

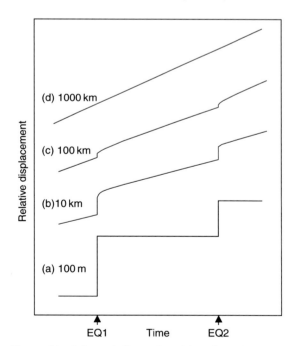

Figure 11   Schematic illustration of the effect of the earthquake cycle on geodetic station positions (see text for explanation).

specific example, the fault is strike slip with two stations either side of the fault located at equal distance normal to the fault strike (**Figure 12**). 'Displacement' is defined as the relative position of the two stations in the direction parallel to the strike of the fault. Each curve represents a different distance from the fault (that is, half the distance between stations). For purposes of illustration, the plot is not to scale. The plot shows the time of characteristic earthquakes EQ1 and EQ2, which are of the largest magnitude that can typically occur on this particular fault, such that smaller earthquakes produce displacements so small that they can be neglected for purposes of illustrating the earthquake cycle. Thus EQ1 and EQ2 represent the start and end of an earthquake cycle. The size of these earthquakes ($M_W \sim 7$) is sufficient that they rupture from seismogenic depth through to the surface, with co-seismic slip approximately constant with depth.

Case (a) at distance 100 m shows a displacement equal to the co-seismic slip on the rupture plane. In between earthquakes, the distance between stations is so small that no deformation is detected. Hence case (a) is effectively equivalent to a geological determination of co-seismic fault slip. In the opposite extreme, case (d) at 1000 km from the fault in the far field shows no detectable co-seismic displacement. (This assumes naively that this is the only active fault in the region of this scale). This displacement represents the far-field driving force transmitted through the crust that ultimately causes the earthquakes. In a sense, the earthquake represents the crust 'catching up' to where it would be if the fault were continuously sliding as a frictionless plane. Thus case (d) shows the

same average displacement per year as would a regression to curve (a) over a sufficient number of earthquake cycles. Thus case (d) also represents: (1) the slip rate at depth below the locked (seismogenic) portion of the crust, assuming the crust behaves perfectly elastically; and (2) the mean slip rate inferred by geological observations of recent Quaternary earthquakes over several earthquake cycles, assuming that the activity of this fault is in steady state equilibrium and is not evolving in time.

Case (b) represents the strain accumulation and release where strain rates are highest in the near-field of the fault. In this case, the co-seismic displacement is slightly damped due to the co-seismic rupture being of finite depth. On the other hand, the time series captures subsequent near-field postseismic effects following each earthquake (Pollitz, 1997). These processes include afterslip (creep) caused by a velocity-hardening rheology, and poroelastic relaxation in response to co-seismic change in pore pressure. Significant aftershock might in some cases be a contributing factor. These processes affect GPS position time series in the days to months following the earthquake (Kreemer *et al.*, 2006a).

Over periods of years to decades, the strain for case (b) is slowly released in the viscoelastic layers beneath the crust which will affect the time series. This occurs because at the time of the earthquake, the crust displaces everywhere and stresses the layers beneath. These layers react instantly as an elastic medium, but as time increases, they start to flow viscously. As the time series ages toward the second earthquake EQ2, most of the viscoelastic response has decayed, and the time series becomes flat. This

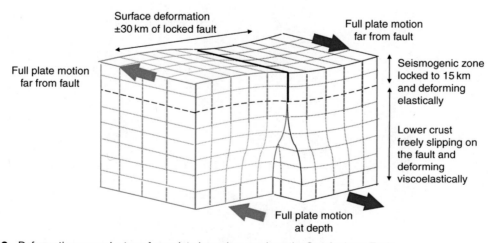

Figure 12   Deformation around a transform plate boundary, such as the San Andreas Fault.

represents the phase of interseismic strain accumulation. However, the slope of this part of the curve is significantly lessened owing to transient postseismic processes earlier in the earthquake cycle. Thus near-field measurements of strain alone can significantly underestimate the seismic hazard unless this is taken into account. This dampening phenomenon also makes it more difficult to pinpoint the location of active faults capable of generating large earthquakes.

Stations in case (c) are still sufficiently close that co-seismic displacements can be detected, but are far enough from the rupture that near-field relaxation do not contribute to the time series. On the other hand, the crust in this intermediate field responds significantly to deep viscoelastic relaxation at the base of the crust and beyond, into the upper mantle. Precisely how the pattern of deformations looks in the years after a large earthquake depends on the relative effective viscosity of these various layers (Hetland and Hager, 2006). Thus GPS networks with stations at various distances about a recently ruptured fault can be used to probe rheology versus depth. Note that in going from EQ1 to EQ2, the relative velocity between the pair of stations decreases in time. Thus the strain rate can depend considerably on the phase of the earthquake cycle (Dixon et al., 2003). Thus in regions of low strain were there are many faults that rupture infrequently (such as the Great Basin, western USA) it is not uncommon to observe strain rates that are almost entirely transient in nature and can exceed interseismic strain by an order of magnitude (Hetland and Hager, 2003; Gourmelen and Amelung, 2005). This makes it more difficult to interpret strain-rate maps in terms of seismic hazard, except in those cases where the strain rates are so large that logically they must be dominated by interseismic strain accumulation on dominant faults (Kreemer et al., 2006b). Such analysis is on the leading edge of research, and requires careful modeling of the earthquake cycle in any given region of interest.

11.3.6 Surface Mass Loading

Earth's time-variable geometrical shape, gravity field, and rotation in space are all connected by Earth's dynamic response to the redistribution of near-surface mass, including mass in the ocean, continental water, ice sheets, and the atmosphere. As a consequence, measurements of the Earth's geometrical shape from GPS can be used to infer surface mass redistribution, and therefore predict changes to the gravity field and

Earth rotation. Thus GPS measurements of Earth's shape can be independently checked by comparison with time-variable gravity as measured in space by geodetic satellites, or residual measurements of Earth rotation (that is, change in Earth rotation driven by change in moment of inertia).

Surface loading therefore represents a unifying theme in geodesy, connecting various types of geodetic measurement and geodynamic models (**Figure 8**). With an assumed structure and rheology of the Earth, it becomes possible to estimate surface mass redistribution from the changing shape of the Earth (Plag et al., 1996). Conversely, with a known source of mass redistribution (e.g., inferred by gravity measurements), it should be possible to invert the measured shape of the Earth to solve for Earth's structure (and rheology, if we include the time-variable response). That is, the ratio of the Earth's gravitational response to geometrical response can be used to infer Earth's structure and rheology.

Loading models have traditionally used Green's functions, as derived by Farrell (1972), and applied in various geodetic investigations (e.g., Van Dam et al., 1994). The Green's function approach is fundamentally based on load Love number theory, in which the Earth's deformation response is a function of the spherical harmonic components of the incremental gravitational potential created by the surface load. To study the interaction between loading dynamics and the terrestrial reference frame, it is convenient to use the spherical harmonic approach (Lambeck, 1988; Mitrovica et al., 1994; Grafarend et al., 1997) (therefore the conclusions must also apply to the use of Green's functions).

The following 'standard model' is based on a spherically symmetric, radially layered, elastic Earth statically loaded by a thin shell on the Earth's surface. Farrell (1972) used such a model to derive Green's functions that are now prevalent in atmospheric and hydrological loading models (van Dam et al., 2001). The Preliminary Reference Earth Model (PREM) (Dziewonski and Anderson, 1981) yields load Love numbers almost identical to Farrell's (Lambeck, 1988; Grafarend et al., 1997).

It is analytically convenient to decompose the Earth system as a spherical solid Earth of radius R_E, plus surface mass that is free to redistribute in a thin surface layer ($<< R_E$) of surface density $\sigma(\Omega)$ which is a function of geographical position Ω (latitude φ, longitude λ). Let us express the total redistributed load as a spherical harmonic expansion:

$$\sigma(\Omega) = \sum_{n=1}^{\infty} \sum_{m=0}^{n} \sum_{\Phi=C}^{S} \sigma_{nm}^{\Phi} Y_{nm}^{\Phi}(\Omega) \qquad [25]$$

where $Y_{nm}^{\Phi}(\Omega)$ are defined in terms of associated Legendre polynomials: $Y_{nm}^{C} = P_{nm}(\sin\varphi)\cos m\lambda$ and $Y_{nm}^{S} = P_{nm}(\sin\varphi)\sin m\lambda$.

The summation begins at degree $n = 1$ assuming that mass is conserved in the Earth system. It is this initial degree one term that relates to the origin of the reference frame. It can be shown (Bomford, 1971) that, for a rigid Earth, such a thin-shell model produces the following incremental gravitational potential at the Earth's surface, which we call the 'load potential':

$$V(\Omega) = \sum_{n} V_n(\Omega)$$
$$= \frac{4\pi R_E^3 g}{M_E} \sum_{n} \sum_{m} \sum_{\Phi} \frac{\sigma_{nm}^{\Phi} Y_{nm}^{\Phi}(\Omega)}{(2n+1)} \qquad [26]$$

where g is acceleration due to gravity at the Earth's surface, and M_E is the mass of the Earth. This load potential results in a displacement of the geoid called the 'equilibrium tide'. As shall be addressed later, the load deforms the solid Earth, and in doing so creates an additional potential.

According to load Love number theory, solutions for surface displacements $\Delta s_h(\Omega)$ in the local height direction, and $\Delta s_l(\Omega)$ in any lateral direction specified by unit vector $\hat{\mathbf{l}}(\Omega)$ are given by (Lambeck, 1988):

$$\Delta s_h(\Omega) = \sum_{n} h_n' V_n(\Omega)/g$$
$$\Delta s_l(\Omega) = \sum_{n} l_n' \hat{\mathbf{l}}(\Omega) \cdot \nabla V_n(\Omega)/g \qquad [27]$$

and the additional potential caused by the resulting deformation is:

$$\Delta V(\Omega) = \sum_{n} k_n' V_n(\Omega) \qquad [28]$$

where h_n', l_n', and k_n' are degree-n load Love numbers, with the prime distinguishing Love numbers used in loading theory from those used in tidal theory. The surface gradient operator is defined $\Delta = \hat{\boldsymbol{\varphi}}\partial_{\phi} + \hat{\boldsymbol{\lambda}}(1/\cos\phi)\partial_{\lambda}$, where $\hat{\varphi}$ and $\hat{\lambda}$ are unit vectors pointing northward and eastward, respectively.

The net loading potential (load plus additional potential) relative to Eulerian observer (the 'space potential' as observed on a geocentric reference surface) is

$$U(\Omega) = V(\Omega) + \Delta V(\Omega)$$
$$= \sum_{n} (1 + k_n') V_n(\Omega) \qquad [29]$$

The net loading potential relative to Lagrangean observer (the 'body potential' as observed on the deforming Earth's surface) must also account for the lowering of Earth's surface due to loading. From equations [28] and [26], the body potential is

$$U'(\Omega) = U(\Omega) - g\Delta s_h(\Omega)$$
$$= \sum_{n} (1 + k_n' - h_n') V_n(\Omega) \qquad [30]$$

Therefore the 'space' and 'body' combinations of load Love number, $(1 + k_n')$ and $(1 + k_n' - h_n')$, are relevant to computing gravity acting on Earth-orbiting satellites and Earth-fixed instruments, respectively.

Solutions for surface deformations of the thin-shell loading model are found by substituting [25] into [26] and [28]:

$$\Delta s_h(\Omega) = \frac{4\pi R_E^3}{M_E} \sum_{n} \sum_{m} \sum_{\Phi} \frac{h_n'}{2n+1} \sigma_{nm}^{\Phi} Y_{nm}^{\Phi}(\Omega)$$

$$\Delta s_l(\Omega) = \frac{4\pi R_E^3}{M_E} \sum_{n} \sum_{m} \sum_{\Phi} \frac{l_n'}{2n+1} \sigma_{nm}^{\Phi} \hat{\mathbf{l}}.\nabla Y_{nm}^{\Phi}(\Omega) \quad [31]$$

$$U(\Omega) = \frac{4\pi R_E^3}{M_E} \sum_{n} \sum_{m} \sum_{\Phi} \frac{1 + k_n'}{2n+1} \sigma_{nm}^{\Phi} Y_{nm}^{\Phi}(\Omega)$$

Thus GPS data on station coordinate variations around the globe can be used to invert eqn [30] for the surface mass coefficients (up to some degree and order n) and hence the surface mass field by substitution into eqn [24] (Blewitt and Clarke, 2003). Truncation of the expansion is of course necessary due to the discrete and finite coverage of GPS data, especially considering the sparcity of data in certain areas such as over the ocean. This implies that the surface mass field will be smoothed. Nevertheless, the long-wavelength information from geodesy is in principle useful to constrain the continental-scale integral of basin-scale hydrological models.

While it is in widespread use, the above standard loading model might be improved by incorporating Earth's ellipticity (Wahr, 1981), mantle heterogeneity (Dziewonski and Anderson, 1981; Su et al., 1994; van Dam et al., 1994; Plag et al., 1996), and Maxwell rheology (Peltier, 1974; Lambeck, 1988; Mitrovica et al., 1994). There is no consensus model to replace PREM yet, however, the general approach to reference frame considerations described here would be applicable to improved models.

To date, surface mass loading has primarily been investigated by gravimetric methods (e.g., GRACE and SLR, see elsewhere in this volume), and the application of geometric measurements from GPS is still in its infancy. The most promising application of GPS in this respect is to the lower degree harmonic components of the global surface mass field, to which satellite missions such as GRACE are least sensitive.

References

Altamimi Z, Angermann D, Argus D, *et al.* (2001) The terrestrial reference frame and the dynamic Earth. *EOS, Transactions of the American Geophysical Union* 82(25): 273–279.

Altamimi Z, Sillard P, and Boucher C (2002) ITRF2000: A new release of the international terrestrial reference frame for Earth science application. *Journal of Geophysical Research* 107(B10): 2214 (doi:10.1029/2001JB000561).

Argus DF, Peltier WR, and Watkins MM (1999) Glacial isostatic adjustment observed using very long baseline interferometry and satellite laser ranging geodesy. *Journal of Geophysical Journal* 104(B12): 29077–29093.

Ashby N and Allan D (1984) Coordinate time on and near the Earth. *Physical Review Letters* 53(19): 1858.

Bar-Sever Y (1996) A new model for GPS yaw attitude. *Journal of Geodesy* 70: 714–723.

Bar-Sever YE, Kroger PM, and Borjesson JA (1998) Estimating horizontal gradients of tropospheric path delay with a single GPS receiver. *Journal of Geophysical Research* 103: 5019–5035.

Bennett RA, Wernicke BP, and Davis JL (1998) Continuous GPS measurements of contemporary deformation across the northern Basin and Range province. *Geophysical Research Letters* 25: 563–566.

Bennett RA, Wernicke BP, Niemi NA, Friedrich AM, and Davis JL (2003) Contemporary strain rates in the northern Basin and Range province from GPS data. *Tectonics.* 22(2): 1008 (doi:10.1029/2001TC001355).

Beutler G, Gurtner W, Bauersima I, and Langley R (1985) Modeling and estimating the orbits of GPS satellites. In: *Proceedings of the 1st International Symposium on Precise Positioning with the Global Positioning System.* US Department of Commerce, Rockville, MD.

Beutler G, Mueller, II, and Neilan RE (1994a) The International GPS Service for Geodynamics: development and start of official service on 1 January 1994. *Bulletin of Geodesique* 68: 39–70.

Beutler G, Brockmann E, Gurtner W, Hugentobler U, Mervart L, and Rothacher M (1994b) Extended orbit modeling techniques at the CODE Processing Center of the International GPS Service for Geodynamics (IGS): Theory and initial results. *Manuscripta Geodaetica* 19: 367–386.

Beutler G, Drewes H, and Verdun A (2004) The new structure of the International Association of Geodesy (IAG) viewed from the perspective of history. Geodesist's Handbook. *Journal of Geodesy* 77: 560–575.

Bevis M, Businger S, Herring TA, Rocken C, Anthes RA, and Ware RH (1992) GPS Meteorology: remote sensing of atmospheric water vapor using the global positioning system. *Journal of Geophysical Research* 97(D14): 15787–15801.

Bianco G, Devoti R, Fermi M, Luceri V, Rutigliano P, and Sciarretta C (1977) Estimation of low degree geopotential coefficients using SLR data. *Planetary and Space Science* 46(11/12): 1633–1638.

Bierman G (1977) *Factorization Methods for Discrete Sequential Estimation.* New York, NY: Academic Press.

Bilham R (1991) Earthquakes and sea level: Space and terrestrial metrology on changing planet. *Reviews of Geophysics* 29(1): 1–29.

Blewitt G (1989) Carrier phase ambiguity resolution for the Global Positioning System applied to geodetic baselines up to 2000 km. *Journal of Geophysical Research* 94(B8): 10187–10283.

Blewitt G (1990) An automatic editing algorithm for GPS data. *Geophysical Research Letters* 17(3): 199–202.

Blewitt G and Lichten SM (1992) Carrier phase ambiguity resolution up to 12000 km: Results from the GIG'91 experiment. In: *Proceedings of the 6th International Symposium on Satellite Positioning.* Columbus: Ohio State University.

Blewitt G (1993) Advances in Global Positioning System technology for geodynamics investigations. In: Smith DE and Turcotte (eds.) *Geodynamics Series Vol. 25: Contributions of Space Geodesy to Geodynamics: Technology,* pp. 195–213. (ISBN 0-87590-526-9), Washington DC: American Geophysical Union.

Blewitt G and Lavallée D (2002) Effect of annual signals on geodetic velocity. *Journal of Geophysical Research* 107(B7): (10.1029/2001JB000570).

Blewitt G and Clarke P (2003) Inversion of Earth's changing shape to weigh sea level in static equilibrium with surface mass redistribution. *Journal of Geophysical Research* 108(B6): 2311 (doi:10.1029/2002JB002290).

Blewitt G, Kreemer C, Hammond WC, Plag H-P, Stein S, and Okal E (2006) Rapid determination of earthquake magnitude using GPS for tsunami warning systems. *Geophysical Research Letters* 33: L11309 (doi:10.1029/2006GL026145).

Blewitt G (2006) The fixed point theorem of ambiguity resolution for precise point positioning of GPS networks: Theory and applications. *EOS, Transactions of the American Geophysical Union* 87(52), Fall Meet. Suppl., Abstract G43A-0977.

Bock Y, Gourevitch SA, Counselman CC, King RW, and Abbot RI (1986) Interferometric analysis of GPS phase observations. *Manuscripta Geodaetica* 11: 282–288.

Bock Y, Nikolaidis RM, de Jonge PJ, and Bevis M (2000) Instantaneous geodetic positioning at medium distances with the Global Positioning System. *Journal of Geophysical Research* 105(B12): 28223–28253.

Bock Y, Prawirodirdjo L, and Melbourne TI (2004) Detection of arbitrarily dynamic ground motions with a dense high-rate GPS network. *Geophysical Research Letters* 31(B10): L06604 (doi:10.1029/2003GL019150).

Boehm J, Niell A, Tregoning P, and Schuh H (2006) Global Mapping Function (GMF): A new empirical mapping function based on numerical weather model data. *Geophysical Research Letters* 33: L07304 (doi:10.1029/2005/GL025546).

Bomford G (1971) *Geodesy, 3rd Rev. edn.,* (ISBN: 978-0198519195). Oxford, UK: Oxford University Press.

Bosworth JM, Coates RJ, and Fischetti TL (1993) The Development of NASA's Crustal Dynamics Project. In: Smith DE and Turcotte DL (eds.) *AGU Geodynamics Series Vol. 25: Contributions of Space Geodesy in Geodynamics: Crustal Dynamics,* pp. 1–20. Washington DC: AGU.

Boucher C and Altamimi Z (1993) Development of a Conventional Terrestrial Reference Frame. In: Smith DE and Turcotte DL (eds.) *Geodynamics Series Vol. 24: Contributions of Space Geodesy in Geodynamics: Crustal Dynamics,* pp. 89–98, Washington, DC: AGU.

Chao BF, O'Connor WP, Chang ATC, Hall DK, and Foster JL (1987) Snow load effect on the Earth's rotation and gravitational field, 1979–1985. *Journal of Geophysical Research* 92: 9415–9422.

Chen G and Herring TA (1997) Effects of atmospheric azimuthal asymmetry on the analysis of space geodetic data. *Journal of Geophysical Research* 102: 20489–20502.

Chen JL, Wilson CR, Eanes RJ, and Nerem RS (1999) Geophysical interpretation of observed geocenter variations. *Journal of Geophysical Research* 104(B2): 2683–2690.

Cheng MK and Tapley BD (1999) Seasonal variations in low degree zonal harmonics of the Earth's gravity field from satellite laser ranging observations. *Journal of Geophysical Research* 104(B2): February 10: 2667–2681.

Cheng M and Tapley BD (2004) Variations in the Earth's oblateness during the past 28 years (2004). *Journal of Geophysical Research (Solid Earth)* 9402.

Choi K, Bilich A, Larson A, and Axelrad P (2004) Modified sidereal filtering: Implications for high-rate GPS positioning. *Geophysical Research Letters* 31(22): L22608.

Ciufolini I and Pavlis EC (2004) A confirmation of the general relativistic prediction of the Lense-Thirring effect. *Nature* 431: 958–960.

Counselman CC, Hinteregger HF, and Shapiro II, (1972) Astronomical applications of differential interferometry. *Science* 178: 607–608.

Counselman CC and Shapiro 11 (1979) Miniature interferometer terminals for Earth surveying. *Bulletin Geodesique* 53(2): 139–163.

Davis JL, Herring TA, Shapiro II, Rogers AEE, and Elgered G (1985) Geodesy by radio interferometry: Effects of atmospheric modeling errors on estimates of baseline length. *Radio Science* 20: 1593–1607.

Davis JL, Murray MH, King RW, and Bock Y (1987) Assessing the effects of atmospheric errors on estimates of relative position obtained from GPS data. *EOS, Transactions of the American Geophysical Union* 68(16): 286.

Davis JL, Elgered G, Niell AE, and Kuehn CE (1993) Ground-based measurement of gradients in the 'wet' radio refractivity of air. *Radio Science* 28: 1003–1018.

Davis JL, Bennett RA, and Wernicke BP (2003) Assessment of GPS velocity accuracy for the Basin and Range Geodetic Network (BARGEN). *Geophysical Research Letters* 30(7): 1411 (doi:10.1029/2003GL016961).

Degnan JJ (1993) Millimeter accuracy satellite laser ranging: A review. In: Smith DE and Turcotte DL (eds.) *Geodynamics Series Vol. 25: Contributions of Space Geodesy to Geodynamics: Technology*, 9, pp. 133–162. (ISBN 0-87590-526), Washington, DC: American Geophysical Union.

DeMets C, Gordon RG, Argus DF, and Stein S (1990) Current plate motions. *Geophysical Journal International* 101: 425–478.

DeMets C, Gordon RG, Argus DF, and Stein S (1994) Effect of recent revisions of the geomagnetic reversal time scale on estimates of current plate motions. *Geophysical Research Letters* 21: 2191–2194.

Devoti R, Luceri V, Sciarretta C, et al. (2001) The SLR secular gravity variations and their impact on the inference of mantle rheology and lithospheric thickness. *Geophysical Research Letters* 28(5): 855–858.

Dickey JO, Bender PL, Faller JE, et al. (1994) Lunar laser ranging. A continuing legacy of the Apollo program. *Science* 265: 482–490.

Dixon TH, Golombek MP, and Thornton CL (1985) Constraints on Pacific plate kinematics and dynamics with Global Positioning System measurements. *IEEE Transactions on Geoscience and Remote Sensing Vol. GE-23*(4): 491–501.

Dixon TH (1991) An introduction to the Global Positioning System and some geological applications. *Reviews of Geophysics* 29(2): 249–276.

Dixon TH, Norabuena E, and Hotaling L (2003) Paleoseismology and GPS: earthquake cycle effects and geodetic vs geologic fault slip rates in the eastern California shear zone. *Geology* 31: 55–58.

Dong D and Bock Y (1989) GPS network analysis with phase ambiguity resolution applied to cmstal deformation studies in California. *Journal of Geophysical Research* 94(B4): 3949–3966.

Dow J, Gurtneer W, and Schluter W (2005a) The IGGOS viewed from the Space Geodetic Services. *Journal of Geodynamics* 40(4–5): 375–386.

Dow JM, Neilan RE, and Gendt G (2005b) The International GPS Service (IGS): Celebrating the 10th anniversary and looking to the next decade. *Advances in Space Research* 36(3): 320–326 (doi:10.1016/j.ast.2005.05.125).

Dunn P, Torrence M, Kolenkiewicz R, and Smith D (1999) Earth scale defined by modern satellite ranging observations. *Journal of Geophysical Research Letters* 26(10): 1489–1492.

Dziewonski AM and Anderson DL (1981) Preliminary reference Earth model. *Physics of the Earth and Planetary Interiors* 25: 297–356.

Evans AG, Hill RW, Blewitt G, et al. (2002) The Global Positioning System geodesy odyssey. *Navigation, Journal of the Institute of Navigation* 49(1): 7–34.

Eubanks TM (1993) Variations in the orientation of the Earth. In: Smith DE and Turcotte DL (eds.) *Geodynamics Series, Vol. 24: Contributions of Space Geodesy in Geodynamics: Crustal Dynamics*, pp. 1–54. Washington, DC: AGU.

Farrell WE (1972) Deformation of the Earth by surface loads. *Reviews of Geophysics* 10: 761–797.

Fliegel HF, Gallini TE, and Swift ER (1992) Global Positioning System radiation force model for geodetic applications. *Journal of Geophysical Research* 97: 559–568.

Fliegel HF and Gallini TE (1996) Solar force modeling of Block IIR Global Positioning System satellites. *Journal of Spacecraft and Rockets* 33(6): 863–866.

Freymueller JT and Kellogg JN (1990) The extended tracking network and indications of baseline precision and accuracy in the North Andes. *Geophysical Research Letters* 17: 207–210.

Ge M, Gendt G, Dick G, Zhang FP, and Reigber Ch (2005) Impact of GPS satellite antenna offsets on scale changes in global network solutions. *Geophysical Research Letters* 32(6): L06310 (doi:10.1029/2004GL022224).

Gegout P and Cazenave A (1993) Temporal variations of the Earth gravity field for 1985–1989 derived from LAGEOS. *Geophysical Journal International* 114: 347–359.

Gourmelen N and Amelung F (2005) Post-seismic mantle relaxation in the Central Nevada Seismic. *Science* 310: 1473–1476 (doi:10.1126/science.1119798).

Grafarend EW, Engels J, and Varga P (1997) The spacetime gravitational field of a deforming body. *Journal of Geodesy* 72: 11–30.

Gross RS, Blewitt G, Clarke PJ, and Lavalleé D (2004) Degree-2 harmonics of the Earth's mass load estimated from GPS and Earth rotation data. *Geophysical Research Letters* 31: L07601, doi:10.1029/2004GL019589.

Gurtner W, Beutler G, Bauersima I, and Schildknecht T (1985) Evaluation of the GPS carrier difference observations: The BERNESE second generation software package. In: Goad C (ed.) *1st International Symposium on Precise Positioning with the Global Positioning System*, U.S. Dept. of Commerce, Rockville, MD.

Hager BH, King RW, and Murray MH (1991) Measurement of crustal deformation using the Global Positioning System. *Annual Review of Earth and Planetary Sciences* 19: 351–382.

Haines AJ (1982) Calculating velocity fields across plate boundaries from observed shear rates. *Geophysical Journal of the Royal Astronomical Society* 68: 203–209.

Haines AJ and Holt WE (1993) A procedure for obtaining the complete horizontal motions within zones of distributed deformation from the inversion of strain rate data. *Journal of Geophysical Research* 98: 12057–12082.

Hammond WC (2005) The ghost of an earthquake. *Science* 310: 1440–1442.

Heflin MB, Bertiger WI, Blewitt G, et al. (1992) Global geodesy using GPS without fiducial sites. *Geophysical Research Letters* 19: 131–134.

Henson DJ, Collier EA, and Schneider KR (1985) Geodetic applications of the Texas Instruments TI-4100 GPS navigator. In: Goad C (ed.) *Proceedings of 1st International Symposium on Precise Positioning with the Global Positioning System*, pp. 191–200, Rockville, MD:US Deptartment of Commerce.

Herring TA, Davis JL, and Shapiro II (1990) Geodesy by radio interferometry: the application of Kalman filtering to the analysis of very long baseline infererometry data. *Journal of Geophysical Research* 95: 12561–12581.

Herring TA and Pearlman MR (1993) Future developments and synergism of space geodetic measurement techniques. In: Smith DE and Turcotte DL (eds.) *Geodynamics Series, Vol. 25: Contributions of Space Geodesy in Geodynamics: Crustal Dynamics*, pp. 21–26. Washington, DC: AGU.

Hetland EA and Hager BH (2003) Postseismic relaxation across the Central Nevada Seismic Belt. *Journal of Geophysical Research* 108: 2394 (doi:10.1029/2002JB002257).

Hetland EA and Hager BH (2006) The effect of rheological layering on postseismic and interseismic displacements. *Geophysical Journal International* 166: 277–292.

Hill EM and Blewitt G (2006) Testing for fault activity at Yucca Mountain, Nevada, using independent GPS results from the BARGEN network. *Geophysical Research Letters* 33: L14302 (doi:10.1029/2006GL026140).

Holt WE, Ni JF, Wallace TC, and Haines AJ (1991) The active tectonics of the Eastern Himalayan Syntaxis and surrounding regions. *Journal of Geophysical Research* 96: 14595–14632.

Holt WE and Haines AJ (1993) Velocity fields in deforming Asia from the inversion of earthquake-released strains. *Tectonics* 12: 1–20.

Holt WE (1995) Flow fields in the Tonga slab determined from the moment tensors of deep earthquakes. *Geophysical Research Letters* 22: 989–992.

Holt WE and Haines AJ (1995) The kinematics of northern South Island New Zealand determined from geologic strain rates. *Journal of Geophysical Research* 100: 17991–18010.

Holt WE, Li M, and Haines AJ (1995) Earthquake strain rates and instantaneous relative motions within central and east Asia. *Geophysical Journal International* 122: 569–593.

Holt WE, Kreemer C, and Haines AJ (2005) Project helps constrain continental dynamics and seismic hazards. *EOS, Transactions of the American Geophysical Union* 86: 383–387.

IERS (eds. Dennis D. McCarthy and Gérard Petit) (2004) IERS Conventions 2003, IERS Technical Note32, Frankfurt am Main: Verlag des Bundesamts für Kartographie and Geodäsie, 127 pp. paperback, ISBN 3-89888-884-3.

Jackson J, Haines AJ, and Holt WE (1992) Determination of the complete horizontal velocity field in the deforming Aegean Sea region from the moment tensors of earthquakes. *Journal of Geophysical Research* 97: 17657–17684.

Jackson JA, Haines AJ, and Holt WE (1994) Combined analysis of strain rate data from satellite laser ranging and seismicity in the Aegean Region. *Geophysical Research Letters* 21: 2849–2852.

Jackson JA, Haines AJ, and Holt WE (1995) The accommodation of Arabia–Eurasia plate convergence in Iran. *Journal of Geophysical Research* 100: 15205–15219.

Kaplan GH (1981) The IAU Resolutions on Astronomical Constants, Time Scale and the Fundamental Reference Frame, U. S. Naval Observatory Circular No. 163.

Kedar S, Hajj GA, Wilson BD, and Heflin MB (2003) The effect of the second order GPS ionospheric correction on receiver positions. *Geophysical Research Letters* 30(16): 1829 (doi:10.1029/2003GL017639).

King RW, Abbot RI, Counselman CC, Gourevitch SA, Rosen BJ, and Bock Y (1984) Interferometric determination of GPS satellite orbits. *EOS, Transactions of the American Geophysical Union* 65: 853.

King RW and Bock Y (2005) *Documentation for the GAMIT GPS processing software Release 10.2*. Cambridge, MA: Massachusetts Institute of Technology.

Kostrov VV (1974) Seismic moment and energy of earthquakes, and seismic flow of rocks (English Translation). *Izvestiya of the Academy of Sciences of the USSR, Physics of the Solid EarthIzvestiya* 1(56): 23–44.

Kreemer C, Haines J, Holt WE, Blewitt G, and Lavallée D (2000) On the determination of a global strain rate model. *Earth, Planets and Space* 52: 765–770.

Kreemer C, Holt WE, and Haines AJ (2003) An integrated global model of present-day plate motions and plate boundary deformation. *Geophysical Journal International* 154: 8–34.

Kreemer C, Blewitt G, and Maerten F (2006a) Co- and postseismic deformation of the 28 March 2005 Nias Mw 8.7 earthquake from continuous GPS data. *Geophysical Research Letters* 33: L07307 (doi:10.1029/2005GL025566).

Kreemer C, Blewitt G, and Hammond WC (2006b) Using geodesy to explore correlations between crustal deformation characteristics and geothermal resources. *Geothermal Resources Council Transactions* 30: 441–446.

Kuang D, Rim HJ, Schutz BE, and Abusali PAM (1996) Modeling GPS satellite attitude variation for precise orbit determination. *Journal of Geodesy* 70: 572–580.

Lambeck K (1988) *Geophysical geodesy:* The slow deformations of the Earth. Oxford, UK: Clarendon Press.

Langbein JL, Wyatt F, Johnson H, Hamann D, and Zimmer P (1995) Improved stability of a deeply anchored geodetic monument for deformation monitoring. *Geophysical Research Letters* 22: 3533–3536.

Larson km and Agnew D (1991) Application of the Global Positioning System to crustal deformation measurement: 1. Precision and accuracy. *Journal of Geophysical Research* 96(B10): 16547–16565.

Larson KM, Webb FH, and Agnew D (1991) Application of the Global Positioning System to crustal deformation measurement: 2. The influence of errors in orbit determination networks. *Journal of Geophysical Research* 96(B10): 16567–16584.

Larson KM, Freymueller JT, and Philipsen S (1997) Global plate velocities from the Global Positioning System. *Journal of Geophysical Research* 102: 9961–9981.

Larson K, Bodin P, and Gomberg J (2003) Using 1 Hz GPS data to measure deformations caused by the denali fault earthquake. *Science* 300: 1421–1424.

Leick A (2004) *Satellite Surveying*, 3rd edn. Hoboken, NJ: Wiley.

Lichten SM and Border JS (1987) Strategies for high-precision Global Positioning System orbit determination. *Journal of Geophysical Research* 92: 12751–12762.

MacDoran PF (1979) Satellite Emission Radio Interferometric Earth Surveying SERIES-GPS geodetic system. *Bulletin of the Green's weekly digest* 53: 117–138.

Mader GL (1999) GPS antenna calibration at the National Geodetic Survey. *GPS Solutions* 3(1): 50–58.

Mader GL and Czopek FM (2002) The Block IIA satellite – Calibrating antenna phase centers. *GPS World* 13(5): 40–46.

Mao A, Harrison (CGA), and Dixon TH (1999) Noise in GPS coordinate time series. *Journal of Geophysical Research* 104: 2797–2816.

McCarthy D (ed.) (1996) *IERS Conventions, IERS Tech. Note 21*. Paris: IERS Cent. Bur., Observatoire de Paris.

McMillan DS (1995) Atmospheric grandients from very long baseline interferometry observations. *Geophysical Research Letters* 22(9): 1041–1044.

Meehan T, Srinivasan J, Spitzmesser D, *et al.* (1992) The TurboRogue GPS Receiver. In: *Proc. of the 6th Intl, Symp. on Satellite Positioning*, vol 1, pp. 209–218. Columbus, Ohio.

Minster JB and Jordan TH (1978) Present-day plate motions. *Journal of Geophysical Research* 83: 5331–5354.

Mitrovica JX, Davis JL, and Shapiro II (1994) A spectral formalism for computing three-dimensional deformations due to surface loads: 1. Theory. *Journal of Geophysical Research* 99: 7057–7073.

Moore AW (2007) The International GNSS Service: Any questions?. *GPS World* 18(1): 58–64.

Nerem RS, Chao BF, Au AY, *et al.* (1993) Time variations of the Earth's gravitational field from satellite laser ranging to LAGEOS. *Geophysical Research Letters* 20(7): 595–598.

Nikolaidis RM, Bock Y, de Jonge PJ, Agnew DC, and Van Domselaar M (2001) Seismic wave observations with the Global Positioning System. *Journal of Geophysical Research* 106(B10): 21897–21916.

Niell A (1996) Global mapping functions for the atmospheric delay at radio wavelengths. *Journal of Geophysical Research* 101(B2): 3227–3246.

Pearlman MR, Degnan JJ, and Bosworth JM (2002) The International Laser Ranging Service. *Advances in Space Research* 30(2): 135–143.

Peltier WR (1974) The impulse response of a Maxwell Earth. *Reviews of Geophysics and Space Physics* 12: 649–669.

Plag H-P, Juttner H-U, and Rautenberg V (1996) On the possibility of global and regional inversion of exogenic deformations for mechanical properties of the Earth's interior. *Journal of Geodynamics* 21: 287–309.

Plag HP, Ambrosius B, Baker TF, *et al.* (1998) Scientific objectives of current and future WEGENER activities. *Tectonophysics* 294: 177–223.

Pollitz FF (1997) Gravitational-viscoelestic postseismic relaxation on a layered spherical Earth. *Journal of Geophysical Research* 102: 17921–17941.

Pollitz FF (2003) Post-seismic relaxation theory on a laterally heterogeneous viscoelastic model. *Geophysical Journal International* 154: 1–22.

Prescott WH, Davis JL, and Svarc JL (1989) Global Positioning System measurements for crustal deformation: Precision and accuracy. *Science* 244: 1337–1340.

Ray J (ed.) (1998) IERS *Analysis Campaign to Investigate Motions of the Geocenter*, International Earth Rotation Service Tech. Note 25., Obs. Paris.

Rogers AEE, Knight CA, Hinteregger HF, *et al.* (1978) Geodesy by radio interferometry: Determination of a 1.24-km base line vector with -5-mm repeatability. *Journal of Geophysical Research* 83: 325–334.

Rothacher M, Beutler G, Gurtner W, Schildknecht T, and Wild U (1990) *BERNESE GPS Software Version 3.2*. Printing Office, University of Berne: Switzerland.

Rothacher M, Springer TA, Schaer S, and Beutler G (1998) Processing strategies for regional networks. In:Brunner FK (ed.) *Advances in Positioning and Reference Frames, IAG Symposium No. 118*, pp. 93–100. Berlin: Springer

Rummel R, Rothacher M, and Beutler G (2005) Integrated Global Geodetic Observing System (IGGOS) – Science rationale. *Journal of Geodynamics* 40: 4–5, 357–362.

Sabadini R, Donato GDi, Vermeersen LLA, Devoti R, Luceri V, and Bianco G (2002) Ice mass loss in Antarctica and stiff lower mantle viscosity inferred from the long wavelength time dependent gravity field. *Geophysical Research Letters* 29(10), doi:10.1029/2001.

Savage JC (1983) Strain accumulation in the Western United States. *Annual Review of Earth and Planetary Sciences* 11: 11–41 (doi:10.1145/annurev.ea.11.0501083.000303).

Schlüter W, Himwich E, Nothnagel A, Vandenberg N, and Whitney A (2002) IVS and its important role in the maintenance of the Global Reference Systems. *Advances in Space Research* 30(2): 145–150.

Segall P and Davis JL (1997) GPS applications for geodynamics and earthquake studies. *Annual Reviews of Earth and Planetary Science* 25: 301–336.

Sella G, Dixon T, and Mao A (2002) REVEL: A model for recent plate velocities from space geodesy. *Journal of Geophysical Research* 107(B4) (doi:10.1029/ 2000JB000033).

Schmid R and Rothacher M (2002) Estimation of elevation-dependent satellite antenna phase center variations of GPS satellites. *Journal of Geodesy* 77: (doi:10.1007/s00190-003-0339-0, 440-446).

Schmid R and Rothacher M (2003) Estimation of elevation-dependent satellite antenna phase center variations of GPS satellites. *Journal of Geodesy* 77: 440–446, doi:10.1007/ s00190-003-0339-0.

Schmid R, Rothacher M, Thaller D, and Steigenberger P (2005) Absolute phase center corrections of satellite and receiver antennas. Impact on GPS solutions and estimation of azimuthal phase center variations of the satellite antenna. *GPS Solutions* 9(4): 283–293.

Schupler BR, Allshouse RL, and Clark TA (1994) Signal characteristics of GPS user antennas. *Journal of the Institute of Navigation* 41: 277–295.

Shen-Tu B, Holt WE, and Haines AJ (1995) Intraplate deformation in the Japanese Islands: A kinematic study of intraplate deformation at a convergent plate margin. *Journal of Geophysical Research* 100: 24275–24293.

Shen-Tu B, Holt WE, and Haines AJ (1999) The kinematics of the western United States estimated from Quaternary rates of slip and space geodetic data. *Journal of Geophysical Research* 104: 28927–28955.

Silver PG, Bock Y, Agnew D, *et al.* (1999) A Plate Boundary Observatory, IRIS Newsletter,16: 3–9.

Smith DE, Kolenkiewicz R, Nerem RS, *et al.* (1994) Contemporary global horizontal crustal motion. *Geophysical Journal International* 119: 511–520.

Smith DE and Turcotte DL (eds.) (1993) *Geodynamics Series Vol. 25: Contributions of Space Geodesy to Geodynamics: Technology*. (ISBN 0-87590-526-9), Washington, DC: American Geophysical Union.

Smith KD, von Seggern D, Blewitt G, *et al.* (2004) Evidence for deep magma injection beneath Lake Tahoe, Nevada-California. *Science* (doi:10.1126/science.1101304).

Sovers OJ and Border JS (1987) Observation model and parameter partials for the JPL geodetic software. GPSOMC, JPL Pub. 87-21, Jet Propulsion Laboratory, Pasadena. CA.

Spakman W and Nyst MCJ (2002) Inversion of relative motion data for fault slip and continuous deformation in crustal blocks. *Earth and Planetary Science Letters* 203: 577–591.

Steigenberger P, Rothacher M, Dietrich R, Fritsche M, Rülke A, and Vey S (2006) Reprocessing of a global GPS network. *Journal of Geophysical Research* 111: B05402 (doi:10.1029/ 2005JB003747).

Stein S (1993) Space geodesy and plate motions. In: Smith DE and Turcotte DL (eds.) *Contributions of Space Geodesy in Geodynamics: Crustal Dynamics*, AGU Geodynamics Series, vol. 23, pp. 5–20. AGU.

Su W, Woodward RL, and Dziewonski AD (1994) Degree 12 model of shear velocity heterogeneity in the mantle. *Journal of Geophysical Research* 99: 6945–6980.

Swift ER (1985) NSWC's GPS orbit/clock determination system. In: *Proceedings of the 1st Symposium on Precise Positioning with the Global Positioning System*. Rockville, MD: US Department of Commerce.

Tapley BD, Schutz BE, Eanes RJ, Ries JC, and Watkins MM (1993) LAGEOS Laser Ranging Contributions to Geodynamics, Geodesy and Orbital Dynamics. In: Smith DE and Turcotte DL (eds.) *Geodynamics Series, vol. 24: Contributions of Space Geodesy in Geodynamics: Crustal Dynamics*, pp. 147–174. Washington, DC: AGU.

Tavernier G, Fagard H, Feissel-Vernier M, *et al.* (2005) The International DORIS Service, IDS. *Advances in Space Research* 36(3): 333–341.

Thomas JB (1988) Functional description of signal processing. in the Rogue GPS Receiver, *JPL Publication 88-15*, Pasadena, California (1988).

Tinnon M, Holt WE, and Haines AJ (1995) Velocity gradients in the northern Indian Ocean inferred from earthquake moment tensors and relative plate velocities. *Journal of Geophysical Research* 100: 24315–24329.

Torge W (2001) *Geodesy*, 3rd edn., 400 pp. Berlin: Walter de Gruyter (ISBN 978-3110170726).

Tralli DM, Dixon TH, and Stephens SA (1988) The effect of wet tropospheric path delays on estimation of geodetic baselines in the Gulf of California using the Global Positioning System. *Journal of Geophysical Research* 93(B6): 6545–6557.

Tralli DM and Lichten SM (1990) Stochastic estimation of tropospheric path delays in Global Positioning System geodetic measurements. *Bulletin Geodesique* 64: 127–159.

Tregoning P and Herring TA (2006) Impact of a priori zenith hydrostatic delay errors on GPS estimates of station heights and zenith total delays. *Geophysical Research Letters* 33: L23303 (doi:10.1029/2006GL027706).

Truehaft RN and Lanyi GE (1987) The effects of the dynamic wet troposphere on radio interferometric measurements. *Radio Science* 22: 251–265.

Van Dam T, Blewitt G, and Heflin MB (1994) Atmospheric pressure loading effects on GPS coordinate determinations. *Journal of Geophysical Research* 99: 23939–23950.

Van Dam TM, Wahr J, Milly PCD, Shmakin AB, Blewitt G, and Larson km (2001) Crustal displacements due to continental water loading. *Geophysical Research Letters* 28: 651–654.

Watkins MM and Eanes RJ (1997) Observations of tidally coherent diurnal and semidiurnal variations in the geocenter. *Geophysical Research Letters* 24: 2231–2234.

Wdowinski S, Bock Y, Zhang J, Fang P, and Genrich J (1997) Southern California Permanent GPS geodetic array; spatial filtering of daily positions for estimating coseismic and postseismic displacements induced by the 1992 Landers earthquake. *Journal of Geophysical Research* B102(8): 18057–18070.

Webb FH and Zumberge JF (1993) *An introduction to GIPSY-OASIS II. JPL Publication D-11088*. Pasadena, CA: Jet Propulsion Laboratory.

Wernicke BP, Friedrich AM, Niemi NA, Bennett RA, and Davis JL (2000) Dynamics of plate boundary fault systems from Basin and Range Geodetic Network (BARGEN) and geologic data. *GSA Today* 10(11): 1–7.

Wahr JM (1981) Body tides on an elliptical, rotating, elastic and oceanless Earth. *Geophysical Journal* 64: 677–703.

Williams JG, Boggs DH, Yoder CF, Ratcliff JT, and Dickey JO (2001) Lunar rotational dissipation in solid body and molten core. *Journal of Geophysical Research* 106(E11): 27933–27968.

Williams JG, Tunyshev S, and Boggs DH (2004) Progress in lunar laser ranging tests of relativistic gravity. *Physical Review Letters* 93 (doi:10.1103/PhysRevLett.93.261101).

Williams SDP (2003) The effect of coloured noise on the uncertainties of rates estimated from geodetic time series. *Journal of Geodesy* 76: 483–494 (doi: 10.1007/s00190-002-0283-4).

Wilson P and Reinhart E (1993) The Wegener-Medlas Project preliminary results on the determination of the geokinematics of the eastern Mediterranean. In: Smith DE and Turcotte DL (eds.) *Geodynamics Series, vol. 23: Contributions of Space Geodesy in Geodynamics: Crustal Dynamics*, pp. 299–310. Washington, DC: AGU.

Wu JT, Wu SC, Hajj GA, Bertiger WI, and Lichten SM (1993) Effects of antenna orientation on GPS carrier phase. *Manuscripta Geodaetica* 18: 91–93.

Wyatt F (1982) Displacement of surface monuments; horizontal motion. *Journal of Geophysical Research* 87: 979–989.

Ziebart M, Cross P, and Adhya S (2002) Modeling photon pressure, the key to high-precision GPS satellite orbits. *GPS World* 13(I): 43–50.

Zumberge JF, Heflin MB, Jefferson DC, Watkins MM, and Webb FH (1997) Precise point positioning for the efficient and robust analysis of GPS data from large networks. *Journal of Geophysical Research* 102(B3): 5005–5018.

Relevant Websites

http://www.ggos.org – Global Geodetic Observing System.

http://www.world-strain-map.org – Global Strain Rate Map Project.

http://igscb.jpl.nasa.gov – International GNSS Service.

http://www.unavco.org – Promoting Earth Science by Advancing High-Precision Techniques for the Measurement of Crustal Deformation.

12 Interferometric Synthetic Aperture Radar Geodesy

M. Simons and P. A. Rosen, California Institute of Technology, Pasadena, CA, USA

12.1 Introduction

12.1.1 Motivation

In 1993, Goldstein *et al.* (1993) presented the first satellite-based interferometric synthetic aperture radar (InSAR) map showing large strains of the Earth's solid surface – in this case, the deforming surface was an ice stream in Antarctica. The same year, Massonnet *et al.* (1993) showed exquisitely detailed and spatially continuous maps of surface deformation associated with the 1992 $M_w 7.3$ Landers earthquake in the Mojave Desert in southern California. These papers heralded a new era in geodetic science, whereby we can potentially measure three-dimensional (3-D) surface displacements with nearly complete spatial continuity, from a plethora of natural and human-induced phenomena. An incomplete list of targets to date includes all forms of deformation on or around faults (interseismic, aseismic, coseismic, and postseismic) aimed at

constraining the rheological properties of the fault and surrounding crust, detection and quantification of changes in active magma chambers aimed at understanding a volcano's plumbing system, the mechanics of glaciers and temporal changes in glacier flow with obvious impacts on assessments of climate change, and the impact of seasonal and anthropogenic changes in aquifers. Beyond detection of coherent surface deformation, InSAR can also provide unique views of surface disruption, through measurements of interferometric decorrelation, which could potentially aid the ability of emergency responders to respond efficiently to many natural disasters.

That InSAR can take advantage of a satellite's perspective of the world permits one to view large areas of Earth's surface quickly and efficiently. In solid Earth geophysics, we are frequently interested in rare and extreme events (e.g., earthquakes, volcanic eruptions, and glacier surges). Therefore, if we want to capture these events and their natural variability, we cannot simply rely on dense instrumentation of a few select areas; instead, we must embrace approaches that allow global access. Given easy access to data (which is not always the case), this inherently global perspective provided by satellite-based InSAR also allows one the luxury of going on geodetic fishing trips, whereby one essentially asks "I wonder if….?", in search of the unexpected (e.g., **Figure 1**). In essence we must not limit

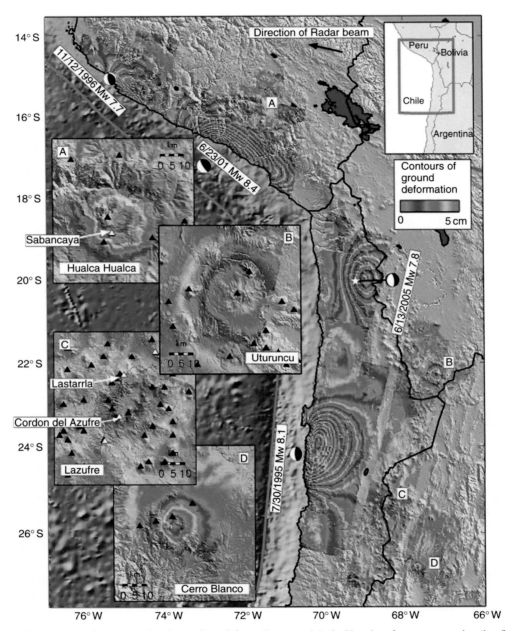

Figure 1   Mosaic of interferograms showing surface deformation associated with subsurface magma migration (inset panels A–D and associated tags in main image) and from three megathrust earthquakes and one intraslab earthquake all with M_w between 7.7 and 8.5. These interferograms are all from right-looking descending orbit C-band radars (*ERS*-1, *ERS*-2, and *ENVISAT*). The phase has been unwrapped from the natural fringe rate and rewrapped at 5 cm per fringe. The subduction trench is indicated by the red line and the centroid moment tensors by beachballs. Note the original survey for volcano deformation is a clear example of the 'fishing trip' approach enabled by InSAR. Figure courtesy of Matt Pritchard.

ourselves to hypothesis testing, but rather we must also tap the inherently exploratory power of InSAR.

12.1.2   History and Overview

Operating at microwave frequencies, synthetic aperture radar (SAR) systems provide unique images representing the electrical and geometrical properties of a surface in nearly all weather conditions. Since they provide their own illumination, SARs can image in daylight or at night. SAR mapping systems typically operate on airborne or spaceborne platforms following a linear flight path, as illustrated in **Figure 2**. Raw image data are collected by transmitting a series of coded pulses from an antenna illuminating a swath offset from the flight track. The echo of each pulse is recorded during a period of reception between the transmission events. When a number of pulses are collected, it is possible to perform 2-D matched-filter compression on a collection of pulse echoes to focus the image. This technique is known as SAR because in the along-track, or azimuth,

direction, a large virtual aperture is formed by coherently combining the collection of radar pulses received as the radar antenna moves along in its flight path (Raney, 1971). Although the typical physical length of a SAR antenna is on the order of meters, the synthesized aperture length can be on the order of kilometers. Because the image is acquired from a side-looking vantage point (to avoid left-side/right-side ambiguities), the radar image is geometrically distorted relative to the ground coordinates (**Figure 2**).

Figure 3 illustrates the InSAR system concept. By coherently combining the signals from two antennas, the interferometric phase difference between the received signals can be formed for each imaged point. In this scenario, the phase difference is essentially related to the geometric path length difference to the image point, which depends on the topography. With knowledge of the interferometer geometry, the phase difference can be converted into an altitude for each image point. In essence, the phase difference provides a third measurement, in

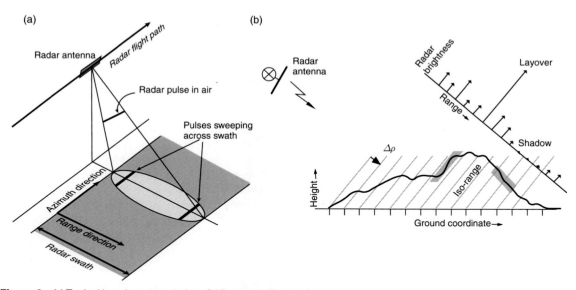

Figure 2   (a) Typical imaging scenario for a SAR system. The platform carrying the SAR instrument follows a curvilinear track known as the 'along-track' or 'azimuth' direction. The radar antenna points to the side, imaging the terrain below. The distance from the aperture to a target on the surface in the look direction is known as the the 'cross-track' or 'range direction' and is terrain dependent. The radar sends a pulse that sweeps through the antenna beam, effectively returning the integrated backscatter over the pulse and azimuth beam extent at any given instant. The azimuth extent can be many kilometers. Matched-filtering creates fine resolution in range. Synthetic aperture processing creates fine resolution in azimuth. (b) The 3-D world is collapsed into two dimensions in conventional SAR imaging. After image formation, the radar return is resolved into an image in range–azimuth coordinates. This panel shows a profile of the terrain at constant azimuth, with the radar flight track into the page. The profile is cut by curves of constant range, spaced by the range resolution of radar, defined as $\Delta\rho = c/2\Delta f_{BW}$, where c is the speed of light and Δf_{BW} is the range bandwidth of the radar. The backscattered energy from all surface scatterers within a range resolution element contribute to the radar return for that element.

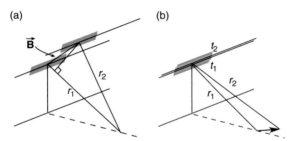

Figure 3 (a) InSAR for topographic mapping uses two apertures separated by a 'baseline', B, to image the surface. The phase difference between the apertures for each image point, along with the range and knowledge of the baseline, can be used to infer the precise shape of the imaging triangle to derive the topographic height of the image point. A range difference exists because the scene is viewed from two different vantage points. This is described by a shift in the point target response as presented in the text. (b) InSAR for deformation mapping uses the same aperture to image the surface at multiple times. A range difference is generated by a change in the position of the scene from one time to the next, imaged from the same vantage point. This range difference is described by a scene shift, not a point-target response shift, the mathematics is the same but for a sign change.

addition to the along- and cross-track location of the image point, or 'target', to allow a reconstruction of the 3-D location of the targets.

The InSAR approach for topographic mapping is similar in principle to the conventional stereoscopic approach. In stereoscopy, a pair of images of the terrain are obtained from two displaced imaging positions. The 'parallax' obtained from the displacement allows the retrieval of topography because targets at different heights are displaced relative to each other in the two images by an amount related to their altitudes (Rosen *et al.*, 2000). The major difference between the InSAR technique and stereoscopy is that, for InSAR, the 'parallax' measurements between the SAR images are obtained by measuring the phase difference between the signals received by two InSAR antennas. These phase differences can be used to determine the angle of the target relative to the baseline of the interferometric SAR directly. The accuracy of the InSAR parallax measurement is typically several millimeters to centimeters, being a fraction of the SAR wavelength, whereas the parallax measurement accuracy of the stereoscopic approach is usually on the order of the resolution of the imagery (several meters or more).

Typically, the postspacing of the InSAR topographic data is comparable to the fine spatial resolution of SAR imagery, while the altitude

measurement accuracy generally exceeds stereoscopic accuracy at comparable resolutions. The registration of the two SAR images for the interferometric measurement, the retrieval of the interferometric phase difference, and subsequent conversion of the results into digital elevation models (DEMs) of the terrain can be highly automated, representing an intrinsic advantage of the InSAR approach. As discussed in later sections, the performance of InSAR systems is largely understood both theoretically and experimentally. These developments have led to airborne and spaceborne InSAR systems for routine topographic mapping.

The InSAR technique just described, using two apertures on a single platform, is often called 'cross-track interferometry' (XTI) in the literature. Other terms are 'single-track' and 'single-pass' interferometry (**Figure 3(a)**).

Another interferometric SAR technique was advanced by Goldstein and Zebker (1987) for measurement of surface motion by imaging the surface at multiple times (**Figure 3(b)**). The time separation between the imaging can be a fraction of a second to years. The multiple images can be thought of as 'time-lapse' imagery. A target movement will be detected by comparing the images. Unlike conventional schemes in which motion is detected only when the targets move more than a significant fraction of the resolution of the imagery, this technique measures the phase differences of the pixels in each pair of the multiple SAR images. If the flight path and imaging geometries of all the SAR observations are identical, any interferometric phase difference is due to changes over time of the SAR system clock, variable propagation delay, or surface motion in the direction of the radar line of sight (LOS).

In the first application of this technique described in the open literature, Goldstein and Zebker (1987) augmented a conventional airborne SAR system with an additional aperture, separated along the length of the aircraft fuselage from the conventional SAR antenna. Given an antenna separation of roughly 20 m and an aircraft speed of about $200 \, \text{m s}^{-1}$, the time between target observations made by the two antennas was about 100 ms. Over this time interval, clock drift and propagation delay variations are negligible. This system measured tidal motions in the San Francisco Bay area with an accuracy of several cm s^{-1} (Goldstein and Zebker, 1987). This technique has been dubbed 'along-track interferometry' (ATI) because of the arrangement of two antennas along the flight track on a single platform.

In the ideal case, there is no cross-track separation of the apertures, and therefore no sensitivity to topography.

ATI is merely a special case of space 'repeat-track interferometry' (RTI), which can be used to generate topography and motion. The orbits of several space-borne SAR satellites have been controlled in such a way that they nearly retrace themselves after several days. Aircraft can also be controlled to accurately repeat flight paths. If the repeat flight paths result in a cross-track separation and the surface has not changed between observations, then the repeat-track observation pair can act as an interferometer for topography measurement. For spaceborne systems, RTI is usually termed 'repeat-pass interferometry' in the literature (**Figure 3**).

If the flight track is repeated perfectly such that there is no cross-track separation, then there is no sensitivity to topography, and radial motions can be measured directly as with an ATI system. However, since the temporal separation between the observations is typically days to many months or years, the ability to detect small radial velocities is substantially better than the ATI system described above. The first demonstration of RTI for velocity mapping was a study of the Rutford ice stream in Antarctica (Goldstein et al., 1993). The radar aboard the ERS-1 satellite obtained several SAR images of the ice stream with near-perfect retracing so that there was no topographic signature in the interferometric phase, permitting measurements of the ice stream flow velocity of the order of $1\,m\,yr^{-1}$ (or $3 \times 10^{-8}\,m\,s^{-1}$) observed over a few days (Goldstein et al., 1993).

Most commonly for repeat-track observations, the track of the sensor does not repeat itself exactly, so the interferometric time-separated measurements generally comprise the signature of topography and of radial motion or surface displacement. The approach for reducing these data into velocity or surface displacement by removing topography is generally referred to as 'differential interferometric SAR.'

Goldstein et al. (1988) conducted the first proof-of-concept experiment for spaceborne InSAR using imagery obtained by the SeaSAT mission. In the latter portion of that mission, the spacecraft was placed into a near-repeat orbit every 3 days. Gabriel et al. (1989) used data obtained in an agricultural region in California, USA, to detect surface elevation changes in some of the agricultural fields of the order of several cm over approximately a 1-month period. By comparing the areas with the detected surface elevation changes with irrigation records, they concluded that these areas were irrigated in between the observations, causing small elevation changes from increased soil moisture. Gabriel et al. (1989) were actually looking for the deformation signature of a small earthquake, but the surface motion was too small to detect. These early studies were then followed by the aforementioned seminal applications to glacier flow and earthquake-induced surface deformation (Goldstein et al., 1993; Massonnet et al., 1993).

All civilian InSAR-capable satellites to date have been right-looking in near-polar sun-synchronous orbits. This gives the opportunity to observe a particular location on the Earth on both ascending and descending orbit passes (**Figure 4**). With a single satellite, it is therefore possible to obtain geodetic measurements from two different directions, allowing vector measurements to be constructed. The variety of available viewing geometries can be increased if a satellite has both left- and right-looking capability. Similarly, neighboring orbital tracks with overlapping beams at different incidence angles can also provide diversity of viewing geometry.

In an ideal mission scenario (see Section 12.5), observations from a given viewing geometry will be acquired frequently and for a long period of time to provide a dense archive for InSAR analysis. The frequency of imaging is key in order to provide optimal time resolution of a given phenomena, as well as to provide the ability to combine multiple images to detect small signals. Of course, many processes of interest are not predictable in time, thus we must continuously image the Earth in a systematic fashion in order to provide recent 'before' images. For a given target, not all acquisitions are necessarily viable for InSAR purposes. The greatest nemesis for InSAR geodesy comes from incoherent phase returns between two image acquisitions. This incoherence can be driven by observing geometry (i.e., the baseline is too large) or by physical changes of the Earth's surface (e.g., snowfall). Thus any InSAR study begins with an assessment of the available image archive. **Figure 5** uses an example from the ERS-1 and ERS-2 image archive to illustrate how one would go about choosing images for InSAR processing, assuming you wanted to make all available pairs that were not decorrelated due to large baselines, snow, or temporal separation that was too large. In theory, a future mission would have sufficiently tight control on the satellite orbit such that baseline selection would not be an issue.

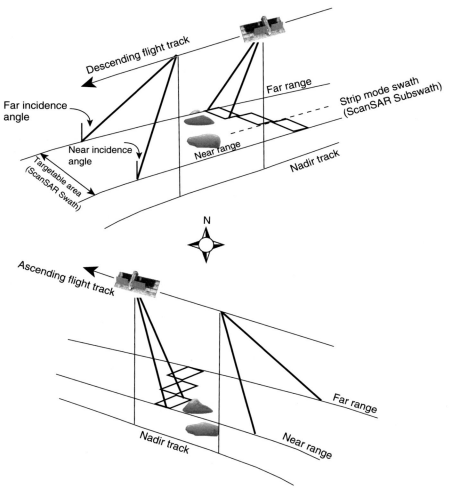

Figure 4   A rendition of ascending and descending tracks for a single right-looking radar system. On successive repeat tracks, geodetic measurements can be made from two different line-of-sight directions, giving possible vector measurements over time. Systems that are capable of imaging in a continuous strip can operate in strip mode with a swath width constrained by the details of the radar system. Systems that can electronically steer their antenna in the cross-track direction are usually capable of ScanSAR operation, and can image a much wider swath, as illustrated.

12.1.3   Scope

Today, spaceborne InSAR enjoys widespread application, in large part because of the availability of suitable globally acquired SAR data from the ERS-1, ERS-2, and ENVISAT satellites operated by the European Space Agency, JERS-1 and ALOS satellites operated by the National Space Development Agency of Japan, RADARSAT-1 operated by the Canadian Space Agency, and SIR-C/X-SAR operated by the United States, German, and Italian space agencies. As more and more radar data become available from international civilian radar satellites, and as scientific demands become greater on the use of these data, including extraction of ever more subtle and well-

calibrated geophysical signals, it is essential to understand the characteristics of the image, how they are processed, and how that processing can affect the interpretation of the image data.

There exist both commercial and freely available software for conventional InSAR processing. While they may differ in detail, they must all follow a basic processing flow. **Figure 6** presents such a flow, derived from the authors' experience in developing the Repeat Orbit Interferometry Package (ROI_PAC) software suite (Rosen *et al.*, 2004). This flow diagram explicitly calls out potential iterative cycles and use of external data and intermediary models. Major phases of this processing are described

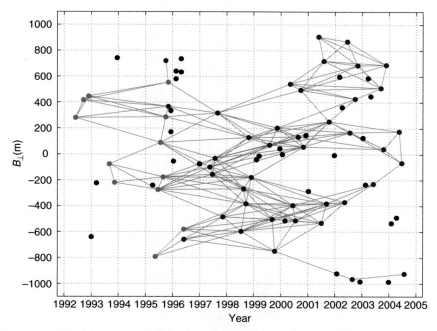

Figure 5 A typical perpendicular component of baseline, $B_\perp$, vs time plot used to select data for InSAR purposes. Here, each dot represents a different acquisition of the same scene by the ERS-1 or ERS-2 satellites. This example is for track 485 and frame 2853 corresponding to the Long Valley Caldera region of California. Lines connecting the dots represents interferograms that are likely to work. We have chosen to indicate only pairs that satisfy the constraints: $B_\perp < 250\,\text{m}$, neither scene occurs in winter when snowfall and rain cause decorrelation, and $\Delta T < 3.5\,\text{years}$. For viable scenes, red and blue dots indicate ERS-1 and ERS-2, respectively.

in the text. Given quality data and metadata, an initial complete processing from raw data to a georeferenced deformation image can now be done automatically and quickly (in under a few hours) on an average laptop computer. Subsequent iterative refinements to maximize the quality and quantity of the observations can require significantly more effort.

In this chapter, we aim to provide a review of the basic theory of InSAR for geodetic applications. Numerous review articles and books on the topic of InSAR already exist (e.g., Massonnet and Feigl, 1995; Rosen *et al.*, 2000; Burgmann *et al.*, 2000; Hanssen, 2001), and our goal is not to repeat these works more than necessary. Instead, we attempt to provide an overview of what data, processing, and analysis schemes are currently used and a glimpse of what the future may hold. As part of this discussion, we present our biased view of what constitutes best practices for use of InSAR observations in geodetic modeling. Finally, we provide a basic primer on the ties between different mission design parameters and their relationship to the character of the resulting observations. In general, this review borrows heavily

from our previous work with many colleagues, and where appropriate, we point the reader to the original sources for a more complete discussion. Much of the SAR processing discussion is derived and simplified from Rosen *et al.* (2000), although here, this discussion is augmented to include a variety of more recent techniques including persistent scatterers, ScanSAR interferometry, and pixel tracking.

12.2 InSAR

Interferometry relies on the constructive and destructive interference of electromagnetic waves from sources at two or more vantage points to infer something about the sources or the relative path length of the interferometer. For InSAR, the interference pattern is constructed from two complex-valued synthetic aperture radar images, and interferometry is the study of the phase difference between two images – acquired from different vantage points, different times, or both.

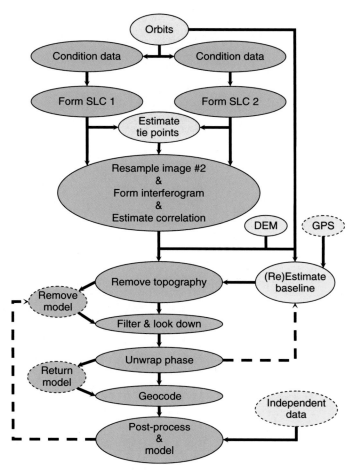

Figure 6 Representative differential InSAR processing flow diagram. Blue bubbles represent image output, yellow ellipses represent nonimage data. Flow is generally down the solid paths, with optional dashed paths indicating potential iteration steps. DEM, digital elevation model; SLC, single look complex image.

Appendix 1 describes the SAR technique, developing a model for the image one would obtain from an idealized surface that consists entirely of a single reflective point, known as a point target, then further considering the image effects of natural surfaces. Fine image resolution is achieved in the cross-track, or range, direction by transmitting a coded waveform with sufficient bandwidth. Matched-filter compression of each received signal pulse then recovers the range resolution. In the along-track, or azimuth, direction, a SAR forms a large synthetic aperture by coherently combining an ensemble of the radar pulses received as the SAR moves along in its flight path. Matched filtering then focuses the image in azimuth.

In Appendix 1, it is shown that there is a phase term $\exp\{-j2kr\}$, where $k = 2\pi/\lambda$ is the wave number and λ is the radar wavelength, that characterizes the two-way propagation distance, $2r$, from the radar

sensor to the point target and back again. For a general surface, there is an additional phase term contributed by each surface scatterer. The net phase of each image point is the sum of these two terms: the intrinsic phase of the surface, which tends to be random, and the propagation phase term.

A resolution element can be represented as a complex phasor of the coherent backscatter from the scattering elements on the ground and the propagation phase delay, as illustrated in **Figure 7**. The backscatter phase delay is the net phase of the coherent sum of the contributions from all elemental scatterers in the resolution element, each with their individual backscatter phases and their differential path delays relative to a reference surface normal to the radar look direction.

Radar images observed from two nearby antenna locations have resolution elements with

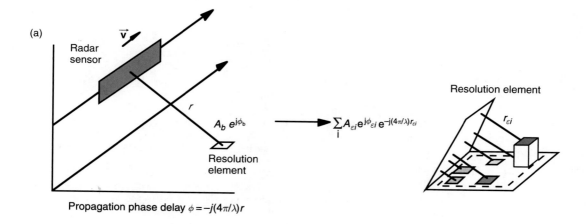

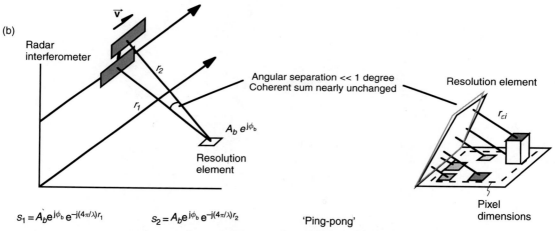

Figure 7   Illustration of the elements of phase in SAR and InSAR observations. (a) Each resolution element in a radar image has a complex value. Its amplitude, A_b, is related to the electrical reflectivity of the surface and its roughness. Its phase is the sum of the propagation phase delay, $-j(4\pi/\lambda)r$, and the surface backscatter phase, ϕ_b. The surface contribution, $A_b e^{j\phi_b}$, can be modeled as the coherent sum of contributions from elemental complex scatterers, $A_{\varepsilon i} e^{j\phi_{\varepsilon i}}$, each with their individual path delays, $r_{\varepsilon i}$, relative to a common reference. Since the arrangement of scatterers within a resolution element is generally random, the phase ϕ_b is also random from element to element. (b) For InSAR, when two observations are made from nearly the same viewing geometry, the backscatter values, $A_b e^{j\phi_b}$, are nearly the same in each, and the phase difference, $\phi_1 - \phi_2$, is essentially the difference in the path delay $-j(4\pi/\lambda)(r_1 - r_2)$.

nearly the same complex phasor return, but with a different propagation phase delay. In interferometry, the complex phasor information of one image is multiplied by the complex conjugate phasor information of the second image to form an interferogram, effectively canceling the common backscatter phase in each resolution element, but leaving a phase term proportional to the differential path delay. Ignoring the slight difference in backscatter phase of the surface observed from two different vantage points treats each resolution element as a point scatterer.

12.2.1   The Interferogram

As shown in Appendix 1, for a fixed point target and a platform moving to synthesize an aperture in azimuth, the range and azimuth-compressed point-target signal, $r_{zcc\delta}$ is

$$r_{zcc\delta}(x', r'; x_0, R_0) = e^{-j4\pi R_0/\lambda} \operatorname{sinc}\left(\frac{\pi}{\Delta R}(r' - R_0)\right)$$
$$\times \operatorname{sinc}\left(\frac{\pi}{\Delta X}(x' - x_0)\right) \qquad [1]$$

where we have explicitly called out the dependence on the location of the point target in the definition of the point-target response. As described in Appendix 1, x_0 represents the location of the fixed point target in the along-track coordinate direction, R_0 represents the 'closest approach' range from the platform to the target at that x_0, ΔR is the range resolution after range matched-filtering, and ΔX is the along-track resolution after synthetic aperture processing. The subscript z indicates that the function is complex valued, cc indicates both range and azimuth compression have been applied, and δ indicates this is a point-target, or delta-function response, for example a small bright reflecting object surrounded by a surface that reflects on energy back to the radar. For a general complex scene $\Gamma(x, r)$, the SAR image after compression is given by convolution of Γ with the point-target response

$$\Gamma_{zcc}(x', r') = \int\int \Gamma(x, r) r_{zcc\delta}(x', r'; x, r) dx dr \quad [2]$$

$$= \int dx \int dr\, \Gamma(x, r) e^{-j4\pi r/\lambda}$$

$$\times \operatorname{sinc}\left(\frac{\pi}{\Delta R}(r'-r)\right) \operatorname{sinc}\left(\frac{\pi}{\Delta X}(x'-x)\right) \quad [3]$$

which can be verified by substituting $\Gamma(x, r) = \delta(x - x_0, r - R_0)$ to recover the impulse response $r_{zcc\delta}$.

The SAR image estimate Γ_{zcc} is the convolution of the actual reflectivity with a 2-D function similar to a delta function but with finite width (the sinc function). This convolution smears out the intrinsic reflectivity as a point-spread function in optics. If the system bandwidths were to become infinite, the sinc functions would become delta functions

$$\operatorname{sinc}\left(\frac{\pi}{\Delta R}(r'-r)\right) \to \delta(r'-r) \quad \text{as} \quad \Delta R \to 0 \quad [4]$$

$$\operatorname{sinc}\left(\frac{\pi}{\Delta X}(x'-x)\right) \to \delta(x'-r) \quad \text{as} \quad \Delta X \to 0 \quad [5]$$

such that

$$\Gamma_{zcc}(x', r') \to \Gamma(x', r') e^{-j4\pi r'/\lambda} \quad [6]$$

Now consider two observations of the reflectivity acquired from slightly different ranges (**Figures 7** and **8**).

For the original observation at r

$$\Gamma_{zcc,1}(x', r') = \int dx \int dr\, \Gamma(x, r) e^{-j4\pi r/\lambda}$$

$$\times \operatorname{sinc}\left(\frac{\pi}{\Delta R}(r'-r)\right) \operatorname{sinc}\left(\frac{\pi}{\Delta X}(x'-x)\right) \quad [7]$$

For an observation at $r + \delta r$, the point-target response is now shifted, but the scene must still be referenced to the original range r:

$$r_{zcc\delta}(x', r'x, r + \delta r) = e^{-j4\pi(r+\delta r)/\lambda} \operatorname{sinc}\left(\frac{\pi}{\Delta R}(r'-(r+\delta r))\right)$$

$$\times \operatorname{sinc}\left(\frac{\pi}{\Delta X}(x'-x_0)\right) \quad [8]$$

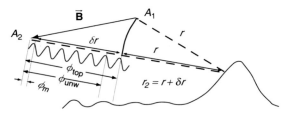

Figure 8 Phase in interferogram depicted as cycles of electromagnetic wave propagating a differential distance δr. Phase in the interferogram is initially known modulo 2π: $\phi_m = W(\phi_{top})$, where ϕ_{top} is the topographically induced phase and $W()$ is an operator that wraps phase values into the range $-\pi < \phi \geq \pi$. After unwrapping, relative phase measurements between all pixels in the interferogram are determined up to a constant multiple of 2π: $\phi_{unw} = \phi_m + 2\pi k_{unw}$, where k_{unw} is a spatially variable integer dependent on the pixel coordinates of the interferogram. Absolute phase determination is the process to determine the overall multiple of $2\pi k_{abs}$ that must be added to the phase measurements so that it is proportional to the range difference. The reconstructed phase is then $\phi_{top} = \phi_m + 2\pi k_{unw} + 2\pi k_{abs}$.

$$\Gamma_{zcc,2}(x', r') = \int dx \int dr\, \Gamma(x, r) e^{-j4\pi(r+\delta r)/\lambda}$$

$$\times \operatorname{sinc}\left(\frac{\pi}{\Delta R}(r'-(r+\delta r))\right)$$

$$\times \operatorname{sinc}\left(\frac{\pi}{\Delta X}(x'-x)\right) \quad [9]$$

Again letting the bandwidths tend to infinity, we get

$$\Gamma_{zcc,1}(x', r') \to \Gamma(x', r') e^{-j4\pi r'/\lambda} \quad [10]$$

$$\Gamma_{zcc,2}(x', r') \to \Gamma(x', r'-\delta r) e^{-j4\pi r'/\lambda} \quad [11]$$

Note a shift in the registration of the true scene reflectivity.

$$\Gamma_{zcc,2}(x', r'+\delta r) \to \Gamma(x', r') e^{-j4\pi(r'+\delta r)/\lambda} \quad [12]$$

When the true scene reflectivity is aligned, there is a phase difference between the reconstructed phase proportional to δr, a geometric term. The true scene reflectivity is a complex number with random phase. However, after we have aligned the scene reflectivity by shifting $\Gamma_{zcc,2}(x', r')$ by δr, or equivalently by looking up the value of $\Gamma_{zcc,2}$ at location $r' + \delta r$, then the scene phase is common to both observations. We can form the product

$$I(x', r') = \Gamma_{zcc,1}(x', r')\Gamma_{zcc,2}^*(x', r'+\delta r) \quad [13]$$

$$= |\Gamma(x', r')|^2 e^{-j4\pi(\delta r)/\lambda} \quad [14]$$

The function $I(x', r')$ is the 'interferogram', a complex quantity, the phase of which is just a geometric term related to the range difference δr (when bandwidths are infinite) between the two images. The range difference can be caused by a vantage point difference as described here, and illustrated in **Figure 3(a)**, or by a shift in the scene location (**Figure 3(b)**) or a combination of the two. If the scene shifts rather than the point target, then the sign of the range difference changes, but the form of the interferogram is the same.

Consider the case where the range difference δr arises from a cross-track separation of two observation points, as illustrated in **Figure 3(a)**.

The phase of the interferogram eqn [14] is the difference in the geometric path length phases of the two images

$$\phi_I = \phi_1 - \phi_2 = \frac{4\pi}{\lambda}(r_2 - r_1) = \frac{4\pi}{\lambda}\delta r \qquad [15]$$

There is a clear dependence on the relative lengths of the two sides of the triangle on the height of the surface, which in general is not known *a priori*. Thus δr is not known exactly to align the reflectivities to form the interferogram. However, in practical systems, one can match the reflectivity estimates in the two SAR observations to within sufficient accuracy (generally much better than the image resolution) to derive a sufficient estimate of δr for alignment. Once formed, the interferogram for the cross-track interferometer then contains a record of the variability of the height of the surface. It is possible to invert the phase to reconstruct the height. It turns out it can be done quickly and efficiently, and is a powerful tool for topographic mapping. Note that the sign of the propagation phase delay is set by the desire for consistency between the Doppler frequency, f_D, and the phase history, $\varphi(t)$ (Rosen *et al.*, 2000).

Only the principal values of the phase, modulo 2π, can be measured from the complex-valued resolution element. The total range difference between the two observation points that the phase represents in general can be many multiples of the radar wavelength, or, expressed in terms of phase, many multiples of 2π. The typical approach for determining the unique phase that is directly proportional to the range difference is to first determine the relative phase between pixels via the so-called 'phase-unwrapping' process. This connected phase field will then be adjusted by an overall constant multiple of 2π. The second step determines this required multiple of 2π, and is referred to as 'absolute phase determination.' **Figure 8** shows the principal value of the phase, the unwrapped phase, and absolute phase for a pixel.

12.2.2 Interferometric Baseline and Height Reconstruction

In order to generate topographic maps or data for other geophysical applications using radar interferometry, we must relate the interferometric phase and other known or measurable parameters to the topographic height. It is also desirable to derive the sensitivity of the interferometrically determined topographic measurements to the interferometric phase and other known parameters. In addition, interferometric observations have certain geometric constraints that preclude valid observations for all possible image geometries.

The interferometric phase as previously defined is proportional to the range difference from two antenna locations to a point on the surface. This range difference can be expressed in terms of the vector separating the two antenna locations, called the interferometric baseline. The range and azimuth position of the sensor associated with imaging a given scatterer depends on the portion of the synthetic aperture used to process the image (see Appendix 1). Therefore the interferometric baseline depends on the processing parameters, and is defined as the difference between the location of the two antenna phase center vectors at the time when a given scatterer is imaged.

The equation relating the scatterer position vector, **T**, a reference position for the platform **P**, and the look vector, **l**, is

$$\mathbf{T} = \mathbf{P} + \mathbf{l} = \mathbf{P} + r\hat{l} \qquad [16]$$

where r is the range to the scatterer and $\hat{l}$ is the unit vector in the direction of **l** (**Figure 9**). The position **P** can be chosen arbitrarily, but is usually taken as the position of one of the interferometer antennas. Interferometric height reconstruction is the

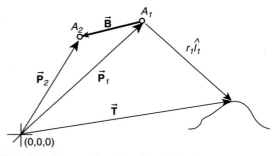

Figure 9 Vectors describing the relationship between the phase centers of the radar antennas defining the interferometer and the surface location, as described in the text.

determination of a target's position vector from known platform ephemeris information, baseline information and the interferometric phase. Assuming $\mathbf{P}$ and r are known, interferometric height reconstruction amounts to the determination of the unit vector $\hat{l}$ from the interferometric phase. Letting $\mathbf{B}$ denote the baseline vector from antenna 1 to antenna 2, setting $\mathbf{P} = \mathbf{P}_1$ and defining

$$\mathbf{B} = \mathbf{P}_2 - \mathbf{P}_1 \quad B = |\mathbf{B}| \equiv \langle \mathbf{B}, \mathbf{B} \rangle^{1/2} \quad [17]$$

we have the following expression for the interferometric phase

$$\phi = \frac{2\pi p}{\lambda}(r_2 - r_1) = \frac{2\pi p}{\lambda}(|\mathbf{l}_2| - |\mathbf{l}_1|) \quad [18]$$

$$= \frac{2\pi p}{\lambda} r_1 \left[\left(1 - \frac{2\langle \hat{l}_1, \mathbf{B} \rangle}{r_1} + \left(\frac{B}{r_1} \right)^2 \right)^{1/2} - 1 \right] \quad [19]$$

where $p = 2$ for repeat-track systems and $p = 1$ for two-aperture systems with a single transmitter and two receivers (Rosen *et al.*, 2000), and the subscripts refer to the antenna number. This expression can be simplified assuming $B \ll r$ by Taylor-expanding eqn [19] to first order to give

$$\phi \approx -\frac{2\pi p}{\lambda} \langle \hat{l}_1, \mathbf{B} \rangle \quad [20]$$

illustrating that the phase is approximately proportional to the projection of the baseline vector on the look direction (Zebker and Goldstein, 1986).

When the baseline lies entirely in the plane of the look vector and the nadir direction, we have $\mathbf{B} = (B \cos(\alpha), B \sin(\alpha))$, where α is the angle the baseline makes with respect to a reference horizontal plane. Then, eqn [20] can be rewritten as

$$\phi = -\frac{2\pi p}{\lambda} B \sin(\theta - \alpha) \quad [21]$$

where θ is the look angle, the angle the LOS vector makes with respect to nadir, shown in **Figure 10**.

The intrinsic fringe frequency in the slant plane interferogram is given by

$$\frac{\partial \phi}{\partial r} = \frac{2\pi p}{\lambda} B \cos(\theta - \alpha)$$
$$\times \frac{1}{r \sin \theta} \left[-\frac{r}{h_p + r_e} + \cos \theta + \frac{h_0 + r_e \sin \tau_c}{h_p + r_e \sin(i - \tau_c)} \right] \quad [22]$$

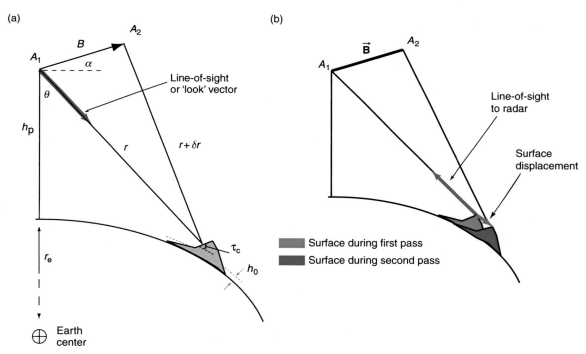

Figure 10 SAR interferometry imaging geometry in the plane normal to the flight direction for topography (a) and deformation (b) mapping.

where

$$\sin i = \frac{h_p + r_e}{h_0 + r_e} \sin \theta \qquad [23]$$

and i is the local incidence angle relative to a spherical surface, h_p is the height of the platform, and τ_c is the surface slope angle in the cross-track direction as defined in **Figure 10** at left. From eqn [22], the fringe frequency is proportional to the perpendicular component of the baseline: $B_\perp = B\cos(\theta - \alpha)$. As $B_\perp$ increases or as the local terrain slope approaches the look angle, the fringe frequency increases. Also from eqn [22], the fringe frequency is inversely proportional to λ, thus longer wavelengths result in lower fringe frequencies. If the phase changes by 2π or more across the range resolution element, Δr, the different contributions within the resolution cell do not add to a well-defined phase, resulting in what is commonly referred to as decorrelation of the interferometric signal. Thus, in interferometry, an important parameter is the critical baseline, defined as the perpendicular baseline at which the phase rate reaches 2π per range resolution element. From eqn [22], the critical baseline satisfies the proportionality relationship

$$B_{\perp, \text{crit}} \propto \frac{\lambda}{\Delta r} \qquad [24]$$

This is a fundamental constraint for interferometric radar applied to natural (distributed scattering) surfaces. Point targets, sometimes called permanent scatterers, can maintain phase coherence beyond this critical baseline, however. Difficulty in phase unwrapping increases (see Section 12.2.4) as the fringe frequency approaches this critical value.

The fringe variation in the interferogram is 'flattened' by subtracting the expected phase from a surface of constant elevation. The resulting fringes follow the natural topography more closely. Letting $\hat{l}_0$ be a unit vector pointing to a surface of constant elevation, h_0, the flattened phase, ϕ_{flat}, is given by

$$\phi_{\text{flat}} = -\frac{2\pi p}{\lambda}\left(\left\langle \hat{l}, \mathbf{B} \right\rangle - \left\langle \hat{l}_0, \mathbf{B} \right\rangle\right) \qquad [25]$$

where

$$\hat{l}_0 = (\sin \theta_0, -\cos \theta_0) \qquad [26]$$

and $\cos \theta_0$ is given by the law of cosines

$$\cos \theta_0 = \frac{r_0^2 + \left(r_e + h_p\right)^2 - \left(r_e + h_0\right)^2}{2\left(r_e + h_p\right) r_0} \qquad [27]$$

assuming a spherical Earth with radius r_e and a slant range to the reference surface r_0.

Equation [25] can be simplified by expanding the look angle $\theta = \theta_0 + \delta\theta$, where $\delta\theta$ is the contribution to the look angle at range r_0 from the topographic relief relative to a reference surface, and θ_0 is the look angle to the reference surface at range r_0. If the topographic relief is represented by δz, then $\delta\theta = \delta z / r_0 \sin \theta_0$, and

$$\phi_{\text{flat}} = -\frac{2\pi p}{\lambda}\left(\left\langle \hat{l}, \mathbf{B} \right\rangle - \left\langle \hat{l}_0, \mathbf{B} \right\rangle\right) \approx -\frac{2\pi p}{\lambda} B_\perp \frac{\delta z}{r_0 \sin \theta_0} \qquad [28]$$

is the component of baseline perpendicular to the LOS. Equation [28] tells us several things about the fringes:

- The flattened fringes are proportional to the topographic height directly. A poorman's topographic map then can be generated by flattening the phase and examining the fringes.
- The flattened fringes are proportional to the perpendicular component of the baseline. For zero baseline, there are no fringes, even if there is a large parallel component of the baseline. For large baselines, there are many cycles of phase change for a given topographic change. From this equation, we can define h_a, the 'ambiguity height', as $h_a = \partial h / \partial \phi = \lambda r_0 \sin \theta_0 / 2\pi p B_\perp$.
- In the absence of topographic variations, there is still an intrinsic variation of fringes across an interferogram given by the flattening phase.

12.2.3 Differential Interferometry

The theory just described assumes that the imaged surface is stationary over time, or that the surface is imaged by the interferometer at a single instant. When there is motion of the surface between radar observations there is an additional contribution to the interferometric phase variation. **Figure 10** at right shows the geometry when a surface displacement occurs between the observation at $\mathbf{P}_1$ (at time t_1) and the observation at $\mathbf{P}_2$ (at $t_2 > t_1$). In this case, $\mathbf{l}_2$ becomes

$$\mathbf{l}_2 = \mathbf{T} + \mathbf{D} - \mathbf{P}_2 = \mathbf{l}_1 + \mathbf{D} - \mathbf{B} \qquad [29]$$

where $\mathbf{D}$ is the displacement vector of the surface from t_1 to t_2. The interferometric phase expressed in terms of this new vector is

$$\phi = \frac{4\pi}{\lambda}\left(\langle \mathbf{l}_1 + \mathbf{D} - \mathbf{B}, \mathbf{l}_1 + \mathbf{D} - \mathbf{B}\rangle^{1/2} - r_1\right) \qquad [30]$$

Assuming, as above, that $|\mathbf{B}|$, $|\mathbf{D}|$, and $|\langle \mathbf{B}, \mathbf{D}\rangle|$ are all much smaller than r_1, the phase reduces to

$$\phi = \frac{4\pi}{\lambda}(-\langle \mathbf{l}_1, \mathbf{B}\rangle + \langle \mathbf{l}_1, \mathbf{D}\rangle) \qquad [31]$$

Typically, for spaceborne geometries $B < 1$ km, and D is of order meters, while $r_1 \approx 600\text{--}800$ km. This justifies the usual formulation in the literature that

$$\phi_{\text{obs}} = \phi_{\text{topography}} + \phi_{\text{displacement}} \qquad [32]$$

In some applications, the displacement phase represents a nearly instantaneous translation of the surface resolution elements, for example, earthquake deformation. In other cases, such as glacier motion, the displacement phase represents a motion tracked over the time between observations. Intermediate cases include slow and/or variable surface motions, such as volcanic inflation or surging glaciers. Equations [31] and [32] highlight that the interferometer measures the projection of the displacement vector in the radar LOS direction. To reconstruct the vector displacement, observations must be made from different LOS directions (see Section 12.4.5).

The topographic phase term is not of interest for displacement mapping, and must be removed. Several techniques have been developed to do this. They all essentially derive the topographic phase from another data source, either a DEM or another set of interferometric data. The selection of a particular method for topography measurement depends heavily on the nature of the motion (steady or episodic), the imaging geometry (baselines and time separations), and the availability of data.

It is important to appreciate the increased precision of the interferometric displacement measurement relative to topographic mapping precision. Consider a discrete displacement event such as an earthquake where the surface moves by a fixed amount $\vec{\mathbf{D}}$ in a short time period. Neither a pair of observations acquired before the event (pair 'a') nor a pair after the event (pair 'b') would measure the displacement directly, but together would measure it through the change in topography. According to eqn [30], and assuming the same imaging geometry for 'a' and 'b' without loss of generality, the phase difference between these two interferograms (i.e., the difference of phase differences) is

$$\phi = \phi_a - \phi_b \qquad [33]$$

$$= \frac{4\pi}{\lambda}\left[\left(\langle \mathbf{l}_1 - \mathbf{B}, \mathbf{l}_1 - \mathbf{B}\rangle^{1/2} - r_1\right)\right. \qquad [34]$$

$$-\left(\langle \mathbf{l}_1 + \mathbf{D} - \mathbf{B}, \mathbf{l}_1 + \mathbf{D} - \mathbf{B}\rangle^{1/2}\right.$$

$$\left.\left. -\langle \mathbf{l}_1 + \mathbf{D}, \mathbf{l}_1 + \mathbf{D}\rangle^{1/2}\right)\right] = 0 \qquad [35]$$

to first order, because $\vec{\mathbf{D}}$ appears in both the expression for $\mathbf{l}_2$ and $\mathbf{l}_1$. The nature of the sensitivity difference inherent between eqns [31] and [35] can be seen in the 'flattened' phase (see eqn [28]) of an interferogram, often written (Rosen et al., 1996)

$$\phi = -\frac{4\pi}{\lambda} B \cos(\theta_0 - \alpha)\frac{z}{r_0 \sin \theta_0} + \frac{4\pi}{\lambda}\delta r_{\text{disp}} \qquad [36]$$

where δr is the surface displacement between imaging times in the LOS direction, and z is the topographic height above the reference surface. In this formulation, the phase difference is far more sensitive to changes in topography (surface displacement) than to the topography itself. From eqn [36] $\delta r = \lambda/2$ gives one cycle of phase difference, while z must change by a substantial amount, essentially r_0/B, to affect the same phase change. For example, for ERS, $\lambda = 5.6$ cm, $r_1 \approx 830$ km, and typically $B \leq 200$ m, implying $\delta r_{\text{disp}} = 2.8$ cm to generate one cycle of phase, $z \geq 450$ cm to have the same effect.

The time interval over which the displacement is measured must be matched to the geophysical signal of interest. For ocean currents, the temporal baseline must be of the order of a fraction of a second because the surface changes quickly and the assumption that the backscatter phase is common to the two images could be violated. At the other extreme, temporal baselines of several years may be required to make accurate measurements of slow deformation processes such as interseismic strain.

Reconstruction of the scatterer position vector depends on knowledge of the platform location, the interferometric baseline length, orientation angle, and the interferometric phase. To generate accurate topographic or displacement maps, radar interferometry places stringent requirements on knowledge of the platform and baseline vectors, as well as the intrinsic accuracy of the phase measurements and, in the case of differential interferometry, supporting topographic data sets. One source of phase noise is the refractivity of the atmosphere, which varies along the radar propagation path in space and time. Refractivity fluctuation due to turbulence in the atmosphere is a minor effect for two-aperture cross-track interferometers, but is much more important for repeat-track systems (Rosen et al., 1996). Sensitivities to these parameters are discussed in detail in Rosen et al. (2000) and Zebker et al. (1994).

12.2.4 Phase Unwrapping

The phase of the interferogram must be unwrapped to remove the modulo-2π ambiguity before estimating topography or surface displacement. The literature describing approaches to phase unwrapping is quite large (Rosen *et al.*, 2000), with initial development of so-called branch-cut techniques for InSAR applications by Goldstein *et al.* (1988), followed by over a decade of exploration of other techniques.

A simple approach to phase unwrapping would be to form the first differences of the phase at each image point in either image dimension as an approximation to the derivative, and then integrate the result. Direct application of this approach, however, allows local errors due to phase noise to propagate, causing errors across the full SAR scene. The unwrapped solution should, to within a constant of integration, be independent of the path of integration. This implies that in the error-free case, the integral of the differenced phase about a closed path is zero. Phase inconsistencies are therefore indicated by non-zero results when the phase difference is summed around the closed paths formed by each mutually neighboring set of four pixels. These points have either a positive or negative integral (by convention performed in clockwise paths). Integration of the differenced phase about any closed path yields a value equal to the sum of the enclosed points of inconsistency. As a result, paths of integration that encircle a non zero sum must be avoided. In branch-cut methods, this is accomplished by connecting oppositely signed points of phase inconsistency with lines that the path of integration cannot cross. Once these barriers have been selected, phase unwrapping is completed by integrating the differenced phase subject to the rule that paths of integration do not cross the barriers.

The phase unwrapping problem becomes particularly difficult when the phase in the interferogram is intrinsically discontinuous, due to layover problems or true shear topography. Most algorithms are based on the assumption that the phases are continuous, and often natural phase discontinuities, often corrupted with inherent phase noise, are difficult to interpret.

A full treatment of phase unwrapping for geodetic imaging applications is beyond the scope of this chapter. There are a number of algorithms available for use, including branch-cut algorithms (Goldstein *et al.*, 1988) and statistical cost network flow techniques (Chen and Zebker, 2001). These techniques yield unwrapped phase images that are multiples of 2π of the original wrapped phase image. In the case of branch-cut algorithms, there are often regions that are blocked off from unwrapping by barriers that form a complete circuit. For network flow, the entire image is unwrapped. In all cases, there will be individual pixels or areas that are placed on the wrong multiple of 2π, and it is often quite difficult to identify these points without additional information.

12.2.5 Correlation

The relationship between the scattered electromagnetic fields seen at the interferometric receivers after image formation is characterized by the complex correlation function, γ, defined as

$$\gamma = \frac{\langle \Gamma_1 \Gamma_2^* \rangle}{\sqrt{\langle |\Gamma_1|^2 \rangle \langle |\Gamma_2|^2 \rangle}} \qquad [37]$$

where Γ_i represents the SAR reflectivity at the i antenna, and angular brackets denote averaging over the ensemble of speckle realizations. For completely coherent scatterers such as point scatterers, we have that $\gamma = 1$, while $\gamma = 0$ when the scattered fields at the antennas are independent. The magnitude of the correlation $|\gamma|$ is often referred to as the 'coherence.' (Several authors distinguish between the "coherence" properties of fields and the correlation functions that characterize them (e.g., Born and Wolf (1989)), whereas others do not make a distinction.)

In general, the correlation will comprise contributions from a number of effects:

$$\gamma = \gamma_N \gamma_G \gamma_Z \gamma_T \qquad [38]$$

where γ_N is the correlation influenced by noise in the radar system and processing approach, γ_G is that influenced by the different observing geometries, γ_Z describes the influence on correlation of the vertical extent of scatterers (e.g., due to vegetation), and γ_T describes the influence of repositioning of scatterers within a resolution element over time (Li and Goldstein, 1990; Zebker and Villasenor, 1992; Rodríguez and Martin, 1992; Bamler and Hartl, 1998; Rosen *et al.*, 2000). It is often more convenient to discuss decorrelation, defined as $\delta_X = 1 - \gamma_X$, where X is N, G, Z, or T.

Geometric decorrelation, δ_G, also called baseline or speckle decorrelation, is due to the fact that, after removing the phase contribution from the center of the resolution cell, the phases from the scatterers located away from the center are slightly different at each antenna (see **Figure 7**). The degree of decorrelation can then be estimated from the differential phase of two points located at the edges of the area obtained by projecting the resolution cell phase from each scatterer within the resolution cell, as shown in **Figure 7**. Using this simple model, one can estimate that the null-to-null angular width of the correlation function, $\Delta\theta$, is given by

$$\Delta\theta \approx \frac{B_\perp}{r} \approx \frac{\lambda}{\Delta r_l} \qquad [39]$$

where $B_\perp$ is the projection of the interferometric baseline onto the direction perpendicular to the look direction, and Δr_l is the projection of the ground resolution cell along the same direction, as illustrated in **Figure 11**.

The geometric correlation term is present for all scattering situations, and depends on the system parameters and the observation geometry. A general expression for it is

$$\gamma_G = \frac{\displaystyle\int \mathrm{d}s\, \mathrm{d}r\, W_1(r,\,x) W_2^*(r+\delta_r,\, x+\delta_x)}{\displaystyle\int \mathrm{d}x\, \mathrm{d}r\, W_1(r,\,x) W_2^*(r,\,x)} \times \exp[jr(p\kappa_r + 2\delta k)] \exp[j\tan\tau_x \kappa_z s] \qquad [40]$$

where $k \equiv 2\pi/\lambda$ is the wave number; δk represents the shift in the wave number corresponding to any difference in the center frequencies between the two interferometric channels; δ_r and δ_x are the misregistration between the two interferometric channels in the range (r) and azimuth (x) directions, respectively; $W_i(r, x)$ is the SAR point-target response in the range and azimuth directions; and τ_x is the surface slope angle in the azimuth direction. In eqn [40], κ_r and κ_z are the interferometric fringe wave numbers in the range and vertical directions, respectively. They are given by

$$\kappa_r = \frac{kB_\perp}{r\tan(\theta-\tau_c)} \qquad [41]$$

$$\kappa_z = \frac{kB_\perp \cos\tau_c}{r\sin(\theta-\tau_c)} = \kappa_r \frac{\cos\tau_c}{\cos(\theta-\tau_c)} \qquad [42]$$

The value of δk can be adjusted to recenter the range spectrum of each interferometric channel. This can be accomplished in principle by bandpass filtering the range spectrum differently in each channel. Under the right conditions, one can adjust the center frequencies to create the condition $2\delta k = -\kappa_r$, which leads to $\gamma_G = 1$ (Gatelli *et al.*, 1994). In other words, the geometric decorrelation term in principle can be eliminated by proper choice of center frequencies for two observations. In practice, as can be seen from the equation above, κ_r depends on the look angle and surface slope, so that adaptive iterative processing is required in order to implement the approach exactly.

The volumetric correlation term can be understood in terms of an effective increase in the size of the projected range cell Δr_l because the scattering elements in a given range cell are now extended not just on a surface but in a volume (Rosen *et al.*, 2000). If the range resolution is infinitesimally small, the volume decorrelation effect can be understood as being due to the geometric decorrelation from a plane cutting through the scattering volume perpendicular to the look direction. It was shown in Rodríguez and Martin (1992) that the volumetric correlation γ_Z can be written as

$$\gamma_Z(\kappa_z) = \int \mathrm{d}z\, f(z)\exp[-j\kappa_z z] \qquad [43]$$

provided the scattering volume could be regarded as homogeneous in the range direction over a distance defined by the range resolution. The function $f(z)$, the 'effective scatterer probability density function (pdf)', is given by

$$f(z) = \frac{\sigma(z)}{\displaystyle\int \mathrm{d}z\, \sigma(z)} \qquad [44]$$

where $\sigma(z)$ is the effective normalized backscatter cross-section per unit height. The term 'effective' is

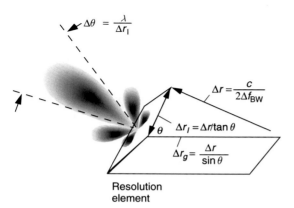

Figure 11 A view of geometric decorrelation showing the effective beam pattern of a ground resolution element 'radiating' to space. The mutual coherence field propagates with radiation beam width in elevation of $\Delta\theta \approx \lambda/\Delta r_l$. These quantities are defined in the figure.

used to indicate that $\sigma(z)$ is the intrinsic cross-section of the medium attenuated by all propagation losses through the medium. The specific form for $\sigma(z)$ depends on the scattering medium. Models for this term, and its use in the remote sensing of vegetation height, will be discussed later.

In repeat-pass systems, there is another source of decorrelation. Temporal decorrelation, δ_T, occurs when the surface changes between the times when the images forming an interferogram are acquired (Zebker and Villasenor, 1992). As scatterers become randomly rearranged over time, the detailed speckle patterns of the image resolution elements differ from one image to the other, so the images no longer correlate.

While it is difficult to describe these effects analytically, this can often be a strong limitation on the accuracy of repeat-pass data, so a few illustrative examples are in order. Open water, where the surface is roughened by wind or turbulence, is constantly changing over time, so two images will completely decorrelate ($\gamma = 0$). Similarly, an agricultural field, where the entire surface has been turned over due to tilling, will have $\gamma = 0$. Standing water with no vegetation present above water will also completely decorrelate because no signal is scattered back toward the radar. However, with vegetation present above the surface, the water serves as a mirror that permits signal return from the water as scattered off the vegetation (Alsdorf et al., 2001; Wdowinski et al., 2004). Rain can change the arrangement of vegetation on a surface (e.g., sagging branches or stalks), reducing the correlation by an amount dependent on the density of altered vegetation. In some cases, as the surface dries, the vegetation bounces back to its original position and correlation is at least partially restored. However, wind typically will alter the positions of scatterers in vegetation canopies over time, so correlation is generally degraded in vegetation. Snow can destroy correlation in the winter months, with correlation restored after the snow is gone. For interferometry applied to geophysical processes, we rely on the block motion of pixels without scatterer rearrangement to provide estimates of the geodetic motion. These effects, and changes due to ground shaking (building collapse, landslides, liquefaction, etc.), can impair the ability to measure displacements. On the other hand, it can also be a means for understanding the nature of the surface and the severity of the geophysical effects on the ground.

In addition to these field correlations, thermal noise in the interferometric channels also introduces phase

noise in the interferometric measurement. The correlation due to thermal noise alone can be written as

$$\gamma_N = \frac{1}{\sqrt{1 + SNR_1^{-1}}\sqrt{1 + SNR_2^{-1}}} \qquad [45]$$

where SNR_i denotes the signal-to-noise ratio for the i channel (Zebker and Villasenor, 1992). In addition to thermal noise, which is additive, SAR returns also have other noise components, due to, for example, range and Doppler ambiguities. An expression for the decorrelation due to this source of error can only be obtained for homogeneous scenes, since, in general, the noise contribution is scene dependent. Typically, for simplicity these ambiguities are treated as additive noise as part of the overall system noise floor.

The effect of decorrelation is the apparent increase in noise of the estimated interferometric phase. Rodríguez and Martin (1992) presented the analytic expression for the Cramer–Rao bound (Sorenson, 1980) on the phase variance

$$\sigma_\phi^2 = \frac{1}{2N_L}\frac{1 - \gamma^2}{\gamma^2} \qquad [46]$$

where N_L is the number of independent samples used to derive the phase, and is usually referred to as the 'number of looks.' The actual phase variance approaches the limit eqn [46] as the number of looks increases, and is a reasonable approximation when the number of looks is greater than four. An exact expression for the phase variance can be obtained starting from the probability density function for the phase when $N_L = 1$, and then extended for an arbitrary number of looks (Goodman, 1985; Joughin et al., 1994; Lee et al., 1992; Touzi and Lopes, 1996). The expressions, however, are quite complicated and must be evaluated numerically in practice.

Note that the estimate of the correlation is usually accomplished by computing the expectation operations in eqn [37] as spatial averages over a number of pixels in an interferometric pair. This leads to biased estimates of the correlation, and care must be exercised in interpreting the estimate. For example, in open water, where actually the coherence of the fields is zero, the correlation estimate will produce decidedly nonzero estimates, in the range of 0.1–0.3, depending on the number of samples used in the estimate, N_L. The estimator is a random variable with a probability distribution shape that depends on the intrinsic coherence, and the number of samples used in the estimate. In the limit, where only one sample is used, the correlation estimate will be 1!

12.3   InSAR-Related Techniques

12.3.1   ScanSAR or Wide Swath Interferometry

For previously flown SAR systems, the width of the swath has been limited to somewhat less than 100 km. As discussed in Section 12.5, SAR antennas must satisfy particular minimum area criteria to ensure noise due to ambiguities below a required level. For wide swath, they must also be quite long, which can be difficult and costly to implement. For example, to achieve a 300 km swath in typical Earth orbits using the typical strip mapping method, the antenna would have to be over 40 m in length. To achieve wide swaths with an antenna sized for a swath smaller than 100 km, the ScanSAR technique has been developed (Tomiyasu, 1981; Moore *et al.*, 1981). ScanSAR requires an antenna with rapid electronic steering capability in the elevation direction. In ScanSAR,

the antenna is electronically steered to a particular elevation angle and radar pulses are transmitted and echoes received for a time period that is a fraction (say one-tenth) of the synthetic aperture time. After that 'dwell period,' also known as the 'burst period,' the antenna is electronically steered to another elevation angle (and other radar parameters such as the PRF, bandwidth, and antenna beam shape are changed), and observations are made for another short dwell period. This process is repeated until each of the elevation directions, needed to observe the entire wide swath is obtained at which point the entire cycle of elevation dwell periods is repeated (**Figure 12**).

For any given elevation direction, or subswath, there are large gaps in the receive echo timeline, yet after processing the data, a continuous strip mode image can be formed. This is true because the extent of the antenna beam in the along-track direction on the ground is equal to the synthetic aperture

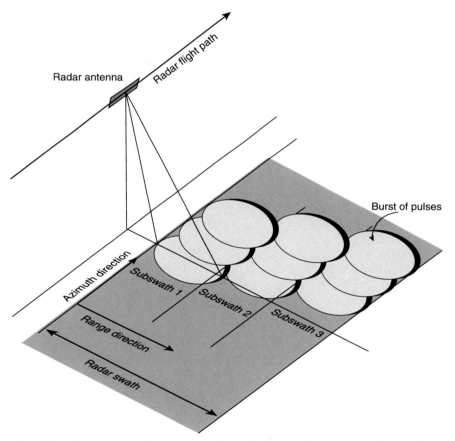

Figure 12   ScanSAR for a three-beam system. The radar transmits a burst of pulses to illuminate subswath 1, then electronically switches to point to subswath 2, then subswath 3. This cycle is repeated for the extent of the data take. The aspect ratio of the beams in this figure is highly stretched in azimuth to illustrate the pulsed behavior for the bursts and the significant beam overlap from burst to burst within a subswath. This overlap allows for construction of continuous maps.

length. As long as the dwell periods for any given subswath occur more than once in the synthetic aperture time, there is guaranteed continuous coverage of all points on the ground.

It is important to understand this method of generating radar data because it has strong implications for the quality of the geodetic data that are derived, and for the constraints that are imposed on the use of the data. First we note that the data contained in any given pulse include the full Doppler spectrum of information. We are transmitting over a broad range of angles (the beam width) and that defines the Doppler frequency content. So each burst of pulses contains the full Doppler spectrum of information. If one were to derive the spectrum in the along-track dimension, the full Doppler bandwidth would be represented. However, we note that any given scattering element within a burst period only contributes a portion of its full Doppler history because it is not observed over the full synthetic aperture time. Thus each scattering element is only resolved commensurate with the burst period relative to the synthetic aperture time: if the resolution in strip mode is $L/2$, then, the resolution in ScanSAR mode is $(L/2) * T_a/T_b$, where T_a is the synthetic aperture time, and T_b is the burst duration. If one were to attempt to achieve maximal resolution possible, one would divide the synthetic aperture time by the number of subswaths needed, and set the burst duration to this time. However, as described following, it is generally better to create several short bursts within one synthetic aperture. This degrades resolution further, but improves the radiometric characteristics of the data.

For interferometry, the most important aspect of ScanSAR is the fact that each scattering element provides in a burst only a small portion of its total Doppler history (Bamler and Eineder, 1996; Guarnieri and Prati, 1996; Lanari et al., 1998). This is equivalent to the statement that each scattering element is observed over a small range of azimuth angles within the beam. In order for interferometry to work, a scattering element must be observed with the same range of Doppler frequencies from one pass to the other. Only then will the images be coherent, with similar speckle patterns. This implies that from pass to pass, each observation must observe from the same group of angles. In the case of strip mode SAR, this implies that the intrinsic pointing of the antenna beam be the same from pass to pass. In this case, the Doppler history of each scattering element will

follow the same course. For ScanSAR, this condition also implies a timing constraint on the bursts. Each burst must occur at the same location in space relative to a scattering element from pass to pass. This constraint makes it much more challenging for mission operators, particularly with short bursts. **Figure 13** illustrates the workflow associated with ScanSAR interferometry as well as required intermediary data objects.

For a satellite in a long repeat period orbit, for example 32 days, the ground swath will be sized to achieve global coverage, around 85 km for the 32-day repeat period orbit. So from an interferometric point of view, using ScanSAR with this period does not improve on the interferometric interval intrinsically. However, ScanSAR will increase the number of times a given scattering element will be observed by a factor of the number of ScanSAR beams. For a four-beam ScanSAR with 340 km swath in a 32-day repeat orbit, a scattering element will be observed roughly every 8 days, so one could make interleaved 32-day repeat interferograms. Alternatively, one could place the radar in a shorter repeat period orbit, for example 8 days, and observe always in ScanSAR mode with four beams. This would allow 8-day interferograms with no interleaving. The advantage of the latter is that decorrelation will be less of an issue in the basic repeat-pass measurement. The advantage of the former is that there will be greater angular diversity in the measurement, potentially resulting in better constrained models of deformation.

12.3.2 Permanent Scatterers and Time-Series Analysis

The two major error sources in InSAR are decorrelation due to temporal and geometric effects and phase errors introduced by the spatially and temporally random variations of the refractive index of the atmosphere and ionosphere. Decorrelation creates areas that are spatially disjoint, and irregularly so over time, leading to difficulty in interpreting the geodetic measurements. Because the repeat period of the orbital satellites is on the order of weeks, temporal sampling of dynamic phenomena is also poor. Continuous GPS measurements share similar characteristics – spatially disjoint measurements that also are subject to atmospheric effects – but have the advantage of dual-frequency measurements for ionospheric correction, and continuous temporal sampling while observing multiple satellites, which

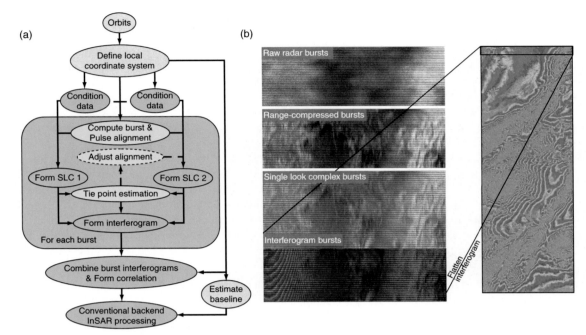

Figure 13 (a) Representative differential InSAR processing flow diagram for ScanSAR interferometry. Blue bubbles represent image output, yellow ellipses represent nonimage data. Flow is generally down the solid paths, with optional dashed paths indicating potential iteration steps. (b) Illustration of the workflow by explicity example. The top three black and white panels show a collection of bursts from one scene for one ScanSAR subswath. The top panel displays raw radar data, with a small artificial gap between bursts to delimit them. The next panel shows the bursts after range compression – features appear to be sharper in the range (across image) dimension. The next panel shows the bursts after azimuth compression – single look complex bursts – features are now sharp in both dimensions, and the existence of surface features in three or four successive bursts is apparent, showing the intrinsic overlap of the data. The bottom center panel shows burst interferograms formed from two SLC burst sequences. After combination and flattening, the bursts form a continuous interferogram as shown on the right, with the left sequence contained in the box at the top as indicated. This interferogram, as well as those of the other subswaths can then be processed as usual.

allows for a fine sampling of atmospheric variability, leading to robust correction algorithms. Given that important areas of the world have significant issues with decorrelation and atmospheric water vapor, interferometry research incorporates methods to reduce the problem to one more similar to continuous GPS. Corner reflectors have long been used as coherent calibration sources for interferometric systems, as they are observable over a wide range of angles and have a well-defined phase center. Usai was the first to note that some man-made structures can behave much like corner reflectors, as though they are continuously reliable coherent scatterers over time and independent of the interferometric baseline (Usai, 1997). By identifying points in a series of radar images that maintain their coherence over time, we can create a network of phase measurements over time and space that are directly analogous to GPS measurements, though sampled once per month

rather than continuously. Ferretti and colleagues (Ferretti *et al.*, 2000, 2001) were the first to systematize and popularize the treatment of these discrete point networks for long-term trend analysis. A number of groups have pursued similar techniques (Wegmuller, 2003; Hooper *et al.*, 2004). **Figure 14** illustrates the way this technique overcomes the atmospheric error issues.

In practice, it is challenging to determine the location of time-coherent scatterers because the phase of each point is comprised of a topographic component, a displacement component, and noise components, all of which are different in each image of a series of images:

$$\phi_{\Delta t_i} = \phi_{\text{topo},\Delta t_i} + \phi_{\text{def},\Delta t_i} + \phi_{\text{atm},\Delta t_i} + \phi_{\text{noise},\Delta t_i} \quad [47]$$

where Δt_i is the time interval for interferogram i, and the interferometric phase, $\phi_{\Delta t_i}$, is broken into its principal components. Though the topography is assumed to be static over time, the phase term

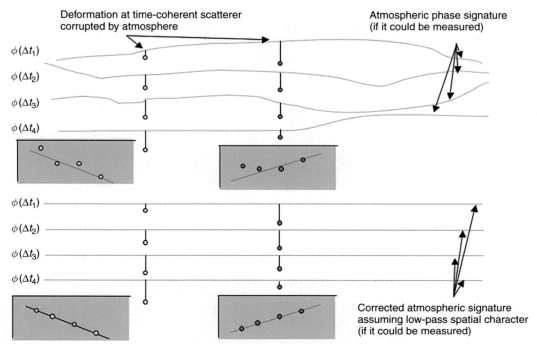

Figure 14 The 'permanent scatterer' technique identifies time-coherent scatterers by estimating the contributions of topography, deformation, and atmospheric delay to the phase under model constraints through correlation maximization. Topography is assumed to be static (with the interferometric phase proportional to baseline), deformation is assumed to follow some functional form (e.g., linear or sinusoidal with time), and atmospheric delay is assumed to vary randomly in time and with long spatial wavelength.

$\phi_{\text{topo},\Delta t_i}$ is variable because each interferogram has a different perpendicular component of the baseline $B_{\perp,i}$. Similarly, the other component terms will change over time. If we simply coalign all available images, then estimate the correlation over time for each point using eqn [37] (with expectation approximated by time averaging), the variability of the phase over time will lead to an estimate of zero correlation. Ferretti *et al.* (2000, 2001) chose to identify time-coherent points by using a brightness threshold, as points with high mean brightness over time have small phase dispersion. This requires that images are radiometrically calibrated, such that variations in the brightness are due to scene variations, not radar system variations (e.g., radar antenna pointing variability).

Once these initial points are identified, it is necessary to find a solution for each phase term for all i that maximizes the correlation estimate. The maximization procedure involves a search over reasonable domains of expected phase values, which can be intensive. For example, if a DEM is available, the topographically induced phases are known, except for errors introduced by the DEM

errors, so the search range for topography should be bounded in a region around the known topography by that error. The deformation is parametrized by a functional form such as linear or sinusoidal, and the solution of the parameters is part of the maximization. The atmospheric error is assumed to be spatially slowly varying, and uncorrelated over time.

This technique has been employed with success in urban areas where man-made time-coherent scattering points abound. Results include the measurement of subsidence rates of individual buildings at the level of less than $1 \, \text{mm} \, \text{yr}^{-1}$, and seasonal effects due to groundwater withdrawal and recharge (Ferretti *et al.*, 2000; Colesanti *et al.*, 2003). Applications to problems of geophysical interest are limited because the areas of interest are generally larger and sufficient time-coherent scatterers frequently do not exist in nonurban environments. Burgmann *et al.* (2006) combined GPS-derived horizontal velocities and permanent scatterer InSAR estimates of uplift in the San Francisco Bay Area to track tectonic uplift in areas not subject to seasonal effects, at an accuracy of better than $1 \, \text{mm} \, \text{yr}^{-1}$.

Other methods have been developed to exploit long time series available for over a decade of SAR observations, but without the restriction of using only time-coherent point scatterers, and without specific assumptions of the model of deformation. By limiting data to only those scene combinations where the baseline is well below the critical baseline, one can expand the area of usable image pixels that are from natural surfaces, but are coherent in time as well (Lundgren *et al.*, 2001; Berardino *et al.*, 2002; Schmidt and Burgmann, 2003; Lanari *et al.*, 2004b). In these methods, each collection of images allows the generation of a dense spatial network of points (for well-correlated areas, potentially every image pixel could be included in the network). For each point, interferometric pairs formed from various combinations of images yield phase differences over available time intervals in the time series where correlation is good, with the shortest sampling interval being the satellite repeat period. From these phase difference measurements for each network point, an inversion is carried out to reconcile all differences. The inversion adjusts each interferogram by a constant to bring all differences into agreement, and attempts to integrate the phase differences in a piecewise linear fashion, minimizing the distance between phase estimates from one time step to the next. The inversion can be carried out with constraints that smooth the estimates of phase. It is also possible to incorporate parametrized model functions in the inversion, for example, for seasonal effects, if desired. These techniques derive directly from GPS network inversion methods, with the substantive difference being the much higher density of samples in space, but much lower density of samples in time, for InSAR. These approaches are further discussed later in the chapter.

12.3.3   Speckle Tracking and Pixel Tracking

InSAR provides the satellite LOS component of relative displacements within an image. Albeit at lower sensitivities, we can also measure the along-track component of the displacement field using cross-correlation techniques. This is frequently referred to as speckle tracking, pixel tracking, or range and azimuth offsets. For crustal deformation purposes, we frequently only consider the along-track or azimuth offsets (AZOs) which are a purely horizontal component of the displacement field and are by construction perpendicular to the LOS phase-based observations. This approach is strictly speaking not

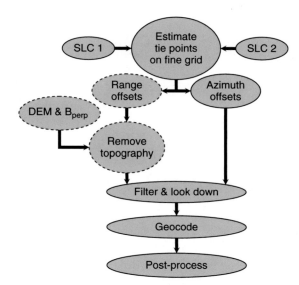

Figure 15   Generic processing flow diagram for generating pixel-tracking measurements. We assume here that the products of the associated differential InSAR processing have already been made.

an interferometric approach, but we include it here since it uses the same single look complex (SLC) images used in forming interferograms (**Figure 15**).

At its most basic level, this approach is simply automatic tie-point estimation and frequently uses the same algorithms and software used to calculate tie points when coregistering two SLCs to form an interferogram or to coregister a DEM-based height map to an interferogram to remove topographic effects (e.g., Michel *et al.*, 1999a, 1999b; Michel and Rignot, 1999; Joughin, 2002; Short and Gray, 2004; Gray *et al.*, 2001). Given a master and slave SLC and a rough guess at their relative gross offsets, a chip of $N \times M$ pixels (typically several tens of pixels on a side) is cross-correlated from both images to provide range and azimuth offset with subpixel accuracy. This process can then be repeated over a grid. Typical sensitivities are about somewhere between 1/10 and 1/30 of a pixel dimension. For the ERS satellites, these numbers correspond to about 10–20 cm sensitivity, with the advantage that they do not need to be phase unwrapped. Since the measurement is done using ensembles of pixels, the spatial resolution is substantially degraded relative to standard interferometric observations. This sensitivity of the measurement is limited by the dimensions of a pixel. While the range offsets provide the same component of the displacement field as the interferometric phase, it is usually much less sensitive and is therefore not usually used. However, there are times, when correlation is low, that the phase-based measurements will not work, but the range offsets can (e.g., Peltzer *et al.*, 1999). In such cases, the range offsets need to be corrected for topographic effects in a way similar to what is done for the phase measurements. The AZO measurement complement the interferometric phase measurements, and therefore always provide useful observations when the expected displacements are large enough to be detected.

Algorithmically, the approach is the same as that used to generate tie points or offset measurements using optical imagery (Michel *et al.*, 1999a, 1999b). With InSAR data, we can achieve extremely accurate estimates of offsets with no inherent contrast in the mean radar backscatter. Since SAR data exhibit high-contrast speckle characteristics from pixel to pixel (Appendix 1) that are the same in interferometrically coherent SAR images, allowing good matching.

Cross-correlation can be performed using the complex images directly or using just the amplitude of the complex image. When complex images are used, we are in essence computing small interferograms and assessing the quality of the fringes therein. Because the phase is used, when the coherence is good, this method can lead to very tight constraints on the pixel offsets. However, when the coherence is poor, the phase contribution leads to extremely poor correlation estimates even when common surface features that are well correlated are present. This is truly a 'speckle-tracking' approach in that if the speckle differs between images, it will not track well. Cross-correlation of amplitude images tracks features, and so would be more accurately called 'feature-tracking' or 'pixel-tracking'. Because speckle appears as high-contrast features from pixel to pixel, amplitude matching also tracks speckle, though it lacks the tightness of the match of complex image correlation when the coherence is good. But it performs very well when images are rich with surface features in areas of poor interferometric correlation. Experimentation with both approaches show that very little matching accuracy is lost by correlating amplitude rather than complex images.

The complex or amplitude image is typically over-sampled by a factor of 2 before the cross-correlation; this is needed to avoid spectral aliasing of the images. The peak of the cross-correlation surface is identified therefore at half-pixel spacing as a first approximation to the offset location. The correlation surface is then interpolated to find the correlation peak with finer granularity. Accuracies have been empirically estimated at about 1/30 of a pixel, on the order of 10–15 cm for typical SAR systems.

We illustrate the use of pixel tracking with an example taken from the 1999 M_w 7.6 Chi–Chi earthquake in Taiwan. This event was a thrust earthquake where the footwall had relatively little topography, while the hanging wall is extremely rugged. Standard InSAR produced clean fringes in the footwall region, but is completely decorrelated in the hanging wall (**Figure 16**). In contrast, the AZO observations can be made for the entire image, with the caveat that strong spatial (median) filtering was required to remove outliers (**Figure 17**).

Instead of using speckle or pixel tracking methods, as described above to infer the along-track component of displacement, Bechor and Zebker (2006) proposed splitting the aperture normally used in forming a single interferogram into separate interferograms using the forward and backward squinting SLCs (relative to the nominal squint angle for the standard SLC). These two interferograms can then in turn be differenced to produce a map of along-track displacements. They showed accuracies on the order of a few centimeters when the interferometric coherence was very good. In areas of lower coherence, the phase difference-derived estimates were comparable to typical AZO estimates.

12.4 A Best-Practices Guide to Using InSAR and Related Observations

Our general goal with geodetic imaging is to discover new crustal deformation processes and to estimate the value and uncertainties of controlling parameters associated with these as well as previously recognized mechanisms. The approaches adopted for using these data vary depending on the extent and type of observations available and the geophysical target. Here, we address some of the approaches that have found some utility in different situations. Beyond the exploratory mode of just looking at the images to discover things, we can consider how to best combine the images, deal with long time series, and how to incorporate these observations in parameter estimation schemes in a computationally efficient manner. In all cases, we must consider the different components of the measurements that act as apparent errors in the geodetic signal. From the perspective of modeling tectonic processes, the primary sources of error come from inadequacies in our knowledge of the satellite orbits and propagation delays accumulated in the

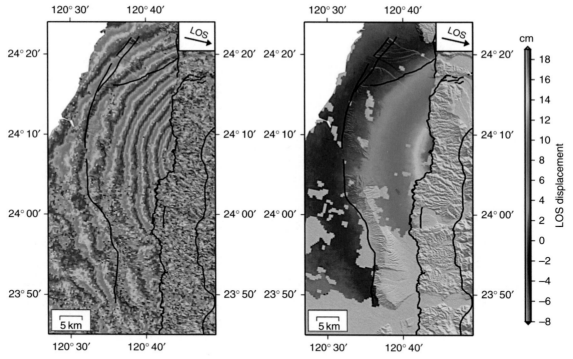

Figure 16 Wrapped at 2.8 cm per fringe (left) and unwrapped (right) interferogram for the 1999 M_W 7.6 Chi-Chi, Taiwan continental thrust earthquake using ERS C-band data from 5/5/99 to 9/23/99. Black lines indicate major fault traces. The Chulengpu Fault is the primary slipping structure for the Chi-Chi earthquake and forms the eastern limit of the unwrapped phase. The phase is successfully unwrapped only on the footwall where the topographic gradients and vegetation cover are dramatically less than in the hanging wall. Adapted from Levy F, Hsu Y, Simons M, LePrince S, and Avouac JP (2005) Distribution of coseismic slip for the 1999 Chi Chi Taiwan earthquake: New data and implications of varying 3D fault geometry. *EOS Transactions of the American Geophysical Union* 86: B1305.

ionosphere and troposphere. Here, we begin with a summary of error sources, since that impacts all uses of the data. At a minimum, knowledge of the error structure (i.e., the covariance structure) is important in order to correctly execute any estimation of geophysical parameters. Furthermore, in certain circumstances, a given component of the error may be reduced using a variety of techniques.

12.4.1   Interferometric Baseline Errors

The primary impact of orbital errors is to induce long-wavelength phase ramps associated with incorrect removal of topographic effects. If $B_\perp$ is long, errors in $B_\perp$ can also lead to short-wavelength errors in regions of rough topography. Typically, apparent long-wavelength deformation gradients associated with errors in $B_\perp$ are dealt with by removing a best-fit bilinear or biquadratic polynomial ramp from both the observations and the models before comparing the two (e.g., Pritchard *et al.*, 2006; Pritchard and

Simons, 2006), or by using independent data, usually GPS observations, to either re-estimate the interferometric baseline or to constrain the longest wavelengths of the physical model (e.g., Peltzer *et al.*, 2001; Simons *et al.*, 2002). Alternatively, if the deformation signal is known to be localized within the interferogram, one can flatten the image by assuming zero deformation in the far field.

In order to merge GPS observations with InSAR data, we must take care how these data are combined. The surface deformation field frequently shows a strong cyclical seasonal component (e.g., Dong *et al.*, 2002). This seasonal signal is exemplified by the GPS time series in **Figure 18**, which shows a greater than 1 cm peak-to-peak seasonal displacement. Of course, with InSAR data, one is sensitive to spatial gradients in these seasonal signals on the scale of the InSAR image, which will usually be smaller than implied by these kind of absolute GPS observations. If one uses GPS to refine estimates of $B_\perp$ then one should use the GPS data before any seasonal signals have been

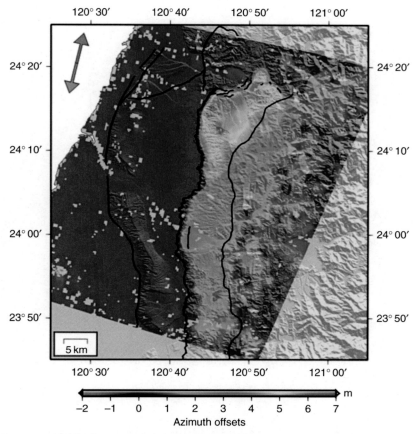

Figure 17 Azimuth offset-based horizontal displacements for the 1999 M_W 7.5 Chi-Chi, Taiwan continental thrust earthquake derived from ERS C-band data from 1/21/99 to 9/23/99. Initial offset measurements are estimated using a correlation window of 23 × 160 pixels equivalent to a resolution of 320 m. A bilinear ramp component of the displacement field is constrained to match that estimated using available GPS data. These offsets are then spatially filtered with a 3-km-wide median filter. Adapted from Levy F, Hsu Y, Simons M, LePrince S, and Avouac JP (2005) Distribution of coseismic slip for the 1999 Chi Chi Taiwan earthquake: New data and implications of varying 3D fault geometry. *EOS Transactions of the American Geophysical Union* 86: B1305.

removed. Conversely, if one assumes the long wavelength in the data are unconstrained and uses the GPS to constrain the model directly, then clearly the seasonal signal should be removed first. This latter approach is safest since the seasonal signal in the GPS is not completely due to surface displacements, and similarly, there are long-wavelength errors in interferograms caused by tropospheric delays and not by errors in $B_\perp$.

12.4.2 Propagation Delays

The atmosphere and ionosphere introduce propagation phase and group delays to the SAR signal. In repeat-track systems, the propagation effects can be more severe. The refractive indices of the atmosphere and ionosphere are not homogeneous in space or time. For a spaceborne SAR, the path delays can be very large, depending on the frequency of the radar (e.g., greater than 50 m ionospheric path delay at L-band), and can be quite substantial in the differenced phase that comprises the interferogram (many centimeters differential tropospheric delay, and meter-level ionospheric contributions at low frequencies). These effects in RTIs were first identified by Massonnet *et al.* (1993) and later by others (Goldstein, 1995; Massonnet and Feigl, 1995; Tarayre and Massonnet, 1996; Rosen *et al.*, 1996; Zebker *et al.*, 1997; Hanssen, 2001; Lohman and Simons, 2005; Gray *et al.*, 2000; Mattar and Gray, 2002). Ionospheric delays are dispersive, so frequency-diverse measurements can potentially

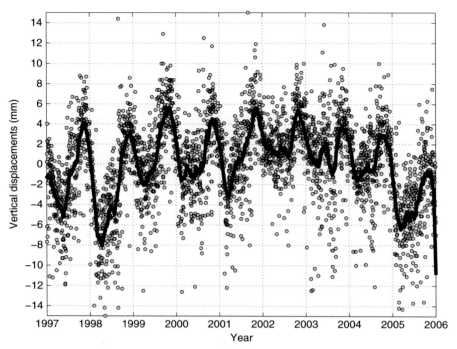

Figure 18 Dots indicate vertical GPS time series from the continuous GPS site TABL located on the central San Andreas Fault, CA. The displacements have been detrended and have the common mode removed. The solid line indicates the running 2-month average. Data from http://sopac.ucsd.edu.

help mitigate the effect, as with two-frequency GPS systems. Tropospheric delays are nondispersive, and can mimic topographic or surface displacement effects. Schemes for distinguishing tropospheric effects from other effects have been proposed (Massonnet and Feigl, 1995), and use of averaging interferograms to reduce atmospheric noise is common (Zebker *et al.*, 1997; Fujiwara *et al.*, 1998). Given even a purely horizontally stratified troposphere, one expects topographically correlated delays due to differences in total water content. Given a simple model of the distribution of water and a few points of calibrations from GPS estimates of zenith wet delay (ZWD) or from radiosondes, one can attempt to correct for this effect (Delacourt *et al.*, 1998); unfortunately, excursions from horizontal stratification are the norm, and in fact, in any given interferogram, it is common to see phase variations that correlate with topography but with amplitude (and even sign) that vary from feature to feature.

There are a variety of more involved approaches to deal with the problem of path delays associated with spatial and temporal variations in tropospheric water vapor content. One can estimate statistically their impact on the estimates of tectonic deformation

and account for these in the data covariances when modeling. Alternatively, one can attempt to explicitly model and remove these delays.

In terms of estimating the appropriate statistics, several studies have attempted to define the variance and distance-dependent covariance of these delays (e.g., Massonnet and Feigl, 1995; Goldstein, 1995; Zebker *et al.*, 1997; Williams *et al.*, 1998; Hanssen, 2001; Emardson *et al.*, 2003; Li *et al.*, 2004, 2006b; Lohman and Simons, 2005). Using estimates of ZWD from GPS data in Southern California, Emardson *et al.* (2003) assume an isotropic (azimuth independent) model and find that the variance, σ, between two locations varies as $\sigma = cL^{\alpha} + kH$, where L and H are the relative horizontal and vertical separation distances between locations. The values of c, α, and k are estimated to be 2.5, 0.5, and 4.8, respectively. The value of α is expected to be more or less region independent, but the value of c may vary between regions. This model must then be scaled to account for the InSAR LOS direction. Limitations of this model include the assumption that the height dependence is independent of absolute height, and that there are no atmospheric variations at scales smaller than the spacing in the

GPS networks, which could influence the estimate of the height dependence (Xu *et al.*, 2006). Lohman and Simons (2005) find reasonable agreement between the estimates from Emardson *et al.* (2003) with empirical estimates from real interferograms (**Figures 19** and **20**). Adopting this statistical approach assumes that one has a large ensemble of interferograms to use in one's analysis. If one has only a few image pairs, then the statistics of small numbers comes into play when interpreting a specific feature in a given interferogram. Furthermore, if one removes the long wavelengths from a given interferogram, then one is also affecting the estimates of σ. These difficulties suggest that at a minimum, one should estimate the full covariance empirically, if necessary, removing an initial model of the phenomena of interest first (Lohman and Simons, 2005).

Since GPS can measure both the ZWD and its spatial gradient, it is possible to use the GPS observations to make a wet delay image to correct individual interferograms (Williams *et al.*, 1998; Xu *et al.*, 2006; Li *et al.*, 2004, 2006b; Onn and Zebker,

2006). This approach requires estimates of the delay made at or near the time of acquisition of the radar images. The viability of this approach clearly increases with density of GPS sites. Onn and Zebker (2006) demonstrate the use of time series of ZWD combined with a frozen flow assumption and observations of wind speed, to increase the spatial resolution of the images of wet delay beyond just the station spacing. Essentially, by assuming that the pattern of the wet delay is changing slowly, they can use estimates of ZWD before and after the image acquisition times to predict variations between GPS stations. To date, these approaches have not taken advantage of the spatial gradient of the ZWD that can also be estimated when processing the GPS delay. While the spacing of the GPS network will always remain a fundamental limitation, since the wet delay has a red spatial spectrum, any correction at medium and long wavelengths should significantly reduce effects due to tropospheric delays. Unfortunately, there are only a few, albeit important, regions of the world where the GPS network is

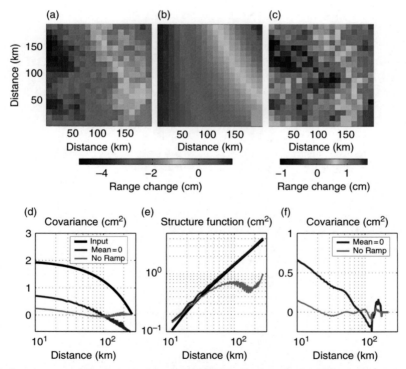

Figure 19 Effect of ramp removal on character of noise. (a) One realization of noise with power-law structure function. (b) Best-fit quadratic ramp to (a). (c) Remaining signal after ramp removal. (d) Input covariance (thick line) and inferred covariance for synthetic noise in (a) after removal of mean (blue) and quadratic ramp (red). (e) Structure function for same scenario. (f) Covariance inferred for a real interferogram, after removal of the mean and with and without the removal of the best-fitting quadratic ramp. Figure from Lohman RB and Simons M (2005) Some thoughts on the use of InSAR data to constrain models of surface deformation. *Geochemistry Geophysics Geosystems* 6 (doi:10.1029/2004GC00084).

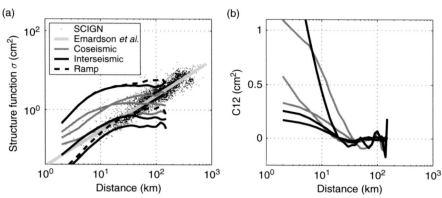

Figure 20 Spatial structure function versus distance for interferograms spanning the Hector Mine earthquake (black), and interferograms spanning time periods before or after the earthquake (gray). Dashed lines indicate the structure function for the interferogram before a quadratic ramp was removed. GPS data (small dots) are from the Southern California Integrated GPS Network (SCIGN) (adapted from Emardson et al. (2003)). The thick gray line indicates the power-law fit from Emardson et al. (2003). (b) Covariance versus distance for the coseismic and interseismic interferograms from (a). (a) Adapted from Emardson TR, Simons M, and Webb FH (2003) Neutral atmospheric delay in interferometric synthetic aperture radar applications: Statistical description and mitigation. *Journal of Geophysical Research* 108 (doi:10.1029/2002JB001781).

sufficiently dense to make any of these methods viable.

The lack of dense GPS networks in all regions of geodetic interest motivates the need for satellite-based estimates of the wet delay. Significant progress has been made to this end using observations from near-infrared imaging radiometers such as NASA's MODIS (Li et al., 2005) and the MERIS instrument on ESA's ENVISAT satellite (Li et al., 2006a; Puyssegur et al., 2007). The latter has the advantage of estimating the water delay in the same viewing geometry and at the same time as the acquisition of the radar data. Unfortunately, imaging radiometers are limited to use in the daytime and cloud-free conditions. However, if clouds are not pervasive, then the resulting holes can be interpolated. Thermal infrared measurements could potentially be used for night-time observations. Ideally, use of image-based estimates would be done in combination with GPS estimates of wet delay where and when available.

Thus far, we have only addressed direct measurements of the wet delay. Recently, progress has been made using high-resolution weather models to estimate the spatial variation in wet delay (Foster et al., 2006; Puyssegur et al., 2007). Foster et al. (2006) demonstrate the use of the MM5 weather model (Grell et al., 1995) to correct interferograms from the big island of Hawaii. In this case, the model is updated every 12 h using local meteorological data, and estimates of wet delay are made every hour and interpolated to the time of acquisition for each radar image. Using this model, they can generally reduce the variance by about a half at wavelengths of 30 km, although they find limitations in the model in regions of high topographic gradient. Puyssegur et al. (2007) go one step further and combine the MM5 model with MERIS observations. In the near future, where available, imaging radiometer and GPS wet delay observations will be combined with physical models to produce the best estimates of wet delay possible. Indeed, one would expect that any future dedicated InSAR mission would produce and distribute estimates of wet delay for each image acquired. Clearly, once the best estimates of wet delay are removed, it is straightforward to empirically estimate the residual covariance structure on an interferogram-by-interferogram basis (Lohman and Simons, 2005).

12.4.3   Stacking Single-Component Images

Independent of whether or not one can correct interferograms for path delay effects, if the primary geophysical target is a single event that occurred quickly (in terms of satellite InSAR, this implies anything lasting less than the orbit repeat time) or alternatively is a gradual process occurring at constant rate, it may be useful to increase the signal-to-noise ratio by stacking (i.e., averaging) multiple interferograms. Such stacks have the advantage of reducing effects due to tropospheric delays since these are uncorrelated on timescales of more than a day (e.g., Hanssen, 2001; Emardson *et al.*, 2003) and thus are useful to discover small signals (e.g., Peltzer *et al.*, 2001; Lyons and Sandwell, 2003; Gourmelen and Amelung, 2005; Pritchard *et al.*, 2006). Stacking images can also reduce computational burden in parameter estimation schemes by reducing the amount of observations (e.g., Pritchard *et al.*, 2006).

These stacks can be made either in radar coordinates (choosing one interferogram as the master) or in geocoded (latitude and longitude) coordinates. Before stacking, one needs to account for long-wavelength errors or phase ramps associated with errors in estimates of the interferometric baseline for each interferogram. The phase ramp is usually parametrized as a bilinear or biquadratic polynomial function of azimuth and range (or geographic coordinates). Neglecting to deal with the ramp can result in biased stacks. If several GPS displacement vectors are available for the region of interest, they can be used to resolve the ramp parameters either before or after forming the stack.

The simplest and most common brute force stack is made by just adding all the interferograms together. In this approach, regions of decorrelation in the resultant stack will be the union of decorrelated regions in the individual interferograms. Therefore, one may choose not to include an individual interferogram if it has an excessively large amount of decorrelation. An example of such a single LOS image stacks for a short-duration event comes from images of the coseismic displacements from deeper and smaller earthquakes in Chile (Pritchard *et al.*, 2006). In most cases, it is advisable to take a more formal approach to stacking.

When considering making stacks, one should separate the case of a single rapid event from that of linear secular deformation. In the case of a single rapid event, the stack is made in terms of displacements. Scenes are averaged together in a straightforward fashion – ideally, using a weighted average, where the weighting is the inverse covariance matrix of the observations.

The full covariance is constructed of three primary parts: (1) the variance of a given pixel on the diagonal, (2) the intra-image pixel-to-pixel covariance primarily due to ionospheric and tropospheric path delays, and (3) intra-image covariances associated with use of a single common image in two separate interferometric pairs (Emardson *et al.*, 2003). Assuming ergodicity, one estimates the pixel variance empirically using estimates of the local phase variance over a small patch of pixels. Doing this for each image is important since different interferograms or offset images may have had different amounts of spatial filtering applied to them, as well as different amount of temporal decorrelation. The intra-image pixel-to-pixel covariance should also be computed empirically since the degree of filtering and long-wavelength ramp removal may vary from image to image.

Usually, this covariance is assumed to depend only on the relative distance between two given pixels (Hanssen, 2001; Emardson *et al.*, 2003; Lohman and Simons, 2005). In contrast to the inter-image pixel covariance, the intra-image covariance can be derived analytically (Emardson *et al.*, 2003). When calculating a displacement stack with the full covariance matrix, one would normally pose the problem as a least-squares problem. The use of a purely least-squares approach relying on an L2 norm, is somewhat debatable since phase noise is not necessarily Gaussian. It may in fact be more robust to use an L1 norm, which is equivalent to asking for the median and not the mean displacement. Regardless of the adopted norm, using a formal approach that includes the local estimate of variance permits one to include all available data regardless of the level of decorrelation.

For a constant rate process, the estimation of the rate at any given pixel is best described using a bit more formality. The underlying equation is simply

$$\rho_i = T_i v + \epsilon_i \qquad [48]$$

where ρ_i, T_i, v, and i are the observed range change, time span, rate, and error, respectively, for a pixel in the ith image. We can rewrite this equation as the linear system for the ensemble of images:

$$d = \theta v + E \qquad [49]$$

where d, θ, and E are the vectors of observations, time spans, and errors, respectively. This system has a general weighted least-squares solution

$$v = (\theta^t W \theta)^{-1} \theta^t W d \qquad [50]$$

where as described earlier, the weight matrix, W, is the inverse of the full data covariance matrix. This approach naturally deals with regions that are decorrelated in some but not all of the images. The same issue described above relating to L2 versus L1 norms applies here.

While not recommended, one could assume that the pixel variances are constant within a given displacement image and between different displacement images, as well as ignore all the covariances. These frequently adopted but unnecessary simplifications lead to the explicit least-squares solution:

$$v = \frac{\sum_{i=1}^{N} \Delta T_i \Delta \rho_i}{\sum_{i=1}^{N} \Delta T_i \Delta T_i} \quad [51]$$

In a study of postseimic deformation from historic earthquakes in the Basin and Range Province of the western US, Gourmelen and Amelung (2005) remove the long-wavelength ramp from individual component interferograms, then add the residual interferograms up and divide the total displacement by the summed total time span of all the interferograms. This is equivalent to converting all the individual interferograms to rates, then averaging the individual rates, weighted by their respective time spans. In other words

$$v = \frac{\sum_{i=1}^{N} \Delta \rho_i}{\sum_{i=1}^{N} \Delta T_i} = \frac{\sum_{i=1}^{N} (\Delta \rho_i / \Delta T_i) \Delta T_i}{\sum_{i=1}^{N} \Delta T_i} \quad [52]$$

Equation [52] is only the same as the least-squares solution shown in eqn [51] if all the interferograms span equal time periods. In that special case, then

$$v = \frac{1}{N \Delta T} \sum_{i=1}^{N} \Delta \rho_i \quad [53]$$

In a study of slow interseismic deformation in the Eastern California Shear Zone, Peltzer *et al.* (2001) effectively adopted eqn [53]. Given that they use interferograms of approximately the same duration (about 4 years), they simply averaged the rates (unweighted by time), which is close to the least-squares solution assuming constant and uncorrelated data errors.

In general, given today's computational resources, there is no reason not to use the full weighted least-squares estimation (e.g., eqn [50]). In particular, this approach allows a more rigorous estimate of both rates and their associated errors. When a final stack is complete, if desirable, it is then possible to create a mask based on the final estimated variances.

12.4.4 InSAR Time Series

We can adopt more complicated models than the linear one just described. For instance, given the potential for seasonal signals (e.g., **Figure 18**; Amelung *et al.*, 1999; Hoffmann *et al.*, 2001; Schmidt and Burgmann, 2003), it may be desirable to augment

eqn [48] by explicitly including a seasonal variation in the estimation process. Indeed, given enough temporal sampling, there exists the possibility for a wide range of time-series approaches, effectively mimicking all the approaches adopted in GPS analysis.

There are a couple of variants to current GPS analysis approaches. The first approach aims to make the cleanest time series possible for later mechanical modeling. This approach can include estimates of site-specific signals including seasonal signals (frequently assumed to be sinusoidal), random walk and flicker noise at each site, and spatially correlated processes such as daily variations in reference frame estimates (e.g., Dong *et al.*, 1998, 2003, 2006). Generally, the temporal and spatial analyses are separated in these approaches. A more complex underlying temporal evolution can also be adopted including a superposition of behaviors including linear rates, coseismic steps, and postseismic exponential or logarithmic decays. At a minimum, these more complicated model terms reduce biases in estimating seasonal and reference frame contributions and can always be added back in for subsequent mechanical modeling. Of course, the inferred model terms, such as coseismic steps, are themselves useful as input into subsequent mechanical models.

An alternative approach uses data from a network of GPS sites to simultaneously solve for time-varying parameters of a specified mechanical model and the aforementioned nontectonic signals. This approach is exemplified by the extended network inversion filter (ENIF) (Segall and Matthews, 1997; McGuire and Segall, 2003; Fukuda *et al.*, 2004).

The parallels between GPS and InSAR time-series analysis are numerous. Obviously, the physical processes of interest are the same, and they are sensitive to similar nontectonic processes such as seasonal effects. Other similarities exist such as the ambiguity in absolute displacements with InSAR data, which is equivalent to reference frame error in GPS. Despite these parallels, InSAR time-series analysis has a few unique challenges. In terms of nontectonic signals, we may want to remove the effects of the troposphere (empirically or by physical model) and we have errors due to inaccurate removal of topographic effects stemming from inaccurate orbits. With typical suites of interferograms from a given orbital track, we may also face incompletely connected chains of dates. For instance, given images acquired at times A, B, C, and D, we may be able to form interferograms I_{AB}, I_{BC}, and I_{CD}, or perhaps we only have I_{AC} and I_{BD}. In the later case, we would like to be able to construct a continuous time record.

To deal with these issues, we have the same variants in modeling approaches for InSAR time series as just described with GPS data. One approach attempts to decompose a suite of interferograms into its component time intervals (e.g., *AB*, *BC*, and *CD*), and to solve for the incremental displacements. Ambiguities in combining unconnected pairs (or collections of pairs) requires some form of regularization of the time series. One approach, based on use of singular value decomposition (SVD), finds a time series that fits the available interferograms while minimizing the implied velocities in any given underlying time interval (Berardino *et al.*, 2002). This approach is equivalent to minimizing the temporal gradient of the deformation field. Alternatively, one could use explicit Laplacian damping to minimize the roughness of the temporal evolution. The approach described by Berardino *et al.* (2002) begins with unwrapped interferograms, and assumes that they have been tied to some stable reference point to define a common phase bias. This approach is a pixel-by-pixel algorithm at its core (important for computational purposes), but they apply subsequent spatial and temporal filtering to provide cleaner time series. An SVD or principal component approach may also be useful to isolate atmospheric signals in large InSAR time series from spatially and temporally coherent deformation fields (Ballatore, 2006). Examples of models using large time series of InSAR data include deformation at volcanoes (Lundgren *et al.*, 2001, 2004), in the Santa Clara Basin (California) (Schmidt and Burgmann, 2003), in the Los Angeles Basin (California) (Lanari *et al.*, 2004a), and faulting in the Asal Rift (Djibouti) (Doubre and Peltzer, 2007).

As with GPS data, an alternative approach to time-series modeling involves explicit use of a physical model. For example, Pritchard and Simons (2006) studied time-dependent postseismic slip and seismic slip on the subduction zone megathrust in northern Chile assuming a simple model of time-dependent slip on a fault plane, and imposing Laplacian smoothing in time in order to tie together unconnected groups of interferograms. Alternatively, they could have used a Kalman filter approach, analogous to the ENIF approach used with GPS data; however, the number of interferograms available did not support the increased complexity of the modeling approach. Of course, the mechanical modeling approach benefits directly from the availability of

any additional temporally continuous data such as can be provided by GPS observations.

12.4.5 Vector Displacements

Another form of stacking can be useful when data from multiple viewing geometries are available. In particular, we may want to (1) compare geodetic imaging data directly with other single-component geodetic data, such as leveling or electronic distance measurements (EDM), without going through a physical model, (2) get a better grasp on what are purely vertical versus horizontal displacements, (3) as in the single-component stacks described above, combine as many images of the geophysical event to reduce noise and look for unexpected (undiscovered) processes, and (4) reduce the total amount of data used in a parameter estimation task. In these cases, it may be useful to combine displacement images from three or more viewing geometries to construct the best estimate of the east, north, and vertical displacement field. The input data need not be homogeneous in type. For instance, one could combine right-looking images from ascending and descending orbits and pixel offset estimates from one or more tracks to resolve the full 3-D displacement field (Fialko *et al.*, 2001a, 2005). Similarly, one could use observations from different overlapping beams of adjacent orbital tracks (Wright *et al.*, 2004). The whole system would then be weighted by the covariance *W* described in the previous section. If sufficient data are available, then one can explicitly include estimation of ramp parameters, which as before, are best constrained if at least a few independent geodetic observations are available. Estimates of the full 3-D coseismic displacement field exist for several earthquakes including the 1999 M_w 7.1 Hector Mine earthquake (Fialko *et al.*, 2001a) and the 2003 M_w 6.5 Bam earthquake in Iran (Fialko *et al.*, 2005; Funning *et al.*, 2005) (**Figure 21**). With respect to future mission design, Wright *et al.* (2004) points out the reduced sensitivity of 3-D deformation maps to north–south motions for missions with near-polar orbits.

While we may not need to explicitly make 3-D decompositions of the displacement fields, having multiple components of the displacement fields are clearly important to constrain physical models such as earthquake and volcano source models (Fialko *et al.*, 2001b; Lohman *et al.*, 2002). For example, for all earthquake models, when only one LOS component is observed, there will be a tradeoff between the amplitude and rake of the fault slip. This tradeoff will

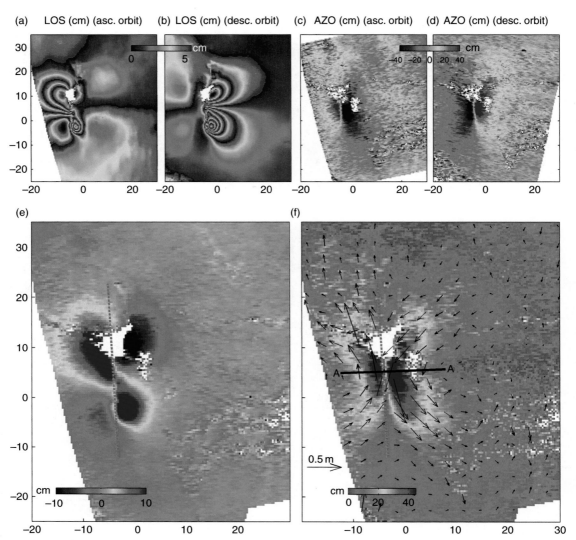

Figure 21   Phase and offset measurements for the 2003 M_W 6.5 Bam, Iran shallow strike-slip earthquake. Images are from the C-band ENVISAT ASAR instrument. The coordinate axes are in kilometers, with the origin at 58.4° E, 29 N. Colors denote displacements in centimeters. (a) Interferogram for 11/16/2003–1/25/2004, ascending orbit. (b) Interferogram for 12/3/2003–2/11/2004, descending orbit. (c) Azimuthal offsets, ascending orbit. (d) Azimuthal offsets, descending orbit. Derived vertical (e) and horizontal (f) components of the surface displacement field. Arrows show the subsampled horizontal displacements. Dashed line shows the surface projection of the fault plane inferred from the inverse modeling of the data. Figure from Fialko Y, Sandwell D, Simons M, and Rosen P (2005) Three-dimensional deformation caused by the Bam, Iran, earthquake and the origin of shallow slip deficit. *Nature* 435: 295–299.

be further exasperated for small earthquakes if one does not know the location well (Lohman *et al.*, 2002). Generally, we note that it may be dangerous to oversimplify the problem by assuming *a priori* that the deformation field is either purely vertical or purely horizontal in nature. For instance, in large strike-slip systems, there is a temptation to assume that the near-fault displacement field is purely fault parallel. However, even the largest strike-slip systems

frequently have vertical or fault normal deformation associated with them that is either tectonic in origin (e.g., Klinger *et al.*, 2005) or due to hydrologic effects associated with faults acting as flow barriers (Bell *et al.*, 2002). The presence of any vertical displacements is particularly problematic, given the high sensitivity to vertical deformation in most InSAR data. Indeed, any ignored vertical displacement will corrupt estimates of the horizontal displacement, where the amplitude of this corruption is amplified by at least $1/\tan\theta$, where θ is the angle of incidence. This effect increases as the fault strike approaches the azimuth of the orbital track.

12.4.6   The Luxury of Sampling – Rationale and Approach

Geodetic imaging can produce a very large number observations – of order 10^6 pixels. In many cases, one wants to use these observations to constrain parameters from a mechanical model, such as distributed fault slip in an elastic space. In practice, the computational expense of the inverse problem can be computationally bound by the number of Green functions one need calculate. In the case of linear problems where the source geometry is known, the Green functions may not be known analytically (e.g., as with 3-D variations in material properties) or one may still want to impose nonlinear constraints, both of these requirements can be expensive to implement. Even worse, for nonlinear problems, such as when we do not know the geometry of the source, then we must recalculate the Green functions at each iteration.

Having spatially continuous observations provides us the opportunity to use a selected subset of these observations. This possibility leads to the question of the optimal subset of observations to pick. It is most precise to view this problem as spatially variable averaging and sampling. Proposed approaches fall in two classes: in the first, the image sampling is done based on properties of the data themselves (Jonsson *et al.*, 2002; Simons *et al.*, 2002), and in the second, the sampling is controlled by the character of the model (Lohman and Simons, 2005). Jonsson *et al.* (2002) and Simons *et al.* (2002) proposed similar approaches that rely on successive subdivision of the deformation image. In both approaches, the image is cut into quadrants and a low-order best-fit surface is removed from the phase field in each quadrant. For a given quadrant, if the residual is greater than a prescribed threshold, the quadrant is further divided into four new quadrants, with the process allowed to continue until a minimum quadrant size is reached. For both approaches, one should then use the mean (or median) of the pixels in the quadrant and assign the value to the center of the quadrant. Jonsson *et al.* (2002) and Simons *et al.* (2002) differ in that the former considers the residual after removing the mean from each quadrant, why as the latter removes a bilinear function. Both approaches work well, but for the same number of points, the Simons *et al.* (2002) approach does a better job at constraining a given fault slip model.

Underlying the Simons *et al.* (2002) approach is the recognition that any physical model can at a minimum produce a linearly varying displacement field, and thus the ability to constrain detailed behavior of a given model lies in the curvature of the deformation field. In essence, we are attempting to choose a data set that includes as many points as possible while maintaining a nearly diagonal data resolution matrix. Both of these approaches are limited by the inherent noise in the observations, and can give spurious regions of high sampling (e.g., far away from a target fault) if there are unwrapping errors or regions of decorrelation.

An alternative approach to image sampling involves using a best-guess initial model parametrization (Lohman and Simons, 2005). A given linear model can be written as $Gm = d$, where G is the design matrix, m is the vector of model parameters, and d is the vector of observations. This has a generalized solution $m_{est} = G^{-g}d$, where G^{-g} is the generalized inverse and the model and data resolution matrices, R and N, can be written as $R = G^{-g} G$ and $N = GG^{-g}$ (e.g., Menke, 1989). Lohman and Simons (2005) propose to find a distribution of samples that has the most points while keeping N nearly diagonal. The algorithm is similar to that of Simons *et al.* (2002) and Jonsson *et al.* (2002) in the use of successive division into quadrants, with the difference being that at each stage, an estimate of N is made, and refinement stops when N becomes sufficiently nondiagonal. As with the other approaches, for any given quadrant, this approach uses the weighted mean (or median) value, where the weight takes into account the pixel variances and covariances. This approach is ideal if one has a reasonable first guess at the model geometry. It is of course also sensitive to the assumed model parametrization – in this case the size of each fault patch. Regardless, all these schemes permit one to constrain models with about 1% of the original data without significant loss of information. It is worth emphasizing that even where the final sample spacing is large, the data variance will be relatively low, since more points go into this estimate than in regions of finer spacing, thus information is not lost.

These spatially variable averaging/sampling approaches are particularly important when modeling shallow sources with the potential of causing complex deformation patterns. A common example of this class of problems comes from shallow earthquakes. **Figure 22** demonstrates the difference in sampling patterns that result from the different approaches described here. This figure also demonstrates the difference in the ensuing model resolution.

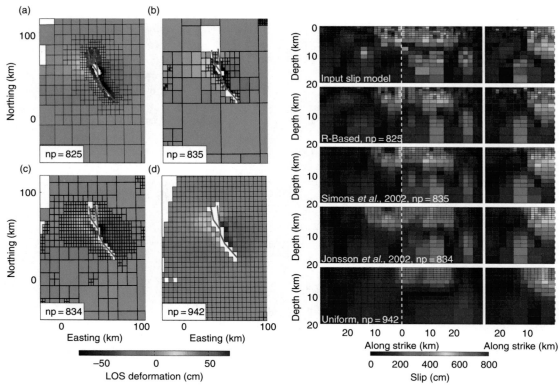

Figure 22 Left: Examples of different interferogram sampling approaches. Results are shown using (a) the data resolution-based algorithm from Lohman and Simons (2005), (b) a variable sampling proportional to data curvature approach from Simons et al. (2002), (c) a variable sampling proportional to data gradient Jonsson et al. (2002), and (d) uniform sampling. Right: Comparison of inferred slip distribution when using the different sampling approaches shown at left. In each case, the total number of points retained is similar. The fault geometry is shown at left by the red lines. The inferred slip is similar for all inversions, but the resolution of shallow slip features increases from bottom to top. Lohman RB and Simons M (2005) Some thoughts on the use of InSAR data to constrain models of surface deformation. *Geochemistry Geophysics Geosystems* 6 (doi: 10.1029/2004GC00084).

The choice of algorithm clearly impacts the ability to distinguish variations in the shallowest parts of the model (Lohman and Simons, 2005).

12.4.7 Decorrelation as Signal

Interferometric temporal decorrelation is usually viewed at least as a nuisance and sometimes as a complete barrier to making useful displacement measurements. However, in some cases, the spatial distribution of decorrelation, as well as its temporal evolution, can serve as useful indicators of geological processes. There are many causes of temporal decorrelation that are not necessarily of interest, among these are the impact of weather-related processes (rain, snow, etc.). Of more interest is decorrelation caused by rearrangement of the scatterers within a pixel associated with intense shaking, damage, etc. Examples of the use of such

decorrelation include mapping of active lava flows (Zebker et al., 1996), as well as near-fault decorrelation near shallow earthquake ruptures in the 1995 Kobe, Japan earthquake (Rosen et al., 2000), the 1999 Hector Mine, California earthquake (**Figure 23**; Simons et al., 2002), and the 2003 Bam, Iran earthquake (Fielding et al., 2005). This latter form of decorrelation has been found to clearly outline fault traces that slip in earthquakes and can serve as an obvious way to constrain surface fault geometry, and so should be used to guide postearthquake field studies. In general, InSAR decorrelation images provide a synoptic view of the spatial extent (along-strike and cross-strike) of faulting that is not easily achieved from the ground.

A more important potential societal benefit of decorrelation measurements is their use immediately after an earthquake, landslide, or eruption to map out regions of damage (e.g., Fielding et al., 2005).

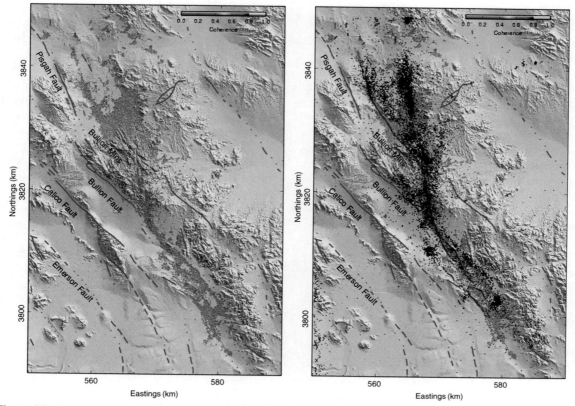

Figure 23 Interferometric decorrelation from the 1999 M_w 7.1 Hector Mine, California earthquake. Only areas with correlation less than 0.8 are shown. Images are from the C-band ERS satellite and span 35 days. Same as left but with aftershocks and the fault trace as mapped in the field. Figure modified from Simons M, Fialko Y, and Rivera L (2002) Coseismic deformation from the 1999 M_w 7.1 Hector mine, California, earthquake as inferred from InSAR and GPS observations. *Bulletin of the Seismological Society of America* 92(4): 1390–1402.

Frequently, existing communication systems are incapacitated by the event, and local inhabitants are not able to report the level of devastation. Rapidly made decorrelation maps produced during or soon after the event could be used to guide emergency response.

The usefulness of a decorrelation map is limited by 'uninteresting' sources of decorrelation not associated by the event. In particular, all forms of temporal decorrelation, especially if not monotonic, can mask the correlation image. Obvious examples include rain, snow fall, and vegetation changes occurring during the time span of the image pair. The most important ways to improve the usefulness of decorrelation measurements include rapid revisit times to limit the possibility of extraneous processes and the use of long-wavelength InSAR (e.g., L-band as opposed to C-band).

12.5 The Link between Science and Mission Design

InSAR and other space geodetic techniques are primarily designed to measure the displacement of Earth's surface over time. The particular characteristics of the measurements are tied to the specific implementation of the InSAR instrument and the mission characteristics. Clearly, there is a close link between the science that can be done with an InSAR mission and the design of that mission. It is important for the research community to understand this link in order to accomplish their goals. A simple example of this would be avoiding the use of data from a mission that provides an image over an area once per month, when the phenomenon of interest has changes that occur on timescales of days. In this section, we describe the flow from scientifically driven measurement needs to the basic parameters of an InSAR mission. With such an understanding, it is then possible to characterize the performance of existing missions and productively discuss the design trades for future systems.

It is worth stating that InSAR measurements are a poor proxy for what scientists really would like to know about geophysical systems. The desired starting point to address larger geophysical questions – for example, what are the mechanics of earthquakes and how do fault systems interact? – would be measurements of the state of the crust, its pressure, temperature, and distribution of material properties throughout medium, to use as input to geophysical models. These *in situ* measurements at depth are impossible to obtain, so scientists model them through observations of the motions of Earth's surface. So it is important initially to understand the sensitivity relationship between model parameters and surface displacements. For example, if any reasonable change in a model parameter changes the modeled surface displacement by 1 mm, there is limited value in measuring deformation to only 1 cm accuracy.

So the question arises: what can we reasonably expect to measure with an InSAR system, and conversely, what is required of that system in order to advance science? A repeat-pass InSAR system measures the range displacement of any image element through a differencing of the phase from one epoch to another. Using a time series of observations, a repeat-pass InSAR system measures a spatial distribution of range displacements in discrete time intervals. To use InSAR displacement measurements in geophysical models, the measurements must have adequate displacement accuracy, both absolute and relative, spatial resolution, spatial coverage, and temporal sampling. These requirements differ for each specific scientific investigation.

For a given system, we have seen that the accuracy of the range displacement measurement is determined by noise induced by decorrelation of the radar echoes, by random phase delays introduced by propagation effects through the time-variable atmosphere, and by systematic knowledge uncertainties in the radar path delays and orbit. In addition, the incidence angle and azimuth angle of the observation can affect the accuracy of the measurement greatly; measuring a horizontal displacement with a system that looks steeply down toward nadir is not desirable.

As described earlier, decorrelation is comprised of principally three components: thermal decorrelation is related to the noise level relative to the signal level

of the radar system; geometric decorrelation is related to the arrangement of scatterers on the surface and how they change with differences in time or viewing geometry; and other decorrelation terms that derive from noise (e.g., quantization, ambiguities, sidelobes) that is dependent on the signal level itself. There is a strong functional dependence of these decorrelation terms on system parameters such as power, antenna size, wavelength, etc., and great interplay among them.

The range resolution of a SAR system is inversely proportional to waveform bandwidth. The required range resolution is usually set by the scale size of the ground feature that is being mapped. For InSAR systems, however, there is a relationship between system bandwidth and desirable interferometric baselines. Finer resolution implies less decorrelation due to nonzero baselines. Thus, even if the final map resolution desired is only 100 m, one might require a system to have 10 m resolution in range so that the constraints on the repeat-orbit accuracy are manageable.

The along-track resolution of a conventional strip map SAR system is equal to half the along-track length of the antenna, independent of range and wavelength. This is illustrated schematically in **Figure 24**, in which the antenna size is grossly exaggerated in size relative to the antenna beam width and the range, so that the salient characteristics can be viewed on a single page. The resolution along-track is determined by the spread in frequency content a given surface element experiences as the SAR beam illuminates it. Because of its beam width, the radar signal experiences Doppler shifts across its beam, such that the received echo contains a spectrum of information. The beam width is given by $\phi_D = \lambda/L$, where λ is the wavelength of the radar tone and L is the length of the antenna along-track. The frequency bandwidth associated with the Doppler shifts within this beam width is given by $\Delta f_{D,t} = 2v\phi_D/\lambda = 2v/L$. In spatial coordinates, the spatial frequency bandwidth is $\Delta f_{D,x} = 2/L$, which implies a resolution of $L/2$. Thus, while wavelength and antenna length determine the beam width of the antenna, the spatial frequency content is not dependent on the wavelength in SAR.

For ScanSAR systems the resolution is determined by the length of the burst of pulses in a given scan, which is generally related to the number of ScanSAR beams. For spotlight SAR systems, the resolution is determined by the length of time over which the observation is made.

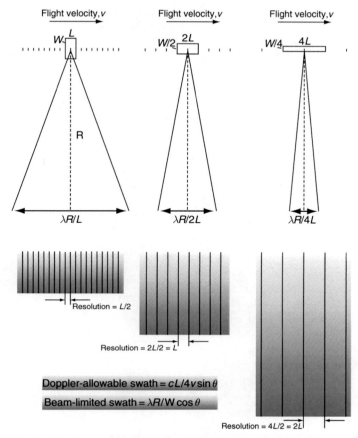

Figure 24 Illustrating the trade-off between swath extent (coverage) and resolution for strip mode SAR systems. In order to image a wide swath in strip mode, the antenna must be long in the along-track dimension. However, the resolution along track is half the antenna dimension, so the resolution is degraded as the swath width increases.

In addition to the thermal noise present in a radar system, there are a number of noise sources that play a significant role in the design of a radar and the accuracy one can expect to achieve for InSAR displacements. SAR systems point to an angle off nadir to avoid echoes from both sides of the nadir track: such echoes would be ambiguously combined in the SAR receiver and could not be distinguished from each other. Even with off-nadir pointing, the transmitted and received energy cannot be completely localized in time as the signal spreads throughout the illumination area, resulting in a wide range of time over which a given echo can return (as well as some energy from the opposite side of nadir in some cases). Because the radar transmits pulses of energy over 1000 times per second for typical spaceborne systems in order to properly sample the Doppler spectrum, it can often occur that energy from time intervals outside the area of interest defined by this sampling rate, for example, from a previous or later

pulse, arrives at the receiving antenna at the same time as the desired echo energy. These corrupting echoes, generally occurring at much lower amplitude, are called range ambiguities. The magnitude of these are controlled by the pulse rate – generally lower pulse rate allows more time to collect all echoes from a pulse – and by shaping the illumination area by manipulating the shape of the antenna pattern.

On the other hand one cannot lower the pulse rate below the point where the along-track pulses become undersampled. As we have seen, to create a narrow Doppler spectrum, we desire a long antenna in the along-track direction. Physical constraints on the size of the flight system, as well as a common desire for reasonably high resolution, limit this size, and therefore limit the minimum pulse rate. The illumination pattern in the along-track direction also has extent beyond the nominal beam width (antenna sidelobes), so the pulse sampling rate will naturally cause aliasing of some energy from beyond the along-track antenna main beam extent.

To first order, then, if a wide swath is desired, then a low pulse rate must be chosen to allow enough time between pulses for the received echo to be unambiguously acquired. This then requires that the Doppler bandwidth, and hence the antenna beam width, be narrow, which then mandates a long antenna in the along-track direction. Furthermore, the antenna must be limited in size in the elevation direction to create a wide-enough beam to receive energy from the wide swath of interest.

For a particular spacecraft altitude, the swath size is determined by these ambiguity constraints for most practical spaceborne systems. As the radar antenna of fixed beam width is pointed at greater distances off nadir, the swath illuminated on the ground becomes broader from projection effects, but the usable swath extent is usually narrower because of ambiguities. This then means that the antenna must be larger in the elevation dimension to limit the swath to maintain performance.

These effects then influence mission design for an InSAR system. Scientists studying deformation want to be able to observe any point on the Earth at some regular interval. They also would like rapid repeat coverage to be able to track rapid changes of the Earth. Suppose the mission requirement is to repeat an orbit every 8 days. For exact repeat, this requires an integer number of orbits in this time. There is an 8-day repeat polar orbit at about 760 km altitude that contains 115 orbits in 8 days. This then implies that the separation of the orbit tracks at the equator will be $2\pi 6378/115 \approx 340$ km. Thus, the SAR must be able to cover 340 km of swath, either all at once with a very wide swath from a very long and skinny antenna (giving very low resolution), or using multiple smaller beams with smaller swaths covering different angles off nadir at different times. ScanSAR, where the radar sends a collection of pulses illuminating one subswath, then electronically steers the antenna to the next subswath off nadir and sends the next collection of pulses, and so on with multiple beams, allows full coverage of the wide swath in one pass again, at the expense of resolution, and somewhat degraded ambiguity performance.

These design space possibilities – frequency, resolution, antenna size, orbit altitude and control, system power, viewing angle, repeat period, observation modality – are the playground of scientists, working with system designers, to optimize a mission to capture meaningful geophysical signals in the presence of the

noise sources that are present in radar measurements. Space faring nations are increasingly relying on SAR and InSAR for scientific discovery and monitoring, with the trend moving away from large multimode systems to simpler instruments that do a few things well. One of these missions will no doubt be dedicated InSAR mission, a true geodetic instrument in the model of GPS, but with global reach and dramatically improved denser coverage of the Earth.

Acknowledgments

Radar images used in this chapter were provided by ESA under multiple proposals (M.S. PI) as well as via the NSF/NASA/USGS-sponsored WInSAR data consortium. P.A.R. thanks Gilles Peltzer for several of the graphics included in this paper, which they jointly developed previously for teaching purposes. This chapter is Contribution number 9166, Geological and Planetary Sciences, California Institute of Technology, Pasadena CA 91125. This research was supported in part by the Gordon and Betty Moore Foundation. This is Caltech Tectonic Observatory Contribution 67. For P.A.R., this paper was prepared at the Jet Propulsion Laboratory, California Institute of Technology, under a contract with the National Aeronautics and Space Administration.

Appendix 1

Radar Imaging

The power of radar interferometry for geodesy stems from its high-resolution images that are generated from a controlled coherent light source. The coherence of the measurements ensures that the phase associated with each complex image element contains both the round-trip geometric path length from the radar sensor to the surface and back, and the ensemble phase associated with the coherent summation of scattering within the image element. The imaging process is quite involved and is described in detail in many other places, for example, Elachi (1988), Raney (1998), and Franceschetti and Lanari (1999). However some basic description of the process will frame the discussion of interferometry. This appendix is a compilation of the basics from a variety of sources, and is designed to cover all the essentials for understanding radar images, and their phase, in one convenient narrative.

The most common radar waveform in civilian spaceborne radars is 'chirp' coding:

$$f(t) = A\cos\left(\omega_0 t + \frac{B_r}{2\tau_p}t^2\right)\text{rect}\left(\frac{t}{\tau_p}\right) \quad [54]$$

This waveform is transmitted typically over 1000 times per second and its echo is received, digitized, and transmitted to a recorder. The presence of a point target on the ground surrounded by a surface with no inherent reflectivity causes the radar system to respond with the following received signal:

$$r(t, R) = p(R)g(R)\cos\left(\omega_0(t - 2R/c) + \frac{1}{2}\frac{B_r}{\tau_p}(t - 2R/c)^2\right)$$

$$\times \text{rect}\left(\frac{t - 2R/c}{\tau_p}\right) \quad [55]$$

where for now, $p(R)$ and $g(R)$ are generic functions to represent $1/R^2$ loss in power and backscatter at R, respectively. They will be defined with greater specificity shortly. Note if we think of the radar as a system, the point target can be thought of as a delta function input, implying that $r(t, R)$ would be the system impulse response.

If numerous point targets are arranged in range, then the total received signal at any given time will be the sum of point-target returns over these ranges:

$$r(t) = \int_R p(R)g(R)\cos\left(\omega_0(t - 2R/c) + \frac{1}{2}\frac{B_r}{\tau_p}(t - 2R/c)^2\right)$$

$$\times \text{rect}\left(\frac{t - 2R/c}{\tau_p}\right)dR \quad [56]$$

This equation states that the received signal is the convolution of the ground backscatter, modulated by the geometric decay in power with range, with the transmitted chirp. Viewing the transmitted chirp as the radar system impulse response, this equation is a typical expression for the output of a linear system driven by $p(R)g(R)$.

The response of the radar to the presence of a line of scatterers in range, represented in its complex-valued form $r_z(t)$, is

$$r_z(t) = K \int_{t'} p(t')g(t')e^{j\left(\omega_0(t - t') + \frac{1}{2}(B_r/\tau_p)(t - t')^2\right)}$$

$$\times \text{rect}\left(\frac{t - t'}{\tau_p}\right)dt' \quad [57]$$

where K is an arbitrary constant and $t' = 2R/c$. The convolution of the surface with the chirp smears out the information from any given scatterer over the extent τ_p. The goal is to recover the original $g(t')$ from $r_z(t)$.

Matched filtering

Suppose

$$p(t')g(t') = g_0 e^{j(\omega_0 t' - 2kR_0)}\delta(t' - t_0) \quad [58]$$

where $R_0 = ct_0/2$, implying there is a single-point scatterer on the ground at range R_0. In the radar receiver, normally one of the first hardware functions is to essentially remove the rapidly varying phase term $\exp(j\omega_0 t')$ by a process called 'heterodyning'. The received signal is multiplied by a reference signal $\exp(-j\omega_0' t')$, with $\omega_0 \approx \omega_0'$ so the narrowband received signal (the bandwidth is $B \ll \omega_0$) is near zero frequency, in the so-called 'video band'. (In practice there can be several stages of filtering and heterodyning to achieve the video signal, and often it is arranged so the spectrum of the real signal has its positive and negative frequencies symmetrically arranged around zero, each centered at $B/2$ and $-B/2$, respectively.) For our purposes, we let $\omega_0 = \omega_0'$, so that

$$p(t')g(t') = g_0 e^{-j2kR_0}\delta(t' - t_0) \quad [59]$$

with the understanding that the 'carrier' has been removed. Then

$$r_{z\delta}(t) = g_0 e^{-j2kR_0}e^{j(1/2)\left(B_r/\tau_p\right)(t - t_0)^2}\text{rect}\left(\frac{t - t_0}{\tau_p}\right) \quad [60]$$

$$= g_0 e^{-j2kR_0}b(t - t_0) \quad [61]$$

Since $r_{z\delta}$ is the response of the radar to an impulsive 'signal' source (i.e., the ground), we recognize

$$b(t) = e^{j(1/2)\left(B_r/\tau_p\right)t^2}\text{rect}\left(\frac{t}{\tau_p}\right) \quad [62]$$

as the impulse response of the radar system. For this system, b is just the transmitted signal. The Fourier Transform of b is known as the system transfer function $H(\omega)$. One of the properties of Fourier transforms is that the convolution of functions in the time domain is equivalent to the product of functions in the frequency domain

$$r_z(t) = \int g(t')b(t - t')dt' \leftrightarrow R_z(\omega) = G(\omega)H(\omega) \quad [63]$$

Since we are interested in recovering $g()$ from $r_z()$, it is clear that if we can multiply $R_z(\omega)$ by the reciprocal of $H(\omega)$, the inverse Fourier transform will retrieve $g()$. It turns out that $H(\omega)$ is of the form

$$H(\omega) \approx e^{j\kappa\omega^2} \quad [64]$$

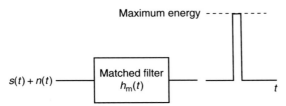

$$s(t) + n(t) \longrightarrow \boxed{\begin{array}{c} \text{Matched filter} \\ h_{\mathrm{m}}(t) \end{array}} \longrightarrow$$

Figure 25   Illustration of the matched filter concept, where a signal in noise is optimally detected by passing it through a filter matched to its characteristics. In our case $s(t)$ is what we called the impulse response of the radar system $h(\,)$.

which implies $R_z(\omega)H^*(\omega) = G(\omega)$, that is, we can filter r_z with the conjugate spectrum of the impulse response to recover g.

This specific result is a special case from matched filter theory, where an optimal filter is designed to best detect a transmitted waveform in the presence of noise (**Figure 25**). The matched filter of a signal $s(t)$ has an impulse response

$$b_{\mathrm{m}}(t) = s^*(-t) \qquad [65]$$

In our case, $s(\,)$ is what we defined as radar impulse response $h(\,)$, so $b_{\mathrm{m}}(t) = b^*(-t)$. Thus the recovered, or 'compressed', signal r_{zc} for our delta function is given by

$$r_{zc}(t) = \int r_z(t')b_{\mathrm{m}}(t-t')\mathrm{d}t' \qquad [66]$$

$$r_{zc\delta}(t) = g_0 e^{-\mathrm{j}2kR_0} \int b(t'-t_0)b_{\mathrm{m}}(t-t')\mathrm{d}t' \qquad [67]$$

$$= g_0 e^{-\mathrm{j}2kR_0} \int b(t'-t_0)b^*(t'-t)\mathrm{d}t' \qquad [68]$$

So we see that convolution with the matched filter is equivalent to correlation with the conjugate of the function itself. This makes some sense intuitively, in that as we slide the filter along our received signal and integrate, they will match best and give the highest integral when they are aligned with lag $t = t_0$ and when all phase terms are cancelled, making the integrand always positive. Since the Fourier transform of $b^*(-t)$ is just $H^*(\omega)$, we see that the result above considering the specific form of b in eqn [64] is actually a more general result. Namely, we can in general filter r_z with the conjugate spectrum of H to retrieve g, or equivalently (**Figure 25**),

$$r_{zc}(t) = \int r_z(t')e^{-\mathrm{j}(1/2)\left(B_r/\tau_{\mathrm{p}}\right)(t'-t)^2}\mathrm{rect}\left(\frac{t'-t}{\tau_{\mathrm{p}}}\right)\mathrm{d}t' \qquad [69]$$

Expanding eqn [68], we have

$$r_{zc\delta}(t) = g_0 e^{-\mathrm{j}2kR_0}\int e^{\mathrm{j}(1/2)\,(B_r/\tau_{\mathrm{p}})\,(t'-t_0)^2}\mathrm{rect}\left(\frac{t'-t_0}{\tau_{\mathrm{p}}}\right)$$
$$\times\, e^{-\mathrm{j}\,(1/2)(B_r/\tau_{\mathrm{p}})(t'-t)^2}\mathrm{rect}\left(\frac{t'-t}{\tau_{\mathrm{p}}}\right)\mathrm{d}t' \qquad [70]$$

If we ignore the rect functions, and assume infinite integration limits, the integration becomes fairly straightforward because the quadratic phase terms involving the integration variable cancel:

$$r_{zc\delta}(t) = g_0 e^{-\mathrm{j}2kR_0}e^{\mathrm{j}(1/2)\left(B_r/\tau_{\mathrm{p}}\right)\left(t_0^2 - t^2\right)}$$
$$\times\int e^{-\mathrm{j}\left(B_r/\tau_{\mathrm{p}}\right)(t'\,t_0)}e^{\mathrm{j}\left(B_r/\tau_{\mathrm{p}}\right)(t'\,t)}\mathrm{d}t' \qquad [71]$$

$$= g_0 e^{-\mathrm{j}2kR_0}e^{\mathrm{j}(1/2)\left(B_r/\tau_{\mathrm{p}}\right)\left(t_0^2 - t^2\right)}\int e^{\mathrm{j}\left(B_r/\tau_{\mathrm{p}}\right)t'(t-t_0)}\mathrm{d}t' \qquad [72]$$

$$= g_0 e^{-\mathrm{j}2kR_0}e^{-\mathrm{j}(1/2)\left(B_r/\tau_{\mathrm{p}}\right)\left(t^2 - t_0^2\right)}\delta(t-t_0) \qquad [73]$$

So we see that we exactly recover the delta function with an infinite bandwidth signal and impulse response. If we were to leave the impulse response bandwidth infinite but apply the windowing of the received signal, then the integration limits above are limited to the pulse extent τ_{p}, and we obtain the usual sinc result:

$$r_{zc\delta}(t) = g_0 e^{-\mathrm{j}2kR_0}e^{-\mathrm{j}(1/2)\left(B_r/\tau_{\mathrm{p}}\right)\left(t^2 - t_0^2\right)}\frac{\sin \pi B(t-t_0)}{\pi B(t-t_0)} \qquad [74]$$

where $B = B_r/2\pi$ (**Figure 26**). The first null is at the expected location of $1/B$, defining the resolution of the compressed signal (**Figure 26**).

When both rect functions are considered in the integral, the quadratic phase term cancels and the sinc function is modified. The integration is carried out in three regimes as shown in **Figure 27**.

Using these limits,

$$r_{zc\delta}(t) = g_0 e^{-\mathrm{j}2kR_0}\mathrm{rect}\left(\frac{t-t_0}{\tau_{\mathrm{p}}}\right)\frac{\sin \pi B(t-t_0)\left(\tau_{\mathrm{p}}-|t-t_0|\right)}{\pi B(t-t_0)} \qquad [75]$$

When $t - t_0 \ll \tau_{\mathrm{p}}$ this reduces to the usual expression without the quadratic phase:

$$r_{zc\delta}(t) \approx g_0 e^{-\mathrm{j}2kR_0}\frac{\sin \pi B(t-t_0)}{\pi B(t-t_0)} \qquad [76]$$

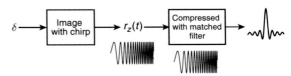

Figure 26   Illustration of the matched filter for the delta-function scene and linear FM chirp. The compressed signal is close to a sinc function.

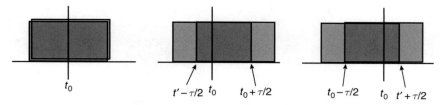

Figure 27 The three regimes of integration as the windowed received signal and matched filter have zero, positve and negative lag. Modified from Franceschetti G and Lanari R (1999) *Synthetic Aperture Radar Processing*. Boca Raton, FL: CRC Press.

Radar signal return from a 3-D surface

We examine what comprises $p()$ and $g()$ in eqn [57] in an imaging radar system looking at the Earth, where we do not necessarily have a neat line of scatterers arranged in range. Instead we have a spatial distribution of scatterers within the beam on a 3-D topographically modulated surface (refer back to **Figure 2** for the SAR imaging geometry). The receive echo will integrate over the azimuth direction and the range pulse at any instant. For a real aperture radar, that is one for which the resolution in azimuth is dictated by the beam width of the antenna, the point-target concept extends to an integrated backscatter value over azimuth $g'(R)$ and the same arguments and equations for the simple point targets described above are obtained.

Figure 28 shows the imaging scenario looking down on the Earth from above the sensor. This more clearly shows that the received echo energy at any given instant within the pulse will effectively be the integrated return from every scatterer within the instantaneous pulse extent on the ground in range and azimuth, weighted by the transmitted chirp waveform. Because the chirp signal is extended in range, the response from any given scatterer is also extended in range. Variations in range relative to a nominal flat surface within the beam induced by topography or surface relief will lead to geometric distortion in the range trace: the measurement is fundamentally an integration (convolution) over phase fronts (i.e., constant range).

Being more specific about the function $g()$, following Raney (1998), we define the geometry of the phase front incident on a elemental surface with reference to **Figure 29**. The surface element is like the facet model, bounded in range by a distance ΔR, and presents a cross-sectional area normal to the incident wave of $da\,dz$, where $da = R\,d\phi$ and $dz = \Delta R/\tan\theta$, where θ is the look angle.

With this definition, we can define the relationship between the normalized back scattering

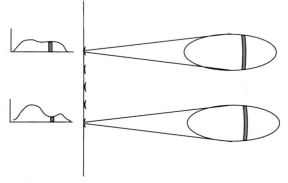

Figure 28 Radar imaging as viewed looking down at the Earth from above the sensor. The sweep of the pulse across the radar beam leads to a return echo trace for each pulse as depicted on the left. Every IPP, a pulse is sent and a trace recorded. The collection of pulses, that is, the along-track history of range profiles, is therefore a 2-D representation of the surface, albeit poorly resolved. There is considerable redundant information for any given surface element in the pulses because the beam samples each point many times as it travels along track (e.g., pulses every 5 m observed over a beam extent of 5 km). Through matched filtering techniques, a fine resolution image of the surface can be generated.

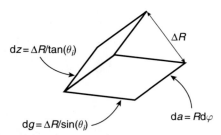

$$dz = \Delta R/\tan(\theta_i)$$

$$\Delta R$$

$$dg = \Delta R/\sin(\theta_i)$$

$$da = R\,d\varphi$$

Figure 29 Depiction of a cell in range defined by two phase fronts separated by range distance ΔR. This could be the transmit pulse length in range units, for example. Modified from Raney RK (1998) Radar fundamentals: Technical perspective. In: Henderson FM and Lewis AG (eds.) *Manual of Remote Sensing Volume 2, Principles and Applications of Imaging Radar*, 3rd edn. Hoboken, NJ: J. Wiley.

cross-subsection σ_0 and the reflectivity normal to the look direction:

$$|\Gamma(a, R)|^2 \, da \, dz = \frac{\sigma_0(a, R)}{\sin\theta} \, da \Delta R \qquad [77]$$

This equation defines Γ. In general, the reflectivity is a complex quantity, taking into account phases from random scatterers and differential path delays within the elemental area.

The received field from this elemental area is then

$$r(t) = \frac{P_T^{1/2} G\lambda}{(4\pi)^{3/2} R^2} \Gamma(R, \phi) R \, d\phi \, dz \, p(\theta(R), \phi)$$

$$\times e^{-j2kR} e^{j(1/2)(B_r/\tau_p)(t - 2R/c)^2} \mathrm{rect}\left(\frac{t - 2R/c}{\tau_p}\right) \qquad [78]$$

At every instant $t - 2R/c$ all the signals that arrive at the receiver add up. We have constructed the elemental area on the phase front such that an integration over these elements is at fixed $2R/c$. Thus, we can define the integration over the phase front to be a function of time that is a generalization of the range-only function $g()$ studied earlier:

$$g_\Gamma(R) = e^{-j2kR} \iint p(\theta(R), \phi) \Gamma(R, \phi) R \, d\phi \, dz \qquad [79]$$

If the scatterers were all arranged on a single phase front at range R, then the received signal would be

$$r(t) = \frac{P_T^{1/2} G\lambda}{(4\pi)^{3/2} R^2} g_\Gamma(R) e^{j(1/2)(B_r/\tau_p)(t - 2R/c)^2} \mathrm{rect}\left(\frac{t - 2R/c}{\tau_p}\right) \qquad [80]$$

For scatterers arranged over the entirety of the surface, we have an additional integration over all phase fronts:

$$r(t) = \frac{P_T^{1/2} G\lambda}{(4\pi)^{3/2} R^2} \int g_\Gamma(R) e^{j(1/2)(B_r/\tau_p)(t - 2R/c)^2}$$

$$\times \mathrm{rect}\left(\frac{t - 2R/c}{\tau_p}\right) dR \qquad [81]$$

This is a convolution of the transmitted pulse with the integrated phase front. The range compression process (matched filtering) described above returns to us a record of the total return from phase fronts resolved in time to $1/B$ or in range to $c/2B$. The next step is to exploit the redundancy in the range signal as a function of pulse number to build resolution in the along-track dimension.

Azimuth signal and aperture synthesis

Real aperture radars are very useful for a variety of applications, generally those which require detection of a bright object in a broad region as is common in military applications, or where a low-resolution averaged measure of surface roughness or dielectric is sufficient. Oceans and ice sheets are generally featureless, so high resolution is not often needed. Large-area estimates of backscatter can be interpreted as wind speed, for example, on global scales, with the sensor dealing with only modest amounts of data.

Often, however, especially for land applications, fine-resolution images are important because the natural spatial variability of the surface is rapid. So it would be very nice if we could overcome the azimuth beam limit on resolution without building an enormously long antenna. It turns out that it is possible using the techniques of aperture synthesis.

Returning momentarily to a point target on the ground, consider a train of pulses transmitted as a spacecraft travels along a straight line path (**Figure 30**) above a flat Earth. The pulse is transmitted and subsequently received at a slightly later time, but for the purpose of understanding the concepts, we can imagine a 'stop-and-shoot' model where the radar to target to radar distance is $2R_i$ for a given pulse i. The transmitted and received pulses are shown schematically in **Figure 30**, illustrating the changing range from pulse to pulse.

In this geometry,

$$R_i = \sqrt{R_0^2 + (x_i - x_0)^2}, \quad x_i = vt_i, \quad t_i = i\,\mathrm{PRI} + t_0 \qquad [82]$$

where R_0 is the shortest range from the flight track to the target, also called 'broadside', and x_0 is the azimuth at that location. Thus the phase factor we have been carrying along for a return from a broadside pulse, $\exp(-j2kR_0)$, must now be updated from pulse to pulse. After demodulation and range compression, the received signal from a point target for pulse i is

$$r_{zc\delta, x_i}(t) = g_0 p(x_i - x_0) e^{-j2kR_i(x_i)} \mathrm{sinc}\{\pi B(t - t_i)\} \qquad [83]$$

where $t_i = 2R_i/c$. This is the range-compressed form eqn [76] with an adjusted range term, and an explicit function $p()$ denoting the antenna pattern of the radar antenna. Writing this entirely in terms of range and azimuth coordinates gives

$$r_{zc\delta, x_i}(R_f) = g_0 p(x_i - x_0) e^{-j2kR_i(x_i)} \mathrm{sinc}\left\{\frac{\pi}{\Delta R}(R_f - R_i)\right\} \qquad [84]$$

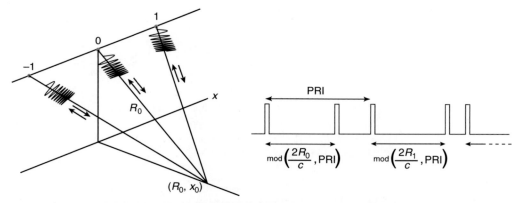

Figure 30 Illustration of range history in azimuth of the pulsed radar. Ignoring the finite speed of light, the radar can be thought of as transmitting and receiving from each point on orbit. Because of the finite extent of the azimuth beam, a single point target on the ground will be seen in a succession of received pulse echoes, with decreasing-to-increasing range location as the spacecraft flies by.

where R_f is the range variable, also called 'fast time' because it represents the time variation of the signal across a single pulse, and x_i is the azimuth location of a given pulse, also called 'slow time' since it characterizes the slower variation of the range trace from pulse to pulse.

If the transmitting antenna is length L in azimuth, then the target will be illuminated for an azimuth distance (3 dB) of

$$X_{ill} = \frac{\lambda}{L} R_0 \qquad [85]$$

The echo field from the point target will be sampled $N = X_{ill}/\Delta x_p$ times while it is in the beam, where $\Delta x_p = v\,\mathrm{PRI}$ is the pulse spacing on the flight track (**Figure 31**). The field is sampled effectively by a linear array of identical antennas spaced by Δx_p. It is shown in the homework (and any number of books) that the antenna pattern of a linear array of this sort is the product of the antenna pattern of the physical antenna and the array factor. The physical antenna is small, leading to a broad sinc-like beam in azimuth. The array factor is also sinc-like, but is much narrower

$$AF = \frac{\sin\big((N/2)k\Delta x_p \sin\theta\big)}{\sin\big((1/2)k\Delta x_p \sin\theta\big)} \qquad [86]$$

In the limit as $\Delta x_p \to 0$, and for small θ

$$AF \to N\frac{\sin((k/2)X_{ill}\theta)}{(k/2)X_{ill}\theta} \qquad [87]$$

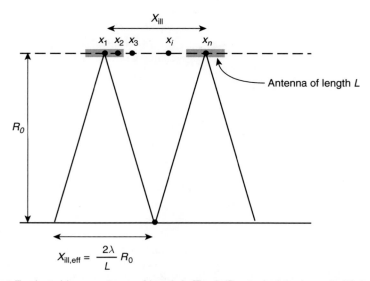

Figure 31 Point target illuminated by an antenna of length L. The 3 dB extent of the beam is $\lambda R_0/L$, but because two-way propagation introduces a doubling of the range and consequently the phase, the effective extent of illumination is $2\lambda R_0/L$.

which has the expected 3 dB beam width of $\theta_{3\,\text{dB}} = \lambda/X_{\text{ill}}$. Thus the sampled linear array in space serves as a very long antenna of length X_{ill}. Without careful deliberation, we might conclude that the resolution in azimuth we could achieve is given by its effective 3 dB beam extent $\Delta X = \theta_{3\,\text{dB}} R_0 = L$; however, we must take into account the two-way propagation of the radar signal. The above arguments are correct for an antenna synthesized to illuminate an area at range R_0. However, since $2R_0$ is the actual distance propagated, we can think of the beam extent for illumination as actually at twice the distance, that is, the phase variation across the aperture is as if the antenna were at twice the distance. Thus,

$$X_{\text{ill, eff}} = \frac{\lambda}{L} 2R_0 \qquad [88]$$

and the resolution allowed by an antenna of this length is

$$\Delta X = L/2 \qquad [89]$$

This remarkable result states that the resolution of a synthetic aperture system is independent of range and velocity, and is just half the physical antenna length. As the range is increased, the synthetic aperture increases in length and the angular extent of synthetic beam narrows in proportion to maintain fixed resolution. If the physical antenna decreases in size for a fixed range, the illuminated area increases, increasing the synthetic aperture and narrowing the synthetic beam width. Since the range is fixed, the resolution becomes finer in proportion to the reduction in L.

Consider R_i as a continuous variable of the azimuth coordinate

$$R(x) = \sqrt{R_0^2 + (x - x_0)^2} \qquad [90]$$

$$\approx R_0 \left(1 + \frac{(x - x_0)^2}{2R_0^2}\right) \qquad [91]$$

$$\approx R_0 + \frac{(x - x_0)^2}{2R_0} \qquad [92]$$

The constant phase term $\exp(-j2kR_0)$ is very difficult to measure absolutely (billions of cycles), so for now we will ignore it and focus on the spatially varying component of the phase

$$e^{-j2k(x-x_0)^2/2R_0} = e^{-j\pi(x-x_0)^2/F^2} \qquad [93]$$

where $F^2 = \lambda R_0/2$. F is known as the Fresnel zone and it is the distance along the synthetic array where the wave front is within $\pi/4$ radians of phase (see **Figure 32**). Typical Fresnel zone size is on the order of 100–200 m. Since the array is coherent over this length, points simply summed together in azimuth will add coherently and form a meaningful, albeit low-resolution representation of the surface in azimuth:

$$\pi\frac{(x-x_0)^2}{F^2} < \frac{\pi}{4} \rightarrow x - x_0 < \frac{F}{2} \qquad [94]$$

The range-compressed pulse history can now be written as

$$r_{zc\delta}(x_i, R_f) = g_0 p(x_i - x_0) e^{-j2kR_0} e^{-j\pi(x-x_0)^2/F^2}$$
$$\times \text{sinc}\left\{\frac{\pi}{\Delta R}\left[R_f - \left(R_0 + \frac{(x_i + x_0)^2}{2F^2}\right)\right]\right\} \qquad [95]$$

where we explicitly call out the functional dependence on x_i in the argument to $r_{zc\delta}$.

The form of the azimuth signal is rather similar to the range signal before compression. Using the same approach to recover the point target in azimuth, we can define a matched filter in azimuth that is a similar conjugate chirp signal:

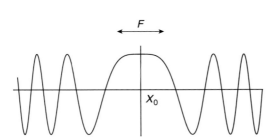

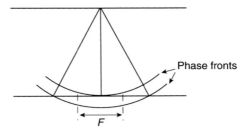

Figure 32   Illustration of the Fresnel zone, where the phase from an incident wave front varies by less than $\pi/4$ radians across the zone (right). Symmetric chirp at left is the real part of the azimuth chirped phasor. It is nearly constant over the Fresnel zone. The synthetic array is coherent over this length and points simply summed in azimuth will add coherently, and will be resolved at this length.

$$b_{am}(x) = b_a^*(-x) = p(-x)e^{j\pi x^2/F^2} \qquad [96]$$

The doubly compressed point-target signal is then

$$r_{zcc\delta}(x, R) = g_0 e^{-j2kR_0} \, \text{sinc}\left\{\frac{\pi}{\Delta R}[R - R_0]\right\} \int p(x' - x_0)$$
$$\times e^{-j\pi(x'-x_0)^2/F^2} p(x' - x) e^{j\pi(x'-x)^2/F^2} \, dx' \qquad [97]$$

where we ignored the x-dependent term in the range sinc function because processors generally take care of this term before performing the matched filter in azimuth. If we allow $p()$ to be a rect-function of extend X_{ill} as we did for range, then we will obtain exactly the same expression as we did for range, and the doubly compressed point-target signal becomes

$$r_{zcc\delta}(x, R) = g_0' e^{-j2kR_0} \, \text{sinc}\left\{\frac{\pi}{\Delta R}[R - R_0]\right\}$$
$$\times \text{sinc}\left\{\frac{\pi(x - x_0)X_{ill}}{F^2}\right\} \qquad [98]$$

Since in this case we explicitly took into account the two-way propagation in the phase, X_{ill} is the actual illuminated extent, not the double-length effective extent used in the heuristic argument earlier. The first null of the azimuth response is where

$$x - x_0 = \frac{F^2}{X_{ill}} \qquad [99]$$

$$= \frac{\lambda R_0/2}{\lambda R_0/L} = \frac{L}{2} \qquad [100]$$

We take this as the 3 dB resolution in azimuth, and we see it is identical to the heuristically derived resolution.

Expressed in terms of the range resolution $\Delta R = c/2B$ and azimuth resolution $\Delta X = L/2$, the overall impulse response of the radar system and matched filtering operations is

$$r_{zcc\delta}(x, R) = g_0' e^{-j2kR_0} \, \text{sinc}\left\{\frac{\pi}{\Delta R}[R - R_0]\right\}$$
$$\times \text{sinc}\left\{\frac{\pi}{\Delta X}[x - x_0]\right\} \qquad [101]$$

For a surface described by a general reflectivity function in range and azimuth $\Gamma(x, R)$ related to the surface backscatter cross-subsection through eqn [77], the matched filter response will be the linear convolution with the impulse response in eqn [101]:

$$\Gamma_{zcc}(x, R) = \int e^{-j2kR'} \Gamma(x', R') \text{sinc}\left\{\frac{\pi}{\Delta R}[R - R']\right\}$$
$$\times \text{sinc}\left\{\frac{\pi}{\Delta X}[x - x']\right\} dx' \, dR' \qquad [102]$$

Range-Doppler images

SAR images generated in this style of filtering are often called 'range-Doppler' images. The quadratic variation of the phase of a point target with time along track leads to a linear frequency variation:

$$\phi(t) = -\frac{\pi}{F^2} v^2 (t - t_0)^2 \qquad [103]$$

$$\omega(t) = -2\pi \frac{v^2}{F^2} (t - t_0) \qquad [104]$$

$$f_D(t) = -\frac{v^2}{F^2} (t - t_0) \qquad [105]$$

Note the Doppler frequency is positive as the sensor approaches the target ($t < t_0$). The Doppler bandwidth is the totality of frequency content for any target. This is limited by the beam width in azimuth, $T_{ill} = X_{ill}/v$, giving

$$f_{D,ill} = \frac{2}{\lambda R} v^2 \frac{X_{ill}}{v} = \frac{2v}{L} \qquad [106]$$

This is in time units (Hz). In spatial units, $f_{Dx,ill} = 2/L$, which is just the reciprocal of the spatial resolution.

Another way of describing this Doppler bandwidth is through the Doppler equation $f_D = -2\vec{v} \cdot \hat{l}/\lambda$, where $\vec{v}$ is the vector velocity of the sensor, and $\mathbf{l}$ is the direction from the sensor to a point on the ground. At any instant of time, there is a one-to-one correspondence between the azimuth position of a target and the Doppler frequency (**Figure 33**). The antenna beam illuminates an area on the ground, with each target at a given angle off boresight having a unique Doppler frequency. The Doppler frequency can be written as $f_D = -2v \cos \theta_{az}/\lambda$, where θ_{az} is the angle formed by the velocity vector $\vec{v}$ and the look vector $\hat{l}$. Defining $\theta_{b.s.}$ as the angle measured relative to broadside (that is, perpendicular to the velocity vector), and assuming that the antenna's boresight is oriented at broadside, $f_D = -2v \sin \theta_{b.s.}/\lambda$. The total Doppler bandwidth at that instant is given by the Doppler frequencies at the edge of the beam. The total beam extent is then

$$2\theta_{b.s.} = X_{ill}/R \qquad [107]$$

$$= \lambda R/(LR) \qquad [108]$$

$$= \lambda/L \qquad [109]$$

as expected. Thus, the Doppler bandwidth $f_{D,ill}$, is also $2v(2 \sin \theta_{b.s.})/\lambda = 2v/L$.

At any given instant, the received signal is comprised of the full complement of Doppler frequencies

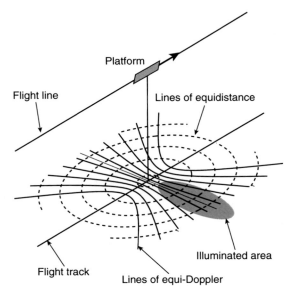

Figure 33 Range-Doppler coordinates of the surface as seen by a sensor at an instant of time. Targets in a band of Doppler frequencies contribute to the signal return. It is up to the azimuth matched filtering operation to sort out the targets into unique Doppler bins. Modified from Elachi C (1988) *Spaceborne Radar Remote Sensing: Applications and Techniques*. New York: IEEE Press.

because targets are distributed throughout the beam, each with their own Doppler frequency. In forming the synthetic aperture at some reference time, we are rearranging each of the targets in the range-Doppler image to lie at the Doppler frequency appropriate to that geometry at that reference time (**Figure 33**).

To be more explicit, we pick a collection of pulses that will be used to form our synthetic aperture, noting that this establishes a position in space at a time instant as a reference for defining Doppler frequencies. We can also model the ground as a collection of point targets arranged in azimuth, such that

$$r_{zc\delta, \text{tot}} = \sum_{k=1}^{N} A_k e^{-j\pi(x-x_k)^2/F^2} \operatorname{sinc} \frac{\pi}{\Delta R} \left(R - R_0 - (x-x_k)^2/2R_0 \right)$$

[110]

For the sake of this discussion, we allow the sinc function to become a delta function in range and ignore its variability with x, which is small and correctable. Thus,

$$r_{zc\delta} \approx \sum_{k=1}^{N} A_k e^{-j\pi(x-x_k)^2/F^2}$$

[111]

The azimuth compression process can be written then as

$$
\begin{aligned}
r_{zcc\delta}(x) &= \int_{x-X_{\text{ill}}/2}^{x+X_{\text{ill}}/2} r_{zc\delta}(x') e^{j\pi(x'-x)^2/F^2} dx' \\
&= \int_{x-X_{\text{ill}}/2}^{x+X_{\text{ill}}/2} \sum_{k=1}^{N} A_k e^{-j\pi(x'-x_k)^2/F^2} e^{j\pi(x'-x)^2/F^2} dx' \\
&= e^{j\pi x^2/F^2} \int_{x-X_{\text{ill}}/2}^{x+X_{\text{ill}}/2} \sum_{k=1}^{N} A_k e^{-j\pi x_k^2/F^2} \\
&\quad \times e^{j2\pi x' x_k/F^2} e^{-j2\pi x' x/F^2} dx' \\
&= \sum_{k=1}^{N} e^{j\pi x^2/F^2} \int_{x-X_{\text{ill}}/2}^{x+X_{\text{ill}}/2} A_k' e^{-j2\pi x'(f_{\text{Dx}} - f_{k,\text{Dx}})} dx' \quad [112]
\end{aligned}
$$

where $f_{\text{Dx}} = x/F^2$ is the Doppler spatial frequency. From this arrangement of terms, we can recognize that the compressed azimuth signal is just the sum of Fourier transforms of the individual complex exponentials of complex amplitude A_k'. Because the argument of the exponential is the Doppler frequency, we see the transform domain is the Doppler domain, and each exponential (being representative of a scatterer at a particular azimuth location x_k relative to the reference location) transforms to an individual Doppler frequency $f_{k\text{Dx}}$. Thus for a general continuous surface, the summation becomes an integration, and we see the compression process consists of focusing the array by applying a quadratic phase correction, followed by a transform that sorts the scatterers into correct Doppler bins for that time reference.

We represent and execute the imaging as a convolution in spatial variables because the processed data sample spacing is independent of range on input and output. In the focussing-Fourier-transform approach, the Doppler sample spacing is controlled by the implementation of Fourier transform, which is usually accomplished by fast Fourier transform (FFT). In this case, if the input data are sampled at a spacing $L/2$, the Doppler spacing will be $2/L/N$, where N is the number of points in the FFT.

Other Doppler considerations

If the antenna pattern was isotropic and we have a continuous time system, the phase and Doppler of a single point target would be as depicted in **Figure 34**. Imposing an antenna pattern where the antenna pattern is pointing 'broadside,' that is orthogonal to the velocity vector, the phase and Doppler history would be truncated according to the extent of the beam (see **Figure 34**).

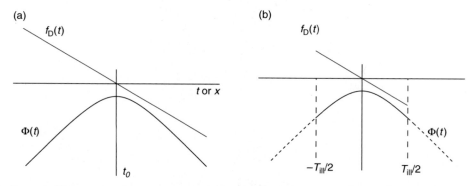

Figure 34   Phase and Doppler history over time for a point target. (a) The trace for a system with an isotropic antenna: the trace keeps going. (b) The finite illumination time of a finite-dimension antenna truncates the phase and Doppler history.

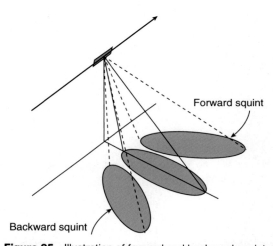

Figure 35   Illustration of forward and backward squint.

It is most commonly the case that the finite extent beam is 'squinted' forward or backward with respect to the velocity vector of the sensor. For an aircraft this is most commonly due to cross-winds causing the fuselage to crab. For spacecraft, there is a natural latitude-dependent angular offset between the velocity vector of the inertial orbit and the rotating Earth below, causing a natural squint (**Figure 35**).

The effect of this squint on the phase and Doppler history of a target is shown in **Figure 36**. Note the shape of the phase or Doppler does not change, just the range of phases or frequencies considered. Thus, there is a unique relationship between the pointing and the location of the scatterers. In processing data, the extent of the antenna beam (Doppler bandwidth) and the 'centroid' of the Doppler, or the degree of squint as represented by the center of the Doppler bandwidth, are specified in one way or another.

Figure 37 shows that the spectra are shifted to be centered on the Doppler spread associated with a particular squint. It is important to know the squint to ensure that the processing is centered on the proper part of the azimuth spectrum. The centroid specified in processing defines the direction from which the ground is imaged after processing.

The squint can be derived from information provided by the spacecraft and radar manufacturer. Often spacecraft have sensitive position and attitude sensors that record the information needed to calculate the direction of the antenna boresight. When this information is not available, the Doppler centroid itself can be estimated from the data. This estimated centroid can be used directly in the processing. The centroid generally varies across range, so often the centroid is specified as a function of range.

To understand the arrangement of spectral energy for a collection of scatterers, consider a collection of point-target scatterers all at the same range, but at different azimuths:

$$
\begin{aligned}
r_{zc\delta,\text{tot}} = {} & A_0 e^{-j\pi(x-x_0)^2/F^2} \\
& \times \operatorname{sinc}\frac{\pi}{\Delta R}\left(R - R_0 - (x-x_0)^2/2R_0\right) \\
& + A_1 e^{-j\pi(x-x_1)^2/F^2} \\
& \times \operatorname{sinc}\frac{\pi}{\Delta R}\left(R - R_0 - (x-x_1)^2/2R_0\right) + \cdots \\
& + A_n e^{-j\pi(x-x_n)^2/F^2} \operatorname{sinc}\frac{\pi}{\Delta R}\left(R - R_0 - (x-x_n)^2/2R_0\right)
\end{aligned}
$$

[113]

If we ignore the antenna pattern modulation of target energy and denote the azimuth spectrum of an individual scatter response as

$$
F(\omega) = \text{F.T.}\left\{ e^{-j\pi x^2/F^2} \operatorname{sinc}\frac{\pi}{\Delta R}\left(R - R_0 - x^2/2R_0\right) \right\} \quad [114]
$$

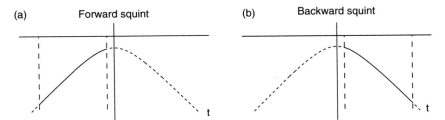

Figure 36 Phase history over time for a point target. (a) The trace for a forward squinted system. (b) The trace for a backward squinted system. Only phase is shown, but with reference to the previous figures, the Doppler history is similarly constrained. Forward (backward) squinted systems see more positive (negative) Doppler frequencies.

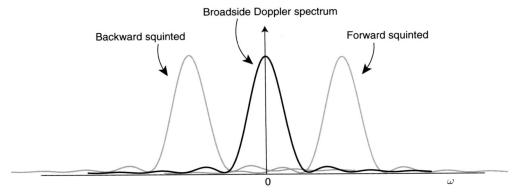

Figure 37 Hypothetical Doppler spectra of broadside, forward and backward squinted signals. For a collection of scatterers over the beam pointed with a particular squint, all scatterers experience the same Doppler frequency spread, so the spectra would be as shown, controlled in extent and magnitude by the antenna pattern in azimuth.

where F.T. denotes the Fourier Transform, the spectrum of $r_{zc\delta, \text{tot}}$ would be

$$R_{zc\delta, \text{tot}}(\omega) = F(\omega)\left(A_0 e^{-j\omega x_0} + A_1 e^{-j\omega x_1} + \cdots + A_n e^{-j\omega x_n}\right)$$

[115]

Thus, except for a phase ramp in azimuth frequency space, all scatterers have the same spectrum. Clearly, the process of applying the matched filter neutralizes this common component, and the linear phase terms in frequency distinguish the targets in position upon inverse transformation. This again shows the spatial-Doppler duality. The x dependence of the energy's localization in range $x^2/2R_0$ is called the 'range migration' and is seen to be part of the function $F(\omega)$ in the frequency domain. Hence any effects this may introduce can be dealt with for all targets simultaneously by manipulating the spectrum.

We ignored the 'range migration' term $(x-x_0)^2/2R_0$ in the argument to the sinc function in range, but frequently this is not advisable. The processing of data

on an orthogonal grid is enforced by the need for high-speed computations, and the most efficient processing methods are those that are separable in range and azimuth and for which FFT-based convolutions can be performed. For broadside imaging, the term can be ignored when $X_{\text{ill}}^2/2R_0$ is a fraction of the range resolution ΔR. In other words, when the migration over the full extent of the azimuth matched filter (equivalent to the beam extent on the ground) is smaller than a range-resolution element, the azimuth matched filter will catch all the energy for a given scatterer in a single range bin.

The sampled azimuth spectrum and range migration

Since the range migration correction terms depend on the Doppler frequency, one must know the correct Doppler frequency to properly compress the imagery. This is also true for computing the correct azimuth matched filter; however, the sampled nature of the azimuth signal adds an interesting wrinkle to the processing that is often confusing. Since the

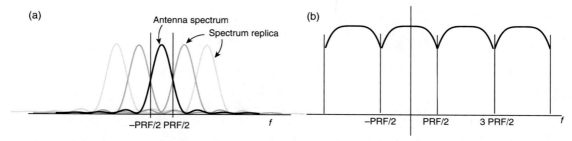

Figure 38 The azimuth signal is sampled at the PRF. The resulting discrete time signal in azimuth has a circular spectrum with all energy at frequency higher than the PRF aliased into the circular bandwidth of the PRF. (a) Illustrates the replicas of the continuous spectrum that are aliased, and (b) the resulting circular spectrum.

azimuth signals are inherently derived from pulses acquired along track, the azimuth signal is a discrete time signal sampled at the PRF. The actual discrete time spectrum is then a circular spectrum, with all energy at frequencies higher than the PRF aliased into the range of one PRF (**Figure 38**).

Thus without knowledge of the geometry of the spacecraft and the observation, it is often difficult to know the actual Doppler centroid, since one usually estimates it from the spectrum only in the range of the PRF. One can process the data with an incorrect Doppler centroid, off by some number of PRFs (also called 'ambiguities') for instance, and the imagery will look almost correct. Since the signal is sampled, the matched filter spectrum for the actual centroid will be the same as for the incorrect centroid as long as the centroid is wrong by an integer number of ambiguities. However, the range migration correction is in terms of absolute Doppler, so if the wrong ambiguity is used, the images will be poorly focused in range (because energy is spread in range for each scatterer) and in azimuth (because the compression uses incomplete information for each scatterer).

If the attitude of the platform is known, then the Doppler can be computed from

$$f_D = -\frac{2\vec{v} \cdot \hat{l}}{\lambda} = -\frac{2v}{\lambda} \sin\theta \sin\phi_{sq} \qquad [116]$$

where $\vec{v}$ is the velocity vector, $\hat{l}$ is the look vector from antenna to target, θ is the look angle, and ϕ_{sq} is the squint angle. If the attitude parameters are not known, then one has to estimate the ambiguity. This can be done by trial and error – examine the focus for a variety of integer ambiguity choices – or by one of several automated techniques. A popular method of ambiguity estimation is to split the azimuth spectrum into two pieces and process each with a particular Doppler ambiguity assumption. If the correction is

the right one, both side bands will be adjusted to the correct range and the two images produced will be properly registered in range. If the correction is not right, there will be a residual shift between the two images in range. From the magnitude of this shift it is often possible to estimate the correct ambiguity. Unfortunately the imagery are lower resolution in azimuth and defocused in range, so estimating the shift is sometimes difficult.

Speckle

Radar images are naturally noisy because of the coherent interaction of the electro-magnetic wave with scatterers within a resolution cell, known as speckle. Speckle is an important part of the radar literature, as it is an important limitation of the imagery that many people have worked hard to minimize. It is also an important concept for interferometry, which exploits the spatially fixed properties of speckle to extract phase differences between images. The normalized backscatter cross-section σ_0 is an average quantity for a naturally varying random surface. For homogenous surfaces made up of a continuum of scatterers, the backscatter cross-section is a measure of the roughness of the surface and its natural reflectivity, as shown in a previous section. Viewing the SAR imaging process as a lens (**Figure 39**), after imaging, the random (complex) surface reflectivity is convolved with the sinc-function impulse response of the imaging system, accounting for the random differential path length introduced by surface roughness.

In a given resolution element, we consider the sum of the collection of small independent scatterers, where each scatterer has its own amplitude and phase

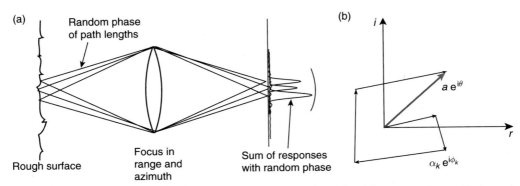

Figure 39   Illustration of the concept of radar imaging of a random surface (a) and the coherent sum of independent scatterers as a phasor (b).

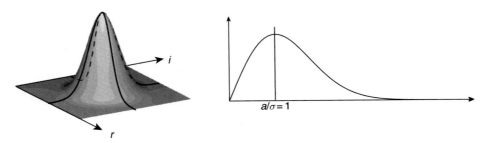

Figure 40   The real and imaginary parts of a SAR image are circularly Gaussian (*left*). The amplitude of the SAR pixel is Rayleigh distributed, with peak at $a = \sigma$.

$$ae^{j\theta} = \sum_{k=1}^{N} \alpha_k e^{j\phi_k} \qquad [117]$$

If α_k and ϕ_k are statistically independent of each other and α_k are identically distributed with mean $\bar{\alpha}$ and second moment $\overline{\alpha^2}$, and ϕ_k are uniformly distributed in the interval $[-\pi, \pi]$, then one can show that for large N

$$r = a \cos \theta \qquad [118]$$

$$i = a \sin \theta \qquad [119]$$

are Gaussian-distributed zero-mean random variables with joint probability density function

$$p_{RI}(r, i) = \frac{1}{2\pi\sigma^2} e^{-(r^2 + i^2)/2\sigma^2} \qquad [120]$$

where $\sigma^2 = \overline{\alpha^2}/2$ (**Figure 40**). The variance of the real and imaginary parts of the SAR image are related to the mean power of the intrinsic scatterers but are zero mean random variables themselves. As such for homogeneous areas, averaging many complex pixels together in a region will eventually reduce the signal to zero on average and not improve the quality of the image. This is a common mistake among beginners. It

is the amplitude of the image, a, or its intensity (power), $I = a^2$, that interests us.

For the amplitude $a = \sqrt{r^2 + i^2}$, we can use the rules of transformation of random variables (multiply the distribution by the Jacobian of the transformation and substitute in the new variables) to get the distribution

$$p_A(a) = \frac{a}{\sigma^2} e^{-a^2/2\sigma^2}, \qquad a \geq 0 \qquad [121]$$

which is known as the Rayleigh distribution (**Figure 40**). The phase is uniformly distributed over the interval $(-\pi, \pi)$

$$p_{\Theta}(\theta) = \frac{1}{2\pi}, \qquad -\pi < \theta < \pi \qquad [122]$$

The mean and variance of the amplitude are

$$\bar{a} = \sqrt{\frac{\pi}{2}} \sigma \qquad [123]$$

$$\sigma_a^2 = \left(2 - \frac{\pi}{2}\right) \sigma^2 \qquad [124]$$

Clearly the mean amplitude of the image scales with the variability in the image, which is related to the mean backscattered power $\overline{\alpha^2}$ of the scatterers comprising the resolution element. The brighter the image, the noisier the image, and vice-versa. Thus, at some signal level such that the signal level returned is sufficiently larger than the thermal noise in an image, having a more powerful radar does not necessarily improve image quality.

The intensity, or power, of the image $I = a^2$ has an exponential distribution

$$p_I(I) = \frac{1}{\bar{I}} e^{-I/\bar{I}}, \quad I \geq 0 \tag{125}$$

with mean and variance

$$\bar{I} = 2\sigma^2, \quad \sigma_I^2 = 4\sigma^4 \tag{126}$$

The mean power is equal to its standard deviation, as expected from the mean amplitude scaling with σ. In the display and analysis of an image with only 'one look,' meaning processed at full resolution using the full synthetic aperture, the SNR of the intensity image is just I since $\bar{I}^2/\sigma_j^2 = 1$; the brighter the image, the noisier. In the design of a radar system, one must have sufficient power to receive a return from the surface above the thermal noise (satisfying the radar equation, where $\overline{\alpha^2}$ is related to the backscattering cross-section), but not so high that we are wasting power. For interferometry, we will soon see however that more power in general is better!

The radiometric resolution is defined as the ability to discriminate surfaces of different brightnesses. It has been defined to be

$$Q = 10 \log \left(\frac{\bar{I} + \sigma_I}{\bar{I} - \sigma_I} \right) \tag{127}$$

which is something like the distance apart in amplitude of discriminable amplitudes. For a single-look image, that distance is infinite. In order to improve this situation, it is often necessary to average many pixels together. The act of averaging N

independent (Gaussian, but also approximately true for non-Gaussian) random variables with identical distributions reduces the variance by a factor of N. Thus for

$$I' = \frac{1}{N} \sum_{k=1}^{N} I_k \tag{128}$$

we have $\bar{I}' = \bar{I}$, $\sigma_{I'}^2 = \sigma_I^2/N$, and the radiometric resolution becomes

$$Q = 10 \log \left(\frac{\bar{I}' + \sigma_{I'}}{\bar{I}' - \sigma_{I'}} \right) = 10 \log \left(\frac{\sqrt{N}+1}{\sqrt{N}-1} \right) \tag{129}$$

The resolution distance approaches 0 for arbitrarily large N. This averaging technique is often called 'taking looks' because it can be shown to be equivalent to segmenting the azimuth spectrum into separate frequency bands, forming a power image from each band, then stacking the images in an average. Since azimuth band corresponds to a particular look direction relative to a target on the ground, this technique is called taking looks.

Of course averaging together a large number of pixels will lower the intrinsic resolution of the image, so designing the system to have a larger bandwidth is desirable if radiometric resolution is important. Note also that there is a limit to how far one can go to improve the statistics. The improvement above assumes that all the pixels come from the same distribution. If the scene is variable in intrinsic brightness, not all the benefits will be realized.

If there is a bright target within a resolution cell, then there is a mean value associated with the Gaussian random variables (**Figure 41**). This complex random variable has a Rician distribution given by

$$p_A(a) = \frac{a}{\sigma^2} e^{-(a^2+s^2)/2\sigma^2} I_0\left(\frac{as}{\sigma^2}\right), \quad a \geq 0 \tag{130}$$

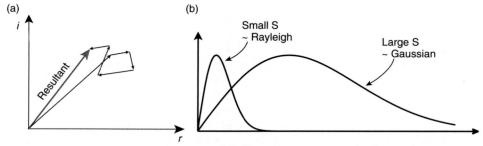

Figure 41 (a) A phasor diagram with a bright specular target (within a scattering cell. (b) The Rician distribution of the amplitude.

where I_0 is the modified Bessel function of the first kind and s is the specular target strength. This distribution approaches a Gaussian distribution for large s relative to σ. The distribution of the phase, which is uniform for $s = 0$, becomes more peaked as s grows, eventually approximating a Gaussian also:

$$p_\Theta(\theta) \to \frac{s/\sigma}{\sqrt{2\pi}} e^{-s^2\theta^2/2\sigma^2} \quad [131]$$

Often specular targets are used for calibration of the radar. If the specular target is not sufficiently bright, then the amplitude and phase will be corrupted by the background random scatterers.

Another way to model the surface is in terms of a general reflectivity function that is statistical in nature. Since

$$\Gamma_{zcc}(x', r') = \iint dr\, dx \Gamma(x, r)$$
$$\times \operatorname{sinc}\left(\frac{\pi}{\Delta X}(x'-x)\right)\operatorname{sinc}\left(\frac{\pi}{\Delta R}(r'-r)\right) [132]$$

if Γ is a statistical function, the power that is meaningful is the expected power scattered back from the resolution cell, which is related by scale factors and geometric normalizations to $E\{\Gamma\Gamma^*\}$. Writing this out explicitly, we have

$$E\{\Gamma_{zcc}\Gamma_{zcc}^*\}(x', r') = \int dx_1 \int dx_2 \int dr_1 \int dr_2$$
$$\times E\{\Gamma(x_1, r_1)\Gamma^*(x_2, r_2)\} \quad [133]$$

$$= \operatorname{sinc}\left(\frac{\pi}{\Delta X}(x'-x_1)\right)\operatorname{sinc}\left(\frac{\pi}{\Delta X}(x'-x_2)\right) \quad [134]$$

$$= \operatorname{sinc}\left(\frac{\pi}{\Delta R}(r'-r_1)\right)\operatorname{sinc}\left(\frac{\pi}{\Delta R}(r'-r_2)\right) \quad [135]$$

If the scatterers that comprise the reflectivity are uncorrelated, then the expected value of the product of the surface with itself when not coaligned would be zero:

$$E\{\Gamma(x_1, r_1)\Gamma^*(x_2, r_2)\} = E\{\Gamma(x_1, r_1)\}E\{\Gamma^*(x_2, r_2)\} = 0$$
$$[136]$$

However, when the surface is self-aligned

$$E\{\Gamma(x_1, r_1)\Gamma^*(x_2, r_2)\} = |\Gamma(x_1, r_1)|^2 \quad [137]$$

More succinctly,

$$E\{\Gamma(x_1, r_1)\Gamma^*(x_2, r_2)\} = E\{|\Gamma(x_1, r_1)|^2\}$$
$$\times \delta(x_1-x_2)\delta(r_1-r_2) \quad [138]$$

Using this expression in eqn [135] simplifies the integrals

$$E\{\Gamma_{zcc}\Gamma_{zcc}^*\}(x', r') = \int dx \int dr\, E\{|\Gamma(x, r)|^2\} \quad [139]$$

$$\operatorname{sinc}^2\left(\frac{\pi}{\Delta X}(x'-x)\right)\operatorname{sinc}^2\left(\frac{\pi}{\Delta R}(r'-r)\right) \quad [140]$$

$$= E\{|\Gamma|^2\}\Delta X\Delta R \quad [141]$$

For homogeneous surfaces, we can expect the mean power to be essentially constant over the resolution cell, so we can take the expectation outside the integration. Also the variables x' and r' can profitably be set to zero since there is no dependence on space. Then the integration over the sinc functions produces a constant, which is proportional to the area under them. This turns out to be ΔX and ΔR. So we see that $E\{\Gamma_{zcc}\Gamma_{zcc}^*\}$ is related to the normalized backscattering cross-section σ_0 through the function Γ.

Doppler and the Interferometric Baseline

The precise definition of interferometric baseline and phase, and consequently the topographic mapping process, depends on how the SAR data comprising the interferometer are processed. Consequently, a brief overview of the salient aspects of SAR processing is in order.

Processed data from SAR systems are sampled images. Each sample, or pixel, represents some aspect of the physical process of radar backscatter. A resolution element of the imagery is defined by the spectral content of the SAR system. Fine resolution in the range direction is achieved typically by transmitting pulses of either short time duration with high peak power, or of a longer time duration with a wide, coded signal bandwidth at lower peak transmit power. Resolution in range is inversely proportional to this bandwidth. In both cases, the received echo for each pulse is sampled at the required radar signal bandwidth.

For ultra-narrow pulsing schemes, the pulse width is chosen at the desired range resolution, and no further data manipulation is required. For coded pulses, the received echoes are typically processed with a matched filter technique to achieve the desired range resolution. Most spaceborne platforms use chirp-encoding to attain the desired bandwidth and consequent range resolution, where the frequency is linearly changed across the pulse.

Resolution in the azimuth, or along-track, direction, parallel to the direction of motion, is achieved by synthesizing a large antenna from the echoes received from the sequence of pulses illuminating a target. The pulses in the synthetic aperture contain an unfocused record of the target's amplitude and phase history. To focus the image in azimuth, a digital 'lens' that mimics the imaging process is constructed, and is applied by matched filtering. Azimuth resolution is limited by the size of the synthetic aperture, which is governed by the amount of time a target remains in the radar beam. The azimuth beam width of an antenna is given by $\theta_{BW} \equiv k\lambda/L$, where λ is the wavelength, L is the antenna length, and k is a constant that depends on the antenna ($k = 1$ is assumed in this chapter). The size of the antenna footprint on the ground in the azimuth direction is approximately given by

$$l_{az} = \rho\theta_{BW} = \rho\frac{\lambda}{L} \qquad [142]$$

where ρ is the range to a point in the footprint.

During the time a target is in the beam, the range and angular direction to the target are changing from pulse to pulse, as shown in **Figure 30**. To generate SAR image, a unique range or angle must be selected from the family of ranges and angles to use as a reference for focusing the image. Once selected, the target's azimuth and range position in the processed image is uniquely established. Specifying an angle for processing is equivalent to choosing a reference Doppler frequency. The bold dashed line from pulse N-2 to the target in **Figure 30** indicates the desired angle or Doppler frequency at which the target will be imaged. This selection implicitly specifies the time of imaging, and therefore the location of the radar antenna. This is an important and often ignored consideration in defining the interferometric baseline. The baseline is the vector connecting the locations of the radar antennas forming the interferometer; since these locations depend on the choice of processing parameters, so does the baseline. For two-aperture cross-track interferometers, this is a subtle point; however, for repeat-track geometries where the antenna pointing can be different from track to track, careful attention to the baseline model is essential for accurate mapping performance.

References

Alsdorf DE, Smith LC, and Melack JM (2001) Amazon floodplain water level changes measured with interferometric SIR-C radar. *Geoscience and Remote Sensing, IEEE Transactions on* 39: 423–431 (ISSN 0196–2892).

Amelung F, Galloway DL, Bell JW, and Zebker H (1999) Sensing the ups and downs of Las Vegas: InSAR reveals structural control of land subsidence and aquifer-system deformation. *Geology* 27: 483–486.

Ballatore P (2006) Synthetic aperture radar interferometry: Separation of atmospheric artifacts from effects due to the topography and the terrain displacements. *Earth Planets Space* 58: 927–935.

Bamler R and Eineder M (1996) ScanSAR processing using standard high-precision SAR algorithms. *IEEE Transactions on Geoscience and Remote Sensing* 34: 212–218.

Bamler R and Hartl P (1998) Synthetic aperture radar interferometry. *Inverse Problems* 14: 1–54.

Bawden GW, Thatcher W, Stein RS, and Hudnut K (2001) Tectonic contraction across Los Angeles after removal of groundwater pumping effects. *Nature* 412: 812–815.

Bechor NBD and Zebker HA (2006) Measuring two-dimensional movements using a single InSAR pair. *Geophysical Research Letters* 33 (doi:10.1029/2006GL026883).

Bell JW, Amelung F, Ramelli AR, and Blewitt G (2002) Land subsidence in Las Vegas, Nevada, 1935–2000: New geodetic data show evolution, revised spatial patterns, and reduced rates. *Environ* 8: 155–174.

Berardino P, Fornaro G, Lanari R, and Sansosti E (2002) A new algorithm for surface deformation monitoring based on small baseline differential SAR interferograms. *IEEE Transactions on Geoscience and Remote Sensing* 40: 2375–2383.

Born M and Wolf E (1989) *Principles of Optics, 6th edn.* Oxford: Pergamon Press.

Burgmann R, Hilley G, Ferretti A, and Novali F (2006) Resolving vertical tectonics in the San Francisco Bay area from permanent scatterer InSAR and GPS analysis. *Geology* 34(3): 221–224.

Burgmann R, Rosen PA, and Fielding EJ (2000) Synthetic aperture radar interferometry to measure Earth's surface topography and its deformation. *Annual Review of Earth and Planetary Science* 28: 169–209.

Chen CW and Zebker HA (2001) Two-dimensional phase unwrapping with use of statistical models for cost functions in nonlinear optimization. *Journal of the Optical Society of America A* 18: 338–351.

Colesanti C, Ferretti A, Novali F, Prati CC, and Rocca F (2003) SAR monitoring of progressive and seasonal ground deformation using the permanent scatterers technique. *IEEE Transactions on Geoscience and Remote Sensing* 41: 1685–1701.

Delacourt C, Briole P, and Achache J (1998) Tropospheric corrections of SAR interferograms with strong topography: Application to Etna. *Geophysical Research Letters* 25: 2849–2852.

Dong D, Fang P, Bock Y, Cheng MK, and Miyazaki S (2002) Anatomy of apparent seasonal variations from GPS-derived site position time series. *Journal of Geophysical Research* 107: 2075.

Dong D, Fang P, Bock Y, Webb F, Prawirodirdjo L, Kedar S, et al. (2006) Spatiotemporal filtering using principal component analysis and Karhunen–Loeve expansion approaches for regional GPS network analysis. *Journal of Geophysical Research* 111: B03405 (doi:10.1029/2005JB003806).

Dong D, Herring TA, and King RW (1998) Estimating regional deformation from a combination of space and terrestrial geodetic data. *Journal of Geodesy* 72: 200–214 (doi:10.1007/s001900050161).

Dong D, Yunck T, and Heflin M (2003) Origin of the international terrestrial reference frame. *Journal of Geophysical Research* 108: 2200.

Doubre C and Peltzer G (2007) Fluid-controlled faulting process in the Asal rift, Djibouti, from 8 yr of radar interferometry observations. *Geology* 35: 69–72.

Elachi C (1988) *Spaceborne Radar Remote Sensing: Applications and Techniques.* New York: IEEE Press.

Emardson TR, Simons M, and Webb FH (2003) Neutral atmospheric delay in interferometric synthetic aperture radar applications: Statistical description and mitigation. *Journal of Geophysical Research* 108 (doi:10.1029/2002JB001781).

Ferretti A, Prati C, and Rocca F (2000) Nonlinear subsidence rate estimation using permanent scatterers in differential SAR interferometry. *IEEE Transactions on Geoscience and Remote Sensing* 38: 2202–2212.

Ferretti A, Prati C, and Rocca F (2001) Permanent scatterers in SAR interferometry. *IEEE Transactions on Geoscience and Remote Sensing* 39: 8–20.

Fialko Y, Simons M, and Agnew D (2001a) The complete (3-D) surface displacement field in the epicentral area of the 1999 M_w 7.1 Hector mine earthquake, California, from space geodetic observations. *Geophysical Research Letters* 28: 3063–3066.

Fialko Y, Simons M, and Khazan Y (2001b) Finite source modeling of magmatic unrest in Socorro, New Mexico, and Long Valley, California. *Geophysical Journal International* 146(1): 191–200.

Fialko Y, Sandwell D, Simons M, and Rosen P (2005) Three-dimensional deformation caused by the Bam, Iran, earthquake and the origin of shallow slip deficit. *Nature* 435: 295–299.

Fielding EJ, Talebian M, Rosen PA, et al. (2005) Surface ruptures and building damage of the 2003 Bam, Iran, earthquake mapped by satellite synthetic aperture radar interferometric correlation. *Journal of Geophysical Research* 110 (doi:10.1029/2004JB003299).

Foster J, Brooks B, Cherubini T, Shacat C, Businger S, and Werner CL (2006) Mitigating atmospheric noise for InSAR using a high resolution weather model. *Geophysical Research Letters* 33 (doi:10.1029/2006GL026781).

Franceschetti G and Lanari R (1999) *Synthetic Aperture Radar Processing.* Boca Raton, FL: CRC Press.

Fujiwara S, Rosen PA, Tobita M, and Murakami M (1998) Crustal deformation measurements using repeat-pass JERS-1 synthetic aperture radar interferometry near the Izu Peninsula, Japan. *Journal of Geophysical Research* 103: 2411–2426.

Fukuda J, Higuchi T, Miyazaki S, and Kato T (2004) A new approach to time-dependent inversion of geodetic data using a Monte Carlo mixture Kalman filter. *Geophysical Journal International* 159(1): 17–39.

Funning GJ, Parsons B, Wright TJ, Jackson JA, and Fielding EJ (2005) Surface displacements and source parameters of the 2003 Bam (Iran) earthquake from Envisat advanced synthetic aperture radar imagery. *Journal of Geophysical Research* 110: B09406.

Gabriel AK, Goldstein RM, and Zebker HA (1989) Mapping small elevation changes over large areas: Differential radar interferometry. *Journal of Geophysical Research* 94: 9183–9191.

Gatelli F, Guarnieri AM, Parizzi F, Pasquali P, Prati C, and Rocca F (1994) The wave-number shift in SAR interferometry. *IEEE Transactions on Geoscience and Remote Sensing* 32: 855–865.

Goldstein RM (1995) Atmospheric limitations to repeat-track radar interferometry. *Geophysical Research Letters* 22: 2517–2520.

Goldstein RM, Engelhardt H, Kamb B, and Frolich RM (1993) Satellite radar interferometry for monitoring ice sheet motion: Application to an Antarctic ice stream. *Science* 262: 1525–1530.

Goldstein RM and Zebker HA (1987) Interferometric radar measurement of ocean surface currents. *Nature* 328: 707–709.

Goldstein RM, Zebker HA, and Werner CL (1988) Satellite radar interferometry: Two-dimensional phase unwrapping. *Radio Science* 23: 713–720.

Goodman JW (1985) *Statistical Optics.* New York: Wiley-Interscience.

Gourmelen N and Amelung F (2005) Post-seismic mantle relaxation in the Central Nevada seismic belt. *Science* 310: 1473–1476.

Gray AL, Mattar KE, and Sofko G (2000) Influence of ionospheric electron density fluctuations on satellite radar interferometry. *Geophysical Research Letters* 27: 1451–1454.

Gray AL, Short N, Mattar KE, and Jezek KC (2001) Velocities and flux of the Filchner ice shelf and its tributaries determined from speckle tracking interferometry. *Canadian Journal of Remote Sensing* 27: 193–206.

Grell GA, Dudhia J, and Stauffer PJ (1995) A description of the fifth generation Penn State/NCAR Mesoscale Model (MM/5). NCAR Tech. Note 398, National Center for Atmospheric Research Boulder, CO.

Guarnieri AM and Prati C (1996) ScanSAR focusing and interferometry. *IEEE Transactions on Geoscience and Remote Sensing* 3: 1029–1038.

Hanssen RF (2001) *Radar Interferometry: Data Interpretation and Error Analysis.* Dordrecht, Netherlands: Kluwer.

Hoffmann J, Zebker HA, Galloway DL, and Amelung F (2001) Seasonal subsidence and rebound in Las Vegas Valley, Nevada, observed by synthetic aperture radar interferometry. *Water Resources Research* 37: 1551–1566.

Hooper A, Zebker H, Segall P, and Kampes B (2004) A new method for measuring deformation on volcanoes and other natural terrains using insar persistent scatterers. *Geophysical Research Letters* 31: L23611.

Jonsson S, Zebker H, Segall P, and Amelung F (2002) Fault slip distribution of the 1999 M_w 7.1 Hector Mine, California, earthquake, estimated from satellite radar and GPS measurements. *Bulletin of the Seismological Society of America* 92: 1377–1389.

Joughin I (2002) Ice-sheet velocity mapping: A combined interferometric and speckle-tracking approach. *Annals of Glaciology* 34: 195–201.

Joughin I, Winebrenner DP, and Percival DB (1994) Probability density functions for multi-look polarimetric signatures. *IEEE Transactions on Geoscience and Remote Sensing* 32: 562–574.

Klinger Y, Xu XW, Tapponnier P, Van der Woerd J, Lasserre C, and King G (2005) High-resolution satellite imagery mapping of the surface rupture and slip distribution of the M_w 7.8, 14 November 2001 Kokoxili earthquake, Kunlun Fault, northern Tibet, China. *Bulletin of the Seismological Society of America* 95: 1970–1987.

Lanari R, Hensley S, and Rosen PA (1998) Chirp-Z transform based SPECAN approach for phase-preserving ScanSAR image generation. *IEEE Proceedings - Radar, Sonar and Navigation* 145: 254–261.

Lanari R, Lundgren P, Manzo M, and Casu F (2004a) Satellite radar interferometry time series analysis of surface deformation for Los Angeles, California. *Geophysical Research Letters* 31: L23613 (doi:10.1029/2004GL021294).

Lanari R, Mora O, Manunta M, Mallorqui JJ, Berardino P, and Sansosti E (2004b) A small baseline approach for investigating deformations on full resolution differential SAR interferograms. *IEEE Transactions on Geoscience and Remote Sensing* 42: 1377–1386.

Lee J-S, Hoppel KW, Mango SA, and Miller AR (1992) Intensity and phase statistics of multilook polarimetric and interferometric SAR imagery. *IEEE Transactions on Geoscience and Remote Sensing* 32: 1017–1028.

Levy F, Hsu Y, Simons M, LePrince S, and Avouac JP (2005) Distribution of coseismic slip for the 1999 Chi Chi Taiwan earthquake: New data and implications of varying 3D fault geometry. *EOS Transactions of the American Geophysical Union* 86: B1305.

Li F and Goldstein RM (1990) Studies of multibaseline spaceborne interferometric synthetic aperture radars. *IEEE Transactions on Geoscience and Remote Sensing* 28: 88–97.

Li ZW, Ding XL, and Liu G (2004) Modeling atmospheric effects on InSAR with meteorological and continuous GPS observations: Algorithms and some test results. *Journal of Atmospheric and Solar-Terrestrial Physics* 66: 907–917.

Li ZH, Fielding EJ, Cross P, and Muller JP (2006b) Interferometric synthetic aperture radar atmospheric correction: GPS topography-dependent turbulence model. *Journal of Geophysical Research* 111: B02404.

Li Z, Muller JP, Cross P, Albert P, Fischer J, and Bennartz R (2006a) Assessment of the potential of MERIS near-infrared water vapour porducts to correct ASAR interferometric measurements. *International Journal of Remote Sensing* 27: 349–365.

Li ZH, Muller JP, Cross P, and Fielding EJ (2005) Interferometric synthetic aperture radar (InSAR) atmospheric correction: GPS, moderate resolution imaging spectroradiometer (MODIS), and InSAR integration. *Journal of Geophysical Research* 110: B03410.

Lohman RB and Simons M (2005) Some thoughts on the use of InSAR data to constrain models of surface deformation. *Geochemistry Geophysics Geosystems* 6 (doi:10.1029/2004GC00084).

Lohman RB, Simons M, and Savage B (2002) Location and mechanism of the Little Skull Mountain earthquake as constrained by satellite radar interferometry and seismic waveform modeling. *Journal of Geophysical Research* 107: 2118.

Lundgren P, Casu F, Manzo M, et al. (2004) Gravity and magma induced spreading of Mount Etna volcano revealed by satellite radar interferometry. *Geophysical Research Letters* 31: L04602 (doi:10.1029/2003GL018736).

Lundgren P, Usai S, Sansosti E, et al. (2001) Modeling surface deformation observed with synthetic aperture radar interferometry at Campi Flegrei caldera. *Journal of Geophysical Research* 106: 19355–19366 (doi:10.1029/2001JB000194).

Lyons S and Sandwell D (2003) Fault creep along the southern San Andreas from interferometric synthetic aperture radar, permanent scatterers, and stacking. *Journal of Geophysical Research* 108: 2047.

Massonnet D and Feigl KL (1995) Discrimination of geophysical phenomena in satellite radar interferograms. *Geophysical Research Letters* 22: 1537–1540.

Massonnet D, Rossi M, Carmona C, et al. (1993) The displacement field of the Landers earthquake mapped by radar interferometry. *Nature* 364: 138–142.

Mattar KE and Gray AL (2002) Reducing ionospheric electron density errors in satellite radar interferometry applications. *Canadian Journal of Remote Sensing* 28: 593–600.

McGuire JJ and Segali P (2003) Imaging of aseismic fault slip transients recorded by dense geodetic networks. *Geophysical Journal International* 155: 778–788 (doi:10.1111/j.1365–246X.2003.02022.x).

Menke W (1989) *International Geophysics Series, vol.45: Geophysical Data Analysis: Discrete Inverse Theory*, Academic Press, rev. edn.

Michel R, Avouac JP, and Taboury J (1999a) Measuring near field coseismic displacements from SAR images: Application to the Landers earthquake. *Geophysical Research Letters* 26: 3017–3020.

Michel R, Avouac JP, and Taboury J (1999b) Measuring ground displacements from SAR amplitude images: Application to the Landers earthquake. *Geophysical Research Letters* 26: 875–978.

Michel R and Rignot E (1999) Flow of Glaciar Moreno, Argentina, from repeat-pass Shuttle Imaging Radar images: Comparison of the phase correlation method with radar interferometry. *Journal of Glaciology* 45: 93–100.

Moore R, Classen J, and Lin YH (1981) Scanning spaceborne synthetic aperture radar with integrated radiometer. *IEEE Transactions on Aerospace and Electronic Systems* 17: 410–420.

Onn F and Zebker HA (2006) Correction for interferometric synthetic aperture radar atmospheric phase artifacts using time series of zenith wet delay observations from a GPS network. *Journal of Geophysical Research* 111: (doi:10.1029/2005JB004012).

Peltzer G, Crampe F, Hensley S, and Rosen P (2001) Transient strain accumulation and fault interaction in the Eastern California Shear Zone. *Geology* 29: 975–978.

Peltzer G, Crampe F, and King G (1999) Evidence of nonlinear elasticity of the crust from the Mw 7.6 Manyi (Tibet) earthquake. *Science* 286: 272–276.

Pritchard ME and Simons M (2006) An aseismic slip pulse in northern Chile and along-strike variations in seismogenic behavior. *Journal of Geophysical Research* 111: (doi:10.1029/2006JB004258).

Pritchard ME, Ji C, and Simons M (2006) Distribution of slip from 11 $M_w > 6$ earthquakes in the northern Chile subduction zone. *Journal of Geophysical Research* 111: (doi:10.1029/2005JB004013).

Puyssegur B, Michel R, and Avouac JP (2007) Tropospheric phase delay in InSAR estimated from meteorological model and multispectral imagery. *Journal of Geophysical Research* (in press).

Raney RK (1971) Synthetic aperture imaging radar and moving targets. *IEEE Transactions on Aerospace and Electronic Systems* AE S-7: 499–505.

Raney RK (1998) Radar fundamentals: Technical perspective. In: Henderson FM and Lewis AG (eds.) *Manual of Remote Sensing Volume 2, Principles and Applications of Imaging Radar*, 3rd edn., Hoboken, NJ: J. Wiley.

Rodríguez E and Martin JM (1992) Theory and design of interferometric synthetic-aperture radars. *Proceedings of the IEEE* 139: 147–159.

Rosen PA, Hensley S, Joughin IR, et al. (2000) Synthetic Aperture Radar Interferometry. *Proceedings of the IEEE* 88: 333–382.

Rosen PA, Hensley S, Peltzer G, and Simons M (2004) Updated repeat orbit interferometry package released. *Eos Transactions of the American Geophysical Union* 85: 47 (doi:10.1029/2004EO050004).

Rosen PA, Hensley S, Zebker HA, Webb FH, and Fielding EJ (1996) Surface deformation and coherence measurements of kilauea volcano, Hawaii, from SIR-C radar interferometry. *Journal of Geophysical Research* 268: 1333–1336.

Schmidt DA and Burgmann R (2003) Time-dependent land uplift and subsidence in the Santa Clara Valley, california, from a large interferometric synthetic aperture radar data set. *Journal of Geophysical Research* 108: 2416–2428.

Segall P and Matthews M (1997) Time dependent inversion of geodetic data. *Journal of Geophysical Research* 102: 22391–22409.

Short NH and Gray AL (2004) Potential for RADARSAT-2 interferometry: glacier monitoring using speckle tracking. *Canadian Journal of Remote Sensing* 30: 504–509.

Simons M, Fialko Y, and Rivera L (2002) Coseismic deformation from the 1999 M_w 7.1 Hector mine, California, earthquake as inferred from InSAR and GPS observations. *Bulletin of the Seismological Society of America* 92(4): 1390–1402.

Sorenson H (1980) *Parameter Estimation*. New York: Marcel Dekker.

Tarayre H and Massonnet D (1996) Atmospheric propagation heterogeneities revealed by ERS-1 interferometry. *Geophysical Research Letters* 23(9): 989–992.

Tomiyasu K (1981) Conceptual performance of a satellite borne wide swath radar. *IEEE Transactions on Geoscience and Remote Sensing* 19: 108–116.

Touzi R and Lopes A (1996) Statistics of the stokes parameters and the complex coherence parameters in one-look and multilook speckle fields. *IEEE Transactions on Geoscience and Remote Sensing* 34: 519–531.

Usai S (1997) The use of man-made features for long time-scale InSAR. In: *Proceedings IEEE International Geoscience and Remote Sensing Symposium* pages, 1542–1544 Singapore, 1997. cdrom.

Wdowinski S, Amelung F, Miralles-Wilhelm F, Dixon T, and Carande R (2004) InSAR-based hydrology of the Everglades, South Florida. In: *Geoscience and Remote Sensing Symposium, 2004. IGARSS '04. In Proceedings, 2004 IEEE International*, vol. 3, pp. 1870–1873.

Wegmuller U (2003) Potential of interferometry point target analysis using small data stacks. *Proceedings of Fringe '03 workshop*, pp. 1–3. Frascati, Italy (cdrom).

Williams S, Bock Y, and Fang P (1998) Integrated satellite interferometry: Tropospheric noise, GPS estimates and implications for interferometric synthetic aperture radar observations. *Journal of Geophysical Research* 103: 27051–27067.

Wright TJ, Parsons BE, and Lu Z (2004) Toward mapping surface deformation in three dimensions using InSAR. *Geophysical Research Letters* 31(1): L01607.

Xu CJ, Wang H, Ge LL, Yonezawa C, and Cheng P (2006) InSAR tropospheric delay mitigation by GPS observations: A case study in Tokyo area. *Journal of Atmospheric and Solar-Terrestrial Physics* 68(6): 629–638.

Zebker HA, Rosen PA, Hensley S, and Mouganis-Mark P (1996) Analysis of active lava flows on Kilauea Volcano, Hawaii, using SIR-C radar correlation measurements. *Geology* 24: 495–498.

Zebker HA, Rosen PA, and Hensley S (1997) Atmospheric effects in interferometric synthetic aperture radar surface deformation and topographic maps. *Journal of Geophysical Research* 102: 7547–7563.

Zebker HA and Goldstein RM (1986) Topographic mapping from interferometric SAR observations. *Journal of Geophysical Research* 91: 4993–4999.

Zebker HA, Rosen PA, Goldstein RM, Gabriel A, and Werner CL (1994) On the derivation of coseismic displacement fields using differential radar interferometry: The Landers earthquake. *Journal of Geophysical Research* 99: 19617–19634.

Zebker HA and Villasenor J (1992) Decorrelation in interferometric radar echoes. *IEEE Transactions on Geoscience and Remote Sensing* 30: 950–959.

Printed in the United States
By Bookmasters